FROGFISHES

FROGFISHES

Biodiversity, Zoogeography, and Behavioral Ecology

Theodore W. Pietsch and Rachel J. Arnold

JOHNS HOPKINS UNIVERSITY PRESS

BALTIMORE

Furthermore:
a program of the J. M. Kaplan Fund

This book was brought to publication through the generous assistance of Foster Bam and the Foster-Davis Foundation, Inc., Greenwich, Connecticut; Joan K. Davidson, Ann Birckmayer, and Furthermore: a program of the J. M. Kaplan Fund, New York, New York; and Thomas Boehm and the Jung Foundation for Science and Research, Hamburg, Germany.

The material in this volume is based in part on work supported by the US National Science Foundation under Grants GB-40700, DEB 76-82279, DEB 78-26540, DEB 03-14637, T. W. Pietsch, principal investigator. Any opinions, findings, conclusions, or recommendations expressed in this work are those of the authors and do not necessarily reflect the views of the National Science Foundation.

Johns Hopkins University Press
2715 North Charles Street
Baltimore, Maryland 21218-4363
www.press.jhu.edu

Library of Congress Cataloging-in-Publication Data
Names: Pietsch, Theodore W., author. | Arnold, Rachel J.
Title: Frogfishes : biodiversity, zoogeography, and behavioral ecology /
    Theodore W. Pietsch and Rachel J. Arnold.
Description: Baltimore : Johns Hopkins University Press, 2020. | Includes
    bibliographical references and index.
Identifiers: LCCN 2018058172 | ISBN 9781421432526 (hardcover : alk. paper) |
    ISBN 1421432528 (hardcover : alk. paper) | ISBN 9781421432533 (electronic) |
    ISBN 1421432536 (electronic)
Subjects: LCSH: Antennariidae—Evolution. | Antennariidae—Ecology. |
    Antennariidae—Geographical distribution.
Classification: LCC QL638.A577 P538 2019 | DDC 597/.62—dc23
LC record available at https://lccn.loc.gov/2018058172

A catalog record for this book is available from the British Library.

Title-page illustration: *Pira Vtoewah, forma monstrosa*, now known as *Antennarius multiocellatus*. The earliest published illustration of a frogfish, reproduced from *Novus Orbis, seu Descriptionis Indiae Occidentalis, Libri XVIII*, by Johannes de Laet, 1633.

Special discounts are available for bulk purchases of this book. For more information, please contact Special Sales at 410-516-6936 or specialsales@press.jhu.edu.

Johns Hopkins University Press uses environmentally friendly book materials, including recycled text paper that is composed of at least 30 percent post-consumer waste, whenever possible.

Contents

5. Evolutionary Relationships   364

6. Zoogeography   387

7. Behavioral Ecology   411

Preface

Once upon a time there was a frog called Mr. Jeremy Fisher; he lived in a little damp house amongst the buttercups at the edge of a pond.

"I will get some worms and go fishing and catch a dish of minnows for my dinner," said Mr. Jeremy Fisher. "If I catch more than five fish, I will invite my friends Mr. Alderman Ptolemy Tortoise and Sir Isaac Newton. The Alderman, however, eats salad."

—Beatrix Potter, *The Tale of Mr. Jeremy Fisher*, 1906

Since *Frogfishes of the World* was published in 1987, an enormous amount of new information about this incredible group of fishes has accumulated: new species have been discovered, described, and formally named; geographic distributions have been expanded as marine habitats and ecosystems have been more thoroughly explored; new molecular techniques and greater phylogenetic understanding have resulted in new and sometimes unexpected perspectives on evolutionary relationships; and new insights into feeding, locomotion, and especially reproductive behavior have been observed and recorded through an ever-expanding, worldwide interest in underwater research by students and professional scientists as well as aquarists and recreational divers. While 30 years ago it was difficult to find high-quality color photographs of frogfishes, state-of-the-art digital camera equipment and talented photographers everywhere have flooded the Internet with a multitude of astonishing images, both stills and videos. But while these are all good developments, there is a downside: the newly discovered interest in these fishes around the world has resulted in heavy exploitation by the aquarium industry in some regions, to the point that local populations are severely threatened. Thirty years ago, it did not occur to us to include a discussion of conservation, but it appears in this volume. So, our purpose here is very much the same as it was 30 years ago—to add new knowledge to what has been previously known about frogfishes; to bring together in a single volume a previously scattered array of facts (as well as to dispel the fiction); and to demonstrate the surprising diversity and beauty of frogfishes by reproducing the very best of the available images of this remarkable assemblage of marine shorefishes.

Frogfishes constitute a family of bony fishes, the Antennariidae, one of 18 families of anglerfishes that together form the order Lophiiformes. Nearly all frogfishes are structurally and chromatically cryptic forms whose feeding strategy consists of maintaining the immobile, inert, and benign appearance of a sponge or coralline-algae-encrusted rock, while wiggling a highly conspicuous lure—a uniquely modified, baited, dorsal-fin spine—to attract their prey. This extraordinary feeding mode, combined with their unusual frog-like appearance and tetrapod-like locomotion, makes them one of the most intriguing groups of aquatic vertebrates for the amateur aquarist as well as the professional ichthyologist.

This volume is the result of a critical review of the systematic literature of the Antennariidae, as well as a detailed examination of 4,916 preserved specimens (including nearly all extant type material) made available by the curators and staffs of numerous institutions located around the world. Following an introduction that emphasizes the history of our knowledge of the Antennariidae, from the earliest mention of their existence to the present, we provide a synopsis of the characters that best differentiate the taxa. This is followed by extensive evidence for the recognition of 15 genera and 52 living species of frogfishes. The known fossil material is reviewed as well. All extant taxa are provided with an annotated synonymy, and each is diagnosed, described, and illustrated. Keys and tables are provided to facilitate their identification. Evolutionary relationships are analyzed using both morphological and molecular approaches. Geographic and vertical distributions are discussed and all verifiable localities are plotted on maps; the ecological and historical zoogeography of the family and its subunits are also analyzed. Finally, we summarize the known facts concerning feeding dynamics, defensive behavior, locomotion, and reproduction, and finish with tips for divers and aquarists, emphasizing the urgent need for conservation. Although this is largely a technical systematic reference work devoted to synonymies, diagnoses, and descriptions, prepared to satisfy the needs and interests of the professional ichthyologist, we hope that we have presented these fascinating creatures in such a way that this volume will be enjoyed by all those who find excitement in the wonders of the natural world.

Acknowledgments

Many people have generously given their help with this study by providing loans and gifts of specimens or by making collection data available. Many others have assisted by making the authors feel welcome during visits to their respective institutions or by working with them in the field. To all of these people we extend our most sincere gratitude. In addition to those named previously in *Frogfishes of the World* (1987), we give special thanks to the following: Belinda Bauer, Tasmanian Museum and Art Gallery, Hobart, Tasmania; Manuel Biscoito, Museu Municipal do Funchal, Madeira, Portugal; Dianne Bray, Museums Victoria, Melbourne, Australia; David Catania, California Academy of Sciences, San Francisco; Alessio Datovo, Museu de Zoologia, Universidade de São Paulo, Brazil; Christopher Davis, Louisiana Department of Wildlife and Fisheries, Baton Rouge, Louisiana; Guy Duhamel, Patrice Pruvost, and Romain Causse, Muséum national d'Histoire naturelle, Paris, France; Richard F. Feeney, Natural History Museum of Los Angeles County, Los Angeles; Ralph Foster, South Australian Museum, Adelaide; Bryan Gim, University of California at Berkeley; Peter W. Glynn, Rosenstiel School of Marine and Atmospheric Science, Miami, Florida; Alastair Graham, Australian National Fish Collection, Commonwealth Scientific Industrial Research Organization, Hobart, Tasmania; Kiyoshi Hagiwara, Yokosuka City Museum of Nature and Human Culture, Yokosuka, Japan; Kiyotaka Hatooka and Shoko Matsui, Osaka Museum of Natural History, Osaka, Japan; Hsuan-Ching Ho, Institute of Marine Biology, National Dong Hwa University, Checheng, Taiwan; Samuel Iglésias, Station de Biologie Marine du Museum national d'Histoire naturelle et du Collège de France, Concarneau; Hisashi Imamura and Toshio Kawai, Laboratory of Marine Zoology, Faculty of Fisheries, Hokkaido University, Hakodate, Hokkaido, Japan; Jeffrey W. Johnson, Queensland Museum, South Brisbane, Australia; Seishi Kimura, Fisheries Research Laboratory, Mie University, Shima, Japan; Luis Alvarez Lajonchere, Felipe Poey Natural History Museum, Havana, Cuba; Shawn Larson, Seattle Aquarium, Seattle, Washington; James Maclaine, The Natural History Museum, London; Keiichi Matsuura and Gento Shinohara, National Science Museum, Tokyo, Japan; Gerald McCormack, Cook Islands Natural Heritage Project, Rarotonga, Cook Islands; Mark McGrouther, Australian Museum, Sydney; Caleb McMahan and Susan Mochel, Field Museum of Natural History, Chicago; Glenn Moore, Sue Morrison, and Mark Allen, Western Australian Museum, Perth; Zenaida Navarro, Center for Marine Research, University of Havana, Havana, Cuba; Yoichi Sato, Tokushima Prefectural Museum, Tokushima, Japan; David Schleser, Dallas World Aquarium, Dallas, Texas; Kwang-Tsao Shao and Shih-Pin Huang, Academia Sinica Institute of Zoology, Taipei, Taiwan; Dirk Steinke and Jeremy deWaard, Centre for Biodiversity Genomics, University of Guelph, Canada; Arnold Suzumoto and Loreen O'Hara, Bernice P. Bishop Museum, Honolulu,

Hawaii; Luke Tornabene and Katherine P. Maslenikov, Burke Museum of Natural History and Culture, University of Washington, Seattle; Jeffrey T. Williams, Diane Pitassy, and Shirleen Smith, National Museum of Natural History, Washington, DC.

Scores of diver-photographers, aquarists, and collection curators from around the world have been overwhelmingly generous in providing images—and describing their underwater observations of frogfishes in the field and in aquaria—asking for nothing more than a note of thanks. The long list includes Graham Abbott, the late Mary Jane Adams (courtesy of John and Suzanne Kelley), Gerry Allen, Shabir Ali Amir, Gilbert Angermann, Calder Atta, Laurent Auer, Georg Bachschmid, Frank Baensch, Stephane Bailliez, Janine Baker, Daniel Barker, Mike Bartick, Fred Bavendam, Danny van Belle, Sergey Bogorodsky, Sheila Bowtle, Chuck Boxwell, Martin Buschenreithner, Janine Cairns-Michael, Kelly and John Casey, Prosanta Chakrabarty, Vincent Chalias, Tony Cherbas, Linda Cline, Brandon Cole, Carol Cox, Helmut Debelius, Robert Delfs, Ned and Anna DeLoach, Gary Dunnett, Anne DuPont, Marcel Eckhardt, Mark Erdmann, Diana Fernie, Salvador Garcia, Jim Garin, Johanna Gawron, Daniel Geary, Scott Gietler, Bryan Gim, Frank de Goey, Andrew Goodson, Bill Goodwin, Dan Gotshall, Gabriel Grimsditch, David Grobecker, Sanne and Rokus Groeneveld, David Hall, Aaron Halstead, David Harasti, Rob Harcourt, Matthew Harrison, Rudy Hayat, Elaine Heemstra, Ronald Holcom, Christa Holdt, John Hoover, Paul Humann, Linda Ianniello, Scott and Jeanette Johnson, Zane Kamat, Shoichi Kato, the late Alex Kerstitch (courtesy of Lloyd Findley and Rick Brusca), Eiji Kozuka, Rudie Kuiter, Anthony Kuntz, Jacob Loyacano, John Lewis, Robert and Cherie van der Loos, Paul Macdonald, Colin Marshall, Susana Martins, Guilherme Mascarenhas, Volker Mattke, Brian Mayes, Michael McKnight, Alicia Hermosillo McKowen, Jonathan Mee, Scott W. Michael, Irene Middleton, Nick Missenden, Glenn Moore, Ellen Muller, Steven Ng, Bronson Nagareda, Junichi Nakamoto, John Neuschwander, Nirupam Nigam, Nobuhiro Ohnishi, Matthew Oldfield, Dan Olsen, Mika Oyry, Seth Parker, Rob Paton, James Peake, Gerson Kim Penetrante, Sarah Pikarski, Brian Potvin, Peter Psomadakis, Jack Randall, Scott Rettig, Ross Robertson, Luiz Rocha, Caroline Rogers, Jeff Rosenfeld, Galina Della Ross, Chip Roy, Frank Schneidewind, Kim Sebo, Hiroshi Senou, Benz Severin, Bruce Shafer, David Shale, Jessica Groomer Smith, Robyn Smith, Mark Snyder, Daniel Stassen, Roger Steene, Keoki Stender, William Stohler, Ann Storrie, Alexius Sutandio, Arnold Suzumoto, Augy Syahailatua, Cindy Tan, Juuyoh Tanaka, Andrew Taylor, Warren Taylor, Chifuku Tetsuro, Heidi Thoricht, Gordon Tillen, Hiroyuki Tokuya, Tom Trnski, Hiroyuki Uchiyama, Roger Uwate, Marli Wakeling, Keri Wilk, Jeff Williams, Rick Winterbottom, Peter Wirtz, Mitsuhiko Yanagita, Marna Zanoff, and Teresa Zuberbühler.

For photos of fossil material, we thank Giorgio Carnevale, Università degli Studi di Torino, Torino, Italy; Mariagabriella Fornasiero, Letizia Del Favero, and Stefano Castelli, Museo di Geologia e Paleontologia, Università di Padova, Italy; Roberto Zorzin and Anna Vaccari, Museo Civico di Storia Naturale, Verona, Italy; Philippe Loubry and Christiane Chancogne, Muséum national d'Histoire naturelle, Paris; and Gary T. Takeuchi and the Department of Vertebrate Paleontology, Natural History Museum of Los Angeles County, Los Angeles, California.

For providing additional illustrative materials, we are greatly indebted to Françoise Antonutti and Olivier Pauwels, Royal Belgian Institute of Natural Sciences, Brussels; Stephan Atkinson, Picture Library, Natural History Museum, London; Tad Bennicoff, Smithsonian Institution Archives, Washington, DC; Sarah Broadhurst and Ann Sylph, Zoological Society of London; Giorgio Carnevale, Università degli Studi di Torino,

Torino, Italy; Isabelle Charmantier and Andrea Deneau, Linnean Society of London; Judy Choi, University of Chicago Press, Chicago; Emmanuelle Choiseau, Bibliothèque Centrale, Muséum national d'Histoire naturelle, Paris; Alessio Datovo, Museum of Zoology, University of São Paulo, São Paulo, Brazil; Joyce Faust, Art Resource, Inc., New York; Rick Feeney, Natural History Museum of Los Angeles County, Los Angeles; Anne-Claire Gallet and Yolande Boulade, Réunion des Musées Nationaux—Grand Palais, Paris; Lea Gardam, South Australian Museum, North Terrace, Adelaide; Martin F. Gomon, Museums Victoria, Melbourne, Australia; Penny Haworth, National Research Foundation, South African Institute for Aquatic Biodiversity, Grahamstown; James E. Hayden, Wistar Institute, Philadelphia, Pennsylvania; Bart Lahr, Het Scheepvaart-museum, Amsterdam; James M. Lawrence, Reef to Rainforest Media, LLC; Mark McGrouther, Patricia Egan, and Vanessa Finney, Australian Museum, Sydney; Martien van Oijen, Naturalis Biodiversity Center, Leiden; Lisa Palmer and Sandra Raredon, National Museum of Natural History, Smithsonian Institution, Washington, DC; Tia Reber, Bernice P. Bishop Museum, Honolulu, Hawaii; Mai Reitmeyer, American Museum of Natural History, New York; Hiroshi Senou, Kanagawa Prefectural Museum of Natural History, Odawara, Kanagawa, Japan; Hiroki Tabo, Sanseido Co. Ltd., Tokyo, Japan; Karin Tucker, University of California Press, Berkeley and Los Angeles; Godard Tweehuysen, Naturalis Bibliotheek, Naturalis Biodiversity Center, Amsterdam; Rosemary Volpe, Peabody Museum of Natural History, Yale University, New Haven, Connecticut; and William Watson, NOAA Fisheries, Southwest Fisheries Science Center, La Jolla, California. Thanks also to Secretary Prosanta Chakrabarty and the American Society of Ichthyologists and Herpetologists for permission to reproduce copyrighted materials previously published in *Copeia*.

We thank David Barnes, James Bell, Cristina Diaz, Eduardo Leal Esteves, Roberto Alcantara Gomes, and Fernando Moraes for help with sponge identification; Mike Bartick and Aaron Halstead for providing information on unpublished sightings of *Fowlerichthys avalonis* off southern California; Lloyd Findley for help in locating images of eastern Pacific frogfishes; Jim Garin for habitat information on *Fowlerichthys ocellatus* in the Gulf of Mexico; Vincent Chalias for aid in procuring specimens through the tropical fish trade; Vincent Chalias, Daniel Geary, David Hall, and Linda Ianniello for habitat descriptions of *Nudiantennarius subteres*; Bryan Gim for behavioral observations of *Histiophryne* in captivity; Samuel Iglesias for examining specimens on our behalf at the Muséum national d'Histoire naturelle, Paris; Ralph Foster and Jeff Johnson for providing information on collections of *Histiophryne* archived in the South Australian Museum, Adelaide, and Queensland Museum, Brisbane, respectively; Chery Kinnick, Bernie Noll, and Alan Presley, University of Washington Libraries, for providing access to rare and obscure literature; Jack Lovell for information on Hawaiian occurrences of *Histrio histrio*; Nobuhiro Ohnishi for Japanese occurrences of *Antennatus flagellatus*; Martien van Oijen for examining Pieter Bleeker types for us at the Naturalis Biodiversity Center, Leiden; James W. Orr for help in every way but specifically for providing radiographs; Diane Pitassy for numerous courtesies but especially an exhaustive search among collections at the Smithsonian for additional specimens of *Nudiantennarius subteres* and for a detailed examination of "warts" in the skin of *Antennarius pardalis*; Abby Simpson and Hans Aili for translations of French and Latin; Peter Wirtz for data on Eastern Atlantic frogfishes; and Scott W. Michael for contributing to Chapter 8. The molecular work was carried out in the Molecular Ecology Research Laboratory, School of Aquatic and Fishery Sciences, University of Washington.

Many people to whom we are hugely indebted have given their time and expertise to provide constructive criticism during the course of this investigation. Early drafts of parts of the manuscript were critically read by Frank Baensch, Giorgio Carnevale, Daniel Gledhill, G. David Johnson, Christopher P. Kenaley, Scott W. Michael, and Masaki Miya. The entire manuscript was reviewed by Gerald R. Allen, Bruce A. Carlson, John H. Caruso, Alastair Graham, Bruce C. Mundy, James W. Orr, and Andrew L. Stewart. From among the latter group, all of whom provided much-appreciated constructive feedback, we must single out Bruce Mundy for the huge effort he made to read nearly every word and comment in exhaustive detail.

At Johns Hopkins University Press, sincere thanks are extended to Tiffany Gasbarrini, Senior Acquisitions Editor, Life Sciences, Mathematics, and Physics, for skillfully directing the publication of this volume, and to the rest of the JHUP publishing team: Esther Rodriguez, Acquisitions Assistant; Julie McCarthy, Managing Editor; Debby Bors and Andre Barnett, Senior Production Editors; Julie Burris, Designer; and Jack Holmes, Publicist. Thanks also go to Liz Radojkovic, freelance copyeditor.

Support for the research presented here has come from numerous sources spread out over a 40-year period beginning in 1967 when the senior author was a first-year graduate student at the University of Southern California. The bulk of the work was generously supported by the U.S. National Science Foundation, most recently through Grant DEB-0314637, T. W. Pietsch, principal investigator. We are extremely grateful to the Foundation and to all its program officers and staff.

Smaller segments of the research were supported by National Geographic Society Grant No. 1826; grants from the William F. Milton Fund, Harvard University, and from the Office of Graduate Studies and Research, California State University at Long Beach; PHS Biomedical Research Support Grant No. RR-07096 administered through the Graduate School Research Fund, University of Washington; the Eppley Foundation for Research, New York; the Lerner-Gray Fund for Marine Research, American Museum of Natural History; and the Ichthyology Fund, Natural History Museum of Los Angeles County. Travel support provided by the School of Aquatic and Fishery Sciences, University of Washington, provided numerous opportunities for the authors to report on research progress. Finally, generous support to the junior author, while a graduate student at the University of Washington, was provided by the Dorothy T. Gilbert Ichthyological Research Fund. To all these agencies, and to those who administer the funds, we are extremely grateful.

Finally, for providing the subsidy needed to publish this volume, we give very special thanks to Foster Bam of the Foster-Davis Foundation, Inc., Greenwich, Connecticut (kindly supported by a recommendation from Eric T. Schultz); to Joan K. Davidson and Ann Birckmayer of Furthermore Grants in Publishing: a program of the J. M. Kaplan Fund, New York, New York; to Thomas Boehm for support through an unrestricted grant from the Jung Foundation for Science and Research, Hamburg, Germany; and finally to all those who responded to our crowd-funding campaign.

FROGFISHES

1. Introduction

They are the most grotesque—we had almost said the most hideous—of all fishes, and, as their vernacular name of frogfish implies, they have nearly as much the appearance of frogs or toads as of fish. . . . The imagination can scarcely conceive more fanciful forms than such as are actually in this group; and the monstrous combinations which painters have represented under the aspect of animals can scarcely surpass the singularity of many of these real fish.

—William Swainson, 1838: 202

Frogfishes of the family Antennariidae are typically small, globose anglerfishes easily distinguished from members of allied families by the presence of three well-developed dorsal-fin spines, laterally directed eyes, a large, anterodorsally directed mouth, and a short, laterally compressed body (Pietsch, 1981b). They share with closely related families a peculiar and unique mode of feeding that is characterized most strikingly by the structure of the first dorsal-fin spine (called the "illicium"), which is placed near the tip of the snout and modified to serve as a luring apparatus.

Frogfishes spend the greater part of their lives squatting on the bottom in shallow to moderately deep water, or, as in the case of *Histrio histrio*, clinging to floating *Sargassum*. Despite their seemingly nonthreatening and rather sedentary nature, all are voracious carnivores that wait patiently for smaller fishes or crustaceans to pass by, while they wriggle their bait to attract prey toward their cavernous mouths. Despite this intriguing lifestyle, their relative abundance, and their wide geographic distribution in relatively accessible waters of nearly all tropical to temperate oceans and seas of the world, the biology of frogfishes has been poorly known and the literature about them widely scattered.

The Antennariidae are part of a much larger assemblage—an order of teleost fishes called the Lophiiformes, containing 18 families, 74 genera, and approximately 371 living species, the monophyletic origin of which seems certain on the basis of at least seven unique and morphologically complex, shared derived features. The families are distributed among five currently recognized suborders (Pietsch, 1984d, 2009; Fig. 1): the Lophioidei (Caruso, 1976, 1981, 1983, 1985, 1986; Caruso and Bullis, 1976; Caruso and Suttkus, 1979), Antennarioidei (Last et al., 1983; Pietsch and Grobecker, 1987; Last and Gledhill, 2009; Pietsch et al., 2009b; Carnevale and Pietsch, 2010; Arnold and Pietsch, 2012), Chaunacoidei (Caruso, 1989a, 1989b; Caruso et al., 2006; Ho and Shao, 2010; Ho and Last, 2013; Ho et al., 2013a, 2015a, 2015b; Ho and McGrouther, 2015), Ogcocephaloidei (Ochiai and Mitani, 1956; Bradbury, 1967, 1980, 1988, 1999; Endo and Shinohara, 1999; Ho et al.,

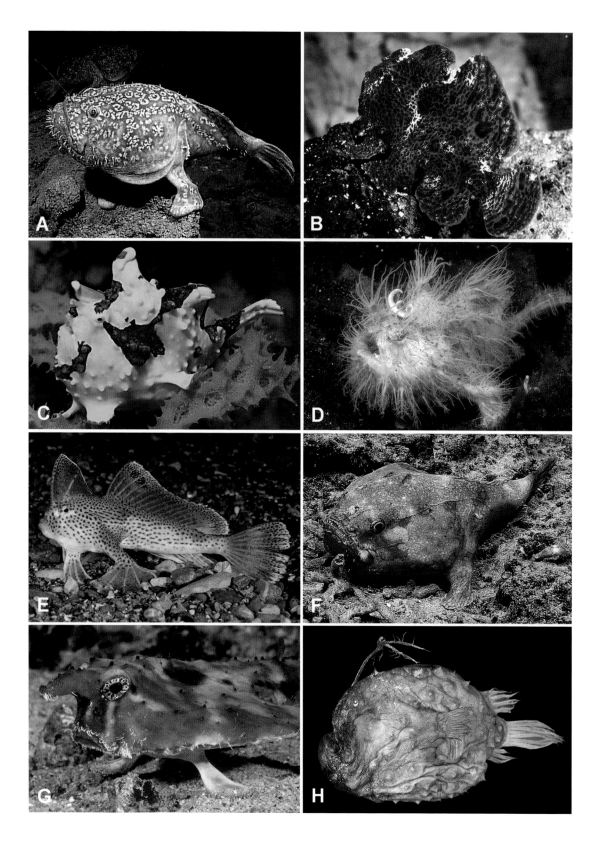

2010, 2013b; Ho, 2013; Derouen et al., 2015), and the deep-sea Ceratioidei (Bertelsen, 1951, 1984; Pietsch and Orr, 2007; Pietsch, 2009; Prokofiev, 2014; Ho et al., 2016).

The family Antennariidae has had a long and complex taxonomic history. Since Carl Linnaeus (1707–1778) published the tenth edition of his *Systema Naturae* in 1758, 185 nominal species have been described and placed in some 34 nominal genera and subgenera. The result of this descriptive proliferation, prior to the revisionary work of Pietsch and Grobecker (1987), was an extremely confusing nomenclature that involved well over 350 different combinations of generic, subgeneric, specific, and varietal names for what we now recognize as 15 genera and 52 species. Except for those dealing with the relatively well-understood Western Atlantic fauna, few authors prior to 1987 were able to apply the correct Latinized binomial to the material on which they reported, a situation largely due to the lack of workable keys to the identification of antennariid genera and species. Finally, and mostly because of these taxonomic problems, virtually nothing useful was available prior to 1987 concerning geographic distributions and depth ranges, nor anything on the ecology or behavior of the species. While many of these difficulties were resolved with the publication of *Frogfishes of the World* (Pietsch and Grobecker, 1987), problems of identification still exist and a wealth of new and exciting information about these animals is yet to be revealed.

The popularity of frogfishes has grown enormously in the past 30 years. Not so long ago they were generally thought to be drab and colorless, and because they hardly ever move they were largely dismissed, providing little interest for divers and aquarists (Burgess and Axelrod, 1972)—not to mention, in the case of the latter, the certainty that most frogfishes will devour anything else that moves in a home aquarium. But, as people have come to know them better, a cadre of frogfish aficionados around the world has grown within the dive community and among aquarists. In so many ways, they are enormously interesting. Here is a fish that through cunning and trickery turns would-be predators into prey, a voracious and sometimes cannibalistic animal that can expand its mouth cavity twelvefold and gulp down prey larger than itself in less than 6 msec (Grobecker and Pietsch, 1979)—less time than is needed by normal vertebrate striated mus-

---

Fig. 1. (*opposite*) The five suborders of anglerfishes that together form the teleost order Lophiiformes. (**A**) Lophioidei, containing the Goosefishes and Monkfishes, a single family, four genera, and 27 living species of shallow- to deep-water forms; represented here by *Sladenia shaefersi*, about 570 mm SL, northern Gulf of Mexico, depth 1,165 m. Courtesy of the NOAA *Okeanos Explorer* Program. (**B–E**) Antennarioidei, the Frogfishes and Handfishes, four families, 23 genera, and about 68 species of laterally compressed, shallow to moderately deep-water, benthic forms; represented by (**B**) *Antennarius commerson*, 111 mm SL, aquarium specimen, UW 20983. Photo by David B. Grobecker. (**C**) *Antennarius maculatus*, about 80 mm SL, Balicasag Island, Bohol, Philippines. Photo by Christa Holdt. (**D**) *Antennarius striatus*, 150 mm SL, Lembeh Strait, Sulawesi, Indonesia. Photo by Frank Schneidewind. (**E**) *Brachionichthys hirsutus*, about 90 mm SL, Tasmania, Australia. Photo by Fred Bavendam. (**F**) Chaunacoidei, the Gapers, Coffinfishes, and Sea Toads, a single family, at least two genera, and perhaps as many as 29 species of globose, deep-water benthic forms; represented by *Chaunax umbrinus*, about 300 mm SL, off Niihau, Hawaiian Islands, depth 328 m. Courtesy of the NOAA *Okeanos Explorer* Program. (**G**) Ogcocephaloidei, the Batfishes, a single family of 10 genera and about 77 species of dorso-ventrally flattened, deep-water benthic forms; represented by *Ogcocephalus darwini*, about 100 mm SL, Cocos Island, Costa Rica. Photo by Fred Bavendam. (**H**) Ceratioidei, the Seadevils, containing 11 families, 35 genera, and 169 currently recognized species of globose to elongate, mesopelagic, bathypelagic, and abyssal-benthic forms; represented by *Himantolophus stewarti*, 124 mm SL, CSIRO H.5652-01, South Tasman Rise, Tasman Sea. Photo by T. W. Pietsch.

cle to complete a single contraction. We are talking about a feeding strategy that is unique to lophiiform fishes, one that has intrigued scientists as well as the casual observer since the days of Aristotle. One cannot look at a frogfish without being struck with the notion that it must be a mimic of substrate and structure, encrusted with marine organisms. What we see is in fact an aggressive mimetic adaptation, which when coupled with a luring mechanism that depends primarily on the resemblance of a bait to small edible marine organisms (Pietsch and Grobecker, 1978), produces one of nature's most highly evolved examples of "lie-in-wait" predation.

This is also a fish that "walks" and "hops" across substrate, and clambers about over rocks and coral like a four-legged terrestrial beast but, at the same time, has the ability to jet-propel itself through open water. Myths concerning its amphibious nature and ability to remain out of water for days on end date back more than 300 years (Renard, 1719). The muscular pectoral fins of these fishes bear a remarkable resemblance to the forelimbs of a tetrapod, and in benthic situations, these structures, together with the anteriorly placed pelvic fins, function to produce two basic tetrapod-like gaits. Alternatively, when in mid-water, frogfishes display three distinct locomotory modes, the most unusual of which is jet propulsion. The unique combination of enlarged, well-muscled oral and opercular cavities and highly specialized, restricted opercular openings located behind the pectoral-fin bases are admirably suited to this form of locomotion.

Many of these fishes lay their eggs encapsulated in a morphologically complex, floating, mucoid mass—often referred to as an "egg raft" or "purple veil" due to its purplish gray or brown coloration—a structure that functions to broadcast a large number of small eggs over great geographic distances, providing for development in relatively productive surface waters. Perhaps more remarkable, a number of antennarioids have evolved an alternative mode of recruitment that involves parental retention and care of a relatively small number of large eggs that hatch into relatively large, advanced young. Some carry their eggs on the side of their body, attached by a complex system of acellular membranes, while others carry them dangling from rays of the dorsal fin. Still others hold and protect a cluster of eggs completely hidden from view within a pocket formed by the body and pectoral, dorsal, and caudal fins.

Far from "colorless," frogfishes are among the most colorful of nature's productions; the images in this volume attest to their brilliance and their complex color patterns. Not only do frogfishes exist in nearly every imaginable color and color pattern, but most have the ability to completely alter their hue and pattern in a matter of seconds or days.

Frogfishes are superb subjects for further enquiry: fascinating aquarium fishes, easily maintained; ideal laboratory animals for scientific investigations of all kinds. They deserve wide attention; and to all who might be still in doubt, we recommend you read on.

Historical Perspective

The history of our knowledge of antennariid fishes begins sometime prior to the year 1630, when an unknown Dutchman supplied Johannes de Laet (Fig. 2), a director of the Dutch West India Company, with an illustration of a frogfish from Maranhão, northeastern Brazil (Whitehead, 1979: 437). A woodcut made from this drawing was first published under the name *Pira Vtoewah, forma monstrosa* (representing a species now known as *Antennarius multiocellatus*) in de Laet's second, Latin edition of *Novus Orbus, seu Descriptionis Indiae Occidentalis* of 1633 (Cuvier, 1828: 41; Pietsch, 1995a: 47). This woodcut (reproduced on the title page of this volume) reappeared in a 1640 French edition of the

Fig. 2. Johannes de Laet (1581–1649). Born at Antwerp, he enrolled in the University of Leiden in 1597 as a student of philosophy, defending his thesis in 1599. In 1606 he again enrolled at Leiden, this time as a student of divinity, and until 1621 his concern was primarily with theology. When, in that same year, he became one of the 19 directors of the Dutch West India Company, his chief occupations, in addition to his business concerns, were geography, history, and language. De Laet was a prolific writer, and many of his publications were connected with his work for the Company (Bekkers, 1970: xv). Of greatest interest to natural historians are his *Nieuwe Wereldt ofte Beschrijvinghe van West-Indien* (New World; or, Description of the West Indies), originally published in Dutch in 1625, later translated into Latin (1633) and French (1640); and his editing of William "Piso's De Medicina Brasiliensicis" and George Marcgrave's "Historia Naturalis Brasiliae," published together in 1648. Engraving by Jan Gerritsz van Bronckhorst; courtesy of Bart Lar and the Collection het Scheepvaartmuseum, Amsterdam.

same work, and later, with a detailed description, as *Guaperua Brasiliensibus*, in George Marcgrave's "Historia Naturalis Brasiliae" of 1648. Until well into the eighteenth century, this remained the only published figure of an antennariid, and like many of the accounts and illustrations in Marcgrave's (1648) "Historia," it was reproduced by numerous subsequent authors (e.g., Jonston, 1650, and many later editions; Willughby, 1686; Petiver, 1702; and Ray, 1713; see synonymy of *Antennarius multiocellatus*, p. 131). Of special interest is a previously unpublished sketch labeled *Guapena*, made by Linnaeus sometime between August 1727 and the beginning of 1730 and clearly modeled after Marcgrave (Fig. 3A).

Somewhat later, and on the other side of the world, another Dutchman, Captain Willem de Vlamingh (1640–1698), and his crew aboard the frigate *Geelvinck* were searching for survivors of a shipwreck off the coast of Western Australia (Pietsch, 1995b, 1: 57; Holthuis and Pietsch, 2006: 33). On the morning of 29 December 1696, amid discoveries of rats as big as common house cats (wallabies, *Notamacropus eugenii*) and tracks of tigers and other ferocious beasts (most probably dingoes, *Canis lupus*), there was observed a remarkable fish, "about two feet long, with a round head and a sort of arms and legs and even something like hands" (Major, 1859: 120; see also Whitley, 1970: 18). No doubt this was a frogfish, and as far as we know this reference is the earliest published mention of an antennariid from the Indo-Pacific region. Judging from the large size it must have been the Giant Frogfish, *Antennarius commerson* (see p. 114).

In 1719, Louis Renard (1678/79–1746), a French book dealer, publisher, and agent to the British Crown, published his *Poissons, Ecrevisses et Crabes* (with subsequent editions in 1754 and 1782), a collection of extraordinary colored drawings of Indo-West Pacific

A

Gvapena.

Natando ventrem inflat ut pila.

Sambia. Loop-visch, ou Poisson courant d'Amboine. Je l'ay atrapé sur le Sable et l'ay gardé trois jours en vie dans ma maison comme un petit chien qui me suivoit par tout fort familiere-ment. M.ʳ Scott en a un à Amsterdam dans l'esprit de vin.

B

Fig. 3. Early representations of frogfishes. (**A**) A rough copy of George Marcgrave's (1648) figure of *Guaperua* made by Carl Linnaeus; the caption reads, "While swimming it blows up its belly like a ball." Folio 166, in *Caroli Linnaei Manuscripta Medica Tom. 1,* dating from August 1727 to the beginning of 1730; courtesy of Isabelle Charmantier, Andrea Deneau, and the Linnean Society of London. (**B**) *Sambia*, the Walking-fish or Common Fish of Ambon. One of hundreds of paintings made in the late 1690s and very early 1700s by Samuel Fallours, an artist in the employ of the Dutch East India Company at Ambon, Indonesia. This bizarre depiction of an antennariid, along with numerous other, equally fantastical illustrations of fishes, crustaceans, grasshoppers, a dugong, and a mermaid, all said to have been modeled by Fallours exactly from nature, were later published by Louis Renard in his *Poissons, Ecrevisses et Crabes* of 1719 (vol. 2, fig. 33; see Pietsch, 1984a, 1995). Fallours's legend to his illustration reads: "I caught it on the sand and kept it alive in my house for three days; it followed me everywhere with great familiarity, much like a little dog. Mr. Scott in Amsterdam has one preserved in spirits of wine."

fishes and other marine organisms produced in the Dutch East Indies during the late seventeenth century (Holthuis, 1959; Pietsch, 1984a, 1995b). Among these drawings are three figures of frogfishes, one called *Kikvorst*, the "Frog of Manipe" (Part 2, pl. 38, fig. 171) and two labeled *Sambia* or *Loop-visch*, the latter name literally meaning "walking fish" (Part 1, pl. 43, fig. 212; and Part 2, pl. 7, fig. 33; Fig. 3B). Although these drawings are impossible to identify to species, they represent the earliest published figures of Indo-Pacific antennariids. Many of the figures in Renard's (1719) book, including *Kikvorst* and *Sambia*, also appeared in the "Collectio Nova Piscium Amboinensium" of Henrici Ruysch (1718), and in the third volume of François Valentijn's (1726) *Oud-en Nieuw Oost-Indiën* (see Pietsch, 1995b).

The next notable contribution to frogfish biology was that of Albertus Seba (Fig. 4), published in the first volume of his *Locupletissimi rerum Naturalium Thesauri* of 1734. Like many of his contemporaries, Seba believed that antennariids were anuran amphibians (or at least, tadpole-like fish stages in frog development). Accordingly, he classified them with the true frogs and referred to them as *Rana piscatrix*—Latin for "fishing frog," a name that also implied his knowledge of the unique feeding mode of these animals. Although rather crudely done, Seba's figures (Fig. 5) clearly indicate two species, *Antennarius scaber*

Fig. 4. Albertus Seba (1665–1736) at the age of 66 years. Born in Etzel, Ostfriesland, northwest Germany, he settled in Amsterdam after 1696 and became a prominent citizen of that town. As a pharmacist he amassed a considerable fortune and devoted much of his time and money to his interests in natural history. Seba brought together a remarkably fine cabinet of natural curiosities that became world famous. Much of his collection was described and figured in his magnificent four-volume publication *Locupletissimi rerum Naturalium Thesauri* (1734–1765), commonly referred to as the "Thesaurus." For details of his life and work, see Engel (1937) and Holthuis (1969). Engraving by Jacobus Houbraken; reproduced from the frontispiece of Volume 1 of the "Thesaurus," published in 1734.

Fig. 5. The earliest published images of two species of frogfishes: Albertus Seba's *Rana Piscatrix*. His figure 3 represents *Antennarius striatus* (Shaw, 1794); figures 4–6 show *Histrio histrio* (Linnaeus, 1758). Figure 7, labeled as juveniles (i.e., *Pullus Ranae Piscatricis, quartae*), is a composite: the two examples on the right represent *Histrio histrio*, but the three on the left depict *Scyllaea pelagica*, a nudibranch that inhabits floating *Sargassum*. After Seba, *Locupletissimi rerum Naturalium Thesauri* (1734, pl. 74).

(his pl. 74, fig. 3) and *Histrio histrio* (his figs. 4–6, and two small drawings on the far right labeled fig. 7). However, three of the five drawings of his figure 7 (those on the far left), indicated by Seba as the young of *Histrio histrio* (i.e., *Pullus Ranae piscatricis, quartae*), show no distinct detail and are quite unidentifiable as fishes. It was Linnaeus (1754b: 56; see also Osbeck, 1757: 306, and Linnaeus, 1758: 656) who realized that these three drawings depict *Scyllaea pelagica* (referred to as *Lepus pelagicus*), a nudibranch syntopic with *Histrio histrio*

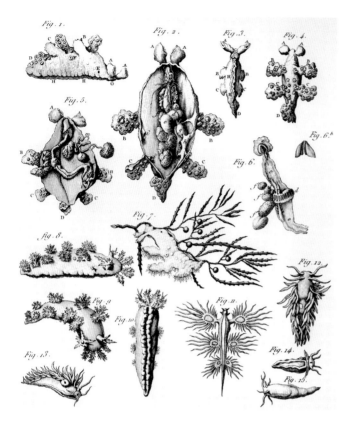

Fig. 6. *Scyllaea pelagica* and its anatomy. Turned upside-down with its gills in a ventral position, this nudibranch, endemic to floating *Sargassum* around the world, was mistaken by Albertus Seba (1734) for the young of a frogfish. After Cuvier (1817a, figs. 1–7).

in floating *Sargassum* weed. Thus, it seems that Seba felt obliged to discover a kind of metamorphosis in his *Rana piscatrix* similar to that of common frogs, in which a mollusk becomes a frogfish (Valenciennes, 1837: 391). "With this singular idea in mind, [Seba] presented a drawing of it [*Scyllaea pelagica*] upside down so that the gills [normally dorsal in position] are found in about the same place as the fins of the fish" (Cuvier, 1817a: 1). This error is not so difficult to comprehend when one compares the drawings of Seba with the illustrations of *Scyllaea pelagica* published by Cuvier (1817a) (Figs. 5, 6).

Several authors (e.g., Cuvier, 1817b: 422; Valenciennes, 1837: 392; Swainson, 1838: 203; Gill, 1879c: 225) have written of the confusion caused by Linnaeus (1758: 237; Fig. 7) when he combined all previous descriptions of antennariids under the single name *Lophius histrio*. A review of Linnaeus's sources (Wheeler, 1979), however, shows that no great nomenclatural problems exist. Only two species are hidden in the Linnaean synonymy. That he clearly had *Histrio histrio* in mind is evidenced not only by his mention of its existence in *Sargassum* (*Habitat in Pelago inter Fucos natantem*), but also by the fact that the only extant Linnaean speci-

Fig. 7. Carolus Linnaeus (1707–1778), known after his ennoblement in 1757 as Carl von Linné, was a Swedish naturalist and physician, who established binomial nomenclature, the modern system of naming organisms, for which he earned the epithet "father of modern taxonomy." Born in the countryside of Småland, in southern Sweden, he received most of his higher education at Uppsala University and began giving lectures in botany there in 1730. He lived in the Netherlands between 1735 and 1738, where he studied and published, among other things, the first edition of his *Systema naturae* (1735). He then returned to Sweden, where he became professor of medicine and botany at Uppsala. In the 1740s, he went on several journeys through Sweden to find and classify plants and animals. In the 1750s and 1760s, he continued to collect and classify animals, plants, and minerals and published a great number of volumes. At the time of his death, he was one of the most acclaimed scientists in Europe. For a full biography, see Blunt, 2001. Portrait by Alexander Roslin, 1775, original in the National Museum, Stockholm, Sweden.

Fig. 8. Philibert Commerson (1727–1773). Born at Châtillon-les-Dombes, Ain, France, he went to Montpellier in 1747, where he received a complete medical education, obtaining his bachelor's degree in 1753 and his doctorate and license to practice medicine in 1754. Primarily interested in botany, he established a notable herbarium. At the request of Linnaeus, he described the ichthyofauna of the Mediterranean; never published, this work was later used by Bernard-Germain-Étienne de La Ville-sur-Illon, comte de Lacepède for his *Histoire naturelle des Poissons* (1798–1803). Following his service as naturalist on Louis-Antoine, Comte de Bougainville's circumnavigation of the globe (1766–1769), Commerson did not return to France but remained at Mauritius, where he died virtually alone. Eight days after his death, the French Academy of Sciences, unaware of his passing, elected him associate botanist. For details of Commerson's life and work, see Oliver (1909).

mens of this family are *H. histrio* (for a more detailed discussion of the taxonomic history of *H. histrio*, see Comments, p. 192).

The first serious study of antennariid fishes was made by the celebrated French naturalist and doctor of medicine Philibert Commerson (Fig. 8). During his service as botanist and naturalist to the king, on Louis Antoine de Bougainville's *Voyage Autour du Monde* (1766–1769), Commerson amassed huge collections of natural objects, and made extensive notes and drawings. Among his surviving manuscripts (Commerson, MSS 889, 891) are Latin descriptions of four species of *Antennarius* that he observed at Mauritius (*in vado littorum nesogalliae*) soon after his arrival on that island in mid-November 1768 (Bougainville, 1772: 389). Although all are named solely by lengthy Latin polynomials (that include the earliest use of the name *Antennarius*), at least two of his species (to be known later as *Antennarius striatus* and *A. commerson*) are precisely described in great detail, and three of the four are represented by illustrations that leave no doubt about their identity (Fig. 9). Unfortunately, however, Commerson did not live to publish his work. Soon after his death at Mauritius in March 1773, his papers, together with most of his collections and numerous drawings of plants and animals, were sent to France, where they were passed on to Georges-Louis Leclerc, Comte de Buffon (1707–1788) and later deposited at the famous Jardin des Plantes in Paris (now in the Bibliothèque Centrale, Muséum national d'Histoire naturelle, Paris). The fishes collected by Commerson, however, were unaccounted for until André Marie Constant Duméril (1774–1860) discovered them still in their original cases in an attic of Buffon's house some twenty years later (Cuvier, 1845: 95; Oliver, 1909: 219). If the material on which Commerson based his descriptions of frogfishes was contained in these collections, none of it has survived.

Commerson's talent as an ichthyologist might have been lost if it were not for the fact that his manuscripts and drawings were later incorporated by Bernard-Germain-Étienne de La Ville-sur-Illon, Comte de Lacepède (Fig. 10) in his *Histoire Naturelle des Poissons*, originally published in five volumes between 1798 and 1803. The frogfish descriptions, together with engravings made from three of the four drawings (and the polynomials of Commerson quoted in footnotes), appeared in Volume 1 (Lacepède, 1798: 321–329; Fig. 11). Despite the fact that many authors have cited Lacepède as the original describer

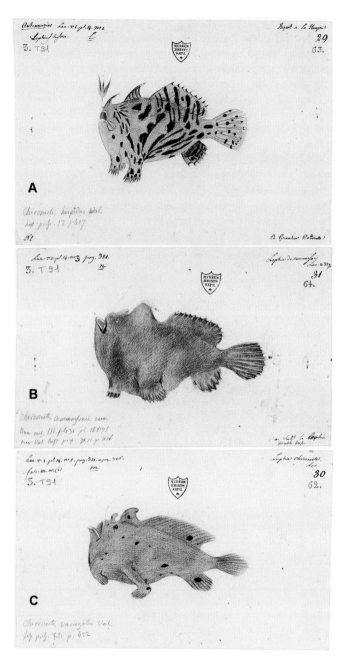

Fig. 9. Frogfishes from the manuscripts of Philibert Commerson, based on material observed by him at Mauritius in 1769 and 1770. (**A**) *Antennarius striatus* (Shaw), *Antennarius antenna tricorni* of Commerson; (**B**) *Antennarius commerson* (Latreille), *Antennarius bivertex, totus ater, puncto mediorum laterum albo* of Commerson; (**C**) *Antennarius pictus* (Shaw), *Antennarius chironectes, obscure rubens, maculis nigris raris inspersus* of Commerson. Drawings by Pierre Sonnerat, Commerson MSS 889, 891, Bibliothèque Centrale, Muséum national d'Histoire naturelle, Paris; courtesy of Joyce Faust, © Art Resource, Inc., New York.

of these species (and have erroneously attributed the name *Antennarius* to him), he provided no Latin binomials, referring to the four species only in the French vernacular (Anon., 1910, 1925).

A few descriptions of antennariids appeared just before and soon after the turn of the eighteenth century. Notable among these were the original descriptions of *Antennarius striatus* and *A. pictus* by George Shaw (1794; Fig. 12), based in part on material discovered at Tahiti by Sir Joseph Banks during the first voyage of Captain James Cook aboard the H.M.S. *Endeavour* (1768–1771), and the original descriptions of *A. hispidus* and *A. ocellatus* published by Marcus Elieser Bloch and Johann Gottlob Schneider (1801). However, a thorough critical review of the frogfishes did not appear until the works of Georges Cuvier (Fig. 13) and his pupil and successor, Achille Valenciennes (Fig. 14). With the first edition of his *Règne Animal Distribué d'après son Organisation* (1816), and "Sur le Genre CHIRONECTES Cuv. (*Antennarius*. Commers.)" (1817b), Cuvier set the stage for a full treatment of the group, published some 20 years later by Valenciennes (1837) in the twelfth volume of the *Histoire Naturelle des Poissons* (see Pietsch, 1985a). In the latter work, Valenciennes recognized 22 species (excluding two species now placed in the antennarioid family Brachionichthyidae; see Evolutionary Relationships, p. 368). Of these 22 species, seven were previously named by Cuvier (1816, 1817b, 1829), and nine were newly described by Valenciennes (1837).

Although only 11 of these forms are recognized today, the contributions of Cuvier and Valenciennes were by far the most important up until this time. They provided the first concise anatomical definition of the group, clearly separating antennarioids from lophioids. They not only included descriptions and figures (Fig. 15) that far exceeded the quality of those of their predecessors, but they also brought together everything possible concerning

Fig. 10. Bernard-Germain-Étienne de La Ville-sur-Illon, comte de Lacepède (1756–1825). Born at Agen and largely educated at home by his father, he took an early interest in natural history while devoting much of his leisure time to music, becoming a good performer on the piano and organ and acquiring considerable mastery of composition. Friendship with Georges-Louis Leclerc, Comte de Buffon (1707–1788) led to his appointment in 1785 as demonstrator at the Jardin du Roi in Paris and a recommendation that he continue Buffon's encyclopedic *Histoire naturelle, générale et particulière*. Subsequent volumes of this great work authored by Lacepède were published under the title *Histoire des quadrupèdes ovipares et des serpens* (1788–1789). In 1798, he published the first volume of *Histoire naturelle des poissons*; the fifth and final volume appeared in 1803. For more on Lacepède, see Adler (1989: 14) and Bornbusch (1989).

the geographic distribution and life history of the antennariids. And because they were among the earliest to realize the importance of maintaining material for future study, many of their species are still represented by type specimens remarkably well-preserved in the collections of the Muséum National d'Histoire Naturelle in Paris. Through such efforts, Cuvier and Valenciennes established a firm basis for all future work on frogfish biology.

With the authoritative review of Valenciennes (1837) in wide circulation, little of significance was published on frogfishes for the next two decades. One exception was the work of Felipe Poey y Aloy (1799–1891), a student of Cuvier who devoted himself solely to the fish fauna of Cuba. Over a 30-year period, beginning in the early 1850s, Poey published a number of papers that contained information on Western Atlantic frogfishes. Of particular significance was his *Quironectos Cubanos* (1853), his 412-page *Synopsis Piscium Cubensium* (1865), and Enumeratio Piscium Cubensium (1876).

In his attempt to identify and classify the relatively large number of frogfishes then present in the collections of the British Museum, Albert Günther (1861a: 184; Fig. 16) wrote:

> The great variability to which . . . [frogfishes] are subject . . . [has given naturalists] ample opportunity of affixing some character to . . . fictitious species. This variability is so great, that scarcely two specimens will be found which are exactly alike; [without the opportunity to examine comparative material] it would have been too hazardous to treat them as merely synonyms, and therefore I have preferred to admit them into the list of species, although I have not the slightest doubt that more than one-half of them will prove to be individual varieties.

Thus, Günther (1861a) was one of the first to realize the problems inherent in frogfish systematics, yet he and others like him, without the means to assess morphological variability, added substantially to the number of nominal species. Thirty are listed (38, if one counts the names in footnotes, pp. 183, 184), along with brief descriptive information, localities, and records of specimens present in the British Museum collections. Günther was certainly correct in his prediction that more than half of these forms would prove to be merely varieties; only nine of his names represent species recognized today.

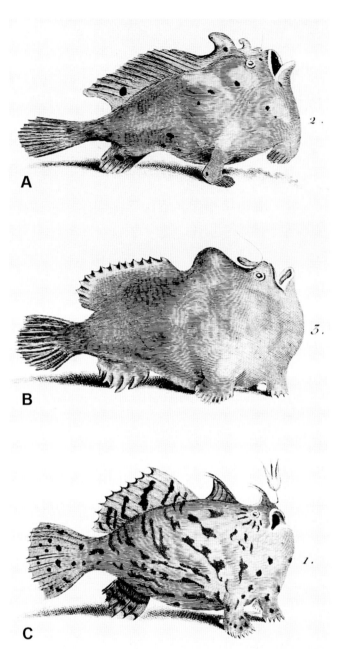

Fig. 11. The frogfishes of Bernard-Germain-Étienne de La Ville-sur-Illon, comte de Lacepède based on the manuscripts of Philibert Commerson. (**A**) *Antennarius pictus* (*La Lophie chironecte* of Lacepède); (**B**) *Antennarius commerson* (*La Lophie commerson* of Lacepède); (**C**) *Antennarius striatus* (*La Lophie histrion* of Lacepède). After Lacepède (1798, pl. 14).

Pieter Bleeker (Fig. 17), a surgeon in the service of the Dutch East India Army, and unsurpassed in the early exploration of the fish fauna of the Indo-West Pacific, published no fewer than 24 papers dealing with frogfishes (see References, p. 525). These studies culminated in his review of the "Antennarii" that appeared in Volume 5 of his great *Atlas Ichthyologique des Indes Orientales Néêrlandaises* published in 1865. In this work, beautifully illustrated with large color plates (many of which are reproduced in this volume; Fig. 18), Bleeker recognized 24 East-Indian species, eight times the number previously recorded from this geographic area. Yet he remained unsatisfied: "I do not doubt that subsequent research may double the number of the already 24 known species, and I think that the twelve new species, that happy circumstances have permitted me to introduce to science, will be followed by many more" (Bleeker, 1865b: 8). But because he was working essentially alone, without the aid of a previously identified reference collection and necessary ichthyological literature, particularly the works of Cuvier and Valenciennes, his publications suffered. Most serious was his propensity for giving to every individual a new name (Günther, 1880: 30), the result among frogfishes being that only six of the 24 species he recognized are considered valid today, and of these six, only one (*Abantennarius dorehensis*) was originally described by him.

Subsequent to the major works of Günther (1861a) and Bleeker (1865b), and prior to the worldwide revisionary study of Leonard P. Schultz (1957), a number of important contributions appeared. Among these were the publications of Theodore Nicholas Gill (Fig. 19). Gill described four new species in 1863, but more important was a series of notes published in 1879 that provided concise definitions of the then known genera and families of lophiiform fishes, as well as a hierarchical classification, with synonymies and keys to all higher taxa. Finally, his "Angler Fishes: Their Kinds and Ways" of 1909 summarized all that was then known about anglerfish natural history. David Starr Jordan (Fig. 20), either alone or with collaborators,

A

B

Fig. 12. Original drawings from the Thomas Watling Collection held by the Zoological Library, Natural History Museum, London, used as models for the engravings that accompanied the original descriptions of (**A**) *Antennarius striatus* and (**B**) *Antennarius pictus* published by George Shaw (1794, pls. 175, 176, upper fig.). Drawings by Frederick Polydore Nodder; courtesy of Stephan Atkinson and the Picture Library, Natural History Museum, London.

Fig. 13. Jean Léopold Nicolas Frédéric Cuvier (1769–1832), better known by his adopted but legally inaccurate name Georges Cuvier, was a colossal figure in biology during the first quarter of the nineteenth century. His primary scientific contributions were in comparative anatomy, zoological classification, and the study of the fossil record. After the completion of the *Règne animal, distribué d'après son organisation* of 1816, a publication said by Jordan (1905: 400) to be of no less importance than the *Systema naturae* itself, Cuvier directed his attention almost entirely to ichthyology. Nearly all that was known about fishes up to that time was summarized by Cuvier, together with his pupil and successor Achille Valenciennes (see Fig. 14), in the monumental *Histoire Naturelle des Poissons*. Consisting of 22 volumes published between 1828 and 1849, the "Natural History" contains descriptions of 4,514 nominal species, about two-thirds written by Valenciennes after the death of Cuvier in 1832 (Gill, 1872; Monod, 1963; Pietsch, 1985a). For details of Cuvier's life and work, see Jardine (1846); for an assessment of the major features of Cuvier's zoological theories and practice, see Coleman (1964). Portrait by Mathieu Ignace van Brée, 1798; courtesy of Emmanuelle Choiseau, © Muséum national d'Histoire naturelle, Paris.

Fig. 14. Achille Valenciennes (1794–1865), protégé of Georges Cuvier and Alexander von Humboldt. Born in the Muséum d'Histoire naturelle in Paris, he spent his entire life associated with that institution, starting out as a *préparateur* in 1812, becoming an *aide-naturaliste* associated with the chair of reptiles and fishes held by Lacepède, and in 1832 gaining the chair of annelids, mollusks, and zoophytes, succeeding Henri-Marie Ducrotay de Blainville (1777–1850). His major scientific achievement was his collaboration with Cuvier on the classic work *Histoire Naturelle des Poissons* (1828–1849). Although best known as an ichthyologist, he also published a number of papers on reptiles, mollusks, and zoophytes; a large work on sponges was planned but never completed. For a complete listing of the ichthyological contributions of Valenciennes, see Blanc (1963); for additional biographical material, see Hallez (1866), Milne-Edwards (1867), and Monod (1963).

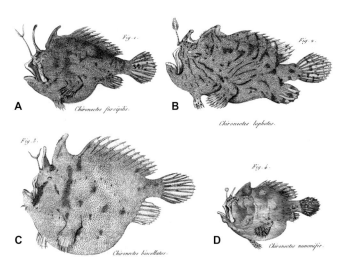

Fig. 15. Frogfishes from Georges Cuvier's 1817 review of the genus *Chironectes*. (**A**) *Kuiterichthys furcipilis* (Cuvier, 1817b), holotype, 63 mm SL, MNHN A.4618. (**B**) *Antennarius hispidus* (Bloch and Schneider, 1801), holotype of *Chironectes lophotes* Cuvier, 1817b, specimen lost. (**C**) *Antennarius biocellatus* (Cuvier, 1817b), holotype, 84 mm SL, MNHN A.4620. (**D**) *Abantennarius nummifer* (Cuvier, 1817b), holotype, 48 mm SL, MNHN A. 181. After Cuvier (1817b, pl. 17).

Fig. 16. Albert Carl Ludwig Gotthilth Günther (1830–1914). Born at Mohringen near Stuttgart, Württemberg, in southern Germany, he was educated at the University of Tübingen. There under the direction of the physician Wilhelm von Rapp (1794–1868) he produced a thesis, *Die Neckar-Fische* (later published as *Die Fische des Neckars*; see Günther, 1853), thus beginning a lifelong systematic study of fishes that placed him among the greatest of ichthyologists. In 1857 he emigrated to England, where he found almost immediate but temporary employment as a cataloger in zoology at the British Museum. He was appointed to a permanent post in 1862 and became Keeper of Zoology in 1875, holding that position until his retirement in 1895. Günther's greatest contribution to ichthyology is his *Catalogue of the fishes of the British Museum*, the last attempt by anyone to cover the entire fish fauna of the world. This monumental work, published in eight volumes between 1859 and 1870 and containing descriptions of 6,843 species, provided an immediate stimulus to the study of fishes and became the basic authority on systematic ichthyology for many years (Jordan, 1905: 402). For more on the life of Günther and other "founders of science" at the British Museum, see Albert E. Günther (1980).

Fig. 17. Pieter Bleeker (1819–1878). Born at Zaandam, Holland, he left for Batavia (Jakarta) in 1842, where over the next 30 years he assembled an immense collection of fishes from Java, Sumatra, and the Molucca Islands. These were described in some 500 papers, culminating in his beautifully illustrated, nine-volume magnum opus, the *Atlas Ichthyologique* (1862–1878). For details of Bleeker's life and an index of his ichthyological papers, see Weber and de Beaufort (1911); for a more recent account, see Boeseman (1983). Courtesy of Godard Tweehuysen and the Naturalis Biodiversity Center, University of Amsterdam.

published between 1883 and 1925 some 15 papers containing information on frogfishes. Of these, the most significant were the section on anglerfishes in Volume 3 of *Fishes of North and Middle America*, by Jordan and Barton Warren Evermann (1898), and "A Review of the Pediculate Fishes or Anglers of Japan," by Jordan (1902). Finally, of significance mainly in the quantity of their publications referring to antennariids, are Henry Weed Fowler (1878–1965), with no less than 18 such works appearing between the years 1903 and 1959, and Gilbert Percy Whitley (1903–1975), with 25 papers published between 1927 and 1970.

Although Whitley certainly must be given some credit, our present-day knowledge of Australian frogfishes is due primarily to the studies of James Douglas Ogilby (1853–1925) and Allan Riverstone McCulloch (Fig. 21). Ogilby's most important contribution was "Some New Pediculate Fishes" of 1907, in which he described two new genera and three new species, and provided one of the earliest keys to the antennariids and brachionichthyids of Australasia. But more significant was a paper by McCulloch and Edgar Ravenswood Waite (1866–1928) in 1918. In addition to descriptions of two new genera and species, these authors provided a revised key to genera, and the first accurate illustrations of Australian frogfishes (some of which are reproduced in this volume); of the five species recognized by McCulloch and Waite (1918), all are considered valid today.

The South African frogfish fauna is known to us almost solely through the studies of James Leonard Brierly Smith (Fig. 22). In *The Sea Fishes of Southern Africa*, first published in 1949, Smith gave brief descriptions and color illustrations of 10 species (Fig. 23). Although the Latinized names applied were accurate for only five species, he correctly hypothesized the existence of at least seven South African antennariids (see Smith and Heemstra,

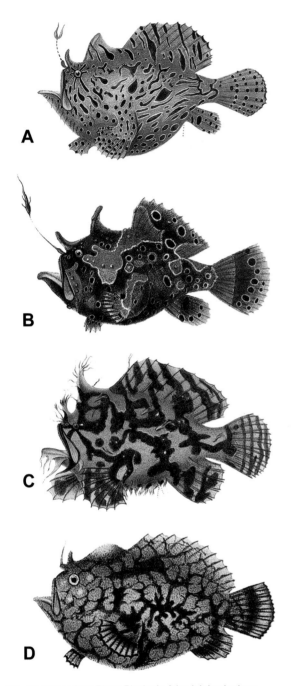

Fig. 18. Frogfishes from Pieter Bleeker's *Atlas Ichthyologique*.
(**A**) *Antennarius striatus* (Shaw, 1794), holotype of *Antennarius lacepedii* Bleeker, RMNH 28017, 80 mm SL, Ambon, Indonesia (after Bleeker, 1865b, pl. 197, fig. 5). (**B**) *Antennarius maculatus* (Desjardins, 1840), *Antennarius phymatodes* of Bleeker, Ambon, Indonesia (after Bleeker, 1865b, pl. 197, fig. 1). (**C**) *Histrio histrio* (Linnaeus, 1758), holotype of *Antennarius lioderma* Bleeker, RMNH 6289, 82 mm SL, Ambon, Indonesia (after Bleeker, 1865b, pl. 199, fig. 8). (**D**) *Antennatus tuberosus* (Cuvier, 1817b), *Antennarius bigibbus* of Bleeker, Ambon, Indonesia (after Bleeker, 1865b, pl. 199, fig. 3). Drawings by Ludwig Speigler under the direction of Pieter Bleeker.

1986). In his "Fishes of Aldabra, Part X," published in 1958, Smith was one of the very few to correctly apply the specific name *tuberosus* (Cuvier, 1817b), realizing that Lacepède (1798: 325) used the name *Antennarius bigibbus* not as a binomen, but only as a footnote to his *La Lophie Double-Bosse* quoted from the manuscripts of Commerson (see Comments, p. 256).

As the only modern attempt to revise the family Antennariidae on a worldwide basis, the monograph of Leonard Peter Schultz (Fig. 24), published in 1957, is as significant as the work of Achille Valenciennes that appeared 120 years earlier. Yet, seen in a contemporary light, Schultz's contribution did little to rectify the tremendous confusion that by then had made it nearly impossible for even the most skilled ichthyologist to correctly identify a frogfish. The problem was chiefly that Schultz lacked the opportunity to examine enough material. His review was based on collections then deposited in only two institutions—the National Museum of Natural History in Washington, DC, and the Field Museum of Natural History, Chicago. He distinguished 13 genera (a different assemblage of taxa than recognized here), three of which were accepted solely on the basis of the literature, two on the examination of a single specimen, and three others on the examination of four specimens or fewer. His *Kanazawaichthys* was later shown to be the prejuvenile form of *Fowlerichthys radiosus* (see Comments, p. 49). No osteological comparisons were made and, at best, his genera are differentiated on superficial characters. At least half are distinguished by combinations of characters that may be found within the 23 species that he assigned to the genus *Antennarius*.

Of the 131 nominal species cited in his paper (185 are cited in this monograph), Schultz recognized 58; of these, material for 15 was unavailable for examination, descriptions of 14 were based on only one or two specimens each, and type material

Fig. 19. Theodore Nicholas Gill (1837–1914), described by David Starr Jordan in the dedication of his *Guide to the Study of Fishes* (1905) as "Ichthyologist, Philosopher, Critic, Master in Taxonomy" and, later in the same volume (p. 405), "the keenest interpreter of taxonomic facts yet known in the history of ichthyology." A prolific writer, Gill published over 500 scientific papers across some 60 years (1853–1912). Although his papers were rarely long and included no great monographs, their influence, especially on the classification of fishes, has been profound. For details of Gill's life and a bibliography of his most significant contributions to systematic zoology and ecology, see Dall (1916).

Fig. 20. David Starr Jordan (1851–1931). Self-described in the subtitle of *Days of a Man* (1922), as a "Naturalist, Teacher, and Minor Prophet of Democracy," Jordan can easily be called the greatest of American ichthyologists. Born in Gainesville, New York, and educated at Cornell University, he began his professional career in 1872 as Instructor of Botany at Lombard University, Galesburg, Illinois. By 1885, at the age of 34, he was President of Indiana University, and six years later he became President of Stanford, a position he held until his appointment as Chancellor of that university in 1913. Jordan was a prolific author on a wide variety of subjects. Alice N. Hays, in her bibliography of Jordan's writings (1952), lists 2,017 titles, 645 of which were ichthyological. But perhaps more significant than his publications was his extraordinary ability as a teacher. Carl Hubbs (1964b: 198) wrote that a "major secret of Jordan's greatness in ichthyology was his ability to inspire others to work in the same field." To his students and biological colleagues, most of whom worked on fishes, he left a legacy that was passed on for several generations and is still felt today. Most of the more active ichthyologists of today are able to trace their lineage back to Jordan. For a rich and thorough documentation of Jordan's life, see his two-volume autobiography *The Days of a Man* (1922); more concise accounts are provided by Myers (1951) and Hubbs (1964a, 1964b).

was unavailable for 43. The characters used to separate these forms are largely unconvincing or nonexistent. For example, within the nominal genus *Phrynelox* (a synonym of *Antennarius*), Schultz recognized nine species; yet, except for what prove to be imagined differences in the lengths of the first and second dorsal-fin spines and in pectoral fin-ray counts, all are described with an identical combination of features (two are here considered synonyms of *A. scaber* and seven are *A. striatus*). Much more could be criticized—such as his impossible key and "presumed phylogeny" in which the most derived

Fig. 21. Allan Riverstone McCulloch (1885–1925) in New Guinea, about 1922. Born in Sydney, he became associated at age 13 with the Australian Museum as an unpaid assistant to Edgar Ravenswood Waite (1866–1928), who was then in charge of vertebrates. Even in those early days young McCulloch was distinguished by his enthusiasm, his ability, and his determination to succeed. In 1906, not quite 21, he succeeded Waite as Assistant in Charge of Vertebrates. By that time, and although he was skilled in all branches of zoology (e.g., a recognized authority on decapod crustaceans), he had decided to devote his life to the study of fishes. He had a gift for taxonomic work, and our knowledge of the Australian ichthyofauna was greatly augmented by his many published papers. At the time of McCulloch's death, following several years of poor health apparently brought on by overwork, Jordan is said to have described him as "unquestionably the greatest authority on fish in the southern hemisphere, and one of the eight men in the world who really knew about fish" (Anderson, 1926: 142). For details of his life and a list of his publications, see Anderson (1926). Photo by Frank Hurley; courtesy of Mark McGrouther, Claire Vince, and the Australian Museum, Sydney.

Fig. 22. James Leonard Brierley Smith (1897–1968), following an initial career in chemistry, became one of the most prolific modern-day contributors to ichthyology. Primarily responsible for our present knowledge of the marine ichthyofauna of South Africa and the islands of the western Indian Ocean, J. L. B. Smith rocketed to fame in 1939, through his astonishing announcement of the discovery of the living coelacanth *Latimeria chalumnae*, a species representative of a group of fishes thought to have been extinct for some 70 million years. For more of his life, see his largely autobiographical book *Old Fourlegs: The Story of the Coelacanth* (1956a). Courtesy of Penny Haworth and the South African Institute for Aquatic Biodiversity, Grahamstown, South Africa.

genera are apparently considered to be the most primitive—but it is enough to say that Schultz (1957) did little more than to bring together a portion of the primary published literature on the Antennariidae and to synonymize some of the more obvious nominal species. In summary, of the 58 nominal species recognized by Schultz, only 27 names were accurately applied, and only 34 of the descriptions represent valid species.

Fig. 23. Frogfishes from James Leonard Brierly Smith's *The Sea Fishes of Southern Africa*. Paintings by Margaret Mary Smith; after Smith (1949, pl. 98), courtesy of Penny Haworth and the South African Institute for Aquatic Biodiversity, Grahamstown, South Africa.

Fig. 24. Leonard Peter Schultz (1901–1986). Born in Albion, Michigan, he received his undergraduate education at Albion College in 1924, a master's degree from the University of Michigan in 1926, and his PhD from the University of Washington in 1932. He taught at the University of Michigan between 1925 and 1927 and the University of Washington from 1928 to 1936. At UW, he, along with his students, undertook a vigorous program of collecting, adding to the fish collection during his nine-year tenure an estimated 4,200 lots of fishes, thus establishing a valuable teaching and research resource. In 1936, he accepted a position at the Smithsonian Institution, serving as assistant curator of fishes until 1938, curator-in-charge of the Division of Fishes until 1965, and eventually senior zoologist, a post he held until his retirement in 1968. Between 1927 and 1969, he produced about 214 scientific publications, most of which were on fishes. For more on Schultz, see Springer (1987). Photo courtesy of Lisa Palmer and the Smithsonian Institution, NMNH, Division of Fishes, Washington, DC.

Fig. 25. Lieven Ferdinand de Beaufort (1879–1968) in 1929. De Beaufort was a distinguished Dutch naturalist, Professor of Zoogeography, and Director of the Amsterdam Zoological Museum from 1922 until his retirement in 1949. An accomplished ornithologist, a biological explorer, an able zoogeographer, and a conservationist, he made his greatest scientific contribution in the field of ichthyology. As collaborator with and successor to Max Weber (1852–1937), another distinguished Dutch naturalist, de Beaufort centered his attention on the fishes of the Indo-Australian Archipelago. In 1911, Weber and de Beaufort published the first of seven coauthored volumes of *The Fishes of the Indo-Australian Archipelago*, a monographic treatment of the ichthyofauna of the East Indies that was continued with vigor by de Beaufort after the death of Weber in 1937. The last volume of this magnum opus, Volume 12, containing a revision of the Indo-Australian antennariids, was published in 1962 in collaboration with John C. Briggs (1920–2018). For details of de Beaufort's life and a list of his publications, see Engel et al. (1969). Courtesy of Godard Tweehuysen and the Naturalis Biodiversity Center, University of Amsterdam.

As part of volume XI of *The Fishes of the Indo-Australian Archipelago*, published in 1962, Lieven Ferdinand de Beaufort (Fig. 25) and John Carmon Briggs (1920–2018) reviewed the antennariids of this region with an approach that far exceeded in quality Schultz's (1957) revision. Although based largely on the literature (but in part on examination of collections housed in European institutions), their interpretation of the work published by their predecessors was superior, and their synonymies, descriptions, and distributional data were much more detailed and precise. Although the names used are correct for only five species, 10 of the 12 descriptions given by Beaufort and Briggs (1962) are accurate accounts of valid species.

In yet another attempt to revise the family Antennariidae, Yseult Le Danois (1964) published a monograph of 162 pages illustrated with 76 line drawings, by far the largest work heretofore published on any lophiiform family: *Étude Anatomique et Systématique des Antennaires, de l'Ordre des Pédiculates*. The work is in the style of her earlier studies on tetraodontiform fishes and her later publications on lophiid and chaunacid anglerfishes (Le Danois, 1974, 1979). Our criticisms of her methodology are largely of the sort expressed by James C. Tyler in his 1963 critique of Le Danois's work on the classification of the Tetraodontiformes. Suffice it to say, Le Danois's (1964) "revision" added significantly to the confusion that by this time characterized the taxonomy of the entire family.

Aside from a few faunal works that include descriptions and figures of antennariids, such as Böhlke and Chaplin's (1968) *Fishes of the Bahamas*, Randall's (1968) *Caribbean Reef Fishes*, and Burgess and Axelrod's (1972–1976) series of volumes on *Pacific Marine Fishes* and *Fishes of the Great Barrier Reef*, little of significance was published on the family in the decade following the work of Le Danois (1964). In the mid-1970s, however, as a logical extension of doctoral dissertation research on deep-sea ceratioid anglerfishes, one of us (TWP) undertook a program of studies on the biology of antennariids, initially working alone but soon collaborating with students, first at California State University, Long Beach, and later, after 1977, at the University of Washington in Seattle. With then graduate student David Brian Grobecker, papers were published on aggressive mimicry (Pietsch and Grobecker, 1978, 1981, 1985), high-speed cinematographic evidence for ultrafast feeding (Grobecker and Pietsch, 1979), parental care as an alternative reproductive strategy (Pietsch and Grobecker, 1980), and much later an article for *Scientific American* entitled "Frogfishes: masters of aggressive mimicry" (Pietsch and Grobecker, 1990a). In addition to a number of smaller papers, including accounts for *FAO Species Identification Sheets for Fishery Purposes* (Pietsch, 1978a, 1981a, 1984e) and chapters in various books (*Ontogeny and Systematics of Fishes*, 1984d; *Smiths' Sea Fishes*, 1986a; and *Fishes of the Northeastern Atlantic and Mediterranean*, 1986b), several papers on the osteology and interrelationships of antennariids were published (Pietsch, 1981b, 1984d). The genera of frogfishes were diagnosed and compared in 1984—of the 28 nominal genera and subgenera recognized at the time, 10 were accepted as valid, plus three additional genera were introduced to accommodate previously described species; a key to the genera was provided, as well as comments on evolutionary relationships and geographic distribution. In that same year, new species of the genera *Rhycherus* (Pietsch, 1984c) and *Echinophryne* were described, the latter in collaboration with Rudie H. Kuiter (Pietsch and Kuiter, 1984). A short paper on the functional morphology of the feeding mechanism was published in 1985 (Pietsch, 1985b) and also, in that year, antennariids described and pictured in the manuscript materials for the *Histoire Naturelle des Poissons* were used as sources for understanding the fishes described by Cuvier and Valenciennes (Pietsch, 1985a; see also Pietsch, 1986c). All of this work finally culminated in 1987, with the publication of *Frogfishes of the World: Systematics,*

Zoogeography, and Behavioral Ecology by Stanford University Press—although a final draft of the manuscript was submitted to the Press in early 1984, it did not appear until July 1987.

Following this monographic treatment of the family, the next decade and a half was marked by few but nevertheless significant contributions: new species of *Antennatus* (Ohnishi et al., 1997; Randall and Holcom, 2001) and *Lophiocharon* (Pietsch, 2004a) were published. Descriptions of the first fossil frogfishes appeared: a new species of *Antennarius* from the Miocene of Raz-el-Aïn, Algeria (Carnevale and Pietsch, 2006); and a new genus and species (*Eophryne barbutii*) from the Eocene of Monte Bolca, Italy (Carnevale and Pietsch, 2009b), representing the earliest known skeletal record for the family.

In 2005, Rachel J. Arnold began graduate studies at the University of Washington, eagerly hoping to work on anglerfishes. Her major professor, (TWP), was admittedly not very encouraging, implying that there was not much left to do. How embarrassing in retrospect—he was sorely mistaken! Eventually convincing him that he was wrong, she launched into a systematic review of frogfishes employing a molecular phylogenetic approach. In a master's thesis that could have easily passed as a doctoral dissertation, her work, published in early 2012 (Arnold and Pietsch, 2012), made dramatic and surprising reassessments of how we view frogfish evolution. Along the way, a number of smaller contributions were made: two new species of *Histiophryne* were described in 2011 and 2012 (Arnold and Pietsch, 2011; Arnold, 2012) and a new species of *Kuiterichthys* was described in 2013 (Arnold, 2013).

In late January 2008, we were startled to see photos of a never-before-seen frogfish, sent to us by employees of a sport-dive operation located on Ambon Island, Indonesia. Excited that something important had been discovered, the University of Washington Office of News and Information issued a press release on 2 April 2008 ("New fish has a face even Dale Chihuly could love") that was picked up instantly and circulated for months by news agencies and various online periodicals around the world. Documented by extraordinary images by well-known underwater photographer David J. Hall, and named the "Psychedelic Frogfish," *Histiophryne psychedelica*, a description was published in early 2009 (Pietsch et al., 2009a).

Specimens of an undescribed genus and species from New South Wales, Australia, which had been recognized since 1980 but set aside for lack of material, made their way to us in 2005 and 2009. Named the Red-fingered Frogfish, *Porophryne erythrodactylus*, a description was published in 2014 (Arnold et al., 2014). The "Lembeh Frogfish" or "Ocellated Frogfish," a distinct antennariid, known for many years, especially among members of the dive community, and suggested by some to represent an undescribed species, was identified as *Nudiantennarius subteres* (Pietsch and Arnold, 2017); and, finally, another new species of *Histiophryne*, described in 2018 from Western and South Australia (Arnold and Pietsch, 2018), brings us up to the present.

Though it is often easy to find fault with the research efforts of our ichthyological predecessors, we acknowledge that this monograph is the culmination of nearly 400 years of work by hundreds of others. Unlike the vast majority of these contributors, and as the most recent revisers, we have benefited enormously from easy access to worldwide collections, the best laboratory facilities and libraries, advanced molecular techniques, generous support in the form of research funds, and substantial cooperation from the diving community in providing specimens and photographs. Acknowledging all of these advantages, and that much is yet to be learned about these animals, we hope that what is presented in this volume is as correct and as complete as possible, and that where we have made decisions, they have been as objective as possible.

Structure of This Volume

This volume is divided into eight chapters. Chapter 1 is an introduction devoted largely to a review of the history of our knowledge of the Antennariidae, from its earliest beginnings in the mid-seventeenth century to the present. The approach and procedures used in our study are outlined in Chapter 2. Chapter 3 is a synopsis of the morphological characters that best describe and differentiate the genera and species of the family, including data that set the stage for the systematic accounts that follow. Chapter 4, devoted to detailed systematic accounts of all antennariid taxa, by necessity makes up the largest section of this volume. An annotated synonymy, including references to all pertinent pre-Linnaean literature, is given for each species. This is followed by a list of material examined, diagnosis, description, summary of geographic distribution and depth preference, and, when applicable, comments on geographic variation, ontogenetic change, habitat preference, problems of identification, and nomenclatural difficulties.

In Chapter 5, we analyze evolutionary relationships, incorporating to a large extent the cladistic philosophy and methods of Hennig (1966). Relationships are examined at all levels. We begin with the establishment of monophyly for the Lophiiformes and discuss its phyletic position among other teleost orders. This is followed by an examination of subordinal and familial relationships and relationships of the antennariid genera, both morphological and molecular, resulting in a fully revised classification of the family.

The geographic distribution and zoogeography of the Antennariidae are analyzed in Chapter 6. Here we begin with a detailed description of the horizontal and vertical ranges of all taxa. This is followed by a partial reconstruction of the ecological and historical factors responsible for the present-day distributional patterns shown by the family and its various subunits. To provide some historical explanation, the distributional data are analyzed using the vicariant biogeographic methodology of Croizat (1958, 1964), following Pietsch and Grobecker (1987). Evidence for dispersal is also discussed.

In Chapter 7, under the term "behavioral ecology," we bring together as much as is known concerning the biology of frogfishes. Much of this information is gathered from the existing scientific literature, but a larger portion is the result of an analysis of data obtained through an examination of preserved material, controlled laboratory experimentation performed on live material, and direct field observations of the organisms in their natural habitat. We begin with a detailed discussion of the various aspects of feeding, including aggressive mimicry, aggressive resemblance and foraging-site selection, feeding behavior and biomechanics, and voracity and cannibalism. Next we discuss defensive behavior, inter- and intraspecific aggression, body inflation, venom, and poison. This is followed by a section on locomotion, with analyses of the functional anatomy of the pectoral girdle and fin, benthic "walking" and the use of tetrapod-like gaits, and the mechanism by which frogfishes jet-propel themselves through the water. We end the chapter with discussions of reproduction, ovarian morphology and egg-raft structure, fecundity, sexual size differences, courtship and spawning behavior, hybridization, egg and larval development, parental care, and the alleged nests of *Histrio histrio*.

In Chapter 8, in collaboration with Scott W. Michael, we provide guidance for divers seeking to observe frogfishes in their natural environment and for aquarists on how to best care for frogfishes in captivity. And, finally, we end with a discussion of the need for conservation of these extraordinary animals.

The reference material in the back of this volume includes an appendix titled Reallocation of Nominal Species of Frogfishes, a glossary, references, illustration credits, and an index.

2. Approach and Procedures

The common names we use for frogfishes are those currently in vogue; when no commonly accepted vernacular was available, we either derived names from a translation of the specific name or chose them for some descriptive attribute. We have tried to provide a complete synonymy for each species, with citations for most of the known literature, but we have referred only to material that could be identified with certainty. Specific locality data are provided only for type material and collections that are of historical importance. For most species, we consulted recently collected specimens, but diagnoses and descriptions are based primarily on material listed by Pietsch and Grobecker (1987); additional available material can be easily found by searching various biodiversity databases (e.g., FishNet2, www.fishnet2.net/search.aspx).

The sets of character states used in diagnoses are in nearly all cases fully comparable at any categorical level; for example, diagnoses of genera are consistently comparative, character for character, among genera, and those of species-groups likewise are fully comparable within genera, etc. For most species, descriptive accounts pertain to specimens greater than 30 mm SL; for those few species characterized by individuals that remain relatively small (e.g., *Abantennarius dorehensis* and members of the *Antennarius pauciradiatus* group, with almost 80% of the known material less than 30 and 20 mm SL, respectively), descriptive accounts pertain to specimens greater than 12 mm SL. Descriptions are presented as concisely as possible; superficial, non-informative features are consistently avoided. Features that pertain to all members of a taxon are listed only in the descriptive account of that taxon and are not repeated for each of its subunits. As with the diagnoses, descriptions are always comparative, character for character, among all taxa of equal rank. For questions regarding characters and character states, and particularly their distribution among antennariid taxa, see Chapter 3, "What Makes a Frogfish?" Keys to the identification of genera, species-groups, and species are most applicable to specimens greater than 30 mm SL; for those few species with individuals that remain relatively small (see above), the keys work well for material greater than about 12 mm SL.

Standard length (SL), used throughout (unless indicated otherwise) for metamorphosed material, is recorded to the nearest half millimeter in specimens less than 100 mm SL, and to the nearest millimeter in larger specimens. Total length (TL) is used in describing early life-history stages. The illicium is the first dorsal-fin spine; the esca is the terminal "bait" of the illicium. Illicium length is the distance from the articulation of the pterygiophore of the illicium and the illicial bone to the base of the esca. To ensure accurate fin-ray counts, we often removed skin from the pectoral fins and made incisions to reveal the rays of the dorsal and anal fins. Terminology used in describing the various parts of the angling apparatus follows Bradbury (1967). Definitions of terms used for the different stages of development are those of Martin and Drewry (1978: 3).

In illustrating frogfishes, we have, whenever possible, included images of historical importance—such as the earliest known depictions of species, and drawings of type specimens—as well as the best available color photographs of these animals taken in their natural environment. When necessary, we have resorted to photos taken in aquaria and in the laboratory. While many of the more commonly observed species are represented by multiple images, included to display as wide a range of chromatic variation as possible, others have rarely or, to the best of our knowledge, never been photographed. Reluctantly, in a few cases, we have included low-resolution photos only because nothing better was available—thinking readers would rather have something than nothing at all.

The identification of any individual frogfish may be made by comparing it to the figures and descriptions herein, but more accurately by using the keys, tables, and distributional charts. The diagnoses and descriptions are based in part on a detailed comparative osteological examination of specimens cleared and stained with alizarin red S following the trypsin digestion technique of Taylor (1967). In many cases, we made dissections of uncleared specimens to confirm observations made on cleared and stained material and to determine ontogenetic changes. Bone terminology follows that of Nybelin (1963), Bradbury (1967), Pietsch (1972, 1981b), and Carnevale and Pietsch (2012). All osteological material examined is listed by Pietsch and Grobecker (1987: 22), except for the batrachoidiform and lophiiform material listed in previous studies of the osteology and relationships of lophiiform fishes (Pietsch, 1972, 1974, 1979, 1981b).

In our attempt to resolve the numerous systematic problems encountered in the family, we made an effort to locate most of the available material in collections around the world. In so doing, we were able to examine (and list in this monograph) 4,916 specimens, some species represented by hundreds of archived specimens (e.g., *Abantennarius sanguineus*, 552 specimens examined), and others being extremely rare (e.g., *Antennatus flagellatus* and *Histiophryne psychedelica* known from only two and three individuals, respectively). Our analysis of this material supports the recognition of 52 more or less well-defined species (a 21% increase in the number recognized by Pietsch and Grobecker, 1987), rendering valid about 28% of the total number of nominally described forms. Of the 34 nominal genera and subgenera, we recognize 15 (modifications of Schultz, 1957; Pietsch, 1984b; Pietsch and Grobecker, 1987; Arnold and Pietsch, 2012). The systematic revision is based on material borrowed from, or examined at, the following institutions:

AIM   Auckland Institute and Museum, Auckland, New Zealand
AMNH   American Museum of Natural History, New York
AMS   Australian Museum, Sydney
ANSP   Academy of Natural Sciences, Philadelphia
ASIZP   Academia Sinica Institute of Zoology, Taipei, Taiwan
BMNH   British Museum (Natural History), London, England
BPBM   Bernice P. Bishop Museum, Honolulu, Hawaii
CAS   California Academy of Sciences, San Francisco
CBG   Center for Biodiversity Genomics, University of Guelph, Guelph, Ontario, Canada
CMA   Cabrillo Marine Aquarium, San Pedro, California
CSIRO   Australian National Fish Collection, Commonwealth Scientific Industrial Research Organization, Hobart, Tasmania
CSULB California State University, Long Beach

FAKU   Fisheries Research Station, Kyoto University, Maizuru, Kyoto, Japan

FMNH   Field Museum of Natural History, Chicago, Illinois

FRLM   Fisheries Research Laboratory, Mie University, Shima, Japan

FSBC   Florida Department of Natural Resources, Marine Research Laboratory, St. Petersburg

FSUT   Florida State University, Tallahassee

GCRL   Gulf Coast Research Laboratory, Ocean Springs, Mississippi

GMBL   Grice Marine Biological Laboratory, Charleston, South Carolina

HUJ   Zoological Museum, Hebrew University of Jerusalem, Israel

HUMZ   Laboratory of Marine Zoology, Faculty of Fisheries, Hokkaido University, Hakodate, Hokkaido, Japan

IFAN   Institut Fondamental d'Afrique Noire, Universite de Dakar, Dakar, Senegal

IU   Indiana University, collections now held at California Academy of Sciences, San Francisco

KFRL   Kanudi Fisheries Research Laboratory, Port Moresby, Papua New Guinea

KPM   Kanagawa Prefectural Museum of Natural History, Odawara, Japan

LACM   Natural History Museum of Los Angeles County, Los Angeles, California

MBUCV Instituto de Zoologia Tropical, Universidad Central de Venezuela, Caracas

MCSNV Museo Civico di Storia Naturale, Verona, Italy

MCZ   Museum of Comparative Zoology, Harvard University, Cambridge, Massachusetts

MGPD   Museo di Geologia e Paleontologia, Università di Padova, Padova, Italy

MMF   Museu Municipal do Funchal, Madeira, Portugal

MNHN Museum national d'Histoire naturelle, Paris, France

MNRJ   Museu Nacional, Departmento de Zoologia, Rio de Janeiro, Brazil

MZUSP Museu de Zoologia, Universidad de São Paulo, Brazil

NCIP   Pusat Penelitian dan Pengembangan Oseanologi, Jakarta, Indonesia

NMNZ Museum of New Zealand Te Papa Tongarewa, Wellington, New Zealand

NMV   National Museum of Victoria, Melbourne, Australia

NRMS   Naturhistoriska Riksmuseet, Stockholm, Sweden

NSMT   National Museum of Nature and Science, Tsukuba, Japan

NTM   Northern Territory Museum of Arts and Sciences, Darwin, Australia

NTUM National Taiwan University, Taipei, Republic of China

OMNH Osaka Museum of Natural History, Osaka, Japan

QMB   Queensland Museum, Brisbane, Queensland, Australia

RMNH   Rijksmuseum van Natuurlijke Historic, Leiden, Netherlands

ROM   Royal Ontario Museum, Toronto, Canada

SAIAB   South African Institute for Aquatic Biodiversity, Grahamstown, South Africa (formerly RUSI: Rhodes University, J. L. B. Smith Institute of Ichthyology)

SAM   South African Museum, Cape Town, South Africa

SAMA   South Australian Museum, Adelaide, Australia

SIO   Scripps Institution of Oceanography, University of California, La Jolla

SMF   Forschungsinstitut Senckenberg, Frankfurt am Main, Germany

SU   Stanford University, collections now held at California Academy of Sciences, San Francisco

TAU   Zoological Museum, Tel-Aviv University, Israel

TKPM   Tokushima Prefectural Museum, Tokushima, Japan

TMH	Tasmanian Museum and Art Gallery, Hobart, Australia
TU	Museum of Natural History, Tulane University, Belle Chasse, Louisiana
UA	University of Arizona, Tucson
UCLA	University of California, Los Angeles
UCR	Museo de Zoología, Universidad de Costa Rica, San Jose, Costa Rica
UF	Florida State Museum, University of Florida, Gainesville
UG	Marine Laboratory, University of Guam, Mangilao
UMMZ	Museum of Zoology, University of Michigan, Ann Arbor
URM	University of the Ryukyus Museum, Okinawa, Japan
USNM	National Museum of Natural History, Washington, DC
USPS	Institute for Marine Sciences, University of the South Pacific, Suva, Fiji
UW	Burke Museum of Natural History and Culture, University of Washington, Seattle
WAM	Western Australian Museum, Perth, Australia
ZIM	Zoologisches Institut und Zoologisches Museum der Universität Hamburg, Germany
ZMA	Instituut voor Taxonomische Zoologie, Zoologisch Museum, Universiteit van Amsterdam, Netherlands
ZMB	Zoologisches Museum der Humboldt-Universität, Berlin, Germany
ZMU	Zoologiska Museum, Uppsala, Sweden
ZMUB	Zoologisk Museum, University of Bergen, Norway
ZMUC	Zoologiske Museum, University of Copenhagen, Denmark
ZUEC	Institute de Biologia, Universidad Estadual de Campinas, São Paulo, Brazil

For a description of procedures used to determine evolutionary relationships, see the introductory paragraphs of Chapter 5. The methodology used in the reconstruction of historical factors responsible for the present-day distributional patterns shown by the family Antennariidae and its subunits is outlined in Chapter 6 (see Zoogeography, p. 387). Terminology used in describing frogfish distributions, particularly in assigning names to large-scale areas of endemism, is as follows (taken primarily from Springer, 1982: 8):

Tropical: That portion of the world's oceans on either side of the equator, bounded on the north and south by the average annual surface isotherms for 20° C (Fig. 273).

Subtropical: That portion of the world's oceans outside of the tropics, bounded on the north and south by the average annual surface isotherm for 12° C.

Western Atlantic: Coastal and island localities of the Gulf of Mexico, Caribbean Sea, and the east coast of the Americas.

Eastern Atlantic: Coastal and island localities of Western Europe and Africa, including the islands of the Azores, Ascension, and St. Helena.

Western Indian Ocean: That part of the Indian Ocean between the African coast (including the Red Sea) and 80° E longitude.

Indo-Pacific (not the same as Indo-West Pacific): The Indian Ocean, including contiguous seas, and the Pacific Ocean as far east as Easter Island, but excluding the area occupied by the coast and offshore islands of the Western Hemisphere (Guadalupe, Revillagigedos, Clipperton, Galápagos, San Félix, San Ambrosio, Juan Fernández, etc.).

Eastern Pacific: The eastern portion of the Pacific Ocean excluding the Indo-Pacific.

Western Pacific: The Pacific Ocean west of the western margin of the Pacific Plate, composed mostly of inland seas (South China Sea, Arafura Sea, Coral Sea, etc.).

Indo-West Pacific: Indian Ocean plus the Western Pacific (not the same as Indo-Pacific).

Indo-Australian Archipelago: An area that extends from Taiwan to Australia (including Tasmania), and comprising of all the inland seas and islands of the Philippines, Malaysia, Indonesia, New Guinea, and the Solomons.

Pacific Plate: The largest of all the earth's lithospheric plates, occupying most of the area that has been referred to as the Pacific Basin (see Springer, 1982: 8, figs. 1–4), and forming a major subunit of the Indo-Pacific biogeographic region. In the tropical zone of the world, the western margin of the Pacific Plate is bounded by, but does not include, the Bonin Islands, the Marianas, Guam, Yap, Palau, the Bismarck Archipelago, the Solomons, the New Hebrides, Fiji, and the Lau, Tonga, and Kermadec Islands; its eastern margin is limited by the San Andreas Fault of North America and by the East Pacific Ridge, where it includes the Revillagigedo Islands and Clipperton Island, but not the Galápagos Islands or Easter Island.

Andesite Line: The boundary between the basalts of the oceanic crust and islands, and the volcanic rocks of the circum-Pacific belt (essentially the boundary between the continental crust and oceanic crust), and for all practical purposes coinciding with the westernmost margin of the Pacific Plate (Holmes, 1978: 664; Springer, 1982: 152).

We obtained living antennariids primarily from local tropical-fish dealers, but brought some back to the laboratory from various localities as a result of our own collecting efforts. These we kept in 150- to 300-liter, closed-system aquaria and maintained on a diet of goldfish (*Carassius auratus*) or Coho and Chinook salmon fingerlings (*Oncorhynchus kisutch* and *O. tshawytscha*) when available from the hatchery of the School of Aquatic and Fishery Sciences, University of Washington. Although prey of all kinds appeared to elicit identical behavioral responses, we offered only natural prey items (damselfishes of the genus *Dascyllus* or *Chromis*) when filming or when recording functional and behavioral data by other means. To assure an uninhibited response, we conditioned experimental animals prior to filming by exposing them to photographic illuminators at each feeding.

High-speed films were made at 800 and 1,000 frames per second with a Hycam K200 HE-15 movie camera equipped with a 75-mm macrolens; a +1 diopter allowed for extremely close work. Regular-speed films (32 and 64 frames per second) were made with a Bolex H-16 camera equipped with a 25-mm macrolens. Kodak EF 7242 film and four 1,000-watt quartz illuminators were used. Films were analyzed with a Moviola M-50 table viewer. All still photography was accomplished with a Canon AE-1, equipped with an Ikelite underwater housing during fieldwork.

Additional analyses of behavior were made using a Panasonic NV-8030 video tape recorder, Panasonic WV-1350 television camera, and Panasonic monitor/receiver. During 24-hour video recordings of vagility and association with structure (see Aggressive Resemblance and Foraging-Site Selection, p. 431), light intensity was altered to simulate day and night (11 hours of bright artificial lighting and 11 hours of low-intensity red light, with a 1-hour intermediate level of light representing dawn and dusk). For these experiments, 227-liter aquaria were used, each with a bottom measuring 43 by 122 cm. "Structure" consisted of two assemblages of rocks and coral, equally spaced from each other and the walls of the aquarium, on an 8- to 10-cm bed of crushed coral; each assemblage occupied approximately 10% of the bottom area of the tank.

In experiments designed to test a hypothesis of chemical attraction via glands in the esca (see Aggressive Mimicry, p. 411), water samples were taken with a 60-cc syringe attached to a clear plastic tube 5 mm in diameter. Experimental aquaria were fully enclosed with black plastic to obstruct the movements of the experimenter while manipulating the plastic tube. We recorded an approach when the prey item passed within a distance of two standard lengths from the tip of the tube. The resulting data were compared using a Mann-Whitney U test (Siegel, 1956).

Escae examined by transmission-electron microscopy were prepared by the methods outlined by Chi et al. (1978). The tissues were fixed in 2% glutaraldehyde and 2% paraformaldehyde in a 0.1-M solution of cacodylate buffer at a pH of 7.4 for 2 hours. After fixation, the tissues were washed three times with buffer, post-fixed in 1% osmium tetroxide, dehydrated in alcohol, and embedded in Epan. Thin sections, stained with uranyl acetate and lead acetate, were examined with a JEOL 100-B electron microscope at 60 kv. Material for the scanning-electron microscope was air dried and coated with 50 A carbon, followed by 300 A of 60/40 Au/Pd, and examined with a Cambridge Stereoscan Mark II.

Unless indicated otherwise in the figure caption, all original line drawings in this volume were penciled by the senior author, with the aid of a Wild M-5 Stereomicroscope equipped with a camera lucida, and inked and stippled by Cathy L. Short of Bothell, Washington.

3. What Makes a Frogfish?

Systematic studies of anglerfishes have suffered primarily from a lack of adequately preserved material available for examination, but also from the paucity of characters that can be used for taxonomic purposes. In frogfishes, nearly all the traditionally utilized morphometric characters are worthless in distinguishing the many morphometrically similar species. Values for head length, width, and depth, length of upper and lower jaw, pectoral lobe, base of the unpaired fins, and caudal peduncle are difficult to establish with accuracy, and so highly variable intraspecifically, and so plastic, depending on the mode of preservation, that they are of little importance. Thus, except for a number of internal osteological and soft-tissue characters, the separation of antennariid taxa is restricted primarily to differences in the morphology of the luring apparatus and associated second and third cephalic spines of the dorsal fin, presence or absence of a caudal peduncle, relative development of the dermal spinules, counts and bifurcation of fin rays, and, to a much smaller extent, color and color pattern. These and other external characters of general importance are discussed in more detail below.

Illicial Apparatus and Associated Cephalic Spines

The spinous dorsal fin of antennariids consists of three cephalic spines, the first (hereafter referred to as the "illicium," plural "illicia," a Latin word meaning lure or inducement, first used by Garman, 1899: 13) and second of which are supported by a single elongate pterygiophore (see Pietsch, 1981b: 396, 410, figs. 13, 36). Unlike some ceratioid and ogcocephalid anglerfishes (Bertelsen, 1951: 18; Bradbury, 1967: 403; Pietsch, 2009: 33), in which the illicial pterygiophore can be protruded and retracted in the longitudinal plane, the pterygiophore of antennariids is relatively immobile, its anterior end usually terminating distinctly posterior to the symphysis of the upper jaw. In members of the *Antennarius striatus* group the illicial pterygiophore extends anteriorly, beyond the symphysis of the upper jaw, slightly overhanging the mouth.

The illicium is highly variable in width, from nearly as thick as the second dorsal spine (e.g., in some members of the genus *Fowlerichthys*) to exceptionally thin and threadlike (as in *Tathicarpus* and members of the *Antennarius pictus* group). Its length varies from minute (*Histiophryne*) to nearly 50% of standard length (*Tathicarpus*). It is usually smooth and naked, but may be more or less covered with dermal spinules (e.g., *Echinophryne* and *Lophiocharon lithinostomus*). Its position on the dorsal surface of the snout is dependent on the position of the underlying supporting pterygiophore.

The illicium of nearly all antennariids terminates in a conspicuous fleshy bait or esca (for exceptions to this rule, see Tables 1, 14). The term "esca" (plural "escae"), a Latin word meaning bait, first used among modern authors by Garman (1899: 13), has been employed since classical times (e.g., Cicero, in his *De Natura Deorum*, Book 2(49), lines 125ff, ca. 45 B.C.; see also Rackham, 1933: 242). Escae vary from simple spherical balls of tissue to highly ornate structures that appear to simulate to a remarkable degree small marine organisms, such as fishes, crustaceans, and worms (see Aggressive Mimicry, p. 411). The morphology of the esca is to a large extent species-specific but unlike ceratioid anglerfishes—in which morphological variation is remarkably small and there are no known exceptions to uniqueness at the species level (see Regan and Trewavas, 1932; Bertelsen, 1951; Pietsch, 1974, 2009)—some antennariids display a considerable amount of intraspecific variation in this structure (e.g., *Antennarius striatus* and *Abantennarius coccineus*; Figs. 74 and 146, respectively). Furthermore, some closely related species (e.g., within *Fowlerichthys* and *Abantennarius*), as well as species of different genera that are otherwise highly distinct (e.g., *Antennarius striatus* and *Rhycherus filamentosus*), may have remarkably similar escae. Nevertheless, the esca provides an extremely useful character complex. In fact, regeneration experiments on captive antennariids show that there is a high degree of genetic control over escal morphology (see Aggressive Mimicry, p. 411).

In most antennariids, the illicium and esca are provided with some degree of protection from the nibblings and attacks of would-be predators. When laid back onto the dorsal surface of the head in a non-feeding situation, the illicium fits into a narrow shallow groove, immediately to the left or right of the second dorsal-fin spine, the esca often coming to lie within a shallow depression between the second and third dorsal-fin spines (see Mowbray, in Barbour, 1942: 25; Böhlke and Chaplin, 1968: 718). The esca is thus capable of being covered and hidden by the second dorsal-fin spine when the latter is fully depressed. The extent to which the esca may be protected in this way reaches its full development in members of the *Antennarius pauciradiatus* group, in which the membrane that extends between the second and third dorsal-fin spines forms a small aperture within which the esca may become totally enveloped (see Description, p. 162).

The second dorsal-fin spine, like the illicium described above, also varies considerably in thickness and length. It may be relatively short and stout (less than 10% SL, as in *Allenichthys*) or extremely long and slender (nearly 30% SL, as in *Nudiantennarius* and *Kuiterichthys*). In most species the second spine is connected posteriorly by membranous skin to the dorsal surface of the cranium; the extent to which this membrane is covered with dermal spinules, and whether or not the membrane reaches posteriorly to the third dorsal-fin spine, varies considerably between genera and species, providing a number of important distinguishing characters.

In nearly all antennariids both the second and third dorsal-fin spines are free and mobile (more or less erect in preserved material), the second widely separated from the third, and the third widely separated from the soft-dorsal fin. However, in *Antennatus* the third spine is relatively immobile, and more or less bound down to the surface of the cranium by thick spinulose skin; in *Histiophryne* both the second and third spines are bound down to the surface of the cranium. These spines are more or less connected to each other and to the soft-dorsal fin by membranous skin in *Lophiocharon* and in members of the *Antennarius pauciradiatus* group.

Dermal Spinules

The skin of the head, body, and fins (except for the medial surface of the pectoral fins) of most antennariids is totally covered with close-set dermal spinules. Except for a tiny crescent-shaped spinule associated with each pore of the acoustico-lateralis system, spinules are consistently absent in only two genera, *Rhycherus* and *Phyllophryne* (Table 14). Each spinule is bifurcate, except for those of the genus *Histiophryne*, which are reduced and simple. In many species, spinules associated with the pores of the acoustico-lateralis system appear to be highly modified, and often highly ornate, fusions of two or more spinules (see Acoustico-Lateralis System, p. 34). On the generic level there is considerable variation in the size and shape of spinules (Fig. 26). In most genera the spines of each spinule are no longer than approximately twice the distance between the tips of the spines. In *Antennatus* the spinules are proportionately smaller than those of other genera, and much more closely spaced, with the length of the two spines of each spinule being about three-and-one-half to six times the distance between the tips of the spines (Fig. 26D). *Echinophryne crassispina* differs from all other antennariids in having the dermal spinules present in at least two distinct size classes (Fig. 26F). Finally, in *Echinophryne mitchelli* the dermal spinules are remarkably long (some in large specimens greater than 4% SL) giving the animal a brushlike texture (Fig. 26G).

Gill Filaments

Schultz (1957: 50, 53) identified the condition of the gill filaments on the first arch as a character that can be used to distinguish *Histrio histrio* from all other antennariids: filaments extending dorsally onto the proximal half of the first epibranchial in *H. histrio*, but present only on the lower half of the first ceratobranchial in other antennariids. This character, however, has proved to be much more complex. For example, a situation identical to that in *H. histrio* occurs in *Antennarius indicus*; in *Lophiocharon*, filaments extend only to the proximal tip of the first epibranchial; in *Fowlerichthys avalonis*, filaments are present along the entire length of the first ceratobranchial. This character is highly variable within genera and species-groups and even within single species—and, for that matter, nearly impossible to quantify—and the anatomical nature of the opercle of frogfishes (see below for a description of the restricted opercular opening) makes it extremely difficult to examine routinely. Therefore, the relative development of gill filaments is not discussed further.

Opercular Opening

Like most other members of the order Lophiiformes, antennariids are characterized by having a greatly restricted opercular opening that consists of a small oval or circular, sometimes tubelike aperture, usually situated immediately behind and slightly below the base of the pectoral fin. The only exceptions to this rule are found in *Abantennarius duescus* and *A. analis*, in which the opercular opening is displaced posteriorly to a point about halfway between the bases of the pectoral and anal fins in *A. duescus*, and further posteriorly to the base of the anal fin in *A. analis*.

Caudal Peduncle

In antennariids there is a clear-cut difference between "caudal peduncle present" and "peduncle absent." In the former, the membranous, posteriormost margin of the soft-dorsal and anal fins is attached to the body distinctly anterior to the base of the outermost rays of the caudal fin. In the latter, the membranous, posteriormost margin of these fins is attached to the body at the base of the outermost rays of the caudal (i.e., at the

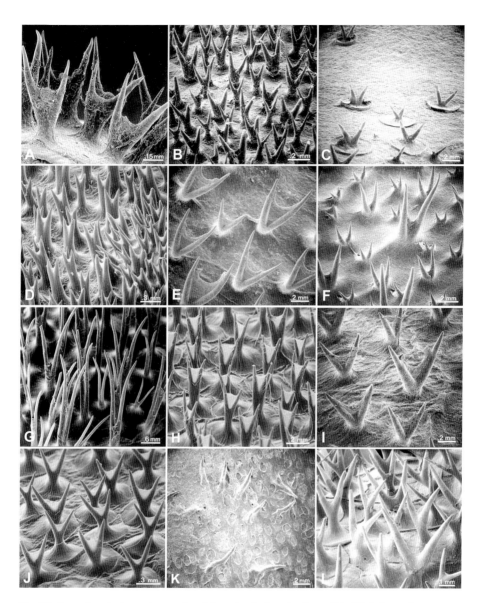

Fig. 26. Scanning electron micrographs of dermal spinules of antennariids. (**A**) *Antennarius hispidus*, UW 20870, 67 mm SL. (**B**) *Antennarius pictus*, UW 20887, 73 mm SL. (**C**) *Nudiantennarius subteres*, BMNH 1866.8.14.108, 64 mm SL. (**D**) *Antennatus tuberosus*, UW 20923, 32.5 mm SL. (**E**) *Kuiterichthys furcipilis*, NMV A.2848, 65 mm SL. (**F**) *Echinophryne crassispina*, NMV A.2901, 55.5 mm SL. (**G**) *Echinophryne mitchellii*, NMV A.530, 80 mm SL. (**H**) *Echinophryne reynoldsi*, NMV A.3064, 31.5 mm SL. (**I**) *Tathicarpus butleri*, AMS IB.3018, 58 mm SL. (**J**) *Lophiocharon trisignatus*, CAS 40369, 61 mm SL. (**K**) *Histiophryne narungga*, AMS I.16170–001, 45 mm SL. (**L**) *Allenichthys glauerti*, WAM P.5965, 103 mm SL. After Pietsch and Grobecker (1987, fig. 16).

posteriormost margin of the hypural plate of the caudal skeleton), or, as in *Histiophryne*, beyond the base of the caudal rays, attaching posteriorly to the outermost rays themselves. This character becomes most critical in distinguishing between the two morphologically similar species *Abantennarius coccineus* and *A. nummifer*. In all the numerous specimens examined of these two species (153 and 458, respectively; some individuals, particularly small specimens, requiring dissection for confirmation), no intermediate condition was found.

Acoustico-Lateralis System

In her review of the anatomy and systematics of the Antennariidae, Le Danois (1964) made extensive use of characters of the acoustico-lateralis system, especially differences in the number of pores in the preopercular series and morphological variability in the modified and often highly ornate dermal spinules that are associated with the pores. Our own attempt to analyze this character complex, made during the early stages of this investigation, revealed no useful information. Not only are counts extremely difficult to make, but the interspecific variability is too small to be taxonomically useful. Likewise, the anatomy of the dermal spinules associated with the pores is often so complex and so highly variable, even within a single specimen, that it is often beyond analysis. Except for the fact that the small naked areas usually present between the pores of the acoustico-lateralis system (in spiny-skinned antennariids) are absent in *Antennatus*, we conclude that no useful data can be obtained from further analysis of this character complex.

Fin Rays

A summary of fin-ray counts shows great variability among antennariid genera and species, especially in the dorsal and pectoral fins (Tables 2, 15). The dorsal fin consists of 10 to 16 rays, the anal from 6 to 10, and the pectoral from 6 to 12. All of these rays are biserial and segmented. On the basis of fin-ray counts alone, only one genus, *Tathicarpus*, can be distinguished from all other genera (Table 15). Although fin-ray counts are certainly useful in differentiating the species of *Antennarius*, no species of this genus can be distinguished on the basis of these counts alone (Table 5).

Almost of greater importance than counts is the variability in the bifurcation of the rays themselves. In the dorsal and anal fins the rays may be all simple to all bifurcate; if any rays are bifurcate, the condition nearly always occurs in the posteriormost rays of the fin (the only exception to this rule is *Tathicarpus*, in which only the anteriormost rays are bifurcate). Three states of bifurcation are found in the rays of the caudal fin: all nine bifurcate, only the innermost seven bifurcate, or all simple (Tables 1, 14). The rays of the pectoral fin are either all simple or all bifurcate; the latter state is present only in some members of the genus *Fowlerichthys*. Finally, the five rays of the pelvic fin are simple in all antennariids except for *Fowlerichthys* and some species of *Antennarius* and *Abantennarius* (Tables 4, 10).

Warts

Julien Desjardins (1840: 3), in describing *Antennarius maculatus* (as *Chironectes maculatus*) from Mauritius, was the first to utilize wartlike structures in the taxonomy of frogfishes: "What is most particular to this fish, and conspicuous at first glance, is the quantity of warts or tubercles placed in a regular manner on each side of the body, face, tail, and fins. . . ." Somewhat later, Bleeker (1857, 1865b) also made reference to the importance of warts when he described three new "warty-skinned" antennariids from Ambon. Schultz (1957) followed Bleeker in recognizing two of these "warty-skinned" forms, but in a later paper he (Schultz, 1964: 172) questioned the value of this character, suggesting that the warts were encysted nematodes. Although encysted parasites certainly produce wartlike structures on frogfishes, the character as used here to describe certain members of the *Antennarius pictus* group is no artifact, the swellings being remarkably symmetrical when both sides of the animal are compared (and not associated with any bilaterally symmetrical structures, such as pores of the acoustico-lateralis system).

Cutaneous Appendages

At first glance, the abundance of appendages that adorn the skin of many frogfishes suggests a wealth of characters that might be taxonomically useful. However, except in a few rare cases (e.g., the two cutaneous appendages on the snout, anterior to the base of the illicium of *Histrio histrio*; the peculiar flattened appendages of *Phyllophryne*; and appendages on the chin and sides of the head of some species of *Histiophryne*), the development of these outgrowths of the skin is extremely variable, even within a single species—some specimens of *Antennarius striatus*, for instance, may be nearly devoid of appendages, whereas others collected from the same general locality may display a remarkable array of such structures. There are also photographically documented observations of a "bald" individual of *A. scaber*, at Lake Worth Lagoon near Blue Heron Bridge in southeast Florida, becoming increasingly "hairier" over a period of slightly more than two weeks (Linda Ianniello, personal communication, 26 April 2017). These observations and the results of our own investigation of this character indicate that nothing would be gained by further analysis.

Maxilla

The relationship between the maxilla and its associated soft tissues provides two character states of taxonomic importance. In *Antennarius*, *Antennatus*, and *Abantennarius* the distal two-thirds of the maxilla are tucked within folds of skin when the mouth is closed; thus, only a small proximal portion of this element is directly covered with spinulose skin. In all other antennariid genera, only the distal tip of the maxilla is tucked, so that (in spiny-skinned forms) a much larger, proximal portion of the bone is directly covered with spinulose skin.

Ovarian Morphology

The eggs of antennariids and, for that matter, those of all lophiiform fishes are spawned embedded within a nonadhesive, gelatinous mucoid mass. While gelatinous egg masses are also produced by some ophidiiforms (Sparta, 1929; Mito, 1962; Gordon et al., 1984; Fahay, 1992; Ambrose, 1996) and scorpaenoids (Barnhart, 1937; Orton, 1955; Phillips, 1957; Pearcy, 1962; Washington et al., 1984; Moser, 1996; Koya and Muñoz, 2007; Morris et al., 2011), the peculiar structure of those of lophiiforms differs considerably from all other ovarian products known in fishes (Rasquin, 1958: 362; Breder and Rosen, 1966: 613). The structure of the mass is a replica of the internal surfaces of the ovaries; that is, it is formed by a mucoid impression of the epithelial surface of the lamellae of the ovarian walls (Rasquin, 1958: 340). The ovaries of all antennariids are paired and fused together on the midline, but they come in at least two forms: (1) a double scroll-shaped structure that produces a buoyant, egg-filled, ribbonlike, sheath of mucous, often referred to as an "egg-raft" or "veil" (Collins, 1880: 8; Gill, 1909: 570; Mosher, 1954: 141; Rasquin, 1958: 333; Pietsch and Grobecker, 1987: 351); and (2) a pair of simple oval-shaped organs, more typical of other teleost fishes, that produces a smaller, globular, non-buoyant, egg-filled mucoid mass. The former structure, found in six antennariid genera (which together form the subfamily Antennarinae), serves as a device for broadcasting a large number of small eggs over great geographical distances, with its positively buoyant quality providing for development in relatively productive surface waters (Gudger, 1905: 842; Pietsch and Grobecker, 1987: 351). The latter structure, characteristic of the remaining nine genera of the family (which together constitute the subfamily Histiophryninae), lends itself to a variety of forms of parental care (Pietsch and Grobecker, 1980: 551; Arnold and Pietsch, 2012: 128; Arnold et al., 2014: 538). For more discussion, see Reproduction and Early Life History, p. 461.

Color and Color Pattern

A misconception concerning the variability of coloration and pigment pattern is the primary cause of the multiplicity of available names and nomenclatural complexity that characterizes the Antennariidae (three out of four nominal antennariid species are junior synonyms; see Randall, 1968: 290; Friese, 1973: 31). Having had the opportunity to examine a large amount of material, we know now that chromatic characters, with only a few exceptions (see below), are extremely variable within species and provide little useful data. Two extremes of coloration are present in many species: a more common light-color phase, consisting of a white to beige, yellow, orange, light-brown, or rust background often overlaid with black, brown, pink, or yellow markings; and a dark-color phase consisting of a green, dark-red, dark-brown, to black background, with or without darker-colored markings showing through. The entire range of coloration between these two extremes, however, can be assumed by living individuals of many species during a time period of a few weeks or months, specimens often changing radically on preservation (Smith, 1898: 109; Gill, 1909: 602; Longley and Hildebrand, 1941: 303; Barbour, 1942: 26; Breder, 1949: 94, 95; Gordon, 1955: 388; Böhlke and Chaplin, 1968: 717; Randall, 2005b: 300; Pietsch et al., 2009a). For example, we have witnessed individuals of *Antennatus tuberosus*, maintained in experimental aquaria, gradually change from dark gray to light cream in a period of about two weeks. Similarly, we have seen *Antennarius commerson* change from lemon-yellow to brick-red in three weeks, and totally black individuals of *A. striatus* taking on their more typical yellow-striped phase in about five weeks (Pietsch and Grobecker, 1987: 37, 1990a: 98). More recently, Randall (2005b: 301) described an individual of *A. commerson* in the Waikiki Aquarium in Honolulu that changed from orange to dark reddish-brown in about two weeks, after being moved to a different tank. In other species, particularly members of the *Antennarius pictus* group, the chromatic variety seems almost infinite.

In contrast to other antennariids, in which color change occurs over a period of weeks or months, the Sargassumfish, *Histrio histrio*, is capable of rapid color change, apparently in response to stimulation (e.g., during courtship and spawning behavior; see Reproduction and Early Life History, p. 461) or to match its surrounding, from a grayish-white, bleached-out appearance to a pattern of streaks and mottling of dark browns, olive, and yellow, and vice versa (Mosher, 1954: 149; Böhlke and Chaplin, 1968: 717; Robins and Ray, 1986: 88; Pietsch and Grobecker, 1987: 199; Pietsch, 2002b: 271). (For more on color change and references to figures, see individual species accounts in Chapter 4.)

At the same time, there are some aspects of pigmentation that do provide important taxonomic characters. These all involve color pattern rather than hue, and include such examples as: (1) the typical striped and spotted pattern of members of the *Antennarius striatus* group; (2) the pattern of dark, roughly circular spots on the body and fins of some members of the *A. pictus* group, and the similar spots on the belly of *A. sanguineus*; (3) the "ocelli" of *Antennarius biocellatus*, *Nudiantennarius subteres*, and members of the genus *Fowlerichthys* (distinguished from simple pigment "spots" by their larger size and by being encircled with a narrow lightly pigmented ring); and (4) the "basidorsal spot," a dark, roughly circular spot (typically smaller than an "ocellus" and only rarely encircled by a lightly pigmented ring) at about the middle of the base of the dorsal fin (sometimes extending onto the fin itself), typically present in, but certainly not confined to, members of the genus *Abantennarius*.

Sexual dichromatism in antennariids was first suggested by Schultz (1957: 72), on the basis of observations made by Louis A. Krumholz on black and "zebralike" individuals of *Antennarius scaber* (*Phrynelox nuttingi* and *P. scaber* of Schultz, respectively). Somewhat later, Le Danois (1964: 118) wrote that "the genus *Phrynelox* [i.e., *Antennarius*] displays a very strong sexual dimorphism in the characters of coloration; the differences between males and females are strong enough that frequently [taxonomists have] established different species for each sex. . . . The general rule is that the males have a light, tiger-like uniform, whereas the females are black or very dark in coloration. . . ." Despite these claims (and a similar suggestion by Randall, 1968: 291), we have not been able to show any sexual differences in coloration in any frogfish (nor could Böhlke and Chaplin, 1968: 718, in their examinations of *Antennarius scaber*). In fact, other than size differences between conspecific males and females—males, when observed with females, generally appearing smaller—we have found no means of differentiating sex by any character, in any species, other than by examining the gonads through simple dissection (see Reproduction and Early Life History, p. 461).

4. Biodiversity

Family Antennariidae Jarocki

Type genus *Antennarius* Daudin, 1816.

Antennarii Jarocki, 1822: 401, 406 (family-group taxon to contain *Lophius*, *Malthe*, and *Antennarius*).

Chironectiides Billberg, 1833: 52 (family-group taxon to contain *Chironectes*).

Chironectidae Swainson, 1838: 32, 190, 201 (family taxon to contain *Chironectes*). Swainson, 1839: 195, 330 (after Swainson, 1838).

Halibatrachi Van der Hoeven, 1855: 343 (family-group taxon to contain *Chironectes*, *Chaunax*, *Lophius*, *Malthe*, and *Batrachus*).

Cheironecteoidei Bleeker, 1859c: 370 (family-group taxon to contain *Antennarius*).

Antennariinae Gill, 1861: 47 (subfamily to contain *Antennarius*, i.e., *Histrio*). Gill, 1863: 90 (subfamily to contain *Pterophryne*, *Antennarius*, *Histiophryne*, and *Saccarius*). Arnold and Pietsch, 2012: 124 (subfamily to contain *Fowlerichthys*, *Antennarius*, *Nudiantennarius*, *Histrio*, *Antennatus*, and *Abantennarius*).

Chironecteoidei Gill, 1863: 89 (equals Chironectidae of Swainson, 1838, 1839; includes *Antennarius*, *Brachionichthys*, *Chaunax*, and *Ceratias*).

Antennarioidae Gill, 1863: 89, 90, 91 (includes subfamilies Antennariinae, Brachion-ichthyinae, and Chaunacinae). Lütken, 1878: 325 (after Gill, 1863).

Chironecteoidei Bleeker, 1865b: 4 (family-group taxon to contain subfamilies Chiro-necteiformes and Ceratiaeformes).

Chironecteiformes Bleeker, 1865b: 5 (subfamily to contain *Antennarius*, *Saccarius*, *Brachionichthys*, and *Chaunax*).

Pediculati Day, 1865: 121 (family to contain *Antennarius*). Day, 1876: 271 (family to contain *Antennarius* and *Halieutaea*).

Antennariidi Poey y Aloy, 1868: 404; 1876: 134 (family-group taxon to contain *Antennarius* and *Pterophryne*).

Antennariidae Cope, 1872: 340 (one of two pediculate families). Gill, 1872: 2 (modified after Gill, 1863; one of four pediculate families).

Chironectae Fitzinger, 1873: 48 (family to contain *Chironectes*).

Histrioninae Schultz, 1957: 103 (subfamily to contain *Histrio*).

Pterophrynidae Le Danois, 1964: 14, 127 (family to contain *Pterophryne*, *Trichophryne*, and *Rhycherus*).

Saccariidae Whitley, 1968a: 88 (family to contain *Saccarius*, a synonym of *Antennarius*).

Histiophryninae Arnold and Pietsch, 2012: 128 (subfamily to contain *Rhycherus*, *Porophryne*, *Kuiterichthys*, *Phyllophryne*, *Echinophryne*, *Tathicarpus*, *Lophiocharon*, *Histiophryne*, and *Allenichthys*).

DIAGNOSIS

A lophiiform family unique and derived in having a greatly enlarged third dorsal-fin spine (and associated pterygiophore), as well as a shortened body involving numerous associated specializations but reflected most strikingly by a sigmoid vertebral column (see Pietsch, 1981b, figs. 12, 33–35, 41; 1984b; Pietsch and Grobecker, 1987: 39). The family is further distinguished from all other families of the order by having the following combination of character states: body laterally compressed; mouth strongly oblique to vertical; eyes lateral; posteromedial process of vomer compressed, keel-like; distance between lateral ethmoids considerably less than that between lateral margins of sphenotics; postmaxillary process of premaxilla spatulate; ectopterygoid triradiate; palatine teeth present; dorsal head of quadrate narrow, narrower than ventral margin of metapterygoid; interhyal with a medial, posterolaterally directed process; basihyal present; branchiostegal rays two plus four; opercle reduced, width equal to or less than 25% length of suspensorium; pharyngobranchial I simple (absent in *Lophiocharon* and *Histiophryne*); spinous dorsal of three cephalic spines; illicial pterygiophore and pterygiophore of third dorsal-fin spine with highly compressed, bladelike dorsal expansions; illicial bone not retractable within an illicial cavity; three pectoral radials; pectoral-fin lobe not membranously attached to rays of pelvic fin; pectoral fin single, the rays not membranously attached to side of body; pelvic fin of one spine and five rays (Pietsch, 1981b, 1984b; Pietsch and Grobecker, 1987; Pietsch et al., 2009b).

DESCRIPTION

Body short, globose, laterally compressed; mouth large, upper and lower jaws with two to four, more or less irregular rows of small villiform teeth; opercular opening restricted to a small oval or circular, sometimes tubelike aperture situated immediately ventral or posterior to base of pectoral fin (displaced posteriorly, situated halfway between base of pectoral-fin lobe and anal fin or reaching base of anal fin in some species of *Abantennarius*); spinous dorsal fin of three cephalic spines, anteriormost spine (illicium) free and modified as a lure (reduced in some species of *Antennatus*, minute or absent in *Histiophryne*); a distinct fleshy bait or esca usually present (reduced, barely differentiated from illicium, or absent in *Antennatus*, *Echinophryne*, *Lophiocharon lithinostomus*, and some species of *Histiophryne*); illicium usually naked (encrusted with dermal spinules in *Echinophryne* and *L. lithinostomus*); second and third dorsal-fin spines usually more or less erect (laid back and bound down to surface of cranium by skin of head in *Histiophryne*), usually widely separated from each other and from soft-dorsal fin (connected to each other and to soft-dorsal fin by skin in *Lophiocharon* and in some species of *Antennarius*); posterior end of pterygiophore of illicium usually cylindrical in cross section (dorsoventrally flattened and expanded laterally in *Tathicarpus*); pectoral-fin lobe elongate, leglike, usually broadly attached to side of body (partially free in *Nudiantennarius*, free for most of its length in *Histrio* and *Tathicarpus*); skin usually covered with close-set dermal spinules (spinules usually bifurcated, simple in *Histiophryne*; reduced, minute, or absent in *Nudiantennarius* and *Histrio*; absent in *Rhycherus* and *Phyllophryne*); cutaneous filaments and appendages usually present; endopterygoid and epural usually absent, present in *Fowlerichthys*, *Antennarius*, *Nudiantennarius*, *Histrio*, *Abantennarius*, and *Antennatus*; pseudobranch usually absent, present in *Fowlerichthys*, *Antennarius*, *Histrio*, *Abantennarius*, and *Tathicarpus*; swimbladder usually present, absent in *Kuiterichthys* and *Tathicarpus*.

Vomer and palatine well-toothed; pharyngobranchial I, if present (absent in *Lophiocharon* and *Histiophryne*), usually toothless (a single tooth-bearing plate present on only right side of 73-mm specimen of *Allenichthys* examined osteologically); pharyngobranchials II and III well-toothed; epibranchials usually toothless (a row of six to 11 teeth present on epibranchial I of *Tathicarpus*; a single tooth-bearing plate, or small remnants of a tooth-plate, present on epibranchial I of some specimens of *Fowlerichthys, Antennarius, Abantennarius, Antennatus, Rhycherus, Kuiterichthys, Phyllophryne, Echinophryne,* and *Allenichthys*); ceratobranchials I to IV usually toothless (a single tooth-plate, often reduced to a remnant, present on ceratobranchial I of *Allenichthys, Rhycherus,* and some species of *Echinophryne*); ceratobranchial V with two to six, more or less irregular rows of teeth; hypobranchials II and III usually bifurcate proximally (hypobranchial II reduced and simple in *Rhycherus, Kuiterichthys, Echinophryne, Histiophryne,* and *Allenichthys*); ossified basibranchial very rarely present (of all material examined osteologically, found only in a single 50-mm specimen of *Phyllophryne scortea*, AMS I.20189-018); vertebrae 18 to 23, caudal centra 13 to 18; dorsal-fin rays 10 to 16, anal-fin rays six to 10, pectoral-fin rays six to 14; usually only seven innermost rays of caudal fin bifurcate (all rays bifurcate in *Fowlerichthys, Abantennarius, Antennarius, Rhycherus, Lophiocharon, Allenichthys*, some specimens of *Nudiantennarius subteres*, and some species of *Antennatus*; all simple in *Tathicarpus*).

Color and color pattern highly variable, with two extremes (the entire range of which can be assumed by living individuals of many species during a period of days or weeks, specimens often changing radically on fixation and preservation; see Color and Color Pattern, p. 36): a more common light-color phase with light-tan to yellow, orange, light-brown, or rust background often overlaid with black, brown, pink, or bright-yellow streaks, bars, and spots on head, body, and fins; and a dark-color phase with green, dark red, dark brown, to black background with streaks, bars, and spots often (but not always) showing through as deeper black, the tips of rays of paired fins often white.

Juveniles and adults benthic from the surface to at least 350 m (the latter record, a 252-mm specimen of *Fowlerichthys scriptissimus* taken by hook and line off Mahina, Tahiti, by a local fisherman in July 1990), with the single exception of *Histrio histrio*, which is pseudopelagic in floating *Sargassum*; adults of some species attaining a standard length of at least 432 mm (*Fowlerichthys scriptissimus*, UW 112642); occurring in all major tropical and subtropical oceans and seas of the world except the Mediterranean; 15 genera and 52 living species.

Classification of the Antennariidae

The following classification of the family, based primarily on the molecular phylogenetic study of Arnold and Pietsch (2012), but modified slightly to include newly described taxa, summarizes the taxonomic conclusions made here:

Subfamily Antennariinae Gill, 1861
 Genus *Fowlerichthys* Barbour, 1941
  Fowlerichthys avalonis (Jordan and Starks, 1907)
  Fowlerichthys ocellatus (Bloch and Schneider, 1801)
  Fowlerichthys radiosus (Garman, 1896)
  Fowlerichthys scriptissimus (Jordan, 1902)
  Fowlerichthys senegalensis (Cadenat, 1959)
 Genus *Antennarius* Daudin, 1816
  Antennarius striatus group Pietsch, 1984b

Antennarius hispidus (Bloch and Schneider, 1801)

Antennarius indicus Schultz, 1964

Antennarius scaber (Cuvier, 1817b)

Antennarius striatus (Shaw, 1794)

Antennarius pictus group Pietsch, 1984b

Antennarius commerson (Latreille, 1804)

Antennarius maculatus (Desjardins, 1840)

Antennarius multiocellatus (Valenciennes, 1837)

Antennarius pardalis (Valenciennes, 1837)

Antennarius pictus (Shaw, 1794)

Antennarius biocellatus group Pietsch, 1984b

Antennarius biocellatus (Cuvier, 1817b)

Antennarius pauciradiatus group Pietsch, 1984b

Antennarius pauciradiatus Schultz, 1957

Antennarius randalli Allen, 1970

Genus *Nudiantennarius* Schultz, 1957

Nudiantennarius subteres (Smith and Radcliffe, in Radcliffe, 1912)

Genus *Histrio* Fischer, 1813

Histrio histrio (Linnaeus, 1758)

Genus *Abantennarius* Schultz, 1957

Abantennarius analis Schultz, 1957

Abantennarius bermudensis (Schultz, 1957), new combination

Abantennarius coccineus (Lesson, 1831), new combination

Abantennarius dorehensis (Bleeker, 1859a), new combination

Abantennarius drombus (Jordan and Evermann, 1903), new combination

Abantennarius duescus (Snyder, 1904), new combination

Abantennarius nummifer (Cuvier, 1817b), new combination

Abantennarius rosaceus (Smith and Radcliffe, in Radcliffe, 1912),

new combination

Abantennarius sanguineus (Gill, 1863), new combination

Genus *Antennatus* Schultz, 1957

Antennatus flagellatus Ohnishi, Iwata, and Hiramatsu, 1997

Antennatus linearis Randall and Holcom, 2001

Antennatus strigatus (Gill, 1863)

Antennatus tuberosus (Cuvier, 1817b)

Subfamily Histiophryninae Arnold and Pietsch, 2012

Genus *Rhycherus* Ogilby, 1907

Rhycherus filamentosus (Castelnau, 1872)

Rhycherus gloveri Pietsch, 1984c

Genus *Porophryne* Arnold, Harcourt, and Pietsch, 2014

Porophryne erythrodactylus Arnold, Harcourt, and Pietsch, 2014

Genus *Kuiterichthys* Pietsch, 1984b

Kuiterichthys furcipilis (Cuvier, 1817b)

Kuiterichthys pietschi Arnold, 2013

Genus *Phyllophryne* Pietsch, 1984b

Phyllophryne scortea (McCulloch and Waite, 1918)

Genus *Echinophryne* McCulloch and Waite, 1918

Echinophryne crassispina McCulloch and Waite, 1918

Echinophryne mitchellii (Morton, 1897)

    Echinophryne reynoldsi Pietsch and Kuiter, 1984

Genus *Tathicarpus* Ogilby, 1907

    Tathicarpus butleri Ogilby, 1907

Genus *Lophiocharon* Whitley, 1933

    Lophiocharon hutchinsi Pietsch, 2004a

    Lophiocharon lithinostomus (Jordan and Richardson, 1908)

    Lophiocharon trisignatus (Richardson, 1844)

Genus *Histiophryne* Gill, 1863

    Histiophryne bougainvilli (Valenciennes, 1837)

    Histiophryne cryptacanthus (Weber, 1913)

    Histiophryne maggiewalker Arnold and Pietsch, 2011

    Histiophryne narungga Arnold and Pietsch, 2018

    Histiophryne pogonius Arnold, 2012

    Histiophryne psychedelica Pietsch, Arnold, and Hall, 2009a

Genus *Allenichthys* Pietsch, 1984b

    Allenichthys glauerti (Whitley, 1944a)

Subfamily Antennariinae Gill

Antennariinae Gill, 1861: 47 (subfamily of family Lophioidae to contain *Antennarius*). Schultz, 1957: 53, 62, fig. 1 (subfamily to contain all antennariid taxa except *Histrio histrio*). Arnold and Pietsch, 2012: 117, 122, 124–128, figs. 4, 6 (molecular phylogeny); Arnold et al., 2014: 538 (diagnosis).

Two subfamilies of the Antennariidae, the Antennariinae and Histiophryninae, have recently been identified, based largely on the work of Arnold and Pietsch (2012; see also Arnold, 2013; Arnold et al., 2014). While no convenient distinguishing characters are available for use in a traditional dichotomous key, the Antennariinae, containing six of the 15 recognized genera of the family (*Fowlerichthys, Antennarius, Nudiantennarius, Histrio, Abantennarius,* and *Antennatus*), may be differentiated as follows (Tables 1, 2):

DIAGNOSIS

A subfamily of the Antennariidae, distinguished from its sister group, the Histiophryninae, by the presence of an endopterygoid and epural (Pietsch, 1984b: 41; Pietsch and Grobecker, 1987: 275), and in having an entirely different ovarian morphology: the Antennariinae has double, scroll-shaped ovaries (Pietsch and Grobecker, 1987, fig. 161, pl. 10), while the Histiophryninae has simple, oval-shaped ovaries (Arnold and Pietsch, 2012: 128). Each ovarian type corresponds to a different life history strategy: members of the Antennariinae are broadcast spawners and go through a distinct larval stage, while those of the Histiophryninae undergo direct development and display various forms of parental care (Pietsch and Grobecker, 1980, 1987; Pietsch et al., 2009a; Arnold et al., 2014). In addition, the two subfamilies are clearly distinguished by molecular analysis (see Evolutionary Relationships, p. 377). Finally, the Antennariinae has a broad geographic distribution, with all known genera (except *Nudiantennarius*) found nearly circumglobally throughout the tropics and subtropics, while the Histiophryninae is restricted geographically to the Indo-Australian Archipelago.

Table 1. Comparison of Distinguishing Character States of Genera of the Subfamily Antennariinae

Character	Fowlerichthys	Antennarius	Nudiantennarius	Histrio	Abantennarius	Antennatus
Dermal spinules	bifurcate	bifurcate	bifurcate reduced	bifurcate minute/absent	bifurcate	bifurcate
Illicium	naked/spinulose	naked	naked	naked	naked	naked
Esca	present	present	present	present	present	minute/absent
Pectoral lobe	attached	attached	partially detached	free	attached	attached
Caudal peduncle	present	present	present	present	present/absent	present/absent
Endopterygoid	present	present	present	present	present	present
Pharyngo-branchial I	present	present	present	present	present	present
Epural	present	present	present	present	present	present
Pseudobranch	present	present	absent	present	present	absent
Swimbladder	present	present	present	present	present	present
Dorsal-fin rays	13 (12–14)	12 (11–13)	12	12 (11–13)	12–13 (11–14)	12 (11–13)
Anal-fin rays	8 (7–10)	7 (6–8)	7	7 (6–8)	7 (6–8)	7 (6–8)
Bifurcate caudal-fin rays	9	9	7	7	9	9 (7)
Pectoral-fin rays	12–13 (11–14)	9–12 (8)	9	10 (9–11)	9–12 (8–13)	10–11 (9–12)
Vertebrae	20	19 (18–20)	19	19 (18)	19	19

Key to the Known Genera of the Subfamily Antennariinae

This key differs from most in that it is not dichotomous. It works by progressively elim-inating the most morphologically distinct taxon and for that reason should always be entered from the beginning. All character states listed for each genus must correspond to the specimen being keyed; if not, proceed to the next set of character states. The il-lustrations accompanying the key are diagrammatic; dermal spinules and cutaneous ap-pendages (with minor exceptions) are not shown. For an additional or alternative means of identification, all genera of the Antennariinae may be compared simultaneously by referring to Table 1.

1. Skin smooth, nearly always without dermal spinules but covered with numerous cutaneous appendages; two con-spicuous cutaneous appendages on mid-dorsal line of snout, situated between symphysis of upper jaw and base of illicium; illicium tiny, considerably shorter than second dorsal-fin spine; pectoral-fin lobe detached from side of body for most of its length; pelvic fins long, greater than 25% SL; outermost rays of caudal fin simple, only the seven innermost bifurcate; pseudopelagic in floating *Sargassum* . . . . **Histrio Fischer, 1813, p. 174**
Atlantic and Indo-Pacific Oceans

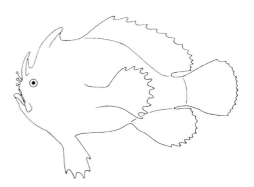

2. Dermal spinules greatly reduced, evident on second and third dorsal-fin spines, anteriormost dorsal-fin rays, on snout and dorsal portion of head, often on chin, but elsewhere few and scattered, usually difficult to detect without microscopic aid; illicium short, about half length of second dorsal-fin spine, the latter long and narrow, 19.5 to 28.1% SL, without posterior membranous connection to head; membranes between rays of paired fins deeply incised; all rays of pelvic fin simple; a large basidorsal ocellus usually present . . . .
. . . . *Nudiantennarius* **Schultz, 1957, p. 168**
Japan, Philippines, and Indonesia

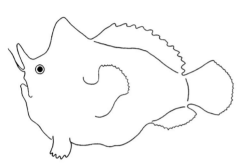

3. Skin thick and firm, everywhere covered with extremely close-set dermal spinules (in some cases, spinules closely clustered to form slightly raised, wartlike patches; naked areas between pores of acoustico-lateralis system absent); illicium more or less tapering to a fine point, esca absent or only barely distinguishable; third dorsal-fin spine more or less immobile, often bound to surface of cranium by skin of head; caudal peduncle usually absent, the membranous posteriormost margin of soft-dorsal and anal fins attached to body at base of outermost rays of caudal fin; all rays of caudal fin usually bifurcate; basidorsal ocellus absent . . . .
. . . . *Antennatus* **Schultz, 1957, p. 238**
Indo-Pacific and Eastern Pacific Oceans

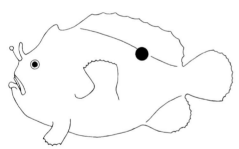

4. Skin covered with dermal spinules; second dorsal-fin spine free, not connected to head by membrane, the posterior surface often naked, devoid of dermal spinules; a darkly pigmented, basidorsal pigment spot usually present, but only rarely encircled by a lightly pigmented ring; anterior end of pterygiophore of illicium terminating distinctly posterior to symphysis of upper jaw; individuals small, never exceeding about 95 mm SL . . . .*Abantennarius* **Schultz, 1957, p. 195**
Atlantic, Pacific, and Indian Oceans

5. Only distal one-fourth of maxilla naked and tucked beneath folds of skin, the remaining proximal portion covered directly with spinulose skin; second dorsal-fin spine membranously attached to head, the membrane nearly always divided into naked dorsal and ventral portions by a dense cluster of dermal spinules; one to three dark ocelli on side of body, each encircled by a lightly pigmented ring; pectoral-fin rays 12 or 13 (rarely 11 or 14), usually all bifurcate; anal-fin rays eight (rarely seven, nine, or 10); all rays of pelvic fin bifurcate . . . .
. . . . *Fowlerichthys* **Barbour, 1941, p. 45**
Atlantic, Pacific, and Indian Oceans

6. Distal two-thirds of maxilla naked and tucked beneath folds of skin, only the extreme proximal end covered directly with spinulose skin; second dorsal-fin spine membranously attached to head, the membrane never divided into naked dorsal and ventral portions by a cluster of dermal spinules; dark ocelli on side of body very rarely present; all rays of caudal fin bifurcate; pectoral-fin rays nine to 12 (rarely eight), all simple, never bifurcate; anal-fin rays seven (rarely six or eight); usually only posteriormost ray of pelvic fin bifurcate . . . . **Antennarius** Daudin, 1816, p. 71

Atlantic, Pacific, and Indian Oceans

Genus *Fowlerichthys* Barbour

Fowlerichthys Barbour, 1941: 12 (type species *Fowlerichthys floridanus* Barbour, 1941 [= *Fowlerichthys radiosus* (Garman, 1896)], by original designation and monotypy).
Kanazawaichthys Schultz, 1957: 62 (type species *Kanazawaichthys scutatus* Schultz, 1957 [= *Fowlerichthys radiosus* (Garman, 1896)], by original designation and monotypy).

DIAGNOSIS

Fowlerichthys is unique among antennariid genera in having the following combination of character states: skin covered with close-set, bifurcate dermal spinules, the length of

Table 2. Frequencies of Fin-Ray and Vertebral Counts of Genera of the Subfamily Antennariinae

Genus	Dorsal-fin rays				Anal-fin rays				
	11	12	13	14	6	7	8	9	10
Fowlerichthys		24	226	4		20	224	9	1
Antennarius	17	419	54		7	431	52		
Nudiantennarius		7				7			
Histrio	3	49	2		1	52	1		
Abantennarius	3	396	266	2	3	633	13		
Antennatus	6	143	4		6	142	5		
TOTALS	29	1,038	552	6	17	1,285	295	9	1

Genus	Pectoral-fin rays (left and right sides combined)							Vertebrae		
	8	9	10	11	12	13	14	18	19	20
Fowlerichthys				10	119	343	8			26
Antennarius	2	163	619	330	16	22		1	77	1
Nudiantennarius		14							7	
Histrio		2	101	5				1	19	
Abantennarius	5	178	426	617	69	1			56	
Antennatus		5	195	102	10				7	
TOTALS	7	362	1,341	1,064	214	366	8	2	166	27

spines of each spinule not more than twice the distance between tips of spines; illicium naked, without dermal spinules (except at extreme base or confined to a narrow row of spinules along anterior margin), about equal to length of second dorsal-fin spine, the latter membranously attached to head, the membrane nearly always divided into naked dorsal and ventral portions by a dense cluster of dermal spinules; esca distinct; only distal 20 to 25% of maxilla naked and tucked beneath folds of skin, the remaining proximal portion covered directly with spinulose skin; pectoral-fin lobe broadly attached to side of body; caudal peduncle present; endopterygoid present; pharyngobranchial I present; epural present; pseudobranch present; swimbladder present; one to three darkly pigmented ocelli on side of body, each encircled by a lightly pigmented ring; dorsal-fin rays 12 to 14 (usually 13); anal-fin rays seven to 10 (usually eight); all rays of caudal fin bifurcate; pectoral-fin rays 11 to 14 (usually 12 or 13); vertebrae 20 (Table 3).

DESCRIPTION

Esca consisting of a tuft of slender filaments, numerous, more or less parallel, vertically aligned folds, or a simple, oval-shaped appendage; distinct, eyelike, escal pigment spots absent; illicium, when laid back onto head, usually fitting into a narrow, naked groove situated on either left or right side of second dorsal-fin spine, the tip of illicium (esca) coming to lie within a shallow depression (sometimes devoid of dermal spinules) between second and third dorsal-fin spines, the esca capable of being covered and protected by

Table 3. Comparison of Distinguishing Character States of Species of the Genus *Fowlerichthys* (measurements as percentage of SL)

Character	F. avalonis	F. ocellatus	F. radiosus	F. scriptissimus	F. senegalensis
Illicium length	6.5–14.1	8.6–15.0	14.6–19.4	6.8–10.6	9.0–13.6
Illicial spinules	absent	present/absent	absent	present/absent	absent
Esca morphology	short appendages	filaments	folds	folds/filaments	filaments
Membrane behind second dorsal-fin spine extending to base of third	yes	yes	yes	yes/no	yes
Membrane behind second dorsal-fin spine divided into dorsal and ventral portions by a cluster of dermal spinules	yes	yes	yes	no	yes
Eye diameter	3.7–7.5	3.9–6.9	7.4–11.4	3.0–6.1	3.9–6.3
Ocelli	1	3	1	1	1
Dorsal-fin rays	13 (12–14)	13 (12)	13 (12)	13	13
Bifurcate dorsal-fin rays	10–13	all	all	3–9	7–13
Anal-fin rays	8 (7–10)	8 (7)	8 (7)	8	8 (7)
Pectoral-fin rays	13 (11–14)	12 (11)	13 (12–14)	13	13 (12)
Pectoral-fin rays bifurcate	yes	yes	no	yes	yes
Maximum standard length	330 mm	320 mm	180 mm	432 mm	285 mm
Distribution	Eastern Pacific	Western Atlantic	Western Atlantic	Indo-Pacific	Eastern Atlantic

second dorsal-fin spine when spine is fully depressed; illicium equal to or slightly longer than second dorsal-fin spine, 6.5 to 19.4% SL; anterior end of pterygiophore of illicium terminating distinctly posterior to symphysis of upper jaw; second dorsal-fin spine straight to strongly curved posteriorly, length 8.0 to 20.6% SL, connected to head by a membrane, the membrane nearly always divided (except in *Fowlerichthys scriptissimus*) into naked dorsal and ventral portions by a dense cluster of dermal spinules, ventral portion usually extending posteriorly (except in some specimens of *F. scriptissimus*), dividing depressed area between bases of second and third dorsal-fin spines and nearly reaching to base of third (Fig. 27); third dorsal-fin spine curved posteriorly, tapering slightly toward distal end, the full length connected to head by membrane, length 15.3 to 29.9% SL; eye not distinctly surrounded by separate clusters of dermal spinules, diameter 3.0 to 11.4% SL; caudal peduncle present, the membranous posteriormost margin of soft-dorsal and anal fins attached to body distinctly anterior to base of outermost rays of caudal fin; dorsal-fin rays 12 to 14 (usually 13), all usually bifurcate (posteriormost seven to nine bifurcate in *F. scriptissimus*, posteriormost seven to 13 bifurcate in *F. senegalensis*); anal-fin rays seven to 10 (usually eight), all bifurcate; pectoral-fin rays 11 to 14 (usually 12 or 13), usually all bifurcate (simple in juveniles less than about 15 mm SL, and in all life stages of *F. radiosus*); all pelvic-fin rays bifurcate (Table 3).

The members of this genus reach a large size compared to other antennariids—all included species are represented by specimens of at least 180 mm SL, and four species are known to exceed 250 mm SL (*F. avalonis*, *F. ocellatus*, *F. scriptissimus*, and *F. senegalensis*; a synapomorphy, see Evolutionary Relationships, p. 381). The average maximum length of material examined of all five species was 252 mm SL.

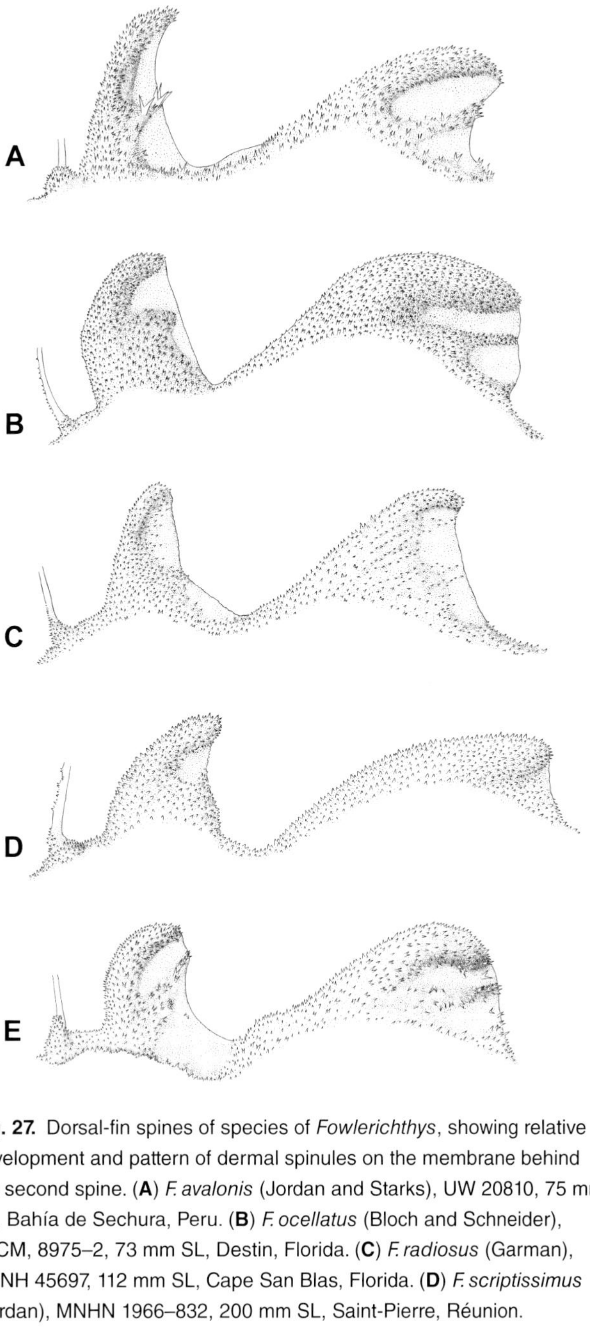

Color and color pattern extremely variable (see species accounts).

Available data indicate that members of this genus are associated with flat, open sand, or mud bottoms; distributed circumtropically, five species.

Fig. 27. Dorsal-fin spines of species of *Fowlerichthys*, showing relative development and pattern of dermal spinules on the membrane behind the second spine. (**A**) *F. avalonis* (Jordan and Starks), UW 20810, 75 mm SL, Bahía de Sechura, Peru. (**B**) *F. ocellatus* (Bloch and Schneider), LACM, 8975–2, 73 mm SL, Destin, Florida. (**C**) *F. radiosus* (Garman), FMNH 45697, 112 mm SL, Cape San Blas, Florida. (**D**) *F. scriptissimus* (Jordan), MNHN 1966–832, 200 mm SL, Saint-Pierre, Réunion. (**E**) *F. senegalensis* (Cadenat), USNM 199254, 88 mm SL, off Liberia. Drawings by Cathy L. Short; after Pietsch and Grobecker (1987, fig. 38).

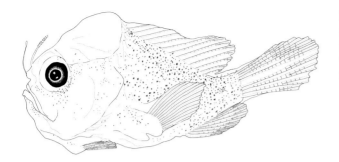

Fig. 28. "Scutatus" postflexion larval stage of *Fowlerichthys radiosus* (Garman): USNM 251937–F21, 16 mm SL. Drawing by Betsy B. Washington; after Pietsch (1984d, fig. 165) and Pietsch and Grobecker (1987, fig. 43).

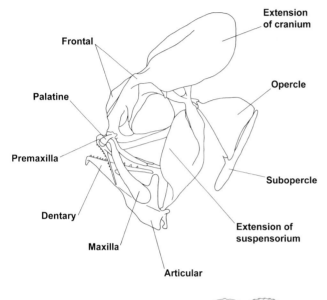

Fig. 29. Lateral view of cranium of "scutatus" postflexion larval stage of *Fowlerichthys radiosus* (Garman): USNM 251937–F21, 17.5 mm SL, with suspensorium, upper and lower jaws, and opercular bones in place. After Pietsch and Grobecker (1987, fig. 44).

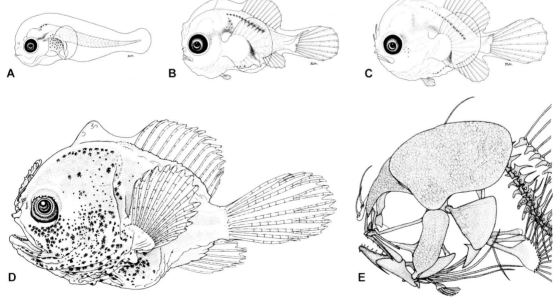

Fig. 30. Larvae of *Fowlerichthys avalonis* (Jordan and Starks). (**A**) preflexion stage, 2.3 mm notochord length. (**B**) flexion stage, 3.3 mm notochord length. (**C**) postflexion "scutatus" larval stage, 3.6 mm SL. (**D**) postflexion "scutatus" larval stage, 7.3 mm SL. (**E**) lateral view of head skeleton, pectoral and pelvic girdles, and anterior part of axial skeleton of postflexion "scutatus" larval stage, 5.3 mm SL. Drawings (**A–C**) by Barbara Sumida McCall, after Watson (1996b, fig. 1); drawings (**D, E**) by William Watson; after Watson (1998, fig. 11).

COMMENTS

The "scutatus" postflexion larval stage was originally described as a new genus and species, *Kanazawaichthys scutatus*, by Schultz (1957: 63, pl. 14, fig. A), but later shown by Hubbs
(1958) to be the prejuvenile of *Fowlerichthys radiosus*. The primary morphological features
that characterize these early life stages are so strikingly different—a pair of shield-like
bony extensions of the cranium that reach beyond the level of the opercular bones, and
an expansion of the anterior margin of the bones of the suspensorium (Figs. 28, 29)—that
their appearance in other antennariids of similar sizes, particularly among members of
the genus *Fowlerichthys*, was expected (Pietsch, 1984d: 322, fig. 165; Pietsch and Grobecker,
1987: 121, figs. 43, 44). Yet, although a few additional antennariids were represented by larvae, no comparable morphological adaptations were found by Pietsch and Grobecker
(1987) despite a thorough search. Since then, however, tiny postflexion larval stages of
Fowlerichthys avalonis (3.4 to 5.3 mm SL, significantly smaller than known "scutatus"
stages of *F. radiosus*, which are typically between about 15 and 24 mm SL; see Hubbs,
1958: 285, table 2), with comparable morphological features, were described and figured
by Watson (1996b: 560, 1998: 234; Fig. 30). The presence of this specialized "scutatus" larval stage in two members of the genus *Fowlerichthys*, while unknown anywhere else in
the family, hints at the possibility that this feature might be unique to *Fowlerichthys*.

Key to the Known Species of the Genus *Fowlerichthys*

This key, like the other keys in this volume, works by progressively eliminating the
most morphologically distinct taxon and for that reason should always be entered from
the beginning. All characters listed for each species must correspond to the specimen
being keyed; if not, proceed to the next set of character states. For an additional or alternative means of identification, all species of *Fowlerichthys* may be compared simultaneously by referring to Table 3.

1. Pectoral-fin rays all simple; a single basidorsal ocellus usually present on side of body;
    illicium relatively long, 14.6 to 19.4% SL; eyes unusually large, diameter greater than
    7% SL; membranes between rays of dorsal and anal fins deeply incised; dorsal-fin rays
    all bifurcate (Table 3) . . . .*Fowlerichthys radiosus* (**Garman, 1896**), **p. 60**
    Atlantic Ocean

2. Pectoral-fin rays all bifurcate; three ocelli on side of body, each surrounded by a narrow
    lightly pigmented ring, one basidorsal in position, one about mid-body, and one cen-
    tered on caudal fin (caudal ocellus very rarely absent); dermal spinules nearly always
    present along anterior margin of illicium (Fig. 27B); pectoral-fin rays never more
    than 12 (Table 3) . . . .*Fowlerichthys ocellatus* (**Bloch and Schneider, 1801**), **p. 55**
    Western Atlantic Ocean

3. Pectoral-fin rays all bifurcate; a single basidorsal ocellus usually present on side of
    body; membrane behind second dorsal-fin spine distinctly divided into naked dorsal
    and ventral portions by a dense cluster of dermal spinules, the ventral portion extend-
    ing across naked depressed area between second and third dorsal-fin spines, nearly
    reaching to base of third (Fig. 27E); esca consisting of a tuft of elongate slender fila-
    ments (Fig. 50) . . . .*Fowlerichthys senegalensis* (**Cadenat, 1959**), **p. 68**
    Eastern Atlantic Ocean

4. Pectoral-fin rays all bifurcate; a single basidorsal ocellus usually present on side of body; membrane behind second dorsal-fin spine distinctly divided into naked dorsal and ventral portions by a dense cluster of dermal spinules, the ventral portion extending across naked depressed area between second and third dorsal-fin spines, nearly reaching to base of third (Fig. 27A); esca consisting of a tight cluster of short, vertically aligned appendages (Fig. 33) . . . .

. . . . *Fowlerichthys avalonis* (**Jordan and Starks, 1907), p. 50**
Eastern Pacific Ocean

5. Pectoral-fin rays all bifurcate; a single basidorsal ocellus usually present on side of body; membrane behind second dorsal-fin spine not divided into naked dorsal and ventral portions, the ventral half thick, densely covered with dermal spinules, and terminating distinctly anterior to base of third spine (Fig. 27D); esca consisting of a simple, oval-shaped appendage with numerous, more for less parallel, vertically aligned folds in small specimens (about 100 mm SL), a dense cluster of fine filaments in larger specimens (greater than 200 mm SL; Fig. 46) . . . .

. . . .*Fowlerichthys scriptissimus* (**Tanaka, 1916), p. 63**
Indo-West Pacific Ocean

Fowlerichthys avalonis (Jordan and Starks)
Roughjaw Frogfish

Figs. 27A, 30–35, 279; Tables 3, 22

Antennarius avalonis Jordan and Starks, 1907: 76, fig. 8 (original description, single specimen, holotype SU 9979, Avalon Bay, Santa Catalina Island, California; see Comments, p. 54). Hiyama, 1937: 66, pl. 101 (Pacific coast of Mexico). Böhlke, 1953: 147 (holotype at CAS). Fitch and Lavenberg, 1975: 16, fig. 4 (distinguishing characters, natural history). Thomson and Lehner, 1976: 9, fig. 3 (ecology, northern Gulf of California). Thomson et al., 1979: 53, fig. 31 (description, ecology; Santa Catalina Island, California, throughout Gulf of California to Peru). Hubbs et al., 1979: 13 (California). Pietsch, 1981b: 419 (osteology). Bartels et al., 1983: 89, 95, tables 1, 6, 10, 11, 15 (Gulf of Nicoya, Costa Rica). Eschmeyer and Herald, 1983: 112 (tropical Eastern Pacific). Pietsch, 1984b: 36 (genera of frogfishes). Pietsch and Grobecker, 1987: 125, figs. 38D, 47, 48, 120, pl. 25 (description, distribution, relationships). Ellis et al., 1988: 175, fig. 2 (Laguna Beach, California). Allen and Robertson, 1994: 82 (tropical Eastern Pacific). Bussing and López, 1994: 68 (Central America). Schneider and Lavenberg, 1995: 856 (tropical Eastern Pacific). Bearez, 1996: 734 (Ecuador). Galván-Magaña et al., 1996: 302 (Cerralvo Island, Baja California, Mexico). Watson, 1996b: 560, fig. 1 (larval development, "scutatus" postflexion stages). De La Cruz Agüero et al., 1997: 52 (Baja California, Mexico). Grove and Lavenberg, 1997: 234, figs. 116c, color fig. 18 (species account, Galápagos Islands). Chirichigno and Vélez, 1998: 221 (Peru). Michael, 1998: 347 (identification, behavior, captive care). Watson, 1998: 219, 220, 234, 235, fig. 11, table 3 ("scutatus" postflexion larval stages, osteology). Beltrán-León and Herrera, 2000: 270, fig. 88 (larvae). Thomson et al., 2000: 24, 55, 296, fig. 16 (Gulf of California). Shedlock et al., 2004: 134, 144, table 1 (molecular phylogeny). Bussing and López, 2005: 46, 47, fig. (Cocos Island). Carnevale and Pietsch, 2006: 452, 453, table 1 (comparison, meristics, first fossil frogfish). Castellanos-Galindo et al., 2006: 199 (Colombia). Kuiter and Debelius, 2007: 123, color fig. (Panama). Mejía-Ladino et al., 2007: 297, figs. 5, 6, tables 1–4, 6 (species account; in key; Colombia). Puentes et al., 2007: 87, 92, tables 1, 3 (Colombia). McCosker and Rosenblatt, 2010: 190 (Galápagos). Sielfeld, 2010: 757, figs. 1–3, tables 1, 2 (Iquique, northern Chile). Carnevale and Pietsch, 2011: 8 (branched pectoral-fin rays). Cruz-Agüero et al., 2012: 2, table 1 (mass stranding, Gulf of California). González-Díaz et al., 2013: 206 (Nayarit, Mexico). Walther-Mendoza et al., 2013: 411, 413, table 1 (Guadalupe Island, northwest Mexico).

Antennarius (*Fowlerichthys*) *avalonis*: Schultz, 1957: 87, fig. 6 (synonymy, description; Mexico, Panama, Peru; in key). Schultz, 1964: 173, table 1 (additional material, Gulf of California).

Antennarius sanguineus (not of Gill): Meek and Hildebrand, 1928: 1013 (description, Panama Bay). Seale 1940: 46 (Costa Rica). Hildebrand, 1946: 501 (description; Isla San Lorenzo, Peru).

Fowlerichthys avalonis: Arnold and Pietsch, 2012: 122, 126, 128 (new combination, molecular phylogeny). Page et al., 2013 (common names). Avendaño-Ibarra et al., 2014: 109, 113, table 3 (larvae; Gulf of California to Colima, Mexico). Murase et al., 2014: 1405, 1410, table 1 (Gulf of Nicoya, Costa Rica). Palacios-Salgado et al., 2014: 480, table 2 (Acapulco, Mexico). Ayala-Bocos et al., 2015: 1 (Socorro Island, Revillagigedo Archipelago). Fuentes et al., 2015: 615, table 3 (off El Salvador, Central America). Del Moral-Flores et al., 2016: 607 (Revillagigedo Archipelago). Martínez-Muñoz et al., 2016: 689, table 1 (Gulf of Tehuantepec).

MATERIAL EXAMINED

One hundred and seventy-five specimens, 12 to 330 mm SL.

Holotype of *Fowlerichthys avalonis*: SU 9979, 252 mm, Avalon Bay, Santa Catalina Island, California; Jordan and Holder, May 1906 (see Comments, p. 54; Fig. 31).

Fig. 31. *Fowlerichthys avalonis* (Jordan and Starks). Holotype, SU 9979, 252 mm SL, collected at Avalon Bay, Santa Catalina Island, California (see Comments, p. 54). Drawing by William S. Atkinson; after Jordan and Starks (1907, fig. 8); courtesy of Lisa Palmer and the Smithsonian Institution, NMNH, Division of Fishes, Washington, DC.

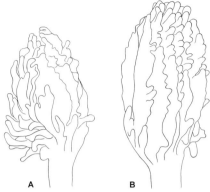

Fig. 32. *Fowlerichthys avalonis* (Jordan and Starks). Pelican Rock, Cabo San Lucas, Baja California. Photo by Salvador Garcia.

Fig. 33. Escae of *Fowlerichthys* avalonis (Jordan and Starks). (**A**) UW 20810, 58 mm SL, Bahía de Sechura. (**B**) UW 20810, 75 mm SL, Bahía de Sechura. Drawings by Cathy L. Short; after Pietsch and Grobecker (1987, fig. 48).

Additional material: *Gulf of California*: CAS, 8 (25–86 mm); LACM, 4 (32–90.5 mm); SIO, 18 (21–117 mm); SU, 5 (14.5–47); UA, 31 (23.5–330 mm); UCLA, 32 (15–138.5 mm). *Mexico outside the Gulf of California*: SIO, 3 (38–54 mm). *Costa Rica*: CAS, 4 (47.5–123 mm); LACM, 1 (44 mm); UCLA, 2 (150–157 mm). *Panama*: FMNH, 4 (35–47 mm); GCRL, 1 (41 mm); MCZ, 1 (13.5 mm); SIO, 4 (12.5–124 mm). *Colombia*: UW, 1 (140 mm). *Ecuador*: UW, 14 (36.5–140 mm). *Galápagos Islands*: LACM, 1 (12 mm). *Peru*: USNM, 32 (54–93 mm); UW 20813, 7 (58–88 mm).

DIAGNOSIS

A member of the genus *Fowlerichthys* unique in having the following combination of character states: a darkly pigmented basidorsal ocellus on each side of body, encircled by a lightly pigmented ring (Figs. 31, 32); illicium without dermal spinules; esca a simple, oval-shaped cluster of numerous short, vertically aligned appendages (Fig. 33); membrane behind second dorsal-fin spine divided into naked dorsal and ventral portions by dense cluster of dermal spinules (Fig. 27A); eye diameter 3.7 to 7.5% SL; dorsal-fin rays 13 (rarely 12 or 14), posteriormost 10 to 13 bifurcate; anal-fin rays eight (rarely seven, nine, or 10); pectoral-fin rays 13 (rarely 11, 12, or 14), all bifurcate; all rays of pelvic fin bifurcate (Table 3).

DESCRIPTION

Esca length not more than 40% illicium length; illicium 6.5 to 14.1% SL, equal to or slightly shorter than second dorsal-fin spine; second dorsal-fin spine more or less straight, hooked at tip, length 10.3 to 20.6% SL; third dorsal-fin spine straight to slightly curved posteriorly, length 15.3 to 29.9% SL; naked ventral portion of membrane behind second dorsal-fin spine extending posteriorly, dividing depressed naked area between second and third dorsal-fin spines and nearly reaching to base of third (Fig. 27A). For description of the "scutatus" postflexion larval stage of this species, see Comments, p. 49.

Coloration in life: "lemon yellow, bright orange and red to shades of brown and black" (Thomson et al., 1979: 53; Fig. 34). Dark brown to greenish-brown on face and body, heavily mottled with black and orange; belly and underside of pectoral fin tan with numerous diffuse brown to yellow-brown spots; unpaired fins tan with large, irregularly shaped brown spots; basidorsal ocellus black surrounded by narrow orange ring (Pietsch and Grobecker, 1987, pl. 24).

Coloration in preservation: gray, brown, dark brown, to yellowish- or pinkish-brown, with irregular, slightly darker spotting and mottling over entire head and body, particularly on belly; three to six dark, irregularly shaped patches usually present on side of body posterior to base of pectoral-fin lobe; more or less interconnected dark blotches usually present across all fins (forming distinct irregular bars when fins are closed), oblique in position on soft-dorsal and anal fins, vertical on caudal; illicium usually lightly banded; esca more or less darkly pigmented at base, the appendages covered with tiny black spots; as many as 10 short, darkly pigmented bars radiating from eye.

One of the largest known antennariids, attaining a standard length of at least 330 mm (UA 75.111.1), exceeded only by its congener *F. scriptissimus*, known to reach 432 mm SL.

DISTRIBUTION

Fowlerichthys avalonis is restricted to the Eastern Pacific Ocean (Fig. 279, Table 22), where it is found throughout the Gulf of California and south along the coasts of Mexico and Central and South America to the port city of Iquique, northern Chile (at about 20° S latitude), with a number of vouchered records from off southern California, at least one

Fig. 34. *Fowlerichthys avalonis* (Jordan and Starks). (**A**) Laguna Beach, Southern California. (**B**) Bahía de los Ángeles, Gulf of California. Photos by Mike Bartick.

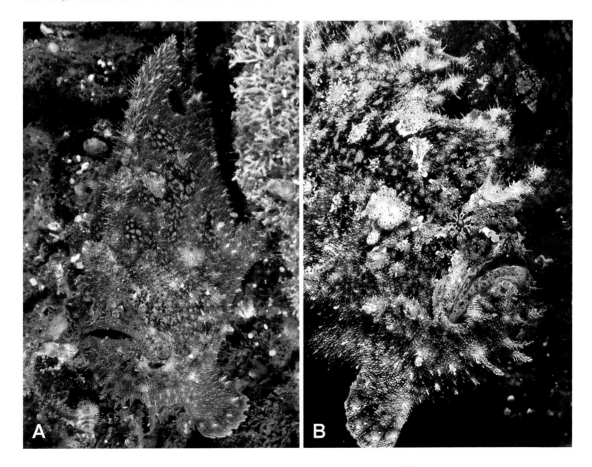

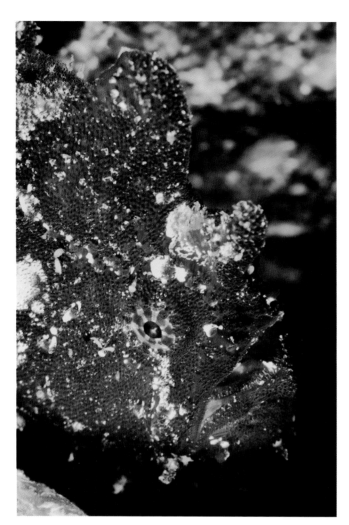

Fig. 35. *Fowlerichthys avalonis* (Jordan and Starks). Sea of Cortez. Photo by Marli Wakeling.

specimen from Guadalupe Island, Mexico (Walther-Mendoza et al., 2013), and several from the Revillagigedo Archipelago (Del Moral-Flores et al., 2016) and the Galápagos Islands (Grove and Lavenberg, 1997; LACM 20898). The type locality is Avalon Bay, Santa Catalina Island, Southern California (see Comments below).

Depths of capture, known for 63 specimens (most of which were taken off flat, sand, or mud bottoms by shrimp or otter trawl), range from the surface to some unknown lower limit that certainly exceeds 200 m and may extend down to at least 300 m (USNM 232094 was trawled somewhere between 192 and 311 m). However, 91% of the material was captured at depths of less than 160 m, 83% at less than 95 m. The average depth for all known captures is 88 m.

COMMENTS

The holotype of *F. avalonis*, said to have been collected in Avalon Bay, Santa Catalina Island, off southern California, was for some 80 years the only known record of this species from outside the Gulf of California, north of the tip of the Baja Peninsula (Fig. 279). Jordan and Starks (1907: 67), in describing this species in their "Notes on Fishes from the Island of Santa Catalina, Southern California," mentioned that "several rare species were obtained" in May 1906, during a visit to the bay of Avalon, "most of which were fishes which had died at the local aquarium." Pietsch and Grobecker (1987: 129) therefore suggested that the specimen of *F. avalonis* might have been brought alive from elsewhere for public display and later, after its death, became mixed with locally collected material. However, a second California specimen (CMA 87.4.1, 275 mm SL), was taken on hook and line off a shallow reef at a depth of 13 m in Scotchmans Cove, Laguna Beach, on 12 April 1986. The specimen was kept alive for four months in a 50-gallon tank at the Cabrillo Marine Aquarium in San Pedro, California, feeding sporadically on free-swimming Topsmelt, *Atherinops affinis* (see Ellis et al., 1988: 175). Since that time, there have been several unpublished sightings of *F. avalonis* off southern California: a specimen photographed by divers "on the wall" at Scripps Canyon, and at least two off Laguna Beach (Aaron Halstead, personal communication, 9 November 2016; Mike Bartick, personal communication, 11 September 2017; Fig. 34A).

Fowlerichthys ocellatus (Bloch and Schneider)
Ocellated Frogfish

Figs. 27B, 36–40, 280; Tables 3, 22

Pescador: Parra, 1787: 1, pl. 1 (Havana, Cuba).

Lophius histrio **var. d. *ocellatus*** Bloch and Schneider, 1801: 142 (original description after Parra, 1787, type material unknown, not in ZMB, iconotype in Parra, 1787, pl. 1; *habitat mare Hauannam alluens*). Cuvier, 1817b: 428 (same as or close to *Chironectes biocellatus*). Gill, 1879c: 225 (after Bloch and Schneider, 1801).

Chironectes ocellatus: Cloquet, 1817: 599 (new combination; binomial for *Pescador* of Parra, 1787; varietal name originally proposed by Bloch and Schneider, 1801, elevated to rank of species; no specimens). Cuvier, 1829: 252 (no additional material). Valenciennes, 1837: 419 (after Parra, 1787, and Bloch and Schneider, 1801). Valenciennes, 1842: 189 (after Valenciennes, 1837). Storer, 1846: 382 (Gulf of Mexico). Poey y Aloy, 1853: 219 (description after Parra, 1787; Cuba). Guichenot, 1853: 215 (after Valenciennes, 1837). Pietsch, 1984b: 36 (genera of frogfishes). Pietsch, 1985a: 80, fig. 18, table 1 (original manuscript sources for *Histoire Naturelle des Poissons* of Cuvier and Valenciennes). Sanders and Anderson, 1999: 42 (off Charleston, South Carolina; based on description and figure from 1822 manuscript of Edmund Ravenel). Stephens, 2000: 62 (reference to 1822 manuscript of Edmund Ravenel).

Chironectes biocellatus (not of Cuvier): Guichenot, 1853: 214 (misidentification, erroneous Western Atlantic locality after Valenciennes, 1837).

Lophius ocellatus: Poey y Aloy, 1861: 382 (new combination; after Bloch and Schneider, 1801). Sanders and Anderson, 1999: 42 (off Charleston, South Carolina; based on description and figure from 1822 manuscript of Edmund Ravenel).

Antennarius ocellatus: Günther, 1861a: 196 (new combination; after Bloch and Schneider, 1801, Valenciennes, 1837). Poey y Aloy, 1868: 405 (after Parra, 1787, Valenciennes, 1837; Cuba). Poey y Aloy, 1876: 134 (after Poey y Aloy, 1868). Jordan, 1885: 45 (Egmont Key, Florida). Garman, 1896: 82 (description, additional material; Bahamas). Jordan and Evermann, 1898: 2722 (Pensacola, Florida; synonymy includes *A. pleurophthalmus* Gill). Evermann and Marsh, 1900: 334 (West Indies; in key). Jordan, 1905: 550 (West Indies). Rea, 1909: 53 (off Charleston, South Carolina). Burton, 1932: 13, fig. ("sargassum fish," off Charleston, South Carolina; drawing reproduced from 1822 manuscript of Edmund Ravenel). Barbour, 1942: 30, 34, 38, pls. 12, 17 (fig. 6) (off South Carolina, Gulf of Mexico; Puerto Rico record erroneous). Caldwell, 1957: 128 (Cedar Key, Florida). Duarte-Bello, 1959: 142 (Cuba). Cervigón, 1966: 873, 877 (Cubagua, Venezuela; in key). Böhlke and Chaplin, 1968: 720, figs. (description, North Carolina to Venezuela). Randall, 1968: 291 (Caribbean). Smith et al., 1975: 5 (Florida). Guitart, 1978: 789, fig. 620 (Cuba). Pietsch, 1978a: 3 (western central Atlantic). Román, 1979: 352, fig. (Venezuela). Pietsch, 1984b: 36 (genera of frogfishes). Pietsch, 1985a: 80, fig. 18, table 1 (original manuscript sources for *Histoire Naturelle des Poissons* of Cuvier and Valenciennes). Robins and Ray, 1986: 88, pl. 14 (North Carolina, Bahamas, northern Gulf of Mexico to South America). Pietsch and Grobecker, 1987: 131, figs. 38F, 51–53, 121 (description, distribution, relationships). Scott and Scott, 1988: 237 (unverified larvae; Scotian Shelf, Canada). McAllister, 1990: 228 (after Scott and Scott, 1988). Cervigón, 1991: 189 (Venezuela). Van Dolah et al., 1994: 10, table 5 (off South Carolina and Georgia). McEachran and Fechhelm, 1998: 819 (Gulf of Mexico).

Michael, 1998: 335, 343, 344, 347, color figs. (identification, behavior, captive care). Smith-Vaniz et al., 1999: 50, 160, pl. 10, fig. 58 (Bermuda; in key). Sanders and Anderson, 1999: 42, 167, 253 (off Charleston, South Carolina; after Burton, 1932). Schmitter-Soto et al., 2000: 153 (Mexico). Stephens, 2000: 62 (reference to 1822 manuscript of Edmund Ravenel). Pietsch, 2002a: 1051 (western central Atlantic). Smith et al., 2003: 13 (Belize). Schwartz, 2005: 145, 146, 148 (North Carolina). Carnevale and Pietsch, 2006: 452, 453, table 1 (comparison, meristics, first fossil frogfish). Jackson, 2006a: 783, 784, tables 1, 2 (reproduction, depth distribution). Walsh et al., 2006: 263, 275, table 4 (southeast United States). Kuiter and Debelius, 2007: 123, color fig. (Caribbean). Mejía-Ladino et al., 2007: 286, figs. 4, 6, tables 1–3, 5 (species account; in key; Colombia). Torruco et al., 2007: 513, 524, table 1 (southwestern Gulf of Mexico). Rohde et al., 2009: xv, xxi (off Charleston, South Carolina; after Burton, 1932). Carnevale and Pietsch, 2011: 8 (branched pectoral-fin rays).

Antennarius pleurophthalmus Gill, 1863: 92 (original description, single specimen, holotype USNM 5886, Key West, Florida). Goode and Bean, 1882: 235 (Gulf of Mexico). Jordan and Gilbert, 1883: 846 (after Gill, 1863). Gill, 1883: 556 (reference to Goode and Bean, 1882). Goode and Bean, 1896: 487 (Florida Keys, 55–154 m). Longley and Hildebrand, 1941: 308 (Florida).

Antennarius astroscopus (not of Nichols): Fowler, 1940b: 19, 21, fig. 37 (west coast of Florida).

Antennarius (Fowlerichthys) ocellatus: Schultz, 1957: 89, pl. 8, fig. 8 (synonymy, description; North Carolina; Puerto Rico record erroneous; in key).

Antennarius ocellatus (not of Bloch and Schneider): Paulin, 1978: 488, fig. 1B (in part, synonymy includes *F. avalonis*; NMNZ P.001510 is *F. scriptissimus*).

Fowlerichthys ocellatus: Arnold and Pietsch, 2012: 126, 128 (new combination, molecular phylogeny). Page et al., 2013 (common names). Smith-Vaniz and Collette, 2013: 169 (Bermuda).

MATERIAL EXAMINED

One hundred and thirty-one specimens, 19 to 320 mm SL.

Iconotype of *Antennarius ocellatus*: in Parra, 1787, pl. 1, Havana, Cuba (see Comments, p. 59; Fig. 36).

Holotype of *Antennarius pleurophthalmus*: USNM 5886, 110 mm, probably Key West, Florida.

Additional material: *Western Atlantic*: AMNH, 1 (157 mm); ANSP, 1 (177 mm); USNM, 3 (56–108 mm). *Florida and Gulf of Mexico*: AMNH, 1 (71 mm); ANSP, 1 (54.5 mm); FMNH, 7 (37.5–190 mm); FSBC, 51 (24.5–320 mm); FSUT, 19 (19–250 mm); LACM, 1 (73 mm); MCZ, 7 (61.5–184 mm); UMMZ, 9 (43–165 mm); USNM, 16 (22–210 mm); UW, 8 (52.5–164 mm). *Caribbean*: FMNH, 1 (85 mm); USNM, 3 (50–165 mm).

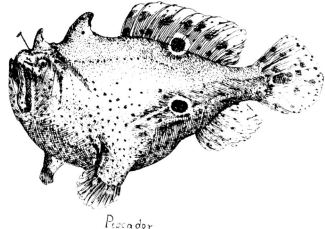

Fig. 36. Antonio Parra's *Pescador.* The iconotype of *Fowlerichthys ocellatus* (Bloch and Schneider), Havana, Cuba. After Parra (1787, pl. 1).

Fig. 37. *Fowlerichthys ocellatus* (Bloch and Schneider). LACM 8975–2, 73 mm SL, near Destin, Florida. Drawing by Cathy L. Short; after Pietsch and Grobecker (1987, fig. 51).

DIAGNOSIS

A member of the genus *Fowlerichthys* unique in having three darkly pigmented ocelli on side of body, each encircled by a lightly pigmented ring (Fig. 37): one basidorsal, one about mid-body, and one centered on the caudal fin (caudal ocellus absent in only one of 131 specimens examined: UW 10592, 52.5 mm SL).

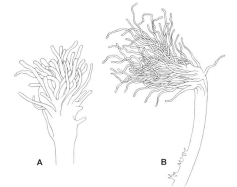

Fig. 38. Escae of *Fowlerichthys ocellatus* (Bloch and Schneider). **(A)** LACM 8975–2, 73 mm SL, vicinity of Destin, Florida. **(B)** UW 10592, 164 mm SL, off Dry Tortugas Light, Florida. Drawings by Cathy L. Short; after Pietsch and Grobecker (1987, fig. 52).

Fowlerichthys ocellatus is further distinguished by having the following combination of character states: illicium with dermal spinules along anterior margin (Fig. 37); esca a dense cluster of elongate, slender filaments (Fig. 38); membrane behind second dorsal-fin spine divided into naked dorsal and ventral portions by dense cluster of dermal spinules (Fig. 27B); eye diameter 3.9 to 6.9% SL; dorsal-fin rays 13 (rarely 12), all bifurcate; anal-fin rays eight (rarely seven); pectoral-fin rays 12 (rarely 11), all bifurcate; all rays of pelvic fin bifurcate (Table 3).

DESCRIPTION

Filaments of esca considerably less than illicium length to more than twice illicium in largest specimens (e.g., FSUT 17078, 205 mm SL; USNM 185983, 210 mm SL); illicium length 8.6 to 15.0% SL, equal to or slightly shorter than second dorsal-fin spine; second dorsal-fin spine curved posteriorly, length 9.5 to 16.5% SL (both spines becoming distinctly shorter in proportion to increasing standard length); third dorsal-fin spine curved posteriorly, length 21.5 to 28.0% SL; naked ventral portion of membrane behind second dorsal-fin spine extending posteriorly, dividing depressed, naked area between second and third dorsal-fin spines and nearly reaching to base of third (Fig. 27B).

Coloration in life: extremely variable; off-white, beige, lemon yellow, orange, red, maroon, pink, green, violet, purple, yellowish-brown, dark brown, gray, with darker mottling,

Fig. 39. *Fowlerichthys ocellatus* (Bloch and Schneider). Destin, Florida. Photo by Jim Garin.

spots and patches (scabby looking patches reminiscent of encrusting coralline algae); ocelli usually black, surrounded by a pale ring, and sometimes an additional dark outer ring (Figs. 39, 40).

Coloration in preservation: beige, light brown, yellowish-brown, brown, or gray to slightly pinkish; lightly pigmented ring of ocelli often surrounded by a more or less diffuse brown ring; numerous small, darkly pigmented spots nearly always present on head, chin, belly, and fins; usually one or two dark, eyelike spots on lateral surface of second and third dorsal-fin spines; illicium nearly always banded; escal filaments darkly pigmented at base; as many as nine short, darkly pigmented bars radiating from eye.

Largest known specimen: 320 mm SL (FSBC 1333); the largest antennariid in the Atlantic Ocean and third in size relative to its congeners, the Eastern Pacific *F. avalonis* reaching 330 mm SL and the Indo-West Pacific *F. scriptissimus*, 432 mm SL.

DISTRIBUTION

Fowlerichthys ocellatus is restricted to the Western Atlantic Ocean. It ranges from off North Carolina to Venezuela, including Bermuda, the Bahamas, the eastern and southwestern coastal margins of the Gulf of Mexico, continental coastal regions of the Caribbean Sea, and also western Cuba (Fig. 280, Table 22). The type locality is Havana, Cuba.

Although occasionally taken by hand in 1 or 2 m in the northern Gulf of Mexico, and often observed by scuba divers at depths of 15 to 25 m (but likely extends down to at least 50 m; Jim Garin, personal communication, 4 June 2017), *F. ocellatus* has most often been collected throughout most of its range by shrimp trawl or scallop dredge on shell rubble, mud, or sand bottoms, ranging in depth from about 20 to 90 m and averaging

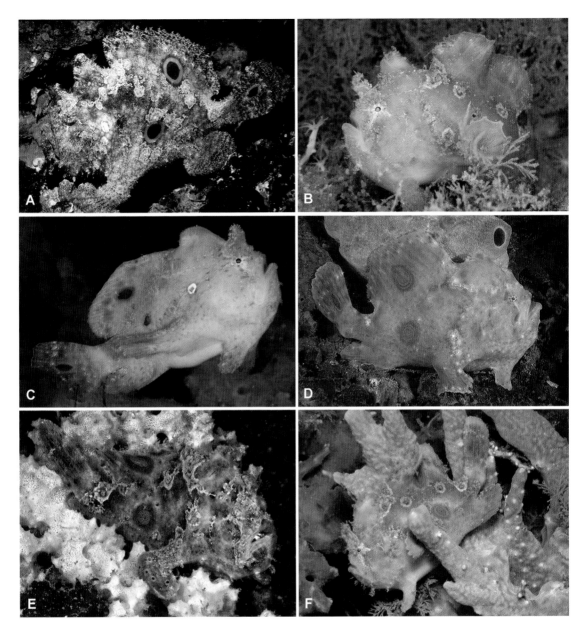

Fig. 40. *Fowlerichthys ocellatus* (Bloch and Schneider). **(A)** Destin, Florida. Photo by Scott W. Michael. **(B)** Blue Heron Bridge, Lake Worth Lagoon, Florida. Photo by Linda Ianniello. **(C)** Blue Heron Bridge, Lake Worth Lagoon, Florida. Photo by Kelly Casey. **(D, E)** Destin, Florida. Photos by Jim Garin. **(F)** Riviera Beach, Lake Worth Lagoon, Florida. Photo by Anne DuPont.

about 50 m. Goode and Bean (1896: 487) reported collections of this species at depths exceeding 150 m (*Blake* haul 242).

COMMENTS

Fowlerichthys ocellatus was first described from Cuban waters by Portuguese naturalist Antonio Parra (1787) under the vernacular name *Pescador*. His figure, reproduced here (Fig. 36), clearly shows the diagnostic triad of ringed ocelli. Bloch and Schneider's (1801) later description, under the name *Lophius histrio* var. d. *ocellatus*, is also undoubtedly that of *F. ocellatus*, but whether it was based on an examination of specimens or founded solely on the authority of Parra (as stated by Günther, 1861a: 197) is unknown. In any case, no type material seems to have survived. Because the species can stand without question on the figure of Parra (1787), no neotype has been designated.

Fowlerichthys radiosus (Garman)
Singlespot Frogfish

Figs. 27C, 28, 29, 41–44, 279; Tables 3, 22

Antennarius **sp.**: Nutting, 1895: 149, pl. 14, fig. 2 (Bahamas).

Antennarius radiosus Garman, 1896: 85, pl. 1 (original description, two specimens, lectotype MCZ 35215, Key West, Florida; whereabouts of paralectotype unknown). Jordan and Evermann, 1898: 2725 (after Garman, 1896). Evermann and Marsh, 1900: 334 (West Indies; in key). Beebe and Tee-Van, 1933: 250, fig. (Bermuda). Longley and Hildebrand, 1941: 305 (additional material, Florida Keys). Barbour, 1942: 31, 34, 38, pls. 16 (fig. 1), 17 (fig. 3) (in part, includes *A. bermudensis*; Pensacola to Key West). Maul, 1959: 15 (Madeira). Palmer, 1960: 149 (first record from Irish waters). Wheeler, 1969: 584, fig. 393 (Irish coast). Monod and Le Danois, 1973: 664 (western central Atlantic). Pietsch, 1978a: 3 (western central Atlantic). Pietsch, 1981a: 2 (eastern central Atlantic). Divita et al., 1983: 404 (Texas coast). Uyeno and Aizawa, 1983: 248 (Suriname, French Guiana). Pietsch, 1984b: 36 (genera of frogfishes). Pietsch, 1984d: 322, fig. 165 ("scutatus" postflexion larval stage). Wheeler and Van Oijen, 1985: 105 (west coast of Ireland). Pietsch, 1986b: 1365, 1366, fig. (eastern North Atlantic). Robins and Ray, 1986: 88, pl. 14 (North Carolina to Cuba, entire Gulf of Mexico). Pietsch and Grobecker, 1987: 118, figs. 38B, 41–44, 120 (description, distribution, relationships). Cervigón, 1991: 190 (Venezuela). Lloris et al., 1991: 220, appendix (Madeira). Boschung, 1992: 79 (Alabama). Van Dolah et al., 1994: 10, table 5 (off South Carolina and Georgia). Arruda, 1997: 125 (Azores). Santos et al., 1997: 125 (Azores record questioned). Aguilera, 1998: 47 (Venezuela). Hoese and Moore, 1998: 164, pl. 103 (Gulf of Mexico). McEachran and Fechhelm, 1998: 821 (Gulf of Mexico). Smith-Vaniz et al., 1999: 160 (Bermuda, records based on misidentifications). Pietsch, 2002a: 1051 (western central Atlantic). Shedlock et al., 2004: 134, 144, table 1 (molecular phylogeny). Ortiz-Ramirez et al., 2005: 23 (early life history). Schwartz, 2005: 145, 146, 148 (North Carolina). Carnevale and Pietsch, 2006: 452, 453, table 1 (comparison, meristics, first fossil frogfish). Jackson, 2006a: 783, 784, tables 1, 2 (reproduction, depth distribution). Walsh et al., 2006: 260, 263, 266, 275, tables 2, 4 (southeast United States). Mejía-Ladino et al., 2007: 288, figs. 4, 6, tables 1–3, 5 (species account; in key; Buritaca and Isla Fuerte, Colombia). Porteiro and Afonso, 2007: 57–59, fig. 1, table 1 (Azores). Wirtz et al., 2008: 8 (Madeira). Hanel et al., 2009: 147 (European seas). O'Connell et al., 2009: 110, appendix (off southeastern Louisiana). Arnold and Pietsch, 2012: 126, 128, figs. 4, 6, table 1 (molecular phylogeny).

Antennarius radiosus (not of Garman): Bean, 1906b: 89 (specimen subsequently made holotype of *A. bermudensis* by Schultz, 1957). Barbour, 1942: 38 (material includes *A. bermudensis*).

Fowlerichthys floridanus Barbour, 1941: 12, pl. 7 (original description, single specimen, holotype ANSP 70170, Palm Beach, Florida).

Kanazawaichthys scutatus Schultz, 1957: 63, pl. 14, fig. A (original description, four spec-imens, holotype USNM 157919, Gulf of Mexico). Hubbs, 1958: 283, tables 1, 2 (preju-venile of *F. radiosus*). Palmer, 1960: 149 (after Hubbs, 1958). Schultz, 1957: 177 (*K. scuta-tus*, synonym of *F. radiosus* after Hubbs, 1958). Pietsch, 1984d: 322 (prejuvenile of *F. radiosus*).

Antennarius (***Fowlerichthys***) ***radiosus***: Schultz, 1957: 87, pl. 7, fig. D (synonymy, descrip-tion; Florida, northern part of Gulf of Mexico; in key). Schultz, 1964: 177 (*K. scutatus* a synonym). Monod and Le Danois, 1973: 664 (synonymy, distribution).

Fowlerichthys radiosus: Arnold and Pietsch, 2012: 126, 128, fig. 6 (new combination, molecu-lar phylogeny). Page et al., 2013 (common names). Derouen et al., 2015: 29, fig. 3, table 1 (molecular phylogeny). Pietsch, 2016: 2053 (eastern central Atlantic; in key).

MATERIAL EXAMINED

Four hundred and fifty-five specimens, 14 to 180 mm SL.

Lectotype of *Antennarius radiosus*: MCZ 35215, 17.5 mm, *Blake* haul 227, Key West, Florida, 66 m.

Holotype of *Fowlerichthys floridanus*: ANSP 70170, 25 mm, off Palm Beach, Florida, 732–915 m, McGinty and McLean, 19 July 1940.

Holotype of *Kanazawaichthys scutatus*: USNM 157919, 21 mm, *Oregon* station 1273, off Mississippi River Delta, on surface (?), 9 March 1955.

Paratypes of *Kanazawaichthys scutatus*: USNM 174946, 1(17 mm), *Oregon* station 1370, east of Chandeleur Island, Louisiana, from stomach of Yellowfin Tuna, 20 August 1955; USNM 157920, 2(16–29.5 mm), data as for holotype.

Additional material: *Western Atlantic*: AMNH, 1 (52.5 mm); FMNH, 1 (64.5 mm); MCZ, 3 (16–117 mm); USNM, 5 (14.5–18.5 mm); ZMUC, 2 (29–30 mm). *Florida*: AMNH, 7 (18–72 mm); ANSP, 3 (23–31 mm); BMNH, 1 (18.5 mm); FMNH, 5 (21–112 mm); FSBC, 13 (17–68 mm); FSUT, 51 (14–56 mm); MCZ, 2 (49–59 mm); UMMZ, 51 (19–25.5 mm); UW, 60 (19–63 mm). *Alabama, Mississippi*: ANSP, 1 (14.5 mm); FMNH, 5 (38–62 mm); FSUT, 27 (19.5–85 mm); UMMZ, 1 (30.5 mm). *Louisiana*: AMNH, 1 (36.5 mm); FMNH, 14 (22–85 mm); FSBC, 4 (52.5–54 mm); LACM, 1 (31.5 mm); TU, 169 (19–38 mm); UMMZ, 1 (24.5 mm). *Texas*: ANSP, 1 (38.5 mm); FMNH, 3 (5986 mm); USNM, 1 (53 mm). *Yuca-tán*: FMNH, 8 (24–41 mm); FSUT, 1 (34 mm); MCZ, 1 (23.5 mm). *Ireland*: BMNH, 1 (11.5 mm). *Madeira*: MMF, 2 (175–180 mm).

DIAGNOSIS

A member of the genus *Fowlerichthys* unique in having the following combination of character states: a single, darkly pigmented, basidorsal ocellus on side of body, encir-cled with a narrow lightly pigmented ring (Fig. 41); illicium relatively long, 14.6 to 19.4% SL, equal to or slightly longer than second dorsal-fin spine, without dermal spinules; esca a simple, oval-shaped appendage with numerous, more or less parallel, vertically aligned folds (Fig. 42); membrane behind second dorsal-fin spine divided into naked dorsal and ventral areas by dense cluster of dermal spinules (Fig. 27C); membranes between rays of unpaired fins deeply incised; eyes unusually large, diameter 7.4 to 11.4% SL; dorsal-fin rays 13 (rarely 12), all bifurcate; anal-fin rays eight (rarely seven); pectoral-fin rays 13 (rarely 12 or 14), all simple; all rays of pelvic fin bifurcate (Table 3).

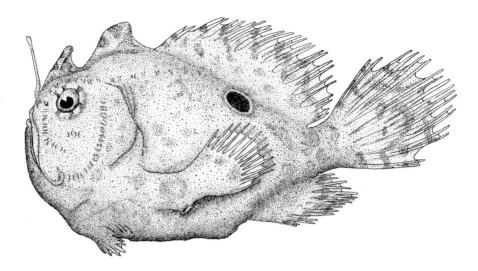

DESCRIPTION

Esca small, length not more than 20% illicium length; second dorsal-fin spine more or less straight, length 11.1 to 18.4% SL (Figs. 27C, 41); third dorsal-fin spine more or less straight, length 17.7 to 25.0% SL; naked ventral portion of membrane behind second dorsal-fin spine extending posteriorly, dividing depressed naked area between second and third dorsal-fin spines and nearly reaching to base of third (Fig. 27C). For description of the "scutatus" postflexion larval stage of this species, see Comments, p. 49.

Coloration in preservation: beige, light brown, reddish brown, to grayish or yellowish brown, occasionally lightly mottled; lightly pigmented ring of basidorsal ocellus occasionally surrounded by narrow brown ring; median fins occasionally lightly barred; illicium usually lightly banded; esca unpigmented; as many as 12 short, darkly pigmented bars radiating from eye (Figs. 43, 44).

Material examined from the Western Atlantic did not exceed 117 mm SL, but the two known specimens from Madeira are considerably larger, measuring 175 and 180 mm SL (MMF 4754 and 23116, respectively).

DISTRIBUTION

In the Western Atlantic, *F. radiosus* ranges from off Long Island, New York (MCZ 38642), to the Florida Keys, throughout the Gulf of Mexico, Cuba, and farther south to Suriname and French Guiana (Fig. 279, Table 22). In the Eastern Atlantic, a single "scutatus" postflexion larval stage (11.5 mm, BMNH 1959.11.25.3) has been reported from off the west coast of Ireland, and two large adults (175 and 180 mm; MMF 4754, 23116) have been reported from Madeira. Its presence at the Azores has been questioned (Santos et al., 1997). The type locality is Key West, Florida.

In the Gulf of Mexico, *F. radiosus* has most often been taken by beach seine and shrimp trawl on muddy bottoms at wide-ranging depths, from just beneath the surface down to a maximum of 275 m and averaging about 90 m. "Scutatus" prejuveniles, less than 25 mm SL, are apparently pelagic, but depths are unknown. There is considerable evidence, at least in shallow inshore waters of the northern Gulf of Mexico subject to Mississippi River outflow, that this species can tolerate brackish conditions. Numerous collections have been made with beach seines at depths of less than 2 m in salinities as low

Fig. 43. *Fowlerichthys radiosus* (Garman). Destin, Florida. Photo by Scott W. Michael.

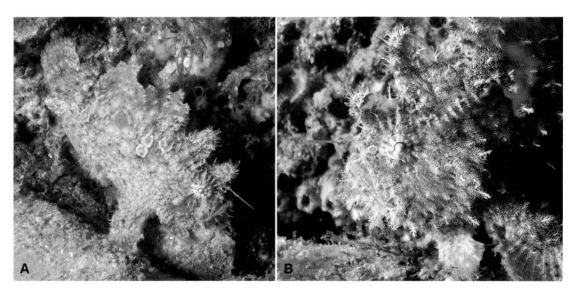

Fig. 44. *Fowlerichthys radiosus* (Garman). (**A**, **B**) Refugia Reef, near Mexico Beach, Florida, about 24 m. Photos by Carol Cox.

as 18.5 ppt; most were made in the winter or early spring months when temperatures and salinities of inshore waters are usually relatively low (John H. Caruso, personal communication, 14 and 19 March 2018).

Fowlerichthys scriptissimus (Jordan)
Calico Frogfish

Figs. 27D, 45–48, 279; Tables 3, 22

Antennarius scriptissimus Jordan, 1902: 373, fig. 4 (original description, single stuffed specimen, holotype originally in Imperial Museum of Tokyo but apparently lost, about 255 mm total length, Boshu, Province of Awa, entrance to Tokyo Bay). Franz, 1910: 87 (synonym of *A. tridens*). Jordan et al., 1913: 425 (after Jordan, 1902; "perhaps a much striped form of *A. tridens*"; Tokyo Bay). Jordan and Hubbs, 1925: 330 (synonym of *A. tridens*; Misaki). Okada et al., 1935: 263, pl. 158 (description, color plate; Tateyama, Boso Peninsula, Japan). Okada, 1938: 274 (Honshu). Tokioka, 1961: 450 (Kyoto, Japan). Yokota and Senou, 1991: 2, fig. 3 (senior synonym of *A. sarasa*; Sagami Sea). Lindberg et al., 1997: 218 (Sea of Japan). Okamura and Amaoka, 1997: 136, fig. (Japan). Pietsch, 1999: 2015 (western central Pacific; in key). Nakabo, 2000: 456 (Japan; in key). Pietsch, 2000: 597 (South China Sea). Nakabo, 2002: 456 (Japan; in key). Randall et al., 2002: 156 (Society Islands). Youn, 2002: 542 (in synonymy of *A. striatus*, Korea). Manilo and Bogorodsky, 2003: S99 (Oman). Senou et al., 2006: 423 (Sagami Sea). Fricke et al., 2009: 28 (Réunion). Nakabo, 2013: 539, figs. (Japan).

Antennarius tridens (not of Temminck and Schlegel): Pietschmann, 1909: 1, pl. 1 (description, Japan).

Antennarius sarasa Tanaka, 1916: 143 (original description, single specimen, holotype originally in Science College Museum, Tokyo, no. 7185, but present whereabouts unknown (mislaid in University Museum, University of Tokyo; Y. Tominaga, personal communication, 6 May 1981), about 200 mm, Tokyo Market). Tanaka, 1918: 432, pl. 119, fig. 346 (after Tanaka, 1916; three specimens collected at Tokyo Market). Okada, 1938: 274 (Honshu, Sikoku). Anonymous, 1981: 162, 490 (*scriptissimusizariuo*, calico frogfish, Japan). Araga, 1984: 103, pl. 88-D (Japan). Pietsch, 1984b: 36 (genera of frogfishes). Pietsch and Grobecker, 1987: 123, figs. 38C, 45, 46, 121 (description, distribution, relationships). Paulin et al., 1989: 134 (New Zealand). Yokota and Senou, 1991: 2 (junior synonym of *A. scriptissimus*; Sagami Sea). Randall, 1995: 84, fig. 170 (Oman). Lindberg et al., 1997: 215 (Sea of Japan). Fricke, 1999: 107 (Réunion). Pietsch, 1999: 2015 (western central Pacific; in key). Pietsch, 2000: 597 (synonym of *A. scriptissimus*). Randall et al., 2002: 156 (Society Islands). Letourneur et al., 2004: 209 (Réunion Island). Carnevale and Pietsch, 2006: 452, 453, table 1 (comparison, meristics, first fossil frogfish). Fricke et al., 2009: 28 (synonym of *A. scriptissimus*). Al-Jufaili et al., 2010: 18, appendix 1 (Oman). Carnevale and Pietsch, 2011: 8 (branched pectoral-fin rays).

Antennarius (*Fowlerichthys*) *sarasa*: Schultz, 1957: 88, pl. 8, fig. A (description after Tanaka, 1918).

Fowlerichthys scriptissimus: Arnold and Pietsch, 2012: 126, 128, figs. 1A, 4, 6, table 1 (new combination, molecular phylogeny). Stewart, 2015: 886 (New Zealand). Han et al., 2017: 1, figs. 1–4, tables 1, 2 (Jejudo Island, Korea).

Antennarius ocellatus (not of Bloch and Schneider): Paulin, 1978: 488, fig. 1B (in part, i.e., NMNZ P.001510, New Zealand; synonymy includes *F. avalonis* and *F. scriptissimus*).

MATERIAL EXAMINED

Twenty-three specimens, 99 to 432 mm SL.

Type specimens apparently lost.

Additional material: BPBM, 1 (252 mm); FMNH, 1 (215 mm); FRLM, 11 (132–208 mm); HUMZ, 1 (202 mm); KPM, 2 (215–221 mm); MNHN, 1 (200 mm); NMNZ, 1 (193 mm); NSMT 1 (228 mm); OMNH, 1 (148 mm); UW, 3 (99–432 mm).

DIAGNOSIS

A member of the genus *Fowlerichthys* unique in having the following combination of character states: a single, darkly pigmented, basidorsal ocellus on each side of body, encircled by a narrow lightly pigmented ring (weakly developed in some specimens); illicium of three largest specimens examined with dermal spinules along anterior margin; esca a simple, oval-shaped appendage with numerous, more or less parallel, vertically aligned folds in smallest known specimen, dividing into dense cluster of short filaments in larger specimens (Fig. 46); membrane behind second dorsal-fin spine not divided into naked dorsal and ventral portions by a cluster of dermal spinules, the dorsal portion weakly developed, the ventral portion thick and usually covered with dermal spinules (Fig. 27D); eye diameter 3.0 to 6.1% SL; dorsal-fin rays 13, only posteriormost three to nine bifurcate; anal-fin rays eight; pectoral-fin rays 13, all bifurcate; all rays of pelvic fin bifurcate (Table 3).

DESCRIPTION

Esca small, length not more than 20% illicium length (Fig. 46); illicium 6.8 to 10.6% SL; second dorsal-fin spine more or less curved posteriorly, length 8.2 to 10.1% SL; depression between second and third dorsal-fin spines naked in 99- and 215-mm specimens, spinulose in remaining specimens examined; third dorsal-fin spine curved posteriorly, length 20.4 to 23.2% SL; membrane behind second dorsal-fin spine extending posteriorly, dividing area between second and third dorsal-fin spines and nearly reaching to base of third in four largest specimens examined (Fig. 27D), but terminating well anterior to base of third spine in 99-mm specimen.

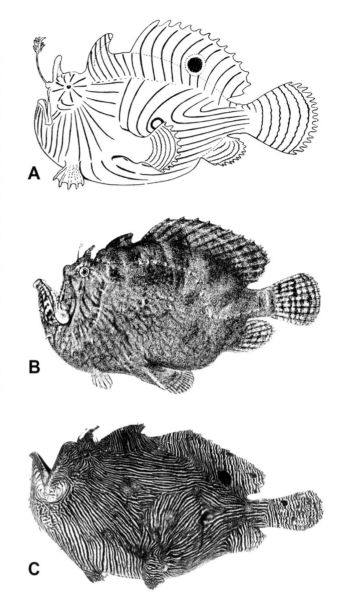

Fig. 45. *Fowlerichthys scriptissimus* (Jordan). (**A**) holotype, a stuffed specimen (presumed lost), about 255 mm in total length, Boshu, Province of Awa, entrance to Tokyo Bay, "a copy of a rough sketch of the type . . . designed only to show the markings . . . not correct as of details of form." After Jordan, 1902: 374, fig. 4. (**B**) holotype of *Antennarius sarasa* Tanaka, specimen presumed lost, about 200 mm SL, Tokyo market. After Tanaka, 1918, pl. 119, fig. 346. (**C**) Tateyama, Boso Peninsula, Chiba Prefecture, Japan, 400 mm total length. After Okada et al., 1935, pl. 158.

Coloration: two distinct color morphs: head, body, and fins everywhere covered with narrow, tightly spaced, dark reddish-brown bars and stripes on a light cream ground

Fig. 46. Esca of *Fowlerichthys scriptissimus* (Jordan). UW 20837, 280 mm SL, Tokyo, Japan. Drawing by Cathy L. Short; after Pietsch and Grobecker (1987, fig. 46).

color (Figs. 47, 48C, E, F); or beige, gray, brown, to dark reddish brown, heavily mottled with darker reticulations over entire head, body, and fins except for inner surface of paired fins (Fig. 48A, D); dark, narrow, irregular banding on fins; as many as 15 oblique bars on soft-dorsal and four oblique bars on anal, five or six vertical bars on caudal, about three bars across outer surface of paired fins; transitional forms showing evidence of both extremes occasionally observed (Fig. 48B). Large basidorsal ocellus usually present (but not indicated by Tanaka, 1918, in his drawing of the holotype of *Antennarius sarasa* reproduced here; see Fig. 45B); illicium banded; esca with dark pigment at base; seven to 16 short, darkly pigmented bars radiating from eye.

The largest known antennariid, reaching a standard length of at least 432 mm (UW 112642).

DISTRIBUTION

Fowlerichthys scriptissimus is known from about 30 specimens collected from off Réunion, Oman, Korea, Japan, Manila Bay in the Philippines, the North Island of New Zealand, and the Cook and Society Islands (Fig. 279, Table 22). Most frequently observed at depths of 19 to 67 m, it extends down to at least 350 m; the latter record (BPBM 31702, 252 mm SL) was made by hook and line off Mahina, Tahiti, by a local fisherman in July 1990. The holotype (see Comments below) was collected at the Tokyo Market.

COMMENTS

In the absence of type material, Pietsch and Grobecker (1987: 58) erroneously placed this species in the synonymy of *Antennarius striatus* following Jordan et al. (1913: 425) who suggested it was "Perhaps a much striped form of *A. tridens*," and Jordan and Hubbs (1925:

Fig. 47. *Fowlerichthys scriptissimus* (Jordan). Osezaki, Izu Peninsula, Suruga Bay, Japan, 35 m. Photo by Hiroyuki Tokuya.

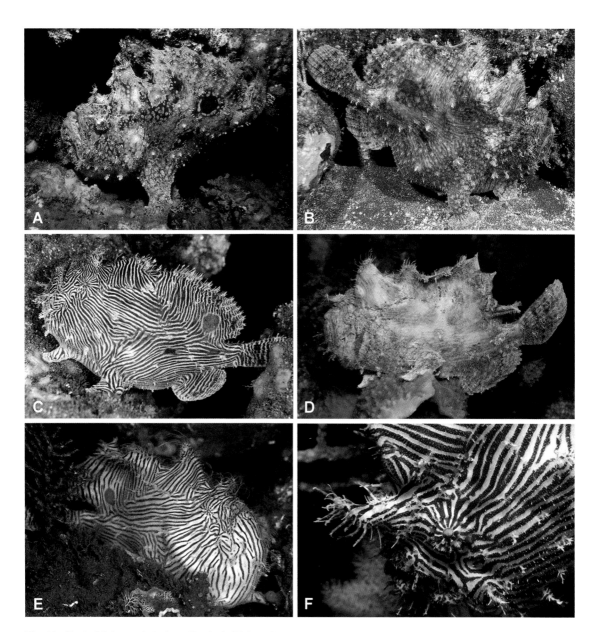

Fig. 48. *Fowlerichthys scriptissimus* (Jordan). (**A**) Hachijo-jima, Izu Islands, Japan. Photo by Shoichi Kato. (**B**) Oshima Island, Sagami Bay, Japan. Photo by Hiroyuki Uchiyama. (**C**) Akinohama, Oshima Island, Sagami Bay, Japan. Photo by Eiji Kozuka. (**D**) Osezaki, Izu Peninsula, Suruga Bay, Japan. Photo by Mika Oyry. (**E**, **F**) Osezaki, Izu Peninsula, Suruga Bay, Japan. Photos by Hiroyuki Tokuya.

330) who placed it in the synonymy of *A. tridens*, along with *A. sanguifluus* and *A. nox*: "We think all these are forms of one highly variant species, as Franz [1910: 87], with Jordan, Tanaka, and Snyder [1913: 425] have already indicated. They are all apparently color-phases of a single species which widely varies in coloration, according to its environment."

Pietsch and Grobecker were influenced further by the figure published by Jordan (1902: 373, fig. 4), despite the warning that it was only "a rough sketch of the type . . .

designed only to show the markings . . . not correct as of details of form." The drawing clearly shows the support for the illicium extending forward beyond the anterior margin of the upper jaw (Fig. 45A), a feature found only in members of the *Antennarius striatus* group. However, the description and color figure published by Okada et al. (1935: 263, pl. 158; see also Okada, 1938: 274) should have been enough to validate *F. scriptissimus* (Fig. 45C) but unfortunately this publication was overlooked by Pietsch and Grobecker. More than half a century later, Yokota and Senou (1991: 2, fig. 3) demonstrated conclusively the distinctiveness of this taxon. That it perhaps contains two species, a striped form and a mottled form, is contradicted by the presence of intermediates that show both aspects of coloration (Fig. 48B).

Fowlerichthys senegalensis (Cadenat)
Senegalese Frogfish

Figs. 27E, 49–51, 280; Tables 3, 22

Antennarius (*Fowlerichthys*) *senegalensis* Cadenat, 1959: 364–67, 382, figs. 1, 18–23, table [original description, eight specimens listed (table on p. 365 and fig. 1), three known: lectotype by subsequent designation of Le Danois (1964) MNHN 1961-971, Senegal; paralectotypes MNHN 1982-131, 1982-133, both from Senegal]. Poll, 1959: 367, fig. 124 (Gabon, Angola). Monod, 1960: 648, figs. 24–45 (anatomy of pectoral lobe and fin). Blache, 1962: 79 (Mauritania to south of Cap Lopez, Gabon). Le Danois, 1964: 15, figs. 2–30, 32–34, 37, 38, 57 (osteology, myology). Schultz, 1964: 177 (after Cadenat, 1959). Blache et al., 1970: 441, fig. 1117 (after Blache, 1962). Monod and Le Danois, 1973: 663 (synonymy, distribution). Pietsch et al., 1986: 138 (in type catalog).

Antennarius pardalis (not of Valenciennes): Navarro, 1943: 164, pl. 38, fig. 13 (Sahara).

Antennarius commersonii (not of Latreille): Poll, 1949: 252, fig. 22 (Mauritania). Cadenat, 1951: 291 (Senegal).

Antennarius **sp.**: Furnestin et al., 1958: 479, fig. 74 (Morocco).

Fowlerichthys senegalensis: Cadenat, 1959: 367, 385, 386, 391–393 (new combination). Arnold and Pietsch, 2012: 126, 128, fig. 6 (molecular phylogeny). Wirtz et al., 2013: 119 (Cape Verde Islands). Pietsch, 2016: 2053 (eastern central Atlantic; in key).

Fowlerichthys ocellatus senegalensis: Le Danois, 1964: 113, 123, figs. 56, 57 (new combination, description, lectotype designated; Senegal, Mauritania).

Antennarius senegalensis: Pietsch, 1981a: 2 (eastern central Atlantic). Pietsch, 1984b: 36 (genera of frogfishes). Pietsch, 1986b: 1365, 1366, fig. (eastern North Atlantic). Pietsch and Grobecker, 1987: 129, figs. 38E, 49, 50, 121 (description, distribution, relationships). Pietsch, 1990: 482 (tropical Eastern Atlantic). Lloris et al., 1991: 220, appendix (West Africa). Paugy, 1992: 572 (West Africa). Azevedo and Heemstra, 1995: 8, fig. 10 (São Miguel, near Ponta Delgada, Azores). Arruda, 1997: 125 (Azores). Debelius, 1997: 91, color fig. (misidentification, figure shows *Antennarius pardalis*; Cape Verde Islands). Santos et al., 1997: 63 (Azores). Calado, 2006: 394, table 4 (Azores; Madeira record in question). Carnevale and Pietsch, 2006: 448, 452, 453, fig. 5b, c, table 1 (comparison, meristics, first fossil frogfish). Kuiter and Debelius, 2007: 124, color fig. (west African coast). Tweddle and Anderson, 2007: 5, table 1 (Congo to Angola). Hanel et al., 2009: 147 (European seas). Carnevale and Pietsch, 2011: 8 (branched pectoral-fin rays). Renous et al., 2011: 97, fig. 4 (amphipedal progression, osteology). Camara et al., 2016: 204, appendix 3 (Guinean shelf).

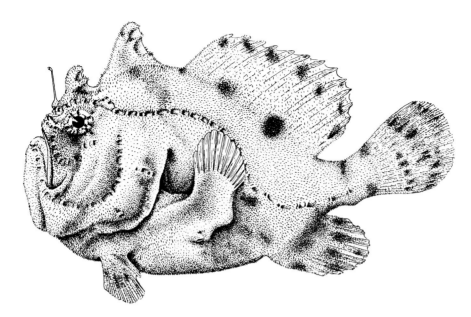

Fig. 49. *Fowlerichthys senegalensis* (Cadenat). 120 mm total length, Gabon, Angola. After Poll (1959, fig. 124); courtesy of Françoise Antonutti, Olivier Pauwels, and the Royal Belgian Institute of Natural Sciences, Brussels.

Fig. 50. Esca of *Fowlerichthys senegalensis* (Cadenat). USNM 199254, 88 mm SL, off Liberia. Drawing by Cathy L. Short; after Pietsch and Grobecker (1987, fig. 50).

MATERIAL EXAMINED

Eight specimens, 80 to 285 mm SL.

Lectotype of *Antennarius senegalensis*: MNHN 1961-971, 252 mm, trawler *Denise*, IFAN 3019, Cap des Biches, Senegal, West Africa, about 45 m, 1949.

Paralectotypes of *Antennarius senegalensis*: MNHN 1982-133, 1 (92.5 mm), trawler *Tréca*, IFAN 1952-175, Cap de Naze, West Africa, about 80 m, 1952; MNHN 1982-131, 1 (265 mm), trawler *Jacqueline*, IFAN 55-105, Cap Rouge, Senegal, West Africa, about 50 m, April 1955.

Additional material: MNHN, 3 (198–285 mm); USNM, 2 (80–88 mm).

DIAGNOSIS

A member of the genus *Fowlerichthys* unique in having the following combination of character states: a large, darkly pigmented, basidorsal ocellus on each side of body, encircled by a lightly pigmented ring (Fig. 49); illicium without dermal spinules; esca a dense cluster of elongate, slender filaments (Fig. 50); membrane behind second dorsal-fin spine divided into naked dorsal and ventral portions by dense cluster of dermal spinules (Fig. 27E); eye diameter 3.9 to 6.3% SL; dorsal-fin rays 13, all or at least seven posteriormost bifurcate; anal-fin rays eight (rarely seven); pectoral-fin rays 12 or 13, all bifurcate; all rays of pelvic fin bifurcate (Table 3).

DESCRIPTION

Escal filaments considerably less than illicium length to more than one and a half times illicium in largest known specimens (e.g., lectotype, MNHN 1961-971, 252 mm SL);

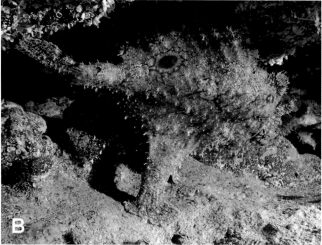

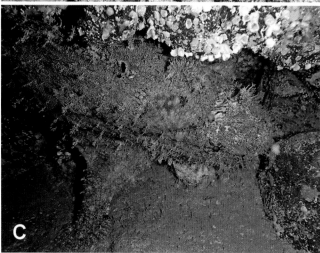

Fig. 51. *Fowlerichthys senegalensis* (Cadenat). (**A**) Santo Antão Island, Cape Verde Islands. Photo by Peter Wirtz. (**B, C**) inside a cave at Tuna Point, Tarrafal, Santiago Island, Cape Verde Islands, said to be at least 40 cm in total length. Photos by Georg Bachschmid.

illicium 9.0 to 13.6% SL, about as long as second dorsal-fin spine; second dorsal-fin spine curved posteriorly, length 8.0 to 13.5% SL; third dorsal-fin spine curved posteriorly, length 17.5 to 25.9% SL; naked ventral portion of membrane behind second dorsal-fin spine extending posteriorly, dividing depressed naked area between second and third dorsal-fin spines and nearly reaching to base of third (Fig. 27E).

Coloration: beige, brown, dark brown, yellow-brown, to pinkish-brown, mottled with darker brown over entire surface of body and fins (except for inner surface of paired fins); as many as seven dark, irregularly shaped patches on body variously located behind or above insertion of pectoral-fin lobe; as many as six dark, evenly spaced and roughly circular patches usually present on anal fin; more or less interconnected dark blotches forming irregular vertical bars across pectoral and caudal fins; illicium usually lightly banded; esca with dark pigment at base; six to 11 short, darkly pigmented bars radiating from eye (Fig. 51).

Largest specimen examined: 285 mm SL (MNHN 1967-2).

DISTRIBUTION

Fowlerichthys senegalensis is restricted to the Eastern Atlantic Ocean where it is most commonly found in coastal waters of tropical West Africa, with several verified observations at the Cape Verde Islands and a least one record from the Azores. Its presence at Madeira, as reported by Calado (2006: 394), has been questioned (Peter Wirtz, personal communication, 29 June 2017; see also Wirtz et al., 2008: 8). The eight specimens examined by us range from Mauritania to Liberia, but material has been reported from as far north as Morocco (Furnestin et al., 1958) and as far south as Cape Morro, Angola (Poll, 1959). The type locality is Cap des Biches, Senegal (Fig. 280, Table 22).

Depths of capture for 10 specimens (unexamined material of Azevedo and Heemstra, 1995, Cadenat, 1959, and Poll, 1959, included) range from 16 to 140 m, with an average depth of 67 m.

Genus *Antennarius* Daudin

There is scarcely another genus of fishes which offers so much difficulty in the discrimination of the species as the present.

—Albert Günther, 1861a: 184

Antennarius Commerson, in Lacepède, 1798: 327, footnote [based on polynomial *Antennarius bivertex, totus ater, puncto mediorum laterum albo* of Commerson, MSS 889(1): 2, 891: 136 (= Lacepède, in Jordan, 1917b: 69); validated by Opinion 24 of the International Commission on Zoological Nomenclature (Anonymous, 1910), but later suspended by Opinion 89 (Anonymous, 1925)].

Antennarius Daudin, 2 October 1816: 193 [type species *Lophius chironectes* Latreille, 1804 (= *Antennarius chironectes, obscure rubens, maculis nigris raris inspersus* Commerson, MSS 889(1), 891, and *Lophius pictus* Shaw, 1794), by subsequent designation of Bleeker, 1865b: 5; dating of Daudin by Mathews and Iredale (1915) and Sherborn (1922: xliv)]. Cuvier, November 1816: 310 (*sensu* Daudin, October 1816, but including *Histrio*).

Antennaria Bory de Saint-Vincent, 1822: 411 (emendation of *Antennarius* Daudin, 1816, therefore taking same type species, *Lophius chironectes* Latreille, 1804).

Artennarius Garthe, 1837, table 4 (misspelling of *Antennarius* Daudin, 1816, therefore taking same type species, *Lophius chironectes* Latreille, 1804).

Saccarius Günther, 1861a: 183 [type species *Saccarius lineatus* Günther, 1861a (= *Lophius striatus* Shaw, 1794), by monotypy].

Antannarius Jones, 1879: 363 (misspelling of *Antennarius* Dauden, 1816, therefore taking same type species, *Lophius chironectes* Latreille, 1804).

Phrynelox Whitley, 1931: 328 (type species *Lophius striatus* Shaw, 1794, by original designation).

Triantennatus Schultz, 1957: 74 [type species *Antennarius* (i.e., *Phrynelox*) *zebrinus* Schultz, 1957 (= *Lophius striatus* Shaw, 1794), by original designation].

Uniantennatus Schultz, 1957: 83 [type species *Antennarius horridus* Bleeker, 1853a (= *Lophius pictus* Shaw, 1794), by original designation]

Chorinectes Bauchot, 1958: 140 (misspelling of *Chironectes* Rafinesque, 1814, preoccupied, therefore taking same type species, *Chironectes variegatus* Rafinesque, 1814; cited as a synonym of *Antennarius* Commerson).

Phymatophryne Le Danois, 1964: 115 (type species *Chironectes maculatus* Desjardins, 1840, by original designation).

Pherynelox Keenleyside, 1979: 32 (misspelling of *Phrynelox* Whitley, 1931, therefore taking same type species, *Lophius striatus* Shaw, 1794, by original designation).

DIAGNOSIS

Antennarius is unique among antennariids in having the following combination of character states: skin covered with close-set, bifurcate dermal spinules, the length of spines of each spinule not more than twice the distance between tips of spines (Fig. 26A, B); ilicium naked, without dermal spinules (except at extreme base), second dorsal-fin spine membranously attached to head; esca distinct; pectoral-fin lobe broadly attached to side of body; caudal peduncle present; endopterygoid present; pharyngobranchial I present;

Table 4. Comparison of Distinguishing Character States of the Species-Groups of the Genus *Antennarius*

Character	A. striatus group	A. pictus group	A. biocellatus group	A. pauciradiatus group
Illicium length in percent of SL	11.2–22.7	18.9–33.8	10.7–16.7	5.2–9.3
Illicial pterygiophore overhanging mouth	present	absent	absent	absent
Membrane behind second dorsal-fin spine	present	present	absent	present*
Ocelli	absent	rare	present	absent
Dark stripes or blotches	present	absent	absent	absent
Dorsal-fin rays	12 (11–13)	12–13 (11)	12	12 (11–13)
Anal-fin rays	7 (6)	7–8 (6)	7 (6)	7 (8)
Pectoral-fin rays	10–13 (9)	10–11 (9)	9 (8–10)	9 (10)
Bifurcate pelvic-fin rays	1	1	1	0
Vertebrae	19 (18)	19 (20)	19	19
Maximum standard length	188 mm	290 mm	120 mm	40 mm

*Forming a unique structure for protection of the esca (see Fig. 112).

epural present; pseudobranch present; swimbladder present; dorsal-fin rays 11 to 13 (usually 13); anal-fin rays six to eight (usually seven); all rays of caudal fin bifurcate; pectoral-fin rays nine to 13 (rarely eight); vertebrae 18 to 20 (usually 19) (Table 4).

DESCRIPTION

Escal morphology highly variable (see species and species-group accounts); illicium, when laid back onto head, usually fitting into a narrow, naked groove situated on either left or right side of second dorsal-fin spine, the tip of illicium (esca) usually (but not in members of the *A. pictus* group) coming to lie within a shallow depression (sometimes devoid of dermal spinules) between second and third dorsal-fin spines, the esca capable of being covered and protected by second dorsal-fin spine when spine is fully depressed (esca of members of the *A. pictus* group perhaps protected in similar way by third dorsal-fin spine); illicium length highly variable, ranging from considerably less than length of second dorsal-fin spine to slightly more than twice its length, 5.2 to 33.8% SL (see species and species-group accounts); anterior end of pterygiophore of illicium terminating considerably posterior to, or extending anteriorly slightly beyond, symphysis of upper jaw; illicium and second dorsal-fin spine relatively closely spaced on pterygiophore, the distance between bases of spines less than 5% SL; second dorsal-fin spine straight to strongly curved posteriorly, free or connected to head by membrane, length 8.5 to 19.4% SL; third dorsal-fin spine curved posteriorly, tapering slightly toward distal end, the full length connected to head by membrane, length 16.7 to 31.7% SL; eye not distinctly surrounded by separate clusters of dermal spinules, diameter 2.6 to 9.1% SL; distal two-thirds of maxilla nearly always naked and tucked beneath folds of skin, only extreme proximal end directly covered with spinulose skin; cutaneous appendages scattered over head, body, and fins, their development highly variable (see species and species-group accounts); wartlike patches of clustered dermal spinules absent; caudal peduncle present; epibranchial I toothless (a few tiny remnants of tooth-plates present in some specimens); ceratobranchial I toothless; vertebrae 19 or 20 (very rarely 18), caudal centra 14 or 15; dorsal-fin rays 11 to 13, all simple to all bifurcate; anal-fin rays 6 to 8, all bifurcate; pectoral-fin rays 8 to 12, all simple or all bifurcate; posteriormost ray of pelvic fin usually bifurcate (simple in members of the *A. pauciradiatus* group; see below).

Color and color pattern extremely variable (see species and species-group accounts).

Twelve species in four species-groups; cosmopolitan in tropical and subtropical waters.

COMMENTS

The name *Antennarius* was originally coined by French naturalist Philibert Commerson (MSS 889, 891) as part of polynomials given to four species of frogfishes collected at Mauritius between 1768 and 1773. Somewhat later, Lacepède (1798) introduced the name to the scientific literature by quoting the polynomials of Commerson in footnotes, but he otherwise provided no Latinized binomials. Opinion 24 of the International Commission on Zoological Nomenclature (Anonymous, 1910) validated the name *Antennarius* as dating from Lacepède (1798), a ruling that was later suspended by Opinion 89 (Anonymous, 1925).

Although fully aware of Commerson's intended name for this genus, Georges Cuvier (1816: 310) proposed the alternative name *Chironectes*, a designation thought by him to be more meaningful than *Antennarius*. It is not clear whether Cuvier knew of Rafinesque's (1814) earlier use of *Chironectes* for a frogfish (see synonymy of *Histrio*, p. 174), but he must have certainly been unaware of Illiger's (1811) still earlier application of this name to a genus of marsupial mammals. In any case, since Cuvier (1816: 310) listed the name *Antennarius* simultaneously with *Chironectes* ["Les Chironectes. (*Antennarius*. Commers.)"], most authors, realizing the unavailability of *Chironectes*, have accepted *Antennarius* as dating from Cuvier (1816; misdated 1817 by many authors). It is evident, however, that François Marie Daudin (2 October 1816; dating by Mathews and Iredale, 1915, and Sherborn, 1922) preceded Cuvier (November 1816; dating by Whitehead, 1967, Cowan, 1969, and Roux, 1976) by at least a month by publishing the name *Antennarius* with a definition and a clear indication of its contents: "We can find in Commerson's manuscripts many species of *lophies* [frogfishes] that this traveler seems to have wanted to place in a different genus. He gave [to this genus] the Latin name *Antennarius* because these *lophies* have one or two filaments [dorsal-fin spines] above the upper lip, terminated by a small mass or two of fleshy appendages" (Daudin, 1816: 193). Daudin then went on to list the four species described in the manuscript by Commerson, together with the French vernacular names provided by Lacepède (1798).

As presently understood, *Antennarius* is an assemblage of 12 species that fall conveniently into four species-groups (Table 5). Whereas monophyly for each of these four groups is supported by one or more synapomorphies, we have not been able to find any convincing morphological feature to establish monophyly for the genus. *Antennarius* is therefore defined by a combination of what appear to be primitive morphological character states. Monophyly is well supported, however, by molecular evidence (see Evolutionary Relationships, p. 377).

Table 5. Frequencies of Fin-Ray and Vertebral Counts of Species of the Genus *Antennarius*

Species	Dorsal-fin rays				Anal-fin rays			Pectoral-fin rays (both sides combined)						Vertebrae		
	11	12	13	14	6	7	8	8	9	10	11	12	13	18	19	20
Antennarius striatus group																
A. hispidus	1	34	1			36				65	7				7	1
A. indicus		14				14						6	22		4	
A. scaber	2	69				71				1	102	5			4	
A. striatus	8	129			1	127			9	171	88	5		1	5	
Antennarius pictus group																
A. commerson	1	3	48				52			1	103				5	
A. maculatus	1	19			1	19				39	1				11	
A. multiocellatus	2	69	1		3	69			4	140					6	
A. pardalis	1	21	2			24			2	41	5				4	
A. pictus	1	70	2		2	71			1	138	1				11	
Antennarius biocellatus group																
A. biocellatus		38			3	35		2	72	1					14	
Antennarius pauciradiatus group																
A. pauciradiatus	1	30				30	1		61	1					5	
A. randalli		5	2			7			14						3	
TOTALS	18	501	56		10	503	53	2	163	598	307	16	22	1	79	1

Key to the Species-Groups of the Genus *Antennarius*

This key, like the other keys in this volume, works by progressively eliminating the most morphologically distinct taxon and for that reason should always be entered from the beginning. All features listed for each species-group must correspond to the specimen being keyed; if not, proceed to the next set of characters. The figures accompanying the key are diagrammatic; dermal spinules and cutaneous appendages (with minor exceptions) are not shown. For an additional or alternative means of identification, all species-groups of *Antennarius* may be compared simultaneously by referring to Tables 4 and 5.

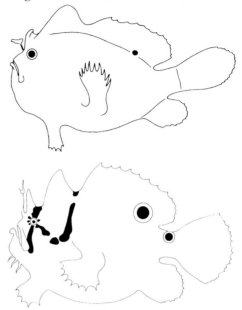

1. Illicium distinctly shorter than second dorsal-fin spine, the latter broadly connected to third spine and the third broadly connected to soft-dorsal fin by thin membranous skin; individuals small, not greater than about 40 mm SL . . . . *Antennarius pauciradiatus* Group, p. 160
   Western Atlantic, Western Pacific to Easter Island

2. Illicium usually slightly shorter than second dorsal-fin spine, the latter free, broad-based, more or less straight and tapering to a point, not connected to head by membrane; chin prominent, extending far forward, well beyond mouth opening, resulting in a bulldog look; a darkly pigmented, basidorsal ocellus encircled by a lightly pigmented ring nearly always present; conspicuous cutaneous appendages on throat; pectoral-fin rays nine (rarely eight or 10) . . . . *Antennarius biocellatus* Group, p. 154
   Western Pacific Ocean

3. Illicium approximately twice length of second dorsal-fin spine, the latter membranously attached to head; anterior end of pterygiophore of illicium terminating distinctly posterior to symphysis of upper jaw; pectoral-fin rays 10 or 11 (rarely nine), never bifurcate; head, body, and fins often with dark, circular spots of pigment (caudal fin often with triad of three large spots), never covered with roughly parallel, darkly pigmented streaks . . . . *Antennarius pictus* Group, p. 111
   Atlantic, Pacific, and Indian Oceans

4. Illicium about as long as second dorsal-fin spine, the latter membranously attached to head; anterior end of pterygiophore of illicium nearly always extending anteriorly beyond symphysis of upper jaw, slightly overhanging the mouth; head, body, and fins usually covered with roughly parallel, darkly pigmented streaks and blotches . . . . *Antennarius striatus* Group, p. 76
   Atlantic and Indo-Pacific Oceans

Antennarius striatus Group Pietsch

DIAGNOSIS

A species-group of *Antennarius* distinguished by having the following combination of character states: head and body usually covered with numerous, darkly pigmented streaks or elongate blotches typically radiating outward from eye and extending posteriorly to level of pectoral fin; similar, more or less parallel, markings extending from pectoral region posterodorsally onto soft-dorsal and anal fins; caudal fin similarly banded (but extremes of light and dark coloration, in which markings are completely absent, also exist); illicium 11.2 to 22.7% SL, about equal to length of second dorsal-fin spine; anterior end of pterygiophore of illicium nearly always extending anteriorly beyond symphysis of upper jaw; esca consisting of a spherical or oval-shaped tuft of numerous, fine filaments (*A. hispidus*), a cluster of flattened, leaflike appendages (*A. indicus*), or two to seven elongate, wormlike appendages (*A. scaber* and *A. striatus*); distinct, eyelike, escal pigment spots absent; second dorsal-fin spine connected to head by a membrane, the membrane not divided into naked dorsal and ventral portions by a cluster of dermal spinules and terminating distinctly anterior to base of third dorsal-fin spine; distal two-thirds of length of maxilla naked and tucked beneath folds of skin, only extreme proximal end directly covered with spinulose skin; caudal peduncle present, the membranous posteriormost margin of soft-dorsal and anal fins attached to body distinctly anterior to base of outermost rays of caudal fin; dorsal-fin rays 11 to 13 (usually 12), as many as four posteriormost rays bifurcate; anal-fin rays six or seven (usually seven), usually all bifurcate but occasionally only posteriormost three to five divided; pectoral-fin rays 10 to 13 (rarely nine, all simple; only posteriormost pelvic-fin ray bifurcate; vertebrae 19 (rarely 18) (Table 6).

Members of this species-group reach a maximum standard length of about 188 mm; all are more or less associated with eel grasses and sponges on flat, open sand or mud bottoms; distributed in tropical and subtropical waters of the Atlantic and Indo-Pacific Oceans as far east as the Hawaiian and Society Islands; four species.

Table 6. Comparison of Distinguishing Character States of Species of the *Antennarius striatus* Group (lengths as percentage of SL)

Character	A. hispidus	A. indicus	A. scaber	A. striatus
Illicium length	11.2–18.1	12.1–14.9	14.1–20.3	13.6–22.7
Second dorsal-fin spine length	11.5–17.4	10.4–14.3	13.8–19.0	11.4–19.0
Esca	tuft of filaments	leaflike appendages	wormlike appendages	wormlike appendages
Ocelli	0	2–3	0	0
Dorsal-fin rays	12 (11–13)	12	11 (12)	11–12
Anal-fin rays	7	7	7	7 (6)
Pectoral-fin rays	10–11 (9–12)	13 (12)	11 (10–12)	10–11 (9–12)
Maximum standard length	148 mm	188 mm	152 mm	155 mm
Distribution	Indo-West Pacific	Western Indian Ocean	Western Atlantic	Eastern Atlantic, Indo-Pacific

This key differs from most in that it is not dichotomous. It works by progressively eliminating the most morphologically distinct taxon and for that reason should always be entered from the beginning. All character states listed for each genus must correspond to the specimen being keyed; if not, proceed to the next set of character states.

1. Esca consisting of a cluster of flattened, leaflike appendages; two or three ocelli on side of body; darkly pigmented streaks and blotches usually more or less confined to fins (Figs. 57, 60); illicium length 12.1 to 14.9% SL; pectoral-fin rays 13 (rarely 12) (Table 6) . . . .*Antennarius indicus* **Schultz, 1964, p. 83**
    Western Indian Ocean

2. Esca consisting of a large spherical or oval tuft of numerous slender filaments; ocelli absent; head, body, and fins usually covered with roughly parallel, darkly pigmented streaks and blotches (Fig. 52); illicium length 11.2 to 18.1% SL; pectoral-fin rays 10 or 11 (rarely nine or 12) (Table 6) . . . .*Antennarius hispidus* **(Bloch and Schneider, 1801), p. 77**
    Indo-West Pacific Ocean

3. Esca consisting of two elongate, wormlike appendages; ocelli absent; head, body, and fins usually covered with roughly parallel, darkly pigmented streaks and blotches (Figs. 63, 64); illicium length 14.1 to 20.3% SL; pectoral-fin rays 11 (rarely 10 or 12) (Table 6) . . . .*Antennarius scaber* **(Cuvier, 1817b), p. 86**
    Western Atlantic Ocean

4. Esca usually consisting of three to seven elongate, wormlike appendages; ocelli absent; head, body, and fins usually covered with roughly parallel, darkly pigmented streaks and blotches (Fig. 69); illicium length 13.6 to 22.7% SL; pectoral-fin rays 10 or 11 (rarely nine or 12) (Table 6) . . . .*Antennarius striatus* **(Shaw, 1794), p. 97**
    Eastern Atlantic and Indo-Pacific Oceans

Antennarius hispidus (Bloch and Schneider)
Hispid Frogfish

Figs. 15, 26A, 52–56, 280, 295B, 296B, 302, 311, 313–316; Tables 5, 6, 22, 24

Lophius, ossiculo frontis tentaculis carnosis, centralibus, terminato: Koelreuter, 1766: 329, pl. 8, fig. 1 (full description, specimens and locality unknown).

Lophius hispidus Bloch and Schneider, 1801: 142 (original description, holotype ZMB 2221, off Coromandel, India). Cuvier, 1816: 310 ["probably same as" *Antennarius antenna tricorni* of Commerson, MSS (= *A. striatus*)].

Lophius histrio (not of Linnaeus): Russell, 1803: 12, pl. 19 (*Kappa-mura-moja*; description and figure represent *A. hispidus*; Vizagapatam, coast of Coromandel).

Chironectes lophotes Cuvier, 1817b: 428, pl. 17, fig. 2 (original description, type apparently lost, not found in MNHN, locality unknown). Bauchot, 1958: 141, 142 (junior synonym of *A. hispidus*). Pietsch et al., 1986: 147 (type catalog).

Antennarius lophotes: Schinz, 1822: 500 (new combination; after Cuvier, 1817b). Gregory, 1933: 389, 394 (description of esca). Randall, 2005a: 71 (synonym of *Antennarius hispidus*).

Lophius hispidus (not of Bloch and Schneider): Bory de Saint-Vincent, 1826: 495 [equated with *L. tricornis* Cloquet, 1817 (= *A. striatus*)].

Chironectes hispidus: Cuvier, 1829: 252 (new combination; after Bloch and Schneider, 1801). Valenciennes, 1837: 401 [in part; i.e., MNHN A.4584 (Ambon), A.4585 (New Guinea), A.4586, A.4634, A.4638 (Pondicherry); A.4544 and A.4588 are *A. striatus*, A.4582 and A.4587 are *A. indicus*]. Valenciennes, 1842: 189 (after Valenciennes, 1837). Jerdon, 1853: 149 ("Kadil madoo"; Madras, India). Bauchot, 1958: 141, 142 (comparison with *"pinniceps"* of Valenciennes, 1837: 410). Pietsch, 1984b: 34 (genera of frogfishes). Pietsch, 1985a: 70, figs. 7–10, table 1 (original manuscript sources for *Histoire Naturelle des Poissons* of Cuvier and Valenciennes).

Riquet-à-la-houppe: Valenciennes, 1837: 410 [in part; vernacular for *Antennarius antenna tricorni* of Commerson, MSS (= *Antennarius striatus*), after Cuvier, 1816; *hispidus* and *striatus* confused].

Antennarius hispidus: Cantor, 1849: 1185 (new combination, Singapore). Cantor, 1850: 203 (after Cantor, 1849). Bleeker, 1852a: 280 (Ambon, Ceram). Bleeker, 1859b: 128 (Indonesia, Ceylon, Mauritius, Réunion). Günther, 1861a: 189 (description, East Indies). Günther, 1861b: 361 (Ceylon). Bleeker, 1865b: 9, 14, pl. 194 (fig. 2), 198 (fig. 1) (description, Singapore to Ceram). Schmeltz, 1869: 19 (Viti Islands). Günther, 1876: 162, pl. 99, fig. A (after Günther, 1861a). Day, 1876: 271, pl. 60, fig. 2 (synonymy includes *C. lophotes* Cuvier, seas of India to Malay Archipelago and beyond). Schmeltz 1879: 48 (Viti Islands). Day, 1889: 231, fig. 84 (after Day, 1876). Jordan and Seale, 1906: 438 (in part, Misool). Jordan and Seale, 1907: 48 (after Jordan and Seale, 1906). Annandale and Jenkins, 1910: 20 (India). Gilchrist and Thompson, 1917: 417 (synonymy includes *C. lophotes* Cuvier, Natal). Hornell, 1921: 80 (aquarium observations, spawning). Hornell, 1923: 28, fig. 14 (aquarium observations, spawning). Whitley, 1927a: 8 (Fiji). Barnard, 1927: 1001, pl. 36, fig. 7 (Natal). Fowler, 1931: 367 (after Schmeltz, 1879). Chevey, 1932: 119 (Indo-China). Fowler, 1934b: 513 (Natal). Smith, 1949: 430, pl. 98, fig. 1234 (South Africa). Munro, 1955: 289, fig. 842 (Ceylon). Schultz, 1957: 90, pl. 8, fig. D (synonymy, description, Philippines; in key). Munro, 1958: 296 (synonymy). Bauchot, 1958: 141, 142 (comparison with *A. scaber* and *"pinniceps"* of Valenciennes, 1837: 410). Fowler, 1959: 561 (Fiji). Beaufort and Briggs, 1962: 214 (synonymy, description, Natal to Hawaii; in key). Le Danois, 1964: 104, fig. 50L (description; India, New Guinea, Ambon). Munro, 1964: 186 (off Kerema Bay, New Guinea). Munro, 1967: 585, pl. 78, fig. 1093 (New Guinea). Le Danois, 1970: 89 (Gulf of Aqaba, Red Sea; but specimen not examined, locality unconfirmed). Kailola, 1975: 47 (Papua New Guinea). Kailola and Wilson, 1978: 26 (Gulf of Papua). Kotthaus, 1979: 46, fig. 494 (Arabian Sea, outside Bay of Kutch). Grobecker and Pietsch, 1979: 1161, figs. 1, 2, table 1 (feeding mechanism). Anonymous, 1982, front and back covers (color photos). Kyushin et al., 1982: 310 (South China Sea). Araga, 1984: 103, pl. 346-F (Chiba Prefecture, Japan). Dor, 1984: 56 (Red Sea). Pietsch, 1984b: 34 (genera of frogfishes). Pietsch, 1984e: 2 (Western Indian Ocean). Thresher, 1984: 41 (egg rafts). Pietsch, 1985a: 70, figs. 7–10, table 1 (original manuscript sources for *Histoire Naturelle des Poissons* of Cuvier and Valenciennes). Pietsch, 1985b: 594, 595, figs. 2, 3 (fast feeding). Pietsch, 1986a: 366, 367, pl. 13 (South Africa). Pietsch and Grobecker, 1987: 71, figs. 16A, 19–21, 117, 142, 152, pl. 6 (description, distribution, relationships). Allen and Swainston, 1988: 38 (northwestern Australia). Paxton et al., 1989: 278 (Australia). Pietsch and Grobecker, 1990a: 99, 100 (aggressive mimicry). Pietsch and Grobecker, 1990b: 79 (after Pietsch and Grobecker, 1990a). Ramaiah and Chandramohan, 1992:

210, table 1 (bioluminescent bacteria). Cunxin, 1993: 117 (South China Sea). Krishnan and Mishra, 1993: 204 (Kakinada-Gopalpur, east coast of India). Goren and Dor, 1994: 14 (Red Sea). Martin et al., 1995: 916 (Gulf of Carpentaria). Mohsin and Ambak, 1996: 633 (Malaysia). Allen, 1997: 62 (tropical Australia). Chen et al., 1997: 188 (Spratly Islands, South China Sea). Kuiter, 1997: 42 (Australia). Lindberg et al., 1997: 216 (Sea of Japan). Michael, 1998: 330, 352, color figs. (identification, behavior, captive care). Fricke, 1999: 105 (Réunion, Mauritius). Mishra et al., 1999: 84 (east coast of India). Pietsch, 1999: 2015 (western central Pacific; in key). Allen, 2000a: 62, pl. 12-12 (plate shows *A. striatus*; Indo-West Pacific). Nakabo, 2000: 455 (Japan; in key). Pereira, 2000: 5 (Mozambique). Pietsch, 2000: 597 (South China Sea). Hutchins, 2001: 22 (Western Australia). Nakabo, 2002: 455 (Japan; in key). Allen and Adrim, 2003: 25 (Indonesia). Allen et al., 2003: 365, color fig. (Indo-West Pacific). Schneidewind, 2003a: 83, 88, figs. (natural history). Heemstra and Heemstra, 2004: 122 (southern Africa). Letourneur et al., 2004: 209 (Réunion Island). Randall et al., 2004: 8 (Tonga). Randall, 2005a: 71, fig. (South Pacific). Allen et al., 2006: 639 (Australia). Carnevale and Pietsch, 2006: 452, table 1 (comparison, meristics, first fossil frogfish). Schneidewind, 2006a: 25, color fig. (luring behavior). Senou et al., 2006: 423 (Sagami Sea). Kuiter and Debelius, 2007: 121, color figs. (Indonesia). Lugendo et al., 2007: 1222, appendix 1 (Zanzibar). Michael, 2007: 57–60, color figs. (possible new species; Bali, Indonesia). Kwang-Tsao et al., 2008: 243 (southern Taiwan). Fricke et al., 2009: 28 (Réunion). Fricke et al., 2011: 367 (New Caledonia). Satapoomin, 2011: 51, appendix A (Andaman Sea). Allen and Erdmann, 2012: 143, 146, color figs. (East Indies). Arnold and Pietsch, 2012: 126–128, figs. 4, 6, table 1 (molecular phylogeny). Larson et al., 2013: 53 (Northern Territory). Nakabo, 2013: 538, figs. (Japan). Yoshida et al., 2013: 54 (Gulf of Thailand). Karplus, 2014: 35, fig. 1.21D (aggressive mimicry; after Pietsch and Grobecker, 1987). Moore et al., 2014: 176 (northwestern Australia). Psomadakis et al., 2015: 174, fig. (Pakistan). Brauwer and Hobbs, 2016 (bioluminescent bacteria, after Ramaiah and Chandramohan, 1992). Joshi et al., 2016: 40 (Gulf of Mannar, Tamil Nadu, India). Wright et al., 2016: 122 (distinguished from *A. striatus*).

Antennarius hispidus (not of Bloch and Schneider): Seale, 1906: 89 (specimen is *Abantennarius coccineus*). Weber, 1913: 562 (material consists of *Abantennarius coccineus*, *A. dorehensis*, *A. randalli*, *Antennatus tuberosus*, and *H. histrio*). Fowler, 1928: 477 (*Antennarius commerson* and *A. pictus*). Burgess and Axelrod, 1973a: 528, color fig. 485 (plate represents *Lophiocharon trisignatus*). Wheeler, 1975, pl. 182 (plate represents *L. trisignatus*).

Antennarius nummifer (not of Cuvier): Schultz, 1957: 102 (in part, FMNH 51864 is *A. hispidus*, North Borneo).

Antennarius hipidus: Allen and Adrim, 2003: 25 (misspelling of specific name; Moluccas to Kalimantan, Indonesia).

Antennarius cf. hispidus: Michael, 2007: 56, 58, color figs. ("pom-pom frogfish," a possible new species; Gilimanuk, West Bali).

MATERIAL EXAMINED

One hundred and fifteen specimens, 11 to 148 mm SL.

Holotype of *Lophius hispidus*: ZMB 2221, 68 mm, off Coromandel coast, Bay of Bengal, India (Fig. 52).

Additional material: *Western Indian Ocean*: ANSP, 2 (103–146 mm); SAIAB, 5 (48–116 mm); SAMA, 4 (40–120 mm); SU, 1 (85 mm); ZIM, 1 (32 mm). *India*: AMS, 1 (97 mm); ANSP, 3 (42–45 mm); BMNH, 4 (57–88 mm); BPBM, 1 (37.5 mm); CAS, 2 (68–71 mm);

Fig. 52. *Antennarius hispidus* (Bloch and Schneider): holotype of *Chironectes lophotes* Cuvier, specimen presumed lost, locality unknown, reproduced from the pencil original that served as the model for the engraving that accompanied the original description of *Chironectes lophotes* published by Cuvier (1817b; after Pietsch, 1985a, fig. 7). After Cuvier and Valenciennes, MS 504, XII.B, Bibliothèque Centrale, Muséum national d'Histoire naturelle, Paris; courtesy of Emmanuelle Choiseau, © Muséum national d'Histoire naturelle, Paris.

Fig. 53. *Antennarius hispidus* (Bloch and Schneider). Gulimanuk, Bali, Indonesia. Photo by Scott W. Michael.

MNHN, 4 (67–84 mm); SU, 8 (47.5–85.5 mm); USNM, 2 (59–80 mm). *Andaman Sea*: USNM, 1 (122 mm); ZMUC, 2 (92.5–94 mm). *Gulf of Thailand*: CAS, 15 (81–148 mm); UMMZ, 1 (130 mm). *Japan*: TKPM, 1 (106 mm). *Vietnam and Taiwan*: CAS, 1 (126 mm); NTUM, 1 (111 mm); ZMUC, 1 (77 mm). *Philippines*: AMNH, 2 (32–67 mm); FMNH, 1 (63 mm); SU, 4 (33–98 mm); USNM, 3 (54–86.5mm); UW, 8 (67–131 mm). *Indonesia and New Guinea*: FMNH, 3 (46–115 mm); KFRL, 13 (39–75 mm); MNHN, 2 (47–64.5 mm); RMNH, 5 (43–130 mm); UW 20875, 1 (36 mm); ZMA, 1 (51 mm). *Australia*: AMS, 2 (43–79 mm); NMV, 1 (11 mm); NTM, 5 (55–100 mm); WAM, 2 (69.5–72 mm). *Fiji*: USNM, 1 (88 mm).

DIAGNOSIS

A member of the *Antennarius striatus* group without ocelli on side of body; esca consisting of a large spherical or oval tuft of numerous slender filaments (Fig. 53); and pectoral-fin rays 10 or 11, rarely nine or 12 (Tables 5, 6). It differs further from its congener, *A. striatus*, in having a slightly shorter illicium (see Pietsch and Grobecker, 1987, fig. 19).

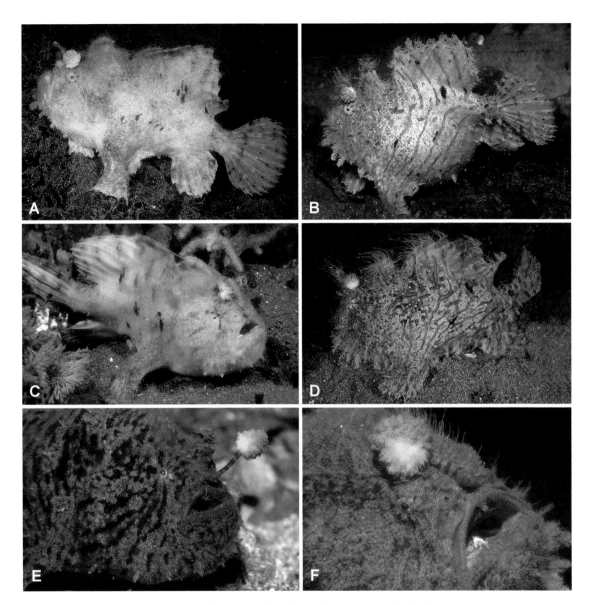

Fig. 54. *Antennarius hispidus* (Bloch and Schneider). (**A**, **B**) Bali, Indonesia. Photos by Rudie H. Kuiter. (**C**) Lembeh Strait, Sulawesi, Indonesia. Photo by Frank Schneidewind. (**D**) Gulimanuk, Bali, Indonesia. Photo by Janine Cairns-Michael. (**E**, **F**) Gulimanuk, Bali, Indonesia. Photos by Frank Schneidewind.

DESCRIPTION

Esca large, highly conspicuous, diameter as great as 10% SL (Fig. 53); illicium 11.2 to 18.1% SL, about equal to length of second dorsal-fin spine; second dorsal-fin spine more or less straight, length 11.5 to 17.4% SL; third dorsal-fin spine curved posteriorly, length 19.9 to 26.4% SL; eye diameter 4.3 to 7.5% SL; dorsal-fin rays 12 (rarely 11 or 13); anal-fin rays seven (Tables 5, 6).

Coloration in life: off white, beige, tan, light yellow, orange, brownish orange, brown, or rarely solid black; typically, everywhere covered with dark stripes or elongate blotches in pattern described for species-group (see Description, p. 76, and Fig. 54).

Coloration in preservation: beige, light yellow, orange, dark yellow-brown, to black; filaments arising from anterior surface of esca more or less darkly pigmented; illicium banded. Lighter-color phases with as many as 10 (usually six to eight) darkly pigmented streaks radiating out from eye; belly without elongate streaks, but occasionally with scattered, dark, circular spots. Solid-black color phase (only one of 115 specimens examined) with tips of pectoral-fin rays white.

Largest specimen examined: 148 mm SL (CAS 50064).

DISTRIBUTION

Antennarius hispidus is widespread throughout the Indo-West Pacific Ocean (Fig. 280, Table 22), ranging from South Africa, Mozambique, the Red Sea, the Pakistani coast, India, and Malaysia to the Molucca Islands, and from Japan and Taiwan to Australia (from the Monte Bello Islands, Western Australia, to Manly Bay, New South Wales), with records from New Caledonia, Fiji, and Tonga (Randall et al., 2004). Although very rarely collected from oceanic islands of the Indian Ocean, Bleeker's (1859b: 128) records for Mauritius and Réunion, questioned by Pietsch and Grobecker (1987: 74), have been confirmed (Fricke, 1999; Letourneur et al., 2004; Fricke et al., 2009). The type locality is Coromandel, India.

The limited information available on depth preference for this species (24 captures) indicates that *A. hispidus* may be taken anywhere between the surface and about 90 m, with an average depth of about 46 m.

COMMENTS

The earliest known account of *A. hispidus*, designated by a lengthy Latin polynomial (see synonymy, p. 77), was published by German naturalist Joseph Theophilus Koelreuter in 1766. His detailed Latin description, including a long list of morphometrics, was accompanied by a remarkably good illustration (Fig. 55). On the other hand, a drawing labeled "Kadil madoo," found in a series of drawings made to illustrate Francis Day's survey of the fishes of India, is rather crude but it clearly represents *A. hispidus* (Fig. 56).

There has been some suggestion that an undescribed species might be hiding within the known material of *A. hispidus*. Underwater photographer and author Scott W. Michael

Fig. 55. Joseph Theophilus Koelreuter's *Lophius, ossiculo frontis tentaculis carnosis, centralibus, terminato*, the earliest known representation of *Antennarius hispidus* (Bloch and Schneider). After Koelreuter (1766, pl. 8, fig. 1).

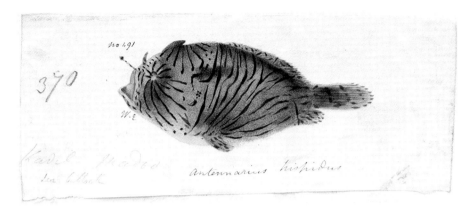

Fig. 56. *Antennarius hispidus* (Bloch and Schneider). "Kadil madoo," one of some 705 drawings from the Francis Day collection that served as the original illustrations for Day's survey of the fishes of India, which he began in about 1861 when living in Cochin (now Kochi; see Whitehead and Talwar, 1976). Drawing by an unknown native artist for Sir Walter Elliot, who presented his collection of drawings to Day sometime prior to 1874; courtesy of Sarah Broadhurst, Ann Sylph, and the Zoological Society of London.

(2007: 58), in a well-reasoned popular article, described *"hispidus*-like" individuals, which he observed most often at Gilimanuk, West Bali, with a "huge, fluffy pom-pom" esca— much larger than usual. These individuals differed also from the standard *hispidus*-look by having a shorter illicium, "almost as short as the esca itself and never appearing to be as long as the second dorsal spine." Calling it the "pom-pom frogfish," he stated further that individuals appeared to be larger than any of the *A. hispidus* he had seen before, some nearly 300 mm in total length—significantly larger than the 175-mm maximum generally cited for *A. hispidus*. While this evidence of a possible unnamed population certainly merits further investigation, and although field observations of this kind are extremely important and always of great interest, collections of specimens as well as tissues for DNA analysis must be made—no matter how reluctant investigators are to disrupt the natural order—before questions of species identification can be adequately addressed.

Antennarius indicus Schultz
Indian Frogfish

Figs. 57–62, 278; Tables 5, 6, 22

Chironectes hispidus (not of Bloch and Schneider): Valenciennes, 1837: 407 (description based in part on MNHN A.4582 and A.4587, *A. indicus* from Sri Lanka and Pondicherry, respectively; see Pietsch, 1985a: 70).

Chironectes nummifer (not of Cuvier): Valenciennes, 1837: 425 (description based in part on MNHN A.4613, *A. indicus*, Mahé, Malabar Coast).

Antennarius indicus Schultz, 1964: 182, pl. 3, tables 1, 2 (original description, single specimen, holotype SU 40090, Vizagapatam, India). Burgess and Axelrod, 1973a: 529, color fig. 490 (in part, plate represents *A. indicus*). Pietsch, 1984b: 36 (genera of frogfishes). Pietsch, 1984e: 2 (Western Indian Ocean). Pietsch and Grobecker, 1987: 115, figs. 38A, 39, 40, 120, pl. 24 (description, distribution, relationships). Randall, 1995: 83, fig. 167 (Oman). Michael, 1998: 343, 344, color fig. (identification, behavior, captive care). Gell and Whittington, 2002: 116, table 1 (Mozambique). Schneidewind,

2003a: 84, fig. (natural history). Carnevale and Pietsch, 2006: 452, 453, table 1 (comparison, meristics, first fossil frogfish). Kuiter and Debelius, 2007: 117, color fig. (misidentification, figure appears to show *Nudiantennarius subteres*; Indonesia). Jawad and Al-Mamry, 2009: 1, 2, fig. 2 (coast of Oman). Al-Jufaili et al., 2010: 18, appendix 1 (Oman). Arnold and Pietsch, 2012: 125–128, figs. 3B, 4, 6, table 1 (molecular phylogeny). Prakash et al., 2012: 943, fig. 5, table 1 (southeast coast of India). Jawad and Hussain, 2014: 1–4, fig. 2, table 1 (10 specimens, 120–260 TL; off Basrah, Iraq, Persian Gulf). Psomadakis et al., 2015: 174, fig. (Pakistan). Joshi et al., 2016: 40 (Gulf of Mannar and Tamil Nadu, India).

Antennarius nummifer (not of Cuvier): Playfair, 1866: 70 (BMNH 1867.3.9.244 is *A. indicus*, Zanzibar). Day, 1876: 272, pl. 59, fig. 2 (description and figure represent *A. indicus*, Madras). Day 1889: 232 (after Day, 1876). Le Danois, 1964: 103 (in part, i.e., MNHN A.4582, A.4587, A.4613). Jones and Kumaran, 1980: 706, fig. 603 (Laccadive Archipelago).

Antennarius oligospilus (not of Bleeker): Wheeler, 1975, pl. 183 (plate represents *A. indicus*).

Antennarius sp.: Wheeler, 1975, pl. 184 (plate shows *A. indicus*).

Antennarius indicus (not of Schultz): Kuiter and Debelius, 2007: 117, color fig. (figure appears to represent *Nudiantennarius subteres*; Indonesia).

Antennarius coccineus (not of Lesson): Jawad and Al-Mamry, 2009: 1, fig. 1 (misidentification, specimen is *A. indicus*; Wadi Shab, Gulf of Oman).

MATERIAL EXAMINED

Twenty-two specimens, 24.5 to 188 mm SL.

Holotype of *Antennarius indicus*: SU 40090, 43.5 mm, Vizagapatam, India, Herre, 25 December 1940 (Fig. 57).

Additional material: BMNH, 3 (52.5–188 mm); CAS, 4 (62–124 mm); MNHN, 3 (33.5–97 mm); USNM, 3 (24.5–82 mm); UW, 8 (63–111 mm).

DIAGNOSIS

A member of the *Antennarius striatus* group unique in having two or three prominent, darkly pigmented ocelli on side of body, each encircled by a lightly pigmented ring—one basidorsal, one slightly smaller on mid-body just posterior to insertion of pectoral-fin

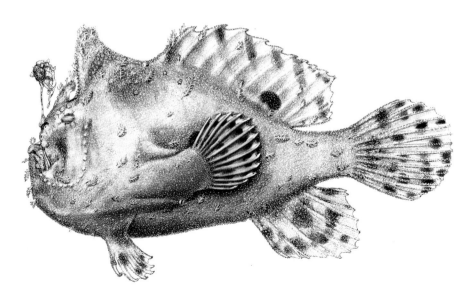

Fig. 57. *Antennarius indicus* Schultz: holotype, SU 40090, 43.5 mm SL, Vizagapatam, India. Drawing by Fanny L. Phillips; courtesy of Lisa Palmer and the Smithsonian Institution, NMNH, Division of Fishes, Washington, DC.

Fig. 58. Esca of *Antennarius indicus* Schultz. CAS 33943, 74.5 mm SL, Madras, India. Drawing by Cathy L. Short; after Pietsch and Grobecker (1987, fig. 40).

Fig. 59. Dorsal-fin spines of *Antennarius indicus* Schultz: CAS 33943, 74.5 mm SL, Madras, India. Drawing by Cathy L. Short; after Pietsch and Grobecker (1987, fig. 38A).

lobe, and occasionally a third, still smaller ocellus placed below basidorsal ocellus somewhat dorsal to base of anal fin (Fig. 57); an esca consisting of a cluster of flattened leaflike appendages (Fig. 58); and pectoral-fin rays 13, rarely 12 (Tables 5, 6).

DESCRIPTION

Esca large, the appendages averaging about 80% the length of illicium (Fig. 58); illicium 12.1 to 14.9% SL, equal to or slightly longer than second dorsal-fin spine; second dorsal-fin spine straight to slightly curved posteriorly, slightly hooked at tip, length 10.4 to 14.3% SL (Fig. 59); third dorsal-fin spine curved posteriorly, length 19.1 to 25.8% SL; eye diameter 3.6 to 6.7% SL; dorsal-fin rays 12; anal-fin rays seven (Tables 5, 6).

Coloration in life: brown to yellow-brown with dark brown to black bars on fins (or interconnected dark blotches forming distinct bars when fins are closed); underside of pectoral fins creamy white with large close-set brown to black spots (Figs. 60, 61).

Coloration in preservation: almost white to light beige on fins; darker beige, brown, yellow-brown, or reddish brown to gray on head and body, with darker mottling, particularly evident on head and belly; dark bars on fins (or more or less interconnected dark blotches forming distinct bars when fins are closed): five or six oblique bars on soft-dorsal, three or four oblique bars on anal, usually four vertical bars on caudal, and three or four bars across outer surface of paired fins; chin and belly nearly always lightly speckled; a light-colored horizontal streak through eye in some specimens; illicium lightly banded; escal appendages with small spots of dark pigment at base; as many as eight short, darkly pigmented bars radiating out from eye.

Largest specimen examined: 188 mm SL (BMNH 1887.11.11.231), but said to attain a total length of at least 260 mm (Jawad and Hussain, 2014; see also Randall, 1995: 83).

DISTRIBUTION

Antennarius indicus is a rarely encountered and very infrequently photographed species, restricted to the Western Indian Ocean, ranging from Mozambique, Zanzibar (BMNH 1867.3.9.244), and the Seychelle Islands (MNHN A.4613) to the Gulf of Mannar, the Coromandel Coast of India, and Sri Lanka, including the Gulf of Aden, coast of Oman, Persian Gulf, Pakistani coast, and the Laccadive Archipelago (Jones and Kumaran, 1980: 706, fig. 603). The type locality is Vizagapatam, India (Fig. 278, Table 22).

Fig. 60. *Antennarius indicus* Schultz: aquarium specimen, locality unknown. Photo by Jessica Groomer Smith.

Fig. 61. *Antennarius indicus* Schultz. (**A**, **B**) locality unknown. Photos taken in the Berlin Aquarium by Gilbert Angermann.

Depth data are available for only two localities: 15 to 22 m off Madras, India (CAS 33943, four specimens), and 22 to 29 m off the Somali Coast, East Africa (USNM 232158 and UW 21269, a single specimen each).

COMMENTS

Recognized by Pietsch (1984b: 36) and Pietsch and Grobecker (1987: 111) as a member of the *Antennarius ocellatus* group (now *Fowlerichthys*), *A. indicus* has since been reallocated to the *A. striatus* group based largely on the molecular analysis of Arnold and Pietsch (2012: 126, figs. 4, 6; see Evolutionary Relationships, p. 377).

It is rather surprising that this species, with its distinctive color pattern, unique escal morphology, and high pectoral-ray count, remained undescribed until 1964 when Leonard P. Schultz recognized it among collections at Stanford University. Although relatively rare, material has been available since the early nineteenth century and a remarkably accurate painting that dates to the 1860s forms part of a collection of drawings made to illustrate Francis Day's survey of the fishes of India (Fig. 62). Valenciennes (1837) confused it with *A. hispidus* (MNHN A.4582, A.4587) and *A. nummifer* (MNHN A.4613); both Playfair (1866) and Day (1876, 1889) likewise confused it with *A. nummifer* (BMNH 1867.3.9.244).

Antennarius scaber (Cuvier)
Splitlure Frogfish

Figs. 63–68, 278, 331; Tables 5–8, 22

Rana Piscatrix, Americana: Seba, 1734: 118, pl. 74, fig. 3 (Western Atlantic).

Batrachus osseum cornu supra nasum gerens: Klein, 1742: 16, no. 5 (after Seba, 1734).

Lophius minor, cute tenuiori rugoso, pinna dorsali majori, cirro nasali bifurco: Browne, 1756: 457, no. 1 [nasal cirrus (i.e., illicium) bifurcate; Jamaica].

Lophius cute scabra: capite cathetoplateo, retuso: Gronovius, 1763: 58, no. 210 (*Inhabitat Oceanum Antillas Americes alluentem*). Gronovius, 1781: 58, no. 210 (after Gronovius, 1763).

Lophius thymelicus: Meuschen, 1781, unpaged [unacceptable binomial for polynomials of Gronovius, 1763, 1781; publication ruled as not consistently binomial by International Commission on Zoological Nomenclature (Anonymous, 1950), rejected by Opinion 261 (Anonymous, 1954)]. Whitley, 1929c: 300 (names of fishes in Meuschen's index to *Zoophylacium Gronovianum*).

Lophius histrio (not of Linnaeus): Bloch, 1785: 13, pl. 111 [*L. corpore scabro*, but plate appears to be a composite showing head and body of *Histrio histrio*, with illicium and bifid esca of *A. scaber* (see Comments, p. 95)]. Bloch, 1787: 10, pl. 111 (after Bloch, 1785). Bonnaterre, 1788: 15, pl. 9, fig. 28 (*L. corpore compresso, scabro*; figure after Bloch, 1785).

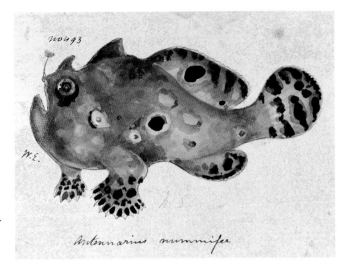

Fig. 62. *Antennarius indicus* Schultz, misidentified as *Antennarius nummifer*. One of some 705 drawings from the Francis Day collection that served as the original illustrations for Day's survey of the fishes of India, which he began in about 1861 when living in Cochin (now Kochi; see Whitehead and Talwar, 1976). Drawing by an unknown native artist for Sir Walter Elliot, who presented his collection of drawings to Day sometime prior to 1874; courtesy of Sarah Broadhurst, Ann Sylph, and the Zoological Society of London.

Gmelin, 1789: 1481 (in part). Sonnini, 1803: 160, 186, pl.10, fig. 1 (*La Lophie Histrion*; figure after Bloch, 1785). Shaw, 1804: 384, pl. 164 (Harlequin Angler, illicium "dividing at top into two dilated oval and pointed appendages"; plate after Bloch, 1785). Bory de Saint-Vincent, 1826: 495 (*sensu* Bloch, 1785; including *Guaperua* of Marcgrave, 1648). Reichenbach, 1840: 102, pl. 37, fig. 229 (*Die gemeine Seekröte*; figure after Bloch, 1785). Bauchot, 1958: 139 (confused with *A. scaber*).

La Baudroie Tachée: Bonnaterre, 1787: 49 [vernacular for *L. histrio* (not of Linnaeus) of Bloch, 1785]. Bonnaterre, 1788: 15, pl. 9, fig. 28 (text after Bonnaterre, 1787; figure after Bloch, 1785). Sonnini, 1803: 160, 186, pl. 10, fig. 1 (after Bonnaterre, 1787, 1788).

Ostracion knorrii Walbaum, 1792: 482 (original description, based solely on a figure and description published by Knorr, 1766: 52, 1767, pl. H IV, fig. 3; type material unknown, Antilles). Parenti and Pietsch, 2003: 187, fig. 1 (invalid, a *nomen oblitum*).

La Lophie Histrion (not of Linnaeus): Sonnini, 1803: 160, 186, pl. 10, fig. 1 (in part; after Lacepède, 1798, but figure after Bloch, 1785). Pietsch, 2001: 53, fig. 31 (*Rana piscatrix minima*, Plumier drawings). Pietsch, 2017: 358, pl. 108A (Plumier drawings, after Pietsch, 2001).

Chironectes scaber Cuvier, 1817b: 425, pl. 16, lower fig. [original description, at least one specimen (Fig. 63A), "provided by Robin from Trinidad, obviously the same species reported from the Antilles by Jacquin and described by Gronovius" (1763, 1781); "Commerson reports to have found it at Mauritius, near the island called Tonneliers"; type material unknown, not found in MNHN; neotype MNHN A.4545 (62 mm SL) hereby designated, one of two specimens cited by Valenciennes (1837: 412), collected by Plée, 1821, Martinique]. Cuvier, 1829: 252 (after Cuvier, 1817b). Valenciennes, 1837: 412 (description; MNHN A.4545, A.4661, Plée, 1821, Martinique).

Valenciennes, 1842: 189, pl. 85, fig. 1 (after Valenciennes, 1837; Fig. 63B). Storer, 1846: 382 (Caribbean). Guichenot, 1853: 214 (Cuba). Bauchot, 1958: 140, 142 (confused with *Antennarius histrio*). Pietsch, 1985a: 71, figs. 11–14, table 1 (original manuscript sources for *Histoire Naturelle des Poissons* of Cuvier and Valenciennes). Pietsch, 2017: 358, pl. 108A (*Rana piscatrix minima*, Plumier drawings).

Antennarius scaber: Schinz, 1822: 499 (new combination; after Cuvier, 1817b). Jordan, 1890: 652 (listed). Garman, 1896: 83 (distinct from *A. tigris*). Jordan and Evermann, 1898: 2722 (St. Lucia). Evermann and Marsh, 1900: 334, 335, pl. 48 (Puerto Rico; in key; Fig. 64). Jordan, 1905: 551, fig. 503 (Puerto Rico). Barbour, 1905: 132 (Bermuda). Bean, 1906b: 87 (Bermuda). Miranda-Ribeiro, 1915: 5, figs. (Brazil; in key). Metzelaar, 1919: 161 (Curaçao). Meek and Hildebrand, 1928: 1012 (synonymy, description, Caribbean coast of Panama). Beebe and Tee-Van, 1933: 251, fig. (Bermuda). Gregory, 1933: 389 (evolution of luring apparatus). Longley and Hildebrand, 1941: 304 (Tortugas). Barbour, 1942: 26, 33, 36, pls. 8, 17 (fig. 1) (South and Central American, Atlantic coasts; West Indies; Bermuda). Gudger, 1945a: 111, figs. 1–3 (luring behavior). Atz, 1950: 110, figs. (luring behavior). Atz, 1951a: 70, figs. (after Atz, 1950). Bauchot, 1958: 139, 142 (confused with *Antennarius histrio* and "*pinniceps*" of Valenciennes, 1837: 410). Herald, 1961: 282, color fig. 143 (New Jersey to Rio de Janeiro). Randall, 1967: 824 (Virgin Islands, Puerto Rico; food habits). Randall, 1968: 293, figs. 322, 323 (description, New Jersey to Rio de Janeiro). Lema, 1976: 40, 47 (Brazil). Lindberg et al., 1980: 342 (common names). Divita et al., 1983: 404 (Texas coast). Molter, 1983: 34, 66, 69, figs. (spawning). Burnett-Herkes, 1986: 587, fig. (Bermuda). Pietsch, 1986a: 368 (synonym of *A. striatus*). Robins and Ray, 1986: 88, pl. 14 (New Jersey, Bermuda, northern Gulf of Mexico to southeastern Brazil). Randall et al., 1990: 55, color fig. (tropical Western Atlantic). Cervigón, 1991: 193 (Venezuela). Van Dolah et al., 1994: 10, table 5 (off South Carolina and Georgia). Aguilera, 1998: 47 (Venezuela). Hoese and Moore, 1998: 164, pl. 102 (Gulf of Mexico). Michael, 1998: 354 (variant of *A. striatus*, perhaps deserving species status). Castro-Aguirre et al., 1999: 175 (Mexico, synonym of *A. striatus*). Smith-Vaniz et al., 1999: 34, 159, 161, pl. 2-23, 2-24 (Bermuda; in key). Schmitter-Soto et al., 2000: 153 (Mexico, synonym of *A. striatus*). Ortiz-Ramirez et al., 2005: 24 (synonym of *A. striatus*). Schwartz, 2005: 145, 146, 148 (synonym of *A. striatus*). Mejía-Ladino et al., 2007: 290 (synonym of *A. striatus*). Michael, 2007: 58 (after Michael, 1998). Fricke et al., 2009: 28 (synonym of *A. striatus*). Smith-Vaniz and Collette,

A

B

Fig. 63. *Antennarius scaber* (Cuvier). (**A**) probably based on the original holotype, a specimen sent to Cuvier from Trinidad by the Abbé Claude C. Robin but now lost. After Cuvier, 1817b, pl. 16, lower fig. (**B**) Cuvier's *Chironectes scaber*, reproduced from the third edition (*Édition des Disciples*) of *Régne Animal*. After Valenciennes, 1842, pl. 85, fig. 1.

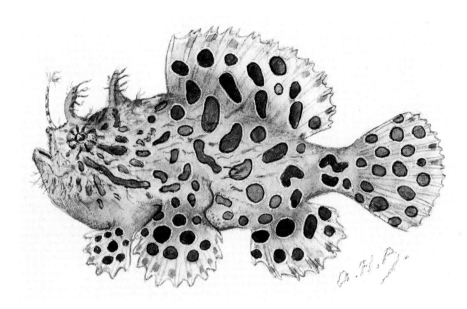

Fig. 64. *Antennarius scaber* (Cuvier). USNM 126076, 54 mm SL, "A fish of small size from the Caribbean Sea; known from St. Lucia, Martinique, and Porto Rico [sic]; only one specimen, 3 inches long, seined on the coral reef at Mayaguez." Drawn by Albertus Hutchinson Baldwin on 20 January 1899 while aboard the *Fish Hawk* during investigations of the United States Fish Commission in Puerto Rican waters, "the fish being placed in an aquarium as soon as caught and the life colors gotten before they had undergone any appreciable change." After Evermann and Marsh (1900: 51, 335, pl. 48); courtesy of Lisa Palmer and the Smithsonian Institution, NMNH, Division of Fishes, Washington, DC.

2013: 169, fig. 21 (Bermuda). Smith-Vaniz and Jelks, 2014: 29 (St. Croix, Virgin Islands). Wright et al., 2016: 121 (synonym of *A. striatus*).

Antennarius histrio (not of Linnaeus): Schinz, 1822: 501 (new combination; after Bloch, 1785). Schinz, 1836, pl. 68, fig. 3 (figure after Bloch, 1785). Günther, 1861a: 188, 189 (in part, includes *L. striatus* Shaw, from "Cook's voyage"). Kner, 1865: 192 (in part, includes *L. histrio* Bloch, 1785, and *C. scaber* Valenciennes, 1837; Riô de Janeiro). Poey y Aloy, 1868: 404 (includes *C. scaber* and *L. spectrum*). Poey y Aloy, 1876: 134 (in part). Bauchot, 1958: 142 (confused with *A. scaber*).

Chironectes mentzelii Valenciennes, 1837: 416 [original description from a drawing (based on material observed in the vicinity of Recife, Brazil) made by German naturalist Christian Mentzel for Johan Maurits of Nassau-Siegen sometime between 1637 and 1644, shown to Valenciennes by M. H. K. Lichtenstein in Berlin between 1826 and 1829 (Whitehead, 1976; 413, 1979: 459; Brienen, 2007: 277); no extant material; iconotype in *Libri Picturati* (Theatri, 1: 23), Biblioteka Jagiellońska, Kraków (copy in Cuvier and Valenciennes, MS 504, XII.B.49; see Pietsch, 1985a: 73, 77, fig. 16)]. Günther, 1861a: 184 (in footnote). Jordan and Evermann, 1898: 2719 (in footnote), 2724 [in synonymy of *A. nuttingi* (=*A. striatus*)]. Miranda Ribeiro, 1915: 8 (misspelled *mentzelli*; in key). Schultz, 1957: 94 (in synonymy of *A. multiocellatus*). Pietsch, 1985a: 73, 77, fig. 16, table 1 (original manuscript sources for *Histoire Naturelle des Poissons* of Cuvier and Valenciennes).

Chironectes tigris Poey y Aloy, 1853: 217, pl. 17, fig. 2 (*Pescador Tigre*, original description, single specimen, holotype MCZ 11619, Cuba). Howell y Rivero, 1938: 219 (MCZ

11611 erroneously listed as holotype; MCZ 11617, 11619, 11621 erroneously listed as paratypes).

Chironectes histrio (not of Linnaeus): Poey y Aloy, 1853: 217 (in part, confused with *Histrio histrio*; Cuba).

Lophius spectrum Gray, 1854: 49 [binomial for polynomials of Browne, 1756 (no. 1), and Gronovius, 1763 (no. 210); constitutes original description, no known type material; Antilles]. Poey y Aloy, 1868: 404 [in synonymy of *A. histrio* (not of Linnaeus)]. Schultz, 1957: 52 ("unidentifiable").

Antennarius tigris: Günther, 1861a: 189 (new combination, description after Poey y Aloy, 1853). Poey y Aloy, 1868: 405 (after Poey y Aloy, 1853). Poey y Aloy, 1876: 134 (after Poey y Aloy, 1853). Jordan and Bollman, 1889: 553 (Green Turtle Cay, Bahamas). Garman, 1896: 83 (distinct from *A. scaber* Cuvier). Jordan and Evermann, 1898: 2723 (after Poey y Aloy, 1853; Garman, 1896). Evermann and Marsh, 1900: 334 (West Indies; in key). Howell y Rivero, 1938: 219 (erroneous type designation). Barbour, 1942: 26, 33, 37, pls. 9 (figs. 1, 2), 17 (fig. 2) (Cuba, Jamaica). Fowler, 1944: 453 (Green Turtle Cay).

Antennarius nuttingii Garman, 1896: 83, pl. 2 (original description, several specimens, Barbour, 1942: 36, lists three; lectotype MCZ 11652 by subsequent designation of Pietsch and Grobecker, 1987, paralectotype MCZ 11660, both from Haiti; existence of additional paralectotype from Great Bahama Bank unknown). Jordan and Evermann, 1898: 2723 (after Garman, 1896). Evermann and Marsh, 1900: 334, 335, pl. 49 (Puerto Rico; in key). Bean, 1906b: 87 (Bermuda). Beebe and Tee-Van, 1933: 251, fig. (Bermuda). Barbour, 1942: 25, 33, 36, pls. 15, 17 (fig. 4) (West Indies, Bermuda). Baughman, 1955: 55 (Texas). Fowler, 1944: 453 (Bahamas). Smith-Vaniz et al., 1999: 161 (synonym of *A. scaber*). (Bermuda; in key). Ortiz-Ramirez et al., 2005: 24 (synonym of *A. striatus*).

Antennarius teleplanus Fowler, 1912: 38, fig. 2 (original description, one specimen, holotype ANSP 38162, New Jersey). Wright et al., 2016: 121 (synonym of *A. striatus*).

Antennarius scaber var. *tigris*: Metzelaar, 1919: 161 (Curaçao).

Antennarius cubensis Borodin, 1928: 3, 24, pl. 3, fig. 1 [original description, one specimen, existence of holotype (40 mm SL) unknown, perhaps mislaid in Vanderbilt Museum, Centerport, New York (D. E. Englot, personal communication, 21 January and 27 May 1981); here placed in synonymy of *A. striatus* on authority of Barbour (1942: 27, 33); Puerto Rico, Cuba]. Barbour, 1942: 27, 33 [synonym of *A. tigris* (= *A. scaber*)]. Schultz, 1957: 73 [in synonymy of *Phrynelox scaber* (= *A. scaber*)].

Antennarius tenebrosus (not of Poey): Barbour, 1942: 28 [specimen of *A. scaber* (USNM 37545) designated neotype of *Chironectes tenebrosus* Poey (= *A. multiocellatus* Valenciennes)].

Phrynelox (Phrynelox) nuttingi: Schultz, 1957: 72, pl. 3, fig. D (new combination, description, Western Atlantic; in key).

Phrynelox (Phrynelox) scaber: Schultz, 1957: 73, pl. 4, figs. C, D (new combination, synonymy, description, Western Atlantic; in key). Schultz, 1964: 176, table 1 (additional specimen, Bahamas).

Phrynelox nuttingi: Duarte-Bello, 1959: 142 (Cuba). Caldwell, 1966: 88 (Montego Bay, Jamaica). Mejía-Ladino et al., 2007: 290 (synonym of *A. striatus*).

Phrynelox scaber: Duarte-Bello, 1959: 142 (Cuba). Le Danois, 1964: 121, fig. 61n (description; MNHN A.4639, A.4661, A.4816 erroneously listed as syntypes, both from Antilles). Caldwell, 1966: 88 (Jamaica). Cervigón, 1966: 873, 875, fig. 374 (Golfo de Cariaco,

Cubagua, Venezuela; in key). Wickler, 1967: 546, 548, fig. 5c (feeding behavior, film no. E 1039/1964). Wickler, 1968: 127, fig. 25 (angling for prey). Böhlke and Chaplin, 1968: 718, figs. (New Jersey to southeastern Brazil). Guitart, 1978: 788, fig. 619 (Cuba). Martin and Drewry, 1978: 380, figs. 199–201 (early developement). Pietsch, 1978b: 3 (western central Atlantic). Román, 1979: 351, fig. (Venezuela; in key). Uyeno and Aizawa, 1983: 247 (Suriname, French Guiana). Mejía-Ladino et al., 2007: 289 (synonym of *A. striatus*).

Antennarius multiocellatus (not of Valenciennes): Le Danois, 1964: 101 (in part, i.e., MNHN A.4619, Bahia, Brazil).

Phrynelox tigris: Le Danois, 1964: 124, fig. 61o (description; MNHN A.4545, A.4703 erroneously listed as paratypes of *Chironectes scaber*; both from Antilles).

Antennarius striatus (not of Shaw): Pietsch, 1984b: 34 (in part; genera of frogfishes). Pietsch, 1985a: 71, 77, figs. 11–14, 16, table 1 (in part; original manuscript sources for *Histoire Naturelle des Poissons* of Cuvier and Valenciennes). Acero and Garzón, 1987: 90 (Santa Marta, Colombia). Pietsch and Grobecker, 1987: 54, figs. 3, 6, 17–19, 116, pls. 2A, 3–5 (in part; description, distribution, relationships). Pietsch and Grobecker, 1990b: 79, figs. (after Pietsch and Grobecker, 1990a). Boschung, 1992: 79 (Alabama). Debelius, 1997: 90, color fig. (Western Atlantic). Myrberg, 1997: 6 (luring behavior). McEachran and Fechhelm, 1998: 822 (Gulf of Mexico). Castro-Aguirre et al., 1999: 175 (Gulf of Mexico). Smith-Vaniz et al., 1999: 161, pl. 2, table 2 (Bermuda). Schmitter-Soto et al., 2000: 153 (Gulf of Mexico). Pietsch, 2001: 53, fig. 31 (*Rana piscatrix minima*, drawings of Plumier). Pietsch, 2002a: 1051 (western central Atlantic). Menezes, 2003: 64 (Brazil). Smith et al., 2003: 13 (Belize). Schwartz, 2005: 145, 146, 148 (North Carolina). Jackson, 2006a: 783, 784, 787, tables 1, 2 (reproduction, depth distribution). Kuiter and Debelius, 2007: 122, 123, color figs. (in part, Brazil). Mejía-Ladino et al., 2007: 289, figs. 4, 6, tables 1–5 (in key; Colombia). Collette and Greenfield, 2009: 79, 80 (Martinique). Pietsch, 2011: 33, fig. (Plumier drawing; after Pietsch, 2001). Monteiro-Neto et al., 2013: 3, table 1 (off Rio de Janeiro). Sampaio and Ostrensky, 2013: 282, table 1 (Brazil). Smith-Vaniz and Collette, 2013: 169 (Bermuda). Garcia et al., 2015: 7 (northeastern Brazil). Ayala et al., 2016: 92, 97, table 3 (larvae, Sargasso Sea). Wright et al., 2016: 121, 122, fig. 2 (Shinnecock Bay, New York).

MATERIAL EXAMINED

One hundred and seventy-two specimens, 26 to 152 mm SL.

Neotype of *Chironectes scaber*: MNHN A.4545, 62 mm, Martinique, Plée, 1821.

Holotype of *Chironectes tigris*: MCZ 11619, 79 mm, Cuba, Poey.

Lectotype of *Antennarius nuttingi*: MCZ 11652, 61 mm, near Jeremie, Haiti, Weinland, April 1865.

Paralectotype of *Antennarius nuttingi*: MCZ 11660, 47.5 mm, data as for lectotype.

Holotype of *Antennarius teleplanus*: ANSP 38162, 72 mm, Corson's Inlet, Cape May Co., New Jersey, Philipps, 30 September 1911.

Neotype of *Antennarius tenebrosus*: USNM 37545, 34 mm, Cuba, Poey.

Additional material: *Bermuda*: ANSP, 2 (71–90 mm); BMNH, 1 (64 mm); MCZ, 3 (97–106 mm); USNM, 1 (81 mm). *Bahamas*: AMNH, 12 (50–109 mm); ANSP, 9 (60–108 mm); CAS, 1 (68 mm); IU, 1 (68 mm). *Florida*: AMNH, 2 (46–99 mm); ANSP, 2 (49–60 mm); FSBC, 2 (59.5–78 mm); LACM, 1 (88 mm); MCZ, 2 (66–96 mm); UF, 1 (26.5 mm). *Louisiana*: UW, 4 (49–75 mm). *Texas*: FSBC, 1 (58.5 mm). *Central America*: FMNH, 1 (98 mm); MCZ, 3 (58–70 mm). *Greater Antilles*: AMNH, 1 (88 mm); ANSP, 3 (42–93 mm); LACM,

Table 7. Geographic Variation in Pectoral-Fin-Ray Counts of *Antennarius scaber* and *Antennarius striatus*

Species/Locality	Pectoral-fin rays (left and right sides combined)			
	9	10	11	12
Antennarius scaber				
Western Atlantic		1	102	5
Antennarius striatus				
Eastern Atlantic	3	90	1	
Australia	5	57	2	
Japan		12	85	5
Hawaiian Islands		12		
TOTALS	8	172	190	10

Fig. 65. *Antennarius scaber* (Cuvier). Blue Heron Bridge, Lake Worth Lagoon, Florida. Photo by Linda Ianniello.

10 (39–92.5 mm); MCZ, 8 (44–94 mm); MNHN, 1 (57 mm); UMMZ, 1 (53 mm); USNM, 2 (26–28 mm); ZMUC, 14 (44–107 mm). *Lesser Antilles*: ANSP, 2 (54–71 mm); BMNH, 6 (42–98 mm); MNHN, 5 (50–152 mm); RMNH, 6 (67–97.5 mm); ROM, 1 (87 mm). *South America*: ANSP, 3 (30–112 mm); BMNH, 2 (55–57 mm); FMNH, 2 (27–31 mm); GCRL, 2 (63.5–73 mm); MBUCV, 2 (40–67 mm); MNHN, 2 (58–63 mm); MNRJ, 8 (65–123 mm); MZUSP, 35 (34–107 mm); RMNH, 1 (60 mm); ZMUC, 1 (112 mm); ZUEC, 1 (75 mm).

DIAGNOSIS

A member of the *Antennarius striatus* group, very similar to *A. striatus*, but distinguished in having two elongate, cylindrical, wormlike escal appendages, and 11 pectoral-fin rays, very rarely 10 or 12 (Tables 6, 7).

DESCRIPTION

Esca bifid, each appendage occasionally bearing smaller, more slender, secondary filaments (among all specimens examined, only one had a trifid esca; ZUEC 741, 75 mm;

Fig. 66. *Antennarius scaber* (Cuvier). (**A**) Blue Heron Bridge, Lake Worth Lagoon, Florida.Photo by Linda Ianniello. (**B**) Riviera Beach, Lake Worth Lagoon, Florida. Photo by Anne DuPont. (**C**) Blue Heron Bridge, Lake Worth Lagoon, Florida. Photo by Linda Ianniello. (**D**) Blue Heron Bridge, Lake Worth Lagoon, Florida. Photo by Kelly Casey. (**E**) Blue Heron Bridge, Lake Worth Lagoon, Florida. Photo by Linda Ianniello. (**F**) Blue Heron Bridge, Lake Worth Lagoon, Florida. Photo by Kelly Casey.

see Comments, p. 109); escal appendages short and stout (less than 4% SL) to extremely long and slender (27.8% SL in MCZ 11621, 79 mm SL); illicium 14.1 to 20.3% SL; second dorsal-fin spine more or less straight, length 13.8 to 19.0% SL; third dorsal-fin spine curved posteriorly, length 18.3 to 25.9% SL; eye diameter 4.5 to 7.4% SL; dorsal-fin rays 11, rarely 12; anal-fin rays seven, rarely six; pectoral-fin rays 11 (rarely 10 or 12; of 108 fins counted, only one had 10 rays and five had 12; Tables 5–7); presence of cutaneous appendages highly variable (Figs. 65, 66); head, body, and fins nearly "bald" to everywhere covered with remarkable array of long, slender, close-set filaments (individuals apparently able

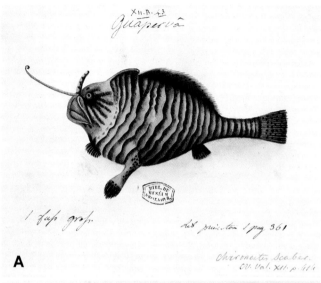

Fig. 67. *Antennarius scaber* (Cuvier). (**A**) *Guaperua*, the earliest known depiction of this species, a watercolor copy of an original made for Count Johan Maurits of Nassau-Siegen sometime between the years 1637 and 1644, based on material observed in the vicinity of Recife, Brazil (Valenciennes, 1837: 414–415); the original painting, examined by Valenciennes in Berlin, by invitation of Martin H. K. Lichtenstein, between 1826 and 1829 (Whitehead, 1979: 459) is now part of the Libri Picturati (Handbook, 1: 361) housed in the Biblioteka Jagiellońska, Kraków, Poland (Whitehead, 1982); although this drawing shows a Brazilian antennariid with an elongate illicium (in this part of the world, characteristic only of *Antennarius multiocellatus*), Valenciennes (1837: 414) associated it with *Chironectes scaber* primarily on the basis of pigment pattern. (**B**) *Rana piscatrix minima*, an early drawing of *Antennarius scaber* (Cuvier) made by French Minimist priest Charles Plumier in the 1690s; observed in the French Antilles, probably Martinique (Pietsch, 2017). After Plumier, MS 25, 90B, Bibliothèque Centrale, Muséum national d'Histoire naturelle, Paris; courtesy of Emmanuelle Choiseau, © Muséum national d'Histoire naturelle, Paris.

to increase or reduce "hairiness" in a matter of several weeks; see Cutaneous Appendages, p. 35).

Coloration in life: typically, light yellow to orange, but often red, green, gray, brown, black, or almost white with dark brown to black stripes or elongate blotches as described for species-group (Figs. 65–66); individuals capable of extreme color change in over a period of several weeks (see Color and Color Pattern, p. 36).

Coloration in preservation: beige, light yellow, orange, dark yellow-brown, to black; wormlike escal appendages occasionally reddish pink (ANSP 10648, 97667); illicium nearly always darkly banded. Lighter-color phases with as many as 12 (usually seven to 10) darkly pigmented streaks radiating out from eye; belly without elongate markings, but usually with scattered, dark circular spots. Solid-black color phase (only about 5% of the material examined) with tips of pectoral-fin rays white.

Largest specimen examined: 152 mm SL (MNHN A.4661).

DISTRIBUTION

As presently understood, *Antennarius scaber* is restricted to the Western Atlantic Ocean, ranging from off the east coast of the United States at Sinnecock Bay, Long Island, New York, to Florida, including Bermuda, the Bahamas, Cuba, the Gulf of Mexico, and throughout the island groups of the Caribbean to the southernmost coast of Brazil at 32° 51′ S latitude (MZUSP 17129; Fig. 278, Table 22). The eastward limit of its distribution has not been established; individuals of this species may be hidden among Eastern Atlantic specimens listed here as *A. striatus* (see Comments below).

This species may be found anywhere between the surface and a maximum depth of about 100 m; for 13 records, the average depth is 40 m.

COMMENTS

Although *Antennarius scaber* was discovered and illustrated as early as 1637 based on material observed in the vicinity of Recife, Brazil, and again in the 1690s in the French Antilles (Fig. 67), the earliest published account appeared in 1734, under the name *Rana Piscatrix, Americana*, in Dutch pharmacist and collector Albertus Seba's *Locupletissimi rerum Naturalium Thesauri*. Although his figure, reproduced here (Fig. 5), is rather crude, the bifurcate wormlike esca and color pattern of the head and body, combined with the locality *"Americana,"* implicit in the name, clearly indicate this species.

A half a century later, with the publication of Marcus Bloch's *Naturgeschichte der Ausländischen Fische*, in 1785, *A. scaber* became so confused with *Histrio histrio* (as *Lophius histrio*) that problems regarding the proper identification of these two species, and the taxonomic distinction between *Antennarius* and *Histrio*, lasted for nearly 200 years (e.g., see Bauchot, 1958: 139; Pietsch and Grobecker, 1987: 69). Although Bloch (1785) clearly referred to a spiny-skinned antennariid (*Lophius corpore scabro*), the figure accompanying the description of his "American Toad-Fish" (his pl. 111, reproduced here as Fig. 68) is clearly a composite, with the head and body of *H. histrio* and the illicium and esca of *A. scaber*. The extent to which this error influenced subsequent work is reflected by no less than 21 publications that make reference to *L. histrio* in the sense of Bloch (1785); seven of these publications reproduced Bloch's illustration (see synonymy, p. 86). Except for Cuvier (1817b: 427), who expressed similar suspicions concerning Bloch's (1785) plate, only Gill (1879c: 225) recognized the error, at least in part, when he wrote that Shaw (1804: 384, pl. 164, after Bloch's pl. 111) "gave, under the term *Lophius histrio*, a quite recognizable figure of that form [*Histrio histrio*], whose only great fault is the delineation of the first spine." For more on confusion related to *H. histrio*, see Comments, p. 192.

Antennarius scaber was formally described by Cuvier (1817b: 427) on the basis of a specimen from Trinidad, provided by French naturalist Claude Robin (see Bauchot et al., 1997: 78), but also on prior published accounts. In the earliest recognition of the striking similarity between *A. scaber* and *A. striatus*, Cuvier wrote, "this is surely the fish described by M. de Lacepède [1798] from a drawing by Commerson [MSS] . . . [in which] the first [dorsal-fin] spine ends in three appendages instead of two, but everything else [about them] is identical. This difference is probably due to *variété accidentelle*. Commerson found [the fish] at Mauritius, near an island called Tonneliers." Cuvier (1817b: 427) went on to say, "it is obviously the same species reported by [Nikolaus Joseph von] Jacquin from the Antilles, and described by Gronovius in the *Zoophylacium*" (1763, 1781; see synonymy, p. 87).

Since that time, the name *"scaber"* had been more or less universally accepted to designate this Western Atlantic member of the *A. striatus* group until Pietsch and Grobecker (1987) synonymized it, along with 25 additional nominal species (Table 8), with the wide-ranging and taxonomically complex *A. striatus*. Consider-

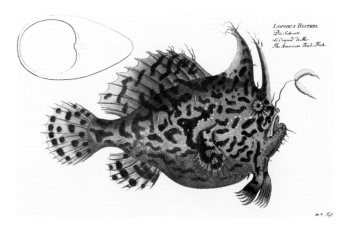

Fig. 68. Marcus Elieser Bloch's "American Toad-Fish," *Lophius histrio*. A composite showing the head and body of *Histrio histrio* (Linnaeus) and the illicium and esca of *Antennarius scaber* (Cuvier). After Bloch (1785, pl. 111).

Table 8. Available Names Currently in the Synonymies of *Antennarius scaber* (Cuvier, 1817b) and *Antennarius striatus* (Shaw, 1794), with Authorship, Date, and Type Locality

Species	Author and Date	Type Locality
Antennarius scaber		
Western Atlantic		
Ostracion knorrii (nomen oblitum)	Walbaum, 1792	Antilles
Chironectes scaber	Cuvier, 1817b	Trinidad, Antilles
Chironectes mentzelii	Valenciennes, 1837	Recife, Brazil
Chironectes tigris	Poey, 1853	Cuba
Lophius spectrum	Gray, 1854	Antilles
Antennarius nuttingii	Garman, 1896	Haiti, Great Bahama Bank
Antennarius teleplanus	Fowler, 1912	New Jersey
Antennarius cubensis	Borodin, 1928	Puerto Rico, Cuba
Antennarius striatus		
Eastern Atlantic		
Antennarius (Triantennatus) delaisi	Cadenat, 1959:374	Senegal
Antennarius (Triantennatus) occidentalis	Cadenat, 1959:381	Dakar
Western Indian Ocean		
Chironectes tricornis	Cloquet, 1817	Mauritius
Lophius commersonianus	Lacepède, in Jordan, 1917b	Mauritius
Antennarius fuliginosus	Smith, 1957	Durban, South Africa
Japan		
Chironectes tridens	Temminck and Schlegel, 1845	Japan
Antennarius nox	Jordan, 1902	Nagasaki
Australia, New Zealand		
Saccarius lineatus	Günther, 1861	New Zealand
Antennarius pinniceps var. *fasciata*	Steindachner, 1866	Port Jackson
Batrachopus insidiator	Whitley, 1934	Sydney
Phrynelox (Triantennatus) zebrinus	Schultz, 1957	Port Jackson
Phrynelox (Triantennatus) atra	Schultz, 1957	Sydney
Western Pacific		
Lophius striatus	Shaw, 1794	Tahiti
Antennarius pinniceps	Bleeker, 1856a	Ambon
Antennarius lacepedii	Bleeker, 1856a	Ambon
Antennarius melas	Bleeker, 1857	Ambon
Antennarius pinniceps Var. β. *bleekeri*	Günther, 1861	Ambon
Antennarius cunninghami	Fowler, 1941	Hawaiian Islands
Phrynelox (Phrynelox) lochites	Schultz, 1964	Mindanao, Philippines

ing the pattern of geographic variation in escal morphology (see below, p. 108) and in the absence of any other significant differences among the more than 690 specimens examined, all of these nominal species were considered by Pietsch and Grobecker (1987: 71) to represent a single, widely distributed species taking the oldest available name, *A. striatus* (Shaw, 1794). Since that time, however, a number of authors (e.g., Williams, 1989: 279; Cervigón, 1991: 193; Aguilera, 1998: 47; Michael, 1998: 354; Smith-Vaniz et al., 1999: 161; Smith-Vaniz and Jelks, 2014: 29) have questioned this decision, some arguing, in the case of *A. scaber*, that the combination of 11 or 12 pectoral rays and a bifid esca distinguishes Western Atlantic specimens from all other "*A. striatus*" and therefore synonymization of *A. scaber* should be reevaluated (Table 8). This suggestion has been followed here (for a more detailed discussion, see Comments, p. 109).

Antennarius striatus (Shaw)
Striated Frogfish

Figs. 1D, 5, 9A, 11C, 12A, 18A, 69–74, 278, 295A, 296A, 297–300, 304, 306B, 317, 320; Tables 5–8, 22–24

Antennarius antenna tricorni: Commerson, MSS 889(3): 19, 891: 123 [observed by Commerson at Tonneliers Island, just outside Port Louis, Mauritius, February 1770; *habitat in vado littorum nesogalliae*; no extant material (see Comments, p. 109)]. Lacepède, 1798: 323, 324 (footnote reference to Commerson, MSS). Cuvier, 1816: 310 ("distinct from *l'histrio* and probably same as *Lophius hispidus*"). Cloquet, 1817: 598 (reference to Commerson, MSS). Bauchot, 1958: 139 [synonym of *"pinniceps"* (= *A. striatus*), authorship given to Valenciennes, 1837].

Lophius striatus Shaw, 1794, pl. 175 [original description, "amongst the number of new animals discovered by Sir Joseph Banks during his first voyage to the South Seas," two specimens from Tahiti listed by Solander (MSS Z1: 41, 127; Z8: 227); lectotype BMNH 1978.3.1.2, paralectotype BMNH 1978.3.1.3 (see Comments, p. 110)]. Shaw, 1804: 385 (Striped Angler; after Shaw, 1794; "observed about the coasts of Tahiti during the first voyage of Captain Cook"). Cuvier, 1816: 310 (after Shaw, 1794). Cuvier, 1817b: 433 [comparison with *Chironectes tuberosus* (= *Antennatus tuberosus*)]. Gill, 1879c: 226 (distinct from *A. scaber*).

La Lophie Histrion (not of Linnaeus): Lacepède, 1798: 302, 321, pl. 14, fig. 1 [description, and Latin polynomial in footnote, based on *Antennarius antenna tricorni* of Commerson, MSS; *voisine des côtes orientales de l'Afrique* (= Mauritius); esca bifid in description, trifid in figure; represents *Antennarius striatus* but confused in part with *Histrio histrio*]. Sonnini, 1803: 160, 186, pl. 10, fig. 1 (in part, description after Lacepède, 1798, but figure after Bloch, 1785).

Lophius histrio (not of Linnaeus) *varietas a*: Bloch and Schneider, 1801: 141 (based on "Striated Lophius" of Shaw, 1794; *habitat ad litora Novae Hollandiae*). Gill, 1879c: 226 (same as *Antennarius striatus*).

Le riquèt à la houppe: Cuvier, 1816: 310 (vernacular for Latin polynomial of Commerson, MSS). Cloquet, 1817: 598 (*Le Riquet a la Houpe*; *sensu* Cuvier, 1816). Bory de Saint-Vincent, 1826: 495 (*Le Riquet a la Houppe*; *sensu* Cuvier, 1816, but including *L. hispidus*). Valenciennes, 1837: 410 (*Riquet-à-la-houppe*; confused with *L. hispidus*).

Lophius hispidus (not of Bloch and Schneider): Cuvier, 1816: 310 ["probably same as" *Antennarius antenna tricorni* of Commerson, MSS (= *Antennarius striatus*)]. Bory de Saint-Vincent, 1826: 495 (*L. striatus* and *L. hispidus* confused).

Chironectes tricornis Cloquet, 1817: 598 (binomial for polynomial of Commerson, MSS; constitutes original description, no known type material; Mauritius)

Lophius tricornis: Bory de Saint-Vincent, 1826: 495 (binomial for polynomial of Commerson, MSS; after Cloquet, 1817).

Chironectes hispidus (not of Bloch and Schneider): Valenciennes, 1837: 407 (description based in part on MNHN A.4544, A.4588, *A. striatus* from Réunion and Bombay, respectively).

Chironectes tridens Temminck and Schlegel, 1845: 159, pl. 81, figs. 2–5 (original description, five specimens, stuffed, lectotype by subsequent designation of Boeseman, 1947, RMNH 1025, paralectotypes RMNH 1021–1024, 36.5–51.5 mm, Japan). Boeseman, 1947: 136 (lectotype designated).

Antennarius tridens: Bleeker, 1853c: 4, 47 (new combination, description, Nagasaki). Bleeker, 1858a: 3 (Japan). Günther, 1861a: 191 (China). Bleeker, 1865b: 9, 14, pl. 195, fig. 3 (description, esca trifid; Ambon). Jordan and Snyder, 1901: 769 ("everywhere common in sandy or muddy bays and inlets"; Yokohama, Japan). Jordan, 1902: 372 (description, Japan). Jordan, 1905: 551 (Japan). Smith and Pope, 1906: 499 (Susaki and Yamagawa, Japan). Franz, 1910: 87, 111, pl. 10, fig. 12 (description, histology of the esca, regeneration; Misaki, Dzushi, Yagoshima). Jordan et al., 1913: 423 (after Temminck and Schlegel, 1845; Tokyo to Nagasaki). Jordan and Thompson, 1914: 312 (Misaki). Jordan and Hubbs, 1925: 330 (*A. scriptissimus, A. sanguifluus,* and *A. nox,* all considered synonyms of *A. tridens*; Misaki). Tanaka, 1930: 929, figs. 508–510 (Japan). Wu, 1931: 60 (coast of Foochow region and Ming River, China). Borodin, 1932: 99 (Southport, Queensland). Okada, 1938: 274 (Honshu to China Sea). Kamohara, 1955: 64, pl. 64, fig. 1 (Japan). Okada, 1955: 430, fig. 387 (description; Japan, China Sea). Suzuki, 1964: 187 (Amami-Oshima). Burgess and Axelrod, 1972: 259, 270, color fig. 468 (natural history). Burgess and Axelrod, 1973a: 532, color fig. 494 (natural history). Paulin, 1978: 488, fig. 1C, 2 (synonymy, New Zealand). Grobecker and Pietsch, 1979: 1162, table 1 (feeding mechanism). Paulin et al., 1989: 134 (New Zealand). Michael, 1998: 354 (variant of *A. striatus,* perhaps deserving species status). Youn, 2002: 542 (in synonymy of *A. striatus,* Korea). Michael, 2007: 58 (after Michael, 1998).

Chironectes scaber (not of Cuvier): Bianconi, 1855: 219 (Mozambique).

Antennarius pinniceps Bleeker, 1856a: 49 [original description, based in part on polynomial of Commerson, MSS; two specimens, lectotype RMNH 6271 by subsequent designation of Pietsch and Grobecker, 1987, paralectotype (98 mm total length) missing, not found at RMNH; Ambon (Valenciennes, 1837: 410, mentioned the name *"pinniceps"* only to point out that Commerson's *"pinniceps"* seemed to be identical to *Chironectes hispidus* Valenciennes, 1837: 407; no valid binomial occurs in literature until Bleeker, 1856a: 49; see Schultz, 1957: 71)]. Bleeker, 1856b: 302 (Bali). Bleeker, 1859b: 129 (Bali, Ambon). Günther, 1861b: 361 (Ceylon). Bleeker, 1865b: 9, 15, pl. 197, fig. 5 (description, esca bifid; Bali, Ambon). Günther, 1869a: 238 (St. Helena). Macleay, 1881: 577 (Sydney). Weber, 1913: 561 (Ujung Pandang, Sulawesi). Gilchrist and Thompson, 1917: 418 (synonymy, Natal). McCulloch, 1929: 407 (New South Wales, Mauritius). Fowler, 1934b: 512 (Durban). Fowler, 1936b: 1133 (West Africa to East Indies). Herre, 1945: 149 (Philippines). Smith, 1949: 431, pl. 98, fig. 1239 (Natal and Delagoa, South Africa). Munro, 1955: 289, pl. 56, fig. 843 (Ceylon). Bauchot, 1958: 139, 142 (a valid species, authorship given to Commerson, in Cuvier and Valenciennes, 1837). Beaufort and Briggs, 1962: 211 (synonymy, description; east coast of Africa to Tahiti; in key). Le Danois, 1964: 105, fig. 50k (description, MNHN A.4544 erroneously listed as type). Edwards and Glass, 1987: 631 (St. Helena). Pietsch and Grobecker, 1987: 57 (in synonymy of *A. striatus*; lectotype designation). Pietsch, 1990: 483 (synonym of *A. striatus*; tropical Eastern Atlantic). Cunxin, 1993: 117 (misidentification, probably intending to refer to *A. striatus*). Youn, 2002: 542 (in synonymy of *A. striatus,* Korea).

Antennarius lacepedii Bleeker, 1856a: 50 (original description, single specimen, with bifid esca; holotype RMNH 28017, Ambon). Bleeker, 1859b: 128 (Ambon).

Antennarius melas Bleeker, 1857: 70 (original description, single specimen, esca bifid, holotype RMNH 6277, Ambon). Bleeker, 1859b: 129 (Ambon). Günther, 1861a: 194 (Ambon). Bleeker, 1865b: 10, 20, pl. 199, fig. 6 (description; after Bleeker, 1857).

Saccarius lineatus Günther, 1861a: 183 (original description, single specimen, holotype BMNH 1979.9.25.1, New Zealand). Hutton, 1872: 30 (after Günther, 1861a). Gill 1879b:

222 (listed). Hutton, 1891: 280 (after Günther, 1861a). Waite, 1912: 197 (synonymy, New Zealand). Phillipps, 1927: 14 (New Zealand). Whitley, 1968a: 88 (after Günther, 1861a).

Antennarius histrio (not of Linnaeus): Günther, 1861a: 188, 189 (in part, synonymy includes *C. scaber* Cuvier). Steindachner, 1895: 24 (Robertsport, Liberia). Bauchot, 1958: 139, 142 (confused with *A. scaber*).

Antennarius pinniceps **Var. α.** *pinniceps*: Günther, 1861a: 190 (BMNH 1853.1.11.14, 1855.9.19.798, both from Mauritius).

Antennarius pinniceps **Var. β.** *Bleekeri* Günther, 1861a: 190 (original description, no known specimens, Ambon). Bleeker, 1865b: 15 (reference to Günther, 1861a).

Antennarius pinniceps **var.** *fasciata* Steindachner, 1866: 457 (original description, no known specimens, Port Jackson).

Antennarius striatus: Günther, 1876: 162, pl. 99 (fig. B) (new combination, description; Mauritius, Australia, Solomon Island). Macleay, 1881: 577 (Port Jackson). McCulloch, 1922: 123, pl. 41, fig. 357a (New South Wales). Barnard, 1927: 1001 (Natal). Fowler, 1928: 477 (Oceania). Griffin, 1929: 387, pl. 65, fig. 10 (New Zealand). McCulloch, 1929: 407 (New South Wales). Fowler, 1934b: 513 (Durban, Zululand). Smith, 1949: 431, pl. 98, fig. 1250 (South Africa as far as Port Alfred). Marshall, 1964: 513, 514, pl. 64, fig. 495 (Queensland, New South Wales; in key). Grant, 1965: 204, fig. ("Striped Angler," Queensland). Smith and Smith, 1966: 151, fig. 1240 (Cape of Good Hope). Kailola, 1975: 47 (Torres Strait). Burgess and Axelrod, 1976: 1910, color fig. 411 (Great Barrier Reef). Pietsch, 1981b: 419 (in part; osteology). Pietsch, 1984b: 34 (in part; genera of frogfishes). Pietsch, 1984d: 321 (in part; early life history). Pietsch, 1984e: 2 (Western Indian Ocean). Pietsch, 1985a: 71, 77, figs. 11–14, 16, table 1 (in part; original manuscript sources for *Histoire Naturelle des Poissons* of Cuvier and Valenciennes). Pietsch, 1985b: 594, 595 (in part; fast feeding). Pietsch, 1986a: 366, 368, fig. 102.7, pl. 13 (South Africa). Edwards and Glass, 1987: 631 (St. Helena). Pietsch and Grobecker, 1987: 54, figs. 3, 6, 17–19, 116, pls. 2A, 3–5 (in part; description, distribution, relationships). Allen and Swainston, 1988: 38 (northwestern Australia). Paxton et al., 1989: 279 (Australia). Pietsch, 1990: 482 (tropical Eastern Atlantic). Pietsch and Grobecker, 1990a: 98, 100, figs. (in part; aggressive mimicry, coloration). Pietsch and Grobecker, 1990b: 79, figs. (after Pietsch and Grobecker, 1990a). Randall et al., 1990: 55, color fig. (in part; circumtropical, excluding Eastern Pacific). Paugy, 1992: 573 (West Africa). Paulin and Roberts, 1992: 100 (New Zealand). Edwards, 1993: 499 (St. Helena). Francis, 1993: 158 (Lord Howe Island). Kuiter, 1993: 48 (southeastern Australia). Mori and Yamato, 1993: 41 (amphipod parasite). Randall et al., 1993a: 223, 227, table 1 (Hawaiian Islands). Bertelsen, 1994: 44, color pls. (aggressive mimicry). Goren and Dor, 1994: 14 (Red Sea). Hutchins, 1994: 42, appendix (Western Australia). Randall, 1996: 46, color fig. (Hawaiian Islands). Allen, 1997: 62 (tropical Australia). Kuiter, 1997: 42 (Australia). Lindberg et al., 1997: 217 (Sea of Japan). Okamura and Amaoka, 1997: 136, figs. (Japan). Randall et al., 1997a: 55, color fig. (in part; circumtropical, excluding Eastern Pacific). Randall et al., 1997b: 11 (Ogasawara Islands). Bavendam, 1998: 41, 48, color figs. (natural history). Bertelsen and Pietsch, 1998: 44, color pls. (aggressive mimicry, after Bertelsen, 1994). Michael, 1998: 329–331, 333, 336, 353, 354, color figs. (identification, behavior, captive care). Fricke, 1999: 107 (Réunion, Mauritius). Johnson, 1999: 724 (Moreton Bay, Queensland). Pietsch, 1999: 2015 (western central Pacific; in key). Allen, 2000a: 62, pl. 12-9 (Indo-West Pacific). Laboute and Grandperrin, 2000: 130 (New Caledonia). Nakabo, 2000: 455 (Japan; in key). Pereira, 2000:

5 (Mozambique). Pietsch, 2000: 597 (South China Sea). Hutchins, 2001: 22 (Western Australia). Shimizu, 2001: 22 (Japan). Shinohara et al., 2001: 309 (Tosa Bay, Japan). Everly, 2002: 393 (early development). Nakabo, 2002: 455 (Japan; in key). Youn, 2002: 214, 541, fig. (Korea). Allen and Adrim, 2003: 25 (Moluccas and Sulawesi, Indonesia). Allen et al., 2003: 366, color figs. (tropical Pacific). Myers and Donaldson, 2003: 610, 614 (Mariana Islands). Schneidewind, 2003a: 83, fig. (natural history). Schneidewind, 2003b: 86, figs. (aggressive mimicry). Yanong et al., 2003: 400, figs. 1–7 (mycobacteriosis). Heemstra and Heemstra, 2004: 122 (southern Africa). Letourneur et al., 2004: 209 (Réunion Island). Shedlock et al., 2004: 134, 139, 144, table 1 (molecular phylogeny). Shimazaki and Nakaya, 2004: 33, figs. 3B, 6D (luring apparatus). Mundy, 2005: 53, 67, 265 (Hawaiian Islands). Ortiz-Ramirez et al., 2005: 23, 24, 26, 29, 30, figs. 2–5, table 1 (early life history, courtship, fertilization). Randall, 2005a: 73, color figs. (South Pacific). Randall, 2005b: 311, fig. 60 (mimicry). Schneidewind, 2005a: 24, 28, figs. (sea urchin mimicry). Schneidewind, 2005b: 14, 17, figs. (after Schneidewind, 2005a). Steene, 2005: 221, 263, 303, color pls. (Lembeh Strait). Allen et al., 2006: 640 (Australia). Carnevale and Pietsch, 2006: 452, table 1 (comparison, meristics, first fossil frogfish). Michael, 2006: 34, 35, 38, 41, color fig. (luring behavior). Schneidewind, 2006b: 28, 32, color figs. (luring behavior, sexual dimorphism). Schneidewind, 2006c: 46, color fig. (luring behavior, list of species). Schneidewind, 2006d: 61, color fig. (sea urchin mimicry). Senou et al., 2006: 423 (Sagami Sea). Kuiter and Debelius, 2007: 122, 123, color figs. (in part; eastern Australia, Japan, but also Brazil). Lugendo et al., 2007: 1222, appendix 1 (Zanzibar). Mejía-Ladino et al., 2007: 289, figs. 4, 6, tables 1–5 (species account; in key; Colombia). Michael, 2007: 57–60, 64, color figs. (possible new species; Lembeh, Sulawesi). Randall, 2007: 130, color fig. (Hawaiian Islands; in key). Kwang-Tsao et al., 2008: 243 (southern Taiwan). Fricke et al., 2009: 28 (Réunion). Pietsch, 2009: 4, fig. 2D (representative antennariid). Pietsch et al., 2009a: 37, 45, fig. 5 (genetics, as an outgroup). Carnevale and Pietsch, 2010: 624, fig. 8C (osteology, phylogeny). Golani and Bogorodsky, 2010: 16 (Red Sea). Miya et al., 2010: 7, 17, table 2 (molecular phylogeny). Randall, 2010: 38 (Hawaii). Fricke et al., 2011: 368 (New Caledonia). Allen and Erdmann, 2012: 142, 151, color figs. (East Indies). Arnold and Pietsch, 2012: 120, 126–128, figs. 3C, 3D, 4, 6, table 1 (in part; molecular phylogeny). Carnevale and Pietsch, 2012: 48 (osteology, phylogeny). Larson et al., 2013: 54 (Northern Territory). Nakabo, 2013: 538, figs. (Japan). Page et al., 2013 (common names). Wirtz et al., 2013: 119 (Cape Verde Islands). Brandl and Bellwood, 2014: 32 (pair-formation). Harasti et al., 2014: 3, 7 (Port Stephens, New South Wales). Karplus, 2014: 35, 37, fig. 1.21B (aggressive mimicry; after Pietsch and Grobecker, 1987). Delrieu-Trottin et al., 2015: 5, table 1 (Marquesas Islands). Derouen et al., 2015: 29, fig. 3, table 1 (molecular phylogeny). Hanel and John, 2015: 138, table 1 (Cape Verde Islands). Stewart, 2015: 884 (northeastern New Zealand). Townsend, 2015: 4 (Sabah, Malaysia). Triay-Portella et al., 2015: 164 (Canary Islands). Brauwer and Hobbs, 2016 (biofluorescent lures). Pietsch, 2016: 2053 (eastern central Atlantic; in key).

Antennarius nox Jordan, 1902: 375, fig. 6 (original description, six specimens, five extant, lectotype by subsequent designation of Böhlke, 1953, SU 7603, Nagasaki, paralectotypes SU 7599 (2), SU 7601, USNM 49819). Jordan, 1905: 551, fig. 501 (after Jordan, 1902). Gill, 1909: 598, fig. 41 (after Jordan, 1902). Jordan et al., 1913: 424, fig. 392 (after Jordan, 1902; "perhaps a black form of *A. tridens*"). Jordan and Thompson, 1914: 312, fig. 87 (Misaki). Jordan and Hubbs, 1925: 330 (synonym of *A. tridens*; Misaki). Okada, 1938: 274 (Honshu to Taiwan). Böhlke, 1953: 148 (lectotype designated).

Antennarius tridens (not of Temminck and Schlegel): Pietschmann, 1909: 1, pl. 1 (specimen is *F. scriptissimus*, Japan).

Lophius commersonianus: Lacepède, in Jordan, 1917b: 104 (designated type species of *Antennarius* Cuvier, 1816; equated with *Antennarius antenna tricorni* Commerson, in Lacepède, 1798).

Antennarius commersoni (not of Latreille): Whitley, 1929b: 137, pl. 31, fig. 5 (New South Wales). Whitley, 1934, unpaged (after Whitley, 1929b).

Phrynelox striatus: Whitley, 1931: 328 (new combination, New South Wales). Whitley, 1935: 302 (natural history). Whitley, 1954: 29 (Wallabi Island, Western Australia). Le Danois, 1964: 120, fig. 61L (Pondicherry, Bombay). Whitley, 1968a: 87 (New South Wales). Allen et al., 1976: 385 (probably senior synonym of *P. zebrinus*; Lord Howe Island). Grant, 1978: 598, color fig. 262 (description, natural history).

Antennarius tridens Var. *striatus*: Borodin, 1932: 99 (Southport, Queensland).

Batrachopus insidiator Whitley, 1934, first page [original description in footnote, single specimen, holotype AMS I.5263 (see Whitley, 1929b: 137) lost (John R. Paxton, personal communication, 11 June 1981), 87.5 mm, Watson's Bay near Sydney]. Whitley, 1935: 302, fig. 5 (after Whitley, 1929b, 1934). Whitley, 1949: 400, caption to fig. (locomotion). Whitley, 1964: 58, 60 (as senior synonym of *Phrynelox atra* Schultz; Australia).

Antennarius scaber (not of Cuvier): Cadenat, 1937: 536, figs. 59, 59 bis (Senegal). Cadenat, 1951: 291, fig. 241 (Senegal). Delais, 1951: 149, figs. 4, 5 (Senegal). Roux and Collignon, 1957: 238, figs. 113, 114 (West Africa).

Antennarius cunninghami Fowler, 1941: 279, fig. 32 (original description, single specimen, holotype ANSP 69892, Hawaiian Islands, Oahu or Maui). Fowler, 1949: 158 (description after Fowler, 1941). Gosline and Brock, 1960: 305, 345 (after Fowler, 1941; in key). Randall et al., 1993a: 223, 227, table 1 (synonym of *A. striatus*, Hawaiian Islands).

Antennarius campylacanthus (not of Valenciennes): Delais, 1951: 147, fig. 3 (Senegal).

Antennarius fuliginosus Smith, 1957: 222, fig. 5 (original description, single specimen, holotype SAIAB 100, Durban, South Africa). Pietsch, 1986a: 368 (synonym of *A. striatus*).

Phrynelox (Phrynelox) striatus: Schultz, 1957: 71, pl. 3, fig. C (synonymy, description; in key). Schultz, 1964: 176, table 1 (additional material; Port Jackson and Lake Illawarra, Australia).

Phrynelox (Phrynelox) melas: Schultz, 1957: 72, pl. 4, fig. A (new combination, synonymy, description, Indo-Pacific; in key).

Antennarius zebrinus: Schultz, 1957: 74 (designated type species of subgenus *Triantennatus*). Thresher, 1984: 41 (egg rafts).

Phrynelox (Triantennatus) cunninghami: Schultz, 1957: 74, pl. 5, fig. A (new combination, description, Hawaii).

Phrynelox (Triantennatus) zebrinus Schultz, 1957: 75, fig. 3, table 2 (original description, six specimens, holotype USNM 47854, Port Jackson). Whitley, 1964: 60 (as synonym of *P. striatus*).

Phrynelox (Triantennatus) atra Schultz, 1957: 76, fig. 4, table 2 (original description, three specimens, holotype FMNH 21705, Sydney). Schultz, 1964: 176, table 1 (additional material; Port Jackson, Honolulu). Whitley, 1964: 60 (as synonym of *Batrachopus insidiator* Whitley, 1934).

Phrynelox (Triantennatus) nox: Schultz, 1957: 78, fig. 5 (new combination, synonymy, description, Japan; in key). Schultz, 1964: 176, table 1 (additional material, Japan).

Phrynelox (Triantennatus) tridens: Schultz, 1957: 79, pl. 5, fig. B [new combination, synonymy, description; Japan, Mauritius (?); in key]. Schultz, 1964: 177, table 1 (additional material, Japan).

Antennarius glauerti (not of Whitley, 1944a) Whitley, 1957a: 207, fig. [original description, single specimen, holotype AMS I.2979, Exmouth Gulf, Western Australia; not same as *Echinophryne glauerti* Whitley, 1944a (= *Allenichthys glauerti*)]. Whitley, 1958: 50 (additional specimen, new record for Queensland).

Antennarius (Triantennatus) delaisi Cadenat, 1959: 374–76, 383, figs. 10, 11 (original description, four specimens, three in MNHN, lectotype MNHN 1982-149, paralectotypes MNHN 1982-134, 1982-135, Senegal). Cadenat, 1960: 1415 (West Africa). Blache, 1962: 79 (after Cadenat, 1959). Schultz, 1964: 177 (after Cadenat, 1959). Blache et al., 1970: 442, fig. 1119 (after Cadenat, 1959). Pietsch et al., 1986: 137 (in type catalog).

Antennarius (Triantennatus) occidentalis Cadenat, 1959: 381, 383, figs. 2, 12–17, 25, 26 (original description, 22 specimens, 14 in MNHN, lectotype MNHN 1982-136 by subsequent designation of Pietsch and Grobecker, 1987, paralectotypes MNHN 1982-137 through 1982-148, Dakar, Gulf of Guinea, tropical West Africa). Cadenat, 1960: 1415 (Los Island, Ghana). Blache, 1962: 79 (after Cadenat, 1959). Schultz, 1964: 177 (after Cadenat, 1959). Blache et al., 1970: 443, fig. 1120 (after Cadenat, 1959). Azevedo, 1971: 92, figs. 2b, c, 5–8 (Angola). Pietsch et al., 1986: 137 (in type catalog).

Phrynelox zebrinus: Mees, 1959: 10 (Shark Bay, Western Australia). Le Danois, 1964: 120, fig. 61m (Madagascar, Australia). Chen et al., 1967: 7, fig. 2 (Taiwan; in key). Allen et al., 1976: 385 (Lord Howe Island). Araga, 1984: 102, pl. 88-A (Japan). Fricke et al., 2011: 368 (synonym of *A. striatus*).

Antennarius occidentalis: Le Danois, 1964: 103, fig. 49h (four specimens, MNHN 56-67, erroneously listed as paratypes; all from Gulf of Guinea). Pietsch, 1986a: 368 (synonym of *A. striatus*).

Phrynelox tridens: Le Danois, 1964: 121, fig. 61p (description, China). Chen et al., 1967: 9, fig. 3 (Taiwan). Allen et al., 1976: 385 (questionable identification, Lord Howe Island). Araga, 1984: 103, pl. 88-C (Japan). Fricke et al., 2011: 368 (synonym of *A. striatus*).

Phrynelox (Phrynelox) lochites Schultz, 1964: 179, pl. 1, tables 1, 2 (original description, four specimens, holotype SU 38194, Dapitan Bay, Mindanao).

Antennarius striatus var. *pinniceps*: Marshall, 1964: 513, 515 (color variant; South Queensland, New South Wales; in key).

Phrynelox nox: Chen et al., 1967: 9, fig. 4 (Taiwan). Araga, 1984: 103, pl. 88-B (Japan).

Triantennatus zebrinus: Munro, 1967: 584, pl. 78, fig. 1090 (new combination, New Guinea). Grant, 1972: 365, fig. (description, natural history). Friese, 1973: 32–36, figs. (natural history). Ortiz-Ramirez et al., 2005: 24 (synonym of *A. striatus*).

Antennarius chironemus (not of Munro): Grant, 1978: 601, pl. 263 (in part, i.e., color plate; origin of specific name unknown).

Phrynelox cunninghami: Tinker, 1978: 508, fig. (description after Fowler, 1941).

Antennarius cf. striatus: Michael, 2007: 60, 62, color figs. (possible new species; Lembeh, Sulawesi).

MATERIAL EXAMINED

Five hundred and eighteen specimens, 16.5 to 155 mm SL.

Lectotype of *Lophius striatus*: BMNH 1978.3.1.2, 77 mm, Tahiti, Society Island, Banks, 13 April–8 August 1769.

Paralectotype of *Lophius striatus*: BMNH 1978.3.1.3, 73 mm, data as for holotype.

Lectotype of *Chironectes tridens*: RMNH 1025, 150 mm, stuffed, vicinity of Nagasaki, Japan, Burger.

Paralectotypes of *Chironectes tridens*: RMNH 1021-24, 4 (51–92 mm), stuffed, data as for lectotype.

Lectotype of *Antennarius pinniceps*: RMNH 6271, 64.5 mm, Ambon, Moluccas, Bleeker.

Holotype of *Antennarius lacepedii*: RMNH 28017, 80 mm, Ambon, Moluccas, Bleeker (Fig. 69).

Holotype of *Antennarius melas*: RMNH 6277, 45 mm, Ambon, Moluccas, Bleeker.

Holotype of *Saccarius lineatus*: BMNH 1979.9.25.1, 102 mm, Bay of Islands, New Zealand, Smith.

Lectotype of *Antennarius nox*: SU 7603, 83.5 mm, Nagasaki, Japan, Jordan and Snyder, summer 1900.

Paralectotypes of *Antennarius nox*: SU 7599, 2 (57.5–60.5 mm), Misaki, Honshu, Japan, Jordan and Snyder, summer 1900; SU 7601, 1 (61.5 mm), Wakanoura, Kii-hanto, Japan, Jordan and Snyder, summer 1900. USNM 49819, 1 (67 mm), Nagasaki, Jordan and Snyder, summer 1900.

Holotype of *Antennarius cunninghami*: ANSP 69892, 85 mm, Oahu or Maui, Hawaiian Islands, 1940.

Holotype of *Antennarius fuliginosus*: SAIAB 100, 124 mm, Durban, South Africa.

Holotype of *Phrynelox zebrinus*: USNM 47854, 98 mm, Port Jackson, New South Wales.

Paratypes of *Antennarius zebrinus*: FMNH 44980, 1 (42 mm), Moreton Bay, Queensland; FMNH 21560, 1 (92 mm), Sydney, Australia. USNM 28659, 1 (84 mm), Australia (?); USNM 47853, 59948, 2 (91–94 mm), Port Jackson, New South Wales.

Holotype of *Phrynelox atra*: FMNH 21705, 73 mm, Sydney, New South Wales.

Paratypes of *Phrynelox atra*: FMNH 21704, 1 (87 mm), Sydney, New South Wales. USNM 164245, 1 (90 mm), Sydney.

Holotype of *Antennarius glauerti* (not of Whitley): AMS IB.2979, 36 mm, Exmouth Gulf, Western Australia, prawn trawl, 3 August 1952.

Lectotype of *Antennarius delaisi*: MNHN 1982-149, 100 mm, IFAN 1429, Gorée, Senegal, on surface, 24 October 1946.

Paralectotypes of *Antennarius delaisi*: MNHN 1982-134, 1 (88 mm), IFAN 52-796, Kayar, Senegal, 50 m, 24 October 1952; MNHN 1982-135, 1 (79 mm), IFAN 53-237, trop. West Africa (Cadenat, 1959, fig. 11).

Lectotype of *Antennarius occidentalis*: MNHN 1982-136, 120 mm, IFAN 51-15, Cap Vert near Dakar, Senegal, January 1951.

Paralectotypes of *Antennarius occidentalis*: MNHN 1982-137, 1 (128 mm), IFAN 51-16, Cap Vert near Dakar, January 1951; MNHN 1982-138, 1 (72 mm), IFAN 51-1097, Dakar, 6 July 1951; MNHN 1982-139, 1 (83 mm), IFAN 52-797, Kayar, 50 m, 14 October 1952; MNHN 1982-140, 1982-141, 2 (67–73 mm), IFAN 49-258, 49-259,

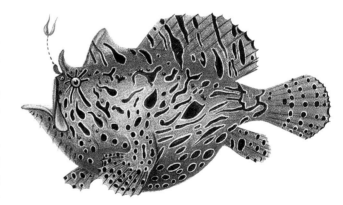

Fig. 69. *Antennarius striatus* (Shaw): holotype of *Antennarius lacepedii* Bleeker, RMNH 28017, 80 mm SL, Ambon, Indonesia. Drawing by Ludwig Speigler under the direction of Pieter Bleeker; after Bleeker (1865b, pl. 197, fig. 5).

south of Dakar, 13 August 1949; MNHN 1982-142, 1 (96 mm), IFAN 55-1418, Region M'Bour, 25–30 m, 24 August 1955; MNHN 1982-143, 2 (68–73 mm), IFAN 55-1616-17, Kayar, 30 m, 25 October 1955; MNHN 1982-144, 1 (83 mm), IFAN 52-176, Dakar, April 1952; MNHN 1982-145, 1982-146, 1982-148, 3 (58–79 mm), IFAN 55-1612, 55-1613, 55-1611, Kayar, 25 October 1955; MNHN 1982-147, 1 (89 mm), IFAN 49-237, south of Dakar, August 1949.

Holotype of *Phrynelox lochites*: SU 38194, 48.5 mm, Dapitan Bay, Mindanao, Philippines, Herre, August 1940.

Paratypes of *Phrynelox lochites*: SU 60459, 2 (33–40 mm), data as for holotype. USNM 197325, 1 (39 mm), data as for holotype.

Additional material: *Eastern Atlantic*: BMNH, 6 (58.5–92 mm); MNHN, 23 (48–118 mm); RMNH, 1 (88 mm); USNM, 7 (47.5–110 mm); UW, 7 (42.5–77 mm); ZMUC, 4 (22.5–50 mm). *Western Indian Ocean*: ANSP, 1 (100 mm); BMNH, 5 (47–104 mm); MNHN, 6 (47–104 mm); SAIAB, 34 (31–129 mm); SAMA, 10 (22–115 mm); USNM, 1 (33 mm); UW, 3 (55–98 mm). *Red Sea*: MNHN, 2 (86–93 mm). *India*: MNHN, 1 (86 mm). *Indonesia and New Guinea*: KFR, 1 (56 mm); RMNH, 5 (52–77 mm); ZMA, 1 (53 mm). *Japan*: ANSP, 4 (49–63 mm); BMNH, 6 (32.5–76 mm); CAS 42839, 4 (48–59 mm); FAKU, 8 (32–83 mm); FMNH, 3 (19–35 mm); HUMZ, 17 (27.5–88.5 mm); IU, 3 (49–53 mm); MCZ, 2 (47.5–57 mm); SU, 34 (25.5–97 mm); UMMZ, 39 (16.5–87 mm); USNM, 25 (20–86 mm); UW, 11 (29–64). *China, Taiwan, and South China Sea*: BMNH, 2 (65–94 mm); BPBM, 1 (58 mm); CAS, 7 (25–96 mm); HUMZ, 3 (42–82.5); MNHN, 1 (38.5 mm); NTUM, 3 (51–103 mm); SU, 3 (43.5–63 mm); USNM, 2 (50.5–81 mm). *Australia*: AMS, 77 (32–141 mm); BMNH, 4 (59–128 mm); MCZ, 1 (59 mm); MNHN, 1 (104 mm); NMV, 2 (86–94 mm); SU, 3 (60–107 mm); USNM, 2 (73–75 mm); UW, 5 (38–128 mm); WAM, 22 (25–126 mm). *Solomon Islands*: AMS, 1 (82 mm); BMNH, 1 (97 mm). *New Zealand*: AIM, 13 (66.5–116.5 mm); NMNZ, 4 (92.5–132 mm). *New Caledonia*: UW, 1 (93 mm). *Fiji*: USPS, 1 (specimen not examined, identification verified from photograph provided by J. Seeto). *Hawaiian Islands*: ANSP, 1 (57 mm); BPBM, 14 (27–70 mm); CAS, 2 (40–69 mm); LACM, 1 (28 mm); MNHN, 2 (55–68 mm); SU, 1 (49 mm). *Tahiti*: BPBM, 1 (81 mm); CAS, 1 (155 mm). *Locality unknown*: UW, 13 (42.5–128 mm, no data).

DIAGNOSIS

A member of the *Antennarius striatus* group, very similar to the Western Atlantic *A. scaber*, but distinguished in having three to seven elongate (very rarely two), cylindrical, wormlike escal appendages, and 10 or 11 pectoral-fin rays, rarely 9 or 12 (Tables 6, 7).

DESCRIPTION

Esca usually trifid, each appendage occasionally bearing much smaller, more slender, secondary filaments (among more than 500 specimens examined, only about 7% had a bifid esca, but including the two type specimens of *A. striatus* from Tahiti; see Geographic Variation, p. 108; Fig. 70); escal appendages short and stout (less than 4% SL) to extremely long and slender (reaching at least 26% SL); illicium 13.6 to 22.7% SL; second dorsal-fin spine more or less straight, length 11.4 to 19.0% SL; third dorsal-fin spine curved posteriorly, length 16.7 to 29.3% SL; eye diameter 3.5 to 7.4% SL; dorsal-fin rays 11 or 12; anal-fin rays seven, rarely six; pectoral-fin rays 9 to 12 (of 268 fins counted, excluding those from Japan, only eight had nine pectoral-fin rays, three had 11, and none had 12; see Tables 6, 7); presence of cutaneous appendages highly variable (Fig. 71); head, body, and fins nearly "bald" to everywhere covered with remarkable array of long, slender, close-

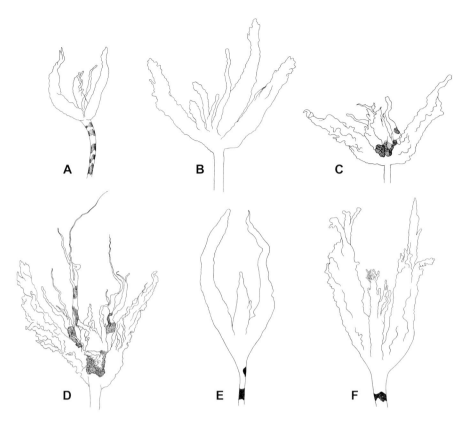

Fig. 70. Escae of *Antennarius striatus* (Shaw). (**A**) BMNH 1868.6.15.20, 83 mm SL, St. Helena. (**B**) UW 20855, 60 mm SL, Cameroun. (**C**) SU 27997, 63 mm SL, Hong Kong. (**D**) USNM 201570, 81 mm SL, Taiwan. (**E**) NMV 54306–7, 94 mm SL, Port Jackson, New South Wales. (**F**) UW 21014, 83 mm SL, Sydney Harbour. Drawings by Cathy L. Short; after Pietsch and Grobecker (1987, fig. 18).

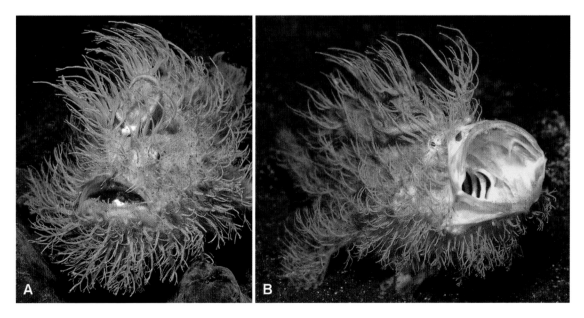

Fig. 71. *Antennarius striatus* (Shaw). (**A**, **B**) Lembeh Strait, Sulawesi, Indonesia. Photos by Roger Steene.

Fig. 72. *Antennarius striatus* (Shaw). Lembeh Strait, Sulawesi, Indonesia. Photo by Mike Bartick.

Fig. 73. *Antennarius striatus* (Shaw). (**A**) Lembeh Strait, Sulawesi, Indonesia. Photo by Frank Schneidewind. (**B**) Lembeh Strait, Sulawesi, Indonesia. Photo by Roger Steene. (**C**) Gulimanuk, Bali, Indonesia. Photo by Scott W. Michael. (**D**) Lembeh Strait, Sulawesi, Indonesia. Photo by Roger Steene. (**E**, **F**) Lembeh Strait, Sulawesi, Indonesia. Photos by Scott W. Michael.

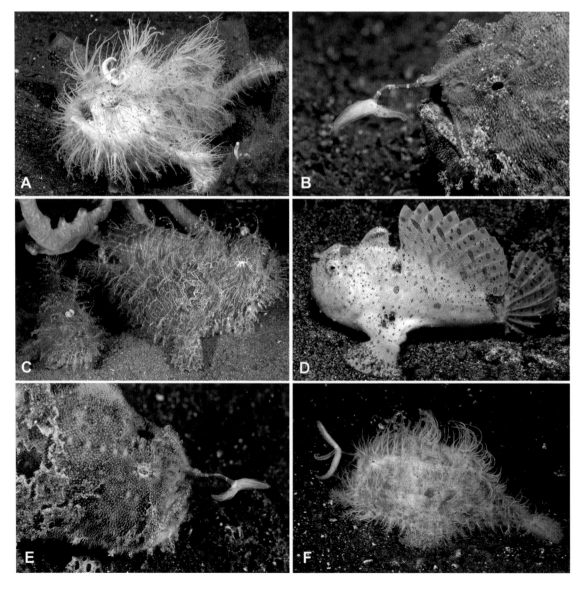

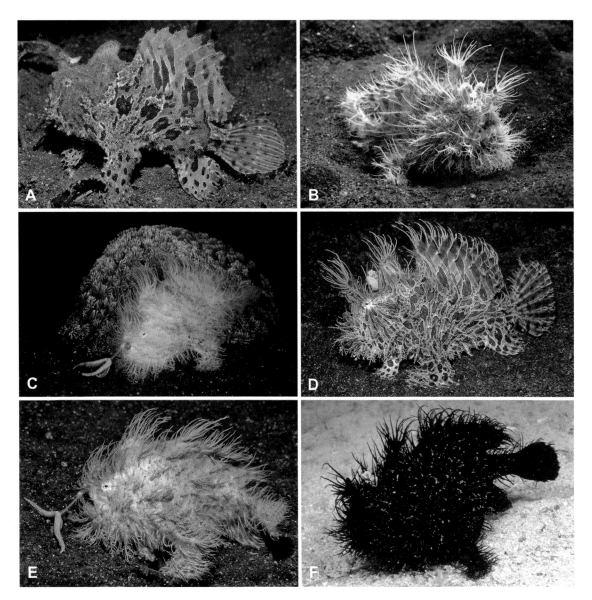

Fig. 74. *Antennarius striatus* (Shaw). (**A**) Lembeh Strait, Sulawesi, Indonesia. Photo by Roger Steene. (**B**) Lembeh Strait, Sulawesi, Indonesia. Photo by Paul Macdonald. (**C**) Lembeh Strait, Sulawesi, Indonesia. Photo by Scott W. Michael. (**D**) Lembeh Strait, Sulawesi, Indonesia. Photo by Roger Steene. (**E**) Lembeh Strait, Sulawesi, Indonesia. Photo by Roger Steene. (**F**) Mindoro, Philippines. Photo by Frank Schneidewind.

set filaments (individuals apparently able to increase or reduce "hairiness" in a matter of several weeks; see Cutaneous Appendages, p. 35).

Coloration in life: extremely variable; typically, light yellow to orange, but often green, gray, brown, black, or almost white with dark brown to black stripes or elongate blotches as described for species-group (see Pietsch and Grobecker, 1987, pl. 4); solid-black individuals known to take on yellow-striped phase in as little as two weeks (Michael, 1998: 354; see Color and Color Pattern, p. 36, and Figs. 72–74).

Coloration in preservation: beige, light yellow, orange, dark yellow-brown, to black; worm-like escal appendages occasionally reddish pink; illicium nearly always darkly banded. Lighter-color phases with as many as 12 (usually seven to 10) darkly pigmented streaks radiating out from eye; belly without elongate markings, but usually with scattered, dark circular spots. Solid black color phase (only about 6% of the material examined) with tips of pectoral-fin rays white.

Largest specimen examined: 155 mm SL (CAS 30166).

GEOGRAPHIC VARIATION

The relatively large amount of material of this geographically wide-ranging species, together with material of the extremely similar *A. scaber*, resurrected here from the synonymy of *A. striatus* (see Comments, p. 95), shows distinct variation in pectoral-ray counts and escal morphology. In the Western Atlantic, 99% of the material of *A. scaber* examined had 11 or 12 pectoral-fin rays. In the Eastern Atlantic, and in the coastal waters of Australia and the Hawaiian Islands, 95% of the material of *A. striatus* examined had nine or 10 pectoral-fin rays. In Japan, however, where this species has been recognized as *A. tridens*, 88% of the specimens examined had 11 or 12 pectoral-fin rays (Table 7). This pattern of geographic variation is somewhat similar to that found in *Abantennarius coccineus* (Table 11), in which slightly higher pectoral-ray counts occur in populations that are peripheral to the center of its range.

The escae of *A. scaber* and *A. striatus* usually consist of either two or three, but sometimes as many as seven, separate, wormlike appendages. This variation, coupled with apparent differences in coloration and color pattern, is at least partially responsible for the 25 additional nominal forms that have been synonymized with these species at various times, dating back more than two centuries (Table 8; for comments on the confusion over bifid versus trifid escae, see Bauchot, 1958; Paulin, 1978; and Pietsch and Grobecker, 1987: 71).

Although escal appendages may often bear slender secondary filaments, all but one of the specimens of *A. scaber* examined had a bifid esca; the exception (ZUEC 741, 75 mm) had three appendages. The esca of the single examined specimen from St. Helena, in the mid-South Atlantic (BMNH 1868.6.15.20), is also distinctly trifid, with an additional, slender filament arising from its base (Fig. 70A). Most Eastern Atlantic specimens (recognized by some as *A. occidentalis*; see synonymy, p. 102) had trifid escae, but several were bifid, and many others had four, five, or even seven separate escal appendages (Fig. 70B). Of the material examined from localities extending from South Africa to the Hawaiian Islands (excluding Japan), only about 7% had a bifid esca (but including the two type specimens of *A. striatus* from Tahiti); all the remaining 93% had a trifid esca, except for an individual from the South China Sea (USNM 232180) with five appendages and three Australian specimens (AMS I.16774, AMS IB.3268, SU 20725) with four or five appendages. All the material examined from Japanese waters (110 specimens) had a trifid esca, except for a single individual (SU 7603) with five escal appendages. Finally, three specimens with a bifid esca (BMNH 1884.2.26.36, China; SU 27997, Hong Kong; USNM 201570, Taiwan) also had five to nine slender, medial filaments, each of which is darkly pigmented at or near its base (Fig. 70C, D). The presence of this cluster of medial filaments and especially the dark pigmentation distinguishes these three individuals from all other specimens of *A. striatus* examined.

DISTRIBUTION

Antennarius striatus, as recognized here, has one of the broadest geographic ranges of any antennariid (Fig. 278, Table 22). In the Eastern Atlantic it is found off the African coast from Senegal to Namibia (BMNH 1866.6.16.13), with records from St. Helena (BMNH 1868.6.15.20), Cape Verde, and the Canary Islands. In the Indo-Pacific it extends from the East African coast (including Zanzibar, Somalia, the Red Sea, and the Western Indian Ocean islands of Réunion and Mauritius) to Japan, Korea, Taiwan, the Philippines, Sulawesi, the Moluccas, Australia, Lord Howe Island, New Caledonia, New Zealand, and eastward to the Hawaiian, Society, and Marquesas Islands, with at least one record from Fiji. The type locality is Tahiti, Society Islands.

Most often found on open sand or mud bottoms, frequently among algae, seagrass, or sponges. Depths of capture, recorded for 79 specimens, range from just beneath the surface to 219 m; however, about 86% of these were taken in 30 m or less. Average depths for known captures are as follows: Eastern Atlantic (30 records), 40 m; Western Indian Ocean (8 records), 43 m; Indo-Australian Archipelago (28 records), 73 m; and the Hawaiian Islands (13 records), 119 m. Although the average depth of known captures made in Australian waters exceeds 70 m, divers have routinely encountered this species in Sydney Harbour, during the months of November to early December and March through April, between about 10 and 25 m. Similar occurences are common at other popular dive sites throughout Indonesia and the Philippines.

COMMENTS

The earliest description of *A. striatus* was provided by French naturalist Philibert Commerson [MSS 889(3): 19, 891: 123] based on material observed by him at Tonneliers Island, Mauritius, in February 1770 (see Introduction, p. 9). An illustration accompanying Commerson's description, made by one of his draftsman, Pierre Sonnerat (*Collection des Vélins du Muséum national d'Histoire naturelle*, Paris, 91: 63), undoubtedly represents this species (Fig. 9A). Commerson's work, however, was never published, but later utilized by Lacepède (1798) in his *Histoire Naturelle des Poissons*. Lacepède (1798, pl. 14, fig. 1) provided a lengthy description of *A. striatus* and published an engraving made from Sonnerat's original drawing, but provided no Latinized binomial. While Sonnerat's drawing and Lacepède's published copy clearly show a trifid esca (Figs. 9A, 11C), Lacepède's text, without explanation, refers to a bifid esca, thus initiating confusion that was to multiply exponentially later on.

The earliest valid description of *A. striatus* is that of George Shaw (1794), illustrated by English copperplate engraver Frederick Polydore Nodder (Fig. 12A) and based on material discovered by Sir Joseph Banks during the first voyage of Captain James Cook aboard H.M.S. *Endeavor* (1768 to 1771). Daniel Solander (MSS Z1: 41, 127; Z8: 227), who accompanied Banks on this expedition, listed two specimens identified as *Lophius histrio* (or *L. histrix*) collected in Tahiti. Günther (1861a: 189) referred to the same two specimens in his *Catalogue of the Fishes in the British Museum*: "b, c. Adult. (? Otaheite.) Cook's Voyage—Types of *Lophius striatus*, Shaw." These specimens are still present in the collections of the Natural History Museum, London, preserved in remarkably good condition. Although neither could have been the specimen used as the model for Nodder's engraving published by Shaw (1794, pl. 175)—both are too small to fit Shaw's claim that the figure depicts the "natural size," and neither is deformed as the specimen depicted (see below)—they are accepted as the syntypes of *A. striatus* as recognized by Günther

(1861b). The larger of the two (BMNH 1978.3.1.2, 77 mm) was designated the lectotype by Pietsch and Grobecker (1987).

The engraving of *A. striatus* that accompanied Shaw's (1794) original description shows a badly deformed specimen (Fig. 12A), nothing like the two well-preserved types in the collections of the Natural History Museum, London. The original drawing from which the engraving was made [together with the original for Shaw's plate 176 (upper fig.) depicting *A. pictus*] is bound in with the drawings made by Thomas Watling (p. 95, no. 317) between 1788 and 1792 (Iredale, 1958; Hindwood, 1970; Gruber, 1983: 13, footnote 13) and now in the Zoological Library of the Natural History Museum. It is well known that the work of two or three additional artists is confused with that of Watling, but how this particular drawing, certainly not painted by Watling, came to be part of this set is unknown. In any case, it seems likely that Shaw, realizing that this drawing represented his new species, simply published it to save the time and expense of having a new drawing made.

From the time of Shaw (1794) and Lacepède (1798) to the revisionary work of Schultz (1957, 1964), frogfishes with a wormlike esca were described under 25 additional specific names (Table 8), allocated to eight nominal genera and involving 60 combinations of generic and subgeneric names (see synonymies, pp. 86, 97). Leonard P. Schultz (1957), in the first attempt to revise the Antennariidae since the works of Cuvier (1816, 1817b) and Valenciennes (1837), recognized two subgenera of frogfishes with wormlike escae within the nominal genus *Phrynelox* Whitley (1931): *Phrynelox*, containing four nominal species, all based on specimens with a bifid esca; and *Triantennatus*, five nominal species, all based on specimens with a trifid esca. This subgeneric distinction (and generic allocation) was later dropped by Paulin (1978: 490), who considered the differences in the esca "to be of specific value only." He recommended that all nominal species based on specimens with bifid escae (*A. pinniceps, A. melas, A. nuttingi, A. scaber,* and *A. lochites*) be synonymized under the name *A. striatus*, and all those based on specimens with trifid escae (*A. nox, A. cunninghami, A. zebrinus,* and *A. atra*) be synonymized under the name *A. tridens*. Considering the pattern of geographic variation in escal morphology described above, and in the absence of any other significant differences among the more than 690 specimens examined, all of these nominal species were considered by Pietsch and Grobecker (1987: 71) to represent a single, widely distributed species taking the oldest available name, *A. striatus* (Shaw, 1794).

Since that time, however, a number of authors have questioned this decision (e.g., Williams, 1989: 279; Cervigón, 1991: 193; Aguilera, 1998: 47; Michael, 1998: 354; Smith-Vaniz et al., 1999: 161; Smith-Vaniz and Jelks, 2014: 29). In the earliest of these criticisms, Williams (1989: 279) argued that, considering the differences in pectoral-fin ray counts and escal morphology (see Geographic Variation, above), at least the Western Atlantic *A. scaber* should be recognized as a valid species, and perhaps, on the same basis, other populations elsewhere, now submerged within "the *A. striatus* complex," should be recognized as well. Smith-Vaniz et al. (1999: 161; see also Smith-Vaniz and Collette, 2013: 169; Smith-Vaniz and Jelks, 2014: 29) agreed, stating that "the combination of 11 or 12 pectoral rays and a bifid esca distinguishes Western Atlantic specimens from all other '*A. striatus*,' and [therefore] synonymization of *A. scaber* should be reevaluated" (note, however, that the type specimens of *A. striatus* from Tahiti both have a bifid esca and 11 pectoral-fin rays). Scott Michael (1998: 354, 2007: 58) also concurred, taking the argument a step further in suggesting that individuals from Japan and adjacent parts of the Western Pacific, which have a trifid esca, should also be recognized as distinct under the name *A. tridens*—although he added that "much more study is required before a definitive conclusion can be made in these cases."

We have certainly pondered these questions at great length, as did Pietsch and Grobecker back in the early 1980s, but resolution of the problem is not all that straightforward. Assigning the name *A. scaber* (Cuvier, 1817b) to the Western Atlantic population—nearly all individuals of which have a bifid esca and 11 pectoral-fin rays— is rather easy and we have done that here. But then what name should be applied to the Eastern Atlantic population, most specimens of which have a trifid esca (but some are bifid and many others have four, five, or even seven escal appendages) and almost invariably 10 pectoral-fin rays? It has been recognized by some as *A. occidentalis* (Cadenat, 1959: 381), but clearly *A. delaisi* (Cadenat, 1959: 374) has priority (see synonymy above).

As many have suggested, the name *A. tridens* (Temminck and Schlegel, 1845) might be resurrected for the Japanese population, but what than are the geographic limits of such a taxon when one considers that 93% of the Indo-Pacific material examined (428 specimens from Japan to temperate Australia and from South Africa to the Hawaiian Islands) had a trifid escae. In this context, it would be perhaps more appropriate to resurrect *A. tricornis* (Cloquet, 1817), the type locality of which is Mauritius. All this, in turn, begs the question, where does this leave *A. striatus*, the types of which are from Tahiti, both with a bifid esca and 11 pectoral-fin rays? Without knowing the locality, one might guess these two specimens came from the Western Atlantic and should therefore be identified as *A. scaber*. Finally, for what it is worth, in regeneration experiments performed by Pietsch and Grobecker (1987: 319), surgical removal of a single wormlike escal appendage of *A. striatus* was followed by regeneration of two similar filaments—the resulting regenerated esca, although different from the original non-regenerated esca, was still well within the expected morphological variation of the species (see Regeneration and the Maintenance of Species-Specific Escae, p. 429).

Clearly, the problem is complex. We agree that *"striatus"* most probably represents a mosaic of populations, each deserving specific recognition, but rather than resurrecting names at this point and risk adding more confusion, we await resolution that will result only from a detailed molecular analysis of material from all pertinent geographic localities. At this point, however, tissues required for DNA extraction are lacking for the Eastern Atlantic population and for most parts of the Indo-Pacific—we strongly encourage the collection of tissues whenever possible (for more discussion, see Intrarelationships of the *Antennarius striatus* Group, p. 383).

Antennarius pictus Group Pietsch

DIAGNOSIS

A species-group of *Antennarius* distinguished by having the following combination of character states: a triad of three, darkly pigmented, circular spots usually present on caudal fin; a row of similar spots on soft-dorsal and anal fins; illicium 18.9 to 33.8% SL, nearly twice length of second dorsal-fin spine; anterior end of pterygiophore of illicium terminating distinctly posterior to symphysis of upper jaw; esca consisting of a broad, laterally compressed appendage with one to several smaller, compressed appendages arising from base (appendages often split longitudinally into two or more separate portions and sometimes appearing filamentous in damaged specimens), and a pair of darkly pigmented, eyelike basal spots usually present; second dorsal-fin spine connected to head

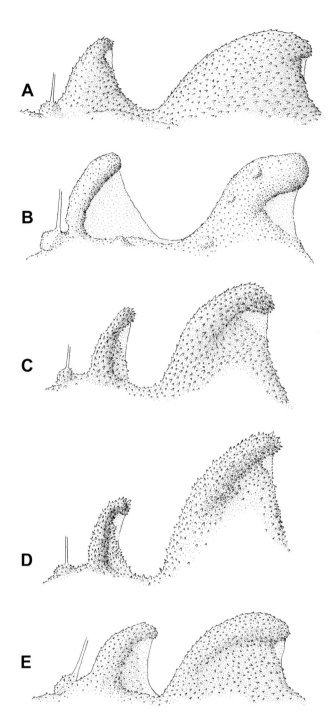

by a membrane, the membrane not divided into naked dorsal and ventral portions by a cluster of dermal spinules, and extending posteriorly across area between second and third dorsal-fin spines, reaching to base of third (in Indo-Pacific species; Fig. 75A, B, E), or terminating distinctly anterior to base of third dorsal-fin spine (in Atlantic species; Fig. 75C, D); distal two-thirds of maxilla naked and tucked beneath folds of skin, only the extreme proximal end directly covered with spinulose skin; caudal peduncle present, the posteriormost margin of soft-dorsal and anal fins attached to body distinctly anterior to base of outermost rays of caudal fin; dorsal-fin rays 12 to 13 (rarely 11), as many as five posteriormost bifurcate; anal-fin rays seven or eight (rarely six), all bifurcate; pectoral-fin rays 10 or 11 (rarely nine), all simple; only posteriormost pelvic-fin ray bifurcate; vertebrae 19 (rarely 20) (Table 9).

Most members do not exceed a standard length of about 160 mm, but *A. commerson* is known to reach 290 mm SL making it one of the largest known antennariids. Limited data indicate that all included species are associated with rocky- or coral-reef communities; distributed in tropical and subtropical waters of all three major oceans of the world; five species.

Key to the Known Species of the *Antennarius pictus* Group

Like the other keys in this volume, the following works by progressively eliminating the most morphologically distinct taxon and for that reason should always be entered from the beginning. All features listed for each species must correspond to the specimen being keyed; if not, proceed to the next set of characters. For an additional or alternative means of identification, all species of the *A. pictus* group may be compared simultaneously by referring to Table 9.

Fig. 75. Dorsal-fin spines of species of the *Antennarius pictus* Group, showing relative development and pattern of dermal spinules on the membrane behind the second spine. (**A**) *A. commerson* (Latreille), UW 20926, 69.5 mm SL, Mauritius. (**B**) *A. maculatus* (Desjardins), UW 20829, 72 mm SL, Philippines. (**C**) *A. multiocellatus* (Valenciennes), LACM 5849, 52 mm SL, Pedro Cays, Jamaica. (**D**) *A. pardalis* (Valenciennes), USNM 199253, 62.5 mm SL, off Bassagos Island, West Africa. (**E**) *A. pictus* (Shaw), UW 20887, 76.5 mm SL, Philippines. Drawings by Cathy L. Short; after Pietsch and Grobecker (1987, fig. 22).

Table 9. Comparison of Distinguishing Character States of Species of the *Antennarius pictus* Group (lengths as percentage of SL)

Character	A. commerson	A. maculatus	A. multiocellatus	A. pardalis	A. pictus
Illicium length	19.3–25.2	26.3–33.8	21.0–31.8	18.9–29.6	19.1–32.0
Second dorsal-fin spine length	10.2–16.9	13.8–18.1	9.2–19.4	10.8–16.3	8.5–17.3
Membrane behind second dorsal-fin spine extending to third	yes	yes	no	no	yes
Membrane behind second dorsal-fin spine thick and spinulose	yes	no	no	no	no
Second dorsal-fin spine tapering from distal end	no	yes	no	no	no
Large warts on head, body, and fins	absent	present	absent	present	absent
Large dark spots on belly	absent	absent	absent	present	absent
Dorsal-fin rays	13 (11–12)	12 (11)	12 (11–13)	12 (11–13)	12 (11–13)
Anal-fin rays	8	7 (6)	7(6)	7	7(6)
Pectoral-fin rays	11 (10)	10 (11)	10 (9)	10 (11)	10 (9–11)
Maximum standard length	291 mm	85 mm	120 mm	97 mm	160 mm
Distribution	Indo-Pacific	Indo-Pacific	Western Atlantic	Eastern Atlantic	Indo-Pacific

1. Dorsal-fin rays 13 (rarely 11 or 12), anal-fin rays eight, pectoral-fin rays 11 (rarely 10); membrane behind second dorsal-fin spine thick, the width at attachment to spine as great as or greater than width of spine itself (Fig. 75A) . . . . *Antennarius commerson* (**Latreille, 1804**), **p. 114**
    Indo-Pacific and Eastern Pacific Oceans

2. Second dorsal-fin spine expanded from base, the posterior membrane naked or only lightly covered with scattered dermal spinules (Fig. 75B); most of head and body covered with large, protruding, wartlike swellings . . . . *Antennarius maculatus* (**Desjardins, 1840**), **p. 123**
    Indo-Pacific Ocean

3. Second dorsal-fin spine slightly tapering from base, the posterior membrane covered with dermal spinules, except along outer margin and usually in a small triangular area just behind tip of spine (Fig. 75E); head and body without large, wartlike swellings . . . . *Antennarius pictus* (**Shaw, 1794**), **p. 145**
    Indo-Pacific Ocean

4. Membrane behind second dorsal-fin spine reduced, terminating distinctly anterior to base of third dorsal-fin spine (Fig. 75C); second dorsal-fin spine short, less than 11% SL in specimens greater than about 90 mm SL (Fig. 91); head and body usually without wartlike swellings, but when present, low, small, and set far apart

(Figs. 93–95); belly without large, darkly pigmented spots . . . .
. . . . *Antennarius multiocellatus* (Valenciennes, 1837), p. 131
Western Atlantic Ocean, Ascension Island, Gulf of Guinea

5. Membrane behind second dorsal-fin spine reduced, terminating distinctly anterior to base of third dorsal-fin spine (Fig. 75D); second dorsal-fin spine long, more than 13% SL in specimens greater than about 90 mm SL (Fig. 91); head and body, especially in specimens greater than about 90 mm SL, covered with numerous, large, close-set, wartlike swellings (Fig. 99); belly of small specimens nearly always covered with large, darkly pigmented spots . . . .*Antennarius pardalis* (Valenciennes, 1837), p. 140
Eastern Atlantic Ocean

Antennarius commerson (Latreille)
Giant Frogfish

Figs. 1B, 9B, 11B, 75A, 76–83, 275, 294, 295D, 296D, 306D, 329F; Tables 5, 9, 22, 23

Antennarius bivertex, totus ater, puncto mediorum laterum albo: Commerson, MSS 889(1): 2, 891: 136 [observed by Commerson at Mauritius, October 1769; *habitat in vado maritimo sub nesogalliae plagis*; no extant material (see Comments, p. 122)]. Lacepède, 1798: 327 (footnote reference to Commerson, MSS).

La Lophie Commerson: Lacepède, 1798: 302, 327, pl. 14, fig. 3 [invalid, not Latinized; description and Latin polynomial in footnote based on Commerson, MSS; *voisine des côtes orientales l'Afrique* (= Mauritius)]. Sonnini, 1803: 161, 192 (after Lacepède, 1798).

Lophius commerson Latreille, March 1804: 73 [binomial for *La Lophie Commerson* of Lacepède, 1798; constitutes original description, no known type material; neotype MNHN A.4623 hereby designated, 90 mm, Mauritius, figured by Cuvier, 1817b, pl. 18, fig. 1; neotype designation of Dor (1984: 56) invalid, not a revisional work].

Lophius commersonii: Shaw, June 1804: 387 ("Commersonian Angler"; no specimens, "observed by Commerson in the Indian seas"). Cuvier, 1816: 310 (after Lacepède, 1798).

Chironectes commersonii: Cloquet, August 1817: 599 (new combination; binomial for *La Lophie Commerson* of Lacepède, 1798). Cuvier, October 1817b: 431, pl. 18, fig. 1 (description, MNHN A.4623, Mauritius). Cuvier, 1829: 252 (after Cuvier, 1817b). Valenciennes, 1837: 426 (description after Cuvier, 1817b). Valenciennes, 1842: 189 (after Cuvier, 1817b). Pietsch, 1985a: 89, figs. 27–29, table 1 (original manuscript sources for *Histoire Naturelle des Poissons* of Cuvier and Valenciennes). Pietsch, 1995b, 1: 98 (*Sambia*; Renard's *Fishes, Crayfishes, and Crabs*).

Antennarius commersonii: Schinz, 1822: 501 (new combination, after Cuvier, 1817b). Bleeker, 1859b: 130 [Ambon, Singapore (after Cantor, 1849), Mauritius]. Günther, 1861b: 361 (Ceylon). Günther, 1876: 163 [in part, i.e., pl. 100 (fig. B), 101, 103 (fig. A)]. Macleay, 1878: 356 (Port Darwin, Australia). Waite, 1901: 47 (Lord Howe Island). McCulloch, 1922: 123 (New South Wales). Fowler, 1927: 32 (Christmas Island). McCulloch, 1929: 407 (in part; Australia to Hawaii). Borodin, 1930: 62 (Maui). Fowler, 1934b: 513 (Natal). Seale, 1935: 378 (Bellona Island, Solomons, but identification not confirmed by us). Smith, 1949: 430, pl. 98, fig. 1236 (South Africa). Le Danois, 1964: 98, fig. 47b (description; MNHN A.4623, described by Cuvier, 1817b, erroneously listed as holotype). Myers, 1989: 68, pl. 13D (Red Sea to Panama; in key). Myers, 1999: 68, pl. 13D (after Myers, 1989). Goldin, 2002: 47, fig. (Samoa). Myers and Donaldson, 2003: 610, 614 (Mariana Islands). Letourneur et al., 2004: 209 (Réunion Island). Castellanos-Galindo et al., 2006: 200 (Colombia). Larson et al., 2013: 53 (Northern Territory).

Chironectes caudimaculatus Rüppell, 1829, pl. 33, fig. 2 (labeled figure only, constitutes original description; dating of Rüppell by Peter J. P. Whitehead, personal communication, 17 March 1983). Rüppell, 1838: 141 (description, one dry specimen, holotype SMF 6783, Red Sea).

Antennarius commersoni (not of Latreille): Cantor, 1849: 1186 (unidentifiable). Cantor, 1850: 204 (after Cantor, 1849). Bleeker, 1865b: 10, 20, pl. 197, fig. 3 (figure represents *A. pictus*). Günther, 1876: 163 [in part; pl. 100 (fig. C), 102 (figs. A, B), 103 (fig. B), 104 (figs. A, B), 105 (fig. A), 106 (fig. B) all represent *A. pictus*]. Schmeltz, 1877: 14 (BMNH 1873.8.1.33 is *A. dorehensis*). Jenkins, 1904: 511 (SU 23259 is *A. pictus*). Jordan and Evermann, 1905: 518 [in part; synonymy includes *A. niger* Garrett (= *A. pictus*)]. Fowler, 1928: 477, pl. 49B (*A. pictus*). Whitley, 1929b: 137, pl. 31, fig. 5 (figure shows *A. striatus*). Fowler, 1931: 367 (*A. pictus*). Whitley, 1934, unpaged (after Whitley, 1929b). Fowler, 1940a: 800 (material is *H. histrio*, USNM 82842). Poll, 1949: 252, fig. 22 (specimen is *F. senegalensis*). Cadenat, 1951: 291 (material is *F. senegalensis*). Munro, 1955: 288, fig. 838 (*A. pictus*). Schultz, 1964: 175, table 1 (*A. pictus*). Gosline and Brock, 1960: 305, 345, fig. 23 (*A. pictus*). Le Danois, 1970: 89 (TAU P.2695 is *A. pictus*). Allen et al., 1976: 385 (Lord Howe Island). Randall et al., 1993b: 365 (Midway Atoll). Sadovy and Cornish, 2000: 44, fig. (figure shows *A. maculatus*).

Antennarius moluccensis Bleeker, 1855c: 424 (original description, single specimen, holotype RMNH 6273, Ambon). Bleeker, 1859b: 129 (Ambon, Goram). Günther, 1861a: 191 (description; Ambon, Goram). Bleeker, 1865b: 9, 17, pl. 196, fig. 2 (description, additional specimen, Celebes). Schultz, 1957: 91, pl. 9, figs. A, C (synonymy, description; California (?), Panama, Oahu; in key). Gosline and Brock, 1960: 305, 345 (in key). Herald, 1961: 282, color fig. 142 (Hawaiian Islands). Beaufort and Briggs, 1962: 206 (synonymy, description; Mauritius, Indonesia, Queensland, Lord Howe Island, Society Islands, Hawaii; in key). Marshall, 1964: 514, pl. 64, fig. 496 (Northern Territory, Queensland, New South Wales, Lord Howe Island; in key). Schultz, 1964: 175, table 1 (additional material, Hawaiian Islands). Tinker, 1978: 511, figs., color pl. 32, figs. 167-7 (description, Hawaii to central Polynesia and East Indies). Araga, 1984: 103, pls. 89-A, 89B (Kii Peninsula, Japan). Randall et al., 1993a: 223, 227, table 1 (synonym of *A. commersoni*, Hawaiian Islands). Michael, 1998: 347 (synonym of *A. commerson*). Fricke et al., 2011: 367 (synonym of *A. commerson*).

Antennarius commersonii Var. α. *Commersonii*: Günther, 1861a: 193 (BMNH 1858.4.21.188, Ambon).

Antennarius caudimaculatus: Günther, 1861a: 197 (after Rüppell, 1838). Klunzinger, 1871: 500 (Red Sea, after Rüppell, 1838). Klunzinger, 1884: 126 (after Rüppell, 1838). Beaufort and Briggs, 1962: 207 (in part, synonymy includes reference to Rüppell, 1838, and Klunzinger, 1871, 1884). Dor, 1984: 55 (Red Sea, after Rüppell, 1838).

Chironectes rubrofuscus Garrett, 1863: 64 [original description, single specimen, holotype lost in San Francisco earthquake, 1906 (William N. Eschmeyer, personal communication, 5 December 1980), 216 mm total length, Hawaiian Islands].

Antennarius goramensis Bleeker, 1865a: 177 (original description, single specimen, holotype RMNH 6288, Goram). Bleeker, 1865b: 9, 17, pl. 195, fig. 2 (description after Bleeker, 1865a). Marshall, 1964: 514, 515 (North Queensland; in key).

Antennarius nummifer (not of Cuvier): Boulenger, 1887: 663 (Muscat; see Beaufort and Briggs, 1962: 219).

Antennarius rubrofuscus: Jenkins, 1904: 511 (new combination, additional specimen, Honolulu). Snyder, 1904: 537 (additional specimen, Honolulu).

Antennarius commersoni: Snyder, 1904: 537 (additional specimen, Honolulu). Dor, 1984: 56 (Red Sea; neotype designation invalid, not being a revisionary work). Pietsch, 1984b: 34 (genera of frogfishes). Pietsch, 1984e: 2 (Western Indian Ocean). Pietsch, 1985a: 89, figs. 27–29, table 1 (original manuscript sources for *Histoire Naturelle des Poissons* of Cuvier and Valenciennes). Pietsch and Grobecker, 1985: 13, fig. (locomotion). Walsh, 1985: 360, table 1 (artificial reefs; Kona, Hawaii). Pietsch, 1986a: 366, 367, fig. 102.2, pl. 13 (South Africa). Pietsch and Grobecker, 1987: 92, figs. 22C, 29–32, 119, pls. 17–20 (description, distribution, relationships). Edwards, 1989: 250, figs. 17, 18 (tetrapod locomotion). Pietsch and Grobecker, 1990a: 96, 98, fig. (aggressive mimicry, color change). Pietsch and Grobecker, 1990b: 74, 76, fig. (after Pietsch and Grobecker, 1990a). Francis, 1993: 158 (Lord Howe Island). Hoover, 1993: 63, 64, color figs. (Hawaiian Islands). Randall et al., 1993a: 223, 227, table 1 (Hawaiian Islands). Strack, 1993: 64 (Ambon). Pietsch, 1995b, 1: 98 (*Sambia*; Renard's *Fishes, Crayfishes, and Crabs*). Randall, 1995: 83 (Oman). Schneider and Lavenberg, 1995: 857 (tropical Eastern Pacific). Larson and Williams, 1997: 349 (Northern Territory). Okamura and Amaoka, 1997: 141, figs. (Japan). Bavendam, 1998: 47, color fig. (natural history). Pietsch, 1999: 2015 (western central Pacific; in key). Nakabo, 2000: 456 (Japan; in key). Pereira, 2000: 5 (Mozambique). Allen and Adrim, 2003: 25 (Papua to Bali, Indonesia). Allen et al., 2003: 363, color figs. (tropical Pacific). Greenfield, 2003: 56, table 1 (Kaneohe Bay, Hawaiian Islands). Hoover, 2003: 63, 64, color figs. (Hawaiian Islands). Schneidewind, 2003a: 82, 85, 87, 88, figs. (natural history). Lieske and Myers, 2004: 46, color fig. (Red Sea to Central America). Bussing and López, 2005: 46, 47, fig. (Cocos Island). Schneidewind, 2005c: 88, fig. (tiny postlarvae). Al-Jufaili et al., 2010: 18, appendix 1 (Oman). Allen and Erdmann, 2013: 32, appendix 3.1 (Bali, Indonesia). Karplus, 2014: 35, fig. 1.21A (aggressive mimicry; after Pietsch and Grobecker, 1987).

Antennarius lutescens Seale, 1906: 89, fig. 2 (original description, single specimen, holotype BPBM 1347, Tahiti). Schultz, 1957: 84 [in synonymy of *Lophiocharon horridus* (= *A. pictus*)].

Antennarius (?*Commersonii*) (not of Latreille): Weber, 1913: 563 (ZMA 116.515 is *Abantennarius nummifer*).

Antennarius lateralis Tanaka, 1917: 200 (original description, single specimen, existence of holotype unknown, 119 mm, Tanabe, Kii Province, Japan). Tanaka, 1918: 494, pl. 135, fig. 378 (after Tanaka, 1917). Okada, 1938: 274 (Honshu, Sikoku).

Lophius commersonianus: Lacepède, in Jordan, 1917b: 69, 96, 104, 130 [on p. 69, equated with *Antennarius bivertex, totus ater, puncto mediorum laterum albo* Commerson (footnote in Lacepède, 1798), and designated type species of *Antennarius* Commerson (in Lacepède, 1798); on p. 96, designated type species of *Chironectes* Cuvier (1817b); but on p. 104, equated with *Antennarius antenna tricorni* Commerson (footnote in Lacepède, 1798) (= *A. striatus*)].

Antennarius hispidus (not of Bloch and Schneider): Fowler, 1928: 477 (in part, includes type of *A. lutescens* Seale).

Lophiocharon goramensis: Whitley, 1941: 45, pl. 2, fig. 29 (new combination, description, additional specimen, AMS IA.5824, off Cairns, northern Queensland).

Antennarius bivertex: Schultz, 1957: 80 (in synonymy; binomial for polynomial of Commerson, MSS, authorship attributed to Lacepède, 1798).

Antennarius chironectes (not of Latreille): Le Danois, 1964: 99 (in part, i.e., MNHN A.4621).

Uniantennatus horridus (not of Bleeker): Le Danois, 1964: 109 (in part, i.e., MNHN 8068).

Antennarius commerson: Randall, 1996: 43, 44, color fig. (corrected spelling of specific name; Hawaiian Islands). Randall et al., 1997b: 11 (Ogasawara Islands). Michael, 1998: 328–330, 334, 336, 338, 345–347, 352, color figs. (identification, behavior, captive care). Fricke, 1999: 104 (Réunion, Mauritius). Laboute and Grandperrin, 2000: 130 (New Caledonia). Pietsch, 2000: 597 (South China Sea). Sadovy and Cornish, 2000: 44 (Hong Kong). Gell and Whittington, 2002: 116, table 1 (Mozambique). Nakabo, 2002: 456 (Japan; in key). Allen and Adrim, 2003: 25 (Indonesia). Robertson et al., 2004: 513, 527, 563, tables 1, 6 (transpacific shore fishes). Mundy, 2005: 263 (Hawaiian Islands). Randall, 2005a: 68, 70, color figs. (South Pacific). Randall, 2005b: 300, figs. 2–7 (mimicry, color change). Carnevale and Pietsch, 2006: 452, table 1 (comparison, meristics, first fossil frogfish). Castellanos-Galindo et al., 2006: 200, 207 (Colombia). Michael, 2006: 37, 41 (behavior). Schneidewind, 2006b: 26, 33, color figs. (mimicry; Sabang Beach, Mindoro). Senou et al., 2006: 423 (Sagami Sea). Kuiter and Debelius, 2007: 118, color figs. (Indonesia). Mejía-Ladino et al., 2007: 296, figs. 5, 6, tables 1–4, 6 (species account; in key; Colombia). Randall, 2007: 125, 127, color figs. (Hawaiian Islands; in key). Senou et al., 2007: 49 (Ryukyu Islands). Kwang-Tsao et al., 2008: 243 (southern Taiwan). Allen and Erdmann, 2009: 594 (West Papua). Fricke et al., 2009: 28, table 6 (Réunion). Pietsch, 2009: 4, fig. 2B, 2C (representative antennariid). Golani and Bogorodsky, 2010: 16 (Red Sea). Miya et al., 2010: 2, fig. 1B (phylogeny, representative frogfish). Motomura et al., 2010: 78 (Yaku-shima Island, southern Japan). Randall, 2010: 36 (Hawaii). Fricke et al., 2011: 367 (New Caledonia). Satapoomin, 2011: 51, appendix A (Andaman Sea). Allen and Erdmann, 2012: 145, color figs. (East Indies). Arnold and Pietsch, 2012: 118, 125, 126, 128, figs. 3A, 4, 6, table 1 (molecular phylogeny). Fricke et al., 2013: 252 (Europa Island, Mozambique Channel). Nakabo, 2013: 540, figs. (Japan). Page et al., 2013: 97 (common names). Pietsch et al., 2013: 664, fig. 4F (sexual dimorphism, spawning). Brandl and Bellwood, 2014: 32 (pair-formation). Moore et al., 2014: 176 (northwestern Australia). Townsend, 2015: 4 (Sabah, Malaysia). Del Moral-Flores et al., 2016: 607 (Revillagigedo Archipelago). Fourriére et al., 2016: 442 (Revillagigedo Archipelago).

MATERIAL EXAMINED

Seventy-three specimens, 14 to 291 mm SL.

Neotype of *Lophius commerson*: MNHN A.4623, 90 mm, Mauritius, Mathieu (Fig. 76).

Holotype of *Chironectes caudimaculatus*: SMF 6783, 206.5 mm, stuffed, Red Sea.

Holotype of *Antennarius moluccensis*: RMNH 6273, 187 mm, Ambon, Bleeker (Fig. 77A).

Holotype of *Antennarius goramensis*: RMNH 6288, 170 mm, Goram Island, south Moluccas, Bleeker (Fig. 77B).

Holotype of *Antennarius lutescens*: BPBM 1347, 96 mm, Tahiti, Society Islands.

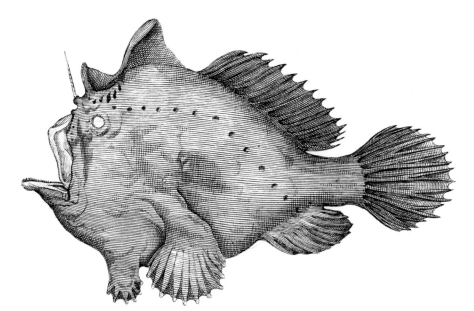

Fig. 76. *Antennarius commerson* (Latreille). Neotype, MNHN A.4623, 90 mm SL, Mauritius. After Cuvier (1817b, pl. 18, fig. 1).

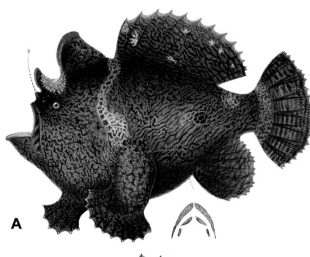

A

B

Fig. 77. *Antennarius commerson* (Latreille). (**A**) holotype of *Antennarius moluccensis* Bleeker, RMNH 6273, 187 mm SL, Ambon, Indonesia (the inset shows the pattern of teeth on the upper jaw and palatines; after Bleeker, 1865b, pl. 196, fig. 2). (**B**) holotype of *Antennarius goramensis* Bleeker, RMNH 6288, 170 mm SL, Goram Island, south Moluccas (after Bleeker, 1865b, pl. 195, fig. 2). Drawings by Ludwig Speigler under the direction of Pieter Bleeker.

Additional material: *Western Indian Ocean*: ANSP, 3 (149–185 mm); BMNH, 1 (99 mm); BPBM, 1 (14 mm); MCZ, 1 (112 mm); MNHN, 6 (44.5–210 mm); RMNH, 1 (62 mm); SAIAB, 7 (75.5–260); USNM, 1 (123 mm); UW, 1 (69.5 mm). *Red Sea*: BMNH, 1 (182 mm); HUJ, 4 (164–230 mm). *Philippines*: UW, 1 (190 mm). *Indonesia*: BMNH, 3 (113–173 mm); RMNH, 3 (99 mm 171 mm). *Australia*: AMS, 3 (50–209 mm); WAM, 1 (291 mm). *Fiji*: USNM, 1 (17.5 mm); USPS, 1 (200 mm). *Hawaiian Islands*: ANSP, 2 (183–210 mm); BPBM, 8 (25–143 mm); USNM, 4 (175–245 mm); UW, 4 (111–192 mm). *Society Islands*: BMNH, 2 (108–165 mm); MCZ, 1 (142 mm). *Eastern Pacific*: GCRL, 1 (197 mm); USNM, 3 (82–195 mm). *Locality unknown*: BMNH, 2

(103–156 mm); FMNH, 1 (185 mm); UW, 2
(68–234 mm).

DIAGNOSIS

A member of the *Antennarius pictus* group
unique in having a greater number of dor-
sal-, anal-, and pectoral-fin rays than its
congeners (Tables 5, 9). *Antennarius commerson* is further distinguished in having
the following combination of character
states: second dorsal-fin spine slightly tapered from base (Fig. 75A); membrane
behind second dorsal-fin spine thick and
swollen (width along attachment to spine
usually as great as spine itself), nearly always completely covered with dermal spinules and extending across area between
second and third dorsal-fin spines, reaching to base of third (Fig. 75A); head and
body sometimes partially or fully covered

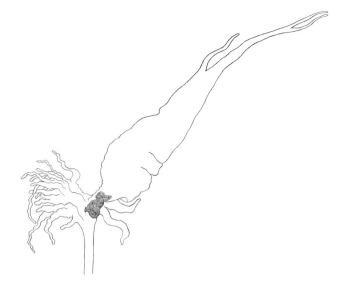

Fig. 78. Esca of *Antennarius commerson* (Latreille). UW 20983, 111 mm
SL, off Makaha Beach, Oahu, Hawaiian Islands. Drawing by Cathy L. Short;
after Pietsch and Grobecker (1987, fig. 30).

with low, rounded, wartlike swellings (but nothing like the warts of *A. maculatus* or *A.
pardalis*); belly without large, darkly pigmented spots; dorsal-fin rays 13 (rarely 11 or 12);
anal-fin rays eight; pectoral-fin rays 11 (rarely 10) (Tables 5, 9).

DESCRIPTION

Appendages of esca short, less than 20% illicium length (Fig. 78); illicium 19.3 to 25.2%
SL; second dorsal-fin spine straight to slightly curved posteriorly, a slight indentation
occasionally present anteriorly at base, length 10.2 to 16.9% SL; naked depression between second and third dorsal-fin spines usually absent; third dorsal-fin spine curved
posteriorly, length 22.4 to 28.1% SL; eye diameter 2.6 to 3.9% SL.

Coloration in life: extremely variable, almost every hue imaginable; white, gray,
yellow, orange, red, green, brown, or solid black, typically peppered with small dark
spots or with large, irregular shaped spots on head, body, and fins; often mottled with
scablike pink patches or irregular white, blue, brown, or black markings (Figs. 79–81).
Bright pink or light yellowish- brown mottled with small brownish spots (USNM 167509);
brown with purple and black mottling; solid bright yellow (USNM 167508). Brownish-
orange with small scattered dark brown spots and some large brown-black patches;
"scabby areas" pink, red, or pale yellow; membranes between premaxillae bright orange (USNM 177856). Lemon-yellow, changing to brick-red in a period of three weeks
(UW 20983, Fig. 82; see Color and Color Pattern, p. 36). Bright red with numerous,
scattered, darker-red, circular spots over entire head, body, and fins; bright-white
saddle on caudal peduncle, on space between second and third dorsal-fin spines (including posterior half of membrane and tip of second dorsal-fin spine), and between
third dorsal-fin spine and soft-dorsal fin; irregularly shaped white patch on dorsal
base of pectoral-fin lobe; delicate cutaneous cirri (particularly abundant on chin and
space between third dorsal-fin spine and soft-dorsal fin) and distal tips of paired fins
white. Green with purple and white mottling. Solid black (SAIAB 5765; UW 20926,
20882).

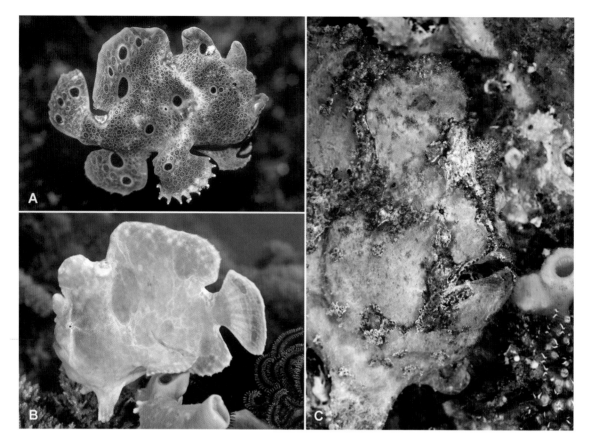

Fig. 79. *Antennarius commerson* (Latreille). (**A**) Bali, Indonesia. Photo by Roger Steene.
(**B**) Sabang, Puerto Galera, Mindoro, Philippines. Photo by Frank Schneidewind. (**C**) Ambon,
Indonesia. Photo by Linda Ianniello.

Coloration in preservation: light beige or tan, green, brown, to black; illicium usually banded; as many as 15 short, darkly pigmented bars radiating from eye. Lighter-
color phases everywhere (except on naked inner surface of paired fins) finely peppered
with small dark spots or mottled with reddish-brown to yellow-brown patches of variable size; dark brown to black circular spots sometimes present (in a pattern characteristic of the *Antennarius pictus* group described above); dark, more or less interconnected
blotches giving appearance of narrow, irregular, vertical bars on caudal fin. Darker-color
phases often with tips of pectoral-fin rays white; body ranging from black with no markings to black or brown with white to pink, green, or blue irregularly shaped patches.

Largest specimen examined: 291 mm SL (WAM P.23603), by far the largest member
of the *A. pictus* group.

DISTRIBUTION

Antennarius commerson is widespread throughout the Western Indian and Indo-Pacific
Oceans, extending into the tropical Eastern Pacific (Fig. 275, Table 22). It has been well-
documented off South Africa, Mozambique, the Red Sea, Oman, Réunion, Mauritius,
Japan, the Ogasawara Islands, Taiwan, Malaysia, the Philippines, Bali to West Papua,
Western Australia and Northern Territory, the Andaman Sea, Lord Howe Island, and
New Caledonia. Farther east it has been reported from Fiji and the Mariana, Hawaiian,

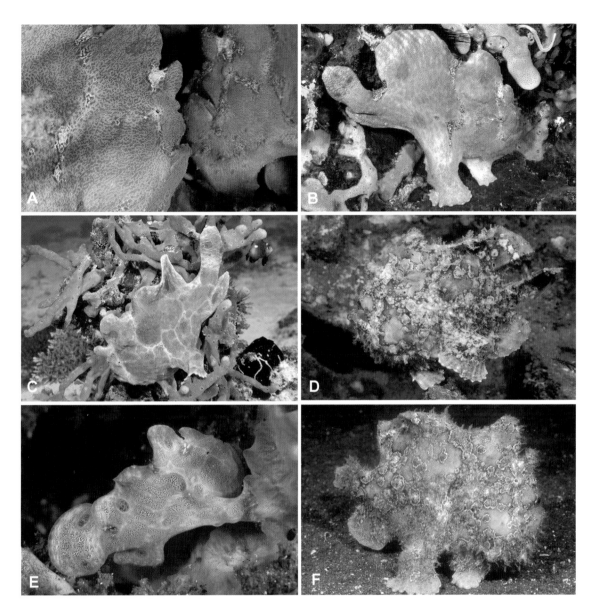

Fig. 80. *Antennarius commerson* (Latreille). (**A**) Anilao, Luzon, Philippines. Photo by Mike Bartick. (**B**) Anilao, Luzon, Philippines. Photo by Janine Cairns-Michael. (**C**) Maldive Islands. Photo by Frank Schneidewind. (**D**) Sabang Beach, Mindoro, Philippines. Photo by Frank Schneidewind. (**E**) Lembeh Strait, Sulawesi, Indonesia (note the damaged dorsal fin). Photo by Scott W. Michael. (**F**) Lembeh Strait, Sulawesi, Indonesia. Photo by Frank Schneidewind.

and Society Islands. Although its presence off Panama (as well as the Cocos and Revillagigedo Islands) and Colombia has now been verified (three adult specimens and an uncollected fourth, photographed by G. M. Wellington at Isla Contadora, Gulf of Panama, in February 1979), the single California record for this species (USNM 32507) remains doubtful (see Schultz, 1957: 92). The type locality is Mauritius, Western Indian Ocean (see Comments, p. 122).

The little information available on depth preference (available for only 12 captures), indicates that *A. commerson* may be taken just below the surface to depths as great as

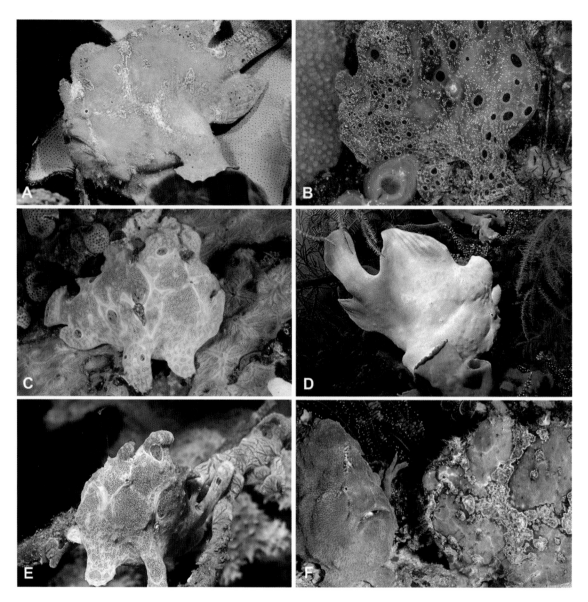

Fig. 81. *Antennarius commerson* (Latreille). (**A**) Pulau Mabul, off the southeastern coast of Sabah, Malaysia. Photo by Frank Schneidewind. (**B**) Bali, Indonesia. Photo by Linda Ianniello. (**C**) Puerto Galero, Philippines. Photo by Linda Ianniello. (**D, E**) Sogod Bay, Lungsodaan, Southern Leyte, Philippines. Photos by Christa Holdt. (**F**) Anilao, Luzon, Philippines. Photo by Linda Ianniello.

45 m, with an average depth of 20 m. It is frequently observed living among sponges (Fig. 294).

COMMENTS

Antennarius commerson was first described and illustrated by Philibert Commerson [MSS 889(1): 2, 891: 136] based on material observed by him at Mauritius in October 1769 (see Introduction, p. 9). Commerson's manuscripts contain a lengthy description and two excellent illustrations (*Collection des Vélins du Muséum national d'Histoire naturelle*, Paris, 91: 64, 65; both probably made by Pierre Sonnerat), one of which is reproduced here (Fig. 9B).

Lacepède (1798: 327), in utilizing Commerson's unpublished notes, failed to provide a Latinized binomial, referring to this species as *La Lophie Commerson* (Fig. 11B). A proper scientific name for the unpublished polynomial of Commerson (MSS) and the French vernacular of Lacepède (1798) was first provided by Latreille (March 1804), preceding Shaw by several months (plates in Shaw dated 1 June 1804). Cuvier (1817b: 431, pl. 18, fig. 1) published a good description and illustration of a specimen collected at Mauritius by a little-known French naturalist named Mathieu (Fig. 76); this specimen (MNHN A.4623, 90 mm), remaining in remarkably good condition, is hereby designated the neotype of *A. commerson* (the designation of Dor, 1984: 56, not being part of a revisional work is invalid).

Of the eight nominal species here considered to be synonyms of *A. commerson*, Rüppell's (1829, 1838) *Chironectes caudimaculatus* is worth mention. Almost immediately following its introduction, this name became firmly associated with what is here recognized as *Lophiocharon trisignatus* (see Comments, p. 318). However, a discrepancy in geographic distribution between the holotype of *C. caudimaculatus* (from the Red Sea) and all known material of *L. trisignatus* (restricted to the Indo-Australian Archipelago) provided the initial clue that this association was in error. A photograph of the holotype (Fig. 83), kindly provided by Wolfgang Klausewitz, Forschungsinstitut Senckenberg, Frankfurt am Main, clearly shows a typical example of *A. commerson*. Radiographs (also provided by Klausewitz) indicate 13 dorsal-fin rays, eight anal-fin rays, and 11 pectoral-fin rays.

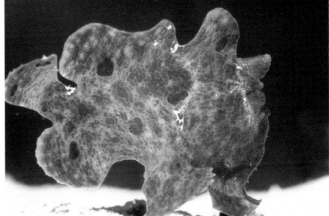

Fig. 82. *Antennarius commerson* (Latreille). UW 20983, 111 mm SL, three color phases assumed by the same aquarium-held individual over a period of three weeks (for full explanation, see Color and Color Pattern, p. 36). Photos by David B. Grobecker; after Pietsch and Grobecker (1987, pl. 18).

Antennarius maculatus (Desjardins)
Warty Frogfish

Figs. 1C, 18B, 75B, 84–89, 275, 295C, 296C, 322, 330A, 333, 337; Tables 5, 9, 22, 24

Chironectes maculatus Desjardins, 1840: 1, pl. 2 (original description, single specimen, holotype MNHN A.4626, Mauritius). Pietsch et al., 1986: 139 (in type catalog).

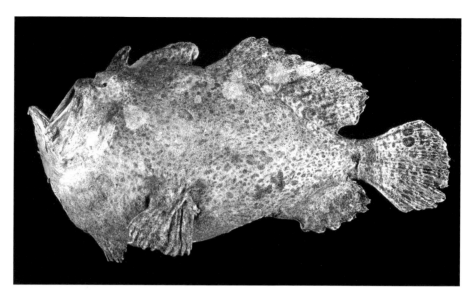

Fig. 83. *Antennarius commerson* (Latreille). Holotype of *Chironectes caudimaculatus* Rüppell, SMF 6783, dried specimen, 206 mm SL, right side reversed, Red Sea. Courtesy of Wolfgang Klausewitz and the Forschungsinstitut Senckenberg, Frankfurt am Main; after Pietsch and Grobecker (1987, fig. 32).

Antennarius leprosus (not of Eydoux and Souleyet): Bleeker, 1857: 68 (description, single specimen subsequently made holotype of *A. guntheri* Bleeker, 1865b). Bleeker, 1859b: 128 (after Bleeker, 1857; Ambon). Günther, 1861a: 198 (description after Bleeker, 1857).

Antennarius phymatodes Bleeker, 1857: 69 (original description, single specimen, holotype RMNH 6285, Ambon). Bleeker, 1859b: 129 (Ambon). Günther, 1861a: 198 (description after Bleeker, 1857; synonymy includes *A. oligospilos* Bleeker). Bleeker, 1865b: 8, 11, pl. 197 (fig. 1), 199 (fig. 5) (after Bleeker, 1857; additional specimen, RMNH 25009, Ambon; pl. 197, fig. 1). Schultz, 1957: 90, pl. 11, fig. A (description after Bleeker, 1857). Beaufort and Briggs, 1962: 204 (in part; synonymy, description, Indonesia). Schultz, 1964: 172, table 1 (synonymy includes *A. oligospilos* Bleeker). Burgess and Axelrod, 1973a: 530, color fig. 491 (plate represents *A. maculatus*). Burgess and Axelrod, 1976: 1910, color fig. 412 (plate represents *A. maculatus*). Grobecker and Pietsch, 1979: 1162, table 1 (feeding mechanism). Araga, 1984: 103, pl. 88-E (Kii Peninsula, Japan). Michael, 1998: 349 (synonym of *A. maculatus*).

Antennarius oligospilos Bleeker, 1857: 70 (original description, single specimen, holotype RMNH 6286, Ambon). Bleeker, 1859b: 129 (Ambon). Fowler, 1928: 479 (East Indies). Schultz, 1957: 95, pl. 11, fig. B (description after Bleeker, 1857). Munro, 1958: 296 (New Guinea). Schultz, 1964: 179 (synonym of *A. phymatodes*).

Antennarius guntheri Bleeker, 1865b: 8, 10, pl. 199, fig. 4 (original description, single specimen, holotype RMNH 6290, Ambon). Schultz 1957: 72 [in synonymy of *Phrynelox melas* (= *A. striatus*)].

Antennarius oligospilus: Bleeker, 1865b: 8, 11, pl. 195 (fig. 1), 200 (fig. 1) (description after Bleeker, 1857; additional material, Ceram and New Guinea). Weber, 1913: 561 (Nusa Island, off Bali).

Antennarius oligospilos (not of Bleeker): Smith, 1949: 431, pl. 98, fig. 1241 (description and figure represent *A. pictus*).

Phymatophryne maculata: Le Danois, 1964: 116, figs. 58, 59 (new combination; in part, MNHN 2107 from Madagascar is *A. pictus*).

Antennarius sp.: Pietsch and Grobecker, 1978: 369, fig. 1 (aggressive mimicry).

Antennarius maculosus (not of Desjardins): Myers and Shepard, 1980: 338 (error for *maculatus*, authorship given to Bleeker; material is *A. pictus*; Guam).

Antennarius maculatus: Grobecker, 1981: 50, figs. (new combination, natural history). Pietsch and Grobecker, 1981: 12 (aggressive mimicry). Pietsch, 1981b: 419 (osteology). Pietsch, 1984b: 36 (genera of frogfishes). Pietsch, 1985a: 85, fig. 23 (original manuscript sources for *Histoire Naturelle des Poissons* of Cuvier and Valenciennes). Pietsch, 1985b: 594, 595, fig. 1 (fast feeding). Pietsch and Grobecker, 1985: 10, figs. (aggressive mimicry, fast feeding). Pietsch and Grobecker, 1987: 87, figs. 22B, 27, 28, 119, pls. 11–16 (description, distribution, relationships). Myers, 1989: 68, pl. 13B (Mauritius to the Solomons, Guam; in key). Pietsch and Grobecker, 1990a: 99, 100, figs. (aggressive mimicry, fast feeding). Pietsch and Grobecker, 1990b: 78, 79, figs. (after Pietsch and Grobecker, 1990a). Amaoka et al., 1994: 23 (aggressive mimicry; after Pietsch and Grobecker, 1978). Helfman et al., 1997: 324, fig. 18.2 (aggressive mimicry). Kuiter, 1997: 44 (Australia). Okamura and Amaoka, 1997: 141, figs. (Japan). Anderson et al., 1998: 22 (Maldive Islands). Bavendam, 1998: 46, color fig. (natural history). Michael, 1998: 329, 330, 332, 333, 334, 347, 348, 350–352, color figs. (identification, behavior, captive care). Fricke, 1999: 105 (Mauritius). Myers, 1999: 68, pl. 13B (after Myers, 1989). Pietsch, 1999: 2015 (western central Pacific; in key). Nakabo, 2000: 457 (Japan; in key). Pietsch, 2000: 597 (South China Sea). Schleichert, 2000: 39, color fig. (natural history). Nakabo, 2002: 457 (Japan; in key). Sazima, 2002: 38 (aggressive mimicry; after Pietsch and Grobecker, 1978). Randall et al., 2002: 155 (Society Islands, Tuamotu Archipelago). Allen and Adrim, 2003: 25 (Bali and Sulawesi to West Papua). Allen et al., 2003: 365, color figs. (tropical Pacific). Myers and Donaldson, 2003: 610, 614 (Mariana Islands). Schneidewind, 2003a: 85, 88, figs. (natural history). Lieske and Myers, 2004: 46, color fig. (Red Sea and Mauritius to Solomon and Mariana islands). Pietsch, 2004b: 235, fig. 236 (natural history). Shedlock et al., 2004: 134, 144, table 1 (molecular phylogeny). Randall, 2005a: 71, color figs. (South Pacific). Schneidewind, 2005a: 20, 22, color figs. (mimicry). Schneidewind, 2005c: 85, 88, figs. (tiny postlarvae). Steene, 2005: frontispiece, 51, 326, color pls. (Anilao, Lembeh Strait). Allen et al., 2006: 640 (Australia). Carnevale and Pietsch, 2006: 452, table 1 (comparison, meristics, first fossil frogfish). Michael, 2006: 34, 36, 38, 40, 41, color figs. (spawning behavior, camouflage). Senou et al., 2006: 423 (Sagami Sea). Kuiter and Debelius, 2007: 118, color figs. (Indonesia, Papua New Guinea). Randall, 2007: 128, color figs. (Kona, Hawaii, based on a video, no known specimens; in key). Senou et al., 2007: 49 (Ryukyu Islands). Allen and Erdmann, 2009: 594 (West Papua). Helfman et al., 2009: 428, fig. 19.2 (aggressive mimicry; after Helfman et al., 1997). Pietsch, 2009: 249, fig. 265 (aggressive mimicry, jet propulsion). Motomura et al., 2010: 78 (Yaku-shima Island, southern Japan). Fricke et al., 2011: 368 (New Caledonia). Satapoomin, 2011: 51, appendix A (Andaman Sea). Allen and Erdmann, 2012: 147, color figs. (East Indies). Arnold and Pietsch, 2012: 124, 128, figs. 1B, 4, 6, table 1 (molecular phylogeny). Nakabo, 2013: 541, figs. (Japan). Brandl and Bellwood, 2014: 32 (pair-formation). Karplus, 2014: 35, 36, figs. 1.21C, 1.22 (aggressive mimicry; after Pietsch and Grobecker, 1987). Moore et al., 2014: 177 (northwestern Australia). Grimsditch et al., 2016: 1, fig. 1 (camouflaged among bleached coral).

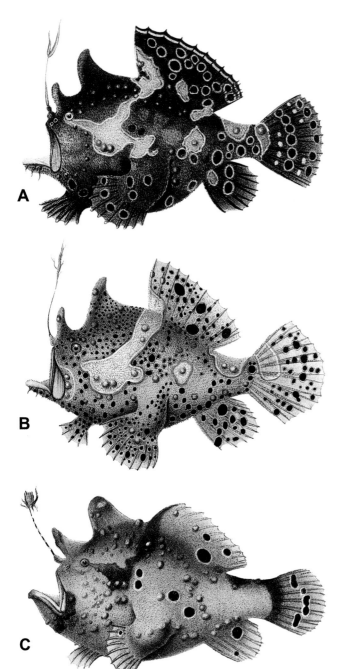

Fig. 84. *Antennarius maculatus* (Desjardins). (**A**) holotype of *Antennarius guntheri* Bleeker, RMNH 6290, 63.5 mm SL, Ambon, Indonesia (after Bleeker, 1865b, pl. 199, fig. 4). (**B**) holotype of *Antennarius phymatodes* Bleeker, RMNH 6285, 67 mm SL, Ambon, Indonesia (after Bleeker, 1865b, pl. 199, fig. 5). (**C**) holotype of *Antennarius oligospilos* Bleeker, RMNH 6286, 55 mm SL, Ambon, Indonesia (after Bleeker, 1865b, pl. 200, fig. 1). Drawings by Ludwig Speigler under the direction of Pieter Bleeker.

Antennarius maculates: Randall et al., 2002: 153 (misspelling of *maculatus*; Society Islands, Tuamotu Archipelago). Prihadi, 2015: 187, figs. 4–6 (Nusa Penida Island, Bali Province).

MATERIAL EXAMINED

Twenty-nine specimens, 30 to 85 mm SL.

Holotype of *Chironectes maculatus*: MNHN A.4626, 72.5 mm, Mauritius, Desjardins, December 1837.

Holotype of *Antennarius guntheri*: RMNH 6290, 63.5 mm, Ambon, Bleeker (Fig. 84A).

Holotype of *Antennarius phymatodes*: RMNH 6285, 67 mm, Ambon, Bleeker (Fig. 84B).

Holotype of *Antennarius oligospilos*: RMNH 6286, 55 mm, Ambon, Bleeker (Fig. 84C).

Additional material: BMNH, 3 (36–65.5 mm); CAS, 1 (55 mm); LACM, 1 (58.5 mm); RMNH, 6 (56.5–85 mm); USNM, 1 (51 mm); UW, 14 (30–72 mm); ZMA, 1 (17 mm).

DIAGNOSIS

A member of the *Antennarius pictus* group unique in having the following combination of character states: second dorsal-fin spine expanded from base, appearing swollen distally (Fig. 75B); membrane behind second dorsal-fin spine thin (width along attachment to spine distinctly less than that of spine itself), nearly always devoid of dermal spinules, and extending across area between second and third dorsal-fin spines reaching to base of third; head, body, and fins covered with numerous, protruding, wartlike swellings (less conspicuous in specimens smaller than about 40 mm SL; Fig. 85); belly without large, darkly pigmented spots; dorsal-fin rays 12 (rarely 11); anal-fin rays seven (rarely six); pectoral-fin rays 10 (rarely 11) (Tables 5, 9).

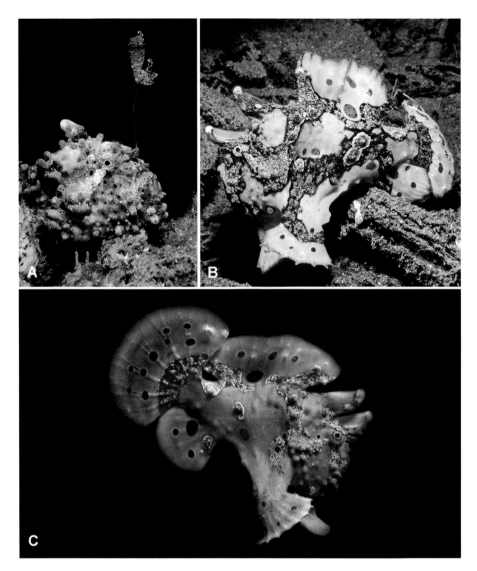

Fig. 85. *Antennarius maculatus* (Desjardins). (**A**) Anilao, Luzon, Philippines. Photo by Mike Bartick. (**B**) Loloata, Papua New Guinea. Photo by Roger Steene. (**C**) Dauin, Negros Island, Philippines. Photo by Daniel Geary.

DESCRIPTION

Esca strongly resembling a small fish (Pietsch and Grobecker, 1978: 370, fig. 1; Fig. 86); broad, laterally compressed appendage of esca as long as 70% of illicium (RMNH 25009, 56.5 mm SL); illicium 26.3 to 33.8% SL; second dorsal-fin spine curved posteriorly, a conspicuous indentation usually present anteriorly at base (Fig. 75B), length 13.8 to 18.1% SL; naked depression between second and third dorsal-fin spines absent; third dorsal-fin spine curved posteriorly, length 24.0 to 26.6% SL; eye diameter 5.8 to 7.8% SL.

Coloration in life: extremely variable; cream, yellow, greenish yellow, bright red, chocolate-brown, to black with numerous, scattered, brown to black circular spots of various sizes; white, pink, red, or rust-brown saddles and irregularly shaped patches located as described below for preserved material (Figs. 87, 88). Esca with a proximal,

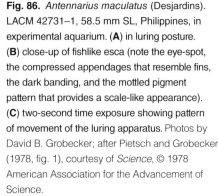

Fig. 86. *Antennarius maculatus* (Desjardins). LACM 42731–1, 58.5 mm SL, Philippines, in experimental aquarium. (**A**) in luring posture. (**B**) close-up of fishlike esca (note the eye-spot, the compressed appendages that resemble fins, the dark banding, and the mottled pigment pattern that provides a scale-like appearance). (**C**) two-second time exposure showing pattern of movement of the luring apparatus. Photos by David B. Grobecker; after Pietsch and Grobecker (1978, fig. 1), courtesy of *Science*, © 1978 American Association for the Advancement of Science.

eyelike, pigment spot on each side; surface of esca mottled with white and chocolate-brown; four or five darkly pigmented vertical bands immediately posterior to "eye-spot" (an area corresponding to the shoulder and pectoral region of a fish), and two lighter-colored bands near distal tip (area corresponding to caudal fin; see Pietsch and Grobecker, 1978; figs. 85A, 86B) (LACM 42731-1; UW 20828, 20829).

Coloration in preservation: creamy white, light pinkish-gray, yellow, brown, dark brown, to almost black; illicium lightly banded in some specimens; as many as nine short, darkly pigmented bars radiating from eye. Lighter-color phases with numerous, scattered, dark brown to black circular spots of various sizes over entire head, body, and fins; brown to pinkish-orange saddles on caudal peduncle and shoulder (usually extending down onto dorsal surface of pectoral-fin lobe); additional brown to pinkish-orange, irregularly shaped patches on second and third dorsal-fin spines, chin, corner of jaw, and side of body posterior to insertion of pectoral-fin lobe. Darker-color phases with numerous, scattered, black circular spots of various sizes over entire head, body, and fins; tan, pinkish-brown to brown saddles on caudal peduncle and shoulder (usually extending down onto opercular region and dorsal surface of pectoral-fin lobe); additional lightly pigmented, irregularly shaped patches located as in lighter-colored phases described above; tips of pectoral- and pelvic-fin rays white.

Largest individual examined: 85 mm SL (RMNH 15590).

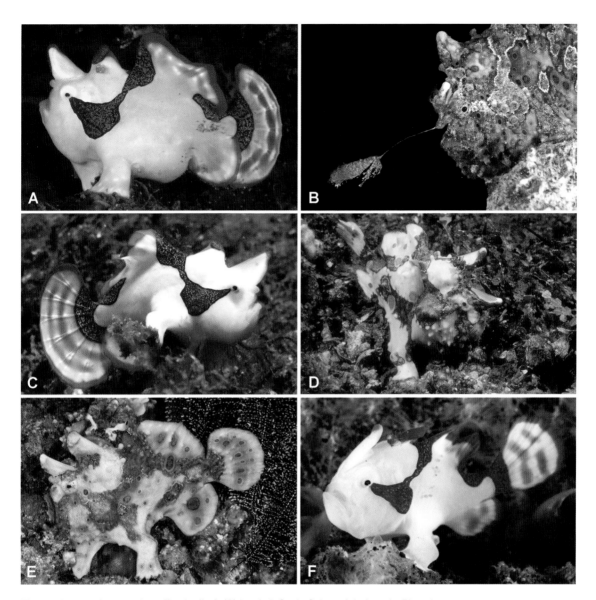

Fig. 87. *Antennarius maculatus* (Desjardins). (**A**) Lembeh Strait, Sulawesi, Indonesia. Photo by Janine Cairns-Michael). (**B**) Anilao, Luzon, Philippines. Photo by Mike Bartick. (**C**) Solitary Islands, New South Wales, Australia. Photo by Matthew Harrison. (**D**) Lembeh Strait, Sulawesi, Indonesia. Photo by Frank Schneidewind. (**E**) Puerto Galera, Mindoro, Philippines. Photo by Linda Ianniello. (**F**) Ambon, Indonesia. Photo by Linda Ianniello.

DISTRIBUTION

Most of the known material of *A. maculatus* has come from the Philippines and Indonesia (from Bali and Sulawesi to West Papua), but scattered records are also known from the Red Sea, Mauritius, and the Maldives to the Society Islands, and from Japan to the Solitary Islands off New South Wales, including northwest Australia, the Andaman Sea, Singapore, the Philippines, Indonesia (Bali, Sulawesi, Ceram, Ambon, and West Papua), Guam, the Marianas, Papua New Guinea, New Britain in the Bismarck Archipelago, Guadalcanal Island in the Solomons, New Caledonia, and the Hawaiian and Society Islands (Fig. 275, Table 22). The type locality is Mauritius.

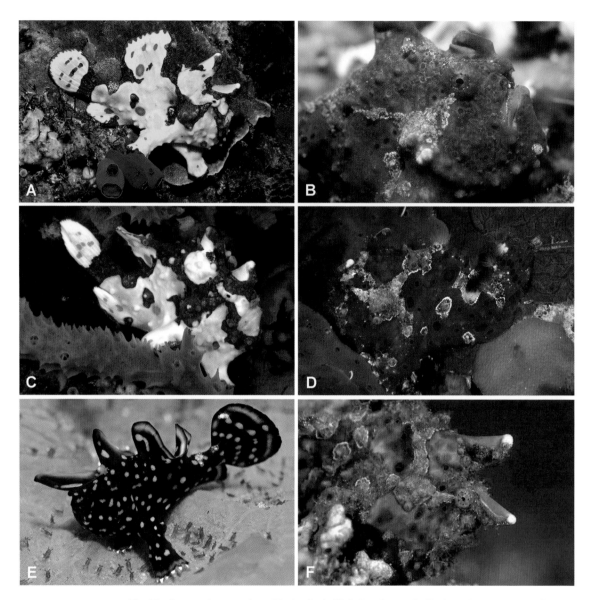

Fig. 88. *Antennarius maculatus* (Desjardins). **(A)** Anilao, Luzon, Philippines. Photo by Roger Steene. **(B)** Tulamben, Bali, Indonesia. Photo by Frank Schneidewind. **(C, D)** Anilao, Luzon, Philippines. Photos by Scott W. Michael. **(E)** Dauin, Negros Island, Philippines. Photo by Daniel Geary. **(F)** Komodo National Park, Rinca Island, Indonesia. Photo by Colin Marshall.

Frequently observed on coastal reefs, associated with algae, sponges, and soft corals, most often at depths ranging from just beneath the surface to 20 m.

COMMENTS

Antennarius maculatus (Desjardins, 1840) is a name that remained hidden in the scientific literature for more than 120 years until Le Danois (1964) reexamined the holotype, referring to it as *Phymatophryne maculata*. Certainly Bleeker (1857, 1865b) was unaware of Desjardins's (1840) excellent description when he described three new "warty-skinned" antennariids from Ambon (*A. phymatodes, A. oligospilos,* and *A. guntheri*; Fig. 84), all based on perceived differences in color and color pattern and now recognized as synonyms of *A. maculatus*. A remarkably accurate pair of images of *A. maculatus* is here reproduced from the manuscript materials for Cuvier and Valenciennes's *Histoire Naturelle des Poissons* (Fig. 89).

Antennarius multiocellatus (Valenciennes) Longlure Frogfish

Title page, Figs. 75C, 90–96, 281, 322B–F, 329A–E, 330B, 332; Tables 5, 9, 22

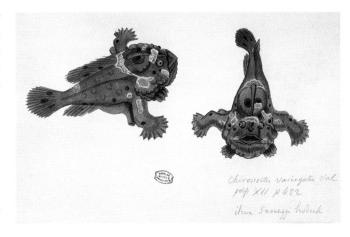

Pira Vtoewah, forma monstrosa: De Laet, 1633: 574, fig. (also spelled *Pira Vtoevvah*, *Vtoehwah piscis* in index; Brazil). De Laet, 1640: 510, fig. (after de Laet, 1633). Valenciennes, 1837: 415, 424 (after de Laet, 1633; confused with *C. scaber* and *C. furcipilis*). De Mello, 1967:

Fig. 89. Two views of *Antennarius maculatus* (Desjardins). From a watercolor original labeled *Chironectes variegatus* Valenciennes (a junior synonym of *Antennarius pictus*), found in the original manuscript sources for the *Histoire Naturelle des Poissons* of Cuvier and Valenciennes (Pietsch, 1985a, fig. 23). After Cuvier and Valenciennes, MS 504, XII.B, Bibliothèque Centrale, Muséum national d'Histoire naturelle, Paris; courtesy of Emmanuelle Choiseau, © Muséum national d'Histoire naturelle, Paris.

156, 158 (review of sources for de Laet, 1633, 1640). Whitehead, 1979: 437, footnote 80 (sources for de Laet, 1633, 1640).

Guaperua Brasiliensibus: Marcgrave, 1648: 150, fig. (description, with figure after de Laet, 1633; spiny-skinned form with illicium about one-third standard length; vicinity of Recife, Brazil; "It feeds on shrimps, while swimming it singularly spreads out its fins and inflates itself so that it appears round like a ball").

Guaperua: Jonston, 1650: 139, 140, pl. 36, fig. 6 (after Marcgrave, 1648). Valenciennes, 1837: 414–416, 424 (after Marcgrave, 1648; confused with *C. scaber* and *C. furcipilis*). De Mello, 1967: 156, 158 (sources for Marcgrave, 1648). Whitehead, 1979: 437–438, footnotes 80, 82 (observations on illustrations in *Libri Picturati*, Biblioteka Jagiellońska, Kraków; see Comments, p. 138).

Guaperua Brasiliensibus Margravii: Willughby, 1686: 90, pl. E.2, fig. 1 ("The American Toad-Fish"; after Marcgrave, 1648).

Piscis Brasilianus cornutus: Petiver, 1702: pl. 20, fig. 6 (after Marcgrave, 1648; "shores of Brazil and several other coasts of the West Indies").

Guaperna Brasiliensibus Marcgr: Ray, 1713: 29, no. 2 (after Marcgrave, 1648; misspelling of *Guaperua*).

Guapena: Linnaeus, 1727–1730: folio 166 ("While swimming it blows up its belly like a ball," a rough sketch after Marcgrave, 1648; Fig. 3A).

Batrachus in fronte corniculum ferens: Klein, 1742: 16, no. 4 (new name for *Guaperua* of Marcgrave, 1648, and Willughby, 1686).

Chironectes principis Valenciennes, 1837: 416 [original description from two drawings made for Count Johan Maurits of Nassau-Siegen between 1637 and 1644, on basis of specimens observed in vicinity of Recife, Brazil; drawings shown to Valenciennes by M. H. K. Lichtenstein in Berlin sometime between 1826 and 1829 (Whitehead, 1976: 417; 1979: 459); no extant material; iconotypes in *Libri Picturati* (Handbooks, 1: 363, 2: 378), Biblioteka Jagiellońska, Kraków (copies of both in Cuvier and Valenciennes, MS 504, XII.B.56, XII.B.50, respectively; see Pietsch, 1985a: 76)]. Pietsch, 1985a: 76, fig. 15, table 1 (original manuscript sources for *Histoire Naturelle des Poissons* of Cuvier and Valenciennes).

Chironectes multiocellatus Valenciennes, 1837: 420 (original description; lectotype MNHN A.4591, established by Pietsch et al., 1986: 140, Cuba; paralectotype MNHN A.4592, Martinique). Storer, 1846: 383 (Caribbean Sea). Guichenot, 1853: 215 (after Valenciennes, 1837; Cuba). Poey y Aloy, 1853: 220 (MNHN A.4591, lectotype, sent to Cuvier). Pietsch, 1985a: 81, table 1 (original manuscript sources for *Histoire Naturelle des Poissons* of Cuvier and Valenciennes). Pietsch et al., 1986: 140 (in type catalog).

Chironectes pavoninus Valenciennes, 1837: 421 [original description, specimen illustrated by Valenciennes at University of Göttingen in 1828 (1829, according to note written on iconotype), accompanied by Alexander von Humboldt; no extant material, not in Göttingen collections (now housed at ZIM, Alfred Post, personal communication, 6 April 1981), iconotype in Cuvier and Valenciennes, MS 504, XII.B.55 (see Pietsch, 1985a: 82); locality unknown]. Pietsch, 1985a: 82, fig. 20, table 1 (original manuscript sources for *Histoire Naturelle des Poissons* of Cuvier and Valenciennes).

Chironectes tenebrosus Poey y Aloy, 1853: 219, pl. 17, fig. 1 [*Pescador negro*, original description, type apparently lost (see Barbour, 1942: 28), Cuba].

Antennarius histrio (not of Linnaeus) **Var. α**: Günther, 1861a: 189 (reference to *C. principis* Valenciennes, 1837, first figure, i.e., *d'un brun très-foncé*).

Antennarius principis: Günther, 1861a: 193 (new combination; description after Valenciennes, 1837; tropical Western Atlantic). Garman, 1896: 84 [comparison with *A. nuttingii* (=*A. scaber*)]. Jordan and Evermann, 1898: 2719 (after Valenciennes, 1837, Günther, 1861a). Evermann and Marsh, 1900: 334 (West Indies; in key). Bean, 1906b: 89 (Bermuda). Miranda-Ribeiro, 1915: 7 (after Valenciennes, 1837; in key). Borodin, 1928: 24 [comparison with *A. cubensis* (=*A. scaber*)]. Meek and Hildebrand, 1928: 1015 (after Günther, 1861a; West Indies, Brazil). Beebe and Tee-Van, 1933: 252 (Bermuda). Barbour, 1942: 27, 28, 29 (synonym of *A. multiocellatus*). Fowler, 1944: 453 (Andros Island, Bahamas). Pietsch, 1985a: 77 (original manuscript sources for *Histoire Naturelle des Poissons* of Cuvier and Valenciennes). Smith-Vaniz et al., 1999: 160 (synonym of *A. multiocellatus*).

Antennarius multiocellatus* Var. α. *multiocellata: Günther, 1861a: 194 (new combination; in part, i.e., specimen "a"; Caribbean).

***Antennarius multiocellatus* Var. β**: Günther, 1861a: 194 (Caribbean).

Antennarius tenebrosus: Günther, 1861a: 197 (new combination; after Poey y Aloy, 1853). Poey y Aloy, 1868: 405 (after Poey y Aloy, 1853, Günther, 1861a). Evermann and Marsh, 1900: 334 (West Indies; in key). Barbour, 1942: 27, 28 (synonym of *A. multiocellatus*). Randall, 1968: 291 (distinct from *A. multiocellatus*, Caribbean). Böhlke and Chaplin, 1968: 721, figs. (description; Florida, Bahamas, Cuba). Guitart, 1978: 791, fig. 622 (Cuba). Pietsch, 1978a: 3 (error). Dennis and Bright, 1988: 3 (synonym of *A. multiocellatus*). Mejía-Ladino et al., 2007: 285 (synonym of *A. multiocellatus*).

Antennarius multiocellatus (not of Valenciennes): Günther, 1861b: 361 (Ceylon). Playfair, 1867: 862 (Var. a, Var. b; material is *A. pictus*, Seychelles). Poey y Aloy, 1876: 135 (in part, synonymy includes *A. pictus*). Poll, 1949: 252, fig. 23 (material is *A. pardalis*, Liberia).

Antennarius annulatus Gill, 1863: 91 (original description, single specimen, holotype USNM 4849, Garden Key, Florida). Goode and Bean, 1882: 235 (Gulf of Mexico). Jordan and Gilbert, 1883: 846 (after Gill, 1863). Gill, 1883: 555 (reference to Goode and Bean, 1882). Longley and Hildebrand, 1941: 307 (description, additional specimen, Tortugas). Barbour, 1942: 27 (synonym of *A. multiocellatus*).

Antennarius corallinus Poey y Aloy, 1865: 188, 413 (original description, single specimen, holotype MCZ 11620, Cuba). Poey y Aloy, 1868: 405 (after Poey y Aloy, 1865).
Poey y Aloy, 1876: 135 (after Poey y Aloy, 1865). Howell y Rivero, 1938: 220 (holotype
in MCZ). Barbour, 1942: 27 (synonym of *A. multiocellatus*).

Antennarius multiocellatus: Poey y Aloy, 1868: 405 (after Valenciennes, 1837). Günther,
1880: 5 (Ascension). Garman, 1896: 82 (Bahamas). Jordan and Evermann, 1898: 2724
(West Indies to Florida Keys). Evermann and Marsh, 1900: 334 (West Indies; in key).
Jordan, 1905: 550 (Cuba). Norman, 1935: 55, 56 (Ascension). Fowler, 1936b: 1133, 1363
(Ascension; St. Helena in error after Günther, 1869a). Howell y Rivero, 1938: 220
(senior synonym of *A. corallinus* Poey). Longley and Hildebrand, 1941: 305 (Tortugas).
Barbour, 1942: 27, 33, 37, pls. 10, 11, 13, 14, 16 (fig. 2), 17 (fig. 5) (Bermuda to Cuba).
Breder, 1949: 95, pl. 9, figs. 1, 2 (color changes). Schultz, 1957: 94, pl. 10, fig. C (synon-
ymy, description; Bermuda, Bahamas, Cuba, Virgin Islands, Tortugas; in key).
Duarte-Bello, 1959: 141 (Cuba). Le Danois, 1964: 101, fig. 48e [in part, MNHN A.4619
is *A. scaber* (Bahia), A.4614 is *A. biocellatus* (locality unknown); A.4590 and A.5264 er-
roneously listed as syntypes]. Schultz, 1964: 175, table 1 (additional specimen, Brazil).
Caldwell, 1966: 88 (Jamaica). Cervigón, 1966: 877 (Cabo Codera, Venezuela; in key).
Randall, 1968: 291, 292, figs. 320, 321 (description, tropical Western Atlantic). Böhlke
and Chaplin, 1968: 719, figs. (description, Bahamas to Central American coast). Bur-
gess, 1976: 58, color photo (luring apparatus). Guitart, 1978: 790, fig. 621 (Cuba).
Pietsch, 1978a: 3 (listed, western central Atlantic). Román, 1979: 352, fig. (Venezuela;
in key). Pietsch, 1984b: 34 (genera of frogfishes). Thresher, 1984: 41 (reproduction).
Pietsch, 1985a: 76, 81, 82, figs. 15, 20, table 1 (original manuscript sources for *Histoire
Naturelle des Poissons* of Cuvier and Valenciennes). Robins and Ray, 1986: 87, pl. 14
(Bermuda, Florida, Bahamas to South America). Acero and Garzón, 1987: 90 (Santa
Marta, Colombia). Pietsch and Grobecker, 1987: 99, figs. 22D, 23, 33–35, 118, 157, pls.
21, 22 (description, distribution, relationships). Dennis and Bright, 1988: 3 (Sabine
Pass, northwestern Gulf of Mexico). Pietsch and Grobecker, 1990a: 101, fig. (aggres-
sive mimicry). Pietsch and Grobecker, 1990b: 78, fig. (after Pietsch and Grobecker,
1990a). Cervigón, 1991: 191 (Venezuela). Burgess et al., 1994: 205 (Cayman Islands).
Hoese and Moore, 1998: 163, pl. 104 (Gulf of Mexico). McEachran and Fechhelm,
1998: 818 (Gulf of Mexico). Michael, 1998: 80, 330, 336, 349, 350, color figs. (identifica-
tion, behavior, captive care). Castro-Aguirre et al., 1999: 175 (Gulf of Mexico). Leão de
Moura et al., 1999: 515 (Brazil). Smith-Vaniz et al., 1999: 159, 160 (Bermuda; in key).
Grace et al., 2000: 47, table 2 (Navassa Island, Caribbean Sea). Ferreira et al., 2001: 357,
table 1 (Arraial do Cabo, southeastern Brazil). Pietsch, 2002a: 1051 (western central
Atlantic). Collette et al., 2003: 100 (Navassa Island, West Indies). Menezes, 2003: 64
(Brazil). Schneidewind, 2003a: 84–86, figs. (natural history). Smith et al., 2003: 13 (Be-
lize). Pietsch, 2004b: 237, fig. (natural history). Dennis et al., 2005: 723, 731, tables 2, 3
(Mona Passage, Greater Antilles). Ortiz and Lalana, 2005: 8 (Cuba). Carnevale and
Pietsch, 2006: 452, table 1 (comparison, meristics, first fossil frogfish). Jackson, 2006a:
783, 784, tables 1, 2 (reproduction, depth distribution). Michael, 2006: 41 (behavior).
Thaler, 2006: 44 (coloration, aquarium care). Kuiter and Debelius, 2007: 123, color
figs. (Caribbean, Brazil). Mejía-Ladino et al., 2007: 284, figs. 4, 6, tables 1–3, 5 (species
account; in key; Colombia). Uyarra and Côté, 2007: 78 (Bonaire). Wirtz et al., 2007: 7,
26, table 1 (São Tomé, Gulf of Guinea). Luiz et al., 2008: 1, 6, 14, fig. 7C, table 1 (Laje
de Santos, Brazil). Williams et al., 2010: 7, figs. 28, 29 (Saba Bank Atoll). Arnold and
Pietsch, 2012: 124, 128, figs. 4, 6, table 1 (molecular phylogeny). Luiz et al., 2013, dataset

S1 (geographic range size). Monteiro-Neto et al., 2013: 3, table 1 (off Rio de Janeiro). Page et al., 2013 (common names). Pietsch et al., 2013: 664, 665, fig. 4A–E (sexual dimorphism, spawning). Sampaio and Ostrensky, 2013: 282, table 1 (Brazil). Smith-Vaniz and Collette, 2013: 169 (Bermuda). Smith-Vaniz and Jelks, 2014: 29 (St. Croix, Virgin Islands). Wirtz et al., 2014: 4 (Ascension Island). Garcia et al., 2015: 7 (northeastern Brazil). Davies and Piontek, 2016: 76 (St. Eustatius).

Antennarius multiocellatus (not of Valenciennes) **Var. β**: Günther, 1869a: 238 (BMNH 1868.6.15.21 from St. Helena is *A. nummifer*). Melliss, 1875: 108 (after Günther, 1869a).

Pterophryne principis: Goode, 1877: 290 (new combination; West Indies).

Antennarius stellifer Barbour, 1905: 132, pl. 4 (original description, single specimen, holotype MCZ 29056, Castle Harbor, Bermuda). Bean, 1906b: 89 (Bermuda). Beebe and Tee-Van, 1933: 253, fig. (Bermuda). Barbour, 1942: 27, 28 (synonym of *A. multiocellatus*). Smith-Vaniz et al., 1999: 160 (synonym of *A. multiocellatus*).

Antennarius verrucosus Bean, 1906a: 31 (original description, single specimen, holotype FMNH 4853, St. George's Island, Bermuda). Bean 1906b: 88, fig. 14 (after Bean, 1906a). Beebe and Tee-Van, 1933: 251, fig. (Bermuda). Barbour, 1942: 27 (synonym of *A. multiocellatus*). Schultz, 1957: 99, pl. 12, fig. D (description; Bermuda, Florida Keys). Pietsch, 1978a: 3 (error). Smith-Vaniz et al., 1999: 160 (synonym of *A. multiocellatus*).

Antennarius astroscopus Nichols, 1912: 109, fig. 1 (original description, single specimen, holotype AMNH 3315, Barbados). Borodin, 1928: 24 (comparison with *A. cubensis* Borodin). Barbour, 1942: 27, 29 (synonym of *A. multiocellatus*).

Antennarius commersoni (not of Latreille) **var. multiocellata**: Lampe, 1914: 253 (Ascension).

Antennarius pardalis (not of Valenciennes): Metzelaar, 1919: 161 (synonymy includes *A. verrucosus* Bean, Bonaire, and St. Martin).

Antennarius tenerosus: Borodin, 1928: 24, 25 [misspelling of *A. tenebrosus* (Poey)].

Antennarius sellifer: Borodin, 1928: 24, 25 [misspelling of *A. stellifer* Barbour; comparison with *A. cubensis* (= *A. scaber*)].

Lophiocharon (**Uniantennatus**) **tenebrosus**: Schultz, 1957: 83, pl. 7 (fig. A), 14 (fig. B) (new combination, description; Miami, Key Biscayne).

Lophiocharon tenebrosus: Duarte-Bello, 1959: 142 (Cuba).

Antennarius mulliocellatus: Gerson, 2001, color fig. (misspelling of specific name; on yellow branching tube sponge).

MATERIAL EXAMINED

One hundred and fifty-nine specimens, 5.0 to 120 mm SL.

Lectotype of *Chironectes multiocellatus*: MNHN A.4591, 65 mm, Havana, Cuba, Poey.

Paralectotype of *C. multiocellatus*: MNHN A.4592, 64.5 mm, Martinique, Lesser Antilles, Garnot.

Holotype of *Antennarius annulatus*: USNM 4849, 58 mm, Garden Key, Florida, Wright.

Holotype of *Antennarius corallinus*: MCZ 11620, 61.5 mm, Cuba, Poey.

Holotype of *Antennarius stellifer*: MCZ 29056, 69 mm, Castle Harbor, Bermuda, Mowbray (Fig. 90).

Holotype of *Antennarius guntheri*: RMNH 6290, 63.5 mm, Ambon, Bleeker.

Holotype of *Antennarius verrucosus*: FMNH 4853, 85 mm, St. George's Island, Bermuda, "in the Reach," Mowbray, 1904.

Holotype of *Antennarius astroscopus*: AMNH 3315, 112.5 mm, Barbados, Lesser Antilles.

Fig. 90. *Antennarius multiocellatus* (Valenciennes). Holotype of *Antennarius stellifer* Barbour, MCZ 29056, 69 mm SL, Castle Harbor, Bermuda. Drawing by Edgar Nathaniel Fischer; after Barbour (1905, pl. 4).

Additional material: *Bermuda*: AMNH, 3 (60–75 mm); ANSP, 2 (78–101 mm); FMNH, 1 (71 mm). *Bahamas*: AMNH, 14 (5–107 mm); ANSP, 13 (9.0–66 mm); BMNH, 1 (25 mm); MCZ, 1 (52 mm); UW, 1 (14.5 mm). *Florida*: AMNH, 3 (17.5–69 mm); MCZ, 1 (15 mm). *Greater Antilles*: ANSP, 5 (13.5–53 mm); BMNH, 1 (39 mm); FMNH, 1 (78 mm); LACM, 5 (12–61 mm); MCZ, 2 (13–109 mm); MNHN, 1 (76 mm); UMMZ, 3 (47–58 mm); ZMUC, 7 (40–110 mm). *Lesser Antilles*: AMNH, 1 (43 mm); ANSP, 9 (9.5–84 mm); BPBM, 1 (88 mm); GCRL, 1 (51 mm); RMNH, 2 (48–54 mm); UF, 1 (13.5 mm). *Central America*: FMNH, 3 (7.5–17 mm); MCZ, 1 (64 mm); UA, 1 (98 mm); ZMUC, 2 (30–31 mm). *Colombia*: MCZ, 1 (43 mm); USNM, 1 (57 mm). *Venezuela*: MBUCV, 5 (29–104 mm). *Brazil*: FMNH, 1 (93 mm); GCRL, 2 (15–30.5 mm); MNRJ, 7 (36–102 mm); MZUSP, 16 (15.5–107 mm); SU, 2 (49.5–93.5 mm); UW, 2 (76–120 mm). *Azores*: MNHN, 1 (98 mm, identification uncertain, see Zoogeography, p. 394). *Ascension Island*: BMNH, 3 (44–56.5 mm); GCRL, 8 (30–52 mm); GMBL, 12 (27–60 mm); USNM, 4 (12–34.5 mm). *Locality unknown*: BMNH, 2 (42–73 mm); ZMB, 1 (59 mm).

DIAGNOSIS

A member of the *Antennarius pictus* group unique in having the following combination of character states: second dorsal-fin spine slightly tapered from base (Fig. 75C); membrane behind second dorsal-fin spine thin (width along attachment to spine less than that of spine itself), covered with dermal spinules (except for extreme outer margin and a small triangular area just behind tip of spine), terminating distinctly anterior to base of third dorsal-fin spine (Fig. 75C); head and body sometimes partially or fully covered with low, rounded, wartlike swellings (but nothing like the warts of *A. maculatus* or *A. pardalis*); belly without large, darkly pigmented spots; dorsal-fin rays 12 (rarely 11 or 13); anal-fin rays seven (rarely six); pectoral-fin rays 10 (rarely nine) (Tables 5, 9).

Antennarius multiocellatus is distinguished from *A. pardalis* by a statistically significant difference in the length of the second dorsal-fin spine (Fig. 91); the absence of large, darkly pigmented spots on the belly that are usually present in *A. pardalis* (especially in

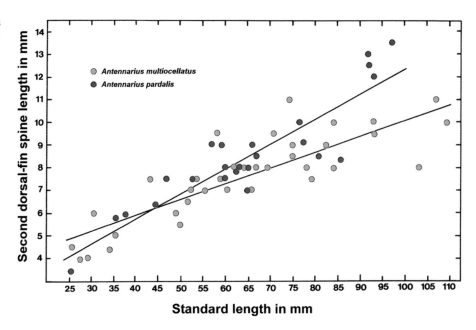

Fig. 91. Regressions ($y = a + bx$) of second dorsal-fin spine length on standard length of the two Atlantic species of the *Antennarius pictus* Group: *A. multiocellatus* (blue), N = 42; *A. pardalis* (red), N = 25. Slopes are significantly different at $p < 0.005$. After Pietsch and Grobecker (1987, fig. 23).

Fig. 92. Esca of *Antennarius multiocellatus* (Valenciennes). USNM 232166, 28.5 mm SL, Southwest Bay, Ascension Island. Drawing by Cathy L. Short; after Pietsch and Grobecker (1987, fig. 34).

smaller specimens); and the lack of large, close-set wartlike swellings that cover the head and body of *A. pardalis* (especially evident in large specimens).

DESCRIPTION

Laterally compressed appendage of esca (Fig. 92) as long as 45% of illicium (USNM 4849, 58 mm SL); illicium 21.0 to 31.8% SL; second dorsal-fin spine straight to slightly curved posteriorly (Fig. 75C), length 9.2 to 19.4% SL; shallow, naked depression between second and third dorsal-fin spines usually present; third dorsal-fin spine curved posteriorly, length 16.8 to 28.2% SL; eye diameter 4.6 to 7.3% SL.

Coloration in life: extremely variable; white, tan, yellow-orange, bright or brick red, pink, green, or solid black; solid red with reddish black spots on body and fins; bright-white saddle on caudal peduncle; rich reddish brown with irregularly shaped gray-green blotches and black circular spots on body and fins; overall dark green to yellow-green with black circular spots; usually a triad of dark spots on caudal fin (Figs. 93–95).

Coloration in preservation: light tan, yellow-brown, dark brown, to black; illicium without banding; as many as nine short, darkly pigmented bars radiating from eye. Lighter-colored phases with slightly darker mottling on head, body, and fins; a more or less conspicuous basidorsal spot, and similar dark circular spots (each sometimes encircled with a lightly pigmented ring) of various sizes, particularly well developed on unpaired fins; usually lightly pigmented saddles on caudal peduncle and shoulder, and irregularly shaped patches (coral-pink in some specimens, e.g., MCZ 11620) on flanks. Darker-colored phases with markings inconspicuous except for slightly lighter-colored saddles on caudal peduncle and shoulder; solid-black phases without markings.

Largest specimen examined: 120 mm SL (UW 117827).

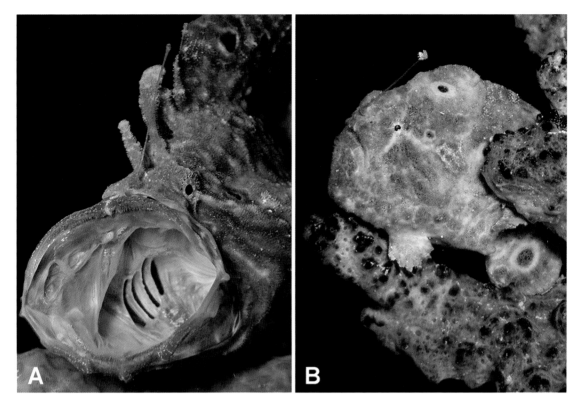

Fig. 93. *Antennarius multiocellatus* (Valenciennes). (**A**, **B**) Bonaire, Netherlands Antilles. Photos by Ellen Muller.

DISTRIBUTION

Antennarius multiocellatus is restricted to the Atlantic Ocean (Fig. 281, Table 22). In the Western Atlantic it ranges from Bermuda to the Bahamas and the eastern and southern tip of Florida, throughout the Caribbean, and along the coasts of Central America, Colombia, Venezuela, and Brazil as far south as Laje de Santos Marine State Park, State of São Paulo (about 24° S latitude; Luiz et al., 2008). In the Eastern Atlantic its presence is well-documented at Ascension Island, and there is at least one record from São Tomé, Gulf of Guinea (Wirtz et al., 2007). Günther's (1869a) record from St. Helena, however, was based on a specimen of *Abantennarius nummifer* (BMNH 1868.6.15.21) collected by J. C. Melliss. A poorly preserved, stuffed specimen from the Azores (MNHN A.5265) may be *A. multiocellatus*, but more likely *A. pardalis*. As indicated below (see Comments, p. 140), these two latter species are difficult to distinguish; it is possible that material of *A. multiocellatus* is hidden among specimens listed here as *A. pardalis*. The type locality is Cuba.

Depths of capture, known for 54 specimens, range between the surface and 91 m. However, 85% of this material was taken at 20 m or less, and 76% in less than 10 m. The average depth for all recorded captures is 12 m.

COMMENTS

Antennarius multiocellatus has been referred to as *A. principis* by nearly a dozen authors, beginning with Günther (1861) and ending with Fowler (1944) (see synonymy, p. 131). The latter name was originally established by Valenciennes (1837: 416) as *Chironectes*

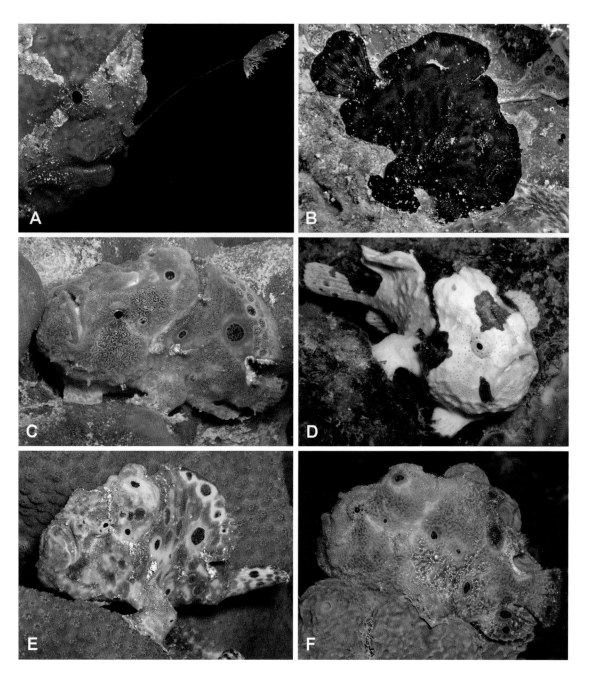

Fig. 94. *Antennarius multiocellatus* (Valenciennes). (**A–F**) Bonaire, Netherlands Antilles.
Photos by Ellen Muller.

principis based solely on two drawings made for Count Johan Maurits of Nassau-Siegen while Maurits was governor-general of northeastern Brazil between 1637 and 1644. These and other illustrations were examined by Valenciennes in Berlin by invitation of M. H. K. Lichtenstein sometime between 1826 and 1829 (Whitehead, 1979: 459). Part of a series of some 800 illustrations depicting Brazilian natural history now contained in the *Libri Picturati* (Handbooks, 1: 363, 2: 378) and preserved in the Biblioteka Jagiellońska, Kraków, Poland (see Whitehead, 1976, 1982; Simonini, 2018), these two drawings remain as the iconotypes of *A. principis* (both reproduced here; Fig. 96).

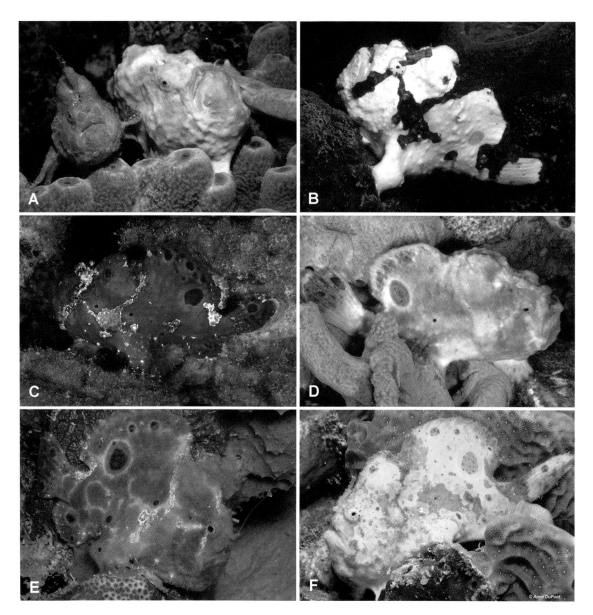

Fig. 95. *Antennarius multiocellatus* (Valenciennes). Bonaire, Netherlands Antilles. **(A–D)** Photos by Scott W. Michael. **(E)** Photo by Ellen Muller. **(F)** Photo by Anne DuPont.

Valenciennes's (1837: 416) brief description of *Chironectes principis* contains primarily characters of color and color pattern that provide little help in identifying the species. The only morphological feature of significance is his statement that "the first dorsal-fin spine is twice as long as the second." This fact, coupled with his mention of Prince Johan Maurits as the source of the drawings (and thus restricting the geographic distribution of the species to the Western Atlantic), makes it certain that Valenciennes was describing what has long been known as *A. multiocellatus*. The two drawings confirm this conclusion by showing an antennariid from Brazil with an illicium that is roughly twice the length of the second dorsal-fin spine (Fig. 96). But because Valenciennes (1837) did not publish the drawings, and because he left no type material to support his

Fig. 96. The earliest known depictions of *Antennarius multiocellatus* (Valenciennes). Watercolor copies of the iconotypes of *Chironectes principis* Valenciennes, made for Count Johan Maurits of Nassau-Siegen sometime between the years 1637 and 1644, based on material observed in the vicinity of Recife, Brazil (Valenciennes, 1837: 414–15); the original paintings, examined by Valenciennes in Berlin by invitation of Martin H. K. Lichtenstein, between 1826 and 1829 (Whitehead, 1979: 459; Pietsch, 1985a: 76), are now part of the *Libri Picturati* (Handbooks, 1: 363, 2: 378) housed in the Biblioteka Jagiellońska, Kraków, Poland (Whitehead, 1982). After Cuvier and Valenciennes, MS 504, XII.B, Bibliothèque Centrale, Muséum national d'Histoire naturelle, Paris; courtesy of Emmanuelle Choiseau, © Muséum national d'Histoire naturelle, Paris.

inadequate description, the name *A. principis* never came into common usage. Instead, the name *A. multiocellatus*, described four pages later in the same publication (Valenciennes, 1837: 420), and for which type material has survived, became the accepted name for the only Western Atlantic member of the *A. pictus* group. Although *A. principis* clearly has page priority over *A. multiocellatus*, the latter was selected by Pietsch and Grobecker (1987: 107) as the name that best ensures stability of nomenclature (see International Code of Zoological Nomenclature, Art. 24a).

Antennarius multiocellatus is often difficult to distinguish from the Eastern Atlantic *A. pardalis*. Both species differ from the Pacific members of the *A. pictus* group in having the membrane behind the second dorsal-fin spine much less developed (not extending posteriorly to the base of the third dorsal-fin spine), but little is available to separate the two. The recognition of the two as separate breeding populations is based on a statistically significant difference in the length of the second dorsal-fin spine (Fig. 91); the absence of large, darkly pigmented spots on the belly that are usually present in *A. pardalis* (especially in smaller specimens); and the lack of large, close-set wartlike swellings that cover the head and body of *A. pardalis* (especially evident in large specimens; see Fig. 99).

Antennarius pardalis (Valenciennes)
Leopard Frogfish

Figs. 75D, 91, 97–100, 281; Tables 5, 9, 22

Chironectes pardalis Valenciennes, 1837: 420, pl. 363 (original description, single specimen, holotype MNHN A.4597, Gorée, Senegal, West Africa). Duméril, 1861: 263 (Gorée). Pietsch, 1985a: 82, fig. 19, table 1 (original manuscript sources for *Histoire Naturelle des Poissons* of Cuvier and Valenciennes). Pietsch et al., 1986: 140 (type catalog).

Antennarius pardalis: Günther, 1861a: 198 (new combination; after Valenciennes, 1837). Capello, 1871: 202 (Guinea-Bissau, West Africa). Peters, 1877: 839 (St. Jago, Cape Verde Islands). Pereira-Guimarães, 1884: 21 (Cape Verde). Tremeau de Rochebrune, 1883: 91 (Gorée, Senegal). Osório, 1894: 183 (Guinea-Bissau). Osório, 1898: 198 (Cape Verde,

Guinea-Bissau). Osório, 1909: 67 (Sal Island, Cape Verde). Pellegrin, 1914: 85 (between Dakar and Cap de Nez). Metzelaar, 1919: 293 (in part; "both sides of Atlantic"). Monod, 1927: 737 (Cameroon). Fowler, 1936b: 1131, 1335, fig. 473 (in part). Cadenat, 1951: 291 (Senegal). Delais, 1951: 146, fig. 2 (Senegal). Cadenat, 1959: 361, 363, 367–373, figs. 3–9, 24, table (description, West Africa). Le Danois, 1964: 102, fig. 49g (Senegal). Azevedo, 1971: 91, figs. 1, 2a, 3, 4 (Angola). Pietsch, 1981a: 2 (eastern central Atlantic). Pietsch, 1984b: 36 (genera of frogfishes). Pietsch, 1985a: 82, fig. 19, table 1 (original manuscript sources for *Histoire Naturelle des Poissons* of Cuvier and Valenciennes). Pietsch and Grobecker, 1987: 107, figs. 22E, 23, 36, 37, 118, pl. 23 (description, distribution, relationships). Pietsch, 1990: 481 (tropical Eastern Atlantic). Paugy, 1992: 572 (West Africa). Afonso et al., 1999: 83 (Gulf of Guinea). Carnevale and Pietsch, 2006: 452, table 1 (comparison, meristics, first fossil frogfish). Wirtz et al., 2007: 7, 27, table 1 (Príncipe, Gulf of Guinea). Wirtz et al., 2013: 119 (Cape Verde). Hanel and John, 2015: 138, table 1 (Cape Verde). Camara et al., 2016: 204, appendix 3 (Guinean shelf). Pietsch, 2016: 2053 (eastern central Atlantic; in key).

Antennarius campylacanthus Bleeker, 1863: 28, pl. 4, fig. 3 (original description, single specimen, holotype RMNH 2098, Ashanti, Ghana).

Antennarius multiocellatus (not of Valenciennes): Buettikofer, 1890: 479 (Liberia). Fowler, 1936b: 1133 (in part; after Buettikofer, 1890). Poll, 1949: 253, fig. 23 (West Africa).

Antennarius vulgaris Osório, 1891: 120 [original description, authorship given to Cuvier and Valenciennes; Ilha de S. Thomé (= São Tomé), Gulf of Guinea; a probable synonym of *A. pardalis*]. Osório, 1898: 198 (after Osório, 1891). Fowler, 1936b: 1129 (possible synonym of *Histrio histrio*). Schultz, 1957: 52 (unidentifiable). Pietsch and Grobecker, 1987: 265 (*nomen nudum*). Afonso et al., 1999: 83 (probable synonym of *A. pardalis*). Wirtz et al., 2007: 27 (after Afonso et al., 1999).

Antennarius commersonii (not of Latreille) **Var. A. campylacanthus**: Steindachner, 1895: 23 (Liberia).

Antennarius pardalis (not of Valenciennes): Metzelaar, 1919: 161 (misidentification, material is *A. multiocellatus*; Bonaire and St. Martin). Poll, 1959: 368, fig. 125 (figure represents *A. striatus*; Congo).

Antennarius commersonii **var. *campylacanthus***: Metzelaar, 1919: 294 (after Steindachner, 1895; Liberia).

Antennarius pardal: Fowler, 1936b: 1131 (in synonymy, misspelling of specific name).

Antennarius commersonii (not of Latreille): Fowler, 1936b: 1132, fig. 474 (in part; description based on either *A. commerson* or *A. pictus* from Hawaii).

Antennarius ampylacanthus: Delais, 1951: 147, fig. 3 (misspelling of specific name; West Africa).

Antennarius **sp.**: Delais, 1951: 149, fig. 6 (Gorée, Senegal).

Lophiocharon (*Uniantennatus*) *campylacanthus*: Schultz, 1957: 85, pl. 7, fig. C (new combination; description after Bleeker, 1863).

Antennarius (*Antennarius*) *pardalis*: Schultz, 1957: 92, pl. 9, fig. D (description after Valenciennes, 1837; in key). Blache, 1962: 79 (coast of West Africa, Mauritania south to Cape Lopez, Gabon). Blache et al., 1970: 442, fig. 1118 (after Blache, 1962).

Uniantennatus tenebrosus (not of Poey): Le Danois, 1964: 109, fig. 55e (Congo).

Antennarius senegalensis (not of Cadenat): Debelius, 1997: 91, color fig. (misidentification, figure shows *Antennarius pardalis*; Cape Verde Islands).

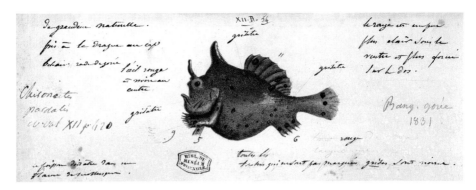

Fig. 97. *Antennarius pardalis* (Valenciennes). A watercolor drawing of the holotype (MNHN A.4597, 47 mm SL) collected off Gorée, Senegal, by French naturalist Paul Charles Rang des Adrets in 1831, found in the original manuscript sources for the *Histoire Naturelle des Poissons* of Cuvier and Valenciennes (Pietsch, 1985a). After Cuvier and Valenciennes, MS 504, XII.B, Bibliothèque Centrale, Muséum national d'Histoire naturelle, Paris; courtesy of Emmanuelle Choiseau, © Muséum national d'Histoire naturelle, Paris.

MATERIAL EXAMINED

Thirty-one specimens, 17.5 to 97 mm SL.

Holotype of *Antennarius pardalis*: MNHN A.4597, 47 mm, Gorée, Senegal, West Africa, Paul Rang (Figs. 97, 98A).

Holotype of *Antennarius campylacanthus*: RMNH 2098, 67 mm, Ashanti, Ghana, West Africa (Fig. 98B).

Additional material: BMNH, 2 (57–68 mm); MNHN, 19 (35.5–97 mm); RMNH, 1 (44.5 mm); USNM, 5 (51–88 mm); ZMUC, 2 (17.5–60.5 mm).

DIAGNOSIS

A member of the *Antennarius pictus* group unique in having the following combination of character states: second dorsal-fin spine slightly tapered from base (Fig. 75D); membrane behind second dorsal-fin spine thin (width along attachment to spine less than that of spine itself), covered with dermal spinules (Fig. 75D) except for extreme outer margin and a small triangular area just behind tip of spine; head and body covered with numerous, large, close-set wartlike swellings (especially evident in specimens greater than about 90 mm SL; Fig. 99); belly usually covered with large, dark brown to black, circular spots, sometimes interconnected by narrow, pigmented bands (Fig. 98); dorsal-fin rays 12 (rarely 11 or 13); anal-fin rays seven; pectoral-fin rays 10 (rarely 11) (Tables 5, 9).

Antennarius pardalis is distinguished from *A. multiocellatus* by a statistically significant difference in the length of the second dorsal-fin spine (Fig. 91); the tendency for *A. pardalis* to have large, darkly pigmented spots on the belly that are consistently absent in *A. multiocellatus*; and the presence of large, close-set wartlike swellings covering the head and body (especially evident in large specimens; Fig. 99) that are also consistently absent in *A. multiocellatus* (Figs. 93–95).

DESCRIPTION

Broad, laterally compressed appendage of esca (Fig. 100) greater than 50% illicium length in some specimens (USNM 199253, 62.5 mm SL); illicium 18.9 to 29.6% SL; second dorsal-fin spine slightly curved posteriorly, length 10.8 to 16.3% SL; shallow naked depression between second and third dorsal-fin spines present or absent; third dorsal-fin spine curved posteriorly, length 21.3 to 26.4% SL; eye diameter 4.6 to 5.8% SL.

Coloration in life: extremely variable; nearly white, lemon yellow, brown, reddish or greenish brown, bright red (Fig. 99), or almost black, usually with slightly darker mottling; scattered, roughly circular dark spots on fins and body, especially in younger specimens, those on belly smaller and more numerous.

Coloration in preservation: light beige, tan, yellow-brown, to dark brown, everywhere mottled with irregularly shaped dark patches; large dark brown to black circular spots (each often encircled by a lightly pigmented ring), particularly well developed on unpaired fins and belly; floor of throat often dark with bright-white blotches; illicium with faint banding in some specimens; as many as seven short, darkly pigmented bars radiating from eye.

Largest specimen examined: 97 mm SL (MNHN 1982-159).

DISTRIBUTION

Antennarius pardalis is restricted to the tropical Eastern Atlantic, ranging from off Senegal southward and eastward to the Congo, with records from Cape Verde, and São Tomé and Príncipe in the Gulf of Guinea (Wirtz et al., 2007; Fig. 281, Table 22). The holotype is from Gorée, Senegal.

Depths of capture, known for 10 specimens, mostly taken by trawl, range from 25 to 200 m, but 70% of the material was taken in less than 60 m; the average depth for all known capture is 70 m. Divers, however, have recorded observations ranging from near the surface to about 25 m.

COMMENTS

Antennarius pardalis is difficult to distinguish from *A. multiocellatus*. Both species differ from the Pacific members of the *A. pictus* group in having the membrane of the second dorsal-fin spine much less

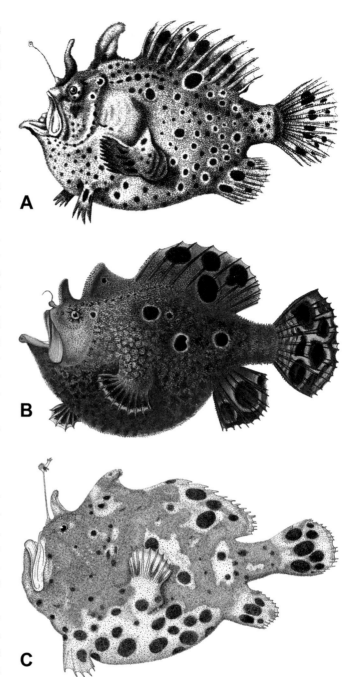

Fig. 98. *Antennarius pardalis* (Valenciennes). (**A**) holotype, MNHN A.4597, 47 mm SL, Gorée, Senegal (after Valenciennes, 1837, pl. 363). (**B**) holotype of *Antennarius campylacanthus* Bleeker, RMNH 2098, 67 mm SL, Ashanti, Ghana, West Africa; note the damaged illicium. Drawing by Ludwig Speigler under the direction of Pieter Bleeker; courtesy of Godard Tweehuysen and the Naturalis Bibliotheek, Naturalis Biodiversity Center, Amsterdam; after Bleeker, 1863, pl. 4, fig. 3). (**C**) USNM 199253, 62.5 mm SL, off Bassagos Island, West Africa. Drawing by Cathy L. Short; after Pietsch and Grobecker, 1987, fig. 36.

Fig. 99. *Antennarius pardalis* (Valenciennes). **(A)** Cape Verde Islands. Photo by John Neuschwander; courtesy of Helmut Debelius. **(B)** Santo Antão, Cape Verde Islands. Photo by Susana Martins. **(C)** Enseada d'Coral, St. Vincent Island, Cape Verde Islands. Photo by Guilherme Mascarenhas. **(D)** Mergulho, Ilha do Sal, Cape Verde Islands. Photo by Susana Martins. **(E)** a male with gravid female, Tarrafal, Santiago Island, Cape Verde Islands. Photo by Georg Bachschmid. **(F)** São Vicente, Cape Verde Islands. Photo by Peter Wirtz.

developed (not extending posteriorly to the base of the third dorsal-fin spine), but very little is available to differentiate the two. The recognition of each as a separate breeding population is based on a statistically significant difference in the length of the second dorsal-fin spine (Fig. 91), the tendency for *A. pardalis* to have large darkly pigmented spots on the belly, a feature consistently absent in *A. multiocellatus* (Figs. 94, 99); and the presence of large, close-set wartlike swellings covering the head and body (especially evident in larger specimens) that are also consistently absent in *A. multiocellatus* (compare Figs. 94 and 99).

Antennarius pictus (Shaw)
Painted Frogfish

Figs. 9C, 11A, 12B, 26B, 75E, 101–105, 281, 301, 306C, 322G, 337; Tables 5, 9, 22, 23

Antennarius chironectes, obscure rubens, maculis nigris raris inspersus: Commerson, MSS 889(1): 13A, 891: 144 [observed by Commerson at Mauritius, about 1770; no extant material (see Comments, p. 153)]. Lacepède, 1798: 325 (footnote reference to Commerson, MSS).

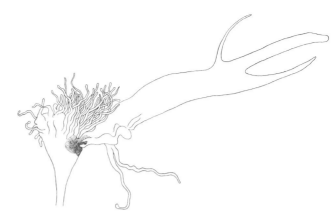

Fig. 100. Esca of *Antennarius pardalis* (Valenciennes). USNM 199253, 62.5 mm SL, off Bassagos Island, West Africa. Drawing by Cathy L. Short; after Pietsch and Grobecker (1987, fig. 37).

Lophius pictus Shaw, 1794, pl. 176, upper fig. ["The Variegated Lophius," original description, "found about the coast of New Holland and the neighbouring isles," no known type material, neotype designated by Pietsch and Grobecker (1987: 87) BPBM 15441, 105 mm, Tahiti; predates *Chironectes pictus* Valenciennes, 1837 (= *Histrio histrio*)]. Shaw, 1804: 386, pl. 165, upper fig. ("Painted Angler," after Shaw, 1794; no specimens, "observed about Otaheitee, New-Holland, etc."). Good and Gregory, 1813, unpaged, pl. 153, upper fig. (after Shaw, 1794). Swain, 1883: 308 (same as *A. multiocellatus* var. *leucosoma* of Günther, 1861a). Whitley, 1931: 328 (confused with *L. histrio* Linnaeus). Schultz, 1957: 104 (in synonymy of *Histrio histrio*).

La Lophie Chironecte: Lacepède, 1798: 302, 325, pl. 14, fig. 2 [invalid, not Latinized; description and Latin polynomial in footnote based on Commerson, MSS; *voisine des côtes orientales de l'Afrique* (= Mauritius)]. Sonnini, 1803: 160, 192 (after Lacepède, 1798).

Lophius histrio (not of Linnaeus) **var. b.** *pictus*: Bloch and Schneider, 1801: 141 (after Shaw, 1794). Gill, 1879c: 225, 226 ("based on species of typical *Antennarius*," not *Histrio*). Schultz, 1957: 104 (in synonymy of *Histrio histrio*).

Lophius chironectes Latreille, 1804: 73 (binomial for *La Lophie Chironecte* of Lacepède, 1798; constitutes original description, no known type material). Cuvier, 1816: 310 (binomial for *La Lophie Chironecte* of Lacepède, 1798).

Lophius variegatus (not of Rafinesque) Cuvier, 1816: 310 [new name for "Variegated Lophius," i.e., *L. pictus* Shaw, 1794; specimens unknown; not same as *Chironectes variegatus* Rafinesque, 1814 (= *Histrio histrio*); see synonymy, p. 181]. Cuvier, 1817b: 433 (possible synonym of *A. chironectes* of Commerson, MSS).

Chironectes verus Cloquet, 1817: 599 (binomial for polynomial of Commerson, MSS, and French vernacular of Lacepède, 1798; constitutes original description, no known type material; includes *L. variegatus* with authorship attributed to Shaw).

Antennarius chironectes: Schinz, 1822: 501 (new combination; after Lacepède, 1798). Bleeker, 1854a: 104 (additional specimen, Banda Neira). Bleeker, 1859b: 128 (Ambon, Banda). Günther, 1861a: 197 (after Bleeker, 1854a). Bleeker, 1865b: 9, 13, pl. 200, fig. 3 (after Bleeker, 1854a, 1859b; RMNH 6283). Schultz, 1957: 93, pl. 10, figs. A, B (in part; synonymy, description; Hawaiian Islands, Samoa; in key). Beaufort and Briggs, 1962: 209 (in part; synonymy, description; Natal, Zanzibar, Seychelles, India, Ceylon, Andamans, Cocos-Keeling, Indonesia, Philippines, Australia, Pacific islands to Hawaii; in key). Smith and Smith, 1963: 61, pl. 56, fig. H (Seychelles). Le Danois, 1964: 99, fig. 47C (in part, MNHN A.4621 is *A. commerson*). Burgess and Axelrod, 1973a: 529, color fig. 488 (plate represents *A. pictus* with illicium lost, esca regenerated; BPBM 5805). Tinker, 1978: 510, fig., color fig. 31 (figs. 167, 167-5) (Hawaii to Philippines and East Indies). Jones and Kumaran, 1980: 705, fig. 602 (Laccadive Archipelago). Randall et al., 1993a: 223, 227, table 1 (synonym of *A. pictus*, Hawaiian Islands). Randall, 1995: 84 (synonym of *A. pictus*, Oman).

Chironectes pictus: Cuvier, 1829: 252 (new combination; var. b. of Bloch and Schneider, 1801, elevated to rank of species-group).

Chironectes variegatus (not of Rafinesque): Cuvier, 1829: 252 (new combination; same as *La Lophie Chironecte* of Lacepède, 1798, and *L. pictus* Shaw, 1794). Valenciennes, 1837: 422 [description based at least in part on MNHN A.4813, locality unknown (specimen is not holotype of *C. variegatus* as stated by Le Danois, 1964: 100); not same as *C. variegatus* Rafinesque, 1814 (= *Histrio histrio*); see synonymy, p. 181]. Valenciennes, 1842: 189 (after Valenciennes, 1837). Jordan, 1917a: 278 ("not the same as *C. variegatus* Rafinesque, 1814"). Pietsch, 1985a: 84, figs. 21–23, table 1 (original manuscript sources for *Histoire Naturelle des Poissons* of Cuvier and Valenciennes).

Chironectes pictus (not of Shaw): Valenciennes, 1837: 393, pl. 364 (material, description, and figure represent *Histrio histrio*; MNHN A.4603, Surinam). Pietsch, 1985a: 84, figs. 21–23, table 1 (original manuscript sources for *Histoire Naturelle des Poissons* of Cuvier and Valenciennes).

Chironectus pictus (not of Valenciennes): Swainson, 1839: 330 (after Shaw, 1804; emendation of generic name).

Lophius sandvicensis Bennett, 1840: 258, fig. (original description, single specimen, existence of holotype unknown, not found in BMNH, 115 mm total length, Oahu).

Cheironectes pictus (not of Shaw) **var. vittatus**: Richardson, 1844: 15, pl. 9, figs. 3, 4 (description and figure represent *Histrio histrio*).

Chironectes leprosus Eydoux and Souleyet, 1850: 187, pl. 5, fig. 3 (original description, single specimen, holotype MNHN A.4595, Hawaii; dating after Bauchot et al., 1982). Montrouzier, 1856: 467 (Woodlark Island, Solomon Sea, off SE coast of New Guinea). Bauchot et al., 1982: 64, 72 (dating of Eydoux and Souleyet). Pietsch et al., 1986: 139 (in type catalog).

Antennarius polyophthalmus Bleeker, 1852b: 644 (original description, single specimen, holotype RMNH 6284, Banda Neira). Bleeker, 1859b: 129 (Banda, Goram). Günther, 1861a: 198 (after Bleeker, 1852b). Bleeker, 1865b: 9, 12, pl. 197, fig. 4 (description after Bleeker, 1852b, 1859b; RMNH 28014). Smith, 1949: 431, pl. 98, fig. 1242 (South Africa).

Antennarius horridus Bleeker, 1853a: 69, 70, 83 (original description, single specimen, holotype RMNH 6276, Solor). Bleeker, 1859b: 128 (Celebes, Flores, Solor, Ambon). Günther, 1861a: 193 (after Bleeker, 1853a). Bleeker, 1865b: 9, 12, pl. 194, fig. 1 (after Bleeker, 1853a, 1859b; RMNH 28012). Jatzow and Lens, 1898: 511, pl. 35, fig. 7 (Zanzibar).

Antennarius leucosoma Bleeker, 1854b: 328 (original description, single specimen, holotype RMNH 6278, Larantuka, Flores). Bleeker, 1859b: 129 (Flores, Ambon). Bleeker, 1865b: 9, 19, pl. 199, fig. 2 (description, additional specimen, Ambon). Schultz, 1957: 92, pl. 9, fig. B (description after Bleeker, 1854b; in key). Beaufort and Briggs, 1962: 214 (synonymy, description; Flores, Ambon, Seychelles).

Chironectes peravok Montrouzier, 1856: 467 (original description, single specimen, whereabouts of holotype unknown, about 75 mm total length, Woodlark Island, Solomon Sea; perhaps the same as *C. leprosus* Eydoux and Souleyet). Schultz, 1957: 93 (in synonymy of *A. chironectes*).

Antennarius marmoratus (not of Shaw) **Var. α. *picta*** (not of Valenciennes): Günther, 1861a: 186 (in part, i.e., *L. histrio* var. b. *pictus* Bloch and Schneider, 1801).

Antennarius multiocellatus (not of Valenciennes) **Var. γ. *leucosoma***: Günther, 1861a: 195 (BMNH 1858.4.21.156, Ambon). Swain, 1883: 308 (listed).

Antennarius multiocellatus (not of Valenciennes) **Var. δ. *leprosa***: Günther, 1861a: 195 (Hawaii).

Chironectes niger Garrett, 1864: 107 [original description, single specimen, holotype lost in San Francisco earthquake, 1906 (William N. Eschmeyer, personal communication, 5 December 1980), 89 mm total length, Hawaiian Islands].

Antennarius commersonii (not of Latreille): Bleeker, 1865b: 10, 20, pl. 197, fig. 3 (RMNH 6275; Ambon, Ceram). Günther, 1876: 163 [in part, i.e., pls. 100 (fig. C), 102, 103 (fig. B), 104, 105 (fig. A), 106 (fig. B)]. Jenkins, 1904: 511 (SU 23259, Hawaii). Jordan and Evermann, 1905: 518 (synonymy includes *A. niger* Garrett). Fowler, 1928: 477, pl. 49B (description, plate represents *A. pictus*). Fowler, 1931: 367 (Oceania). Munro, 1955: 288, fig. 838 (Ceylon). Gosline and Brock, 1960: 305, 345, fig. 23 (Hawaii; in key). Schultz, 1964: 175, table 1 (additional material, Hawaiian Islands and Philippines). Le Danois, 1970: 89 (Gulf of Aqaba).

Antennarius multiocellatus (not of Valenciennes): Playfair, 1867: 862 (BMNH 1867.8.16.76, Seychelles). Whitmee, 1875: 545 (aquarium observations, Samoa). Poey y Aloy, 1876: 135 (in part). Meyer, 1885: 28 (Celebes).

Antennarius nigromaculatus Playfair, 1869: 239 (original description, single specimen, holotype BMNH 1869.1.29.29, Zanzibar).

Pterophryne picta (not of Valenciennes): Goode, 1876: 20 (new combination; in part, synonymy includes *L. histrio* var. b. *pictus* Bloch and Schneider, 1801).

Antennarius pictus (not of Valenciennes): Gill, 1879c: 226 (new combination, "a senior synonym of *A. phymatodes* Bleeker or *A. commersoni* Günther," resurrected *sensu* Shaw, 1794). Whitley, 1957b: 69 ("a species of *Antennarius*," New South Wales). Whitley, 1968b: 39, fig. 3 (AMS IB.548, Ballina, New South Wales). Myers and Shepard, 1980: 338 (Guam). Bauchot et al., 1982: 72 (same as *Chironectes leprosus* Eydoux and Souleyet). Pietsch, 1984b: 34, fig. 1A (genera of frogfishes). Pietsch, 1984e: 2 (Western Indian Ocean). Pietsch, 1985a: 84, figs. 21–23, table 1 (original manuscript sources for *Histoire Naturelle des Poissons* of Cuvier and Valenciennes). Pietsch, 1986a: 366, 368, pl. 13 (South Africa). Randall, 1986: 180 (Marshall Islands). Pietsch and Grobecker, 1987: 79, figs. 16B, 22A, 24–26, 118, 141, pls. 2B, 7–10 (description, distribution, relationships). Allen and Swainston, 1988: 38 (Western Australia). Edwards, 1989: 249, fig. 16 (tetrapod locomotion). Myers, 1989: 68, pl. 13E (East Africa to Hawaiian Islands, Marianas and Marshalls; in key). Paxton et al., 1989: 278 (Australia). Pietsch and Grobecker, 1990a: 96, 99, figs. (aggressive mimicry). Pietsch and Grobecker, 1990b: 74, 77, figs. (after Pietsch and Grobecker, 1990a). Randall et al., 1990: 55, color fig. (East Africa to Hawaiian and Society Islands). Francis, 1993: 158 (Lord Howwe Island). Kuiter, 1993: 47 (southeastern Australia). Randall et al., 1993a: 223, 227, table 1 (Hawaiian Islands). Randall et al., 1993b: 365 (Midway Atoll). Goren and Dor, 1994: 14 (Red Sea). Randall, 1995: 82, 84, figs. 165, 169 (Oman). Randall, 1996: 45, color fig. (Hawaiian Islands). Allen, 1997: 62 (tropical Australia). Kuiter, 1997: 44 (Australia). Lindberg et al., 1997: 217 (Sea of Japan). Okamura and Amaoka, 1997: 141, figs. (Japan). Randall et al., 1997a: 55, color fig. (East Africa to Hawaiian and Society Islands). Anderson et al., 1998: 22 (Maldive Islands). Bavendam, 1998: 42, 43, color figs. (natural history). Michael, 1998: 329, 332, 349–352, color figs. (identification, behavior, captive care). Fricke, 1999: 106 (Réunion, Mauritius). Johnson, 1999: 724 (Moreton Bay, Queensland). Myers, 1999: 69, pl. 13E (after Myers, 1989). Pietsch, 1999: 2015 (western central Pacific; in key). Allen, 2000a: 62, pl. 12-13 (Indo-West Pacific). Nakabo, 2000: 457 (Japan; in key). Pereira, 2000: 5 (Mozambique). Pietsch, 2000: 597 (South China Sea). Schleichert, 2000: 34, 35, 37, color figs. (natural history). Almeida et al., 2001: 114, table 3 (Inhaca Island, Mozambique). Hutchins, 2001: 22 (Western Australia). Nakabo, 2002: 457 (Japan; in key). Randall et al., 2002: 155 (Society Islands, Tuamotu Archipelago). Allen and Adrim, 2003: 25 (Bali to Papua). Allen et al., 2003: 364, color figs. (tropical Pacific). Manilo and Bogorodsky, 2003: S99 (Arabian Sea). Myers and Donaldson, 2003: 610, 614 (Mariana Islands). Schneidewind, 2003a: 84, fig. (natural history). Letourneur et al., 2004: 209 (Réunion Island). Lieske and Myers, 2004: 45, 46, color fig. (South Africa and Red Sea to Hawaii, southern Japan to Great Barrier Reef). Randall et al., 2004: 8 (Tonga). Mundy, 2005: 264 (Hawaiian Islands). Randall, 2005a: 72, color figs. (South Pacific). Randall, 2005b: 305 (mimicry). Schneidewind, 2005c: 88, fig. (tiny postlarvae). Steene, 2005: 264, 327, color pls. (Anilao, Lembeh Strait). Allen et al., 2006: 640 (Australia). Carnevale and Pietsch, 2006: 452, table 1 (comparison, meristics, first fossil frogfish). Michael, 2006: 35, 36, 39, 41, color figs. (spawning behavior, "yawning"). Schneidewind, 2006a: 18, 24, color figs. (luring behavior). Schneidewind, 2006b: 27, 30, color figs. (mimicry). Senou et al., 2006: 423 (Sagami Sea). Kuiter and Debelius, 2007: 119, color figs. (South Africa, Indonesia, Japan). Lugendo et al., 2007: 1222, appendix 1 (Zanzibar). Randall, 2007: 129, color fig. (Hawaiian Islands; in key). Kwang-Tsao et al., 2008: 243 (southern Taiwan). Allen and Erdmann, 2009: 594 (West Papua). Fricke et al., 2009: 28 (Réunion). Al-Jufaili et al., 2010: 18, appendix 1 (Oman). Arnold, 2010a: 68 (Bare Island, Botany Bay). Golani and Bogorodsky, 2010: 16 (Red Sea). Motomura et al., 2010: 78 (Yaku-shima Island, southern Japan). Randall, 2010: 37 (Hawaii). Fricke et al., 2011: 368 (New Caledonia). Satapoomin, 2011: 51, appendix A (Andaman Sea). Allen and Erdmann, 2012: 148, 149, color figs. (East Indies). Arnold and Pietsch, 2012: 118, 120, 124, 127, 128, figs. 1F, 4, 6, table 1 (molecular phylogeny). Larson et al., 2013: 54 (Northern Territory). Nakabo, 2013: 540, figs. (Japan). Brandl and Bellwood, 2014: 32 (pair-formation). Fricke et al., 2014: 35 (Papua New Guinea). Hylleberg and Aungtonya, 2014: 101 (Phuket, Thailand). Moore et al., 2014: 177 (northwestern Australia). Arai, 2015: 94, table 2 (Malaysia). Townsend, 2015: 4 (Sabah, Malaysia).

Antennarius argus Fowler, 1903: 172, pl. 8 (original description, single specimen, holotype ANSP 24208, Zanzibar).

Antennarius laysanius Jordan and Snyder, 1904: 947 (original description, single specimen, holotype SU 8439, Laysan Island, Hawaiian Islands). Jordan and Evermann, 1905: 520, pl. 63 (after Jordan and Snyder, 1904). Böhlke, 1953: 148 (holotype in CAS).

Antennarius sandvicensis: Jordan and Snyder, 1904: 948 (new combination, Honolulu). Jordan and Evermann, 1905: 518 (synonymy includes *A. horridus*, Honolulu). Fowler, 1928: 478 (Honolulu).

Antennarius leprosus: Jordan and Evermann, 1905: 519, fig. 228 (new combination, description, synonymy includes *A. rubrofuscus* Garrett, Honolulu). Fowler, 1927: 32 (Pearl Harbor). Fowler, 1928: 478 (Society Islands). Munro, 1955: 288, fig. 839 (Ceylon). Munro, 1958: 296 (synonymy, New Guinea). Jones and Kumaran, 1980: 704, fig. 601 (Laccadive Archipelago).

Antennarius commersoni (not of Latreille) **var. *nigromaculatus***: Steindachner, 1906: 1413 (after Günther, 1876, pl. 102, fig. A; Samoa).

Antennarius hispidus (not of Bloch and Schneider): Fowler, 1928: 477 (in part).

Antennarius phymatodes (not of Bleeker): Fowler, 1928: 478 (Society Islands, MCZ 11626). Whitley, 1959: 321 (Heron Island, Queensland). Myers and Donaldson, 2003: 610 (synonym of *A. pictus*). Fricke et al., 2011: 368 (synonym of *A. pictus*).

Pterophrynoides histrio (not of Linnaeus) **var. *pictus***: Whitley, 1931: 328 (different from *L. histrio* Linnaeus; New South Wales).

Antennarius oligospilos (not of Bleeker): Smith, 1949: 431, pl. 98, fig. 1241 (South Africa). Herald, 1961: 283, fig. (tropical Indo-Pacific). Smith and Smith, 1963: 61, pl. 56, fig. G (Seychelles).

Lophiocharon (Uniantennatus) horridus: Schultz, 1957: 84, pl. 7, fig. B (new combination, synonymy, Mauritius).

Uniantennatus leprosus: Le Danois, 1964: 108, fig. 55c (new combination, description, Hawaii).

Uniantennatus horridus: Le Danois, 1964: 109, fig. 55d (in part, i.e., MNHN A.4581, Trincomalee).

Phymatophryne maculata (not of Desjardins): Le Danois, 1964: 116 (new combination; in part, i.e., MNHN 2107, Madagascar).

Phrynelox polyophthalmus (not of Bleeker): Le Danois, 1964: 118, fig. 61k (new combination; Hawaii, Réunion).

Antennarius chironemus Munro, 1967: 584, pl. 78, fig. 1092 [origin of specific name unknown, authorship given to Lacepède, 1802 (= 1798) in reference to *La Lophie Chironecte*; synonymy includes *A. chironectes* (Lacepède), *A. commersoni* Bleeker, *A. laysanius* Jordan and Evermann; figure represents *A. pictus* after Jordan and Evermann, 1905; New Guinea]. Grant, 1972: 366, fig. (after Munro, 1967; Queensland coast). Friese, 1973: 32, 33 [after Munro, 1967; supposed hybridization with *Triantennatus zebrinus* (=*A. striatus*)]. Friese, 1974: 8, 10 (after Munro, 1967; Friese, 1973). Kailola, 1975: 47 (Gulf of Papua). Grant, 1978: 601 (in part, color plate 263 represents *A. striatus*). Kailola and Wilson, 1978: 26 (after Munro, 1967; Gulf of Papua).

Antennarius sp.: Burgess and Axelrod, 1973a: 528, color fig. 487 (plate represents *A. pictus*).

Antennarius spilopterus: Friese, 1974: 8, fig. (specific name taken "from some remote source in the aquarium literature," U. E. Friese, personal communication, 10 October 1979; figure shows *A. pictus*).

Lophiocharon horridus: Araga, 1984: 103, pl. 346-E (Japan).

MATERIAL EXAMINED

One hundred and fifty-two specimens, 12 to 160 mm SL.

Neotype of *Lophius pictus*: BPBM 15441, 105 mm, Tahiti, Society Islands, 19 March 1964.

Holotype of *Chironectes leprosus*: MNHN A.4595, 84 mm, *Voyage de la Bonite*, Hawaiian Islands, Eydoux and Souleyet, 1837 (Fig. 101).

Holotype of *Antennarius horridus*: RMNH 6276, 95.5 mm, Lawajong, Solor, Bleeker (Fig. 102A).

Holotype of *Antennarius leucosoma*: RMNH 6278, 44 mm, Larantuka, Flores, Bleeker (Fig. 102B).

Holotype of *Antennarius polyophthalmus*: RMNH 6284, 73 mm, Banda Neira Island, Indonesia, Bleeker (Fig. 102C).

Holotype of *Antennarius nigromaculatus*: BMNH 1869.1.29.29, 69 mm, Zanzibar, Kirk.

Holotype of *Antennarius argus*: ANSP 24208, 63.5 mm, Zanzibar Island, Eliot.

Fig. 101. *Antennarius pictus* (Shaw). Holotype of *Chironectes leprosus* Eydoux and Souleyet, MNHN A.4595, 84 mm SL, Hawaiian Islands. After Eydoux and Souleyet (1850, pl. 5, fig. 3).

Holotype of *Antennarius laysanius*: SU 8439, 68.5 mm, Laysan Island, Hawaiian Islands, Schlemmer.

Additional material: *Western Indian Ocean*: BMNH, 3 (50–82 mm); MNHN, 2 (58.5–109 mm); SAIAB, 10 (22.5–117 mm); UW, 1 (110 mm). *Red Sea*: TAU, 1 (95 mm). *Sri Lanka*: MNHN, 1 (104 mm). *Cocos-Keeling, Indonesia, and New Guinea*: BMNH, 2 (50–60 mm); KFRL, 1 (23 mm); RMNH, 12 (12–91 mm); USNM, 1 (32.5 mm); UW, 2 (71–96.5 mm). *Philippines*: MCZ, 1 (67 mm); SU, 1 (63 mm); UW, 7 (12.5–85 mm). *Australia and New Caledonia*: AMS, 10 (17–158 mm); UW, 1 (71 mm); WAM, 2 (87.5–91 mm). *Guam, Saipan, Marshalls, Tonga, and Samoa*: BMNH, 3 (92–102 mm); CAS, 1 (59 mm); UG, 3 (58–91 mm); USNM, 1 (107 mm); ZMUC, 1 (82 mm). *Midway and Hawaiian Islands*: AMNH, 10 (55–135 mm); BPBM, 28 (25.5–133 mm); CAS, 7 (78–103 mm); FMNH, 2 (24–78 mm); MCZ, 2 (48–98 mm); MNHN, 5 (62–107 mm); SU, 5 (51–98 mm); USNM 6 (32–108 mm); UW, 1 (99 mm). *Society Islands*: ANSP, 1 (112 mm); MCZ, 2 (108–126 mm); USNM, 1 (41 mm). *Locality unknown*: BMNH, 1 (44 mm); MNHN, 1 (125 mm); UW, 5 (67–155 mm).

DIAGNOSIS

A member of the *Antennarius pictus* group unique in having the following combination of character states: second dorsal-fin spine slightly tapered from base (Fig. 75E); membrane behind second dorsal-fin spine thin (width along attachment to spine less than that of spine itself), covered with dermal spinules (except along extreme outer margin and in small triangular area just behind tip of spine) and extending across area between second and third spines reaching to base of third; head and body sometimes partially or fully covered with low, rounded, wartlike swellings; belly without large, darkly pigmented spots; dorsal-fin rays 12 (rarely 11 or 13); anal-fin rays seven (rarely six); pectoral-fin rays 10 (rarely nine or 11) (Tables 5, 9).

DESCRIPTION

Broad, laterally compressed appendage of esca (Fig. 103) as long as 70% of illicium (WAM P.19215, 87.5 mm SL); illicium length 19.1 to 32.0% SL; second dorsal-fin spine curved posteriorly, length 8.5 to 17.3% SL; shallow, naked depression between second and third dorsal-fin spines present or absent; third dorsal-fin spine curved posteriorly, length 19.5 to 28.9% SL; eye diameter 3.8 to 7.2% SL.

Coloration in life: extremely variable; off-white, tan, yellow, orange, bright red, brick-red, yellowish-green, dark green, brown, grey, or black, usually with numerous round spots or ocelli of variable size, and often with three dark spots on caudal fin; dark blue to black with lemon-yellow or orange spots scattered over head, body, and fins; light blue band across distal margin of unpaired fins; illicium and tips of paired fins often white (Figs. 104, 105). Yellow-green with black spots and irregularly shaped patches of pink. Orange with red spots. A deep rust-red with whitish pectorals; head, body, and fins with numerous, scattered, dark brown to black circular spots of various sizes, and mottled with irregularly shaped pink and white patches. Black and white dendritic pattern in mouth (BPBM 5805, 10029).

Coloration in preservation: light tan, yellow-brown, dark brown to black; illicium without banding; as many as 12 short, darkly pigmented bars radiating from eye. Lighter-color phases with numerous, scattered, dark circular spots of various sizes over entire head, body, and fins; light-colored saddles on caudal peduncle and shoulder, and scattered irregularly shaped patches, particularly on nape of neck, cheek, and pectoral regions. Darker-color phases with tips of pectoral-fin rays white; solid black individuals with no conspicuous markings.

Largest specimen examined: 160 mm SL (BPBM 12219).

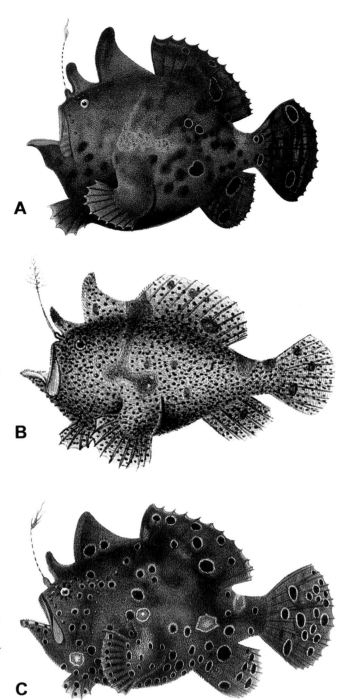

Fig. 102. *Antennarius pictus* (Shaw). (**A**) holotype of *Antennarius horridus* Bleeker, RMNH 6276, 98.5 mm SL, Lawajong, Solor, Indonesia (after Bleeker, 1865b, pl. 194, fig. 1). (**B**) holotype of *Antennarius leucosoma* Bleeker, RMNH 6278, 44 mm SL, Larantuka, Flores, Indonesia (after Bleeker, 1865b, pl. 199, fig. 2). (**C**) holotype of *Antennarius polyophthalmus* Bleeker, RMNH 6284, 73 mm SL, Banda Neira, Indonesia (after Bleeker, 1865b, pl. 197, fig. 4). Drawings by Ludwig Speigler under the direction of Pieter Bleeker.

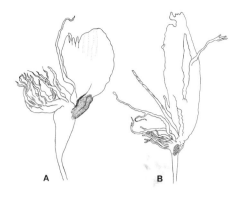

Fig. 103. Escae of *Antennarius pictus* (Shaw). (**A**) BPBM 22602, 40 mm SL, Kaneohe Bay, Oahu, Hawaiian Islands. (**B**) BPBM 13324, 80 mm SL, Coconut Island Reef, Oahu, Hawaiian Islands. Drawings by Cathy L. Short; after Pietsch and Grobecker (1987, fig. 25).

Fig. 104. *Antennarius pictus* (Shaw). (**A**) Lembeh Strait, Sulawesi, Indonesia. Photo by Christa Holdt. (**B**) Milne Bay, Papua New Guinea. Photo by Roger Steene. **C**) Lembeh Strait, Sulawesi, Indonesia. Photos by Scott W. Michael. (**D**) Lembeh Strait, Sulawesi, Indonesia. Photo by David J. Hall. (**E**) Mactan, Cebu. Photo by Juuyoh Tanaka; courtesy of Wikimedia Commons. (**F**) Lembeh Strait, Sulawesi, Indonesia. Photo by Roger Steene.

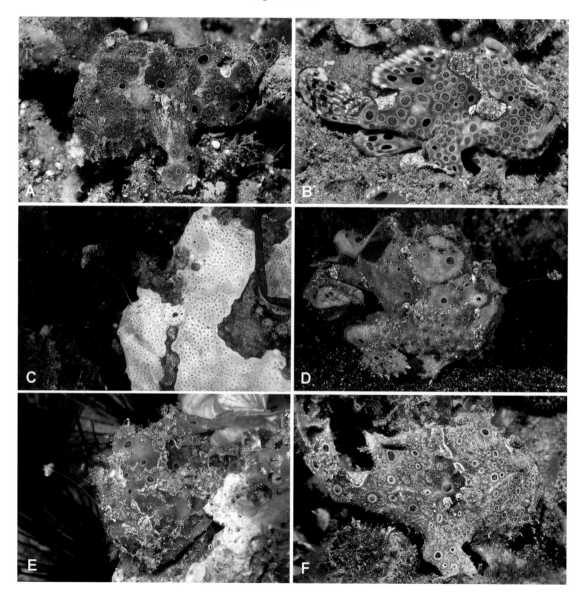

DISTRIBUTION

Antennarius pictus is widespread throughout the Indo-Pacific from South Africa, Mozambique, Zanzibar, the Red Sea, and Oman to the Hawaiian and Society Islands, including Réunion and Mauritius, Japan, Taiwan, Thailand, the Philippines and Indonesia (Lesser Sunda Islands, Sulawesi, Moluccas, Halmahera, West Papua), Australia (Western Australia, Queensland, New South Wales), New Caledonia, Guam, the Marshalls and Marianas, Samoa, and Tonga (Fig. 281, Table 22). The type locality is Tahiti (see Comments, p. 153).

Although often observed near the surface, this species extends to depths of at least 75 m. However, 92% of the material was taken in less than 40 m and 72% in less than 20 m. The average depth for all known captures (25 records) is 14 m.

COMMENTS

Antennarius pictus was first described and illustrated by Philibert Commerson [MSS 889(1): 13A, 891: 144] based on material observed at Mauritius sometime after his arrival there in mid-November 1768 (see Introduction, p. 9). Although what Commerson wrote concerning this spe-

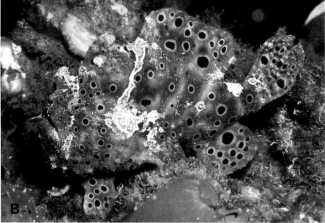

Fig. 105. *Antennarius pictus* (Shaw). (**A**) Anilao, Luzon, Philippines. Photo by Roger Steene. (**B**) Secret Bay, Gilimanuk, North Bali, Indonesia. Photo by Mark V. Erdmann.

cies consists of only a brief descriptive phrase—*Antennarius (Chironectes) obscure rubens maculis nigris nesogallibus raris inspersus. nob.*—the illustration, probably made for him by Pierre Sonnerat, one of his draftsmen (Collection des Vélins du Muséum national d'Histoire naturelle, Paris, 91: 62), clearly represents this species (Fig. 9C). Lacepède (1798: 325, pl. 14), in utilizing Commerson's unpublished work (but without the benefit of a specimen), added further description of what he called *La Lophie Chironecte* and produced an engraving based solely on Commerson's drawing (Fig. 11A), but he failed to provide a Latinized binomial. In the meantime, Shaw (1794, pl. 176, upper figure), along with a copper engraving by Frederick Polydore Nodder (Fig. 12B), provided the first valid description of this species based on material said to inhabit "the coast of New Holland and the neighboring isles." It is not clear whether Shaw actually had a specimen of *A. pictus* in hand, or whether he founded his description solely on the original drawing that was the basis for Nodder's engraving. In any case, a thorough search of the collections of the Natural History Museum, London, failed to produce any original material. A neotype (BPBM 15441, 105 mm, Tahiti) was designated by Pietsch and Grobecker (1987: 87).

The original drawing from which the engraving of Shaw's *A. pictus* was made (reproduced here, Fig. 12B) is bound in with the drawings made by Thomas Watling (p. 95,

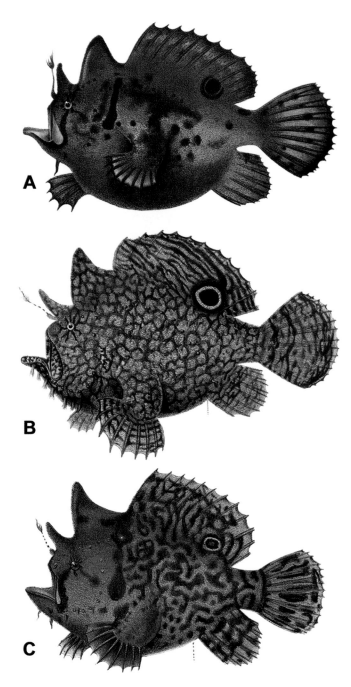

Fig. 106. *Antennarius biocellatus* (Cuvier). (**A**) RMNH 6281, 111 mm SL, Ambon, Indonesia (after Bleeker, 1865b, pl. 194, fig. 3). (**B**) holotype of *Antennarius notophthalmus* Bleeker, RMNH 6280, 95 mm SL, Djungku-lon, West Java, Indonesia (after Bleeker, 1865b, pl. 196, fig. 1). (**C**) RMNH 6281, 70 mm SL, Ambon, Indonesia (after Bleeker, 1865b, pl. 198, fig. 5). Drawings by Ludwig Speigler under the direction of Pieter Bleeker.

no. 318) between 1788 and 1792 (Iredale, 1958; Hindwood, 1970; Gruber, 1983: 13, footnote 13) and now in the Zoological Library of the Natural History Museum, London (for further discussion, see Comments, p. 110).

Antennarius biocellatus Group Pietsch

DIAGNOSIS

A species-group of *Antennarius* distinguished by having the following combination of character states: a darkly pigmented basidorsal ocellus encircled by a lightly pigmented ring (absent in only one of 48 specimens examined; Fig. 106); illicium 10.7 to 16.7% SL, usually slightly shorter than second dorsal-fin spine (equal to or slightly longer in some specimens); anterior end of pterygiophore of illicium terminating distinctly posterior to symphysis of upper jaw; esca a conical, longitudinally folded appendage with numerous, slender filaments arising from base (Fig. 107); second dorsal-fin spine free, not connected to head by membrane, nearly always straight and tapering distally to a point, its posterior surface covered with close-set dermal spinules (Fig. 107); distal two-thirds of maxilla naked and tucked beneath folds of skin, only the extreme proximal end covered directly with spinulose skin; chin prominent, extending far forward, well beyond mouth opening, resulting in a bulldog look; caudal peduncle present, the posteriormost margin of soft-dorsal and anal fins attached to body distinctly anterior to base of outermost rays of caudal fin; dorsal-fin rays 12, as many as four posteriormost rays bifurcate; anal-fin rays seven (rarely six, only two of 33 specimens examined), all bifurcate; pectoral-fin rays nine, rarely eight or 10 (two and one, respectively, of 66 pectoral fins examined), all simple; only posteriormost ray of pelvic fin bifurcate; vertebrae 19 (Tables 4, 5).

A single species, apparently not exceeding a maximum of about 120 mm SL.

Antennarius biocellatus (Cuvier)
Brackish-Water Frogfish

Figs. 15C, 106–111, 282; Tables 4, 5, 22

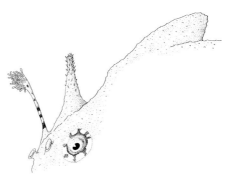

Chironectes biocellatus Cuvier, 1817b: 427, pl. 17, fig. 3 (original description, single specimen, holotype MNHN A.4620, locality unknown). Cuvier, 1829: 252 (after Cuvier, 1817b). Valenciennes, 1837: 417 (erroneous locality, not "American"). Valenciennes, 1842: 189 (after Valenciennes, 1837). Poey y Aloy, 1861: 400 (a rejected species). Schultz, 1957: 73 (in synonymy of *Phrynelox scaber*). Pietsch, 1985a: 79, fig. 17, table 1 (original manuscript sources for *Histoire Naturelle des Poissons* of Cuvier and Valenciennes). Pietsch et al., 1986: 139 (in type catalog).

Antennarius biocellatus: Schinz, 1822: 500 (new combination; after Cuvier, 1817b). Bleeker, 1860b: 4 (Ambon). Günther, 1861a: 196 (after Bleeker, 1860b). Bleeker, 1865b: 9, 18, pl. 194 (fig. 3), pl. 198 (fig. 5) (Ambon). Bleeker, 1873: 123 (China). Pietsch, 1984b: 36 (genera of frogfishes). Pietsch, 1985a: 79, fig. 17, table 1 (original manuscript sources for *Histoire Naturelle des Poissons* of Cuvier and Valenciennes). Pietsch and Grobecker, 1987: 174, figs. 68, 69, 126, pls. 31, 32 (description, distribution, relationships). Allen, 1991: 68 (New Guinea). Kottelat et al., 1993: 83 (freshwater; Sulawesi and western Indonesia). Michael, 1998: 334, 339, color figs. (identification, behavior, captive care). Ni and Kwok, 1999: 136, table 1 (Hong Kong). Pietsch, 1999: 2015 (western central Pacific; in key). Pietsch, 2000: 597 (South China Sea). Donaldson and Myers, 2002: 141 (Palau). Allen and Adrim, 2003: 25 (Papua to Sumatra). Allen et al., 2003: 363, color fig. (Asian Pacific). Carnevale and Pietsch, 2006: 452, table 1 (comparison, meristics, first fossil frogfish). Kuiter and Debelius, 2007: 117, color figs. (Indonesia). Kwang-Tsao et al., 2008: 243 (southern Taiwan). Allen and Erdmann, 2012: 144, color figs. (East Indies). Arnold and Pietsch, 2012: 128, fig. 3E (molecular phylogeny). Kottelat, 2013: 274 (freshwater, southeast Asia). Brandl and Bellwood, 2014: 32 (pair-formation). Fricke et al., 2014: 35 (Papua New Guinea). Joshi et al., 2016: 40 (Gulf of Mannar, Tamil Nadu, India).

Chironectes biocellatus (not of Cuvier): Guichenot, 1853: 214 (misidentification, confused with *Fowlerichthys ocellatus*).

Antennarius notophthalmus Bleeker, 1853b: 543, 544 (original description, single specimen, holotype RMNH 6280, Djungkulon, West Java). Bleeker, 1859b: 129 (listed). Günther, 1861a: 196 (description after Bleeker, 1853b). Bleeker, 1865b: 9, 16, pl. 196, fig. 1 (description, additional material; Java to Ceram, *in mari et ostiis fluviorum*). Bleeker, 1865c: 275 (Ambon). Meyer, 1885 (Celebes). Cott, 1940: 87, 373, fig. 77 (adaptive coloration). Schultz, 1957: 99, pl. 13, figs. A, B (description, Philippines; in key). Beaufort and Briggs, 1962: 216, fig. 48 [in part; synonymy, description; east coast of Africa (in error after Smith, 1949), Malabar coast? (after Day, 1865), China (after Bleeker, 1873), Indonesia, New Guinea, Philippines; in key]. Le Danois, 1964: 101, fig. 48f (description; MNHN A.4549, 145 mm total length, from Nias, erroneously listed as paratype). Schultz, 1964: 175, table 1 (additional specimens, Luzon). Goren and Dor, 1994: 14 (Red Sea). Golani and Bogorodsky, 2010: 62 (synonym of *A. biocellatus*).

Antennarius multiocellatus (not of Valenciennes): Le Danois, 1964: 101 (in part, i.e., MNHN A.4614).

Fig. 107. *Antennarius biocellatus* (Cuvier). UW 20883, 58 mm SL, Philippines, showing details of the illicium and esca and shape of the second and third dorsal-fin spines. Drawing by Cathy L. Short; after Pietsch and Grobecker (1987, fig. 69).

Uniantennatus biocellatus: Le Danois, 1964: 110, fig. 55f (new combination, description; locality in error after Valenciennes, 1837).

Antennarius biocellatus (not of Cuvier): Smith, 1949: 431, pl. 98, fig. 1237 (figure represents *Histrio histrio*). Burgess and Axelrod, 1973b: 828, color pls. 342, 343 (plates represent *Abantennarius coccineus*). Moore et al., 2014: 176 (Ashmore Reef, Timor Sea; specimen is probably *Abantennarius coccineus*).

Antennarius nummifer (not of Cuvier): Burgess and Axelrod, 1974: 1365, color pls. 453, 454 (plates represent *A. biocellatus*; Taiwan).

Antennarius sp.: Kailola, 1975: 47 (Papua New Guinea).

Antennarius notophthalmus (not of Bleeker): Kotthaus, 1979: 46, fig. 494 (ZIM 6007 is *Abantennarius nummifer*).

MATERIAL EXAMINED

Forty-eight specimens, 30 to 118 mm SL.

Holotype of *Chironectes biocellatus*: MNHN A.4620, 84 mm, locality unknown (Figs. 15, 108).

Holotype of *Antennarius notophthalmus*: RMNH 6280, 95 mm, Djungkulon, West Java, Bleeker (Fig. 106B).

Additional material: *Philippines*: CAS, 3 (30–51 mm); SU, 2 (38.5–50 mm); USNM, 1 (45 mm); UW, 15 (31.5–102 mm). *Indonesia*: BMNH, 4 (88–112 mm); MNHN, 1 (104 mm); RMNH, 11 (59–118 mm); ZMUC, 1 (67 mm). *New Guinea, Palau, and Solomon Islands*: BMNH, 1 (76 mm); BPBM, 1 (105 mm); KFRL, 2 (90–113 mm); USNM, 2 (56–88 mm); WAM, 1 (91 mm). *Locality unknown*: MNHN, 1 (66 mm).

DIAGNOSIS

The only member of the *Antennarius biocellatus* group, unique in having the combination of character states given for that species-group (see p. 154).

Fig. 108. *Antennarius biocellatus* (Cuvier). Holotype, MNHN A.4620, 84 mm SL, locality unknown. After Cuvier (1817b, pl. 17, fig. 3).

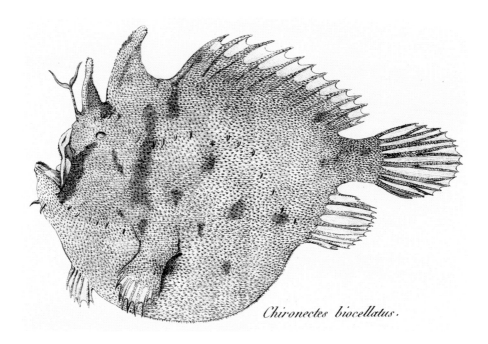

Chironectes biocellatus.

DESCRIPTION

Illicium 10.7 to 16.7% SL; second dorsal-fin spine straight, more or less tapering to a point, length 11.0 to 18.7% SL (Fig. 107); shallow naked depression between second and third dorsal-fin spines present or absent; third dorsal-fin spine curved posteriorly, length 23.3 to 30.5% SL; eye small, diameter 2.9 to 5.3% SL; conspicuous cutaneous appendages present on throat (Tables 4, 5).

Coloration in life: extremely variable; tan, yellow, bright orange, brick-red, chocolate brown, gray, or nearly black, with darker markings; basidorsal ocellus black, surrounded by a narrow yellow ring, which is in turn often surrounded by a narrow black ring (Figs. 109–111).

Coloration in preservation: tan, light gray (sometimes almost purplish), charcoal gray, or dark brownish-gray, to almost black; illicium darkly banded; escal appendage and associated filaments grayish, occasionally darkly pigmented at base; six to eight short, darkly pigmented bars radiating from eye; a second, slightly smaller ocellus present at base of caudal fin in seven of 48 specimens examined (but usually this ocellus on only one side of body). Light-color phase slightly darker on upper part of head, body, and fins; a dark brown to black bar extending from area between second and third dorsal-fin spines through eye to corner of mouth; a second, roughly parallel bar extending from area between third dorsal-fin spine and soft-dorsal fin onto cheek and terminating anterior to base of pectoral-fin lobe; usually a short, dark bar extending posteroventrally from eye;

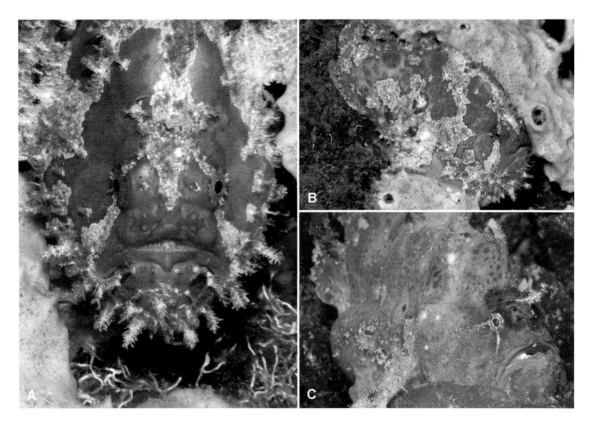

Fig. 109. *Antennarius biocellatus* (Cuvier). (**A**, **B**) Alotau, Milne Bay Province, Papua New Guinea. Photo by Gerry Allen. (**C**) Bali, Indonesia. Photo by Johanna Gawron.

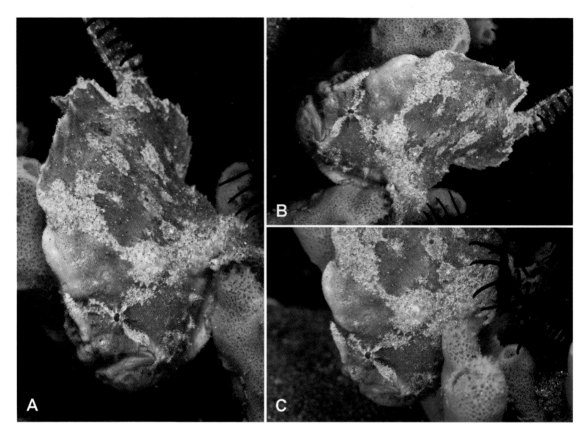

Fig. 110. *Antennarius biocellatus* (Cuvier). (**A–C**) Secret Bay, Gilimanuk, North Bali. Photos by Daniel Stassen.

second dorsal-fin spine and anterior margin of third spine often darkly pigmented; additional dark mottling and scattered spots present, especially on opercular region and upper half of body above pectoral fin; oblique bars on fins faintly visible in some specimens; tips of rays of paired fins banded; cutaneous appendages on throat darkly pigmented. Dark-color phase with belly somewhat lighter in coloration and sometimes faintly mottled; dark, oblique, roughly parallel bars usually present on all fins (as many as 12 bars across dorsal, three or four across anal, four to eight across caudal, two or three across paired fins); tips of all fin rays white. Intermediate-color phases usually covered with a network of heavy dark mottling, particularly on body and fins posterior to opercular region.

Largest specimen examined: 118 mm SL (RMNH 28020).

DISTRIBUTION

Antennarius biocellatus has a relatively narrow geographic distribution restricted to the shallow waters of Taiwan, the Philippines, Palau, Indonesia (Sumatra to West Papua), Papua New Guinea, and the Solomon Islands (Fig. 282, Table 22), with at least one occurrence off Hong Kong and another from the Gulf of Mannar, India (Joshi et al., 2016). The type locality is unknown.

Depths of capture, known for 10 specimens, range from near the surface to approximately 80 m, but eight were taken in 10 m or less; the average depth is 11 m.

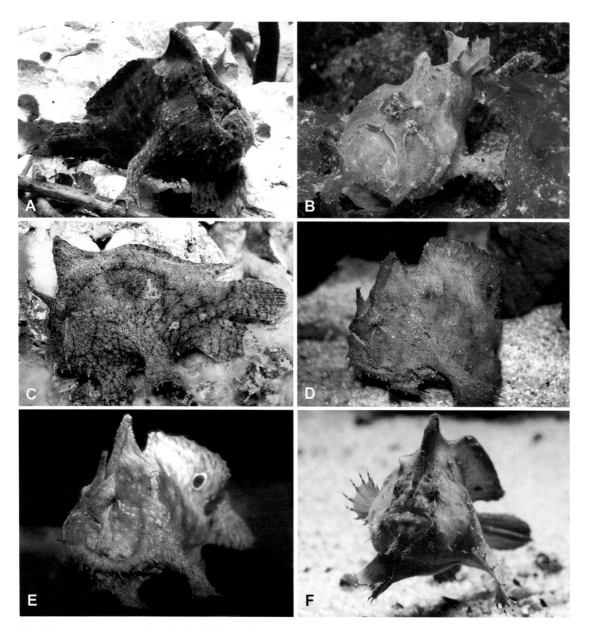

Fig. 111. *Antennarius biocellatus* (Cuvier). (**A**) aquarium specimen, probably Indonesia. Photo by Laurent Auer. (**B**) Bali, Indonesia. Photo by Johanna Gawron. (**C**) aquarium specimen, probably Indonesia. Photo by Scott W. Michael. (**D, E**) aquarium specimens, Indonesia. Photos by Frank de Goey, Ruinemans Aquarium B.V. (**F**) aquarium specimen, probably Indonesia. Photo by Dan Olsen.

COMMENTS

In his original description of *Antennarius biocellatus* (as *Chironectes biocellatus*), Cuvier (1817b: 428) made it clear that the source of his single specimen was unknown but, at the same time, he indicated a strong similarity (if not identical) to Bloch and Schneider's (1801) *Lophius histrio ocellatus*, first introduced by Antonio Parra (1787) as *Pescador* from Cuba and recognized here as *Fowlerichthys ocellatus* (see synonymy, p. 55). Thus, it is not so surprising that Valenciennes (1837: 417) referred to *A. biocellatus*, a species restricted to the Western

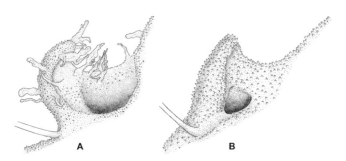

Fig. 112. Left lateral views of illicium and second dorsal-fin spine of species of the *Antennarius pauciradiatus* Group, showing details of aperture formed by membrane between second and third dorsal spines, used to protect the esca (esca not shown). (**A**) *A. pauciradiatus* Schultz, AMNH 26175, 38 mm SL, Bimini, Bahamas. (**B**) *A. randalli* Allen, USNM 232176, 18.5 mm SL, Ch'uan-fan-shih, Taiwan. Drawings by Cathy L. Short; after Pietsch and Grobecker (1987, fig. 70).

Pacific, as "another American chironecte with rough skin introduced by Cuvier." This assumption caused some early confusion with the Western Atlantic *F. ocellatus*.

Antennarius biocellatus is unusual among antennariids in often occupying brackish or even totally freshwater habitats. Bleeker (1865b: 16) was aware of this fact when he wrote *"in mari et ostiis fluviorum."* Twelve of the 48 specimens examined were captured in fresh or nearly freshwater.

Antennarius pauciradiatus Group Pietsch

DIAGNOSIS

A species-group of *Antennarius* distinguished by having the following combination of character states: illicium unusually short, 5.2 to 9.3% SL, considerably less than half length of second dorsal-fin spine; anterior end of pterygiophore of illicium terminating distinctly posterior to symphysis of upper jaw; esca a stout, oblong appendage usually bearing numerous, tentacle-like filaments; second dorsal-fin spine connected to head and to proximal portion of third dorsal-fin spine by a membrane, the membrane not divided into naked dorsal and ventral portions by a cluster of dermal spinules (Fig. 112); distal two-thirds of maxilla naked and tucked beneath folds of skin, only the extreme proximal end directly covered with spinulose skin; caudal peduncle present, the membranous posteriormost margin of soft-dorsal and anal fins attached to body distinctly anterior to base of outermost rays of caudal; dorsal-fin rays 11 to 13 (usually 12), as many as three posteriormost bifurcate; anal-fin rays seven (rarely eight), all bifurcate; pectoral-fin rays nine (rarely 10), all simple; all rays of pelvic fin simple; vertebrae 19 (Tables 4, 5); a small basidorsal pigment spot usually present.

Members of this species-group appear to be neotenic, cryptobenthic forms that reach a maximum standard length of only 40 mm; more than 80% of the known material measured less than 20 mm, and the average standard length was less than 16 mm (see Brandl et al., 2018).

Distributed in tropical waters of the Western Atlantic and Indo-Pacific oceans; two species.

Key to the Known Species of the *Antennarius pauciradiatus* Group

1. Two or more elongate cutaneous appendages arising from distal end of second dorsal-fin spine; membrane connecting second and third dorsal-fin spines with a number of similar appendages arising from distal margin (Fig. 112A) . . . .
. . . . *Antennarius pauciradiatus* **Schultz, 1957, p. 161**
Western Atlantic Ocean

2. Distal end of second dorsal-fin spine without cutaneous appendages; membrane connecting second and third dorsal-fin spines without cutaneous appendages (Fig. 112B) . . . . *Antennarius randalli* **Allen, 1970, p. 163**
    Western Pacific Ocean to Easter Island

Antennarius pauciradiatus Schultz
Dwarf Frogfish

Figs. 112A, 113–117, 282, 322H; Tables 4, 5, 22

Antennarius pleurophthalmus (not of Gill): Longley and Hildebrand, 1941: 308 (two specimens, USNM 116764, Tortugas).

Antennarius pauciradiatus Schultz, 1957: 100, fig. 7 (original description, 13 specimens, holotype USNM 153226, Palm Beach, Florida). Schultz, 1958: 147 ("no pelvic-fin ray divided"). Randall and Randall, 1960: 473 (resemblance to dead *Udotea*; St. John, Virgin Islands). Randall, 1968: 291 (description; Caribbean). Böhlke and Chaplin, 1968: 722, figs. (description; Florida, Bahamas, Cuba). Allen, 1970: 520, fig. 2b, table 2 (comparison with *A. randalli*, esca figured). Pietsch, 1978a: 3 (western central Atlantic). Pietsch, 1984b: 36 (genera of frogfishes). Robins and Ray, 1986: 88, pl. 14 (south Florida, Bahamas, Cuba). Acero and Garzón, 1987: 90 (Santa Marta, Colombia). Ibarra and Stewart, 1987: 9 (type catalog). Pietsch and Grobecker, 1987: 180, figs. 70B, 73, 74, 127 (description, distribution, relationships). Cervigón, 1991: 192 (Venezuela). Bertelsen, 1994: 137 (smallest frogfish). Bertelsen and Pietsch, 1998: 137 (smallest frogfish; after Bertelsen, 1994). McEachran and Fechhelm, 1998: 820 (Gulf of Mexico). Michael, 1998: 345 (identification, behavior, captive care). Smith-Vaniz et al., 1999: 159, 160 (Bermuda; in key). Pietsch, 2002a: 1051 (western central Atlantic). Collette et al., 2003: 100 (Navassa Island, West Indies). Dulvy et al., 2003: 29, table 1 (local extinction, Bermuda). Smith et al., 2003: 13 (Belize). Schneidewind, 2005a: 23 (mimic of green algae). Schneidewind, 2005b: 14 (after Schneidewind, 2005a). Schwartz, 2005: 147 (Florida and farther south). Carnevale and Pietsch, 2006: 452, table 1 (comparison, meristics, first fossil frogfish). Jackson, 2006a: 783, 784, tables 1, 2 (reproduction, depth distribution). Mejía-Ladino et al., 2007: 287, figs. 4, 6, tables 1–3, 5 (species account; in key; Colombia). Williams et al., 2010: 7, figs. 28, 29 (Saba Bank Atoll, Netherlands Antilles). Arnold and Pietsch, 2012: 128, fig. 3F (molecular phylogeny). Baldwin, 2013: 523, fig. 27A (larval coloration). Page et al., 2013 (common names). Smith-Vaniz and Jelks, 2014: 29 (St. Croix, Virgin Islands). Ayala et al., 2016: 92, 97, table 3 (larvae, Sargasso Sea).

MATERIAL EXAMINED

Sixty-eight specimens, 5.5 to 40 mm SL.

Holotype of *Antennarius pauciradiatus*: USNM 153226, 20 mm, off Palm Beach, Florida, Thompson and McGinty, August 1950 (Fig. 113A).

Paratypes of *Antennarius pauciradiatus*: FMNH 50249, 2 (35–40 mm), south end of Biscayne Bay, Grey, July–August 1949. USNM 153146, 2 (14–15.5 mm), off Palm Beach, McGinty, 1950; USNM 153147, 2 (12.5–15.5 mm), off Palm Beach, 37–55 m, Thompson and McGinty, April 1950; USNM 153148, 1 (12 mm), off Palm Beach, 37 m, McGinty, March 1950; USNM 153223, 2 (21–26.5 mm), off Palm Beach, Thompson and McGinty, 2 August 1950; USNM 116764, 2 (30–31 mm), Tortugas, Florida, Longley; USNM 82583, 1 (16.5 mm), off Cape San Antonio, Cuba, Bartsch and Henderson, 24 May 1914.

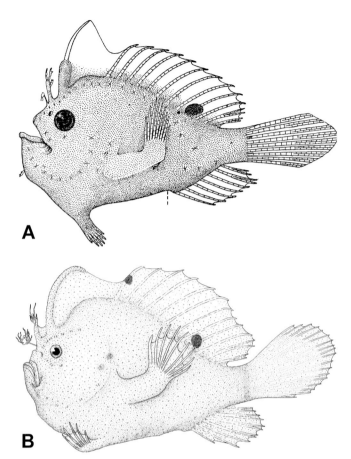

A

B

Fig. 113. *Antennarius pauciradiatus* Schultz. **(A)** holotype, USNM 153226, 20 mm SL, off Palm Beach, Florida. Drawing by A. M. Awl; after Schultz, 1957, fig. 7; courtesy of Lisa Palmer and the Smithsonian Institution, NMNH, Division of Fishes, Washington, DC. **(B)** UF 32425, 17 mm SL, Straits of Florida. Drawing by Cathy L. Short; after Pietsch and Grobecker (1987, fig. 73).

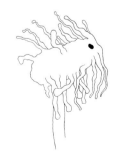

Fig. 114. Esca of *Antennarius pauciradiatus* Schultz. AMNH 26175, 38 mm SL, Bimini, Bahamas. Drawing by Cathy L. Short; after Pietsch and Grobecker (1987, fig. 74).

Additional material: *Bermuda*: USNM, 1 (16 mm); *Bahamas*: AMNH, 23 (8–16 mm); ANSP, 15 (5.5–35 mm); UF, 3 (15.5–16 mm). *Florida*: FMNH, 1 (34.5 mm); FSBC, 2 (8.5–18.5 mm); MCZ, 1 (17.5 mm); UF, 1 (17 mm). *Belize*: FMNH, 3 (12–17.5 mm). *Colombia*: UF, 2 (14.5–15.5 mm). *West Indies*: BPBM, 1 (14 mm); UF, 1 (9.5 mm).

DIAGNOSIS

A member of the *Antennarius pauciradiatus* group unique in having two or more elongate, branched, cutaneous appendages arising from distal end of second dorsal-fin spine; membrane connecting second and third dorsal-fin spines with a number of similar cutaneous appendages arising from distal margin (Fig. 112A). The length of the caudal peduncle of *A. pauciradiatus* appears to be slightly greater than that of *A. randalli*, but the known material of the latter species is insufficient to establish this with certainty.

DESCRIPTION

Esca oval in shape, bearing numerous elongate filaments along dorsal and ventral margins, the distal tips of filaments usually slightly swollen, a conspicuous darkly pigmented "eye-spot" on distal end (Fig. 114); illicium 5.2 to 8.2% SL; second dorsal-fin spine strongly curved posteriorly, length 10.3 to 13.2% SL; membrane between second and third dorsal-fin spines thin, translucent, slightly off-center to left side, forming shallow pocket on right to receive and protect esca, the anterodorsal portion of membrane naked, the remaining area lightly covered with small dermal spinules (Fig. 112A); shallow naked depression between second and third dorsal-fin spines absent; third dorsal-fin spine curved posteriorly, length 24.5 to 31.7% SL, the proximal one-third enveloped by thick muscle (often making distal two-thirds of spine appear much more narrow; see Schultz, 1957: 101, fig. 7); membrane behind third dorsal-fin spine continuous with soft-dorsal fin; eye diameter 5.3 to 9.1% SL; dorsal-fin rays 12 (rarely 11); anal-fin rays seven (rarely eight); pectoral-fin rays nine (rarely 10) (Tables 4, 5).

Coloration in life: nearly white, beige, yellow, tan, or brown, with darker mottling and small scattered, light and dark spots on head, body, and fins; a small dark basidorsal spot often present; dark bars or streaks occasionally present on caudal fin (Figs. 115–117).

Coloration in preservation: pale gray, light beige, to brown (slightly pinkish in some specimens); oblong appendage of esca usually with one to three dark eye-like spots, the escal filaments faintly pigmented; illicium without banding; darkly pigmented bars radiating from eye absent. Light-color phase with tiny scattered dark spots; an irregularly shaped patch above pectoral fin, a band across chin, and spots at base of caudal fin and on outer margin of caudal rays in some specimens; basidorsal spot dark brown. Dark-color phase unknown.

Largest specimen examined: 40 mm SL (FMNH 50249) but about 80% of the material examined was less than 20 mm SL. The average standard length for the 68 known specimens was 15.5 mm. The ovaries of a 24.5 mm SL specimen (ANSP 103017) contained numerous developing eggs.

DISTRIBUTION

Antennarius pauciradiatus is narrowly restricted to the tropical Western Atlantic (Fig. 282, Table 22). The bulk of the material (42 of 68 examined specimens) was collected in the Bahamas, but additional localities include Bermuda, the Atlantic coast of Florida (as far north as about 30° N latitude), Cuba, Puerto Rico, Antigua in

Fig. 115. *Antennarius pauciradiatus* Schultz. Blue Heron Bridge, Lake Worth Lagoon, Florida. Photo by Linda Ianniello.

the Lesser Antilles, Belize, Saba Bank Atoll in the Netherlands Antilles, and San Andrés y Providencia and Santa Marta, Colombia. The holotype was collected off Palm Beach, Florida.

Depths of capture, known for 59 specimens, range from just beneath the surface to 73 m; however, 93% of the material was taken in less than 40 m, and 71% in 20 m or less. The average depth for all known captures is 15 m.

Antennarius randalli Allen
Randall's Frogfish

Figs. 112B, 118–121, 282; Tables 4, 5, 22

Antennarius hispidus (not of Bloch and Schneider): Weber, 1913: 562 (in part, i.e., smaller specimen from Station 240).
Antennarius randalli Allen, 1970: 518, figs. 1, 2a, tables 1, 2 (original description, four specimens, holotype BPBM 6554, Easter Island). Pietsch, 1984b: 36 (genera of frogfishes). Randall, 1986: 180 (Marshall Islands). Pietsch and Grobecker, 1987: 178, figs. 70A, 71, 72, 126, pl. 33 (description, distribution, relationships; Fig. 118). Myers, 1989:

Fig. 116. *Antennarius pauciradiatus* Schultz. (**A–C**) St. Vincent, Lesser Antilles. Photos by Keri Wilk / ReefNet.

68, figs. 1c, 2 (Taiwan, Philippines, Moluccas, Fiji, Marshalls, Easter Island; in key). Pequeño, 1989: 43 (Easter Island). Randall et al., 1993a: 229 (Hawaiian Islands). Randall, 1996: 45, color fig. (Hawaiian Islands). Michael, 1998: 345, color fig. (identification, behavior, captive care). Randall et al., 1993a: 222, 229, pl. 1C (description; Makua, Oahu). Myers, 1999: 69, figs. 1c, 2 (after Myers, 1989). Pietsch, 1999: 2015 (western central Pacific; in key). Pietsch, 2000: 597 (South China Sea). Senou and Kawamoto, 2002: 2 (Japan). Allen and Adrim, 2003: 25 (Bali, Indonesia). Allen et al., 2003: 367, color figs. (tropical Pacific). Mundy, 2005: 265 (Hawaiian Islands). Randall, 2005a: 72, color fig. (South Pacific). Schneidewind, 2005c: 87, fig. (tiny postlarvae). Carnevale and Pietsch, 2006: 452, table 1 (comparison, meristics, first fossil frogfish). Senou et al., 2006: 423 (Sagami Sea). Kuiter and Debelius, 2007: 121, color figs. (Indonesia, Japan). Randall, 2007: 130, color fig. (Hawaiian Islands; in key). Senou et al., 2007: 49 (Ryukyu Islands). Kwang-Tsao et al., 2008: 243 (southern Taiwan). Allen and Erdmann, 2009: 594 (West Papua). Randall, 2010: 38 (Hawaii). Allen and Erdmann, 2012: 150, color figs. (East Indies). Arnold and Pietsch, 2012: 128 (molecular phylogeny, revised classification). Nakabo, 2013: 539, figs. (Okinawa, Ryukyu Islands). Brandl and Bellwood, 2014: 32 (pair formation). Fricke et al., 2014: 35 (Papua New Guinea). Moore et al., 2014: 206 (Hibernia Reef, Timor Sea).

MATERIAL EXAMINED

Twenty-three specimens, 8.0 to 31.5 mm SL.

Holotype of *Antennarius randalli*: BPBM 6554, 17 mm, off Motu Tautara, west coast of Easter Island, rotenone in 20 m, Randall and Allen, 7 February 1969.

Paratypes of *Antennarius randalli*: BPBM 6553, 1 (18.5 mm), offshore between Hanga Roa and Hanga Piko, Easter Island, rotenone in 13 m, Randall and Allen,

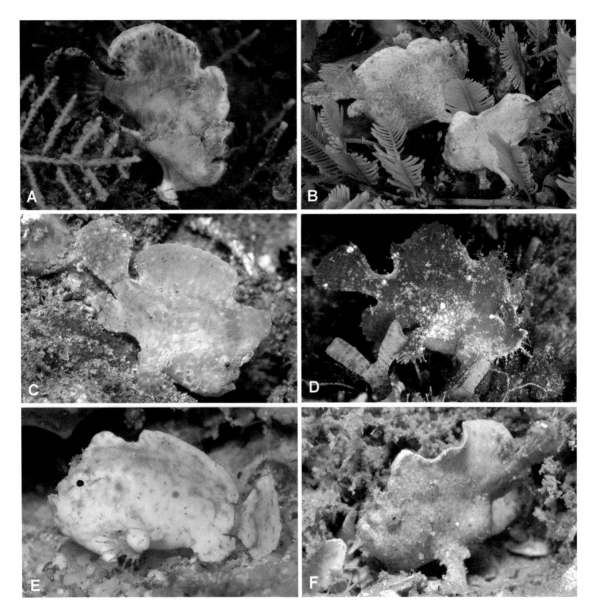

Fig. 117. *Antennarius pauciradiatus* Schultz. (**A**, **B**) Blue Heron Bridge, Lake Worth Lagoon, Florida. Photos by Kelly Casey. (**C**) Destin, Florida. Photo by Scott W. Michael. (**D, E**) Bonaire, Netherlands Antilles. Photos by Ellen Muller. (**F**) Riviera Beach, Lake Worth Lagoon, Florida. Photo by Anne DuPont.

10 February 1969; CAS 24417, 1 (15.5 mm), data as above; USNM 204310, 1 (20.5 mm), data as above.

    Additional material: BPBM, 6 (17–31.5 mm); USNM, 10 (9.5–18.5 mm); UW, 1 (12 mm); WAM, 1 (8 mm); ZMA, 1 (11 mm).

DIAGNOSIS

A member of the *Antennarius pauciradiatus* group unique in having the following character states: distal end of second dorsal-fin spine and membrane connecting second and

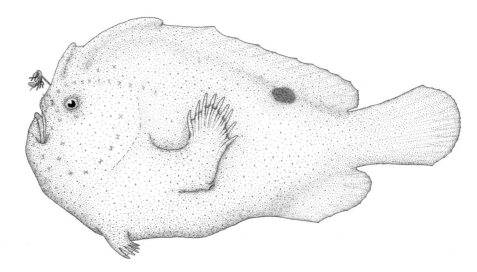

Fig. 118. *Antennarius randalli* Allen. USNM 232176, 18.5 mm SL, Ch'uan-fan-shih, Taiwan. Drawing by Cathy L. Short; after Pietsch and Grobecker (1987, fig. 71).

Fig. 119. Esca of *Antennarius randalli* Allen. USNM 232176, 18.5 mm SL, Ch'uan-fan-shih, Taiwan. Drawing by Cathy L. Short; after Pietsch and Grobecker (1987, fig. 72).

third dorsal-fin spines without cutaneous appendages (Fig. 112B). The length of the caudal peduncle of this species appears to be slightly less than that of its congener, *A. pauciradiatus*, but the known material is insufficient to establish this with certainty.

DESCRIPTION

Esca bilobed, each lobe bearing numerous, slender elongate filaments (Fig. 119); illicium length in 20.5-mm specimen 9.3% SL; second dorsal-fin spine strongly curved posteriorly, length in 20.5-mm specimen 11.2% SL; membrane between second and third dorsal-fin spines thin, translucent, slightly off-center to left side, forming small aperture on right to receive and protect esca, the entire area (except within escal aperture) lightly covered with small dermal spinules (Fig. 112B); shallow, naked depression between second and third dorsal-fin spines absent; third dorsal-fin spine curved posteriorly, length in 20.5-mm specimen 29.8% SL, the proximal one-third enveloped by thick muscle (usually making distal two-thirds of spine appear much more narrow); membrane behind third dorsal-fin spine continuous with soft-dorsal fin; eye diameter in 20.5-mm specimen 8.8% SL; dorsal-fin rays 12 or 13; anal-fin rays seven; pectoral-fin rays nine (Tables 4, 5).

Coloration in life: red or pink to mottled beige, yellow, brown or nearly black with scattered white spots; a rich red-brown fading to orange on chin, with faint spotting of darker brown; a blackish basidorsal spot; two white spots on caudal fin, one near upper

Fig. 120. *Antennarius randalli* Allen: Lembeh Strait, Sulawesi, Indonesia. Photo by Matthew Oldfield.

edge of fin and one directly below near lower edge; a white spot on side anterior to and above pectoral-fin base; a smaller white spot just behind eye and another below origin of soft-dorsal; a few smaller white flecks scattered over body; esca often white (Figs. 120, 121).

Coloration in preservation: light beige to yellow-brown (slightly pinkish in some specimens), brown, dark red-brown, to almost black; illicium without banding; pigmented bars radiating from eye absent. Lighter-color phases with tiny scattered spots and roughly circular or irregularly shaped brown blotches or mottling overall; outer edge of dorsal and anal fins brown; basidorsal spot dark brown; tips of rays of paired fins banded.

Fig. 121. *Antennarius randalli* Allen. (**A**) Lembeh Strait, Sulawesi, Indonesia. Photo by Roger Steene. (**B, C**) Dauin, Negros Oriental, Philippines. Photos by Daniel Geary. (**D**) Ambon, Indonesia. Photo by Linda Ianniello. (**E**) Lembeh Strait, Sulawesi, Indonesia. Photo by Scott W. Michael. (**F**) Nuweiba, Gulf of Aqaba, Red Sea. Photo by Sarah Pikarski.

Darker-color phase (UW 20886) covered with slightly darker spotting or mottling over-all, with tips of all fin rays white; rays of paired fins darkly banded.

Largest specimen examined: 31.5 mm SL (BPBM 32842).

DISTRIBUTION

Antennarius randalli has been observed at numerous, widely separated localities through-out the Indo-Pacific (Fig. 282, Table 22), extending from Cocos-Keeling to the Ryukyu Islands, including Taiwan, Cebu Island in the Philippines, Bali, the Moluccas, Hibernia Reef (Timor Sea), West Papua, and Papua New Guinea, to the Marshall Islands, Fiji, the Marquesas and Hawaiian Islands, and Easter Island. Recent photographs of a pair of in-dividuals (probably male and female) taken at Nuweiba, Gulf of Aqaba, in the Red Sea (Sarah L. Pikarski, personal communication, 5 November 2017; Fig. 120F), vastly extend the known distribution of this species. The type material was collected at Easter Island.

Depths of capture, known for 18 specimens, ranged from just beneath the surface to 54 m. However, all but one of these were taken in 24 m or less; the average depth for all known captures is 17 m.

Genus *Nudiantennarius* Schultz

Nudiantennarius Schultz, 1957: 66 (type species *Antennarius subteres* Smith and Radcliffe, in Radcliffe, 1912, by original designation).

DIAGNOSIS

Nudiantennarius is unique among antennariids in having the following combination of character states: dermal spinules reduced, skin only partially covered with bifurcate der-mal spinules, body often appearing naked, length of spines of each spinule not more than twice the distance between tips of spines (Fig. 26C); esca distinct (Figs. 122, 123); illicium naked, without dermal spinules, about half length of second dorsal-fin spine; second dorsal-fin spine unusually long, narrow, without posterior membrane; pectoral-fin lobe narrow, somewhat detached from side of body; caudal peduncle present, the membranous posteriormost margin of soft-dorsal and anal fins attached to body dis-tinctly anterior to base of outermost rays of caudal fin; endopterygoid present; pharyn-gobranchial I present; epural present; pseudobranch absent; swimbladder present; dorsal-fin rays 12; anal-fin rays seven; all rays of caudal fin usually bifurcate (outermost caudal-fin rays simple, seven innermost bifurcate in UW 117643 and CBG 13028); pectoral-fin rays nine; pelvic-fin rays five, all simple; membranes between rays of paired fins deeply incised (Pietsch and Grobecker, 1987, fig. 77; Fig. 122); vertebrae 19; one or more large basidorsal ocelli usually present (Tables 1, 2).

DESCRIPTION

Dermal spinules greatly reduced, evident on second and third dorsal-fin spines, anteri-ormost dorsal-fin rays, on snout and dorsal portion of head, often on chin, coverage sometimes extending to pectoral lobe, but elsewhere few and scattered, usually diffi-cult to detect without microscopic aid (Fig. 26C); esca a rounded clump of folded tissue (in smallest specimen examined) or an oval-shaped tuft of short, more or less flattened appendages, sometimes with a few short filaments (especially in larger specimens; Fig. 123), attached to illicium at an angle, usually directed slightly ventrally; length of esca 3.9 to 7.2% SL; illicium, when laid back onto head, fitting into a tiny, narrow groove situated alongside of second dorsal-fin spine, tip of illicium (esca) coming to lie within a

Fig. 122. *Nudiantennarius subteres* (Smith and Radcliffe, in Radcliffe). Holotype, USNM 70268, 42 mm SL, collected off Luzon, Philippines (pectoral and pelvic fins retouched). After Radcliffe (1912, pl. 17, fig. 1).

shallow depression between second and third dorsal-fin spines, esca probably capable of being covered and protected by second dorsal-fin spine when spine is fully depressed; illicium about half length of second dorsal-fin spine, 5.9 to 11.3% SL; anterior end of pterygiophore of illicium terminating distinctly posterior to symphysis of upper jaw; illicium and second dorsal-fin spine relatively closely spaced on pterygiophore, distance between bases of spines less than 5% SL; second dorsal-fin

Fig. 123. Esca of *Nudiantennarius subteres* (Smith and Radcliffe, in Radcliffe). BMNH 1866.8.14.108, 64 mm SL, locality unknown. Drawing by Cathy L. Short; after Pietsch and Grobecker (1987, fig. 76).

spine long, narrow, straight to slightly curved posteriorly, not connected to head by membrane (Fig. 122), sometimes with numerous slender, cutaneous filaments; length of second dorsal-fin spine 17.4 to 28.1% SL; third dorsal-fin spine curved posteriorly, connected to head by thick membrane, length 23.0 to 31.3% SL; eye diameter 5.5 to 11.3% SL; only distal tip (about 20 to 25% of length) of maxilla tucked beneath folds of skin; scattered cutaneous appendages often present on head (especially on chin), body, and second and third dorsal-fin spines; epibranchial I toothless; ceratobranchial I toothless; vertebrae 19, caudal centra 14; dorsal-fin rays 12, posteriormost two to five bifurcate; anal-fin rays seven, all bifurcate; pectoral-fin rays nine, all simple; all rays of pelvic fin simple; distal third to half of pectoral- and pelvic-fin rays free, not connected by membrane (Pietsch and Grobecker, 1987, fig. 77; Tables 1, 2).

Coloration in life: extremely variable; typically, overall dark purplish-brown, chocolate brown, to black, with a large brown to black basidorsal ocellus, usually surrounded by a light brown, yellow, orange, or red ring; similar but much smaller ocelli occasionally present on membrane behind third dorsal-fin spine, anterior half of soft-dorsal fin, and upper margin of caudal peduncle; tiny scattered white spots often present on head, body, and fins, especially on caudal fin; illicium, esca, and all or part of second dorsal-fin spine light brown to white; free distal tips of paired fins light brown to almost white. Variations on this general theme are seemingly endless (Figs. 124–127): background coloration may range from pink, red, orange, yellow, green, or gray to off-white; head, body, and fins often monochromatic, but sometimes mottled or covered with small dark, close-set spots; ring of ocellus sometimes surrounded by a series of dark close-set spots;

Fig. 124. *Nudiantennarius subteres* (Smith and Radcliffe, in Radcliffe). Dauin, Negros Oriental, Philippines. Photo by Daniel Geary; after Pietsch and Arnold (2017).

ocellus sometimes absent; illicium without banding; small dark bars of pigment often radiating from eye.

A single species.

Nudiantennarius subteres (Smith and Radcliffe)
Lembeh Frogfish

Figs. 26C, 122–127, 274; Tables 1, 2, 22

Antennarius subteres Smith and Radcliffe, in Radcliffe, 1912: 205, pl. 17, fig. 1 (original description, single specimen, holotype USNM 70268, Lingayen Gulf, west coast of Luzon, Philippines).

Nudiantennarius subteres: Schultz, 1957: 66, pl. 1, fig. D (new combination; after Smith and Radcliffe, in Radcliffe, 1912). Pietsch, 1984b: 36 (genera of frogfishes). Pietsch and Grobecker, 1987: 184, figs. 16C, 75–77, 129 (description, distribution, relationships; new records from Luzon and Ambon). Lindberg et al., 1997: 213 (Sea of Japan). Pietsch, 1999: 2015 (western central Pacific; in key). Pietsch, 2000: 597 (South China Sea). Allen and Adrim, 2003: 25 (Moluccas; Flores locality based on misidentification: NMV A.9676, 11 mm SL, is *Antennarius hispidus*). Kuiter and Debelius, 2007: 117, color fig. (misidentification; after Allen and Adrim, 2003). Allen and Erdmann, 2012: 155 (Indonesia). Arnold and Pietsch, 2012: 128, fig. 1E (molecular phylogeny, *incertae sedis*). Pietsch and Arnold, 2017: 658, figs. 1–5 (redescription, additional material).

Antennarius sp.: Allen et al., 2003: 363, color fig. ("Ocellated Frogfish," "unidentified, possibly undescribed species, known only from Lembeh Strait, Sulawesi").

Fig. 125. *Nudiantennarius subteres* (Smith and Radcliffe, in Radcliffe). (**A**, **B**) Lembeh Strait, Sulawesi, Indonesia. Photos by Johanna Gawron.

Fig. 126. *Nudiantennarius subteres* (Smith and Radcliffe, in Radcliffe). (**A**) Secret Bay, Anilao, Luzon, Philippines. Photo by Stephane Bailliez. (**B**) Pantar Island, Alor Archipelago, Indonesia. Photo by David J. Hall. (**C**) Dauin, Negros Island, Philippines. Photo by Daniel Geary. (**D**) juvenile, Banka, North Sulawesi, Indonesia. Photo by Johanna Gawron. (**E**) hypothesized mated pair, male on the left, female on the right; Seraya Bay, northeast Bali, Indonesia. Photo by Vincent Chalias. (**F**) Secret Bay, Anilao, Luzon, Philippines. Photo by Stephane Bailliez.

MATERIAL EXAMINED

Seven specimens, 17 to 64 mm SL.

Holotype of *Antennarius subteres*: USNM 70268, 42 mm, *Albatross* station 5442, Lingayen Gulf, west coast of Luzon, Philippines, 16° 30′ 36″ N, 120° 11′ 06″ E, beam trawl, 82 m, on a bottom of coral sand, 10 May 1909 (Fig. 122).

Fig. 127. *Nudiantennarius subteres* (Smith and Radcliffe, in Radcliffe). (**A**) Dauin, Negros Island, Philippines. Photo by Daniel Geary. (**B**) Ambon, Indonesia, October 1996. Photo by Scott W. Michael. (**C**) Lembeh Strait, Sulawesi, Indonesia. Photo by Teresa Zuberbühler. (**D**) Dauin, Negros Oriental, Philippines. Photo by Daniel Geary. (**E**) juvenile, Dauin, Negros Oriental, Philippines. Photo by Daniel Geary. (**F**) Lembeh Strait, Sulawesi, Indonesia. Photo by Ned DeLoach.

Additional material: BMNH, 1 (64 mm); CAS, 1 (32 mm); CBG Barcode of Life voucher 13028, 1 (42 mm). UW, 2 (30.5–38 mm). ZMUC, 1 (17 mm).

DIAGNOSIS AND DESCRIPTION

As given for the genus.

Largest known specimen: 64 mm SL (BMNH 1866.8.14.108), but the species is apparently a relatively small frogfish. Six of the seven known preserved specimens range from

30.5 to 64 mm but most estimates of uncollected individuals are considerably smaller, no more than about 22 mm (Pietsch and Arnold, 2017). Accordingly, some observers have hypothesized that the numerous sightings of living specimens are based on juveniles and that the adults, which are rarely seen, exist at much greater depths (Vincent Chalias, personal communication, 11 April 2017). This notion is contradicted, however, by observations of small females through multiple reproductive cycles; in several cases, a heavily gravid female, no more than about 15 mm, has been observed in the company of two slightly smaller individuals, apparently males, foretelling a reproductive event (Daniel Geary, personal communication, 4 May 2017).

DISTRIBUTION

Originally described from a single specimen collected in Lingayen Gulf, on the west coast of Luzon Island, Philippines, the additional preserved specimens of *Nudiantennarius subteres* are from Manila and off Bataan, west of Talaga, Luzon Island; Secret Bay, Gilimanuk, on the west end of Bali; and Ambon, Moluccas Islands (Fig. 274, Table 22). Photographs, however, expand the distribution to include Lembeh Strait and Bangka Island, North Sulawesi; Pantar and Alor Islands in the Alor Archipelago; Dauin and Dumaguete, Negros Island, Philippines; and off the Izu Peninsula, Suruga Bay, Japan (Pietsch and Arnold, 2017).

Early collection data indicate a deep-water existence, at depths between 64 and 90 m (CAS 32765), 82 m (the holotype, USNM 70268), and 128 m (ZMUC P922045)—the reason why Pietsch and Grobecker (1987) called it the "Deepwater Frogfish"—but numerous recent observations by divers indicate a much shallower existence, from 3 to 30 m, although some hypothesize that most observations are based on juveniles, while adults occur at greater depths (Vincent Chalias, personal communication, 11 April 2017).

HABITAT

Lembeh Strait is a kilometer and a half wide and several kilometers long; it is the body of water between Lembeh Island and the northeastern corner of Sulawesi, Indonesia. Most dive sites are located along the Sulawesi side and close to shore; depths vary, but most range from 3 to 9 m, with no sites deeper than about 18 m. Bottoms on which the Lembeh Frogfish has been observed are typically sand, silt, or brown mud, with some soft corals, gorgonians, and sponges, but very little hard coral. There is also tree litter that has fallen into the water in the shallows, and a lot of trash, rubbish, human refuse of all kinds, at many locations, especially those near native villages and towns (David Hall, personal communication, 7 April 2017).

The Pantar dive site, located on the southeast margin of the island in the Alor Archipelago, Indonesia, where David Hall photographed the Lembeh Frogfish in 2006 (personal communication, 7 and 9 April 2017), is a much smaller area—a small bay adjacent to a small native village. Most of the diving was no deeper than 6 m and the bottom was largely made up of very coarse sand or fine coral rubble, with some hard corals and gorgonians; but, unlike Lembeh, there was no rubbish strewn over the bottom. There were numerous small cephalopods, many shrimps, crabs, other crustaceans, and lots of fishes, especially juveniles: small shark species, burrowing snake eels, two species of *Rhinopias* (*R. eschmeyeri* and *R. frondosa*), and several other species of scorpionfishes and waspfishes. Only two species of frogfishes were observed, *Antennarius striatus* and a single individual of the Lembeh Frogfish.

Most of the dive sites in Ambon Bay where the Lembeh Frogfish has been observed offer a typical "muck diving" habitat—course sand- and rubble-covered slopes, with random solitary and encrusting sponges, hydroids, mixed with lots of human refuse, the latter more or less covered with organic growth, which the animals use for shelter (Linda Ianniello, personal communication, 5 June 2017). At Bali, the species occurs most often on black sandy slopes on the northeast margin of the island, and most commonly seen during night dives (Vincent Chalias, personal communication, 11 April 2017). At Dauin on Negros Island, Philippines, the habitat is similar: dark-colored individuals are most commonly found on course sand or gravel, often within patches of green algae, in 4 to 20 m; lighter, more colorful individuals are usually found associated with small, similarly colored sponges, at somewhat greater depths of 12 to 30 m (Daniel Geary, personal communication, 4 May 2017).

COMMENTS

For more than a century since its original description, *Nudiantennarius subteres* remained somewhat of an enigma, until Pietsch and Arnold (2017) provided evidence of its conspecificity with the so-called "Lembeh Frogfish" or "Ocellated Frogfish." The latter had also posed an enigma—its distinctive morphology had been recognized among members of the dive community for almost two decades and among professional ichthyologists at least since the early 2000s (Allen et al., 2003). It was therefore a bit of a mystery that no formal description and scientific name had been applied to it. The many images attributable to this species (posted on the Internet and provided by many underwater photographers; Pietsch and Arnold, 2017) have been variously identified as *Antennarius biocellatus, Antennatus nummifer, Antennatus dorehensis*, and *Antennatus rosaceus* but clearly it cannot be assigned to any of these species—the sum of its characters points to *Nudiantennarius subteres*. The combination of reduced dermal spinules, unique escal morphology, short illicium, and long, slender second dorsal-fin spine, unattached to the head by a membrane, is enough to separate the species from all other frogfishes, except *Histrio histrio* (see below)—the latter remains distinct in a host of ways, including exceptionally large pelvic fins and a pseudopelagic existence in floating *Sargassum*. Finally, *Nudiantennarius subteres* is also genetically distinct, forming a sister-group relationship with *Histrio histrio* (see Evolutionary Relationships, p. 385).

Genus *Histrio* Fischer

Histrio Fischer, 1813: 70, 78 (type species *Lophius histrio* Linnaeus, 1758, by absolute tautonymy; no species listed).

Chironectes Rafinesque, 1814: 19 [type species *Chironectes variegatus* Rafinesque, 1814 (= *Lophius histrio* Linnaeus, 1758), by monotypy; preoccupied by *Chironectes* Illiger, 1811, a genus of marsupial mammals].

Chironectes Rafinesque, 1815: 92 (*nomen nudum*). Cloquet, August 1817: 597 (in part, includes *Antennarius*; type species *Chironectes* variegatus Rafinesque, 1814; first mentioned species, *Chironectes histrio*). Cuvier, October 1817b: 418 [*sensu* Cloquet, August 1817; first mentioned species, *Chironectes laevigatus* (= *Lophius histrio* Linnaeus)].

Batrachopus Goldfuss, 1820: 110 (in part, includes *Antennarius*; substitute name for *Chironectes* Rafinesque, 1814, preoccupied, therefore taking same type species, *Chironectes variegatus* Rafinesque, 1814; no species listed).

Cheironectes Lowe, 1839: 84 (emendation of *Chironectes* Rafinesque, 1814, preoccupied, therefore taking same type species, *Chironectes variegates* Rafinesque, 1814).

Chironectus Swainson, 1839: 330 (emendation of *Chironectes* Rafinesque, 1814, preoccupied, therefore taking same type species, *Chironectes variegatus* Rafinesque, 1814; "body naked, always destitute of scales").

Capellaria Gistel, 1848: viii (in part, includes *Antennarius*; substitute name for *Chironectes* Rafinesque, 1814, preoccupied, therefore taking same type species, *Chironectes variegatus* Rafinesque, 1814; no species listed).

Pterophryne Gill, 1863: 90 [type species *Lophius laevigatus* Bosc, in Cuvier, 1816 (= *Lophius histrio* Linnaeus, 1758, and *Chironectes laevigatus* Cuvier, 1817b), by original designation].

Pterophrynoides Gill, 1879a: 216 (tentative substitute name for *Pterophryne* Gill, 1863, therefore taking same type species, *Lophius laevigatus* Bosc, in Cuvier, 1816; "if considered to be" preoccupied by *Pterophrynus* Lütken, 1863, a genus of amphibians).

DIAGNOSIS

Histrio is unique among antennariids in having exceptionally large pelvic fins (length greater than 25% SL; Fig. 128), greatly reduced dermal spinules, two cutaneous cirri on mid-dorsal line of snout between symphysis of premaxillae and base of illicium (Fig. 129), and a pseudopelagic lifestyle in floating *Sargassum*. *Histrio* is further distinguished from other antennariids by having the following combination of character states: skin nearly always without dermal spinules (tiny, simple spinules rarely present, e.g., UG 5329; a single, tiny, crescent-shaped spinule associated with each pore of acoustico-lateralis system), but everywhere covered with tiny, rounded, close-set papillae; illicium naked, without dermal spinules; esca small but distinct (Fig. 130); pectoral-fin lobe detached from side of body for most of length; caudal peduncle present, the membranous posteriormost margin of soft-dorsal and anal fins attached to body considerably anterior to base of outermost rays of caudal fin; endopterygoid present; pharyngobranchial I present; epural present; pseudobranch present; swimbladder present; dorsal-fin rays 11 to 13 (usually

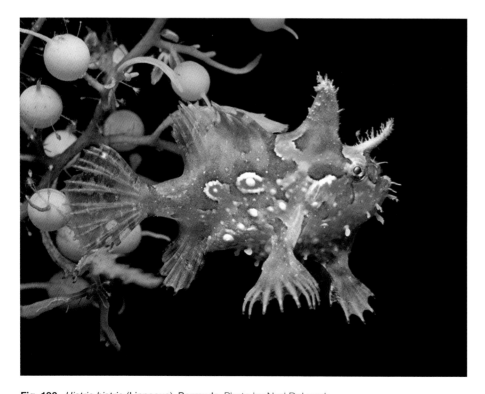

Fig. 128. *Histrio histrio* (Linnaeus). Bermuda. Photo by Ned DeLoach.

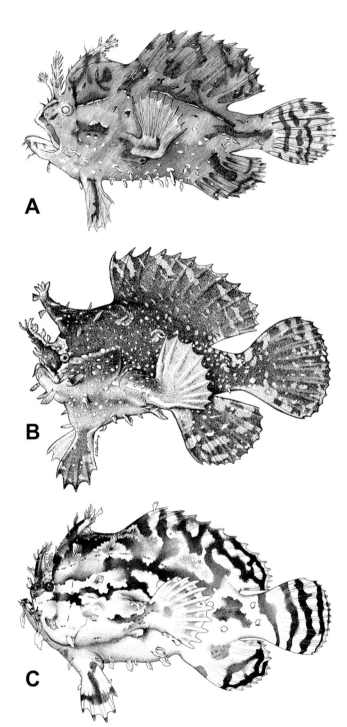

Fig. 130. Esca of *Histrio histrio* (Linnaeus). UW 20959, 52.5 mm SL, Mactan, Cebu, Philippines. Drawing by Cathy L. Short; after Pietsch and Grobecker (1987, fig. 84).

12); anal-fin rays six to eight (usually seven); outermost rays of caudal fin simple, the seven innermost bifurcate; pectoral-fin rays nine to 11 (usually 10); vertebrae 19 (rarely 18) (Tables l, 2).

DESCRIPTION

Esca an oval-shaped or tapering appendage with numerous, more or less parallel folds running from base to tip, and a cluster of short, slender filaments arising from base (Fig. 130); illicium unprotected, no distinct groove alongside second dorsal-fin spine, no shallow depression between second and third dorsal-fin spines (but esca perhaps protected by second dorsal-fin spine and associated cutaneous appendages when spine is fully depressed); illicium less than one-half length of second dorsal-fin spine, 4.5 to 9.3% SL; anterior end of pterygiophore of illicium terminating distinctly posterior to symphysis of upper jaw; illicium and second dorsal-fin spine very closely spaced, the illicium appearing to emerge from base of second; second dorsal-fin spine narrow, straight to slightly curved posteriorly, not connected to head by membrane, the posterior margin densely covered with cutaneous papillae and appendages; second dorsal-fin spine 13.3 to 19.3% SL; third dorsal-fin spine narrow (about equal to width of

Fig. 129. *Histrio histrio* (Linnaeus). (**A**) lectotype of *Lophius laevigatus* Bosc, in Cuvier, MNHN A.3829, 49 mm SL, Carolina, collected by Louis Bosc the Elder. After Cuvier, 1817b, pl. 16, upper figure. (**B**) West Indies, Gulf Stream, identified as *Pterophryne tumida*. Drawing by Sherman Foote Denton; after Jordan (1905, fig. 500); (see also Agassiz, 1888, fig. 210). (**C**) specimen unknown, 44 mm SL, Misaki, Japan, identified as *Pterophryne histrio.* Drawing by Chloe Lesley Starks; after Jordan (1902, fig. 2).

second dorsal-fin spine), slightly curved posteriorly, connected to head by membrane, the membrane densely covered with cutaneous papillae and appendages; third dorsal-fin spine 19.3 to 29.3% SL; eye diameter 5.0 to 8.7% SL; only distal tip (about 20 to 25% of length) of maxilla tucked beneath folds of skin; head, body, and fins covered with numerous, compressed, cutaneous appendages; epibranchial I toothless; ceratobranchial I toothless; vertebrae 18 or 19, caudal centra 13 or 14; dorsal-fin rays 12 (rarely 11 or 13), the posteriormost two or three bifurcate; anal-fin rays seven (rarely six or eight), the posteriormost two to six bifurcate; pectoral-fin rays 10 (rarely nine or 11), all simple; all rays of pelvic fin simple (Tables 1, 2).

Color and color pattern highly variable, capable of sudden change, fading or intensifying, apparently to match the background or in response to stimulation, between a gray-white, bleached-out appearance and a pattern of dark streaks and mottling of browns, olive, and yellow,

Fig. 131. *Histrio histrio* (Linnaeus). Gulf of Mexico. Photo by David Shale.

with numerous small brown to black spots and sometimes narrow, irregular white lines; pectoral and pelvic fins sometimes edged with orange; cutaneous filaments white; basidorsal spot rarely present; illicium without banding; dark bars or streaks radiating from eye usually continuous with mottling of head and body; some juveniles entirely black, with tips of pectoral-fin rays white (Figs. 131, 132).

A single species, Atlantic and Indo-West Pacific.

COMMENTS

The name *Histrio* was established by German-Russian naturalist Johann Gotthelf Fischer von Waldheim (1813) in a series of analytical keys that lead only to genera. Its validity may be questioned because of an error in the original publication that transposes the generic character of *Lophius* and *Histrio* (Jordan, 1917b: 84; 1923: 242). "It is true that 'corpus depressum' is attributed to *Histrio* and 'corpus compressum' to *Lophius* but the *lapsus calami* appears so obvious that no nomenclatorial consequences should be involved" (Monod and Le Danois, 1973: 661).

Histrio histrio (Linnaeus)
Sargassumfish

Figs. 5, 18C, 68, 128–136, 274, 325, 326, 345; Tables 1, 2, 22

A description of this species in its protean manifestations of form and color seems scarcely necessary here, since its characters are well known to every tyro in ichthyology.

—George Brown Goode and Tarleton Hoffman Bean, 1896: 486

Fig. 132. *Histrio histrio* (Linnaeus). (**A, B**) Blue Heron Bridge, Lake Worth Lagoon, Florida. Photos by Linda Ianniello. (**C**) Riviera Beach, Lake Worth Lagoon, Florida. Photo by Anne DuPont. (**D**) Papua New Guinea. Photo by Scott W. Michael. (**E**) Gulf of Mexico. Photo by David Shale. (**F**) Blue Heron Bridge, Lake Worth Lagoon, Florida. Photo by Kelly Casey.

Rana Piscatrix, Americana, cornuta, spinosa: Seba, 1734: 118, pl. 74, fig. 4 (Western Atlantic). Cuvier, 1817b: 425 (same as *Chironectes laevigatus* Bosc, in Cuvier, 1816). Walbaum, 1792: 496 (listed).

Pullus Ranae Piscatricis, spinosae: Seba, 1734: 118, pl. 74, fig. 5 (thought to represent a juvenile of form shown in Seba's fig. 4).

Rana Piscatrix, Americana, quarta: Seba, 1734: 119, p. 74, fig. 6 (Western Atlantic).

Pullus Ranae Piscatricis, quartae: Seba, 1734: 119, p. 74, fig. 7 (in part, i.e., three small drawings on left, thought to represent juveniles of forms shown in fig. 6, are of nudibranch *Scyllaea pelagica*; see Introduction, p. 6).

Batrachus; sive rana piscatrix americana cornuta spinosa: Klein, 1742: 16, no. 3 (after Seba, 1734: 118, pl. 74, figs. 4, 5).

Batrachus capite product: Klein, 1742: 16, no. 6 (after Seba, 1734: 119, pl. 74, fig. 6).

Batrachus mollis: Klein, 1742: 16, No. 7, pl. 3, fig. 4 (figure shows *H. histrio*).

Balistes, quae Guaperua Chinensis: Linnaeus, 1747: 137, pl. 3, fig. 5a, b [seen at Gothenburg (Göteborg, Sweden) on 10 July 1746 in collection of Apothecary Lut during journey through West Gothland; specimen most likely lost, origin probably China].

Lophius (tumidus) pinnis dorsalibus tribus: Linnaeus, 1754a: 20, no. 21 (probably China).

Lophius tumidus: Linnaeus, 1754b: 56 [*L. pinnis dorsalibus tribus*; includes *Balistes, quae Guaperua Chinensis* Linnaeus (1747), *L. (tumidus) pinnis dorsalibus tribus* Linnaeus (1754a), and *Guaperua* of Marcgrave (1648) and Willughby (1686); *Habitat in Pelago inter FUCOS natantes*]. Osbeck, 1757: 305 ["skin smooth, covered with small, tag-like appendages"; found in "The Grass-Sea," probably at Ascension Island in 1752 (Forster, 1771: 112); early mention of mimicry, see Aggressive Mimicry, p. 411]. Osbeck, 1765: 400 (validation of *L. tumidus* as binomial; German translation of original Osbeck, 1757). Linnaeus, 1766: 403 (in synonymy of *L. histrio*). Forster, 1771: 112 (English translation of Osbeck, 1765). Bloch, 1785: 13 (after Osbeck, 1757). Jordan, 1917a: 278 (same as *Chironectes variegatus* Rafinesque, 1814). Holm, 1957: 39 (Linnaean types).

Balistes alepidotus, cirratus, polyodon, dorso diacantho, pinnisventralibus distinctis inermibus: Gronovius, 1754: 53, no. 116 (*Mare Atlanticum, Hispanicum atque Indicum, ubi inter Algas multoties reperitur*).

Lophius histrio Linnaeus, 1758: 283 [original description, three extant syntypes, ZMU Linnesamling nr. 100 (Chin. Lagerstr. nr. 21), 74-mm specimen designated lectotype by Pietsch and Grobecker, 1987, paralectotypes 18 and 26 mm; synonymy includes Linnaeus (1747, 1754a, b), Osbeck (1757), Marcgrave (1648), and Willughby (1686); type locality probably China, *Habitat in Pelago inter Fucum natantem*]. Linnaeus, 1766: 403 (*L. compressus*; synonymy includes *Piscis Brasilianus cornutus* of Petiver, 1702). Hjortberg, 1768: 353, pl. 10 (*in den Seegewächse Sargazo gefangen worden*). Braam-Houckgeest, 1774: 20 (*Land der Eendragt*, south of Java). Meuschen, 1781, unpaged [binomial for polynomial of Gronovius, 1763; publication ruled as not consistently binomial by International Commission on Zoological Nomenclature (Anonymous, 1950) and rejected by Opinion 261 (Anonymous, 1954)]. Gmelin, 1789: 1481 (in part; *in fucis aut pone lapides latens*). Latreille, 1804: 73 (after Linnaeus, 1758). Gray, 1854: 48 (Gronovius collection; BMNH 1854.11.13.92). Lönnberg, 1896: 30 (type catalog). Whitley, 1929c: 300 [names of fishes in Meuschen's (1781) index to *Zoophylacium Gronovianum*]. Holm, 1957: 39 (Linnaean types). Fernholm and Wheeler, 1983: 227–228 (Linnean fish specimens). Wheeler, 1991: 166, fig. 8 (Linnean fish collection).

Lophius (Histrio) pinnis dorsalibus tribus: Linnaeus, 1759: 246, no. 21 (after Linnaeus, 1754a).

Lophius cute alepidota laevi, cathetoplateus: Gronovius, 1763: 58, no. 208 (*Habitat in mari Fucoso*).

Guaperua (not of Marcgrave): Hjortberg, 1768: 353 (*in den Seegewächse Sargazo gefanden worden*). Walbaum, 1792: 495 (after Hjortberg, 1768).

Lophius histrio (not of Linnaeus): Bloch, 1785: 13, pl. 111 (*L. corpore scabro*; plate is a composite showing head and body of *H. histrio*, with illicium and esca of *Antennarius scaber*). Bloch, 1787: 10, pl. 111 (after Bloch, 1785). Bonnaterre, 1788: 15, pl. 9, fig. 28 (*La Baudroie Tachée; L. corpore compresso, scabro*; figure after Bloch, 1785). Gmelin, 1789: 1481 (in part, confused with *Antennarius scaber*, after Bloch, 1785). Walbaum,

1792: 494 (*corpore scabro*, confused with *A. scaber*). Bloch and Schneider, 1801: 141, 142 [*varietas* a. is *Antennarius scaber*; *var.* b. is *A. pictus*; *var.* c. *marmoratus* is a *nomen dubium* (see p. 354); *var.* d. is *Fowlerichthys ocellatus*)]. Russell, 1803: 12, pl. 19 (description and plate represent *A. hispidus*). Sonnini, 1803: 160, 186, pl. 10, fig. 1 (figure after Bloch, 1785). Shaw, 1804: 384, pl. 164 (in part; "Harlequin Angler," illicium "dividing at top into two dilated oval and pointed appendages"; plate after Bloch, 1785). Reichenbach, 1840: 102, pl. 37, fig. 229 (figure after Bloch, 1785). Bauchot, 1958: 139 (confused with *A. scaber*).

La Lophie Histrion (not of Linnaeus): Lacepède, 1798: 302, 321, pl. 14, fig. 1 [description, and Latin polynomial in footnote, based on *Antennarius antenna tricorni* of Commerson, MSS; *voisine des côtes orientales de l'Afrique* (= Mauritius); esca bifid in description, trifid in figure; represents *Antennarius striatus* but confused in part with *Histrio histrio*]. Sonnini, 1803: 160, 186, pl. 10, fig. 1 (after Lacepède, 1798, and Bloch, 1785).

La Lophie Unie: Bosc, 1803: 313, pl. E30 (*Lophie variee* in caption of plate; two specimens, MNHN A.3829 and B.2949, Carolina; on floating weed between Europe and America). Bosc, 1817: 183, pl. E30 (after Bosc, 1803).

Lophius laevis (not of Lacepède): Latreille, March 1804: 73 (Latin binomial for *La Lophie Unie* of Bosc, 1803; no known type material; senior primary homonym of *Lophius laevis* Lacepède, May 1804, a brachionichthyid; see Whitley, 1936, and Pietsch, 1985a: 95).

Lophius raninus Tilesius, 1809: 245 (249), pls. 16, 17 (original description, no known type material, Sea of Japan). Tilesius, 1812: 299, pl. 17, fig. 5 (after Tilesius, 1809).

Lophius cocinsinensis Shaw, 1811, pl. 1012 ("Cochinchina Lophius," original description, no known type material, observed near Bay of Turon, coast of Cochin China; Fig. 133). Valenciennes, 1837: 404 (in discussion; misspelled *concincinensis*). Günther, 1861a: 185

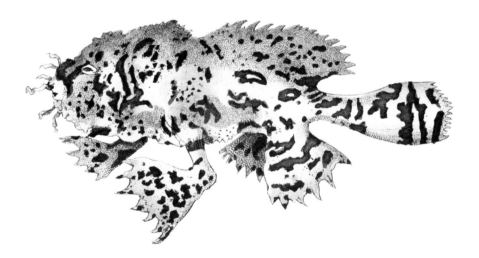

Fig. 133. *Histrio histrio* (Linnaeus). Holotype of *Lophius cocinsinensis* Shaw, about 60 mm SL, specimen lost, said to have been "observed near the Bay of Turon, on the coast of Cochinchina, adhering to a piece of salted meat, which, according to the usual custom of sailors, had been lowered for some time in the water . . . copied from the original drawing of the ingenious Mr. Alexder of the British Museum, who accompanied the voyage of Lord Macartney to China" (Shaw, 1811, pl. 1012). Of this drawing, Valenciennes (1837: 404) wrote: "If it is possible to decipher the monstrous figure provided, it is necessary to [assume that it] was made from nature by a simple amateur. It is believable that the author, as it often happens, represented what he saw in a vague manner, without care." Courtesy of Stephan Atkinson and the Picture Library, Natural History Museum, London.

("Cochinchina Lophius" in synonymy of *Antennarius marmoratus*). Whitley, 1931: 328 (synonym of *Pterophrynoides histrio*).

Chironectes variegatus Rafinesque, 1814: 19 [original description, "compressed; gray olive above with large spots and streaks of black and small white spots; two appendages above eye; between Azores and U.S.," translation of Jordan (1917a: 278); no known type material; not same as *Lophius variegatus* Cuvier (1816) and *Chironectes variegatus* Cuvier (1829) and Valenciennes (1837) (= *Antennarius pictus*)].

Lophius gibbus Mitchill, 1815, pl. 4, fig. 9 ("mouse-fish," original description, no known type material, New York). Cuvier, 1817b: 424 (same as *Chironectes laevigatus*). Mitchill, 1818: 325 (between St. Croix and New York, in gulfweed). DeKay, 1842: 164, pl. 24, fig. 74 (after Mitchill, 1815).

Lophius laevigatus: Bosc, in Cuvier, 1816: 310 (Latin binomial for *La Lophie Unie* of Bosc, 1803; two known syntypes, lectotype MNHN A.3829 by subsequent designation of Pietsch and Grobecker, 1987, paralectotype MNHN B.2949, both from Carolina; Fig. 129A). Bory de Saint-Vincent, 1826: 495 (after Bosc, 1803, and Bosc, in Cuvier, 1816).

Chironectes histrio: Cloquet, 1817: 598 (new combination; includes *L. histrio* Linnaeus and *L. tumidus* Osbeck). Swainson, 1838: 201–202, fig. 34 (fig. after Bloch, 1785).

Chironectes laevigatus: Cloquet, August 1817: 598 (new combination, based on *La Lophie Unie* of Bosc, 1803). Cuvier, October 1817b: 423, pl. 16, upper fig. (additional material, MNHN A.3563, A.3564, Mauritius, MNHN A.4603, Surinam; "skin smooth as a frog"). Cuvier, 1829: 252 (same as "*L. gibba* Mitch."). Valenciennes, 1837: 399 (description; same as *L. gibbus* Mitchill; Carolina, Surinam, Mauritius). Storer, 1839a: 383 (Massachusetts). Storer, 1839b: 73 (after Storer, 1839a). Valenciennes, 1842: 189 (after Valenciennes, 1837). DeKay, 1842: 165, pl. 27, fig. 83 (Charleston to Boston). Storer, 1855: 270, pl. 18, fig. 3 (after Storer, 1839a, 1839b). Pietsch, 1985a: 66, figs. 3, 4, table 1 (original manuscript sources for *Histoire Naturelle des Poissons* of Cuvier and Valenciennes).

Lophius calico Mitchill, 1818: 326 (original description, "skin smooth and scaleless"; type material unknown, New York).

Antennarius laevigatus: Schinz, 1822: 499 (new combination, after Cuvier, 1817b). Bleeker, 1859b: 256 (synonymy, tropical Atlantic). Bleeker, 1860a: 55, 76 [in part, synonymy includes *L. histrio* var. b *pictus* Bloch and Schneider, 1801 (= *Antennarius pictus*); east coast of U.S., Surinam, Cape of Good Hope, Mauritius].

Antennarius histrio (not of Linnaeus): Schinz, 1822: 501 [new combination; in part; reference to *L. histrio* of Bloch, 1785, based on composite figure of *H. histrio* and *A. scaber*]. Schinz, 1836, pl. 68, fig. 3 (after Schinz, 1822). Günther, 1861a: 188 ("skin very rough, covered with small spines"; synonymy, description, and material refer to *A. scaber* and *A. multiocellatus*). Kner, 1865: 192 (in part; synonymy includes *L. histrio*, with reference to Bloch, 1785, but also *C. scaber*; Rio de Janeiro). Poey y Aloy, 1868: 404 (includes *C. scaber* and *L. spectrum*). Poey y Aloy, 1876: 134 (in part, confused with *A. scaber*; Trinidad, Martinique). Steindachner, 1895: 24 [confused with *A. tigris* (= *A. scaber*); Liberia]. Bauchot, 1958: 142 (in part, confused with *A. scaber*).

Lophius geographicus Quoy and Gaimard, 1825: 335, pl. 65, fig. 3 (original description, type material unknown, not found in MNHN, New Guinea; dating after Sherborn and Woodward, 1901: 392).

Antennarius nitidus Bennett, 1827: 375, pl. 9, fig. 2 (original description based on four specimens in "collection of the Zoological Society," but none found in BMNH; locality uncertain, perhaps Cape of Good Hope). Günther, 1861a: 187 (in synonymy of *A. marmoratus* Var. γ. *ranina*).

Chironectes tumidus: Cuvier, 1829: 252 (new combination; after Linnaeus, 1754b). Valenciennes, 1837: 397 (description of additional material, MNHN 171, 172, from masses of floating seaweed, Atlantic Ocean; same as *L. laevigatus* Bosc, in Cuvier, and *L. tumidus* of Osbeck and Linnaeus). Valenciennes, 1842: 189 (after Valenciennes, 1837). Pietsch, 1985a: 63, 65, table 1 (original manuscript sources for *Histoire Naturelle des Poissons* of Cuvier and Valenciennes).

Lophius gibba: Cuvier, 1829: 252 (after Mitchill, 1815; same as *C. laevigatus*).

Chironectes marmoratus (not of Shaw): Cuvier, 1829: 252 [in part based on *Lophius raninus* Tilesius, 1809 (=*H. histrio*), but also includes *L. histrio* var. c. *marmoratus* Bloch and Schneider, 1801 (*nomen dubium*); not same as *L. marmoratus* Shaw, 1794 (*nomen dubium*), see p. 354)]. Lesson, 1831: 144, pl. 16, fig. 2 [description, authorship given to Cuvier; MNHN A.4593 and A.4594 erroneously listed as types by Le Danois (1964: 131); New Guinea]. Valenciennes, 1837: 402 (description; three additional specimens, MNHN 1616, Malabar). Valenciennes 1842: 189 (in part; after Cuvier, 1829). Temminck and Schlegel, 1845: 159, pl. 81, fig. 1 (Japan). Jerdon, 1853: 149 (Tellicherry, India). Boeseman, 1947: 135 (Burger and Von Siebold collection, Japan). Pietsch, 1985a: 68, figs. 5, 6, table 1 (original manuscript sources for *Histoire Naturelle des Poissons* of Cuvier and Valenciennes).

Batracoide: Bibron, 1833: 399, pl. 45, fig. 3 (figure shows *H. histrio*).

Chironectes pictus (not of Shaw) Valenciennes, 1837: 393, pl. 364 [original description, skin smooth, without spines, cutaneous appendages over entire body; lectotype by subsequent designation of Le Danois (1964: 130), MNHN A.4603; paralectotypes lost, not in MNHN; Surinam, central Atlantic; not same as *Lophius pictus* Shaw, 1794 (=*Antennarius pictus*)]. Agassiz, 1872: 155 (description of "nest"). Gudger, 1937: 368 (reference to Agassiz, 1872). Pietsch, 1985a: 63, fig. 2, table 1 (original manuscript sources for *Histoire Naturelle des Poissons* of Cuvier and Valenciennes). Pietsch et al., 1986: 140 (in type catalog).

Guaperua sinensis: Valenciennes, 1837: 397 (a reference to *Balistes, quae Guaperua Chinensis* Linnaeus, 1747; see synonymy, above).

Chironectes nesogallicus Valenciennes, 1837: 401 [original description, three specimens, lectotype MNHN A.4612 by subsequent designation of Le Danois (1964: 130); paralectotypes MNHN A.3563 and A.3564; all from Mauritius]. Pietsch, 1985a: 63, 66, table 1 (original manuscript sources for *Histoire Naturelle des Poissons* of Cuvier and Valenciennes). Pietsch et al., 1986: 140 (in type catalog). Fricke et al., 2009: 29 (synonym of *Histrio histrio*, Réunion).

Chironectus histrio: Swainson, 1839: 330 [emendation of generic name; in part, "body naked," but reference also given to *L. histrio* of Bloch, 1785 (=*Antennarius scaber*), and *L. pictus* Shaw, 1804 (=*A. pictus*)].

Chironectes gibbus: DeKay, 1842: 164, pl. 24, fig. 74 (new combination; after Mitchill, 1815). Storer, 1846: 382 (after Mitchill, 1815; DeKay, 1842).

Antennarius nesogallicus: Guérin-Méneville, 1844: 26, pl. 41, fig. 2 (new combination; *Mer des Indes*, Mauritius). Bleeker, 1853a: 69, 84 (Solor). Bleeker, 1859b: 129 (Flores, Solor).

Cheironectes pictus (not of Shaw) **Var. *vittatus*** Richardson, 1844: 15, pl. 9, figs. 3, 4 (original description, single specimen, not found in BMNH, 71.5 mm total length, Atlantic, in *Sargassum*; emendation of generic name).

Chironectes sp.: Düben, 1845: 111 (Norway).

Cheironectes raninus: Richardson, 1846: 203 (new combination, Japan).

Chironectes arcticus Düben and Koren, 1846: 72, pl. 3, figs. 4, 5 (original description, holotype ZMUB 430, White Sea, Vardø, Norway; a second smaller specimen collected simultaneously but lost prior to description). Nilsson, 1855: 257 (after Düben and Koren, 1846). Fries et al., 1893: 146 (after Düben and Koren, 1846).

Chironectes barbatulus Eydoux and Souleyet, 1850: 184, pl. 5, fig. 1 (original description, single specimen, holotype MNHN A.4596, locality unknown). Schultz, 1957: 52 ("unidentifiable"). Bauchot et al., 1982: 64, 72 (dating of Eydoux and Souleyet; a synonym of *Histrio histrio*; specific name on page 72 misspelled *barbatulatus*). Pietsch et al., 1986: 138 (in type catalog).

Antennarius raninus: Cantor, 1849: 1184 (new combination, description, Sea of Pinang). Cantor, 1850: 202 (after Cantor, 1849). Bleeker, 1853a: 69 (Solor). Bleeker, 1858b: 12 (synonymy, description, Celebes). Bleeker, 1859a: 3 (New Guinea). Bleeker, 1859b: 129 (Indonesia). Bleeker, 1863: 29 (tropical West Africa).

Chironectes histrio (not of Linnaeus): Poey y Aloy, 1853: 217 (in part, confused with *A. scaber*).

Antennarius marmoratus (not of Shaw) **Var. α. picta** (not of Shaw): Günther, 1861a: 186 [new combination; in part, BMNH 1853.11.13.92, 1858.7.24.6-7, 1859.6.3.11-12, 1860.5.26.40; synonymy includes *L. histrio* var. b. *pictus* Bloch and Schneider (= *A. pictus*); Atlantic]. Collett, 1875: 69 (reference to *C. arcticus* Düben and Koren, 1846).

Antennarius marmoratus (not of Shaw) **Var. β**: Günther, 1861a: 186 (BMNH 1848.3.16.171, 1852.11.4.11, Atlantic and Indian Ocean).

Antennarius marmoratus (not of Shaw) **Var. γ. ranina**: Günther, 1861a: 187 (after *L. raninus* Tilesius; BMNH 1860.3.19.957, Atlantic and Indian Oceans, Chinese and Japanese seas).

Antennarius marmoratus (not of Shaw) **Var. δ. gibba**: Günther, 1861a: 187 (after *L. gibbus* Mitchill; BMNH 1847.7.14.10, 1857.12.18.3, Atlantic, Australia).

Antennarius marmoratus (not of Shaw) **Var. ε. marmorata**: Günther, 1861a: 187 (in part, i.e., BMNH 1858.6.1.7, 1860.3.19.427, Madagascar and East Indies; includes *L. histrio* var. c. *marmoratus* Bloch and Schneider, *nomen dubium*).

Antennarius marmoratus (not of Shaw) **Var. ζ**: Günther, 1861a: 188 (BMNH 1852.9.13.135, 1855.9.19.796, Atlantic).

Antennarius barbatulus: Günther, 1861a: 185, 188 (new combination; after Eydoux and Souyelet, 1850).

Antennarius marmoratus (not of Shaw): Günther, 1861b: 361 (Ceylon). Bleeker, 1865c: 275 (Ambon). Day, 1865: 121 (Malabar). Kner, 1865: 192 (after Valenciennes, 1837). Playfair, 1866: 70 (Zanzibar). Playfair, 1867: 861 (Seychelles). Schmeltz, 1869: 19 (Kandavu Island, Fiji; Atlantic). Cope, 1871: 480 (St. Martin). Day, 1876: 272 (Red Sea, East African coast to Malay Archipelago and beyond). Günther, 1876: 162, pl. 100, fig. A (Indian Ocean). Thomson, 1878(1): 188, fig. 44, (2): 17 (in gulfweed, "nests"). Schmeltz, 1879: 48 (Atlantic, Südsee). Günther, 1880: 44 (BMNH 1879.5.14.525, Kai Islands, Moluccas). Tremeau de Rochebrune, 1883: 91 (Gorée, Dakar). Klunzinger, 1884: 125 (Red Sea). Steindachner and Döderlein, 1885: 194 (Japan). Meyer, 1885: 28 (Ternate Island, Moluccas). Vaillant, 1887: 732 (construction of "nest"). Day, 1889: 233 (BMNH 1889.2.1.3570-1, Ceylon). Steindachner, 1895: 23 (Liberia). Gilchrist, 1902: 130 (South Africa). Steindachner, 1903: 432 (Ternate Island, Moluccas). Boulenger, 1904: 720 (among *Sargassum*). Annandale and Jenkins, 1910: 20 (in key). Murray and Hjort, 1912: 611, 615, 633, 671, fig. 471, pl. 5 (figs. 1, 2) (Sargasso Sea). Weber, 1913: 563 (Indonesia). Phisalix, 1922: 582, 609 (ciguatoxic). Duncker and Mohr, 1929: 71 (New Guinea, *aus treibenden Tangmassen*). Maass-Berlin, 1937: 198 (ciguatoxic). Wagner, 1939: 285, figs. 1–7 (aquarium observations). Cott, 1940: 340, 341, fig. 70 (adaptive coloration). Padmanabhan, 1958: 85 (early development). Halstead, 1978: 348

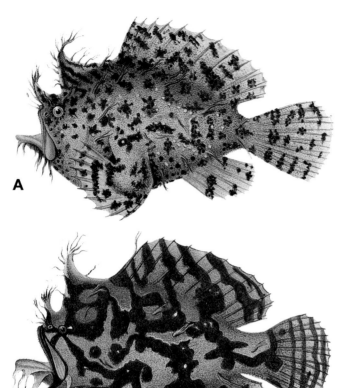

A

B

Fig. 134. *Histrio histrio* (Linnaeus). (**A**) holotype of *Antennarius lioderma* Bleeker, RMNH 6289, 82 mm SL, Ambon, Indonesia (after Bleeker, 1865b, pl. 199, fig. 8). (**B**) Bleeker's *Antennarius marmoratus*, authorship attributed to Günther, specimen unknown (after Bleeker, 1865b, pl. 198, fig. 4). Drawings by Ludwig Speigler under the direction of Pieter Bleeker.

(ciguatoxic). Thresher, 1984: 41 (egg rafts). Fricke et al., 2009: 29 (synonym of *Histrio histrio*).

Chironectes mesogallicus: Guichenot, 1863: 27 (misspelling of *nesogallicus*, Réunion). Fricke et al., 2009: 29 (after Guichenot, 1863; synonym of *Histrio histrio*).

Cheironectes laevigatus: Gill, 1863: 90 (type species of *Pterophryne* Gill).

Chironectes sonntagii Müller, 1864: 180 (original description, type material unknown, Gulf Stream in floating seaweed).

Antennarius lioderma Bleeker, 1865a: 178 (original description; one specimen, holotype RMNH 6289, Ambon; Fig. 134A). Bleeker, 1865b: 24, pl. 199, fig. 8 (after Bleeker, 1865a).

Antennarius marmoratus (not of Shaw) **Var. marmorata**: Bleeker, 1865b: 23, pls. 198 (fig. 4), 199 (fig. 1) (synonymy includes all varieties of *A. marmoratus* Günther, 1861a; Indonesia, New Guinea; Fig. 134B).

Pterophryne variegatus (not of Cuvier, not of Valenciennes): Poey y Aloy, 1868: 405 (new combination; after Rafinesque, 1814; *llevado por las algas flotantes*).

Pterophryne picta (not of Shaw): Goode, 1876: 20 [new combination; in part, synonymy includes *L. histrio* var. b *pictus* Bloch and Schneider, 1801 (= *Antennarius pictus*); warmer parts of Atlantic, pelagic in gulfweed]. Goode, 1877: 290 (West Indies).

Pterophryne laevigatus: Poey y Aloy, 1876: 135 (new combination; includes *Pterophryne variegatus* of Poey y Aloy, 1868; Cuba).

Pterophryne histrio: Gill, 1879a: 216 (new combination, synonymy, description; *Pterophrynoides* for *Pterophryne*). Gill, 1879c: 226 (taxonomic history). Gill, 1883: 556 (reference to Cuvier, 1817b). Ives, 1890: 344 (mimicry). Goode and Bean, 1896: 486 (St. Vincent, Atlantic). Bean, 1897: 373 (Long Island, New York). Smith, 1898: 109 (Woods Hole). Jordan and Evermann, 1898: 2716 (synonymy, description, tropical Atlantic; in key). Jordan, 1902: 368, fig. 2 (synonymy includes *L. histrio* var. c. *marmoratus* Bloch and Schneider, 1801, *nomen dubium*; Japan; Fig. 129C). Bean, 1903: 736 ("mousefish," New York). Gill, 1905b: 841 (not a "nest-maker"). Gudger, 1905: 841 (eggs, egg laying). Jordan, 1905: 550 (East Indies). Bean, 1906b: 89 (Bermuda). Jordan and Seale, 1906: 438 (New Guinea, East Indies). Gill, 1907: 63 (not a "nest-maker"). Jordan and Seale, 1907: 48 (Manila, Luzon). Gill, 1909: 570, 575, 601, 614, figs. 9, 45 (description, natural history, so-called "nests" of *Pterophryne* made by flying fish).

Gudger, 1912: 175 (North Carolina). Snyder, 1912: 450 (Japan). Jordan et al., 1913: 425, fig. 394 (Japan). Miranda-Ribeiro, 1915: 8 (Brazil). Thompson, 1918: 156 (synonymy, South Africa). Barnard, 1927: 1002 (South Africa). Gregory, 1928: 408 (locomotion). Borodin, 1928: 25 (Cuba). Gudger, 1937: 363, fig. 1 ("nests"). Norman, 1939: 111 (Arabian coast). Koefoed, 1944: 3, pl. 3, figs. 1, 2 (North Atlantic in *Sargassum*). Tokioka, 1961: 448 (Kyoto, Japan). Le Danois, 1964: 127, figs. 39–41, 64–66 (synonymy, description, tropical zones of all oceans; MNHN A.4610, A.4637 erroneously listed as types of *C. laevigatus* Cuvier; MNHN A.4603 erroneously listed as type of *C. pictus* Valenciennes; MNHN 171, 172 erroneously listed as types of *C. tumidus* Valenciennes). Pietsch, 2009: 260 (jet propulsion).

Pterophrynoides histrio: Gill, 1879a: 216 (new combination; *Pterophrynoides* for *Pterophryne*). Goode and Bean, 1882: 235 (after Gill, 1879a; Gulf of Mexico). Whitley, 1929b: 137, pl. 31, fig. 4 (New South Wales). Whitley, 1935: 302, fig. 4 (natural history). Whitley, 1948: 31 (Western Australia). Whitley, 1954: 28 (near Cooktown, Queensland). Whitley, 1968b: 38, fig. 2 (New South Wales). Fricke et al., 2011: 368 (synonym of *H. histrio*).

Antannarius marmoratus: Jones, 1879: 363 (adapted for life in *Sargassum*; nests).

Chironectes vittatus: Macleay, 1881: 576 (infra-subspecific name originally proposed by Richardson, 1844, elevated to rank of species-group; Port Jackson).

Antennarius inops Poey y Aloy, 1881: 340 (original description, single specimen, holotype USNM 37434, Puerto Rico). Evermann and Marsh, 1900: 334, 335 (Puerto Rico; in key). Beebe and Tee-Van, 1928: 271, fig. (Port-au-Prince, Haiti).

Pterophryne laevigata: Poey y Aloy, 1881: 340 (after Poey y Aloy, 1876; Puerto Rico). Stahl, 1882: 79, 165 (after Poey y Aloy, 1881).

Antennarius histrio: Jordan and Gilbert, 1883: 845 (tropical Atlantic). Lilljeborg, 1884: 775 (synonymy, Scandinavia). Fries et al., 1893: 145, text-fig. 41, pl. 10, fig. 3 (reference to *C. arcticus* of Düben and Koren, 1846). Collett, 1896: 38 (in *Sargassum*, Azores). Lönnberg, 1896: 30 (Linnaean type specimens). Ehrenbaum, 1901: 75 (reference to Düben and Koren, 1846). Zugmayer, 1911: 116 (*surface dans les sargasses*, central North Atlantic). Ehrenbaum, 1936: 163, fig. 136 (reference to *A. marmoratus* of Murray and Hjort, 1912). Koumans, 1953: 267 (Red Sea, Indonesia). Bauchot, 1958: 142 (in part, confused with *A. scaber*). Santos et al., 1997: 62 (synonym of *Histrio histrio*). Fricke et al., 2009: 29 (synonym of *Histrio histrio*).

Pterophrynoides gibbus: Garman, 1896: 81 (new combination, Tortugas "in Gulf Weed").

Pterophryne gibba: Jordan and Evermann, 1898: 2717 (new combination, synonymy, description, West Indies to Key West and Tortugas; in key). Evermann and Marsh, 1900: 334 (Puerto Rico). Barbour, 1905: 131 (60 specimens, Bermuda). Jordan, 1905: 550 (West Indies, Gulf Stream). Bean, 1906b: 89 (Bermuda). Gudger, 1929: 203 (Tortugas). Gudger, 1937: 364 (egg-rafts).

Pterophryne ranina: Jordan, 1902: 369, fig. 3 (new combination, synonymy includes *Antennarius nitidus* Bennett). Barbour, 1905: 132 (Bermuda). Jordan, 1905: 550 (East Indies). Gill, 1909: 602, fig. 42 (after Jordan, 1902). Jordan et al., 1913: 426, fig. 395 (after Tilesius, 1809; Wakanoura, Japan). Okada, 1955: 431, fig. 388 (Honshu to South China Sea).

Pterophryne tumida: Jordan, 1905: 549, fig. 500 (new combination; West Indies, Gulf Stream; Fig. 129B). Cott, 1940: 340, 341, fig. 71 (adaptive coloration).

Pterophryne marmoratus (not of Shaw): Franz, 1910: 86 (new combination, Japan).

Antennarius hispidus (not of Bloch and Schneider): Weber, 1913: 562 (in part, i.e., station 45, ZMA 116.505).

Antennarius gibbus: Metzelaar, 1919: 160 (new combination, Curaçao).

Peterophryne gibbo: Connolly, 1920: 7 (Gulf Stream in *Sargassum*).

Histrio jagua Nichols, 1920: 62 (original description, single specimen, holotype AMNH 7316, Bermuda). Beebe and Tee-Van, 1933: 249, fig. (Bermuda).

Histrio histrio: McCulloch, 1922: 122, pl. 41, fig. 356a (new combination, New South Wales). Whitley, 1927a: 8 (Fiji, after Schmeltz, 1869). Fowler, 1928: 476, figs. 81, 82 ("nests," Oceania). McCulloch, 1929: 406 (synonymy, Australia). Fowler, 1931: 367 (Oceania). Borodin, 1934: 121 (Fisher's Island, Florida). Fowler, 1936a: 408 (South Africa). Bigelow and Schroeder, 1936: 341 (Gulf of Maine). Fowler, 1936b: 1129, 1335 (synonymy, description, West Africa; in key). Fowler, 1938: 241 (Malaya). Barbour, 1942: 22, 35 ("nests," spawning; synonymy includes *Antennarius inops* Poey and *H. jagua* Nichols). Smith, 1949: 431, pl. 98, fig. 1243 (South Africa). Cadenat, 1951: 291, fig. 240 (Senegal). Delais, 1951: 145, fig. 1 (West Africa). Herre, 1953: 855 (synonymy, Philippines). Mosher, 1954: 141, pls. 1–3 (spawning behavior, early larval development). Gordon, 1955: 387, figs. (natural history). Munro, 1955: 288, pl. 56, fig. 837 (Ceylon; in key). Albuquerque, 1956: 1052 (Portugal). Schultz, 1957: 103, pl. 13, fig. D (sole member of newly proposed subfamily Histrioninae; synonymy; material from Atlantic, Indian, and Pacific Oceans compared, no significant morphological differences throughout range). Breder and Campbell, 1958: 135 (pigmentation). Wheeler, 1958: 246 (Gronovius collection, BMNH 1853.11.13.92, locality unknown). Rasquin, 1958: 333, text fig. 1, pls. 47–74 (ovarian morphology, early development). Duarte-Bello, 1959: 142 (Cuba). Poll, 1959: 365, fig. 123 (tropical West Africa). Fowler, 1959: 560 (Fiji). Adams, 1960: 55, figs. 1–5 (postlarval development). Ray, 1961: 230 (egg-raft morphology). Herald, 1961: 282, pl. 144 (natural history). Beaufort and Briggs, 1962: 197, fig. 47 (synonymy, description; Indo-Australian Archipelago). Smith and Smith, 1963: 61, pl. 56, fig. I (Seychelles). Marshall, 1964: 512, pl. 72, fig. 493 (Queensland, New South Wales). Ochiai, 1964: 41, fig. 1 (aberrant specimen, Nagasaki). Caldwell, 1966: 88 (Jamaica). Cervigón, 1966: 873, 874, fig. 373 (Caigüire, Venezuela; in key). Smith and Smith, 1966: 151, color fig. 1243 (Cape of Good Hope). Ueno and Abe, 1966: 235, fig. 11 (Hokkaido). Ueno, 1966: 535 (after Ueno and Abe, 1966). Chen et al., 1967: 11, fig. 5 (Taiwan). Munro, 1967: 583, pl. 78, fig. 1088 (New Guinea). Wickler, 1967: 546, 547 (feeding and locomotion, film no. E 60/1962). Böhlke and Chaplin, 1968: 715, 717, figs. (New England, Gulf of Mexico; in key). Weis, 1968: 554 (*Sargassum* community). Parin, 1968: 29, fig. 12 (*Sargassum* community). Wheeler, 1969: 583, fig. 392 (after Düben and Koren, 1846; Norway). Bagnis et al., 1970: 88 (ciguatoxic). Blache et al., 1970: 441, fig. 1116 (tropical West Africa). Le Danois, 1970: 83 (Red Sea). Ueno, 1971: 101 (Hokkaido and adjacent waters). Dooley, 1972: 7, 17, 21, 23, fig. 10, tables 3–4 (*Sargassum* community). Friese, 1973: 31 (natural history). Monod and Le Danois, 1973: 661 (synonymy, distribution). Smith, 1973: 219, fig. 1 (energy transformation). Friese, 1974: 7 (natural history). Winterbottom, 1974: 284, fig. 44 (musculature of spinous dorsal fin). Kailola, 1975: 48 (Fairfax Harbour, Papua). Sisson, 1976: 188, color figs. (natural history). Guitart, 1978: 786, fig. 618 (Cuba). Halstead, 1978: 348 (ciguatoxic). Martin and Drewry, 1978: 370, figs. 193–198 (early developement). Paulin, 1978: 485, fig. 1A (New Zealand). Pietsch, 1978a: 3 (western central Atlantic). Román, 1979: 351, fig. (Venezuela; in key). Yamakawa, 1979: 7 (Nansei Islands, Japan). Jones and Kumaran, 1980: 703, fig. 599 (Laccadive Archipelago). Myers and Shepard, 1980: 315 (Guam). Lindberg et al., 1980: 342, fig. (common names). Pietsch, 1981a: 2 (eastern central Atlantic). Pietsch, 1981b: 419 (osteology). Pietsch and Grobecker, 1981: 14 (aggressive mimicry). Bauchot et al., 1982: 72 (same as *Chironectes barbatulus* Eydoux and Souleyet; misspelled *barbatulatus*). Divita et al., 1983: 404 (Texas coast). Araga, 1984: 103, pls. 89-G, 89-H (Japan). Dor, 1984: 57 (Red Sea). Pietsch, 1984b: 37 (genera of frogfishes). Pietsch, 1984d: 321, fig. 164B (early life history). Pietsch, 1984e: 2 (Western Indian Ocean). Thresher, 1984: 40 (reproduction). Pietsch, 1985a: 63, 65, 66, 68, figs. 2–6, table 1 (original manuscript sources for *Histoire Naturelle des Poissons* of Cuvier and Valenciennes). Pietsch and Grobecker, 1985: 16 (cannibalism). Wheeler and Van Oijen, 1985: 105 (off Norway, Barents Sea; after Düben and Koren, 1846). Burnett-Herkes, 1986: 587, fig. (Bermuda). Pietsch, 1986a: 366, 369, fig. 102.9, pl. 13 (South Africa). Pietsch, 1986b: 1367, fig. (eastern North Atlantic). Allen and Swainston, 1988: 38 (northwestern Australia). Robins and Ray, 1986: 88, pl. 14 (circumtropical). Fish, 1987: 1046 (jet propulsion). Kuiter, 1987: 28, color figs. (camouflage, Flores). Pietsch and Grobecker, 1987: 199, figs. 3, 83–85, 128, pls. 37–40 (description, distribution, relationships). Scott and Scott, 1988: 238 (Canada). Edwards, 1989: 249 (tetrapod locomotion). Myers, 1989: 66, 69 (Indo-West Pacific, tropical Atlantic; in key). Paulin et al., 1989: 134 (New Zealand). Paxton et al., 1989: 279 (Australia). McAllister, 1990: 229 (Canada). Pietsch, 1990: 485 (tropical Eastern Atlantic). Pietsch and Grobecker, 1990a: 98 (aggressive mimicry). Pietsch and Grobecker, 1990b: 77 (after Pietsch and Grobecker, 1990a). Randall et al., 1990: 56, color fig. (Indo-Pacific and tropical Atlantic). Boschung, 1992: 79 (Alabama). Paugy, 1992: 570 (West Africa). Pietsch et al., 1992: 247 (Pacific Plate). Cunxin, 1993: 117 (South China Sea). Kuiter, 1993: 48 (southeastern Australia). Burgess et al., 1994: 205 (Cayman Islands). Goren and Dor, 1994: 14 (Red Sea). Randall, 1995: 82, 84, fig. 171 (Oman). Allen, 1997: 62 (tropical Australia). Arruda, 1997: 125 (Azores). Debelius, 1997: 92, color figs. (Azores). Grove and Lavenberg, 1997: 231 (Eastern Pacific records invalid). Kuiter, 1997: 40 (Australia). Lindberg et al., 1997: 222 (Sea of Japan). Okamura and Amaoka, 1997: 136, fig. (Japan). Randall et al., 1997a: 56, color fig. (Indo-Pacific and tropical Atlantic). Randall et al., 1997b: 11 (Ogasawara Islands). Santos et al., 1997: 63 (Azores). Hoese and Moore, 1998: 163, pl. 101 (Gulf of Mexico). McEachran and Fechhelm, 1998: 823 (Gulf of Mexico). Michael, 1998: 335, 353, 355, 356, color figs. (identification, behavior, captive care). Randall, 1998: 228 (zoogeography, offshore pelagic). Castro-Aguirre et al., 1999: 176 (Gulf of Mexico). Fricke, 1999: 109 (Réunion, Mauritius). Johnson, 1999: 724 (Moreton Bay, Queensland). Myers, 1999: 69 (after Myers, 1989). Ni and Kwok, 1999: 136, table 1 (Hong Kong). Pietsch, 1999: 2015 (western central Pacific; in key). Smith-Vaniz et al., 1999: 159, 161 (Bermuda; in key). Allen, 2000a: 62, pl. 12-15 (temperate and tropical seas worldwide). Laboute and Grandperrin, 2000: 132 (New Caledonia). Nakabo, 2000: 454 (Japan; in key). Pietsch, 2000: 597 (South China Sea). Sadovy and Cornish, 2000: 46, fig. (Hong Kong). Schmitter-Soto et al., 2000: 153 (Gulf of Mexico). Watson et al., 2000: 120 (spawning, early development). Yoneda et al., 2000: 315 (gelatinous egg masses). Hutchins, 2001: 22 (Western Australia). Shimizu, 2001: 22 (Japan). Briggs and Waldman, 2002: 57 (New York). Nakabo, 2002: 454 (Japan; in key). Pietsch, 2002a: 1051 (western central Atlantic). Pietsch, 2002b: 270–272, fig. 146 (species account, Gulf of Maine). Youn, 2002: 214, 542, fig. (Korea). Allen and Adrim, 2003: 25 (Papua to Sumatra, Indonesia). Allen et al., 2003: 367, color fig. (circumtropical). Collette et al., 2003: 100 (Navassa Island, West Indies). Greenfield, 2003: 56 (Kaneohe Bay, Hawaiian Islands). Menezes, 2003: 64 (Brazil). Myers and Donaldson, 2003: 614 (Mariana Islands). Schneidewind, 2003a: 84 (natural history). Heemstra and Heemstra, 2004: 122

(southern Africa). Letourneur et al., 2004: 209 (Réunion). Lieske and Myers, 2004: 47, color fig. (all tropical and temperate seas, except Eastern Pacific). Pietsch, 2004b: 236 (natural history). Randall et al., 2004: 8 (Tonga). Shedlock et al., 2004: 134, 144, table 1 (molecular phylogeny). Hemphil, 2005: 50, 51, color fig. (conservation). Mundy, 2005: 266 (Hawaiian Islands). Ortiz-Ramirez et al., 2005: 23 (early life history). Randall, 2005a: 73, color fig. (South Pacific). Randall, 2005b: 300, fig. 1 (mimicry). Schneidewind, 2005a: 23 (locomotion). Schneidewind, 2005b: 14 (after Schneidewind, 2005a). Schwartz, 2005: 145, 146, 148 (North Carolina). Allen et al., 2006: 642 (Australia). Calado, 2006: 394, 395, tables 4, 6 (Azores). Holthuis and Pietsch, 2006: 216, 217, color fig. 65-182 (*Klaauw Vis, Poisson-grenouille des Sargasses*; Lamotius paintings of Indo-West Pacific fishes). Jackson, 2006a: 783, 784, 786, tables 1, 2 (reproduction, early life history, depth distribution). Michael, 2006: 42 (behavior). Senou et al., 2006: 423 (Sagami Sea). Kuiter and Debelius, 2007: 116, 125, color figs. (Indonesia, Australia). Lugendo et al., 2007: 1222, appendix 1 (Zanzibar). Mejía-Ladino et al., 2007: 278, figs. 4, 6, tables 1–5 (species account; in key; Colombia). Porteiro and Afonso, 2007: 57, 58 (Azores). Randall, 2007: 131, color fig. (Hawaiian Islands; in key). Kwang-Tsao et al., 2008: 243 (southern Taiwan). Wirtz et al., 2008: 8 (Madeira). Allen and Erdmann, 2009: 594 (West Papua). Fricke et al., 2009: 28 (Réunion). Hanel et al., 2009: 148 (European seas). O'Connell et al., 2009: 110, appendix (off southeastern Louisiana). Pietsch, 2009: 260 (jet propulsion). Al-Jufaili et al., 2010: 18, appendix 1 (Oman). Carnevale and Pietsch, 2010: 622 (senior synonym of *Lophius laevis* Latreille, 1804). Golani and Bogorodsky, 2010: 16 (Red Sea). Miya et al., 2010: 7, 17, table 2 (molecular phylogeny). Motomura et al., 2010: 78 (Yaku-shima Island, southern Japan). Rogers et al., 2010: 577 (in mangroves, St. John, Virgin Islands). Fricke et al., 2011: 368 (New Caledonia). Obura et al., 2011: 11, table 8 (Phoenix Islands). Renous et al., 2011: 98 (locomotion). Allen and Erdmann, 2012: 154, color fig. (East Indies). Arnold and Pietsch, 2012: 120, 126, 128, figs. 1C, 4, 6, table 1 (molecular phylogeny). Diamond and Bond, 2013: 20, fig. (protective resemblance). Nakabo, 2013: 537, figs. (Japan). Page et al., 2013 (common names). Fricke et al., 2014: 35 (Papua New Guinea). Moore et al., 2014: 177 (northwestern Australia). Smith-Vaniz and Jelks, 2014: 30 (St. Croix, Virgin Islands). Garcia et al., 2015: 7 (northeastern Brazil). Stewart, 2015: 887 (New Zealand). Ayala et al., 2016: 92, 97, table 3 (larvae, Sargasso Sea). Christie et al., 2016: 31, 32 (egg rafts). Davies and Piontek, 2016: 76 (St. Eustatius). Farina and Bemis, 2016: 209, 212, 214 (gill ventilation). Joshi et al., 2016: 40 (Gulf of Mannar, Tamil Nadu, India). Pietsch, 2016: 2052, 2053 (eastern central Atlantic; in key). Wright et al., 2016: 120 (New York). Wirtz et al., 2017: 51 (Cape Verde).

Pterophryne historio: Gregory, 1928: 408 (locomotion; misspelling of specific name).

Histrio gibbus: Meek and Hildebrand, 1928: 1010 (new combination; Colon and Porto Bello, Panama). Beebe and Tee-Van, 1928: 270, fig. (Port-au-Prince, Haiti). Poll, 1949: 268 (Isla Margarita, Venezuela; San Salvador Island, Bahamas).

Histrio pictus (not of Shaw): Beebe and Tee-Van, 1933: 249, fig. (Bermuda). Roule and Angel, 1933: 59 (new combination; mid-Atlantic, on surface). Bigelow and Schroeder, 1953: 541, fig. 287 (Gulf of Maine). Straughan, 1954: 277, figs. (natural history).

Histrio raninus: Fowler, 1936b: 1130 (new combination; description after Günther, 1861a; West Africa; in key).

Antennarius sp.: Norman, 1939: 112 (surface, association with *Sargassum*).

Antennarius commersonii (not of Latreille): Fowler, 1940a: 800 (USNM 82842, Maui?).

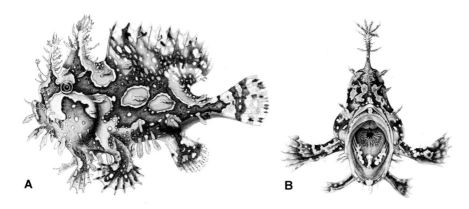

A　　　**B**

Fig. 135. *Histrio histrio* (Linnaeus). (**A**, **B**) lateral and anterior views of Longley and Hildebrand's *Histrio gibba*, USNM 88097, 87 mm TL, "carnivorous, voracious, and peculiar in form and structure," Tortugas, Florida. Painted from life by Manson Valentine; after Longley and Hildebrand (1941: 303, pl. 34); courtesy of Lisa Palmer and the Smithsonian Institution, NMNH, Division of Fishes, Washington, DC.

Histrio gibba: Beebe and Tee-Van, 1933: 248, fig. (Bermuda). Longley and Hildebrand, 1941: 303, pl. 34 (Tortugas; Fig. 135). Fowler, 1944: 67 (Gulf Stream, Caribbean). Breder, 1949: 94, pl. 8, figs. 1–3 (camouflage, spawning). Atz, 1951b: 135 (protective resemblance).

Antennarius biocellatus (not of Cuvier): Smith, 1949: 431, pl. 98, fig. 1237 (figure represents *H. histrio* with illicium added; East London to Delagoa Bay, South Africa).

Lophius pelagicus Banks, in Beaglehole, 1962: 272 (original description, no known type material, from "Gulph-weed," Eastern Atlantic).

Histrio ranina: Ueno, 1966: 536 (Hokkaido).

Histiophryne bigibba (not of Latreille): Le Danois, 1970: 84 (material is *H. histrio*).

Histrio histrio (not of Linnaeus): Grant, 1972, color fig. 87 (plate represents *Abantennarius coccineus*). Grant, 1978: 601, color fig. 264 (after Grant, 1972).

Chironectes barbatulatus: Bauchot et al., 1982: 72 (error for *Chironectes barbatulus* Eydoux and Souleyet).

MATERIAL EXAMINED

Three hundred and seventy-seven specimens, 7.0 to 141 mm SL.

Lectotype of *Lophius histrio*: ZMU Linnesamling nr. 100 (Chin. Lagerstr. nr. 21), 74 mm, probably China, Magnus Lagerström, donations to Academy at Uppsala (see Pietsch and Grobecker, 1987, fig. 85).

Paralectotypes of *Lophius histrio*: ZMU Linnesamling nr. 100 (Chin. Lagerstr. nr. 21), 2 (18–26 mm), data as for lectotype.

Lectotype of *Lophius laevigatus*: MNHN A.3829, 49 mm, Carolina, Bosc (Fig. 129A).

Paralectotype of *Lophius laevigatus*: MNHN B.2949, 27 mm, data as for lectotype.

Lectotype of *Chironectes pictus*: MNHN A.4603, 42.5 mm, Surinam, Levaillant.

Lectotype of *Chironectes nesogallicus*: MNHN A.4612, 30 mm, Mauritius, Péron and Lesueur.

Paralectotypes of *Chironectes nesogallicus*: MNHN A.3563, A.3564, 2 (33–50 mm), Mauritius, Mathieu.

Holotype of *Chironectes arcticus*: ZMUB 430, 28.5 mm, Vardø, Norway.

Holotype of *Chironectes barbatulus*: MNHN A.4596, 64.5 mm, *Voyage de la Bonite*, locality unknown, Eydoux and Souleyet.

Holotype of *Antennarius lioderma*: RMNH 6289, 82 mm, Ambon, Moluccas, Bleeker (Fig. 134A).

Holotype of *Antennarius inops*: USNM 37434, 47.5 mm, Puerto Rico, Poey.

Holotype of *Histrio jagua*: AMNH 7316, 110 mm, Bermuda, Mowbray.

Additional material: *Atlantic*: AMS, 1 (30 mm); ANSP, 10 (13.5–49.5 mm); BMNH, 11 (9.5–89 mm); CAS, 1 (33.5 mm); FSBC, 15 (12–55 mm); LACM, 1 (91 mm); MCZ, 20 (7.0–12.5 mm); MNHN, 17 (23–85.5 mm); SU, 6 (41.5–73 mm); UF, 6 (8.0–14 mm); UMMZ, 47 (10.5–88 mm); UW, 11 (8.0–78 mm). *Indian Ocean and Red Sea*: AMS, 1 (45 mm); BMNH, 6 (11.5–72 mm); HUJ, 1 (59.5 mm); MNHN, 10 (33–107 mm); SAIAB, 24 (11.5–110.5 mm); UMMZ, 1 (45 mm). *Indonesia and New Guinea*: AMS, 2 (37.5–67 mm); BMNH, 4 (50–53 mm); BPBM, 1 (49 mm); KFRL, 2 (16–37 mm); MNHN, 2 (61–65 mm); SIO, 1 (24 mm); USNM, 1 (13.5 mm); ZMA, 4 (11.5–15.5 mm). *South China Sea and Philippines to Japan*: BMNH, 4 (43–94 mm); FAKU, 5 (73–132 mm); FMNH, 7 (16–83.5 mm); SU, 6 (53.5–89.5 mm); UMMZ, 7 (62–141 mm); USNM, 1 (7.5 mm); UW, 5 (12–53 mm). *Australia*: AMS, 31 (32–123 mm); BMNH, 1 (77 mm); MNHN, 2 (19–60 mm); NMV, 1 (33 mm); NTM, 2 (70–80 mm); WAM, 35 (7.5–112 mm). *New Zealand*: AIM, 1 (39 mm); BMNH, 1 (68 mm). *Marianas, New Caledonia, and Tonga*: AMS, 1 (68 mm); BPBM, 1 (48 mm); SIO, 1 (33 mm); UG, 1 (55 mm). *Polynesia*: MNHN, 2 (18–58 mm). *Hawaiian Islands*: USNM, 2 (21.5–23 mm); locality data questioned by Springer, 1982: 13; see Zoogeography, p. 389). *Galápagos Islands*: USNM, 9 (12.5–90 mm), all nine said to have been collected by a Captain Herendeen, first cataloged at USNM in 1877; locality data questioned by Springer, 1982: 13 (see Zoogeography, p. 389). *Locality unknown*: BMNH, 4 (36–76 mm); MNHN, 4 (23–80.5 mm); NRMS, 4 (32–116 mm; "old collection" possibly containing Linnaean type specimens; see Fernholm and Wheeler, 1983: 227); UW, 13 (11.5–59).

DIAGNOSIS AND DESCRIPTION

As given for the genus.

Largest known specimen: 141 mm SL (UMML 204179).

DISTRIBUTION

Histrio histrio has the broadest longitudinal and latitudinal range of any antennariid (Pietsch and Grobecker, 1987; Pietsch, et al., 1992; Fig. 274, Table 22). Its distribution coincides largely with that of floating *Sargassum* (Adams, 1960: 79; Dooley, 1972: 21; Bortone et al., 1977: 62; Coston-Clements et al., 1991: 7), but where *Sargassum* is uncommon, it is found with other kinds of flotsam, including anthropogenic debris. In the Western Atlantic, this species extends from the Gulf of Maine and Georges Bank to the mouth of the Rio de la Plata, Uruguay, including the Gulf of Mexico and throughout the Caribbean. On the eastern side of the Atlantic, it is apparently quite rare; we have seen specimens only from Madeira, the Azores, and off the West African coast, and there is a recent record of five individuals observed at Cape Verde (Wirtz et al., 2017). The record from Vardø, northern Norway (Düben and Koren, 1846), is no doubt based on a straggler taken northward by the North Atlantic and Norwegian Currents.

In the Indian Ocean, *H. histrio* is known from the tip of South Africa eastward to India and Sri Lanka, with verified records from Somalia, Oman, the Red Sea, Madagascar, Zanzibar, Réunion, and Mauritius. In the Western Pacific and along the western margin of the Pacific Plate it occurs from Hokkaido, Japan, to tropical Australia (about as far south as Perth in the west and Sydney in the east), including Hong Kong, Taiwan, the Philippines and Moluccas, and Papua New Guinea, as well as rare but verified occurrences at Guam, the Marianas, Tonga, New Caledonia, and the North Island of New Zealand (Fig. 274).

The presence of *H. histrio* farther east in the Pacific has been questioned. Pietsch and Grobecker (1987: 210) agreed with Springer (1982: 13) that the records of Fowler (1928, 1931), as well as those of Whitley (1927a), from Fiji (Kandavu) and the Society Islands (Raiatea and Huahine), all based on listings of *Antennarius marmoratus* and *A. multiocellatus* in Schmeltz (1869, 1877, 1879), are most likely erroneous. But the records listed by Schultz (1957: 105) from the Hawaiian and Galápagos Islands cannot be so easily dismissed. The first of these is a single lot of two specimens (USNM 82842) purported to have been collected at Maui (field no. 148) during the United States Exploring Expedition of 1838 to 1842, led by United States Navy Lieutenant Charles Wilkes. Fowler (1940a: 742, 800), who listed these same two specimens under the name *Antennarius commersonii* in his report on "The Fishes Obtained by the Wilkes Expedition," wrote "that many of these old specimens have reached into our day in a comparatively fair condition of preservation in alcohol, perishable as they are. . . . Less fortunate have been the labels and data accompanying them, as many were apparently lost or in other cases mixed, so that their origin is now conjecture." Despite the possibility of error regarding the Wilkes material, and with the knowledge that no subsequent reports of *H. histrio* from the Hawaiian Islands had ever been made (despite many years of intensive collecting), Pietsch and Grobecker (1987: 210) concluded that this species might occasionally reach these islands as stragglers brought in with *Sargassum* by ocean currents, particularly the Kuroshio Extension from Japan (Springer, 1982: 13).

Since that time, however, several independent reports of *H. histrio* in the Hawaiian Islands have come to light. Pietsch et al. (1992) described four collections of this species all from off the Kona Coast of Hawaii. The first of these was made on 12 January 1990 by John Spenser, a commercial fisherman, who discovered eight to ten specimens associated with an abandoned cargo net found floating approximately 8 km off Kawili Point; three of these were placed in a holding tank after which the largest (95 mm SL, now registered UW 21449, a female, with ovaries containing thousands of tiny, undeveloped eggs, each measuring approximately 0.2 mm in diameter) promptly ate the other two, each measuring about 50 mm SL (UW 21449 was subsequently maintained alive for about a month; thus, the two smaller specimens were completely digested by the time stomach contents could be examined).

The second collection was made on 23 June 1990 by Seth Sandler: a single juvenile specimen (approximately 10 mm SL) found associated with a plastic soft drink container about 170 m from shore off Kala'au O Kalakoni Point. The third and fourth collections were made on 4 January 1991 off Puhili Point by Jack Lovell: two adult specimens (41 and 54 mm SL) found together under a plastic mattress in a current line, now registered BPBM 34523, and a single juvenile (about 10 mm SL) associated with a piece of rope in the same current line. According to Lovell (personal communication, 4 January 1991), these were only three of some 10 specimens of *H. histrio* that had been collected by him off Kona within the last two years. In addition to these occurrences, Greenfield (2003) reported specimens of this species "rafting with *Sargassum*" in Kaneohe Bay, Oahu. The presence of *H. histrio* in the Hawaiian Islands is thus confirmed. The discovery of two small juveniles (both about 10 mm SL) perhaps indicates further that the recently collected specimens are the products of a breeding population rather than expatriates from the west.

Schultz's (1957: 105) listing of *H. histrio* from the Galápagos Islands is based on three lots and nine specimens (USNM 20403, 84612, and 92698) listed in the catalogs of the Fish Division of the United States National Museum as having been collected by a Captain Herendeen of Woods Hole, Massachusetts, and initially registered at the Museum in 1877. This Captain Herendeen was apparently Edward P. Herendeen, a captain of a San Francisco whaling vessel who occasionally sent to the National Museum specimens (mostly ethnological in nature) that he collected on voyages to the Arctic and the South Pacific. His association with the USNM was probably a result of his acquaintance with William H. Dall, Curator of Mollusks. Herendeen and Dall maintained an extensive correspondence during the 1870s and 1880s, but an examination of these letters, housed in the Smithsonian Institution Archives, failed to document Herendeen's travel to the Galápagos (William Cox, personal communication, Smithsonian Institution Archives, 24 August 1983). Finally, a thorough search for logbooks or journals that might document a Herendeen voyage that reached the Galápagos was unsuccessful. With no additional source of information, and in light of the fact that there are no other records of *H. histrio* from the Galápagos Islands (or for that matter from the entire Eastern Pacific), despite intensive collecting in this region, we conclude that the Herendeen material is highly suspect and probably in error.

The *Sargassum* community has been described as a worldwide, circumtropical phenomenon (see Dooley, 1972: 2), but this may not be strictly true (Bruce C. Mundy, personal communication, 11 July 2018). Best known and most common in the Atlantic Ocean, masses of floating *Sargassum* are rare or absent in the central Pacific, particularly in regions of low atolls (Tsuda, 1976; Phillips, 1995). This may help to explain the absence of *H. histrio* in areas to the east of the Eastern Pacific biogeographic barrier.

The type locality of *H. histrio* is probably China, based on the name *Guaperua Chinensis* applied by Linnaeus (1747) in his earliest mention of the species, and in his latter references (1754a, 1759) to the Chinese collections of Magnus Lagerström (1691–1759) sent to Linnaeus in the 1740s and later donated to the Academy at Uppsala (see the synonymies below; for Lagerström, see Franzén, 1977).

Adams (1960) found that larval and postlarval individuals of this species (up to 4 mm SL) occur at depths of 50 to 600 m, but no specimens larger than 4 mm SL were taken below 50 m. Dooley (1972) did not collect any specimens in *Sargassum* that were smaller than 6 mm SL. All of the material examined here (for which data were available) was taken essentially at the surface. For evidence of the relatively deep occurrence of frogfish larvae, see Boehlert and Mundy (1996).

COMMENTS

The basis for Linnaeus's (1758) description of *Lophius histrio* is complex, but worth reviewing in light of the fact that several authors (e.g., Cuvier, 1817b: 422; Valenciennes, 1837: 392; Gill, 1879c: 223) have written of the confusion caused when he combined all previous descriptions of antennariids under a single name. Linnaeus's earliest mention of *H. histrio* appeared in *Västgöta-resa förrättad år 1746* under the name *Balistes, quae Guaperua Chinensis* (Linnaeus, 1747: 137, pl. 3, fig. 5a, b; Fig. 136). Although he indicated his belief that *Guaperua* of Jonston (1650: 139), a name that originated from Marcgrave (1648: 150), represented a different species (*"Diversissima species a Guaperua* Jonst. Tab. 36. f. 6. *quam ex America possideo"*), no literary source is listed. This indicates that Linnaeus's

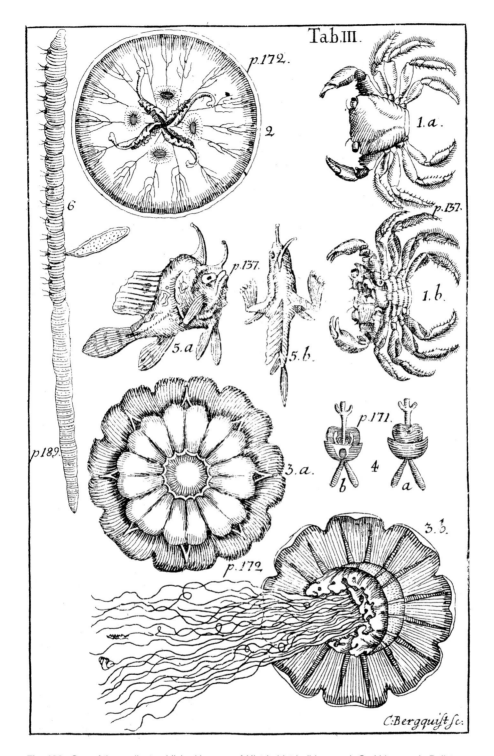

Fig. 136. One of the earliest published images of *Histrio histrio* (Linnaeus). Carl Linnaeus's *Balistes, quae Guaperua Chinensis,* lateral and dorsal views, shown surrounded by a lugworm, two views of a crab, and three of a medusa. After Linnaeus, *Västgöta-resa förrättad år 1746* (1747, pl. 3, fig. 5a, b).

description and accompanying figures, both clearly representing *H. histrio*, were based solely on one or more specimens.

In his catalogue of the museum of King Adolphus Frederick, Linnaeus (1754b: 56) again described *H. histrio*, but this time he called it *Lophius tumidus*, following the earlier use of this name in the published dissertation of Johann Laurentius Odhelius (Linnaeus, 1754a: 20):

> *tumidus. LOPHIUS pinnis dorsalibus tribus.*
>
> *Balistes quae Guaperua chinensis.* It. W-goth. 137. t. 3. f. 5 [i.e., Linnaeus, 1747]
>
>   *Guaperua.* Marcgr. bras. 150. Will. icht. 50. t. E. 2. f. 1. [i.e., Marcgrave, 1648, and Willughby, 1686]
>
>   *Habitat in Pelago inter Fucos natantes.*

Thus, in this later paper, Linnaeus (1754b) synonymized his *Balistes*, no doubt realizing by this time the distinction between *Balistes* and *Lophius*; and further, in a surprising reversal of his earlier opinion that this species was not the same as Jonston's (1650) *Guaperua*, he synonymized *Guaperua* of Marcgrave (1648), the sole basis for *Guaperua* of Jonston (1650) and Willughby (1686).

In the 10th Edition of the *Systema Naturae*, Linnaeus (1758: 237) introduced the name *Lophius histrio* as follows:

> *histrio.* 3. L. *compressus.*
>
>   Chin. Lagerstr. 21. *Lophius pinnis dorsalibus tribus.* [i.e., Linnaeus, 1754a]
>
>   Mus. Ad. Fr. 1. p. 56. *Balistes s. Guaperua chinensis.* [i.e., Linnaeus, 1754b]
>
>   It. Wgoth. 137. t. 3. f. 5. *Balistes s. Guaperua.* [i.e., Linnaeus, 1754b]
>
>   Marcgr, bras. 150. *Guaperua.* [i.e., Marcgrave, 1648]
>
>   Will. icht. 50. t. E. 2. f. 1. *Guaperua.* [i.e., Willughby, 1686]
>
>   Osb. iter. 305. *Lophius tumidus.* [i.e., Osbeck, 1757]
>
>   *Habitat in Pelago inter Fucum natantem.*
>
>   Pinnae D. 1, 1, 12. P. 10. V. 5. A. 7. C. 10.

The descriptions of Linnaeus (1747, 1754a, 1754b) that are cited in the 10th Edition are unequivocal and as far as can be determined based solely on specimens of *H. histrio* (Gill, 1879c: 225). The description of Osbeck (1757), made from one or more specimens collected in "The Grass-Sea" (most likely at Ascension Island in the year 1752; see Forster, 1771: 112), is also certainly of *H. histrio*. Osbeck's discussion of the possible function of the *foliaceous fulcra* that cover the body of *Histrio* is one of the earliest references to protective resemblance in animals (Forster, 1771: 112; see also Aggressive Mimicry, p. 411).

Guaperua, however, as used by Marcgrave (1648) and followed by Willughby (1686), is certainly not synonymous with *H. histrio*. In describing fishes observed in the vicinity of Recife, Brazil, Marcgrave wrote that his "specimen measures a little more than two fingers long from its mouth to its tail," and "between the eyes, . . . it has a fine fiber, about half a finger long. . . . It does not have scales, but is covered with a rough skin as in sharks and rays" (translated from the Portuguese edition of Marcgrave's "Historia," 1942: 150). The condition of the skin indicates that Marcgrave certainly had a member of the genus *Antennarius* rather than *Histrio*; the length of the "fine fiber" or illicium, approximately 30% of standard length, further indicates that Marcgrave's *Guaperua* is *Antennarius multiocellatus* (see Synonymy, p. 131).

Thus, although Linnaeus (1758) may have combined all earlier descriptions of frogfishes under the name *Lophius histrio*, only two species are confused. All evidence indicates that he personally examined only specimens of *H. histrio*. That he clearly had *H. histrio* in mind is further evidenced by his mention of its existence in *Sargassum* weed (*Habitat in Pelago inter Fucos natantem*), and also by the fact that the only extant Linnaean specimens of this family are *H. histrio*. To have been "unprepared for the polymorphous character of the type, his confusion under the synonymy is not at all to be wondered at" (Gill, 1879c: 225).

Following its original description by Linnaeus (1758), *H. histrio* was described under at least 20 additional specific names allocated to six nominal genera and involving 28 combinations of generic and subgeneric names (see Synonymy, p. 178). All evidence indicates, however, that these nominal forms represent a single, morphologically variable and widely distributed species. Schultz (1957: 104), in his revision of the Antennariidae, found no significant differences among material (452 specimens) compared from throughout its "entire range in tropical marine waters," and concluded that only one species was represented. Carl L. Hubbs, in an unpublished letter written to Leonard Schultz, dated 19 November 1957, and now in the Smithsonian Institution Archives, concurred: "I long ago came to the same conclusion regarding the monotypy of *Histrio*, made comparisons, and planned a paper." Results from a detailed molecular analysis may prove contradictory but so far no significant differences in populations have been discovered. Although COI sequences available to date indicate two distinct clusters (Atlantic versus Indo-Pacific, including Japan; Rachel J. Arnold, unpublished data), they are all still greater than 97% similar, providing further evidence that *Histrio* is probably monotypic.

Genus *Abantennarius* Schultz

Abantennarius Schultz, 1957: 66 (type species *Antennarius duescus* Snyder, 1904, by original designation).

DIAGNOSIS

Abantennarius is unique among antennariids in having the following combination of character states: darkly pigmented basidorsal spot usually present; illicium usually about equal to length of second dorsal-fin spine (distinctly shorter than second dorsal-fin spine in *A. dorehensis*; about one-and-a-half to two times the length of second spine in *A. rosaceus* and *A. analis*); anterior end of pterygiophore of illicium terminating distinctly posterior to symphysis of upper jaw; esca ranging in morphology from a simple, unpigmented sphere of folded tissue to an elongate, usually tapering appendage with several slender filaments and a cluster of darkly pigmented, spherical swellings at base, each swelling usually terminating in an elongate, slender filament; second dorsal-fin spine free, not connected to head by membrane, the posterior surface usually devoid of dermal spinules (especially in specimens greater than 30 mm SL); dermal spinules of second dorsal-fin spine often tightly clustered, leaving small naked areas between (most obvious in the smallest known individuals of each species, and probably best developed in *A. rosaceus*); distal two-thirds of maxilla naked and tucked beneath folds of skin, only the extreme proximal end covered directly with spinulose skin; caudal peduncle present or absent; endopterygoid present; pharyngobranchial I present; epural present; pseudobranch present; swimbladder present; dorsal-fin rays 11 to 14 (usually 12 or 13), as many as six (usually only three) posteriormost bifurcate; anal-fin rays six to eight (usually seven), all bifurcate; all rays of caudal fin bifurcate; pectoral-fin rays eight to 13 (usually nine to 12), all simple; usually only posteriormost ray of pelvic fin bifurcate (all simple in *A. duescus* and *A. analis*); vertebrae 19 (Tables 1, 2).

Table 10. Comparison of Distinguishing Character States of Species of the Genus *Abantennarius* (lengths as percentage of SL)

Character	A. analis	A. bermudensis	A. coccineus	A. dorehensis	A. drombus
Illicium length	20.2–27.1	11.1–11.6	8.8–13.2	5.0–9.5	8.6–11.4
Illicial spinules	absent	absent	absent	absent	absent
Second dorsal-fin spine length	10.7–13.7	10.2–12.8	7.1–15.2	6.8–13.0	6.6–11.4
Opercular opening	anal	pectoral	pectoral	pectoral	pectoral
Caudal peduncle	absent	absent	absent	present	absent
Darkly pigmented basidorsal spot	weak	strong	weak	weak	weak
Large dark spots on belly	absent	absent	absent	absent	absent
Dorsal-fin rays	12 (13)	12	12 (13)	12 (11–13)	12 (13)
Pectoral-fin rays	10 (9)	10 (9)	10–11 (9–12)	8–9 (10)	12 (11–13)
Bifurcate pelvic-fin rays	0	1	1	1	1
Maximum standard length	78 mm	61 mm	91 mm	51 mm	86 mm
Distribution	Indo-Pacific	Western Atlantic	Indo-Pacific, Eastern Pacific	Indo-Pacific	Johnston Atoll, Hawaiian Islands

Character	A. duescus	A. nummifer	A. rosaceus	A. sanguineus
Illicium length	9.7–13.8	8.8–14.7	23.3–30.3	8.6–13.3
Illicial spinules	absent	absent	absent	present
Second dorsal-fin spine length	10.0–14.5	8.7–15.2	15.6–18.9	8.5–14.1
Opercular opening	post-pectoral	pectoral	pectoral	pectoral
Caudal peduncle	absent	present	present	absent
Darkly pigmented basidorsal spot	weak	unusually strong	strong	weak
Large dark spots on belly	absent	absent	absent	present
Dorsal-fin rays	12	12 (13)	12 (13)	13–14 (12)
Pectoral-fin rays	9	10–11 (12)	10 (9)	11 (10–12)
Bifurcate pelvic-fin rays	0	1	1	1
Maximum standard length	30 mm	90 mm	42 mm	82 mm
Distribution	Western Pacific	Eastern Atlantic, Indo-Pacific	Indo-Pacific	Eastern Pacific

Members of this genus reach a smaller maximum body size than most other antennariids (members of the *Antennarius pauciradiatus* group being the only notable exception); the average maximum size of the known material of all species combined was 66 mm SL. Most species appear to be associated with rocky- and coral-reef habitats.

Distributed circumtropically; nine species.

COMMENTS

Although diagnosed above as if monophyletic, *Abantennarius*, as recognized here, is probably paraphyletic. This paraphyly is not apparent morphologically, but our preliminary molecular analysis, based on only four of the nine contained species (*A. coccineus*, *A. nummifer*, *A. rosaceus*, and *A. sanguineus*), divides the genus into two clades and places the genus *Antennatus* in between. *Antennatus*, however, is clearly distinct and well-defined—at least based on morphology—and to consider the contents of both genera as one, in an expanded *Antennatus*, seems unwarranted. On the other hand, to erect a new genus to accommodate a portion of the species of *Abantennarius*, appears equally ill-advised considering how little we know about the interrelationships of the latter. It therefore seems best at this point to maintain *Abantennarius* as recognized here until further molecular analysis becomes possible through additional collections of tissues and voucher specimens. For further discussion, see Evolutionary Relationships, p. 383.

Key to the Known Species of *Abantennarius*

This key, like the other keys in this volume, works by progressively eliminating the most morphologically distinct taxon and for that reason should always be entered from the beginning. All characters listed for each species must correspond to the specimen being keyed; if not, proceed to the next set of character states. For an additional or alternative means of identification, all species of *Abantennarius* may be compared simultaneously by referring to Table 10.

1. Opercular opening adjacent to origin of anal fin; illicium long, about twice length of second dorsal-fin spine. . . . . **Abantennarius analis Schultz, 1957, p. 198**
    Indo-Pacific Ocean

2. Opercular opening about halfway between base of pectoral-fin lobe and origin of anal fin; illicium short, about equal to length of second dorsal-fin spine. . . . .
    . . . . **Abantennarius duescus (Snyder, 1904), p. 220**
    Indonesia to Fiji and the Hawaiian Islands

3. Opercular opening situated on or adjacent to pectoral-fin lobe; illicium considerably longer than second dorsal-fin spine; second dorsal-fin spine elongate and slender with discrete clusters of dermal spinules along entire length; caudal peduncle present, the membranous posteriormost margin of soft-dorsal and anal fins connected to body distinctly anterior to base of outermost rays of caudal fin . . . .
    . . . . *Abantennarius rosaceus* **(Smith and Radcliffe, in Radcliffe, 1912), p. 230**
    Western Pacific to Samoa

4. Illicium distinctly shorter than second dorsal-fin spine; opercular opening situated on or adjacent to pectoral-fin lobe; caudal peduncle present, the membranous

posteriormost margin of soft-dorsal and anal fins connected to body distinctly anterior to base of outermost rays of caudal fin; pectoral-fin rays eight or nine (rarely 10) (Table 10) . . . .*Abantennarius dorehensis* (**Bleeker, 1859a**), **p. 212**
Indo-Pacific Ocean

5. Caudal peduncle present, the membranous posteriormost margin of soft-dorsal and anal fins connected to body distinctly anterior to base of outermost rays of caudal fin; a darkly pigmented basidorsal spot nearly always present; belly without large, darkly pigmented spots; dorsal-fin rays 12 (rarely 13) (Table 10) . . . .
. . . .*Abantennarius nummifer* (**Cuvier, 1817b**), **p. 222**
Eastern Atlantic and Indo-Pacific Oceans

6. Caudal peduncle absent, the membranous posteriormost margin of soft-dorsal and anal fins connected to body at base of outermost rays of caudal fin; basidorsal pigment spot present or absent, but when present usually only weakly developed; belly covered with large, darkly pigmented spots; dorsal-fin rays 13 or 14 (rarely 12) (Table 10) . . . . .
. . . .*Abantennarius sanguineus* (**Gill, 1863**), **p. 232**
Eastern Pacific Ocean

7. Caudal peduncle absent, the membranous posteriormost margin of soft-dorsal and anal fins connected to body at base of outermost rays of caudal fin; basidorsal pigment spot well developed; belly without large, darkly pigmented spots; dorsal-fin rays 12 (Table 10) . . . .*Abantennarius bermudensis* (**Schultz, 1957**), **p. 201**
Western Atlantic Ocean

8. Caudal peduncle absent, the membranous posteriormost margin of soft-dorsal and anal fins connected to body at base of outermost rays of caudal fin; body and especially fins densely peppered with small dark spots; esca usually with darkly pigmented spherical swellings (Fig. 151); pectoral-fin rays 12 (rarely 11 or 13) (Table 10) . . . .
. . . . *Abantennarius drombus* (**Jordan and Evermann, 1903**), **p. 215**
Johnston Atoll and Hawaiian Islands

9. Caudal peduncle absent, the membranous posteriormost margin of soft-dorsal and anal fins connected to body at base of outermost rays of caudal fin; body and fins usually without dense peppering of small dark spots; esca usually without darkly pigmented spherical swellings (Fig. 144); pectoral-fin rays 10 or 11 (rarely nine or 12) (Table 10) . . . . *Abantennarius coccineus* (**Lesson, 1831**), **p. 204**
Indo-Pacific and Eastern Pacific Oceans

Abantennarius analis Schultz
Tail-Jet Frogfish

Figs. 137–139, 284; Tables 10, 22

Abantennarius analis Schultz, 1957: 67, fig. 2 (original description, single specimen, holotype USNM 164419, Oahu). Gosline and Brock, 1960: 305, 345 (description; in key). Tinker, 1978: 506, fig. (description; figure after Schultz, 1957).
Antennarius analis: Pietsch, 1984b: 36 (new combination; genera of frogfishes). Randall, 1986: 180 (Marshall Islands). Pietsch and Grobecker, 1987: 163, figs. 62, 63, 124 (description, distribution, relationships). Myers, 1989: 67, fig. 1a (Christmas Island to Hawaiian and Society Islands, Marshalls and Marianas; in key). Paxton et al., 1989:

278 (Australia). Randall et al., 1993a: 223, 227, table 1 (Hawaiian Islands). Kon and Yoshino, 1999: 101, fig. 1 (Okinawa, Ryukyu Islands). Myers, 1999: 68, fig. 1a (after Myers, 1989). Pietsch, 1999: 2015 (western central Pacific; in key). Nakabo, 2000: 455 (Japan; in key). Hutchins, 2001: 22 (Western Australia). Nakabo, 2002: 455 (Japan; in key). Greenfield, 2003: 56, table 1 (Kaneohe Bay, Hawaiian Islands). Randall et al., 2004: 8 (Tonga). Mundy, 2005: 262 (Hawaiian Islands). Randall, 2005a: 69, fig. (South Pacific). Allen et al., 2006: 638 (Australia). Carnevale and Pietsch, 2006: 452, table 1 (comparison, meristics, first fossil frogfish). Randall, 2007: 127, fig. (Hawaiian Islands; in key). Allen and Erdmann, 2012: 144, color fig. (Christmas Island). Nakabo, 2013: 538, figs. (Okinawa, Ryukyu Islands). Hobbs et al., 2014a: 192, table 1 (Christmas Island). Moore et al., 2014: 176 (northwestern Australia). Farina and Bemis, 2016: 214 (gill ventilation, jet propulsion).

Antennatus analis: Arnold and Pietsch, 2012: 128 (new combination, molecular phylogeny, revised classification).

MATERIAL EXAMINED

Fifteen specimens, 8.5 to 78 mm SL.

Holotype of *Abantennarius analis*: USNM 164419, 42 mm, Waikiki, Oahu, Hawaiian Islands, Gosline and Randall, 31 December 1952.

Additional material: BPBM, 3 (15.5–51 mm); CAS, 3 (57–78 mm); ROM, 2 (8.5–27.5 mm); URM, 1 (56.5 mm); USNM, 1 (49.5 mm); UW, 2 (12–30 mm); WAM, 2 (16.5–20.5 mm).

DIAGNOSIS

A member of the genus *Abantennarius* unique in having the opercular opening adjacent to the anal fin (Figs. 137, 138). It is further distinguished in having the following combination of character states: basidorsal pigment spot usually absent (small but darkly pigmented in BPBM 9551, 51 mm SL); illicium 20.2 to 27.1% SL, about twice length of second dorsal-fin spine; esca an elongate, tapering appendage with several slender filaments and a cluster of darkly pigmented, spherical swellings at base (Fig. 139); second dorsal-fin spine more or less straight, length 10.7 to 13.7% SL; caudal peduncle absent, the membranous posteriormost margin of soft-dorsal and anal fins connected to body at base of outermost rays of caudal fin; belly without large, darkly pigmented spots; dorsal-fin rays 12 (rarely 13); anal-

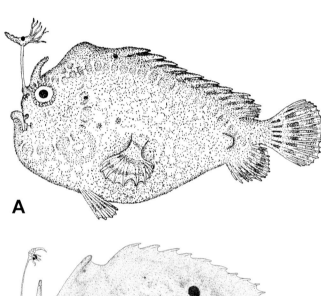

A

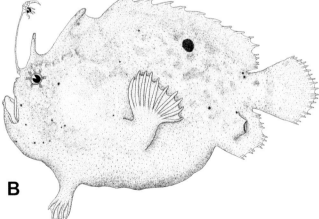

B

Fig. 137. *Abantennarius analis* Schultz. (**A**) holotype, USNM 164419, 42 mm SL, Waikiki, Oahu, Hawaiian Islands. Drawing by Beatrice Jennings Schultz; after Schultz (1957, fig. 2); courtesy of Lisa Palmer and the Smithsonian Institution, NMNH, Division of Fishes, Washington, DC. (**B**) BPBM 9551, 51 mm SL, Malakal Pass, Palau, Caroline Islands. Drawing by Cathy L. Short; after Pietsch and Grobecker (1987, fig. 62).

Fig. 138. *Abantennarius analis* Schultz. UW 20874, 30 mm SL, Larsen Cove, Tutuila Island, American Samoa. Photo by Luke Tornabene.

Fig. 139. Esca of *Abantennarius analis* Schultz. BPBM 11378, 49.5 mm SL, Kawaihae, Hawaii Island. Drawing by Cathy L. Short; after Pietsch and Grobecker (1987, fig. 63).

fin rays seven (rarely six); pectoral-fin rays 10 (rarely nine); all rays of pelvic fin simple (Table 10).

DESCRIPTION

Darkly pigmented escal swellings (as many as five) always present (Fig. 139); anterior and lateral surfaces of second dorsal-fin spine completely covered with close-set dermal spinules; shallow naked depression between second and third dorsal-fin spines present or absent; third dorsal-fin spine straight to slightly curved posteriorly, length 19.0 to 25.7% SL; eye diameter 6.4 to 9.0% SL; distance between opercular opening and base of lowermost ray of caudal fin 7.8 to 13.1% SL.

Coloration in life: light yellowish-gray to brown with faint mottling, the fins more yellowish than body (BPBM 9551).

Coloration in preservation: gray, yellow-brown, pinkish-brown, to brown (belly usually somewhat lighter), with fine, slightly darker mottling over entire head and body, and a few small scattered spots in some specimens; dusky-brown spots on dorsal and anal fins in some specimens; usually two or three dark parallel bars across caudal fin; banding on illicium absent or only faintly present; short, darkly pigmented bars radiating from eye faintly present to absent (Fig. 138).

Largest specimen examined: 78 mm SL (CAS 45576).

DISTRIBUTION

Abantennarius analis ranges from Christmas Island in the Eastern Indian Ocean, Rowley Shoals off northwestern Australia, and Okinawa, Japan, to the Society and Hawaiian Islands, with vouchered records at the Marianas, Marshalls, and Solomons, and the islands of Palau, Yap, Enewetak, Guadalcanal, Fiji, Tonga, Wallis, and Samoa (Fig. 284, Table 22). The type locality is Oahu, Hawaiian Islands. Depths of capture, known for 13 specimens, range from just beneath the surface to 21 m; the average depth for all known captures is 11 m.

COMMENTS

Abantennarius analis and the apparently closely related *A. duescus* differ from all other lophiiform fishes in having the opercular opening displaced posteriorly, adjacent to the anal fin in *A. analis*, and about halfway between the base of the pectoral-fin lobe and the origin of the anal fin in *A. duescus* (compare Figs. 137 and 155). The shared presence of this single derived feature prompted Schultz (1957: 66) to establish the genus *Abantennarius* to contain these two species: "this new genus is characterized by having the gill opening remote from the "elbo" [sic] of the pectoral fin." Both *A. duescus* and *A. analis* share a number of derived features with members of the former *A. nummifer* group (see Diagnosis above, and Evolutionary Relationships, p. 384)—they are all included here within the now expanded genus *Abantennarius*.

Abantennarius bermudensis (Schultz)
Island Frogfish

Figs. 140–142, 276; Tables 10, 22

Antennarius radiosus (not of Garman): Bean, 1906b: 89 (misidentification, USNM 50000, Bermuda). Barbour, 1942: 38 (misidentification, USNM 50000).

Antennarius bermudensis Schultz, 1957: 98, pl. 12, fig. C (original description, two specimens, holotype USNM 50000, Hungry Bay, Bermuda). Cervigón, 1966: 873, 876 (Venezuela; in key). Böhlke and Chaplin, 1968: 723, figs. (description, figure erroneously shows distinct caudal peduncle; Bermuda, Bahamas). Randall, 1968: 291 (description, Caribbean). Pietsch, 1978a: 3 (listed, western central Atlantic). Román, 1979: 353, fig. (Venezuela; in key). Pietsch, 1984b: 36 (genera of frogfishes). Pietsch, 1986b: 1365, fig. (eastern North Atlantic). Acero and Garzón, 1987: 90 (Santa Marta, Colombia). Ibarra and Stewart, 1987: 9 (type catalog). Pietsch and Grobecker, 1987: 153, figs. 58, 59, 123 (description, distribution, relationships). Smith-Vaniz et al., 1999: 159 (Bermuda; in key). Pietsch, 2002a: 1051 (western central Atlantic). Dennis et al., 2005: 723, table 2 (Mona Passage, Greater Antilles). Schwartz, 2005: 147 (Florida and farther south). Carnevale and Pietsch, 2006: 452, table 1 (comparison, meristics, first fossil frogfish). Jackson, 2006a: 783, 784, tables 1, 2 (reproduction, depth distribution). Mejía-Ladino et al., 2007: 282, figs. 4, 6, tables 1–3, 5 (species account; in key; Colombia).

Antennarius bermudensis (not of Schultz): Maul, 1959: 15 (material is *A. nummifer*, MMF 5254, Azores; MMF 2461, 2681, Madeira).

Antennatus bermudensis: Arnold and Pietsch, 2012: 128 (new combination, molecular phylogeny, revised classification). Smith-Vaniz and Collette, 2013: 169 (Bermuda).

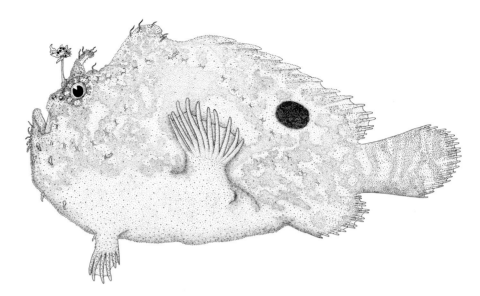

Fig. 140. *Abantennarius bermudensis* (Schultz). AMNH 28454, 43.5 mm SL, West Plana Cay, Bahamas. Drawing by Cathy L. Short; after Pietsch and Grobecker (1987, fig. 58).

Fig. 141. Esca of *Abantennarius bermudensis* (Schultz). AMNH 28454, 43.5 mm SL, West Plana Cay, Bahamas. Drawing by Cathy L. Short; after Pietsch and Grobecker (1987, fig. 59).

MATERIAL EXAMINED

Twenty specimens, 12.5 to 65 mm SL.

Holotype of *Antennarius bermudensis*: USNM 50000, 50 mm, Hungry Bay, Bermuda, Gosling.

Paratype of *Antennarius bermudensis*: FMNH 48862, 61 mm, Harrington Sound, Bermuda, Mobray, April 1931.

Additional material: AMNH, 1 (43.5 mm); ANSP, 8 (12.5–43 mm); MBUCV, 1 (65 mm); MCZ, 2 (20–47 mm); UF, 4 (28–48 mm); USNM, 2 (17–33.5 mm).

DIAGNOSIS

A member of the genus *Abantennarius* unique in having the following combination of character states: a conspicuous, darkly pigmented, basidorsal spot (Fig. 140); illicium 11.1 to 11.6% SL, about as long as second dorsal-fin spine; esca an elongate, tapering appendage with several slender filaments and a cluster of darkly pigmented, spherical swellings at base (Fig. 141); second dorsal-fin spine strongly curved posteriorly, length 10.2 to 12.8% SL; opercular opening adjacent to or on pectoral-fin lobe; caudal peduncle absent, the membranous posteriormost margin of soft-dorsal and anal fins connected to body at base of outermost rays of caudal fin; belly without large, darkly pigmented spots; dorsal-fin rays 12; anal-fin rays seven; pectoral-fin rays 10 (rarely nine); last ray of pelvic fin bifurcate (Table 10).

DESCRIPTION

Dark, spherical escal swellings (as many as seven; ANSP 121747) always present (Fig. 141); anterior and lateral surfaces of second dorsal-fin spines usually completely covered with dermal spinules (but spinules occasionally forming discrete clusters leaving naked spaces between); shallow naked depression between second and third dorsal-fin spines; third dorsal-fin spine curved posteriorly, length 18.2 to 24.4% SL; eye diameter 6.8 to 7.8% SL;

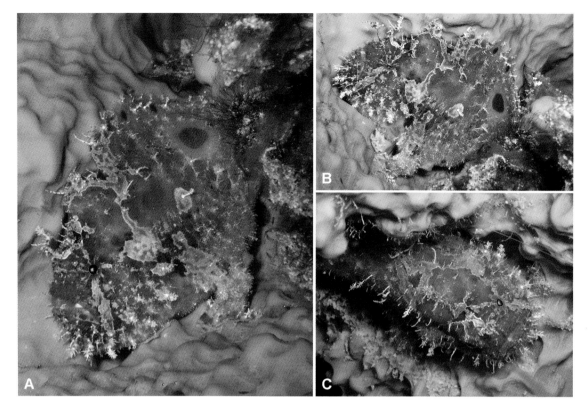

Fig. 142. *Abantennarius bermudensis* (Schultz). (**A–C**) Bonaire, Netherlands Antilles. Photos by Ellen Muller.

numerous, darkly banded, cutaneous filaments usually present on head and upper body, particularly above eye and on second and third dorsal-fin spines.

Coloration in preservation: yellow-brown, brown, to rust above, light tan below, with darker spots, mottling, or irregularly shaped blotches on head and body; fins more or less peppered with brown to black spots, occasionally forming roughly parallel bands across caudal fin; basidorsal spot dark brown; illicium distinctly banded; as many as seven short, darkly pigmented bars radiating from eye (Fig. 142).

Largest specimen examined: 65 mm SL (MBUCV 2070).

DISTRIBUTION

Antennarius bermudensis is restricted to the Western Atlantic Ocean, ranging from Bermuda and the Bahamas to the coastal waters of Colombia and Venezuela as far east as Tobago in the Windward Islands, including the Yucatan Peninsula, Belize, Haiti, Puerto Rico, and Bonaire in the Leeward Antilles (Fig. 276, Table 22). The type locality is Hungry Bay, Bermuda.

Depths of capture, known for 14 specimens, range from just beneath the surface to 65 m, but 78% of the material was taken in less than 30 m. The average depth for all recorded captures is 22 m.

COMMENTS

Abantennarius bermudensis is very similar to the Indo-Pacific *A. coccineus*, and might be synonymized with the latter if it were not for the well-developed basidorsal spot consistently present in the available material of *A. bermudensis*. This spot is weakly developed or absent in *A. coccineus*.

Abantennarius coccineus (Lesson)
Scarlet Frogfish

Figs. 143–146, 276, 318; Tables 10, 11, 22

Chironectes coccineus Lesson, 1831: 143, pl. 16, fig. 1 (original description, single specimen, holotype MNHN A.4599, Mauritius; dating of Lesson after Sherborn and Woodward, 1906: 336; Fig. 143A). Valenciennes, 1837: 430 (description after Lesson, 1831). Pietsch, 1985a: 93, fig. 31, table 1 (original manuscript sources for *Histoire Naturelle des Poissons* of Cuvier and Valenciennes). Pietsch et al., 1986: 139 (in type catalog). Dyer and Westneat, 2010: 605 (synonym of *A. coccineus*).

Antennarius nummifer (not of Cuvier): Bleeker, 1854c: 497 (Ambon). Günther, 1876: 164 (in part; Samoa, Raiatea). Schmeltz, 1879: 48 (after Schmeltz, 1877). Boulenger, 1897: 373 (BMNH 1897.8.23.79-80, Rotuma). Pietschmann, 1930: 23 (Waikiki). Fowler, 1934a: 450 (after Pietschmann, 1930). Koumans, 1953: 268 (RMNH 20110, Morotai Island, Indonesia). Burgess and Axelrod, 1973a: 528, color fig. 486 (plate represents *A. coccineus*).

Antennarius coccineus: Günther, 1861a: 191 (new combination, description, Mauritius). Bleeker, 1865b: 22, pl. 197, fig. 2 (description, additional material, Indonesia; Fig. 143B). Schmeltz, 1869: 19 (Viti Islands). Klunzinger, 1871: 499 (Red Sea). Whitmee, 1875: 544 (aquarium observations, BMNH 1876.7.25.15, Samoa). Macleay, 1881: 578 (Port Jackson). Klunzinger, 1884: 126 (Red Sea). Whitley, 1927a: 8 (Fiji). Smith, 1949: 431, pl. 98, fig. 1238 (Natal). Koumans, 1953: 267 (RMNH 21020, Kera near Timor). Schultz, 1957: 97, pl. 12, fig. A (synonymy, description; Caroline, Marshall, Gilbert Islands; Samoa, Tonga, Moorea; in key). Whitley, 1958: 49 (Noumea, New Caledonia). Beaufort and Briggs, 1962: 203 (synonymy, description, discussion of past confusion with *A. nummifer*; in key). Le Danois, 1964: 97, fig. 47a (description; Madagascar, Fiji; MNHN A.4500 erroneously listed as holotype).

A

B

Fig. 143. *Abantennarius coccineus* (Lesson). (**A**) holotype of *Chironectes coccineus*, MNHN 4599, 91 mm SL, Mauritius, collected by French naturalists René Primevère Lesson and Prosper Garnot (after Lesson, 1831, pl. 16, fig. 1), erroneously depicted with a distinct caudal peduncle (see Comments, p. 211). Courtesy of Emmanuelle Choiseau, © Muséum national d'Histoire naturelle, Paris. (**B**) RMNH 6274, 72 mm SL, probably Ambon, Indonesia. Drawing by Ludwig Speigler under the direction of Pieter Bleeker; after Bleeker (1865b, pl. 197, fig. 2).

Schultz, 1964: 175, table 1 (additional material, Samoa, Phoenix Islands). Schultz et al., 1966: 145, pl. 146, fig. B (synonymy, description after Schultz, 1957). Le Danois, 1970: 87 (description, Gulf of Aqaba). Burgess and Axelrod, 1973a: 529, color fig. 489 (Tahiti). Allen et al., 1976: 384 (Lord Howe Island). Jones and Kumaran, 1980: 703, fig. 599 (Laccadive Archipelago). Araga, 1984: 103 (Japan). Dor, 1984: 56 (Red Sea). Pietsch, 1984b: 36 (genera of frogfishes). Pietsch, 1984e: 2 (Western Indian Ocean). Pietsch, 1985a: 93, fig. 31, table 1 (original manuscript sources for *Histoire Naturelle des Poissons* of Cuvier and Valenciennes). Pietsch, 1986a: 366, 366, fig. 102.1, pl. 13 (South Africa). Pietsch and Grobecker, 1987: 144, figs. 56, 57, 123, pl. 27 (description, distribution, relationships). Allen and Swainston, 1988: 38 (northwestern Australia). Zug et al., 1988: 6 (Rotuma Islands). Myers, 1989: 68, fig. 1b (Red Sea to Costa Rica, Galápagos, and San Felix Island; in key). Paxton et al., 1989: 278 (Australia). Pequeño, 1989: 43 (Desventuradas Islands, Chile). Winterbottom et al., 1989: 14 (Chagos Archipelago). Pietsch and Grobecker, 1990a: 99 (aggressive mimicry). Pietsch and Grobecker, 1990b: 80 (after Pietsch and Grobecker, 1990a). Randall et al., 1990: 54, color fig. (Indo-Pacific to the Americas). Kosaki et al., 1991: 189, fig. 5 (Johnston Atoll). Francis, 1993: 158 (Lord Howe Island). Randall et al., 1993a: 223, 227, table 1 (Hawaiian Islands). Goren and Dor, 1994: 14 (Red Sea). Randall, 1995: 83, fig. 166 (Oman). Schneider and Lavenberg, 1995: 856 (tropical Eastern Pacific). Robertson and Allen, 1996: 122, 128, tables 1, 3 (Clipperton Atoll). Allen, 1997: 62 (tropical Australia). Allen and Robertson, 1997: 820 (Clipperton Atoll). Grove and Lavenberg, 1997: 231, 233 (Galápagos records questioned). Kulbicki and Williams, 1997: 12 (New Caledonia). Lindberg et al., 1997: 216 (Sea of Japan). Okamura and Amaoka, 1997: 141, figs. (Japan). Randall et al., 1997a: 54, color fig. (Indo-Pacific to the Americas). Michael, 1998: 340, 341, 343, color figs. (identification, behavior, captive care). Fricke, 1999: 103 (Réunion, Mauritius). Johnson, 1999: 724 (Moreton Bay, Queensland). Myers, 1999: 68, fig. 1b (after Myers, 1989). Pietsch, 1999: 2015 (western central Pacific; in key). Randall, 1999: 8 (Pitcairn Islands). Allen, 2000a: 62, pl. 12-11 (Indo-Pacific). Mora and Zapata, 2000: 477, table 1 (Gorgona Island, Colombia). Nakabo, 2000: 457 (Japan; in key). Pequeño and Lamilla, 2000: 434 (Desventuradas Islands, Chile). Pereira, 2000: 5 (Mozambique). Pietsch, 2000: 597 (South China Sea). Randall and Earle, 2000: 8 (Marquesas Islands). Hutchins, 2001: 22 (Western Australia). Goldin, 2002: 47, fig. (Samoa). Nakabo, 2002: 457 (Japan; in key). Allen and Adrim, 2003: 25 (Sumatra to West Papua, Indonesia). Allen et al., 2003: 366, color figs. (tropical Pacific). Lozano and Zapata, 2003: 402, table 1 (Gorgona Island, Colombia). Myers and Donaldson, 2003: 610 (Mariana Islands). Schneidewind, 2003a: 84, fig. (natural history). Heemstra et al., 2004: 3314, table 1 (Rodrigues Island). Letourneur et al., 2004: 209 (Réunion Island). Pereira et al., 2004: 4, appendix 1 (Mozambique). Randall et al., 2004: 8 (Tonga). Robertson et al., 2004: 513, 527, 563, tables 1, 6 (transpacific shore fishes). Bussing and López, 2005: 46, 47, fig. (Cocos Island). Randall, 2005a: 69, color fig. (South Pacific). Allen et al., 2006: 639 (Australia). Carnevale and Pietsch, 2006: 452, table 1 (comparison, meristics, first fossil frogfish). Alwany et al., 2007: 85, 86, table 2 (Red Sea). Kuiter and Debelius, 2007: 120, color figs. (Indonesia). Mejía-Ladino et al., 2007: 295, figs. 5, 6, tables 1–4, 6 (species account; in key; Eastern Pacific). Randall, 2007: 128 (Hawaiian Islands). Kwang-Tsao et al., 2008: 243 (southern Taiwan). Allen and Erdmann, 2009: 594 (West Papua). Fricke et al., 2009: 28, table 6 (Réunion). Jawad and Al-Mamry, 2009: 1 (Gulf of Oman). Dyer and Westneat, 2010: 599, 605 (Desventuradas Islands, Chile). Golani and Bogorodsky, 2010: 16 (Red Sea). McCosker and Rosenblatt, 2010: 190 (Galápagos records confirmed). Miya et al., 2010: 7, 17, table 2 (molecular phylogeny). Mundy et al., 2010: 23 (Palmyra Atoll). Fricke et al., 2011: 367 (New Caledonia). Satapoomin, 2011: 51, appendix A (Andaman Sea). Allen and Erdmann, 2012: 144, color fig. (East Indies). Prakash et al., 2012: 943, fig. 5, table 1 (southeast coast of India). Allen and Erdmann, 2013: 32, appendix 3.1 (Bali, Indonesia). Larson et al., 2013: 53 (Northern Territory). Nakabo, 2013: 541, figs. (Okinawa, Ryukyu Islands). Fricke et al., 2014: 35 (Papua New Guinea). Hobbs et al., 2014a: 192, table 1 (Christmas Island). Hobbs et al., 2014b: 209, table 1 (Cocos-Keeling Islands). Hylleberg and Aungtonya, 2014: 101 (Phuket, Thailand). Moore et al., 2014: 176 (northwestern Australia). Delrieu-Trottin et al., 2015: 5, table 1 (Marquesas Islands).

Antennarius hispidus (not of Bloch and Schneider): Seale, 1906: 89 (Society Islands). Weber, 1913: 562 (in part, material includes *A. coccineus*, *A. dorehensis*, *Antennarius randalli*, *Antennatus tuberosus*, and *Histrio histrio*).

Antennarius stigmaticus Ogilby, 1912: 63, pl. 14, fig. 2 (original description, single specimen, holotype QMB I.363, Moreton Bay, Queensland). McCulloch, 1929: 408 (after Ogilby, 1912). Marshall, 1964: 514, 515 (after Ogilby, 1912; in key). Dyer and Westneat, 2010: 605 (synonym of *A. coccineus*).

Antennarius leucus Fowler, 1934b: 512, fig. 53 (original description, single specimen, holotype ANSP 54955, Durban, Natal). Böhlke, 1984: 20 (type catalog). Dyer and Westneat, 2010: 605 (synonym of *A. coccineus*).

Lophiocharon caudimaculatus (not of Rüppell): Le Danois, 1964: 111 (in part, i.e., MNHN 54-25, Aldabra Island).

Abantennarius neocaledoniensis Le Danois, 1964: 126, fig. 63 (original description, single specimen, holotype MNHN A.4589, New Caledonia). Pietsch et al., 1986: 137 (in type catalog). Dyer and Westneat, 2010: 605 (synonym of *A. coccineus*). Fricke et al., 2011: 367 (synonym of *A. coccineus*).

Antennarius chironectes (not of Latreille): Le Danois, 1970: 88 (Gulf of Aqaba, Red Sea).

Uniantennatus caudimaculatus (not of Rüppell): Le Danois, 1970: 90, fig. 3D (Gulf of Aqaba, Red Sea).

Antennarius immaculatus Le Danois, 1970: 91, figs. 2, 3E (original description, four specimens, holotype TAU P.254, Elat, Gulf of Aqaba). Dor, 1984: 56 (after Le Danois, 1970). Golani, 2006: 23 (type catalog). Dyer and Westneat, 2010: 605 (synonym of *A. coccineus*).

Antennarius moai Allen, 1970: 521, figs, 2c, 3, table 1 (original description, two specimens, holotype BPBM 6556, Easter Island). Dyer and Westneat, 2010: 605 (synonym of *A. coccineus*).

Histrio histrio (not of Linnaeus): Grant, 1972, color fig. 87 (misidentification). Grant, 1978: 601, color fig. 264 (after Grant, 1972).

Antennarius biocellatus (not of Cuvier): Burgess and Axelrod, 1973b: 828, color pls. 342, 343 (plates represent *A. coccineus*).

Antennarius sp. cf. coccineus: Myers and Shepard, 1980: 315 (Guam).

Antennarius mohai: Pequeño, 1989: 43 (misspelling of *A. moai* Allen, 1970; Easter Island).

Antennarius coccineus (not of Lesson): Jawad and Al-Mamry, 2009: 1, fig. 1 (misidentification, specimen is *A. indicus*; Wadi Shab, Gulf of Oman).

Antennatus coccineus: Arnold and Pietsch, 2012: 127, 128, figs. 4, 6, table 1 (new combination; molecular phylogeny). Page et al., 2013 (common names). Brandl and Bellwood, 2014: 32 (pair-formation). Galván-Villa et al., 2016: 144 (Bahía Chamela, Mexico).

MATERIAL EXAMINED

Four hundred and fifty-eight specimens, 5.0 to 91 mm SL.

Holotype of *Chironectes coccineus*: MNHN A.4599, 91 mm, Mauritius, Lesson and Garnot (Fig. 143A).

Holotype of *Antennarius stigmaticus*: QMB I.363, 66 mm, Moreton Bay, Queensland.

Holotype of *Antennarius leucus*: ANSP 54955, 59 mm, Durban, Natal, Bell-Marley, 1932.

Holotype of *Antennarius neocaledoniensis*: MNHN A.4589, 52 mm, New Caledonia, Pancher.

Holotype of *Antennarius immaculatus*: TAU P.254, 31 mm, Elat, Gulf of Aqaba, 7 January 1965.

Paratypes of *Antennarius immaculatus*: HUJ F.348, 1 (29 mm), Elat, Theodor, October 1951; HUJ F.9806(605), 1 (19.5 mm), El Arqana, Gulf of Aqaba, 2 June 1969; HUJ F.9808(805), 2 (28–38.5 mm), Dahab, Sinai Peninsula, Gulf of Aqaba, 15 September 1967.

Holotype of *Antennarius moai*: BPBM 6555, 24 mm, Easter Island, large tidepool between Hanga Roa and Hanga Piko, rotenone, 0–1 m, Allen and Randall, 26 January 1969.

Paratype of *Antennarius moai*: LACM 6560-12, 57.5 mm, Easter Island, 85 m NE of sand beach on east side of Anakena Cove, 3–5 m, Parks et al., 1 October 1958.

Additional material: *Western Indian Ocean*: ANSP, 15 (15–77 mm); BMNH, 2 (32.5–35.5 mm); BPBM, 2 (17–29 mm); CAS, 5 (27–64 mm); FMNH, 3 (40–63 mm); MNHN, 2 (20.5–35 mm); SAIAB, 29 (13–71 mm); UMMZ, 2 (37–40 mm); USNM, 10 (17–54.5 mm); UW, 1 (84 mm). *Chagos Archipelago*: BMNH, 1 (58.5 mm); ROM, 17 (5–67 mm); USNM, 4 (13.5–40 mm). *Red Sea*: BPBM, 5 (19–72 mm); HUJ, 13 (8–50 mm); MNHN, 1 (55.5 mm); USNM, 10 (16–61 mm). *Cocos-Keeling and Christmas Islands*: ANSP, 4 (21–50.5 mm); WAM, 2 (43–54 mm). *Indonesia and New Guinea*: AMNH, 1 (35.5 mm); AMS, 1 (12.5 mm); BMNH, 1 (40.5 mm); BPBM, 1 (52 mm); RMNH, 15 (17.5–72 mm); USNM, 15 (10–65 mm); WAM, 1 (23 mm); ZMA, 9 (6.5–45 mm). *Philippines*: AMS, 1 (46 mm); USNM, 21 (10–46 mm); UW, 25 (8.5–53 mm). *Thailand, China, Taiwan, and Ryukyu Islands*: BMNH, 1 (24 mm); BPBM, 1 (25 mm); CAS, 2 (47–60 mm); SU, 1 (40 mm); USNM, 7 (23–70.5 mm); WAM, 1 (23.5 mm); ZMUC, 1 (38 mm). *Australia*: AMNH, 1 (36 mm); AMS, 19 5 (16.5–91 mm); ANSP, 1 (19 mm); FMNH, 1 (60 mm); USNM, 3 (28.5–35.5 mm); WAM, 6 (23–62 mm). *Palau, Carolines, Marshalls, and Marianas*: BPBM, 3 (16–70 mm); CAS, 6 (7–52 mm); FMNH, 1 (39.5 mm); MCZ, 1 (32.5 mm); UG, 3 (9–14.5 mm); USNM, 10 (14.5–63.5 mm); UW, 1 (40 mm). *Bismarcks, Solomons, New Hebrides, and New Caledonia*: AMS, 5 (16.5–50.5 mm); ANSP, 3 (33.5–41 mm); BPBM, 1 (53 mm); ROM, 1 (23 mm); USNM, 2 (58–71 mm). *Gilbert and Ellice Islands*: BMNH, 2 (20.5–45 mm); USNM, 1 (57 mm). *Fiji, Tonga, and Samoa*: AMNH, 1 (36.5 mm); AMS, 2 (32–48.5 mm); BMNH, 3 (46.5–55 mm); MNHN, 1 (68.5 mm); NMNZ, 1 (61 mm); ROM, 26 (8.5–75 mm); USNM, 18 (7.8–58.5 mm); UW, 5 (12–53.5 mm). *Line Islands*: ANSP, 1 (25 mm); BPBM, 4 (17–53.5 mm). *Society Islands*: ANSP, 1 (16.5 mm); BMNH, 2 (39–60 mm); BPBM, 15 (10–68 mm); CAS, 8 (24–66 mm); MCZ, 2 (18–28 mm); RMNH, 1 (49.5 mm); USNM, 1 (65 mm). *Marquesas Islands*: BPBM, 4 (10–52 mm). *Pitcairn*: BPBM, 3 (25–30 mm). *Eastern Pacific*: LACM, 11 (13–68 mm); SIO, 1 (39.5 mm); UCLA, 20 (11.5–64 mm); UCR, 20 (6.5–58 mm); USNM, 1 (23 mm). *Chile*: SIO, 1 (17 mm).

DIAGNOSIS

A member of the genus *Abantennarius* unique in having the following combination of character states: basidorsal pigment spot present or absent, but if present only weakly developed; illicium 8.8 to 13.2% SL, about as long as second dorsal-fin spine; esca ranging

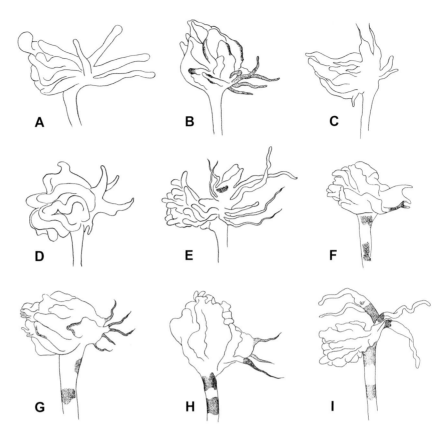

Fig. 144. Escae of *Abantennarius coccineus* (Lesson). (**A**) ANSP 140250, 22 mm SL, Cocos-Keeling. (**B**) UW 20864, 53 mm SL, Philippines. (**C**) BPBM 11439, 53 mm SL, New Caledonia. (**D**) BPBM 22557, 53.5 mm SL, Palmyra. (**E**) BPBM 7830, 18 mm SL, Tahiti. (**F**) BPBM 8354, 44 mm SL, Tahiti. (**G**) BPBM 6092, 68 mm SL, Tahiti. (**H**) BPBM 6168, 66 mm SL, Moorea. (**I**) BPBM 11142, 47.5 mm SL, Marquesas. Drawings by Cathy L. Short; after Pietsch and Grobecker (1987, fig. 57).

in morphology from a simple, unpigmented ball of folded tissue (usually associated with a few short, basal filaments) to an elongate, tapering appendage with a variable number of slender filaments; darkly pigmented, spherical escal swellings only rarely present (Fig. 144); second dorsal-fin spine strongly curved posteriorly, length 7.1 to 15.2% SL; opercular opening adjacent to or on pectoral-fin lobe; caudal peduncle absent, the membranous posteriormost margin of soft-dorsal and anal fins connected to body at base of outermost rays of caudal fin; belly without large, darkly pigmented spots; dorsal-fin rays 12 (rarely 13); anal-fin rays seven (rarely eight); pectoral-fin rays 10 or 11 (rarely nine or 12); last ray of pelvic fin bifurcate (Table 10).

DESCRIPTION

Esca often a bright white ball of tissue, usually without darkly pigmented, spherical swellings (dark swellings present in only about 7% of material examined; see Geographic Variation below; Fig. 146); anterior and lateral surfaces of second dorsal-fin spine usually completely covered with close-set dermal spinules (but the spinules occasionally forming discrete clusters leaving naked spaces between); shallow, naked depression be-

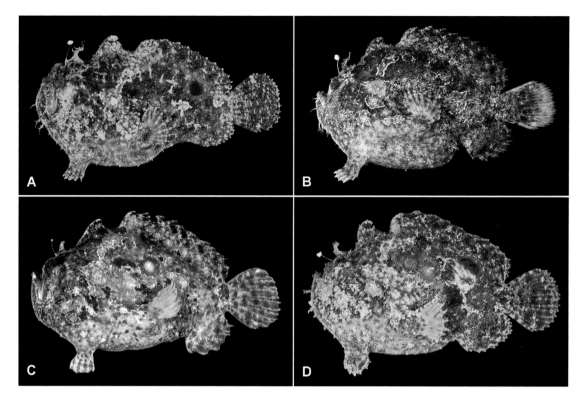

Fig. 145. *Abantennarius coccineus* (Lesson). (**A**) BPBM 29214, 29 mm SL, Enewetak, Marshall Islands. (**B**) BPBM 6118, 48 mm SL, Taravao, Tahiti, Society Islands. (**C**) BPBM 18246, Ras Muhammad, Sinai Peninsula, Red Sea, 73 mm SL. (**D**) Maldives Government Ministry of Fisheries, North Male Atoll, Maldive Islands, 55 mm SL. Photos by John E. Randall; courtesy of Lisa Palmer and the Smithsonian Institution, NMNH, Division of Fishes, Washington, DC.

tween second and third dorsal-fin spines nearly always present; third dorsal-fin spine curved posteriorly, length 16.8 to 26.5% SL; eye diameter 5.3 to 8.2% SL.

Coloration in life: highly variable; tan, brown, green, yellow, orange, bright red, gray or black; mottled yellowish- or reddish-brown (Society Islands, BPBM 11613); mostly red (Marquesas, BPBM 12588); heavily mottled with dark brown over entire body including belly, five bands of spots on tail, dark basidorsal spot present (Red Sea, BPBM 18246; Fig. 144C); yellow with a tinge of orange when caught, changing through orange to brownish-red while in captivity for five days (Enewetak, BPBM 29214; Fig. 144A); reddish, covered with darker red circular spots (Samoa, UW 20803) (Figs. 145, 146).

Coloration in preservation: highly variable (see Geographic Variation below); illicium usually lightly banded; as many as 13 short, darkly pigmented bars radiating from eye.

Largest specimen examined: 91 mm SL (AMS I.16570-001; MNHN A.4599).

GEOGRAPHIC VARIATION

The relatively large amount of available material of this geographically wide-ranging species shows distinct variation in coloration. In the center of its distribution, the Indo-Australian Archipelago, coloration in preservation tends to be an overall tan, yellow, to yellow-brown (occasionally pinkish) with darker brown to black mottling or spotting,

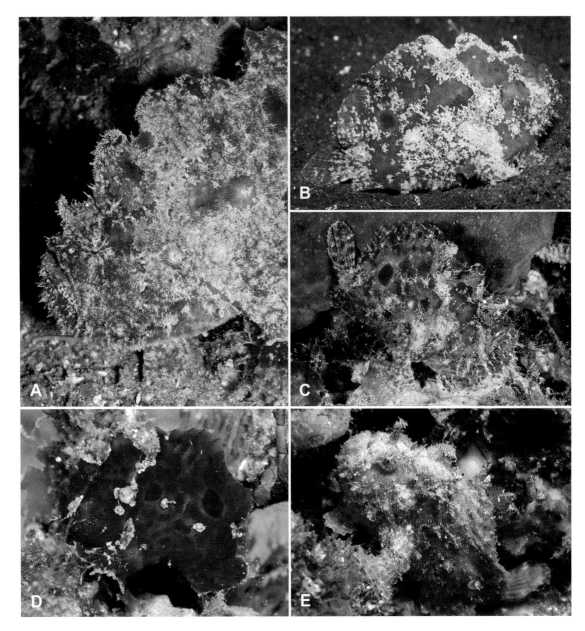

Fig. 146. *Abantennarius coccineus* (Lesson). (**A**) Lembeh Strait, Sulawesi, Indonesia. Photo by Johanna Gawron. (**B**) Lembeh Strait, Sulawesi, Indonesia. Photo by Mark V. Erdmann. (**C**) Bunaken Island, North Sulawesi, Indonesia. Photo by Fred Bavendam); (**D**) Kwajalein Atoll, Marshall Islands (photo by Jeanette Johnson); (**E**) Bahía de Banderas, Puerto Vallarta, Mexico (photo by Alicia Hermosillo McKowen).

especially on the cheeks and opercular region, and sometimes roughly circular or irregularly shaped blotches on the flanks above or behind the pectoral-fin lobes. On the periphery of its range, particularly the Red Sea, Line and Marquesas Islands, and Eastern Pacific populations, individuals tend to be light to dark gray and peppered with brown to black spots; the unpaired fins are usually darker and much more heavily peppered than the body (Fig. 276).

This pattern of geographic variation correlates somewhat with variation in pectoral-fin ray counts. Indo-Australian populations of *Abantennarius coccineus* nearly always have nine or ten pectoral-fin rays, whereas in peripheral populations the count tends to increase slightly. In the Western Indian Ocean and Red Sea, 6% of the pectoral fins counted had 11 rays; in the easternmost island groups of the Pacific Plate, and in the Eastern Pacific, the bulk of the material examined had 11 or 12 pectoral-fin rays (Table 11).

The esca of most individuals of *Abantennarius coccineus* consists of an unpigmented, usually spherical appendage sometimes associated with a few slender, basal filaments (Fig. 144D). This relatively simple morphology grades into a more complex esca consisting of a cluster of elongate, tapering appendages (Fig. 144E), and rarely associated with a cluster of darkly pigmented, spherical swellings. The latter condition was fully developed in only about 7% of the specimens examined, and more or less evenly distributed throughout most of the geographic range of *A. coccineus* (Fig. 276). In the tropical Eastern Pacific, where *A. coccineus* occurs sympatrically with the morphologically similar *A. sanguineus*, none of the specimens examined (53 individuals, 11.5 to 68 mm SL) had the more complex escal morphology of darkly pigmented swellings (see Comments, p. 236).

Table 11. Geographic Variation in Pectoral Fin-Ray Counts of *Abantennarius coccineus*

Geographical locality	Pectoral-fin rays*			
	9	10	11	12
Western Indian Ocean		86	4	
Red Sea		33	3	
Indo-Pacific (excluding localities below)	5	154	1	
Line Islands		3	5	
Marquesas Islands			8	
Pitcairn Island		4	2	
Easter Island			3	1
Eastern Pacific		14	70	
TOTALS	5	294	96	1

*Left and right sides combined.

DISTRIBUTION

Abantennarius coccineus is a wide-ranging species found throughout the Indo-Pacific Ocean from the East African coast, the Red Sea, Oman, and all major insular localities of the Western Indian Ocean, to the easternmost island groups of the Pacific Plate, including the Ryukyus, Taiwan, Christmas Island, Cocos-Keeling, Sumatra to West Papua, tropical and subtropical Australia, Lord Howe Island, New Caledonia, and Tonga to Pitcairn and Easter Islands (Fig. 276, Table 22). But it is replaced at Johnston Atoll and the Hawaiian Islands by the morphologically similar *Abantennarius drombus* (see Comments, p. 218).

It extends farther east as well, to insular and coastal waters of the tropical Eastern Pacific, including Clipperton Atoll, off Puerto Vallarta, Mexico, coastal Costa Rica and Panama, Cocos Island, and south to the Desventuradas Islands, Chile, and there is at least one, perhaps questionable, observation made off the southern tip of Baja California, Mexico. Its presence at the Galápagos Islands, questioned by Grove and Lavenberg (1997: 231), has been confirmed by McCosker and Rosenblatt (2010: 190). The type locality is Mauritius.

Data for depth preference, known for 142 captures, indicates that *A. coccineus* may be taken anywhere between the surface and 75 m. However, 95% of these captures were made in less than 30 m, and 70% in less than 10 m.

COMMENTS

Abantennarius coccineus has often been confused with *A. nummifer* (see Synonymy, p. 204). These two species, however, are easily distinguished from each other by the absence of a caudal peduncle in *A. coccineus* (see Caudal Peduncle, p. 32). The importance of this character was not appreciated by Lesson (1831, pl. 16, fig. 1), who in the original description of *A. coccineus* published a figure that shows a distinct caudal peduncle (Fig. 143A). The inaccuracy of Lesson's drawing was first suspected by Bleeker (1865b: 23) and later verified by Schultz (1957: 97) and Beaufort and Briggs (1962: 204).

Abantennarius dorehensis (Bleeker)
New Guinean Frogfish

Figs. 147–149, 283; Tables 10, 22

Antennarius dorehensis Bleeker, 1859a: 5, 21 [original description, three specimens, lectotype RMNH 6268 designated by Pietsch and Grobecker, 1987, paralectotypes RMNH 28010, Dorei (= Doreh), New Guinea; Fig. 147A]. Günther, 1861a: 184 (footnote reference to Bleeker, 1859a). Bleeker, 1865b: 9, 19, pl. 199, fig. 7 (description, after Bleeker, 1859a). Fowler, 1928: 478 (after Bleeker, 1859a; East Indies). Herre, 1945: 149 (Philippines). Schultz, 1957: 97, pl. 12, fig. B (after Bleeker, 1859a; in key). Munro, 1958: 296 (synonymy). Whitley, 1959: 321 (description; Bay Island, Purdy Archipelago, New Guinea). Beaufort and Briggs, 1962: 201, 207 (description; New Guinea, Philippines; in key). Munro, 1967: 585, pl. 78, fig. 1095 (New Guinea). Myers and Shepard, 1980: 338 (Guam). Collette, 1983: 97 (Umboi Island, Papua New Guinea. Pietsch, 1984b: 36 (genera of frogfishes). Pietsch, 1984e: 2 (Western Indian Ocean). Pietsch, 1986a: 366, 367, fig. 102.3 (South Africa). Pietsch and Grobecker, 1987: 166, figs. 64, 65, 125, pl. 29 (description, distribution, relationships). Myers, 1989: 68, pl. 13A (East Africa to Line and Society Islands; in key). Paxton et al., 1989: 278 (Australia). Pietsch and Grobecker, 1990a: 99, fig. (aggressive mimicry). Pietsch and Grobecker, 1990b: 77, fig. (after Pietsch and Grobecker, 1990a). Cunxin, 1993: 117 (South China Sea). Lindberg et al., 1997: 216 (Sea of Japan). Michael, 1998: 341, color fig. (identification, behavior, captive care). Myers, 1999: 68, pl. 13A (after Myers, 1989). Pietsch, 1999: 2015 (western central Pacific; in key). Allen, 2000b: 96 (Calamianes Islands, Philippines). Nakabo, 2000: 458 (Japan; in key). Pereira, 2000: 5 (Mozambique). Pietsch, 2000: 597 (South China Sea). Hutchins, 2001: 22 (Western Australia). Nakabo, 2002: 458 (Japan; in key). Allen and Adrim, 2003: 25 (Papua to Kalimantan, Indonesia). Allen et al., 2003: 365, color fig. (Indo-Pacific). Myers and Donaldson, 2003: 610, 614 (Mariana Islands). Allen et al., 2006: 639 (Australia). Kuiter and Debelius, 2007: 124, color fig. (Indonesia). Kwang-Tsao et al., 2008: 243 (southern Taiwan). Allen and Erdmann, 2009: 594 (West Papua). Chen et al., 2009: 23 (Green Island, Taiwan). Allen and Erdmann, 2012: 146, fig. (East Indies). Nakabo, 2013: 542, figs. (Okinawa, Ryukyu Islands). Hobbs et al., 2014b: 209, table 1 (Cocos-Keeling Islands). Moore et al., 2014: 176 (northwestern Australia). Arai, 2015: 94, table 2 (Malaysia).

Antennarius nummifer (not of Cuvier): Bleeker, 1865b: 18 (in part, three of four specimens, RMNH 28013). Fowler, 1959: 561 (Fiji).

Antennarius commersoni (not of Latreille): Schmeltz, 1877: 14 (BMNH 1873.8.1.33, Tahiti).

Antennarius altipinnis Smith and Radcliffe, in Radcliffe, 1912: 204, fig. 3 (original description, single specimen, holotype USNM 70267, Nogas Point, Panay, Philippines). Schultz, 1957: 99, pl. 13, fig. C (synonymy, description; Philippines, Palau, Guam, Gilbert Islands, Fiji; in key). Schultz, 1964: 173, table 1 (additional material, Philippines). Schultz et al., 1966: 145, pl. 146, fig. C (after Schultz, 1957). Palmer, 1970: 233 (additional material, Gilbert Islands). Araga, 1984: 103, pl. 89-E (Okinawa, Japan). Myers and Donaldson, 2003: 610 (synonym of *A. dorehensis*).

Antennarius hispidus (not of Bloch and Schneider): Weber, 1913: 562 (in part, i.e., SIBOGA Station 193, Sanana Island, ZMA 101.878).

Antennarius chironectes (not of Latreille): Weber, 1913: 562 (in part, i.e., smaller specimen, Sanguisiapo, Sulu Archipelago, ZMA 116.584).

Antennarius albomarginatus Fowler, 1945: 72, 74, fig. 19 (original description, single specimen, holotype ANSP 71609, Saipan). Myers and Donaldson, 2003: 610 (synonym of *A. dorehensis*).

Antennarius niveus Fowler, 1946: 215, fig. 75 (original description, single specimen, holotype ANSP 72088, Ryukyu Islands).

Antennarius punctatissimus Fowler, 1946: 216, fig. 76 (original description, single specimen, holotype ANSP 72089, Ryukyu Islands).

Antennarius albomaculatus: Fowler, 1949: 158 (error for *A. albomarginatus* Fowler, 1945).

Antennarius doherensis: Carnevale and Pietsch, 2006: 452, table 1 (misspelling of specific name; comparison, meristics, first fossil frogfish).

Antennatus dorehensis: Arnold and Pietsch, 2012: 128 (new combination, molecular phylogeny, revised classification). Brandl and Bellwood, 2014: 32 (pair-formation).

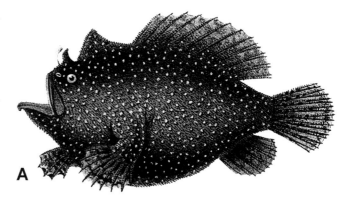

Fig. 147. *Abantennarius dorehensis* (Bleeker). (**A**) lectotype, RMNH 6268, 35 mm SL, Dorei, Irian Barat. Drawing by Ludwig Speigler under the direction of Pieter Bleeker; after Bleeker (1865b, pl. 199, fig. 7). (**B**) USNM 150939, 35 mm SL, Tataan Island, Tawitawi Group, Sulu Archipelago, Philippines. Drawing by Kumataro Ito; courtesy of Lisa Palmer and the Smithsonian Institution, NMNH, Division of Fishes, Washington, DC.

MATERIAL EXAMINED

One hundred and seventy-six specimens, 5.0 to 51 mm SL.

Lectotype of *Antennarius dorehensis*: RMNH 6268, 35 mm, Dorei, Irian Barat, Bleeker (Fig. 147A).

Paralectotypes of *Antennarius dorehensis*: RMNH 28010, 2 (26–51 mm), data as for lectotype.

Holotype of *Antennarius altipinnis*: USNM 70267, 14.5 mm, *Albatross*, Nogas Point, Panay, Philippines, copper sulfate in tidepool.

Holotype of *Antennarius albomarginatus*: ANSP 71609, 34 mm, Saipan Island, Marianas, Tinkham, March–May 1945.

Holotype of *Antennarius niveus*: ANSP 72088, 30.5 mm, Aguni Shima, Ryukyu Islands, Japan, Tinkham, 27 July 1945.

Holotype of *Antennarius punctatissimus*: ANSP 72089, 40 mm, Aguni Shima, Ryukyu Islands, Japan, Tinkham, 27 July 1945.

Additional material: *Western Indian Ocean*: ANSP, 1 (36 mm); BMNH, 4 (23.5–35.5 mm); SAIAB, 41 (16–47 mm); USNM, 5 (12.5–33 mm). *Cocos-Keeling Islands*: ANSP, 5 (13.5–22 mm).

Indonesia, New Guinea, and Ninigo-Hermit Islands: AMS, 1 (35 mm); FMNH, 1 (18 mm); NTM, 6 (12–38 mm); RMNH, 3 (26–26.5 mm); SU, 3 (20–31.5 mm); USNM, 7 (10.5–35 mm); ZMA, 2 (16.5–19 mm). *Australia*: WAM, 1 (16 mm). *Philippines*: CAS, 13 (9–34.5 mm); SU, 8 (13.5–37 mm); USNM, 16 (5–32.5 mm); UW, 19 (7.5–37 mm). *Taiwan and Okinawa*: NTUM, 1 (30.5 mm); USNM, 3 (20–25 mm). *Palau, Marianas, Guam, and Carolines*: ANSP, 1 (27.5 mm); BPBM, 3 (18.5–33 mm); CAS, 7 (21–38 mm); NSMT, 1 (33.5 mm); UG, 3 (7.5–37.5 mm); UMMZ, 1 (22 mm); USNM, 5 (12.5–30 mm). *Solomons and New Hebrides*: AMS, 3 (24.5–29 mm); ANSP, 1 (28.5 mm); USNM, 1 (13.5 mm). *Gilberts*: AMS, 2 (38.5–51 mm); BMNH, 3 (26–37 mm); BPBM, 2 (28.5–36 mm). *Fiji*: ANSP, 2 (21.5–26 mm); BPBM, 2 (16–22 mm); ROM, 2 (13–29.5 mm); USNM, 6 (8–21 mm). *Society Islands*: BMNH, 1 (23.5 mm).

DIAGNOSIS

A member of the genus *Abantennarius* unique in having a short illicium (5.0 to 9.5% SL, distinctly less than length of the second dorsal-fin spine; Figs. 147, 148), a unique escal morphology, and a reduced number of pectoral-fin rays (rarely more than nine; see Table 10).

Abantennarius dorehensis is further distinguished by having the following combination of character states: basidorsal pigment spot present or absent, but if present usually only weakly developed; esca a simple oval-shaped appendage or a more elongate, tapering appendage, nearly always ventrolaterally directed, basal filaments sometimes present, pigmented swellings absent (Fig. 148); second dorsal-fin spine unusually short, length 6.8 to 13.0% SL, slightly curved posteriorly; opercular opening adjacent to or on pectoral-fin lobe; caudal peduncle present, the membranous posteriormost margin of soft-dorsal and anal fins attached to body distinctly anterior to base of outermost rays of caudal fin; belly without large, darkly pigmented spots; dorsal-fin rays 12 (rarely 11 or 13); anal-fin rays seven (rarely eight); pectoral-fin rays eight or nine (rarely 10); last ray of pelvic fin bifurcate (Table 10).

DESCRIPTION

Anterior, lateral, and usually proximal half of posterior surface of second dorsal-fin spine densely covered with close-set dermal spinules; shallow naked depression between second and third dorsal-fin spines small to absent; third dorsal-fin spine curved posteriorly, length 18.7 to 29.3% SL; eye diameter 3.9 to 8.0% SL.

Coloration in life: overall white, maroon, dark violet to brownish gray, with numerous, small white and black spots on head, body, and fins (Figs. 147–149).

Fig. 148. Head of *Abantennarius dorehensis* (Bleeker) showing details of the esca and shape of the second and third dorsal-fin spines, Lembeh Strait, Sulawesi, Indonesia. Photo by Teresa Zuberbühler.

Coloration in preservation: light gray, brown, to black, covered with numerous small white to gray spots or irregularly shaped patches; tips of soft-dorsal and anal-fin rays white; distal half of rays of pectoral fin white, peppered with dark spots; caudal fin with a conspicuous pale bar; illicium without banding; short, pigmented bars radiating from eye faintly developed to absent.

Largest specimen examined: 51 mm SL (AMS I.18055-013; RMNH 28010), but 78% of material examined measured 30 mm SL or less, 60% measured 25 mm SL or less.

DISTRIBUTION

The known range of *Abantennarius dorehensis* extends from the East African coast and Aldabra Islands, east to the Solomon, Caroline, Line, and Society Islands, including the Cocos-Keeling and Ryukyu Islands, Taiwan, throughout the Philippines, Indonesia (Kalimantan, Bali, Molucca Islands, Sangihe Islands, and Samana Island), northwestern Australia (Rowley Shoals), and Papua New Guinea (Fig. 283, Table 22). The type locality is Dorei, Irian Barat, New Guinea.

Compared to other members of the genus, this is an extremely shallow-living species. Depths of capture, known for 46 specimens, range from the surface to 2.5 m; 76% of these, however, were taken in 1 m or less.

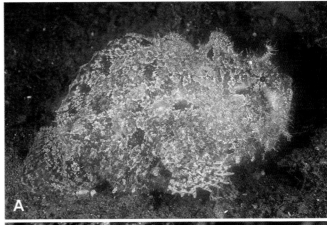

Fig. 149. *Abantennarius dorehensis* (Bleeker). (**A**) Papua New Guinea. Photo by Colin Marshall. (**B**) Lembeh Strait, Sulawesi, Indonesia. Photo by Teresa Zuberbühler.

COMMENTS

Probably because of its extremely shallow distribution, *A. dorehensis* is more commonly collected than most other antennariids. Although it has been correctly identified by most workers, Bleeker (1865b) and Fowler (1959) confused it with *Abantennarius nummifer*, Schmeltz (1877) with *Antennarius commerson*, and Weber (1913) with *A. hispidus* and *A. chironectes* (the latter a synonym of *A. pictus*; see Synonymy, p. 212).

Abantennarius drombus (Jordan and Evermann)
Freckled Frogfish

Figs. 150–154, 276, 303; Tables 10, 22

Antennarius drombus Jordan and Evermann, 1903: 207 (original description, two specimens, holotype USNM 50659, Waikiki, Oahu; Fig. 150A). Jordan and Evermann, 1905: 521, pl. 64 (after Jordan and Evermann, 1903). Jordan and Seale, 1906: 438

(Hilo, Hawaii; Samoa). Böhlke, 1953: 147 (paratype SU 7472, Oahu). Schultz, 1957: 96, pl. 11, figs. C, D (in part, includes *A. coccineus*; synonymy, description; Cocos Island, Eastern Pacific; Hawaiian Islands; in key). Gosline and Brock, 1960: 305, 345 (description; in key). Schultz, 1964: 175, table 1 (additional material, Oahu). Rosenblatt et al., 1972: 5 (misidentification, material is *A. coccineus*; Panama). Tinker, 1978: 511, figs. (in part, includes *A. coccineus*; description; Hawaii to Cocos Island, Eastern Pacific). Parrish et al., 1986 (piscivory). Randall et al., 1993a: 223, 227, table 1 (synonym of *A. coccineus*, Hawaiian Islands). Randall et al., 1993b: 364, fig. 16 (figure shows *A. drombus*; Midway Atoll). Randall, 1996: 44, color fig. (resurrection from synonymy of *A. coccineus*, Hawaiian Islands). Michael, 1998: 349 (variant of *A. coccineus*, perhaps distinct). Greenfield, 2003: 56, table 1 (Kaneohe Bay, Hawaiian Islands). Robertson et al., 2004: 563 (sister species of *A. coccineus*, Hawaiian Islands). Mundy, 2005: 18, 39, 263 (Johnston Atoll, Hawaiian Islands, including Midway). Randall, 2005a: 69 (Hawaiian Islands). Randall, 2007: 128, color fig. (Hawaiian Islands; in key). Dyer and Westneat, 2010: 605 (synonym of *A. coccineus*). Randall, 2010: 37 (Hawaiian Islands).

Antennarius nexilis Snyder, 1904: 537, pl. 13, fig. 23 (original description, two specimens, holotype USNM 50883, Honolulu; Fig. 150B). Jordan and Evermann, 1905: 523, pl. 65, fig. 2 (after Snyder, 1904). Fowler, 1928: 479 (description, additional material, Society Islands). Böhlke, 1953: 148 (paratype SU 7735). Dyer and Westneat, 2010: 605 (synonym of *A. coccineus*).

Antennarius leucus: Mundy, 2005: 263 (misplaced in synonymy of *A. drombus* with reference to Fowler, 1934b).

Antennarius coccineus (not of Lesson): Pietsch and Grobecker, 1987: 144, figs. 56, 57, 123, pl. 27 (in part, failure to recognize *A. drombus*; description, distribution, relationships). Randall et al., 1993b: 364, fig. 16 (figure shows *A. drombus*; Midway Atoll). Norris and Parrish, 1988: 109, table (predator-prey relationships, Hawaiian Islands). Kosaki et al., 1991: 189, fig. 5 (Johnston Atoll). Randall et al., 1993a: 223, 227, table 1 (Hawaiian Islands).

MATERIAL EXAMINED

Forty-eight specimens, 20 to 86 mm SL.

Holotype of *Antennarius drombus*: USNM 50659, 27 mm, Waikiki Reef near Honolulu, Oahu, 19 August 1901 (Fig. 150A).

Paratype of *Antennarius drombus*: SU 7472, 20 mm, data as for holotype.

Fig. 150. *Abantennarius drombus* (Jordan and Evermann). (**A**) holotype, USNM 50659, 27 mm SL, Waikiki Reef near Honolulu, Oahu, Hawaiian Islands. (**B**) holotype of *Antennarius nexilis* Snyder, USNM 50883, 78 mm SL, Honolulu, Oahu, Hawaiian Islands. Drawings by Kako H. Morita; courtesy of Lisa Palmer and the Smithsonian Institution, NMNH, Division of Fishes, Washington, DC.

Holotype of *Antennarius nexilis*: USNM 50883, 78 mm, *Albatross*, Honolulu, Oahu, 1902 (Fig. 150B).

Paratype of *Antennarius nexilis*: SU 7735, 71 mm, data as for holotype.

Additional material: *Hawaiian Islands and Johnston Atoll*: ANSP, 4 (54.5–86 mm); BPBM, 22 (22.5–86 mm); CAS, 6 (35–69 mm); FMNH, 4 (50–63 mm); LACM, 1 (52 mm); NMNZ, 1 (35 mm); ROM, 1 (54 mm); USNM, 5 (30–58 mm).

DIAGNOSIS

A member of the genus *Abantennarius*, very similar to *A. coccineus*, but distinguished in having 12 pectoral-fin rays (rarely 11 or 13) (Table 10); two or more darkly pigmented, spherical escal swellings (very rarely absent; Fig. 151); and the body and especially fins nearly always densely peppered with small dark brown to black spots (Fig. 150).

Abantennarius drombus is further distinguished in having the following combination of character states: basidorsal pigment spot present or absent, if present only weakly developed; illicium 8.6 to 11.4% SL, about as long as second dorsal-fin spine; esca an unpigmented cluster of folded tissue, with a few short, tapering filaments, and a variable number of darkly pigmented, spherical swellings, each terminating in a slender, tapering filament (Fig. 151); second dorsal-fin spine strongly curved posteriorly, length 6.6 to 11.4% SL; opercular opening adjacent to or on pectoral-fin lobe; caudal peduncle absent, the membranous posteriormost margin of soft-dorsal and anal fins connected to body at base of outermost rays of caudal fin; belly without large, darkly pigmented spots; dorsal-fin rays 12, rarely 11 or 13 (of 82 fins counted, 14 had 11 rays, 67 had 12, and one had 13; Table 10); anal-fin rays seven; last ray of pelvic fin bifurcate.

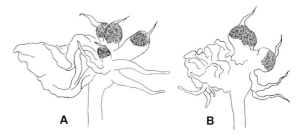

Fig. 151. Escae of *Abantennarius drombus* (Jordan and Evermann). (**A**) FMNH 63616, 54 mm SL, Oahu, Hawaiian Islands. (**B**) BPBM 8912, 62.5 mm SL, Oahu, Hawaiian Islands. Drawings by Cathy L. Short; after Pietsch and Grobecker (1987, fig. 57D, E).

DESCRIPTION

Darkly pigmented, spherical escal swellings nearly always present (much reduced or absent in about 12% of material examined (Fig. 151); anterior and lateral surfaces of second dorsal-fin spine usually completely covered with close-set dermal spinules (but spinules occasionally forming discrete clusters leaving naked spaces between); shallow, naked depression between second and third dorsal-fin spines nearly always present; third dorsal-fin spine curved posteriorly, length 19.2 to 24.5% SL; eye diameter 5.3 to 8.0% SL.

Coloration: overall yellowish- to reddish-brown, tan, or gray; body densely peppered with small dark brown to black spots; more heavily peppered on fins, sometimes forming irregular dark bands on caudal fin; belly somewhat lighter, but equally peppered with small dark spots; illicium usually lightly banded; six to eight short, darkly pigmented bars radiating from eye (Figs. 152–154).

Largest specimen examined: 86 mm SL (BPBM 13334, Johnston Atoll).

DISTRIBUTION

Abantennarius drombus is known only from the Hawaiian Islands (including Midway) and Johnston Atoll (Fig. 276, Table 22). The type locality is Waikiki Reef near Honolulu, Oahu. Depths of capture range from near the surface to 104 m but most of the specimens for which depth is recorded were taken in less than 10 m.

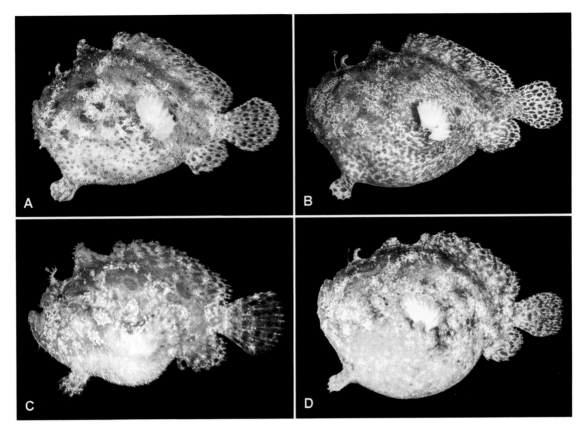

Fig. 152. *Abantennarius drombus* (Jordan and Evermann). (**A**) BPBM 8471, 74 mm SL, Moku Manu, Oahu, Hawaiian Islands. (**B**) BPBM 7927, 93 mm SL, Kaneohe Bay, Oahu, Hawaiian Islands. (**C**) BPBM 34812, 42 mm SL, Midway Atoll. (**D**) BPBM 33971, 72 mm SL, Johnston Atoll. Photos by John E. Randall; courtesy of Lisa Palmer and the Smithsonian Institution, NMNH, Division of Fishes, Washington, DC.

COMMENTS

Pietsch and Grobecker (1987) recognized *Abantennarius drombus* as a junior synonym of the widely distributed Indo-Pacific *A. coccineus*, based on apparent geographic variation in coloration, pectoral-fin ray counts, and escal morphology. In the center of its distribution, the Indo-Australian Archipelago, *A. coccineus* tends to be an overall tan, yellow, to yellow-brown, with darker brown to black mottling or spotting, especially on the cheeks and opercular region. On the periphery of its range, particularly in the Red Sea, Line and Marquesas Islands, and Eastern Pacific populations, individuals tend to be dark brown or gray and densely peppered with brown to black spots; the unpaired fins are usually darker and much more heavily peppered than the body (Fig. 276).

This pattern of geographic variation correlates somewhat with variation in pectoral-ray counts. Indo-Australian populations of *Abantennarius coccineus* nearly always have 10 (rarely nine) pectoral-fin rays, whereas in peripheral populations the count tends to increase slightly. In the Western Indian Ocean and Red Sea, 6% of the pectoral fins counted had 11 rays; in the easternmost island groups of the Pacific Plate, and in the Eastern Pacific, the bulk of the material examined had 11 or rarely 12 pectoral-fin rays.

Fig. 153. *Abantennarius drombus* (Jordan and Evermann). Black Rock, Kaanapali, Maui, Hawaiian Islands. Photo by Scott Rettig.

The esca of most individuals of *Abantennarius coccineus* consists of an unpigmented, usually spherical appendage sometimes associated with a few slender appendages (Fig. 144D). This relatively simple morphology grades into a more complex esca consisting of a cluster of elongate, tapering filaments, and only rarely associated with a cluster of darkly pigmented, spherical swellings (Fig. 144E). The latter condition was fully developed in about 23% of the combined "coccineus-like" specimens examined, and whereas these were more or less evenly distributed throughout the geographic range of *A. coccineus* (Fig. 276), more than half were collected in the Hawaiian Islands, where nearly all the individuals had the more complex escal morphology.

Considering that the Hawaiian material consisted of darkly pigmented, densely peppered individuals, usually with 12 pectoral-fin rays (17% of the specimens examined had 11 pectoral-fin rays) and darkly pigmented escal swellings (present in 88% of specimens examined), Pietsch and Grobecker (1987) concluded that this population represented an extension of the trends observed in *A. coccineus* and synonymized *A. drombus* with the latter. In retrospect, however, considering that a combination of these three characters, coupled with geographic distribution, always distinguishes *A. drombus* from *A. coccineus*, it is here considered valid, following resurrection of this species by Randall (1996: 44) and now recognized as such by many others (e.g., Michael, 1998; Greenfield, 2003; Robertson et al., 2004; Mundy, 2005).

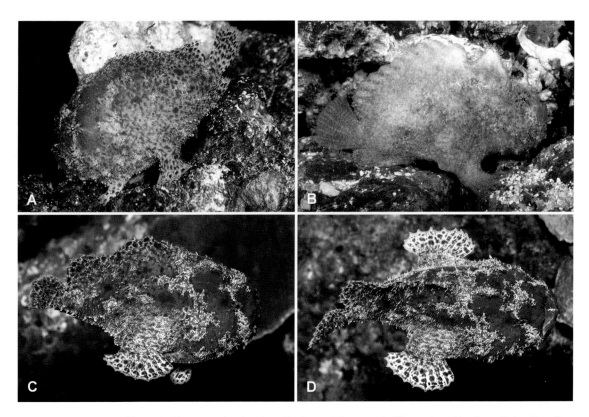

Fig. 154. *Abantennarius drombus* (Jordan and Evermann). **(A)** near Lanai Lookout, Oahu, Hawaiian Islands. Photo by John P. Hoover. **(B)** Oahu, Hawaiian Islands. Photo by Scott W. Michael. **(C, D)** Black Rock, Kaanapali, Maui, Hawaiian Islands. Photos by Robyn Smith.

Abantennarius duescus (Snyder)
Side-Jet Frogfish

Figs. 155, 284; Tables 10, 22

Antennarius duescus Snyder, 1904: 537, pl. 13, fig. 24 (original description, three speci-
    mens, holotype USNM 50884, Hawaiian Islands between Maui and Lanai; Fig. 155).
    Jordan and Evermann, 1905: 522, pl. 65, fig. 2 (after Snyder, 1904). Fowler, 1928: 479
    (after Snyder, 1904). Böhlke, 1953: 147 (paratype in CAS). Gosline and Brock, 1960:
    305, 345 (after Snyder, 1904; in key). Pietsch, 1984b: 36 (genera of frogfishes). Pietsch
    and Grobecker, 1987: 160, figs. 61, 124 (description, distribution, relationships). Ran-
    dall et al., 1993a: 223, 227, table 1 (Hawaiian Islands). Kulbicki and Williams, 1997: 1,
    4, 12 (Ouvea Atoll, Loyalty Islands, New Caledonia). Allen and Adrim, 2003: 25
    (Flores, Indonesia). Mundy, 2005: 263 (Hawaiian Islands). Carnevale and Pietsch,
    2006: 452, table 1 (comparison, meristics, first fossil frogfish). Randall, 2007: 128,
    color fig. (Hawaiian Islands; in key). Fricke et al., 2011: 367 (New Caledonia). Allen
    and Erdmann, 2012: 146, color fig. (East Indies). Fricke et al., 2014: 35 (Papua New
    Guinea). Farina and Bemis, 2016: 214 (gill ventilation, jet propulsion).
Abantennarius duescus: Schultz, 1957: 66, pl. 2, fig. A (new combination; description and fig.
    after Snyder, 1904; in key). Tinker, 1978: 507, fig. (description; figure after Snyder, 1904).
Antennatus duescus: Arnold and Pietsch, 2012: 128 (new combination, molecular phy-
    logeny, revised classification).

Fig. 155. *Abantennarius duescus* (Snyder). Holotype, USNM 50884, 30 mm SL, between Maui and Lanai, Hawaiian Islands. Drawing by William S. Atkinson; courtesy of Lisa Palmer and the Smithsonian Institution, NMNH, Division of Fishes, Washington, DC.

MATERIAL EXAMINED

Twelve specimens, 10 to 30 mm SL.

Holotype of *Antennarius duescus*: USNM 50884, 30 mm, *Albatross* station 3872, Auau Channel between Maui and Lanai, Hawaiian Islands, 58.6–78.7 m (Fig. 155).

Paratypes of *Antennarius duescus*: SU 7736, 1 (14.5 mm), data as for holotype. USNM 126597, 1 (15 mm), *Albatross* station 4128, vicinity of Kauai, Hawaiian Islands, 137 m.

Additional material: USNM, 1 (21.5 mm); WAM, 8 (10–20 mm).

DIAGNOSIS

A member of the genus *Abantennarius* unique in having the opercular opening situated about halfway between base of pectoral-fin lobe and origin of anal fin (Fig. 155).

Abantennarius duescus is further distinguished by having the following combination of character states: basidorsal pigment spot absent (in three known specimens); illicium length 9.7% SL in holotype (13.3 and 13.8% SL in paratypes), about as long as second dorsal-fin spine; escal morphology uncertain (but appears to conform to that of most other members of the genus *Abantennarius*, consisting of a tapering appendage with a few slender filaments and a cluster of pigmented swellings at base); second dorsal-fin spine slightly curved posteriorly, length 10.0% SL in holotype (12.0 and 14.5% SL in paratypes); caudal peduncle absent, the membranous posteriormost margin of soft-dorsal and anal fins connected to body at base of outermost rays of caudal fin; belly without large, darkly pigmented spots; dorsal-fin rays 12; anal-fin rays seven; pectoral-fin rays nine; all rays of pelvic fin simple (Table 10).

DESCRIPTION

Esca of holotype with perhaps a single, small, pigmented swelling; second dorsal-fin spine slender, the anterior and lateral surfaces covered with scattered dermal spinules, the spinules forming a discrete cluster at tip; shallow naked depression between second and third dorsal-fin spines; third dorsal-fin spine curved posteriorly, length 17.7% SL in holotype (23.7 and 26.9% SL in paratypes); eye diameter of holotype 8.7% SL; distance between opercular opening and base of lowermost ray of caudal fin 39.3% SL in holotype (34.5 and 37.2% SL in paratypes).

Coloration in life: extremely variable; an overall purplish-lilac except for a few pinkish spots on body and whitish tips of pectoral- and pelvic-fin rays (holotype); or light bronze above, yellowish-bronze below, with a wide pinkish crescent on upper part of opercles (paratype) (after Snyder, 1904: 538); often reddish to mottled grayish brown.

Coloration in preservation: light grayish-brown with slightly darker-brown, irregularly shaped blotches on head and body and a rounded, pinkish blotch on side about halfway between eye and base of pectoral fin; but, since the known material is badly bleached, the following is extracted from Snyder's (1904: 537) original description: holotype pale brick-red, the unpaired fins darker on edges; rayed portion of paired fins gray below, dusky above; head and body sparsely clouded and spotted with dusky and gray; a large, irregular cross band on chin, extending upward slightly beyond mouth; dusky cloud above pectoral; large, gray spot, bordered with dusky, on head between snout and pectoral; small, ocellated gray spot below pectoral, and a similar spot on body midway between opercular opening and dorsal fin; caudal peduncle with a narrow, vertical, gray band bordered with dusky; mouth immaculate within; paratype brownish-black except for small reddish cloud on nape; fins narrowly edged with red.

The holotype is the largest known specimen at 30 mm SL (USNM 50884).

DISTRIBUTION

Abantennarius duescus is extremely rare, known for more than 90 years only from the three type specimens, all collected near Maui, Lanai, and Kauai in the Hawaiian Islands (Fig. 284, Table 22). There are now, however, records from Karaban Island off Kalimantan; Flores and Komodo in the Lesser Sunda Islands, Indonesia (Allen and Adrim, 2003); Madang and Milne Bay, Papua New Guinea; Ouvea Atoll in the Loyalty Islands, territory of New Caledonia (Kulbicki and Williams, 1997; Fricke et al., 2011); Fiji; and Midway and Johnston atolls.

The non-type material was captured at depths ranging from less than 12 to 60 m, averaging about 23 m; the holotype and one paratype were trawled in 58.6 to 78.7 m, the remaining paratype in 137 m.

COMMENTS

Abantennarius duescus and the apparently closely related species *A. analis* differ from all other lophiiforms in having the opercular opening displaced posteriorly, about halfway between the base of the pectoral-fin lobe and the origin of the anal fin in *A. duescus*, and adjacent to the anal fin in *A. analis* (see Comments, p. 201).

Abantennarius nummifer (Cuvier)
Spotfin Frogfish

Figs. 15D, 156–160, 277, 328; Tables 10, 22

Chironectes nummifer Cuvier, 1817b: 430, pl. 17, fig. 4 (original description, single specimen, holotype MNHN 181, locality unknown; Fig. 156). Cuvier, 1829: 252 (after Cuvier, 1817b). Valenciennes, 1837: 425 (in part, MNHN A.4613 is *F. indicus*). Rüppell, 1838: 141 (Red Sea). Valenciennes, 1842: 189 (after Valenciennes, 1837). Pietsch, 1985a: 87, fig. 25, table 1 (original manuscript sources for *Histoire Naturelle des Poissons* of Cuvier and Valenciennes). Pietsch et al., 1986: 140 (in type catalog).

Antennarius nummifer: Schinz, 1822: 500 (new combination; after Cuvier, 1817b). Bleeker, 1859a: 5 (Doreh, New Guinea). Bleeker, 1859b: 129 (Indonesia). Günther, 1861a: 195 (description; East Indies, Red Sea). Day, 1865: 121 (Malabar). Bleeker, 1865b: 9, 18, pl. 198, fig. 2 (in part, three of four specimens are *A. dorehensis*; Ambon, Ceram; Fig. 157A). Klunzinger, 1871: 499 (synonymy includes, with question, *Chironectes chlorostigma* Ehrenberg, in Valenciennes, 1837; Red Sea). Günther, 1876: 164 (in part, includes *A. coccineus*; Samoa, Raiatea). Klunzinger, 1884: 125 (Red Sea). Jordan and Seale, 1906: 438 (in part, includes *A. coccineus* after Günther, 1876). Annandale and Jenkins, 1910: 20, 21, pl. 1, fig. 5 (India). Weber, 1913: 561 (Sumbawa). McCulloch, 1922: 123 (New South Wales). Fowler, 1928: 478 (Red Sea, Oceania). McCulloch, 1929: 408 (New South Wales, Lord Howe Island). Fowler, 1931: 367 (Raiatea). Fowler, 1949: 158 (Papeete, Tahiti). Munro, 1955: 289, fig. 840 (Ceylon). Schultz, 1957: 102, fig. 8 (in part, USNM 167505 is *A. rosaceus*, FMNH 51864 is *A. hispidus*; Persian Gulf; in key). Munro, 1958: 296 (synonymy, New Guinea). Mees, 1959: 9 (new to Western Australia). Ray, 1961: 230,

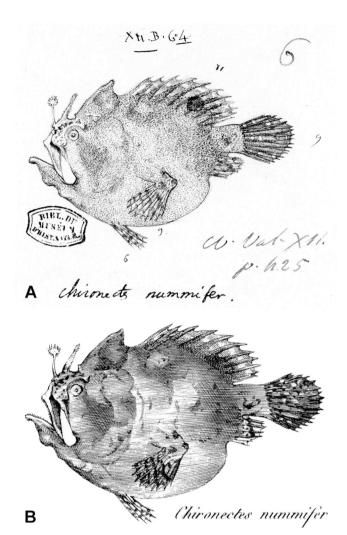

Fig. 156. *Abantennarius nummifer* (Cuvier). Holotype, MNHN 181, 48 mm SL, locality unknown. (**A**) reproduced from the pencil original that served as the model for the engraving that accompanied the original description published by Cuvier (1817b, pl. 17, fig. 4) (after Cuvier and Valenciennes, MS 504, XII.B, Bibliothèque Centrale, Muséum national d'Histoire naturelle, Paris). The conspicuous, dark basidorsal spot, so characteristic of this species—present on the holotype and mentioned by Cuvier (1817b: 430) but not shown in his published version (**B**) of the figure—has been added by an unknown hand. Courtesy of Emmanuelle Choiseau, © Muséum national d'Histoire naturelle, Paris.

fig. 1 (egg raft morphology; identification unconfirmed). Beaufort and Briggs, 1962: 217 (synonymy, description, distinct from *A. coccineus*; Red Sea, India, Ceylon, China, Indonesia, Japan, Philippines, Australia, New Guinea, Lord Howe Island, Gilberts, Samoa, Society Islands, Hawaii; in key). Le Danois, 1964: 103, fig. 50j (in part, MNHN A.4582, A.4587, A.4613 are *A. nummifer*). Marshall, 1964: 514, 515 (Queensland, New South Wales; in key). Schultz, 1964: 175, table 1 (holotype of *A. sanguifluus* examined). Munro, 1967: 585, pl. 78, fig. 1094 (New Guinea). Paulin, 1978: 490, fig. 1D (description, New Zealand). Araga, 1984: 103, pl. 89-F (Chiba Prefecture, Japan). Dor, 1984: 56 (Red Sea).

Pietsch, 1984b: 36 (genera of frogfishes). Pietsch, 1984e: 2 (Western Indian Ocean). Thresher, 1984: 41 (reproduction). Dooley et al., 1985: 45 (Canary Islands). Pietsch, 1985a: 87, 88, figs. 25, 26, table 1 (original manuscript sources for *Histoire Naturelle des Poissons* of Cuvier and Valenciennes). Pietsch, 1986a: 366, 368, fig. 102.5 (South Africa). Edwards and Glass, 1987: 631 (St. Helena). Pietsch and Grobecker, 1987: 138, figs. 54, 55, 122, pl. 26 (description, distribution, relationships). Allen and Swainston, 1988: 38 (northwestern Australia). Myers, 1989: 68, pl. 13E (Red Sea to Society Islands, Marianas and Marshalls; in key). Paulin et al., 1989: 134 (New Zealand). Paxton et al., 1989: 278 (Australia). Randall et al., 1990: 54, color fig. (East Africa and Red Sea to Society Islands and north to Japan). Lloris et al., 1991: 220, appendix (Azores, Madeira, and Canary Islands). Cunxin, 1993: 117 (South China Sea). Francis, 1993: 158 (Lord Howe Island). Goren and Dor, 1994: 14 (Red Sea). Hutchins, 1994: 42, appendix (Western Australia). Martin et al., 1995: 916 (Gulf of Carpentaria). Randall, 1995: 83, fig. 168 (Oman). Allen, 1997: 62 (tropical Australia). Arruda, 1997: 125 (Azores). Carpenter et al., 1997: 126 (Persian Gulf). Debelius, 1997: 91, color figs. (Madeira). Kuiter, 1997: 44 (Australia). Kulbicki and Williams, 1997: 12 (New Caledonia). Lindberg et al., 1997: 216 (Sea of Japan). Okamura and Amaoka, 1997: 141, figs. (Japan). Randall et al., 1997a: 54, color fig. (East Africa and Red Sea to Society Islands and north to Japan). Randall et al., 1997b: 11 (Ogasawara Islands). Santos et al., 1997: 62 (Azores). Michael, 1998: 329, 330, 334, 342, 343, color figs. (identification, behavior, captive care). Fricke, 1999: 106 (Mauritius). Myers, 1999: 69, pl. 13E (after Myers, 1989). Ni and Kwok, 1999: 136, table 1 (Hong Kong). Pietsch, 1999: 2015 (western central Pacific; in key). Allen, 2000a: 62, pl. 12-10 (Indo-West Pacific). Nakabo, 2000: 458 (Japan; in key). Pereira, 2000: 5 (Mozambique). Pietsch, 2000: 597 (South China Sea). Sadovy and Cornish, 2000: 45, fig. (Hong Kong). Schleichert, 2000: 36, color figs. (natural history). Hutchins, 2001: 22 (Western Australia). Goldin, 2002: 48, fig. (Samoa). Nakabo, 2002: 458 (Japan; in key). Allen and Adrim, 2003: 25 (Papua to Sumatra, Indonesia). Allen et al., 2003: 366, color figs. (Indo-Pacific). Greenfield, 2003: 56, table 1 (Kaneohe Bay, Hawaiian Islands). Mishra and Krishnan, 2003: 23 (Pondichery and Karaikal). Myers and Donaldson, 2003: 610, 614 (Mariana Islands). Heemstra et al., 2004: 3314, 3317, table 1 (Rodrigues Island). Letourneur et al., 2004: 209 (Réunion Island). Lieske and Myers, 2004: 46 (Red Sea). Pereira et al., 2004: 4, appendix 1 (Mozambique). Shedlock et al., 2004: 134, 144, table 1 (molecular phylogeny). Mundy, 2005: 264 (Hawaiian Islands). Randall, 2005a: 71, color figs. (South Pacific). Allen et al., 2006: 640 (Australia). Calado, 2006: 394, 395, tables 4, 6 (Azores, Madeira). Carnevale and Pietsch, 2006: 452, table 1 (comparison, meristics, first fossil frogfish). Michael, 2006: 36, 43 (behavior). Senou et al., 2006: 423 (Sagami Sea). Kuiter and Debelius, 2007: 120, color figs. (Indonesia). Randall, 2007: 129, color fig. (Hawaiian Islands; in key). Porteiro and Afonso, 2007: 57, 58, table 1 (Azores). Kwang-Tsao et al., 2008: 243 (southern Taiwan). Wirtz et al., 2008: 8 (Madeira). Allen and Erdmann, 2009: 594 (West Papua). Fricke et al., 2009: 28 (Réunion). Hanel et al., 2009: 147 (European seas). Al-Jufaili et al., 2010: 18, appendix 1 (Oman). Golani and Bogorodsky, 2010: 16 (Red Sea). Fricke et al., 2011: 368 (New Caledonia). Motomura and Aizawa, 2011: 450, fig. 2A, table 1 (Yaku-shima Island, southern Japan). Obura et al., 2011: 11, table 8 (Phoenix Islands). Satapoomin, 2011: 51, appendix A (Andaman Sea). Allen and Erdmann, 2012: 147, color figs. (East Indies). Grandcourt, 2012: 137 (Persian Gulf). Nakabo, 2013: 542, figs. (Japan). Yoshida et al., 2013: 55 (Gulf of Thailand). Espino et al., 2014: 21 (Canary Islands). Fricke et al., 2014: 35 (Papua New Guinea). Hobbs et al., 2014a: 192, table 1 (Christmas Island). Hylleberg and Aungtonya,

2014: 101 (Phuket, Thailand). Moore et al., 2014: 177 (northwestern Australia). Delrieu-Trottin et al., 2015: 5, table 1 (Marquesas Islands). Hanel and John, 2015: 138, table 1 (Cape Verde Islands). Psomadakis et al., 2015: 174, fig. (Pakistan). Christie et al., 2016: 31, 32 (egg rafts).

Chironectes chlorostygma: Ehrenberg, in Valenciennes, 1837: 426 [said to have been originally described from a drawing made for C. G. Ehrenberg, examined by Valenciennes in Berlin in 1828, iconotype in Cuvier and Valenciennes, MS 504, XII.B.51 (Pietsch, 1985a), but two extant specimens, lectotype ZMB 2233 (the larger specimen), designated by Pietsch and Grobecker, 1987, Massuah (= Mesewa, Ethiopia)]. Schultz, 1957: 52 (unidentifiable). Pietsch, 1985a: 88, fig. 26, table 1 (original manuscript sources for *Histoire Naturelle des Poissons* of Cuvier and Valenciennes).

Chironectes chlorostigma: Rüppell, 1838: 141 (emended spelling of specific name; possible synonym of *A. nummifer*). Günther, 1861a: 183 (footnote reference to Valenciennes, 1837). Day, 1876: 272 (in synonymy of *A. nummifer*). Klunzinger, 1871: 499 (in synonymy of *A. nummifer*).

Cheironectes bicornis Lowe, 1839: 84 (original description, single specimen, holotype BMNH 1865.6.20.2, Madeira).

Fig. 157. *Abantennarius nummifer* (Cuvier). **(A)** Bleeker's *Antennarius nummifer*, authorship attributed to Günther, RMNH 6279, 65 mm SL, Ambon or Ceram, Indonesia. Drawing by Ludwig Speigler under the direction of Pieter Bleeker; after Bleeker (1865b, pl. 198, fig. 2). **(B)** about 33 mm SL, Cairns, Queensland, Australia. Drawing labeled *Antennarius stigmaticus*, a synonym of *Abantennarius coccineus*. Watercolor by George James Coates; courtesy of Rick Feeney and the Los Angeles County Museum of Natural History.

Antennarius nummifer (not of Cuvier): Bleeker, 1854c: 497 (material is *A. coccineus*). Bleeker, 1865b: 18 (in part, three of four specimens are *A. dorehensis*, RMNH 6279 is *A. nummifer*). Playfair, 1866: 70 (material is *A. indicus*). Günther, 1876: 164 (material is, in part, *A. coccineus*). Day, 1876: 272, pl. 59, fig. 2 (description and figure represent *A. indicus*). Schmeltz, 1877: 15 (material is *A. coccineus*). Schmeltz, 1879: 48 (after Schmeltz, 1877). Day, 1889: 232 (after Day, 1876). Boulenger, 1897: 373 (BMNH 1897.8.23.79-80 are *A. coccineus*). Pietschmann, 1930: 23 (material is *A. coccineus*). Fowler, 1934a: 450 (after Pietschmann, 1930). Koumans, 1953: 268 (RMNH 20110 is *A. coccineus*). Fowler, 1959: 561 (material is *A. dorehensis*). Burgess and Axelrod, 1973a: 528, color fig. 486 (in part, plate represents *A. coccineus*). Burgess and Axelrod, 1974: 1365, pls. 453, 454 (plate represents *A. biocellatus*).

Chironectes bicornis: Günther, 1861a: 184 (footnote reference to Lowe, 1839; emended spelling of generic name).

Antennarius multiocellatus (not of Valenciennes): Günther, 1869a: 238 (BMNH 1868.6.15.21 from St. Helena is *A. nummifer*). Melliss, 1875: 108 (after Günther, 1869a).

Antennarius sanguifluus Jordan, 1902: 374, fig. 5 (original description, two specimens, lectotype SU 7600 by subsequent designation of Böhlke, 1953, paralectotype USNM 49820, both from Misaki, Japan). Jordan, 1905: 551 ("spotted with blood-red in imitation of coralline patches"). Jordan et al., 1913: 424, fig. 393 (after Jordan, 1902; "perhaps a red form of *A. tridens*"). Jordan and Hubbs, 1925: 330 ("probably the same as *A. tridens*," Misaki). Okada, 1938: 274 (Honshu). Okada and Matsubara, 1938: 457, pl. 112, fig. 2 (after Jordan, 1902). Böhlke, 1953: 148 (lectotype designated). Okada, 1955: 429, fig. 386 (description; Misaki, Boshu). Tokioka, 1961: 448 (Kyoto, Japan).

Antennarius (?*Commersonii*) (not of Latreille): Weber, 1913: 563 (ZMA 116.515, New Guinea).

Antennarius mummifer: Gregory, 1933: 389 (misspelling of specific name).

Antennarius bermudensis (not of Schultz): Maul, 1959: 15 (first Atlantic records of *A. nummifer*; MMF 5254, Azores; MMF 2461, 2681, Madeira).

Antennarius japonicus Schultz, 1964: 181, pl. 2, tables 1, 2 (original description, single specimen, holotype SU 26796, Sagami Bay, Japan). Araga, 1984: 103, pl. 346-G (after Schultz, 1964).

Antennarius chlorostigma: Le Danois, 1970: 84, fig. 1, 3C (new combination; Dahlak Islands, Red Sea). Dor, 1984: 55 (types in ZMB).

Antennarius notophthalmus (not of Bleeker): Kotthaus, 1979: 46, fig. 494 (ZIM 6007, Red Sea). Dor, 1984: 56 (after Kotthaus, 1979).

Antennatus nummifer: Arnold and Pietsch, 2012: 127, 128, figs. 4, 6, table 1 (new combination; molecular phylogeny). Stewart, 2015: 885 (New Zealand). Pietsch, 2016: 2053 (eastern central Atlantic; in key).

MATERIAL EXAMINED

One hundred and fifty-three specimens, 13 to 90 mm SL.

Holotype of *Chironectes nummifer*: MNHN 181, 48 mm, locality unknown (Figs. 15, 156).

Lectotype of *Chironectes chlorostygma*: ZMB 2233, 25 mm, Mesewa, Ethiopia, Hemprich and Ehrenberg.

Paralectotype of *Chironectes chlorostygma*: ZMB 2233, 21.5 mm, Mesewa, Ethiopia, Hemprich and Ehrenberg.

Holotype of *Cheironectes bicornis*: BMNH 1865.5.20.2, 42 mm, Madeira, Lowe.

Lectotype of *Antennarius sanguifluus*: SU 7600, 44 mm, Misaki, Japan, Jordan and Mitsukuri, 1900.

Paralectotype of *Antennarius sanguifluus*: USNM 49820, 24 mm, data as for lectotype.

Holotype of *Antennarius japonicus*: SU 26796, 49 mm, Sagami Sea, Japan, Owston, 11 March 1903.

Additional material: *Eastern Atlantic*: AMNH, 1 (60 mm); BMNH, 1 (68 mm); MMF, 4 (49–90 mm); UW, 4 (42–50 mm); ZMUC, 1 (24.5 mm). *Western Indian Ocean*: BMNH, 33 (12.5–61.5 mm); CAS, 3 (42–60 mm); FMNH, 6 (26.5–68 mm); SAIAB, 7 (26–74 mm); SAMA, 1 (22.5 mm); USNM, 3 (25.5–33 mm). *Red Sea*: BPBM, 1 (40 mm); HUJ, 6 (18–41 mm); MNHN, 1 (37 mm); USNM, 6 (22–44 mm); UW, 1 (20 mm); ZIM, 1 (29.5 mm). *Thailand, Indonesia, and New Guinea*: BMNH, 4 (20–41 mm); CAS, 3 (37–57 mm); RMNH, 1 (65 mm); USNM, 1 (43 mm); UW, 2 (20–52 mm); ZMA, 7 (18–48 mm); ZMUC, 2 (32–37.5 mm). *Philippines*: UW, 2 (22–42 mm). *China and Japan*: ANSP, 4 (33–58 mm); BMNH, 1 (35 mm); BPBM, 4 (39.5–64 mm); SU, 2 (43.5–44 mm); ZMUC, 2 (30.5–39 mm). *Australia, Tasmania, and New Zealand*: AIM, 1 (61 mm); AMS, 8 (34–81 mm); BMNH, 1 (43 mm); CSIRO, 6 (31–60 mm); NMNZ, 1 (66 mm); WAM, 11 (13–43 mm). *Hawaiian Islands*: BPBM, 3 (24.5–42.5 mm). *Society Islands*: MCZ, 1 (17 mm). *Locality unknown*: UW, 2 (37–55 mm).

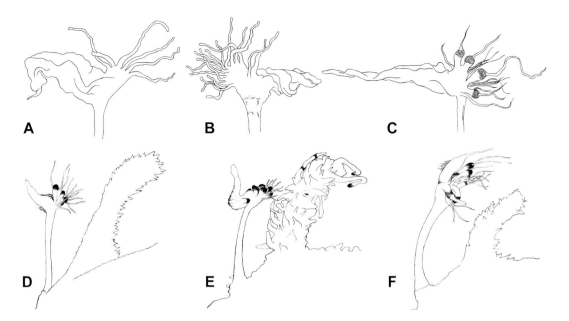

Fig. 158. Illicium, escae, and second dorsal-fin spine of Indo-West Pacific and Eastern Atlantic specimens of *Antennarius nummifer* (Cuvier). (**A**) CAS 35469, 42 mm SL, Grande Comore Island, western Indian Ocean. (**B**) CAS 35469, 45 mm SL, Grande Comore Island, western Indian Ocean. (**C**) CAS 42851, 53.5 mm SL, Goh Chuang, Gulf of Thailand. (**D**) MMF 2461, 49 mm SL, Madeira. (**E**) MMF 2681, 57 mm SL, Madeira. (**F**) USNM 195979 (originally MMF 3898), 65 mm SL, Madeira. Drawings **A**–**C** by Cathy L. Short; after Pietsch and Grobecker (1987, fig. 55); **D**–**F** courtesy of the late Günther Edmund Maul, Museu Municipal do Funchal, Madeira.

DIAGNOSIS

A member of the genus *Abantennarius* unique in having the following combination of character states: a darkly pigmented basidorsal spot nearly always present; illicium length 8.8 to 14.7% SL, about equal to length of second dorsal-fin spine; esca ranging from a simple, unpigmented sphere of irregularly folded tissue (usually associated with a few, short, basal filaments) to an elongate, tapering appendage with a variable number of slender filaments and a cluster of darkly pigmented, spherical swellings at base (Fig. 158); second dorsal-fin spine more or less straight, length 8.7 to 15.2% SL; opercular opening adjacent to or on pectoral-fin lobe; caudal peduncle present, the membranous posteriormost margins of soft-dorsal and anal fins attached to body distinctly anterior to base of outermost rays of caudal fin (see Caudal Peduncle, p. 32); belly without large, darkly pigmented spots; dorsal-fin rays 12 (rarely 13); anal-fin rays seven (rarely eight); pectoral-fin rays 10 or 11 (rarely 12); last ray of pelvic fin bifurcate (Table 10).

DESCRIPTION

Esca with as many as eight darkly pigmented basal swellings, absent in only about 7% of material examined (Fig. 158); anterior and lateral surfaces of second dorsal-fin spine usually completely covered with close-set dermal spinules (but the spinules often forming discrete clusters leaving naked spaces between; e.g., BPBM 21512); shallow naked depression between second and third dorsal-fin spines nearly always present; third dorsal-fin spine curved posteriorly, length 18.1 to 27.9% SL; eye diameter 5.1 to 8.8% SL.

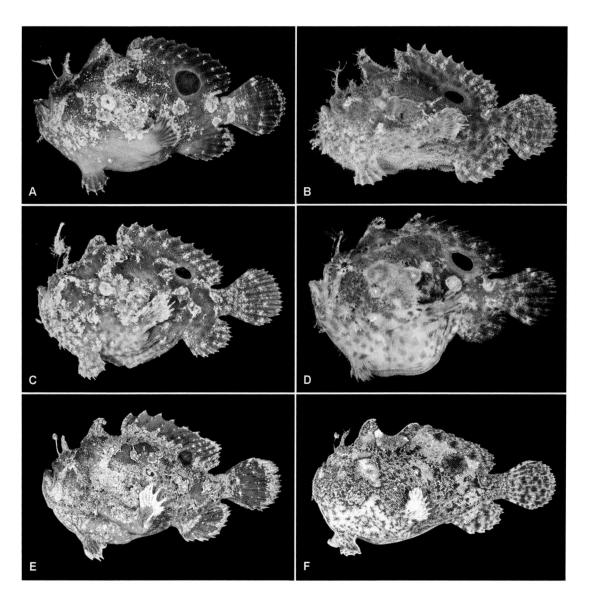

Fig. 159. *Abantennarius nummifer* (Cuvier). (**A**) BPBM 33338, 32 mm SL, Jana Island, Saudi Arabia. (**B**) BPBM 9851, 26 mm SL, off Lahilahi Point, Oahu, Hawaiian Islands. (**C**) BPBM 21512, 42 mm SL, El Hamira, Gulf of Aqaba, Egypt. (**D**) BPBM 16337, 23 mm SL, Mauritius. (**E**) BPBM 18645, 67 mm SL, Kau Yi Chau, Hong Kong. (**F**) BPBM 18994, 48 mm SL, Igaya Bay, Miyake-jima, Izu Islands, Japan. Photos by John E. Randall; courtesy of Lisa Palmer and the Smithsonian Institution, NMNH, Division of Fishes, Washington, DC.

Coloration in life: extremely variable; off-white, yellow, orange, brownish-orange, pink, or blood-red (BPBM 9851, 18645, 18980, 18994). Brown to reddish-brown on head, gradually becoming greenish-brown to brown on body and fins; belly somewhat lighter; scattered irregularly shaped greenish-white blotches on head, body, and fins; basidorsal spot black encircled with diffuse greenish ring (Figs. 157B, 159, 160).

Coloration in preservation: beige, pink, brown, to black; illicium usually lightly banded; as many as 10 short, darkly pigmented bars radiating from eye. Lighter-color phases often with somewhat darker, irregularly shaped patches scattered over entire

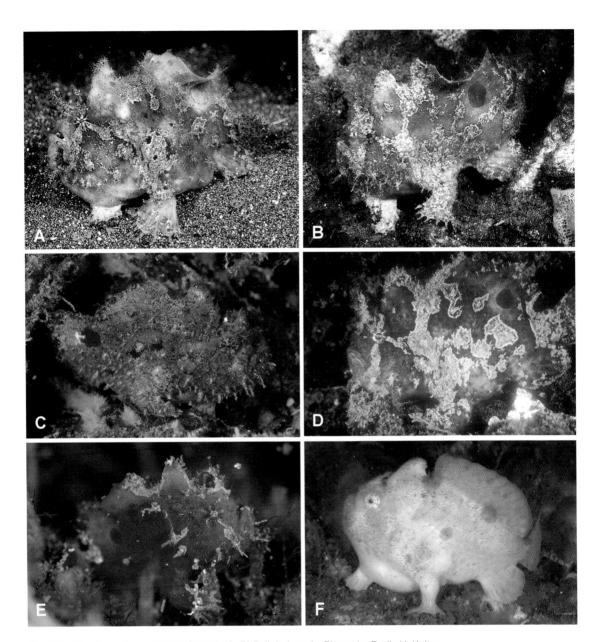

Fig. 160. *Abantennarius nummifer* (Cuvier). (**A**, **B**) Bali, Indonesia. Photos by Rudie H. Kuiter.
(**C**, **D**) Lembeh Strait, Sulawesi, Indonesia. Photos by Colin Marshall. (**E**) Bali, Indonesia. Photo by
Colin Marshall. (**F**) Osezaki, Izu Peninsula, Suruga Bay, Japan. Photo by Scott W. Michael.

body (bright pink-red blotches in some specimens; e.g., BMNH 1868.6.15.21); soft-dorsal, anal, and caudal fins often peppered with numerous dark spots. Darker-color phases with tips of pectoral-fin rays white.

Largest specimen examined: 90 mm SL (MMF 5254).

DISTRIBUTION

Abantennarius nummifer has a typical Indo-Pacific distribution, ranging from the East African coast, the Red Sea, the Persian Gulf, and Pakistani coast, including Mauritius,

Réunion, and Rodrigues, to India and Stri Lanka and farther east, with confirmed records from Japan, Taiwan, the Marshalls and Marianas, Christmas Island, tropical and subtropical Australia, Lord Howe Island, New Caledonia, the north Island of New Zealand, and eastward to the Hawaiian, Society, and Marquesas Islands (Fig. 277, Table 22). Rather surprisingly, it is found as well at insular localities of the Eastern Atlantic, including the Azores, Madeira, Canaries, Cape Verde, and St. Helena. The type locality is unknown.

In the Indo-Pacific, *A. nummifer* is a rather shallow-living frogfish. Although specimens may be taken anywhere between the surface and approximately 176 m, 75% of the 33 known captures were collected in less than 50 m; the average depth for all known Indo-Pacific captures is 19 m. In contrast, the Eastern Atlantic population appears to occupy significantly deeper waters; the six specimens for which data are available were taken between 44 and 293 m, with an average depth of 107 m.

COMMENTS

Abantennarius nummifer has often been confused with *A. coccineus*, due in part to an inaccurate illustration of the latter species published by Lesson, 1831 (see Comments, p. 212). These two species, however, are easily distinguished from each other by the presence of a distinct caudal peduncle in *A. nummifer* and the consistent absence of this feature in *A. coccineus* (see Caudal Peduncle, p. 32).

Abantennarius rosaceus (Smith and Radcliffe)
Rosy Frogfish

Figs. 161–164, 283; Tables 10, 22

Antennarius rosaceus Smith and Radcliffe, in Radcliffe, 1912: 203, pl. 17, fig. 2 (original description, single specimen, holotype USNM 70266, Romblon, Philippines; Fig. 161). Pietsch, 1984b: 36 (genera of frogfishes). Pietsch, 1984e: 2 (Western Indian Ocean). Pietsch and Grobecker, 1987: 170, figs. 66, 67, 125, pl. 30 (description, distribution, relationships). Myers, 1989: 68, fig. 1d (Red Sea, Philippines, Timor, Samoa; Gilbert, Marshalls, and Lord Howe Island; in key). Paxton et al., 1989: 278 (Australia). Yokota et al., 1992: 4 (Japan). Francis, 1993: 158 (Lord Howe Island). Goren and Dor, 1994: 14 (Red Sea). Michael, 1998: 343 (identification, behavior, captive care). Myers, 1999: 69, fig. 1d (after Myers, 1989). Pietsch, 1999: 2015 (western central Pacific; in key). Nakabo, 2000: 458 (Japan; in key). Hutchins, 2001: 22 (Western Australia). Nakabo, 2002: 458 (Japan; in key). Allen and Adrim, 2003: 25 (Java, Timor, Togian and Molucca Islands). Randall, 2005a: 72, color fig. (South Pacific). Allen et al., 2006: 640 (Australia). Carnevale and Pietsch, 2006: 452, table 1 (comparison, meristics, first fossil frogfish). Kuiter and Debelius, 2007: 117, color figs. (Indonesia). Allen and Erdmann, 2009: 594 (West Papua). Golani and Bogorodsky, 2010: 16 (Red Sea). Fricke et al., 2011: 368 (New Caledonia). Allen and Erdmann, 2012: 150, color fig. (East Indies). Allen and Erdmann, 2013: 32, appendix 3.1 (Bali, Indonesia). Nakabo, 2013: 541, figs. (Okinawa, Ryukyu Islands). Fricke et al., 2014: 35 (Papua New Guinea).

Trichophryne rosaceus: Schultz, 1957: 65, pl. 1, fig. B (new combination, description, Bikini). Schultz et al., 1966: 144, pl. 146, fig. A (description after Schultz, 1957).

Antennarius nummifer (not of Cuvier): Schultz, 1957: 102 (in part, i.e., USNM 167505, Gilbert Islands).

Antennatus rosaceus: Arnold and Pietsch, 2012: 127, 128, figs. 4, 6, table 1 (new combination; molecular phylogeny). Brandl and Bellwood, 2014: 32 (pair formation).

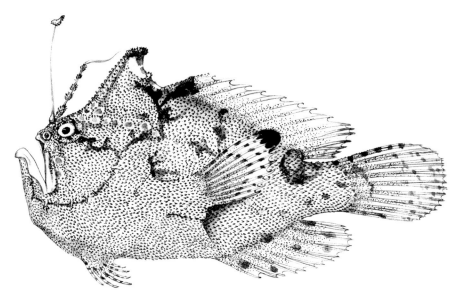

Fig. 161. *Abantennarius rosaceus* (Smith and Radcliffe). Holotype, USNM 70266, 26.5 mm SL, Romblon, Philippines, "taken from ship's side at night, attracted by electric light suspended in water." Drawing by an unknown artist, after Smith and Radcliffe, in Radcliffe (1912: 204, pl. 17, fig. 2); courtesy of Lisa Palmer and the Smithsonian Institution, NMNH, Division of Fishes, Washington, DC.

MATERIAL EXAMINED

Thirteen specimens, 11 to 42 mm SL.

Holotype of *Antennarius rosaceus*: USNM 70266, 26.5 mm, *Albatross*, Romblon, Philippines, "ship's side at night, attracted by electric light suspended in water" (Fig. 161).

Additional material: AMS, 1 (42 mm); BPBM, 1 (33 mm); USNM, 6 (11–21.5 mm); UW, 3 (14–38 mm); WAM, 1 (30 mm).

DIAGNOSIS

A member of the genus *Abantennarius* unique in having an exceptionally long illicium (23.3 to 30.3% SL) and second dorsal-fin spine (15.6 to 18.9% SL) (Fig. 161).

Fig. 162. Esca of *Abantennarius rosaceus* (Smith and Radcliffe). BPBM 17497, 33 mm SL, Fagatele Bay, Tutuila Island, American Samoa. Drawing by Cathy L. Short; after Pietsch and Grobecker (1987, fig. 67).

Abantennarius rosaceus is further distinguished by having the following combination of character states: a conspicuous, darkly pigmented, basidorsal spot; illicium distinctly longer than second dorsal-fin spine; esca an elongate, tapering appendage with several slender filaments and a cluster of darkly pigmented, spherical swellings at base (Fig. 162); second dorsal-fin spine straight, narrow, and tapering (Fig. 161); opercular opening adjacent to or on pectoral-fin lobe; caudal peduncle present, the membranous posterior-most margin of soft-dorsal and anal fins attached to body distinctly anterior to base of outermost rays of caudal fin; belly without large, darkly pigmented spots; dorsal-fin rays 12 (rarely 13); anal-fin rays seven (rarely eight); pectoral-fin rays 10 (rarely nine); last ray of pelvic fin bifurcate (Table 10).

DESCRIPTION

Darkly pigmented escal swellings present (except in some smallest known specimens); second dorsal-fin spine elongate, slender, the dermal spinules forming discrete clusters

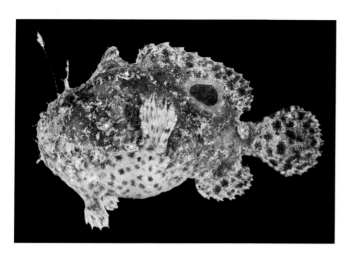

Fig. 163. *Abantennarius rosaceus* (Smith and Radcliffe). BPBM 17497, 37 mm SL, Tutuila, American Samoa. Photo by John E. Randall; courtesy of Lisa Palmer and the Smithsonian Institution, NMNH, Division of Fishes, Washington, DC.

leaving naked spaces between; shallow naked depression between second and third dorsal-fin spines nearly always present; third dorsal-fin spine curved posteriorly, length 22.7 to 26.2% SL; eye diameter 8.2 to 10.2% SL (Table 10).

Coloration in life: highly variable; tan, brown, black, or reddish with darker mottling; purple on upper part of head and body; tan to yellowish-brown below mouth, on belly and fins, with small reddish-brown to brown spots; basidorsal spot black, encircled with yellowish-brown ring (Figs. 163, 164).

Coloration in preservation: light tan, gray, light yellow, to brown (usually somewhat lighter on belly), often with scattered, dark spots, irregularly shaped blotches, or fine mottling over entire head and body; fins usually with a few small scattered spots, but heavily peppered with dark spots in some specimens (e.g., UW 20880); basidorsal spot dark brown to black; illicium usually banded; short, darkly pigmented bars radiating from eye present or absent.

Largest specimen examined: 42 mm SL (AMS I.5399).

DISTRIBUTION

Most of the known material of *Abantennarius rosaceus* was collected in the Philippines and Indonesia (Java to West Papua, including Bali, Timor, and the Togian and Molucca Islands). Additional localities include the Red Sea, the Ryukyu Islands, Papua New Guinea, Lord Howe Island, New Caledonia, American Samoa, the Marshalls (including Bikini Atoll), and Onotoa Atoll in the Gilberts. The type locality is Romblon, Philippines (Fig. 283, Table 22).

Depths of capture range from the surface to 130 m, but 92% of the specimens were taken in less than 55 m and 75% in 21 m or less.

Abantennarius sanguineus (Gill)
Sanguine Frogfish

Figs. 165–168, 277; Tables 10, 22

Antennarius sanguineus Gill, 1863: 91 (original description, two specimens, lectotype USNM 6393 designated by Pietsch and Grobecker, 1987, paralectotype USNM 224077, both from Cabo San Lucas, Baja California, Mexico). Gill, 1883: 555 (priority over *A. leopardinus* Günther, 1864). Jordan and Evermann, 1898: 2721 (after Gill, 1863; synonymy includes *A. leopardinus* Günther). Gilbert and Starks, 1904: 204 (Panama Bay). Schultz, 1957: 95, pl. 10, fig. D (synonymy, description; Cabo San Lucas, Acapulco, Costa Rica, Galápagos, Colombia; in key). Schultz, 1964: 175, table 1 (additional material, Islands Tres Marías). Le Danois, 1964: 100, fig. 48d (description, Gulf of California). McCosker and Rosenblatt, 1975: 93 (Malpelo Island). Thomson et al., 1979: 53, pl. 3a (comparison with *F. avalonis*; central Gulf of California to Peru and Galápagos

Islands). Pietsch, 1981b: 419, figs. 15, 20A, 21, 26A, 28, 33, 36A, 40A (osteology). Pietsch, 1984b: 36 (genera of frogfishes). Pietsch and Grobecker, 1987: 157, figs. 60, 122, pl. 28 (description, distribution, relationships). Pietsch and Grobecker, 1990a: 99 (aggressive mimicry). Pietsch and Grobecker, 1990b: 80 (after Pietsch and Grobecker, 1990a). Allen and Robertson, 1994: 82 (tropical Eastern Pacific). Schneider and Lavenberg, 1995: 857 (tropical Eastern Pacific). Bearez, 1996: 734 (Ecuador). Galván-Magaña et al., 1996: 302 (Cerralvo Island, Baja California, Mexico). Robertson and Allen, 1996: 122, table 1 (Clipperton Atoll). Allen and Robertson, 1997: 820 (Clipperton Atoll). De La Cruz Agüero et al., 1997: 53 (Gulf of California). Helfman et al., 1997: 242, fig. 14.15 (aggressive mimicry). Chirichigno and Vélez, 1998: 221 (Peru). Gotshall, 1998: 22, fig. (central Gulf of California to Peru and Galápagos Islands). Michael, 1998: 331, 342, 343, color figs. (identification, behavior, captive care). Watson, 1998: 219, 220, 234, 235, figs. 1–8, tables 1, 3 (early life history, osteology). Beltrán-León and Herrera, 2000: 272, fig. 89 (larvae). Mora and Zapata, 2000: 477, table 1 (Gorgona Island, Colombia). Thomson et al., 2000: 24, 55, 296, fig. 18, pl. 3a (Gulf of California). Villarreal-Cavazos et al., 2000: 416 (Cabo Pulmo, Gulf of California). Lozano and Zapata, 2003: 402, table 1 (Gorgona Island, Colombia). Bussing and López, 2005: 46, 47, fig. (Cocos Island). Carnevale and Pietsch, 2006: 450, 452, table 1 (comparison, meristics, first fossil frogfish). Castellanos-Galindo et al., 2006: 200, 207 (Colombia). Mejía-Ladino et al., 2007: 292, figs. 5, 6, tables 1, 3, 4, 6

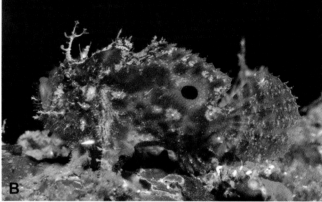

Fig. 164. *Abantennarius rosaceus* (Smith and Radcliffe). (**A**) Penemu Island, Raja Ampat, West Papua, Indonesia. Photo by Gerry Allen. (**B**) Lawadi Beach, Milne Bay, Papua New Guinea. Photo by Gerry Allen. (**C**) juvenile specimen, Sorong, West Papua, Indonesia. Photo by Teresa Zuberbühler.

(species account; in key; Colombia). Tobón-López et al., 2008: 98, table 1 (Gulf of Tribugá, northern Colombia). Helfman et al., 2009: 288, fig. 14.26 (aggressive mimicry; after Helfman et al., 1997). Dyer and Westneat, 2010: 599, 605 (Desventuradas Islands, Chile). McCosker and Rosenblatt, 2010: 190 (Galápagos Archipelago). Erisman et al., 2011: 29, table 1 (Islas Marías Archipelago, Mexico). González-Díaz et al., 2013: 206

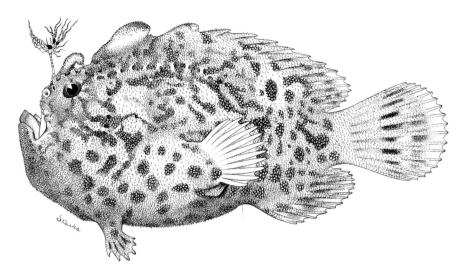

Fig. 165. *Abantennarius sanguineus* (Gill). Holotype of *Abantennarius tagus* Heller and Snodgrass, SU 6351, 54 mm SL, Albermarle Island, Galápagos. Drawing by Chloe Lesley Starks; after Heller and Snodgrass (1903, pl. 20).

(Nayarit, Mexico). Salas et al., 2014: 107, table 2 (Caño Island, Costa Rica). Del Moral-Flores et al., 2016: 607 (Revillagigedo Archipelago). Fourriére et al., 2016: 442 (Revillagigedo Archipelago).

Antennarius leopardinus Günther, 1864: 151 (original description, single specimen, holotype BMNH 1864.1.26.417, Pacific coast of Panama). Günther, 1869b: 388, 439, pl. 69, fig. 3 (description after Günther, 1864). Gill, 1883: 555 (synonym of *A. sanguineus* Gill). Dyer and Westneat, 2010: 605 (synonym of *A. sanguineus*).

Antennarius tagus Heller and Snodgrass, 1903: 226, pl. 20 (original description, single specimen, holotype SU 6351, Albermarle Island, Galápagos; Fig. 165). Nichols, 1939: 48 (Ecuador). Nichols and Murphy, 1944: 259 (Ecuador). Böhlke, 1953: 148 (type in CAS). Dyer and Westneat, 2010: 605 (synonym of *A. sanguineus*).

Antennarius sanguineus (not of Gill): Meek and Hildebrand, 1928: 1013 (description is of *F. avalonis*). Seale, 1940: 46 (description is of *F. avalonis*). Hildebrand, 1946: 501 (description is of *F. avalonis*).

Antennatus sanguineus: Grove and Lavenberg, 1997: 233, figs. 116a, 117, color pl. 17 (new combination; Galápagos Islands). Arnold and Pietsch, 2012: 127, 128, figs. 4, 6, table 1 (molecular phylogeny). Page et al., 2013 (common names). Avendaño-Ibarra et al., 2014: 109, 113, table 3 (larvae; Gulf of California to Colima, Mexico). Palacios-Salgado et al., 2014: 480, table 2 (Acapulco, Mexico). Del Moral-Flores et al., 2016: 607 (Revillagigedo). Galván-Villa et al., 2016: 144 (Bahía Chamela, Mexico).

Antennarius sanguinea: Kuiter and Debelius, 2007: 124, color fig. (emendation of specific name, Panama).

MATERIAL EXAMINED

Five hundred and fifty-two specimens, 7.0 to 82 mm SL.

Lectotype of *Antennarius sanguineus*: USNM 6393, 58 mm, Cabo San Lucas, Baja California, Mexico, Xantus.

Paralectotype of *Antennarius sanguineus*: USNM 224077, 66 mm, data as for lectotype.

Holotype of *Antennarius leopardinus*: BMNH 1864.1.26.417, 46 mm, Pacific coast of Panama, Dow.

Holotype of *Antennarius tagus*: SU 6351, 54 mm, Tagus Cove, Albemarle Island, Galápagos, Heller and Snodgrass, 1898–1899 (Fig. 165).

Additional material: *Gulf of California*: CAS, 3 (29–53 mm); CSULB, 2 (46–51.5 mm); LACM, 6 (17–55.5 mm); MNHN, 1 (79 mm); SIO, 11 (35–57 mm); CAS, 7 (40–74.5 mm); SIO, 48 (8–71.5 mm); UA, 21 (10–68 mm); UCLA, 9 (15.5–69 mm); SIO, 41 (7.5–75 mm); SU, 1 (31.5 mm). *Islas de Revillagigedo*: LACM, 1 (21.5 mm); SIO, 1 (49 mm). *Mexico outside the Gulf*: CAS, 4 (23–42 mm); FMNH 60278, 21 mm, Pie de la Cuesta, Acapulco. LACM 32097-15, 26 mm, Island Clarion; LACM, 2 (35–38 mm); SIO, 26 (10.5–77 mm); UA, 17 (7–77 mm); UCLA, 2 (24–49 mm). *Clipperton Island*: LACM, 1 (19 mm); UCLA, 1 (43 mm). *Nicaragua*: UW, 2 (38–39.5 mm). *Costa Rica*: FNMN, 2 (20–25 mm); LACM, 219 (7.5–61 mm); SU, 3 (42–51 mm); UA, 1 (45 mm); UCR, 3 (27.5–40 mm); UMMZ, 4 (17–28.5 mm). *Panama*: AMNH, 2 (17.5–42.5 mm); BMNH, 4 (22–33 mm); CAS, 8 (15.5–46 mm); GCRL, 5 (28–54 mm); MCZ, 1 (10.5 mm); SIO, 52 (9–57 mm); UW, 1 (53 mm); WAM, 4 (45–52 mm); ZMUC, 1 (29 mm). *Colombia*: SIO, 2 (12–44.5 mm); UA, 2 (37–42.5 mm); ZMUC, 1 (29 mm). *Ecuador and Galápagos Islands*: AMNH, 2 (25.5–61 mm); CAS, 2 (52.5–55.5 mm); FMNH, 3 (32–41.5 mm); LACM, 2 (17–76 mm); SIO, 1 (19 mm); UCLA, 4 (56–82 mm); UW, 8 (12–62.5 mm). *Chile*: SIO, 1 (14 mm).

DIAGNOSIS

A member of the genus *Abantennarius* unique in having a greater number of dorsal-fin rays than other species of the genus (rarely less than 13; Table 10) and the belly covered with large, darkly pigmented spots (in specimens greater than about 25 mm SL; Fig. 165).

Abantennarius sanguineus is further distinguished in having the following combination of character states: basidorsal spot present or absent, but if present usually weakly developed; illicium 8.6 to 13.3% SL, about as long as second dorsal-fin spine; esca an elongate, tapering appendage with several slender filaments and a cluster of darkly pigmented, spherical swellings at base (Fig. 165); second dorsal-fin spine strongly curved posteriorly, length 8.5 to 14.1% SL; opercular opening adjacent to or on pectoral-fin lobe; caudal peduncle absent, the membranous posteriormost margin of soft-dorsal and anal fins connected to body at base of outermost rays of caudal fin; dorsal-fin rays 13 or 14 (rarely 12); anal-fin rays seven (rarely six or eight); pectoral-fin rays 11 (rarely 10 or 12); last ray of pelvic fin bifurcate (Table 10).

DESCRIPTION

Dermal spinules sometimes present along anterior surface of illicium (GCRL 11647; along only proximal half of illicium in GCRL 11447); dark, spherical escal swellings always present in specimens 25 mm SL and larger (Fig. 165); anterior and lateral surfaces of second dorsal-fin spine usually completely covered with close-set dermal spinules (but spinules occasionally forming discrete clusters leaving naked spaces between); shallow, naked depression between second and third dorsal-fin spines always present; third dorsal-fin spine curved posteriorly, length 16.3 to 28.4% SL; eye diameter 5.6 to 8.0% SL.

Coloration in life: highly variable; overall cream, yellow, yellow-brown, orange, red, reddish-brown, brown, lavender, or purple, peppered with dark brown spots, particularly evident on belly; reddish-brown spots on dorsal fin and caudal peduncle; irregularly shaped brownish-black to black patches on flank; usually black peppering on caudal fin (WAM P.25511-001) (Figs. 166–168).

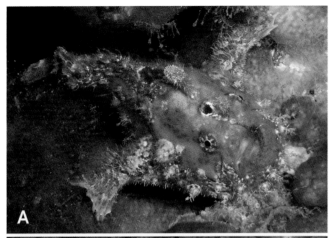

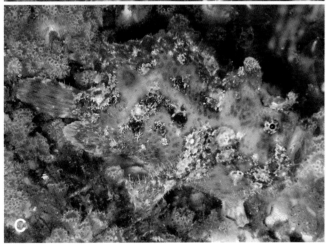

Fig. 166. *Abantennarius sanguineus* (Gill). (**A–C**) Isabela Island, Galápagos. Photos by Martin Buschenreithner.

Coloration in preservation: yellow, yellow-brown, reddish-brown to dark brown (occasionally pinkish) with darker brown spots or irregularly shaped blotches over entire head and body; fins usually more or less covered with small, darkly pigmented spots (spots often forming one to several parallel bands across caudal fin); belly (of specimens approximately 25 mm SL and larger) covered with more or less evenly spaced, reddish-brown to dark brown spots; illicium usually banded; as many as nine short, darkly pigmented bars radiating out from eye.

Largest specimen examined: 82 mm SL (UCLA W64-10).

DISTRIBUTION

Abantennarius sanguineus, restricted to coastal waters of the tropical and subtropical Eastern Pacific (Fig. 277, Table 22), is common throughout the lower half of the Gulf of California (below about 27° N) and south to Isla San Félix, Chile, including Islas Tres Marías and Revillagigedo, Clipperton Island, off Nicaragua, Cocos and Caño Islands off Cost Rica, Panama, Gorgona Island, Colombia, and Ecuador and the Galápagos Islands. The type locality is Cabo San Lucas.

Depths of capture, known for 45 collections, range from just beneath the surface to about 40 m; 80% of these, however, were made in less than 20 m, and the average depth for all known captures is 9 m.

COMMENTS

Abantennarius sanguineus and the morphologically similar *A. coccineus* are sympatric in the tropical Eastern Pacific south of about 10° N (compare Figs. 282 and 284). When specimens are greater than about 25 mm, *A. sanguineus* is easily distinguished by a combination of character states that includes the presence of darkly pigmented escal swellings (absent in the Eastern Pacific population of *A. coccineus*), dark rounded spots on the belly (absent in *A. coccineus*), and 13 or more dorsal-fin rays (less than 4% of the material of *A. sanguineus* counted had 12 dorsal-fin rays, whereas less

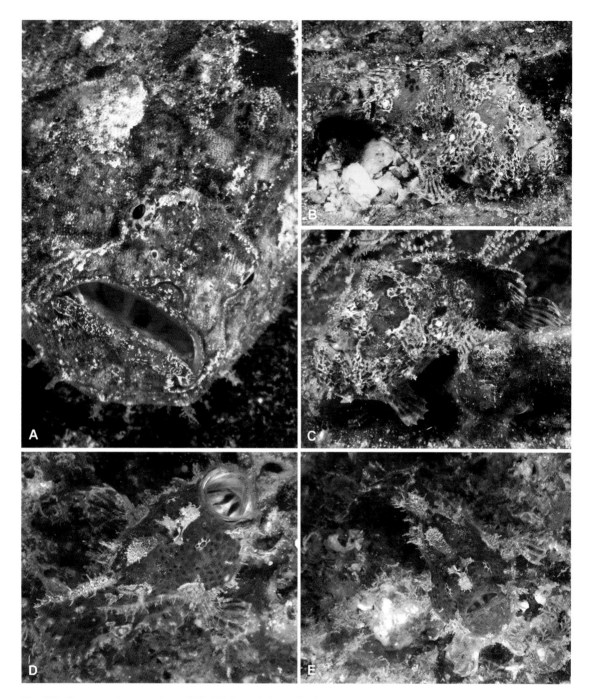

Fig. 167. *Abantennarius sanguineus* (Gill). (**A**) Cousin's Rock, Galápagos. Photo by Brandon Cole.
(**B**, **C**) Punta Vicente Roca, Isabela Island, Galápagos. Photos by Brandon Cole. (**D**, **E**) Bahía de
Banderas, Puerto Vallarta, Mexico. Photos by Alicia Hermosillo McKowen.

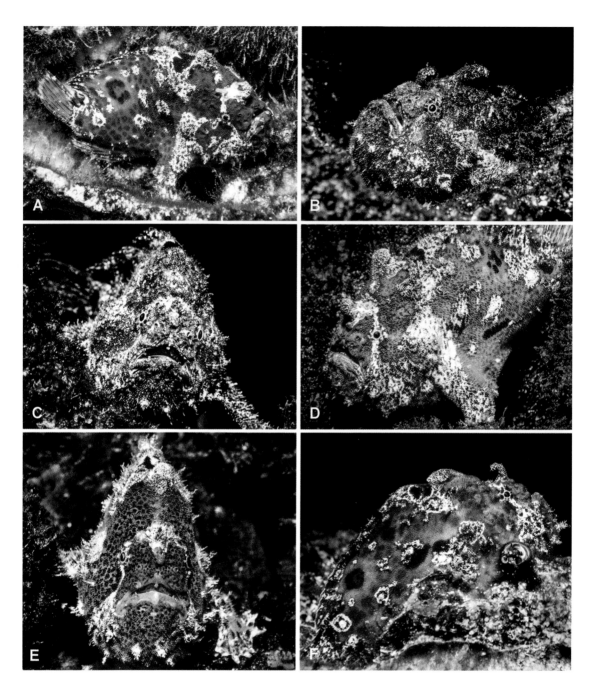

Fig. 168. *Abantennarius sanguineus* (Gill). (**A**) Cape Marshall, Isabela Island, Galápagos Islands. (**B–E**) Punta Vicente Roca, Galápagos Islands. (**F**) Cousin's Rock, near Santiago Island, Galápagos Islands. Photos by David J. Hall.

than 2% of that of *A. coccineus* had 13; Table 11). However, when specimens are less than about 25 mm SL, the pigmentation of the esca and belly are not yet developed, and identification must be based solely on the differences in dorsal-fin ray counts.

Genus *Antennatus* Schultz

Antennatus Schultz, 1957: 80 (type species *Antennarius strigatus* Gill, 1863, by original designation).

DIAGNOSIS

Antennatus is unique among the spinulose genera of the Antennariidae in lacking small naked areas between the pores of the acoustico-lateralis system (e.g., compare with *Fowlerichthys senegalensis*, Fig. 49; Table 1). *Antennatus* is further distinguished by having the following combination of character states: skin covered with extremely close-set, bifurcate dermal spinules, the length of the spines of each spinule three-and-a-half to six times distance between tips of spines (Fig. 26D); illicium naked, without dermal spinules; esca tiny (hardly differentiated from illicium) or absent (illicium more or less tapering to a fine point); pectoral-fin lobe broadly attached to side of body; caudal peduncle present or absent; endopterygoid present; pharyngobranchial I present; epural present; pseudobranch absent; swimbladder present; dorsal-fin rays 11 to 13 (usually 12); anal-fin rays six to eight (usually seven); all rays of caudal fin usually bifurcate (only seven innermost caudal rays bifurcate in *A. linearis*; Randall and Holcom, 2001: 140); pectoral-fin rays nine to 12 (usually 10 or 11); vertebrae 18 or 19 (Tables 12, 13).

DESCRIPTION

Esca, if present, represented by a slender, tapering filament or a small, usually lobed appendage; illicium, when laid back onto head, fitting into a narrow, naked groove on either left or right side of second dorsal-fin spine, the tip of illicium (esca) usually coming to lie within a shallow, longitudinal depression (usually devoid of dermal spinules) between second and third dorsal-fin spines (except in *A. flagellatus*), the esca capable of being covered and protected by second dorsal-fin spine when spine is fully depressed; illicium about equal to or nearly four times length of second dorsal-fin spine, 12.5 to 50.0% SL (Fig. 169); anterior end of pterygiophore of illicium terminating distinctly posterior to symphysis of upper jaw; illicium and second dorsal-fin spine relatively closely spaced on pterygiophore, the distance between bases of spines less than 5% SL; second dorsal-fin spine conical or cylindrical, straight to slightly curved posteriorly, not connected to head by membrane (a tiny transparent membrane is present in some specimens,

Table 12. Comparison of Distinguishing Character States of Species of the Genus *Antennatus* (lengths as percentage of SL)

Character	A. flagellatus	A. linearis	A. strigatus	A. tuberosus
Illicium length	40.2–50.0	15.8–20.7	12.5–21.4	15.4–28.3
Second dorsal-fin length	11.0–13.0	10.4–12.5	10.8–18.8	7.3–11.8
Second dorsal-fin spine	conical/cylindrical	conical, pointed	conical, pointed	cylindrical, not tapered
Esca	minute/absent	absent	minute	absent
Caudal peduncle	present	present/absent	absent	absent
Dorsal-fin rays	11	12	12 (11–13)	12 (11–13)
Anal-fin rays	6, all bifurcate	7, 4–7 bifurcate	7 (6–8), all bifurcate	7 (8), usually all simple
Bifurcate caudal-fin rays	9	7	9	9
Pectoral-fin rays	10	10 (9)	10 (9–11)	11 (9–12)
Coloration	plain, blotched, mottled	covered with closely spaced parallel lines	plain, blotched, mottled	plain, blotched, mottled
Distribution	Japan	Indo-Pacific	Eastern Pacific	Indo-Pacific

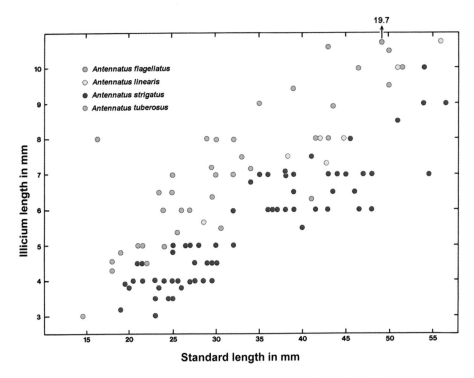

Fig. 169. Relationship between illicium length and standard length for species of *Antennatus*. *A. flagellatus* (green), N = 2; *A. linearis* (yellow), N = 6; *A. strigatus* (red), N = 64; *A. tuberosus* (blue), N = 37. Modified after Pietsch and Grobecker (1987, fig. 78).

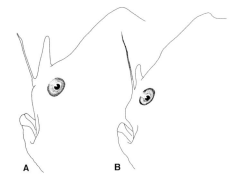

Fig. 170. Lateral view of face of two species of *Antennatus*, showing differences in length of the illicium and length and shape of the second dorsal-fin spine. (**A**) *Antennatus strigatus* (Gill), UCR 140–18, 38 mm SL, Playa del Coco, Costa Rica. (**B**) *Antennatus tuberosus* (Cuvier), USNM 223411, 43 mm SL, Pohnpei, Caroline Islands. Drawings by Cathy L. Short; after Pietsch and Grobecker (1987, fig. 79).

especially in small individuals, but apparent only after the spine has been rotated forward), completely covered with close-set dermal spinules, length 7.3 to 18.8% SL (Fig. 170); third dorsal-fin spine bound down to head and nape of neck by thick, spinulose skin forming a rounded protuberance, length 21.0 to 30.2% SL; eye not surrounded by ring of distinct clusters of dermal spinules, diameter 5.2 to 8.6% SL; distal two-thirds of maxilla naked and tucked beneath folds of skin, only the extreme proximal end covered directly with spinulose skin; head, body, and fins without cutaneous appendage; head and body often with slightly raised, wartlike patches of tightly clustered dermal spinules (Fig. 26D); epibranchial I toothless (remnants of tooth-plate present in some specimens); ceratobranchial I toothless; vertebrae 18 or 19, caudal centra 13 or 14; dorsal-fin rays 11 or 12 (rarely 13), as many as seven (usually only two or three) posteriormost bifurcate; anal-fin rays six or seven (rarely eight), simple or bifurcate; pectoral-fin rays 10 or 11 (rarely nine or 12), all simple; all rays of pelvic fin simple (Tables 12, 13).

Coloration in preservation: white, beige, yellow, to brown, everywhere (but especially on belly and trunk) covered with a network of dark brown, more or less interconnecting blotches (or fine, more or less parallel, closely spaced lines; see description of

Table 13. Frequencies of Fin-Ray Counts of Species of the Genus *Antennatus*

Species	Dorsal-fin rays			Anal-fin rays			Pectoral-fin rays*			
	11	12	13	6	7	8	9	10	11	12
A. flagellatus	2			2				4		
A. linearis		9			9		2	16		
A. strigatus	3	71	3	4	70	3	2	149	3	
A. tuberosus	1	57	1		57	2	1	14	99	4
TOTALS	6	137	4	6	136	5	5	183	102	4

*Left and right sides combined.

A. linearis, p. 245); conspicuous dark brown bar across all fins; basidorsal spot absent; illicium without banding; bars of pigment radiating from eye absent.

Four species: Southern Japan and Indo-Pacific and Eastern Pacific Oceans.

Key to the Known Species of the Genus *Antennatus*

This key, like the other keys in this volume, works by progressively eliminating the most morphologically distinct taxon and for that reason should always be entered from the beginning. All characters listed for each species must correspond to the specimen being keyed; if not, proceed to the next set of character states. For an additional or alternative means of identification, all species of *Antennatus* may be compared simultaneously by referring to Tables 12 and 13.

1. Illicium more than three times length of second dorsal-fin spine, 40.2 to 50.0% SL (Fig. 169); second dorsal-fin spine conical or cylindrical, length 11.0 to 13.0% SL (Fig. 171); caudal peduncle present, membranous posteriormost margin of soft-dorsal and anal fins connected to body distinctly anterior to base of outermost rays of caudal fin; all rays of anal fin bifurcate; pectoral-fin rays 10 (Table 13); all nine caudal-fin rays bifurcate; head and body yellow to creamy white, with brown and pinkish blotches (Fig. 171); more or less parallel, closely spaced lines absent . . . .
. . . . *Antennatus flagellatus* **Ohnishi, Iwata, and Hiramatsu, 1997, p. 242**
Southern Japan

2. Illicium one and a half to nearly two times length of second dorsal-fin spine, 15.8 to 20.7% SL (Fig. 169); second dorsal-fin spine conical, more or less tapering to a point, length 10.4 to 12.5% SL (Fig. 172); caudal peduncle present, membranous posteriormost margin of soft-dorsal and anal fins connected to body distinctly anterior to base of outermost rays of caudal fin; four to seven rays of anal fin bifurcate; pectoral-fin rays 10 (rarely nine) (Table 13); outermost rays of caudal fin simple, seven innermost rays bifurcate; head and body covered with fine, more or less parallel, closely spaced lines (Figs. 172–174) . . . .
. . . . *Antennatus linearis* **Randall and Holcom, 2001, p. 243**
South Africa to Hawaiian Islands

3. Illicium equal to or slightly longer than second dorsal-fin spine, 12.5 to 21.4% SL, emerging far forward on snout, just behind upper lip (Figs. 169, 170A); second dorsal-fin

spine conical, more or less tapering to a point, relatively long, length 10.8 to 18.8% SL (Fig. 170A); caudal peduncle absent, the membranous posteriormost margin of soft-dorsal and anal fins connected to body at base of outermost rays of caudal fin; all rays of anal fin bifurcate; pectoral-fin rays 10 (rarely nine or 11) (Table 13); all nine caudal-fin rays bifurcate; head and body overall pale but more often yellow-brown, with dark brown to black reticulations, marbling, and more or less interconnecting blotches (Figs. 178, 179) . . . . *Antennatus strigatus* (**Gill, 1863**), **p. 247**
Eastern Pacific Ocean

4. Illicium one-and-a-half to two times length of second dorsal-fin spine, 15.4 to 28.3% SL, emerging on snout considerably behind upper lip (Figs. 169, 170B); second dorsal-fin spine cylindrical, not tapering, relatively short, 7.3 to 11.8% SL (Fig. 170B); caudal peduncle absent, the membranous posteriormost margin of soft-dorsal and anal fins connected to body at base of outermost rays of caudal fin; rays of anal fin usually all simple (but occasionally two or three posteriormost, or at most five innermost, bifurcate); pectoral-fin rays usually 11, rarely nine, 10, or 12 (Table 13); all nine caudal-fin rays bifurcate; head and body yellow, brown, or red, with dark brown to black reticulations, marbling, and more or less interconnecting blotches (Figs. 181, 182) . . . .
. . . . *Antennatus tuberosus* (**Cuvier, 1817b**), **p. 250**
Indo-Pacific Ocean

Antennatus flagellatus Ohnishi, Iwata, and Hiramatsu
Japanese Frogfish

Figs. 169, 171, 285; Tables 12, 13, 22

Antennatus flagellatus Ohnishi, Iwata, and Hiramatsu, 1997: 213, figs. 1–3, table 1 (original description, two specimens, holotype NSMT-P 49487, paratype OMNH-P 9999, Kashiwajima Island, Kochi, Prefecture, southern Japan; Fig. 171). Michael, 1998: 354 (identification, behavior, captive care). Nakabo, 2000: 454 (Japan; in key). Randall and Holcom, 2001: 137, 143 (comparison with congeners). Nakabo, 2002: 454 (Japan; in key). Randall, 2005a: 73 (Japan). Arnold and Pietsch, 2012: 127, 128 (molecular phylogeny, revised classification). Nakabo, 2013: 537, figs. (southern Japan; after Ohnishi et al., 1997).

MATERIAL EXAMINED

Two specimens, 16 to 49 mm SL.

Holotype of *Antennatus flagellatus*: NSMT-P 49487, 49 mm, Kashiwajima Island, Kochi, Prefecture, southern Japan, 45 m, 13 December 1992 (Fig. 171A).

Paratype of *Antennatus flagellatus*: OMNH-P 9999, 16 mm, Kashiwajima Island, Kochi, Prefecture, southern Japan, 45 m, 2 May 1993 (Fig. 171B).

DIAGNOSIS

A member of the genus *Antennatus* unique in having an exceptionally long illicium, 40.2 to 50.0% SL, about 3.5 times the length of the second dorsal-fin spine (Figs. 169, 171). It is further distinguished in having the following combination of character states: caudal peduncle present, membranous posteriormost margin of soft-dorsal and anal fins connected to body distinctly anterior to base of outermost rays of caudal fin; esca, when present, a tiny lobed appendage, barely differentiated from illicium; second dorsal-fin spine conical or cylindrical, more or less tapering to a blunt point, length 11.0 to 13.0% SL (Fig. 171); dorsal-fin rays 11; anal-fin rays 6, all bifurcate; all rays of caudal fin bifurcate; pectoral-fin rays 10 (Tables 12, 13); head and body yellow to creamy white, with brown and pinkish blotches, more or less parallel, closely spaced lines absent.

DESCRIPTION

Coloration in life: head, body, and fins creamy yellow, with brown and pinkish mottling and diffuse blotches; four dark spots on dorsal fin (holotype only); two dark bars across width of anal, pectoral, and pelvic fins; a dark, indistinct bar across caudal peduncle; three dark bars on caudal fin (after Ohnishi et al., 1997: 215).

Coloration in preservation: head, body, and fins creamy white, with some faint brown blotches; dark markings on fins faded but clearly apparent (after Ohnishi et al., 1997: 216).

Remaining description as given for the genus.

Largest known specimen: 49 mm SL (NSMT-P 49487).

DISTRIBUTION

Antennatus flagellatus is currently known only from the type material (Nobuhiro Ohnishi, personal communication, 1 September 2016): two specimens collected by scuba at Kashiwajima Island, Kochi Prefecture, southern Japan, from a small rock ledge on a fine sandy slope at a depth of 45 m (Fig. 285, Table 22).

Fig. 171. *Antennatus flagellatus* Ohnishi, Iwata, and Hiramatsu. (**A**) holotype, NSMT-P 49487, 49 mm SL, Kashiwajima Island, Kochi Prefecture, Japan. (**B**) paratype, OMNH-P 9999, 16 mm SL, Kashiwajima Island, Kochi Prefecture, Japan. Photos by Nobuhiro Ohnishi; after Ohnishi et al. (1997, fig. 1).

Antennatus linearis Randall and Holcom
Lined Frogfish

Figs. 169, 172–175, 285; Tables 12, 13, 22

Antennarius tuberosus: Smith, 1958: 60, pl. 1, fig. H (in part, misidentification, "curious spiderweb-like lines on the body"; Aldabra). Smith and Smith, 1963: 61, pl. 56, fig. F (after Smith, 1958).

Antennatus tuberosus: Pietsch and Grobecker, 1987: 192, 195 (in part, misidentifications, 11 unique specimens set aside, "everywhere covered with narrow, more or less parallel, closely set lines").

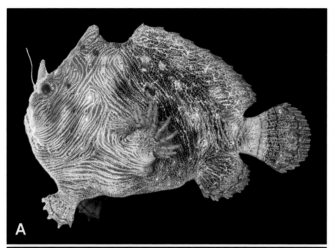

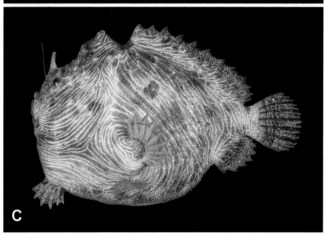

Fig. 172. *Antennatus linearis* Randall and Holcom. **(A)** holotype, BPBM 38704, 41.5 MM SL, off Kahe Point, Oahu, Hawaiian Islands. **(B)** paratype, USNM 361066, 52 mm SL, off Waianae, Oahu, Hawaiian Islands. **(C)** paratype, BPBM 37232, 42 mm SL, Makua, Oahu, Hawaiian Islands. Photos by John E. Randall; courtesy of Lisa Palmer and the Smithsonian Institution, NMNH, Division of Fishes, Washington, DC.

Antennatus linearis Randall and Holcom, 2001: 137, figs. 1–4, tables 1, 2 (original description, eight specimens, holotype BPBM 38704, Oahu, Hawaiian Islands; Fig. 172). Allen and Adrim, 2003: 25 (Moluccas, Indonesia). Mundy, 2005: 265 (Hawaiian Islands). Randall, 2005a: 73 (South Pacific). Schneidewind, 2005c: 88 (tiny postlarvae). Randall, 2007: 130, color fig. (Hawaiian Islands; in key). Randall, 2010: 39 (Hawaii, Indonesia, Mozambique). Allen and Erdmann, 2012: 152, color fig. (East Indies). Arnold and Pietsch, 2012: 127, 128 (molecular phylogeny, revised classification).

MATERIAL EXAMINED

Eighteen specimens, 17.5 to 69.5 mm SL.

Holotype of *Antennatus linearis*: BPBM 38704, 41.5 mm, off Kahe Point, Oahu, Hawaiian Islands, 14 m, early 1997 (Fig. 172A).

Paratypes of *Antennatus linearis*: BPBM 37232, 1 (42 mm), Makua, Oahu, Hawaiian Islands, on sparse rubble and sand, 21.5 m, 14 March 1991; SAIAB 4605, 1 (43 mm), Pinda Islands, Mozambique, 20 September 1956; SAIAB 4607, 1 (51.5 mm), Durban, Natal, South Africa, 7 May 1963; SAIAB 4609, 1 (56.5 mm), Maputo, Mozambique, 6 May 1967; SAIAB 56816, 1 (38 mm), Five Mile Reef, Sodwana Bay, KwaZulu-Natal, 13 August 1997; USNM 209596, 1 (20.5 mm), Haruku Island, point east of Tandjung Naira, Molucca Islands, Indonesia, surge channel, 4.5 m, rotenone, 15 January 1973; USNM 361066, 1 (52 mm), off Waianae, Oahu, Hawaiian Islands, died in aquarium, 20 December 1996 (Fig. 172B, C).

Additional material: BMNH, 1 (43 mm); CAS, 1 (33.5 mm); SAIAB, 7 (17.5–69.5 mm); USNM, 1 (18 mm).

DIAGNOSIS

A member of the genus *Antennatus* unique in having the following combination of character states: caudal peduncle present, membranous posteriormost margin of soft-dorsal and anal fins connected to body

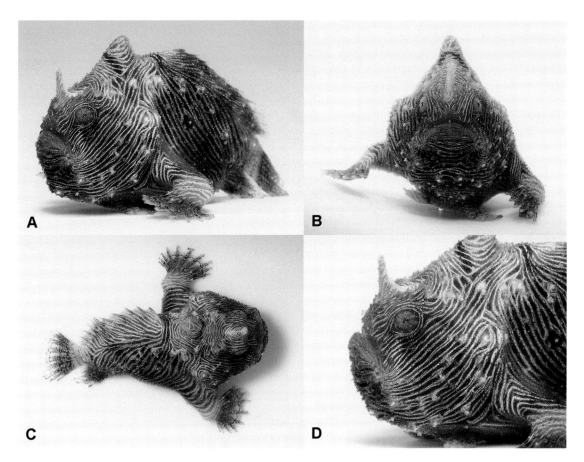

Fig. 173. *Antennatus linearis* Randall and Holcom. (**A–D**) Oahu, Hawaiian Islands. Photos by Keoki Stender.

distinctly anterior to base of outermost rays of caudal fin; esca absent; illicium length 15.8 to 20.7% SL (Fig. 169); second dorsal-fin spine conical, more or less tapering to a point, length 10.4 to 12.5% SL (Fig. 172); dorsal-fin rays 12; anal-fin rays 7, four to seven rays bifurcate; outermost caudal-fin rays simple, seven innermost rays bifurcate (determined by radiographs); pectoral-fin rays 10, very rarely nine (Tables 12, 13); head and body covered with fine, more or less parallel, closely spaced lines (Figs. 172–174).

DESCRIPTION

Coloration in life: head and body of holotype pale, with three sets of numerous, dark, more or less parallel, close-set curving lines, one on body extending downward and posteriorly, then curving forward on abdomen, one on head curving upward and posteriorly from chin, and a third in a small triangular area behind head, with lines converging onto pectoral-fin base; large, obscure, dark blotches on body (the result of dark lines within the blotches being broader and more darkly pigmented); dorsal fin blackish, with whitish blotches except for translucent, naked outer membranes; anal fin with an irregular, dark, double submarginal line; caudal fin pale with a transverse blackish bar across basal fifth of fin, a broad blackish bar in middle of fin, and a narrow, submarginal zone of dark lines forming a reticulum; pectoral fins pale, with a reticulum of fine dark

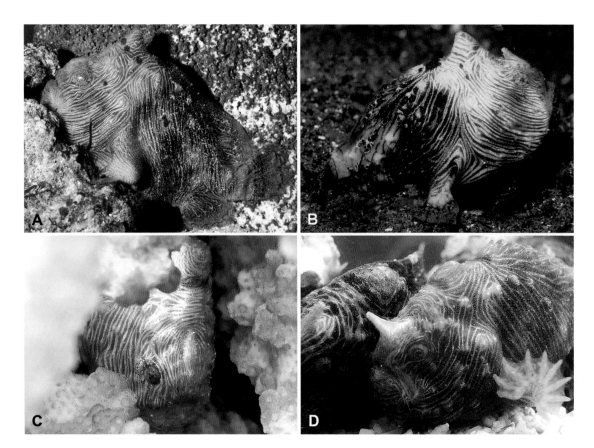

Fig. 174. *Antennatus linearis* Randall and Holcom. (**A**) Oahu, Hawaiian Islands. Photo by Ronald Holcom. (**B**) Milne Bay, Papua New Guinea. Photo by Rob van der Loos. (**C**) Koloa Landing, south shore Kauai, Hawaiian Islands. Photo by Anthony J. Kuntz. (**D**) aquarium specimen, locality unknown. Photo by Bronson Nagareda.

lines; pelvic fins with a distal zone of dark lines forming a reticulum (after Randall and Holcom, 2001: 141) (Figs. 172–174).

Coloration in preservation: lines on head and anterior part of body of holotype brownish-orange, but mostly blackish on posterior part of body; a complex lace-like pattern of fine white lines, superimposed on linear pattern of posterodorsal part of body and on dorsal and anal fins (Figs. 172–174); lines on body of paratypes lighter than those of holotype, but with variation in intensity of brown lines such that some large darker blotches evident; scattered orange-red spots of various sizes on head and body, most associated with wartlike clusters of dense dermal spicules (after Randall and Holcom, 2001: 141).

Remaining description as given for the genus.

Largest known specimen: 69.5 mm SL (SAIAB 4610).

DISTRIBUTION

As presently understood, the distribution of *Antennatus linearis* extends from Natal on the southeast coast of South Africa, to Mozambique, the Comoro Islands, Aldabra, the Moluccas, Papua New Guinea, and the Hawaiian Islands (Fig. 285, Table 22). The holotype was collected off Kahe Point, Oahu, Hawaiian Islands. Depths of capture range from 4.5 to 33 m.

COMMENTS

This species was originally recognized by Pietsch and Grobecker (1987: 195) as distinct from *A. tuberosus*, with which it had been confused since at least 1958 when J. L. B. Smith published his "Fishes of Aldabra, Part X" (see Synonymy, above). In their account of *A. tuberosus*, Pietsch and Grobecker (1987: 195) identified 11 specimens (nearly all of which were eventually designated types of *A. linearis* by Randall and Holcom, 2001) that differed strikingly—"everywhere covered with narrow, more or less parallel, closely set lines"—from the 90 specimens of *A. tuberosus* then examined. Reluctant to describe them as new at the time, they were set aside for future consideration.

The specimens recognized here as *A. linearis* all display the tight linear pigment pattern diagnostic for the species but it should be pointed out that intermediates appear to exist between this species and *A. tuberosus*. Photographs of several uncollected specimens show a mixed pigment pattern, with a head like that of *A. linearis* and the body of *A. tuberosus* (Fig. 175). While this might indicate that these two species are one and the same, significant differences remain: among a few minor distinguishing features (see the key above), a distinct caudal peduncle is present in *A. linearis* and only the innermost seven caudal-fin rays are bifurcate, whereas the caudal peduncle is absent in *A. tuberosus* and all nine caudal-fin rays are bifurcate. Still, without specimens to examine, these intermediates based only on photographs cannot be properly identified to species.

Antennatus strigatus (Gill)
Bandtail Frogfish

Figs. 169, 170A, 176–179, 285; Tables 12, 13, 22

Antennarius strigatus Gill, 1863: 92 (original description, two specimens, lectotype USNM 6267 designated by Pietsch and Grobecker, 1987, paralectotype USNM 224078, Cabo San Lucas, Baja California, Mexico). Jordan and Gilbert, 1883: 650 (Panama). Gill, 1883: 556 (= *A. tenuifilis* Günther, 1869b). Jordan and Evermann, 1898: 2720 (Panama). Gilbert and Starks, 1904: 204 (Panama Bay). Jordan, 1905: 550 (Pacific coast of Mexico). Meek and Hildebrand, 1928: 1014 (description, Panama City Market).

Antennarius tenuifilis Günther, 1869b: 388, 440 (original description, single specimen, holotype BMNH 1866.8.19.1, off Panama City). Gill, 1883: 556 (synonym of *Antennarius strigatus*).

Fig. 175. Uncollected specimens of *Antennatus* with a mixed pigment pattern—a head like that of *A. linearis* and a body of *A. tuberosus* (see Comments, p. 247). (**A**) Milne Bay, Papua New Guinea. Photo by Rob van der Loos. (**B**) Bali, Indonesia. Photo by Zane Kamat.

Antennarius reticularis Gilbert, 1892: 566 (original description, single specimen, holotype USNM 48260, Gulf of California). Jordan and Evermann, 1898: 2719 (after Gilbert, 1892).

Antennarius ziesenhennei Myers and Wade, 1946: 168, pl. 23, fig. 7 (original description, single specimen, holotype LACM 21555, Sulivan Bay, James Island, Galápagos). Grove and Lavenberg, 1997: 235 (synonym of *A. strigatus*).

Antennatus (Antennatus) strigatus: Schultz, 1957: 81, pl. 6, fig. C (new combination, description; Panama, Colombia, Galápagos; in key).

Antennatus bigibbus: Ricker, 1959: 7 (new combination, Revillagigedo Islands).

Antennatus reticularis: Briggs, 1962: 440 (new combination, resurrection from synonymy of *Antennatus strigatus*). Rosenblatt, 1963: 462 (synonym of *Antennatus strigatus*).

Antennatus strigatus: Briggs, 1962: 440 (comparison with *Antennatus reticularis* Gilbert). Rosenblatt, 1963: 462 (*Antennatus reticularis*, a synonym). Schultz, 1964: 176, table 1 (additional material, Clipperton Island and Galápagos). Rosenblatt et al., 1972: 7 (endemic to Eastern Pacific). Thomson et al., 1979: 54, pl. 3b (comparison with *Antennarius avalonis* and *Antennarius sanguineus*; central Gulf of California to Colombia and Galápagos). Pietsch, 1984b: 37 (genera of frogfishes). Pietsch and Grobecker, 1987: 189, figs. 78, 79A, 80, 81, 127, pl. 34 (description, distribution, relationships). Allen and Robertson, 1994: 83 (tropical Eastern Pacific). Schneider and Lavenberg, 1995: 857 (tropical Eastern Pacific). Bearez, 1996: 734 (Ecuador). Robertson and Allen, 1996: 122, table 1 (Clipperton Island). Allen and Robertson, 1997: 821 (Clipperton). De La Cruz Agüero et al., 1997: 52 (Gulf of California). Grove and Lavenberg, 1997: 235, figs. 116b (species account, Galápagos Islands). Ohnishi et al., 1997: 216, 217, table 1 (comparison with congeners; in key). Michael, 1998: 354, color fig. (identification, behavior, captive care). Watson, 1998: 219, 231, 232, 235, figs. 9, 10, tables 2, 3 (early life history). Mora and Zapata, 2000: 477, table 1 (Gorgona Island, Colombia). Thomson et al., 2000: 54, 56, 296, fig. 31A, pl. 3b (Gulf of California). Villarreal-Cavazos et al., 2000: 416 (Cabo Pulmo, Gulf of California). Watson et al., 2000: 123 (early development). Randall and Holcom, 2001: 137, 142, 143 (comparison with congeners). Lozano and Zapata, 2003: 402, table 1 (Gorgona Island, Colombia). Bussing and López, 2005: 46, 47, fig. (Cocos Island). Castellanos-Galindo et al., 2006: 200, 207 (Colombia). Kuiter and Debelius, 2007: 117, color fig. (Gulf of California). Mejía-Ladino et al., 2007: 280, figs. 5, 6, tables 2, 4, 6 (species account; in key; Colombia). McCosker and Rosenblatt, 2010: 190 (Galápagos Islands). Erisman et al., 2011: 29, table 1 (Islas Marías Archipelago, Mexico). Arnold and Pietsch, 2012: 127, 128 (molecular phylogeny, revised classification). González-Díaz et al., 2013: 206 (Nayarit, Mexico). Page et al., 2013 (common names). Salas et al., 2014: 107, table 2 (Caño Island, Costa Rica). Del Moral-Flores et al., 2016: 607 (Revillagigedo Archipelago). Fourriére et al., 2016: 442 (Revillagigedo Archipelago).

Histiophryne bigibba: Le Danois, 1964: 106 (new combination; in part, i.e., MNHN 01-304, Gulf of California).

Histiophryne bigibbus: Dor, 1984: 56 (after Le Danois, 1964).

MATERIAL EXAMINED

One hundred and forty-nine specimens, 12 to 78 mm SL.

Lectotype of *Antennarius strigatus*: USNM 6267, 62 mm, Cabo San Lucas, Baja California, Mexico, Xantus.

Paralectotype of *Antennarius strigatus*: USNM 224078, 56.5 mm, data as for lectotype.

Holotype of *Antennarius tenuifilis*: BMNH 1866.8.19.1, 51 mm, on reefs outside Panama City, Seemann.

Holotype of *Antennarius reticularis*: USNM 48260, 24 mm, *Albatross* station 2825, Gulf of California, 1888.

Holotype of *Antennarius ziesenhennei*: LACM 21555, 46 mm, Sulivan Bay, James Island, Galápagos, tidepool, Ziesenhenne, 21 January 1938.

Additional material: *Gulf of California*: CAS, 3 (21–66 mm); MNHN 1901-304, 2 (30–47 mm); SIO, 27 (12–54.5 mm); SU, 2 (24–25 mm); UA, 2 (42–58 mm); UCLA, 1 (24 mm). *Clipperton Island*: CAS, 3 (19.5–34 mm); UCLA, 32 (21–66 mm). *Costa Rica*: LACM, 32 (13.5–34 mm); SU, 1 (25 mm); UCR, 11 (7.5–38 mm); UMMZ, 1 (20 mm). *Panama*: BMNH, 3 (22.5–30 mm); FMNH, 1 (19 mm); SIO, 14 (12–40 mm); SU, 2 (35–36 mm). *Colombia*: BMNH, 1 (25 mm); FMNH, 1 (27 mm); LACM, 1 (41.5 mm). *Galápagos Islands*: CAS, 1 (67.5 mm); LACM, 1 (78 mm); UCLA, 1 (42 mm).

DIAGNOSIS

A member of the genus *Antennatus* unique in having the following combination of character states: caudal peduncle absent, the membranous posteriormost margin of soft-dorsal and anal fins connected to body at base of outermost rays of caudal fin; esca consisting of a tiny, elongate, more or less lobed appendage, barely differentiated from illicium (Figs. 170A, 176, 177); illicium equal to or slightly longer than second dorsal-fin spine, length 12.5 to 21.4% SL (Figs. 169, 170A); second dorsal-fin spine conical, more or less tapering to a blunt point, relatively long, 10.8 to 18.8% SL (Fig. 170A); dorsal-fin rays 12 (rarely 11 or 13); anal-fin 7 (rarely six or eight), all bifurcate; all rays of caudal fin bifurcate; pectoral-fin rays 10 (rarely nine or 11) (Tables 12, 13).

DESCRIPTION

Coloration in life: highly variable; nearly white, yellow, yellow-brown, brown, red, or purple, with dark brown to black reticulations, marbling, and more or less interconnecting blotches; some individuals with lavender patches reminiscent of encrusting coralline algae; base of dorsal and anal fins often dark-brown to black; a wide dark band usually

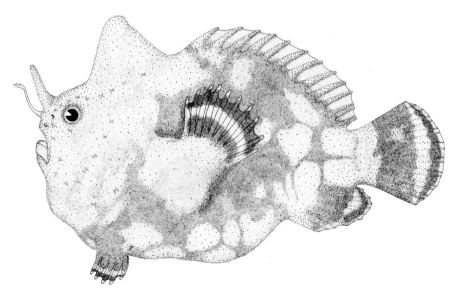

Fig. 176. *Antennatus strigatus* (Gill). UCR 35–1, 26.5 mm SL, Playa del Coco, Costa Rica. Drawing by Cathy L. Short; after Pietsch and Grobecker (1987, fig. 80).

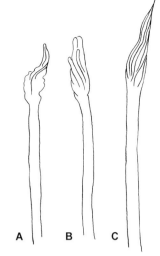

Fig. 177. Escae of *Antennatus strigatus* (Gill). (**A**) UCR 423–14, 18 mm SL, Isla del Caño, Costa Rica. (**B**) UCR 423–14, 27 mm SL, Isla del Caño, Costa Rica. (**C**) UCR 140–18, 38 mm SL, Playa del Coco, Costa Rica. Drawings by Cathy L. Short; after Pietsch and Grobecker (1987, fig. 81).

A   B   C

present across caudal fin; tips of pectoral and pelvic rays often dark-brown to black (Figs. 178, 179).

Remaining description as given for the genus.

Largest specimen examined: 78 mm SL (LACM 20677).

DISTRIBUTION

Antennatus strigatus is restricted to the tropical Eastern Pacific Ocean (Fig. 285, Table 22) where it extends from the southern half of the Gulf of California (below about 26° N latitude) and south to Colombia (including Gorgona Island), coastal Ecuador, and the Galápagos Islands, including the Revillagigedo and Islas Marias Archipelagos, Clipperton Atoll, Cocos, and the Caño Islands off the coast of southern Costa Rica. The holotype is from Cabo San Lucas, Baja California, Mexico.

Depths of capture, known for 36 specimens, range from just beneath the surface to 38 m; 92% of this material, however, was taken in 15 m or less, and 61% in less than 10 m. The average depth for all known captures is 10 m.

Antennatus tuberosus (Cuvier)
Tuberculated Frogfish

Figs. 18D, 26D, 169, 170B, 175, 180–182, 285, 306A; Tables 12, 13, 22, 23

Chironectes tuberosus Cuvier, October 1817b: 432 (original description, three specimens, lectotype MNHN 1104, established by Pietsch et al., 1986: 141, paralectotypes

Fig. 178. *Antennatus strigatus* (Gill). Sea of Cortez. Photo by John Neuschwander; courtesy of Helmut Debelius.

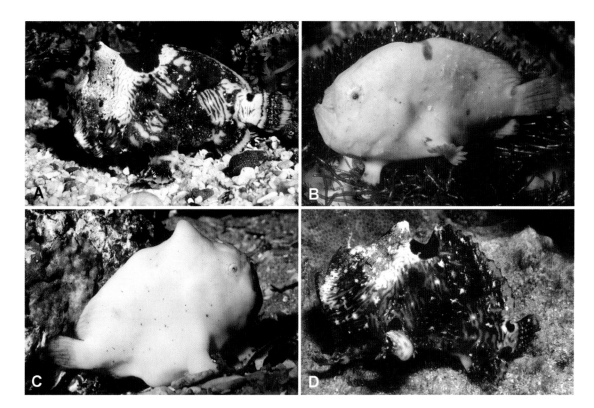

Fig. 179. *Antennatus strigatus* (Gill). (**A**) Sonora, Mexico. Photo by Alex Kerstitch. (**B**) Isla San Pedro Nolasco, Sonora, Mexico. Photo by Alex Kerstitch. (**C**) Sea of Cortez. Photo by Alex Kerstitch. (**D**) Cocos Island, Costa Rica. Photo by Ross Robertson.

MNHN 182 (2), Mauritius). Cuvier, 1829: 252 (after Cuvier, 1817b). Valenciennes, 1837: 428 (description, MNHN A.4615, New Ireland, Bismarck Archipelago). Valenciennes, 1842: 189 (after Valenciennes, 1837). Pietsch, 1985a: 91, fig. 30, table 1 (original manuscript sources for *Histoire Naturelle des Poissons* of Cuvier and Valenciennes). Pietsch et al., 1986: 141 (in type catalog).

Antennarius tuberosus: Schinz, 1822: 501 (new combination; after Cuvier, 1817b). Bleeker, 1858c: 463 (RMNH 6287, Cocos). Bleeker, 1859b: 129 (Cocos, Madagascar, Mauritius). Fowler, 1903: 174 (description, additional specimen, Zanzibar). McCulloch, 1929: 408 (Western Australia). Whitley, 1954: 28 (near Terrigal, New South Wales). Smith, 1958: 60, pl. 1, figs. G, H (in part, misidentification; at least one specimen, SAIAB 4609, is *A. linearis*; description, additional material; Aldabra, Mozambique, Seychelles). Smith and Smith, 1963: 61, pl. F (after Smith, 1958). Pietsch, 1986a: 366, 369, fig. 102.8, pl. 13 (in part, includes *A. linearis*; South Africa). Pereira, 2000: 5 (Mozambique). Luiz et al., 2013, dataset S1 (geographic range size).

Antennarius unicornis Bennett, 1827: 374, pl. 9, fig. 1 (original description, two specimens, lectotype BMNH 1855.12.26.572, designated by Pietsch and Grobecker, 1987, paralectotype BMNH 1981.5.27.7, Madagascar; Fig. 180A).

Chironectes reticulatus Eydoux and Souleyet, 1850: 186, pl. 5, fig. 2 (original description, single specimen, holotype MNHN A.4616, Hawaii; dating, in part, after Bauchot et al., 1982; Fig. 180B). Bauchot et al., 1982: 64, 72 (dating of Eydoux and Souleyet). Pietsch et al., 1986: 141 (in type catalog).

A

B

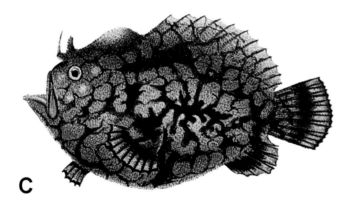

C

Fig. 180. *Antennatus tuberosus* (Cuvier). (**A**) lectotype of *Antennarius unicornis* Bennett, BMNH 1855.12.26.572, 45 mm SL, Madagascar. After Bennett (1828, pl. 9, fig. 1). (**B**) holotype of *Chironectes reticulatus* Eydoux and Souleyet, MNHN A.4616, 30 mm SL, Hawaiian Islands. After Eydoux and Souleyet (1850, pl. 5, fig. 2). (**C**) Bleeker's *Antennarius bigibbus*, authorship attributed to Commerson, Ambon, Indonesia. Drawing by Ludwig Speigler under the direction of Pieter Bleeker; after Bleeker (1865b, pl. 199, fig. 3).

Antennarius bigibbus: Günther, 1861a: 199 (new combination, description; synonymy includes *A. unicornis* Bennett). Günther, 1861b: 361 (Ceylon). Bleeker, 1865b: 10, 21, pl. 199, fig. 3 (description, additional material, Ambon; Fig. 180C). Günther, 1876: 165, pl. 105, fig. B (Huahine). Schmeltz, 1877: 15 (BMNH 1876.5.1.43; Tuamotu Archipelago). Jenkins, 1904: 511 (Honolulu). Snyder, 1904: 537 (Honolulu). Jordan and Evermann, 1905: 520 (synonymy, description, Honolulu). Regan, 1918: 77 (new to coast of Natal). Barnard, 1927: 1001 (Natal). Fowler, 1928: 479 (synonymy, description, Madagascar to Hawaii). Fowler, 1931: 367 (Paumotu, Society Islands). Fowler, 1934a: 450 (New Hebrides). Fowler, 1949: 158 (Oahu). Smith, 1949: 430, pl. 98, fig. 1235 (Natal). Munro, 1955: 289, fig. 841 (Ceylon). Munro, 1958: 296 (synonymy, New Guinea). Beaufort and Briggs, 1962: 202 (synonymy, description, coast of Natal to Hawaii; in key). Bauchot et al., 1982: 72 (same as *Chironectes reticulatus* Eydoux and Souleyet). Randall et al., 1993a: 223, 227, table 1 (synonym of *A. tuberosus*, Hawaiian Islands). Randall et al., 1997a: 55, color fig. (synonym of *A. tuberosus*, East Africa to Samoa and Hawaiian Islands).

Antennarius hispidus (not of Bloch and Schneider): Weber, 1913: 562 (in part, i.e., smaller specimen from station 282).

Antennarius reticulatus: Borodin, 1930: 62 (new combination; authorship erroneously attributed to Gilbert; Hawaii).

Antennatus (*Antennatus*) *bigibbus*: Schultz, 1957: 80, pl. 5, figs. C, D (new combination, description; Tuamotu Islands, Palmyra, Solomons, Hawaiian Islands; in key). Schultz, 1964: 176, table 1 (additional material, Philippines and Hawaiian Islands).

Antennatus bigibbus (not of Latreille): Ricker, 1959: 7 (specimen is *A. strigatus*).

Antennatus bigibbus: Gosline and Brock, 1960: 306, 345 (description, Hawaii; in key). Rosenblatt, 1963: 462 (comparison with *A. strigatus*). Munro, 1967: 584, pl. 78, fig. 1091 (New Guinea). Rosenblatt et al., 1972: 7 (unknown in Eastern Pacific). Tinker, 1978: 509, fig., color fig. 31, fig. 167-4 (description, Hawaiian and Tuamotu Islands to East African coast). Pietsch, 1981b: 419 (osteology).

Histiophryne bigibba: Le Danois, 1964: 106, fig. 51, 52b (new combination; in part, i.e., MNHN 01-304 is *A. strigatus*; MNHN A.4615 erroneously listed as syntype).

Histiophryne bigibba (not of Latreille): Le Danois, 1970: 84, fig. 3B (material, TAU 1687, is *Histrio histrio*).

Histiophryne tuberosa: Le Danois, 1970: 86, fig. 3A (new combination, specimen not examined by us, Gulf of Aqaba).

Antennatus tuberosus: Myers and Shepard, 1980: 315 (new combination, Guam). Pietsch, 1984b: 37, fig. 1B (genera of frogfishes). Pietsch, 1984e: 2 (Western Indian Ocean). Pietsch, 1985a: 91, fig. 30, table 1 (original manuscript sources for *Histoire Naturelle des Poissons* of Cuvier and Valenciennes). Randall, 1986: 180 (Marshall Islands). Pietsch and Grobecker, 1987: 192, figs. 16D, 78, 79B, 82, 127, pls. 35, 36 (description, distribution, relationships). Myers, 1989: 69, fig. 1e (East Africa to Line and Pitcairn Islands, Marshalls, and Hawaiian Islands, Samoa; in key). Pietsch and Grobecker, 1990a: 98 (aggressive mimicry, color change). Pietsch and Grobecker, 1990b: 79 (after Pietsch and Grobecker, 1990a). Randall et al., 1990: 55, color fig. (East Africa to Samoa and Hawaiian Islands). Randall et al., 1993a: 223, 227, table 1 (Hawaiian Islands). Randall et al., 1993b: 365 (Midway Atoll). Goren and Dor, 1994: 14 (Red Sea). Senou et al., 1994: 2 (Ryukyu Islands). Randall, 1996: 46, color fig. (Hawaiian Islands). Ohnishi et al., 1997: 216, 217, table 1 (comparison with congeners; in key). Randall et al., 1997a: 55, color fig. (East Africa to Samoa and Hawaiian Islands). Michael, 1998: 334, 354, 355, color fig. (identification, behavior, captive care). Fricke, 1999: 108 (Mauritius). Myers, 1999: 69, fig. 1e (after Myers, 1989). Pietsch, 1999: 2015 (western central Pacific; in key). Randall, 1999: 8 (Pitcairn Islands). Nakabo, 2000: 454 (Japan; in key). Pietsch, 2000: 597 (South China Sea). Randall and Holcom, 2001: 137, 138, 142, 143, table 1 (comparison with congeners). Nakabo, 2002: 454 (Japan; in key). Allen and Adrim, 2003: 25 (Timor, Moluccas to Bali, Indonesia). Allen et al., 2003: 367, color fig. (Indo-Pacific). Myers and Donaldson, 2003: 614 (Mariana Islands). Randall et al., 2004: 8 (Tonga). Mundy, 2005: 265 (Hawaiian Islands). Randall, 2005a: 73, color fig. (South Pacific). Allen et al., 2006: 641 (Australia). Kuiter and Debelius, 2007: 117, color fig. (Indonesia). Randall, 2007: 131, color fig. (Hawaiian Islands; in key). Senou et al., 2007: 49 (Ryukyu Islands). Allen and Erdmann, 2009: 594 (West Papua). Mundy et al., 2010: 23 (Howland Island, Palmyra Atoll). Randall, 2010: 39 (Hawaii). Fricke et al., 2011: 368 (New Caledonia). Allen and Erdmann, 2012: 152, color figs. (East Indies). Arnold and Pietsch, 2012: 120, 121, 127, 128, figs. 1D, 4, 6, table 1 (molecular phylogeny). Allen and Erdmann, 2013: 32, appendix 3.1 (Bali, Indonesia). Brandl and Bellwood, 2014: 32 (pair-formation). Fricke et al., 2014: 35 (Papua New Guinea). Delrieu-Trottin et al., 2015: 5, table 1(Marquesas Islands).

Histiophryne tuberosus: Dor, 1984: 57 (after Le Danois, 1970; Red Sea).

MATERIAL EXAMINED

Ninety specimens, 9.0 to 69.5 mm SL.

Lectotype of *Chironectes tuberosus*: MNHN 1104, 39.5 mm, Mauritius, Mathieu.

Paralectotypes of *Chironectes tuberosus*: MNHN 182, 2 (15.5–27.5 mm), data as for lectotype.

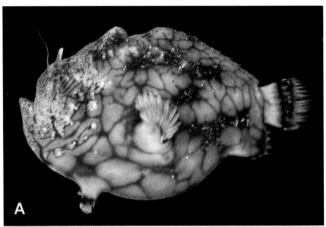

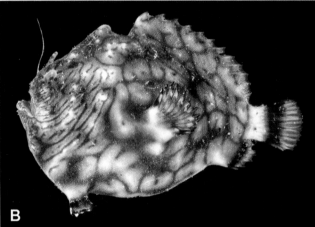

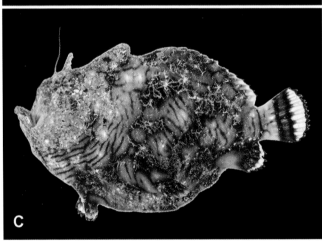

Fig. 181. *Antennatus tuberosus* (Cuvier). (**A**) BPBM 11176, 40.5 mm SL, Oeno, Pitcairn Islands. Photo by John E. Randall. (**B**) BPBM 6924, 32 mm SL, Tahiti, Society Islands. Photo by John E. Randall. (**C**) USNM 408188, 33 mm SL, Mururoa Atoll, French Polynesia. Photo by Jeffrey T. Williams. Courtesy of Lisa Palmer and the Smithsonian Institution, NMNH, Division of Fishes, Washington, DC.

Lectotype of *Antennarius unicornis*: BMNH 1855.12.26.572, 45 mm, Madagascar, Barclay (Fig. 180A).

Paralectotype of *Antennarius unicornis*: BMNH 1981.5.27.7, 32.5 mm, data as for lectotype.

Holotype of *Chironectes reticulatus*: MNHN A.4616, 30 mm, Voyage de la *Bonite*, Hawaiian Islands, Eydoux and Souleyet (Fig. 180B).

Additional material: *Western Indian Ocean*: BPBM, 1 (20 mm); CAS, 2 (11–26 mm); USNM, 1 (32 mm). *Indo-Australian Archipelago*: AMNH, 1 (27 mm); RMNH, 2 (23.5–46.5 mm); SU, 3 (23.5–35 mm); USNM, 2 (18–25 mm); UW, 2 (25–32.5 mm); ZMA, 1 (61 mm). *Guam, Carolines, and Marshalls*: BPBM, 1 (16 mm); CAS, 3 (24–29 mm); MCZ, 1 (27 mm); UG, 1 (19.5 mm); USNM, 2 (30–43 mm); AMS, 1 (44.5 mm); USNM, 2 (34–41 mm). *Gilberts, Howland Island, Tonga, and Samoa*: BPBM, 1 (24.5 mm); MCZ, 1 (25 mm); USNM, 1 (32 mm); UW, 1 (17 mm). *Johnston and Hawaiian Islands*: BPBM, 12 (17.5–41 mm); FMNH, 3 (30–41.5 mm); MNHN, 1 (50 mm); SU, 8 (14.5–25 mm); USNM, 1 (40 mm); UW, 8 (27.5–65 mm); ZMUC, 1 (15 mm). *Cook and Society Islands*: BMNH, 1 (42.5 mm); BPBM, 1 (32 mm); CAS, 1 (30 mm); MCZ, 1 (39 mm); NMNZ, 2 (12.5–28 mm). *Line Islands, Tuamotu Archipelago, and Pitcairn*: BMNH, 5 (10.5–40.5 mm); CAS, 3 (9.0–50 mm); USNM, 3 (17.5–47 mm). *Locality unknown*: UW, 3 (32–48 mm).

DIAGNOSIS

A member of the genus *Antennatus* unique in having the following combination of character states: caudal peduncle absent, the membranous posteriormost margin of soft-dorsal and anal fins connected to body at base of outermost rays of caudal fin; distinct esca absent, the illicium more or less tapering to a fine point (Fig. 170B); illicium one-and-a-half to two times length of second dorsal-fin spine, length

15.4 to 28.3% SL (Figs. 169, 170B); second dorsal-fin spine cylindrical, not tapering, straight to slightly curved posteriorly, relatively short, length 7.3 to 11.8% SL (Fig. 170B); dorsal-fin rays 12 (rarely 11 or 13); anal-fin rays 7 (rarely eight), usually all simple (but sometimes two or three posteriormost, or at most five innermost, bifurcate); all rays of caudal fin bifurcate; pectoral-fin rays 11, (rarely nine, 10, or 12) (Tables 12, 13); head and body yellow-brown with dark brown to black reticulations, marbling, and more or less interconnecting blotches (Figs. 181–182).

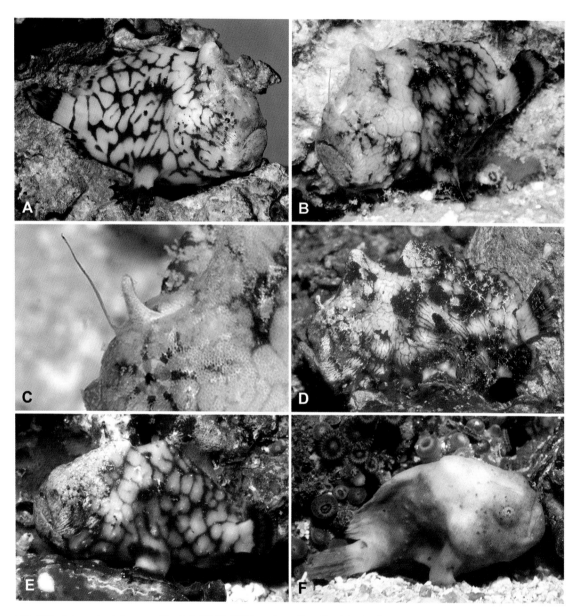

Fig. 182. *Antennatus tuberosus* (Cuvier). (**A**) Waikiki Aquarium, Hawaiian Islands. Photo by Keoki Stender. (**B, C**) Kwajalein Atoll, Marshall Islands. Photos by Jeanette Johnson. (**D**) Anilao, Luzon, Philippines. Photo by Roger Steene. (**E**) Oahu, Hawaiian Islands. Photo by Scott W. Michael. (**F**) juvenile, aquarium specimen, locality unknown. Photo by Scott W. Michael.

DESCRIPTION

Coloration in life: an overall pale or slate gray, but more typically, cream or yellow with reddish-brown or dark brown reticulations and marbling (Figs. 180–182); whitish, scab-like pattern sometimes present on face; a conspicuous dark brown band on tail (BPBM 6924, 11176). Individuals maintained in experimental aquaria observed to change from dark gray to light cream within a period of about two weeks (Pietsch and Grobecker, 1987: 195; see Color and Color Pattern, p. 36).

Remaining description as given for the genus.

Largest specimen examined: 69.5 mm SL (SAIAB 4610).

DISTRIBUTION

Antennatus tuberosus is widely distributed in the Indo-Pacific Ocean from the East African coast and Red Sea (including Mauritius but apparently absent north of the Chagos Archipelago in the Indian Ocean), throughout the Philippines and Indonesia (Bali to the Moluccas and West Papua), north to the Ryukyu Islands, south to New Caledonia and Tonga, and east to the island groups of the Pacific Plate, including the Marshalls and Marianas, Samoa, and the Hawaiian, Marquesas, Line, and Pitcairn Islands (Fig. 285, Table 22). The type locality is Mauritius.

Depths of capture, known for 29 specimens, range from the surface to 73 m; 83% of these, however, were taken in less than 20 m, and 66% in less than 6 m. The average depth for all known captures is 11.2 m.

COMMENTS

This species has long been recognized as *Antennarius bigibbus*, a name that originated in the manuscripts of Commerson [MSS 889(1): 13, 891: 135] based on specimens examined by him at Mauritius sometime between 1768 and 1770. The unpublished materials of Commerson that relate to this form, however, contain only a single, brief descriptive phrase: *"Antennarius (Bigibbus) Nigro et griseo variegatus nobis,"* a reference to body coloration that could easily apply to many Indo-Pacific antennariids. Of the four drawings of antennariids left by Commerson (see Introduction, p. 9), none can be correlated with this description. Lacepède (1798: 326), in utilizing Commerson's unpublished materials for his *Histoire Naturelle des Poissons*, but without examining specimens, recognized Commerson's species under the French vernacular name *La Lophie Double-Bosse*. While admitting that only scant evidence was available for its validity, Lacepède (1798) stated, in effect, that because Commerson was such a clever naturalist, and because he so clearly indicated that *La Lophie Double-Bosse* represented a distinct species, it must be recognized. Six years later, and again without reference to specimens, Latreille (1804) fixed the name *bigibbus* in the scientific literature by being the first to provide a Latinized binomial (*Lophius bigibbus*) for Lacepède's *La Lophie Double-Bosse*. It was Cuvier (1817b: 433), however, who initiated the nomenclatural confusion by hinting at the possible conspecificity of his *Chironectes tuberosus* and *La Lophie Double-Bosse*. Despite Valenciennes's (1837: 429) later statement that *C. tuberosus* seemed to be different from the "double-humped lophius" of Commerson (MSS) and Lacepède (1798), Günther (1861: 199) synonymized the former, recognizing *Antennarius bigibbus* and giving authorship to Lacepède. Since that time, most authors have followed Günther (1861). Although there is every reason to believe that Commerson did indeed have a distinct form in mind, it is now impossible with existing sources to determine just what that entity might have been. For this reason, the name *Lophius bigibbus* is considered a *nomen dubium* (see Species *Incertae Sedis*, p. 352) and *Antennatus tuberosus*, resurrected from synonymy by Pietsch and Grobecker (1987), is recognized as valid.

Subfamily Histiophryninae Arnold and Pietsch

Histiophryninae: Arnold and Pietsch, 2012: 117, 124–128, figs. 5, 7 (molecular phylogeny); Arnold, 2013: 496 (additional species); Arnold et al., 2014: 534, 538 (diagnosis; new genus and species; key to genera).

Two subfamilies of the Antennariidae, the Antennariinae and Histiophryninae, have recently been identified, based largely on the work of Arnold and Pietsch (2012; Arnold et al., 2014). While no convenient distinguishing characters are available for use in a traditional dichotomous key, the Histiophryninae, containing nine of the 15 recognized antennariid genera (*Rhycherus, Porophryne, Kuiterichthys, Phyllophryne, Echinophryne, Tathicarpus, Lophiocharon, Histiophryne,* and *Allenichthys*), is defined as follows:

DIAGNOSIS

A subfamily of the Antennariidae, distinguished from its sister-group, the Antennariinae, in lacking an endopterygoid and epural (Pietsch, 1984b: 41; Pietsch and Grobecker, 1987: 275), and in having an entirely different ovarian morphology: the Histiophryninae has simple, oval-shaped ovaries (Arnold and Pietsch, 2012: 128), while the Antennariinae has double, scroll-shaped ovaries (Pietsch and Grobecker, 1987, pl. 10, fig. 161). Each ovarian type corresponds to a different life history strategy: members of the Histiophryninae undergo direct development and display various degrees of parental care (Pietsch and Grobecker, 1980, 1987; Pietsch et al., 2009a; Arnold et al., 2014), while those of the Antennariinae are broadcast spawners and go through a distinct larval stage. In addition, the two subfamilies are clearly distinguished by molecular analysis (see Evolutionary Relationships, p. 378). Finally, the Histiophryninae is restricted geographically to the Indo-Australian Archipelago, while the Antennariinae has a broad geographic distribution, with all known genera found circumglobally throughout the tropics and subtropics (Tables 14, 15).

Key to the Known Genera of the Subfamily Histiophryninae

This key differs from most in that it is not dichotomous. It works instead by progressively eliminating the most morphologically distinct taxon and for that reason should always be entered from the beginning. All character states listed for each genus must correspond to the specimen being keyed; if not, proceed to the next set of character states. The illustrations accompanying the key are diagrammatic; dermal spinules and cutaneous appendages (with minor exceptions) are not shown. For an additional or alternative means of identification, all genera of the Histiophryninae may be compared simultaneously by referring to Table 14.

Table 14. Comparison of Distinguishing Character States of Genera of the Subfamily Histiophryninae

Character	Rhycherus	Porophryne	Kuiterichthys	Phyllophryne	Echinophryne	Tathicarpus	Lophiocharon	Histiophryne	Allenichthys
Dermal spinules	absent	bifurcate	bifurcate	absent	bifurcate	bifurcate	bifurcate	present/absent simple	bifurcate
Illicium	naked	naked	naked	naked	spinulose	naked	naked/spinulose	naked	naked
Esca	present	present	present	present	absent	present	present/absent	present/absent	present
Pectoral lobe	attached	attached	attached	attached	attached	free	attached	attached	attached
Caudal peduncle	present	present	present	absent	present	present	absent	absent	absent
Endopterygoid	absent	absent	absent	absent	absent	absent	absent	absent	absent
Pharyngobranchial I	present	present	present	present	present	present	absent	absent	present
Epural	absent	absent	absent	absent	absent	absent	absent	absent	absent
Pseudobranch	absent	absent	absent	absent	absent	present	absent	absent	absent
Swimbladder	present	present	absent	present	present	absent	present	present	present
Dorsal-fin rays	13 (12)	13	11–13 (10–14)	15 (14–16)	14–15 (13–16)	11 (10)	13 (12)	13–16	15–16
Anal-fin rays	8 (7)	7	7–8	8 (7–9)	8–9 (10)	7	7 (6–8)	7–9	8
Bifurcate caudal-fin rays	9	9	7	7	7	0	9	7	9
Pectoral-fin rays	11 (9–10)	10–11	8–10 (7–11)	11 (10–12)	10–11	7 (6)	9 (8)	8–9	10–11 (9)
Vertebrae	20	20	20–21	21 (20–22)	22 (21–23)	18	19	20–23 (19)	19–23

1. Illicium extremely long, greater than 24% SL; second and third dorsal-fin spines unusually long, greater than 15% and 24% SL, respectively; skin covered with close-set dermal spinules; pectoral-fin lobe detached from side of body for most of its length; all fin rays exceptionally long; distal one-third to one-half of pectoral-fin rays free, not interconnected by membrane; all rays of caudal fin simple (extreme distal tip of dorsalmost caudal-fin ray bifurcate in some specimens); dorsal-fin rays 11 (rarely 10); pectoral-fin rays seven (rarely six). . . . . *Tathicarpus* **Ogilby, 1907, p. 300**
   Tropical Australia, New Guinea

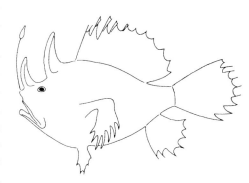

2. Skin smooth, without dermal spinules but covered with close-set, elongate, cutaneous appendages (easily lost through inadequate preservation); third dorsal-fin spine free, only the proximal 20 to 25% connected to nape of neck by membrane; caudal peduncle present, the membranous posteriormost margin of soft-dorsal and anal fins attached to body distinctly anterior to base of outermost rays of caudal fin; all rays of caudal fin bifurcate . . . .
   . . . . *Rhycherus* **Ogilby, 1907, p. 261**
   Subtropical Australia

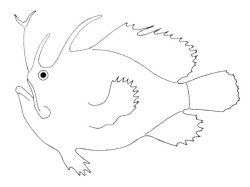

3. Rows of translucent to transparent ocelli on membranes of caudal fin, one row between each two rays (best observed by spreading rays); illicium distinctly curved anteriorly; anterior tip of pterygiophore of illicium upturned; second dorsal-fin spine connected by membrane to base of third spine, the third spine connected to soft-dorsal fin by skin; all rays of soft-dorsal and caudal fins bifurcate; caudal peduncle absent, the membranous posteriormost margin of soft-dorsal and anal fins attached to body at base of outermost rays of caudal fin; pectoral-fin rays nine (rarely eight) . . . . .
   . . . . *Lophiocharon* **Whitley, 1933, p. 305**
   Philippines to Tropical Australia]

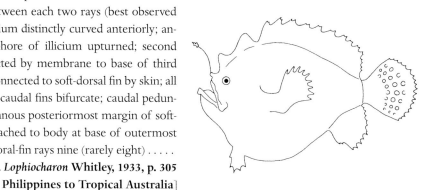

4. Illicium thick, completely covered with close-set dermal spinules, no bulbous esca present; second dorsal-fin spine short and stout (as shown here) or relatively long and slender; skin covered with dermal spinules; outermost rays of caudal fin simple, only the seven innermost rays bifurcate; caudal peduncle present, the membranous posteriormost margin of soft-dorsal and anal fins attached to body distinctly anterior to base of outermost rays of caudal fin; dorsal-fin rays 14 or 15 (rarely 13 or 16) . . . .
   . . . . *Echinophryne* **McCulloch and Waite, 1918, p. 289**
   Victoria, Tasmania

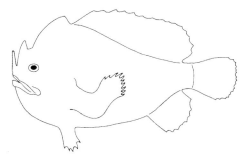

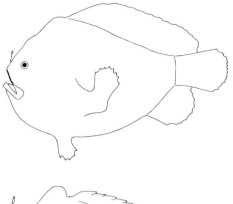

5. Illicium short (less than 10% SL), reduced (often hidden in a narrow groove on snout), minute (evident only with aid of a microscope) or absent; second and third dorsal-fin spines hidden, laid back and bound to surface of cranium by skin of head; skin naked, or covered with tiny, simple spinules (Fig. 26K); caudal peduncle absent, the membranous posteriormost margin of soft-dorsal and anal fins extending posteriorly beyond base of caudal fin and connecting to outermost caudal rays . . . . *Histiophryne* Gill, 1879b, p. 318
Taiwan to Australia

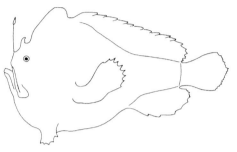

6. Illicium long, more than twice length of second dorsal-fin spine, the latter reduced (length less than 9% SL), recurved with a narrow tapering distal tip; caudal peduncle absent, the membranous posteriormost margin of soft-dorsal and anal fins attached to body at base of outermost rays of caudal fin; all rays of caudal fin bifurcate; dorsal-fin rays 15 or 16, anal-fin rays eight . . . . *Allenichthys* Pietsch, 1984b, p. 348
Western and South Australia

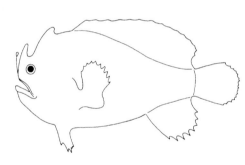

7. Skin smooth, without dermal spinules but with scattered, usually flattened, cutaneous appendages; skin surrounding second and third dorsal-fin spines usually loose, appearing swollen; caudal peduncle absent, the membranous posteriormost margin of soft-dorsal and anal fins attached to body at base of outermost rays of caudal fin; outermost rays of caudal fin simple, only seven innermost bifurcate. . . . .
. . . .*Phyllophryne* Pietsch, 1984b, p. 284
Subtropical Australia, Tasmania

8. Second dorsal-fin spine long (greater than 17% SL), slender, strongly curved posteriorly, not connected to head by membrane; skin covered with close-set dermal spinules; caudal peduncle present, the membranous posteriormost margin of soft-dorsal and anal fins attached to body distinctly anterior to base of caudal fin; outermost rays of caudal fin simple, the seven innermost bifurcate . . . .
. . . .*Kuiterichthys* Pietsch, 1984b, p. 277
Subtropical Australia

9. Second dorsal-fin spine club-shaped, narrow proximally, abruptly expanded distally, not connected to head by membrane; skin covered with close-set dermal spinules; pectoral fins unusually large; caudal peduncle present, the membranous posteriormost margin of soft-dorsal and anal fins attached to body distinctly anterior to base of caudal fin; all rays of caudal fin bifurcate; a conspicuous tuft of filaments, usually brightly colored, on midline of lip of lower jaw; distal margin of some or all fins red . . . .

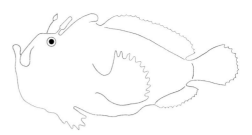

. *Porophryne* **Arnold, Harcourt, and Pietsch, 2014, p. 271**

New South Wales

Genus *Rhycherus* Ogilby

Rhycherus Ogilby, 1907: 17 [type species *Rhycherus wildii* Ogilby, 1907 (= *Chironectes filamentosus* Castelnau, 1872), by monotypy].

DIAGNOSIS

Rhycherus is unique among antennariids in having the head, body, and fins entirely covered with close-set, tapering, cutaneous appendages (length greater than 20% SL in some well-preserved specimens) and in having the third dorsal-fin spine free, only the proximal 20 to 25% of its length connected to nape of neck by membrane (Table 14).

Rhycherus is distinguished further from other antennariid genera by having the following combination of character states: skin without dermal spinules (except for a single, tiny crescent-shaped spinule associated with each pore of acoustico-lateralis system); illicium naked, without dermal spinules; esca distinct; pectoral-fin lobe broadly attached to side of body; caudal peduncle present, the membranous posteriormost margin of soft-dorsal and anal fins attached to body distinctly anterior to base of outermost rays of caudal fin; endopterygoid absent; pharyngobranchial I present; epural absent; pseudobranch absent; swimbladder present; dorsal-fin rays 12 or 13 (usually 13); anal-fin rays seven or eight (usually eight); all rays of caudal fin bifurcate; pectoral-fin rays nine to 11 (usually 11); vertebrae 20 (Tables 15, 16).

DESCRIPTION

Esca a pair of large, wormlike appendages with a small, compressed, medial flap (*R. filamentosus*), or a single, large, tapering appendage with numerous cylindrical filaments arising from base (*R. gloveri*); illicium unprotected, without groove alongside second dorsal-fin spine, the shallow depression between second and third dorsal-fin spines present (but incapable of providing protection for esca); illicium slightly more than one-half to nearly as long as second dorsal-fin spine, 9.9 to 31.6% SL (Fig. 183); anterior end of pterygiophore of illicium terminating slightly posterior to symphysis of upper jaw; illicium and second dorsal-fin spine relatively closely spaced on pterygiophore, the distance between bases of spines less than 7% SL; second dorsal-fin spine narrow (about twice width of illicium), straight to only slightly curved posteriorly, not connected to head by membrane, usually terminating in tight cluster of small, spherical swellings; length of second dorsal-fin spine 18.0 to 32.9% SL; third dorsal-fin spine narrow (about equal to width of second dorsal-fin spine), slightly curved posteriorly, usually terminating in tight cluster of small, spherical swellings, length 19.8 to 34.2% SL; eye protruding conspicuously from surface of head, diameter 3.8 to 8.5% SL; only distal tip (about 20 to

Table 15. Frequencies of Fin-Ray and Vertebral Counts of Genera of the Subfamily Histiophryninae

Genus	Dorsal-fin rays							Anal-fin rays					Pectoral-fin rays (left and right sides combined)							Vertebrae					
	10	11	12	13	14	15	16	6	7	8	9	10	6	7	8	9	10	11	12	18	19	20	21	22	23
Rhycherus			4	40					7	37						10	8	68				14			
Porophryne		14		3					3						1		3	2				3			
Kuiterichthys	2		8	33	2				25	34				2	30	16	68	2				20	21		
Phyllophryne					1	23	1		4	19	2						11	40	1			1	12	3	
Echinophryne				1	6	25	3			16	17	2					25	52					1	17	3
Tathicarpus	1	57						6	58				6	115						12					
Lophiocharon			4	65				2	65	2					3	135					20				
Histiophryne				37	14	58	15		41	66	19				152	86						30	23	29	21
Allenichthys						9	4			13						2	11	14						5	1
TOTALS	3	71	16	179	23	115	23	2	203	187	38	2	6	117	186	249	126	178	1	12	20	68	57	54	25

Table 16. Frequencies of Fin-Ray Counts of Species of the Genus *Rhycherus*

Species	Dorsal-fin rays		Anal-fin rays		Pectoral-fin rays*		
	12	13	7	8	9	10	11
R. filamentosus	3	30	4	29		8	56
R. gloveri	1	10	3	8	10		12
TOTALS	4	40	7	37	10	8	68

*Left and right sides combined.

25% of length) of maxilla tucked beneath folds of skin; epibranchial I with single tooth-bearing plate; ceratobranchial I with single, reduced tooth-plate; vertebrae 20, caudal centra 15; dorsal-fin rays 13 (rarely 12), as many as eight posteriormost bifurcate; anal-fin rays seven or eight, as many as seven posteriormost bifurcate; pectoral-fin rays nine to 11 (usually 11), all simple; all rays of pelvic fin simple (Tables 15, 16).

Coloration in preservation: tan, yellow-brown, to dark brown on upper part of head and body, with three broad, dark brown to black bars (sometimes continuous with dorsal pigmentation), one on cheek, one just above and behind pectoral fin, and one just anterior to caudal peduncle; some additional scattered dark spots on flanks; lower part of head and body, especially belly, light tan to cream; unpaired fins tan to cream with series of tiny black flecks on membrane between rays; basidorsal spot absent; illicium usually darkly banded; small dark spots on periphery of eye in some specimens; some specimens (e.g., NMV A.538) with silvery white, irregularly shaped blotches inside mouth, on dorsal surface of head, on nape of neck, on periphery of eye, at corner of mouth, and on body dorsal to pectoral-fin lobe.

Two species, both restricted to coastal, subtropical waters of Australia.

Key to the Known Species of the Genus *Rhycherus*

1. Illicium long, 19.2 to 31.6% SL (Fig. 183); esca a pair of tapering wormlike appendages, with a low, broad medial flap . . . .
    . . . . ***Rhycherus filamentosus*** (Castelnau, 1872), p. 263
    South Australia and Victoria

2. Illicium short, 9.9 to 17.6% SL (Fig. 183); esca a single, tapering appendage bearing short filaments within a "V-shaped" depression along its inner margin and a tight cluster of more elongate filaments arising from its base . . . .
    . . . . ***Rhycherus gloveri*** Pietsch, 1984c, p. 267
    Western and South Australia

Rhycherus filamentosus (Castelnau)
Tasselled Frogfish

Figs. 183–188, 291, 334–336; Tables 16, 22

Chironectes filamentosus Castelnau, 1872: 244 (original description, single specimen, holotype MNHN A.4617, Gulf St Vincent, South Australia). Castelnau, 1873: 65 (after Castelnau, 1872). Pietsch et al., 1986: 139 (type catalog).

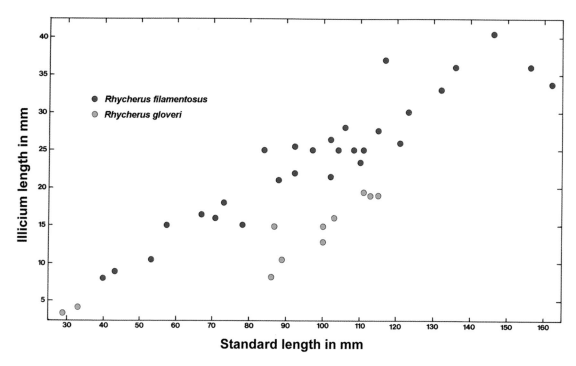

Fig. 183. Relationship between illicium length and standard length of species of *Rhycherus*. *R. filamentosus* (red), N = 30; *R. gloveri* (blue), N = 11. After Pietsch and Grobecker (1987, fig. 101).

Antennarius filamentosus: Macleay, 1881: 579 (new combination, South Australia).

Chironectes bifurcatus McCoy, 1886: 87, pl. 123 (original description, three specimens, lectotype by subsequent designation of Pietsch, 1984c, NMV A.521, Victoria; existence of paralectotypes unknown; Fig. 184). Lucas, 1890: 27 (Port Phillip, Victoria). McCulloch, 1916: 68 (synonym of *R. filamentosus*).

Rhycherus wildii Ogilby, 1907: 18 (original description, single specimen, holotype QMB I.12/778, South Australia). McCulloch, 1916: 68 (synonym of *R. filamentosus*).

Rhycherus filamentosus: McCulloch, 1916: 68 (new combination; in part, i.e., WAM specimen is *R. gloveri*). McCulloch and Waite, 1918: 70, pl. 6, fig. 3 (in part, text-fig. 31 shows *R. gloveri*; Kangaroo Island, Corny Point and Palmerston, South Australia; Fig. 185). Waite, 1921: 180, fig. 299 (in part; after McCulloch and Waite, 1918). Waite, 1923: 208, fig. (in part; after McCulloch and Waite, 1918). McCulloch, 1929: 406 (Brighton, Gulf St Vincent, South Australia). Schultz, 1957: 68, pl. 2, fig. B (synonymy, description after McCulloch and Waite, 1918). Le Danois, 1964: 134, figs. 69, 70b (description after McCulloch and Waite, 1918). Scott et al., 1974: 295, fig. (description, includes *R. gloveri*; Western Australia, South Australia, Victoria, Tasmania; in key). Pietsch, 1981b: 419 (osteology). Kuiter, 1982: 39, color pl. (escal morphology; Portsea, Victoria). Last et al., 1983: 255, fig. 22.7 (Tasmania; in key). Pietsch, 1984b: 40 (genera of frogfishes). Pietsch, 1984c: 70, figs. 1–4 (review based on all known material). Pietsch and Grobecker, 1987: 249, figs. 101, 102B, 103, 132, pls. 48–50 (description, distribution, relationships). Paxton et al., 1989: 280 (southern Australia). Pietsch and Grobecker, 1990a: 98, fig. (aggressive mimicry). Pietsch and Grobecker, 1990b: 76, fig. (after Pietsch and Grobecker, 1990a). Kuiter, 1993: 49 (southeastern Australia). Bertelsen, 1994: 137, color pl. (reproduction, parental care). Pietsch, 1994:

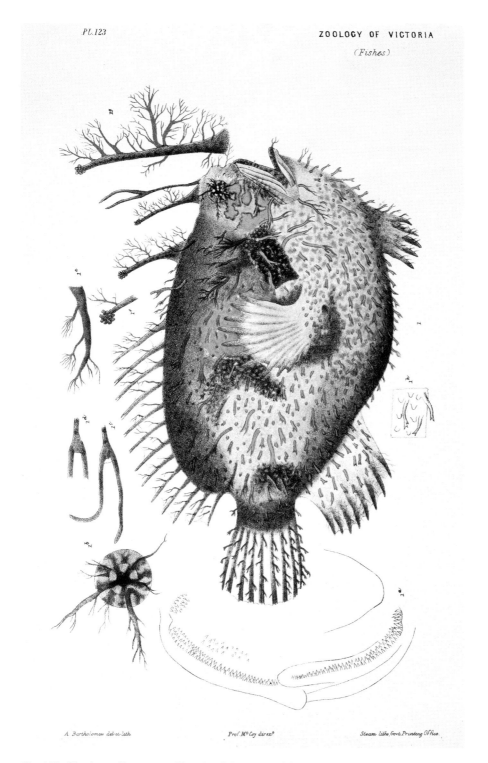

A. Bartholomew del et lith.   Prof. McCoy dir ex.º   Steam litho Govt Printing Office.

Fig. 184. *Rhycherus filamentosus* (Castelnau). Lectotype of *Chironectes bifurcatus* McCoy, NMV A.52l, 110 mm SL, Victoria, Australia, showing details of teeth, eye, esca, dorsal spines, and cutaneous appendages. Drawing and engraving by Arthur Bartholomew; after McCoy (1886, pl. 123); courtesy of Martin F. Gomon and the Museum of Victoria, Melbourne.

296, figs. 262A, 263 (southern Australia). Kuiter, 1997: 40 (southern Australia). Bavendam, 1998: 44, color figs. (natural history). Bertelsen and Pietsch, 1998: 137, color pl. (reproduction, parental care; after Bertelsen, 1994). Michael, 1998: 334 (identification, behavior, captive care). Schleichert, 2000: 38, color fig. (natural history). Allen et al., 2006: 645 Kuiter and Debelius, 2007: 126, color fig. (South Australia, Victoria). Pietsch, 2008: 372, fig. (after Pietsch, 1994). Baker et al., 2009: 18, appendices 2, 3 (South Australia). Arnold and Pietsch, 2012: 120, 128, figs. 1F, 5, 7, table 1 (molecular phylogeny). Saunders, 2012: 122 (reference to original description).

MATERIAL EXAMINED

Thirty-four specimens, 22 to 162 mm SL.

Holotype of *Chironectes filamentosus*: MNHN A.4617, 73 mm, Gulf St Vincent, South Australia, Castelnau.

Lectotype of *Chironectes bifurcatus*: NMV A.521, 110 mm, Brighton shore, Victoria, June 1879 (Fig. 184).

Holotype of *Rhycherus wildii*: QMB I.12/778, 77 mm, South Australia.

Additional material: AMS, 4 (22–115 mm); NMV, 11 (43–132 mm); SAMA, 16 (40–162 mm).

DIAGNOSIS

A member of the genus *Rhycherus* unique in having a long illicium, 19.2 to 31.6% SL (Fig. 183), and an esca consisting of a pair of tapering, wormlike appendages (as long as 25% SL, SAMA F.773), with a low, compressed medial flap (Pietsch and Grobecker, 1987, fig. 102B).

Coloration in life: face and margins of body (as seen in lateral view) blue, fading to light blue on flank; yellow reticulations on face under eye and a yellow patch on flank dorsal to pectoral-fin lobe; a large brown patch posterior to cheek, above and behind pectoral-fin lobe and just anterior to caudal peduncle; cutaneous appendages of head and body yellow; dorsal-fin spines and rays of unpaired fins, and distal portion of rays of paired fins, reddish-brown. Pinkish-yellow to yellow, fading to nearly white on belly, with white and dark brown to black blotches on head and body (Figs. 186–188).

Remaining description as given for the genus.

Largest specimen examined: 162 mm SL (SAMA F.3386).

DISTRIBUTION

Rhycherus filamentosus is restricted to the coastal waters of South Australia (at least as far west as Point Sinclair), Victoria, and northern Tasmania. The type locality is Gulf St Vincent, South Australia (Fig. 291, Table 22). Depths of capture, recorded for nine specimens, range from 3 to 46 m, with an average depth of 23 m.

COMMENTS

Ogilby (1907) established the genus *Rhycherus* with the description of *R. wildii*, within which he also included McCoy's (1886) *Chironectes bifurcatus* (Fig. 184). McCulloch (1916) was apparently the first to recognize the conspecificity of *R. wildii* and *R. bifurcatus*, and further that both of these are synonyms of the much older name *R. filamentosus* (Castelnau, 1872).

Fig. 185. *Rhycherus filamentosus* (Castelnau). 162 mm total length, Kingscote, Kangaroo Island, South Australia. Drawing by Phyllis Clarke; after McCulloch and Waite (1918, pl. 6, fig. 3); courtesy of Lea Gardam and the South Australian Museum, North Terrace, Adelaide.

Rhycherus gloveri Pietsch
Glover's Frogfish

Figs. 183, 189, 190, 291; Tables 16, 22

Rhycherus filamentosus (not of Castelnau): McCulloch, 1916: 68 (new combination; in part, only WAM specimen is *R. gloveri*). McCulloch and Waite, 1918: 70, text-fig. 31 (in part, i.e., "young specimen" shown in text-fig. 31; Wallaroo, Gulf St Vincent; Fig. 189). Waite, 1921: 180 (in part; after McCulloch and Waite, 1918). Waite, 1923: 208 (in part). Whitley, 1945: 42 (WAM P.2315, Cottesloe Beach, Western Australia).

Rhycherus gloveri Pietsch, 1984c: 70, figs. 2, 3B, 4 (original description, 11 specimens, holotype USNM 231754, Bunbuly, Western Australia). Pietsch and Grobecker, 1987: 248, figs. 101, 102A, 132 (description, distribution, relationships). Paxton et al., 1989: 281 (Australia). Hutchins and Smith, 1991: 10 (type catalog). Pietsch, 1994: 297, figs. 262B, 264 (southern Australia). Hutchins, 2001: 22 (Western Australia). Allen et al., 2006: 645 (southern Australia). Moore et al., 2009: 10 (type catalog). Pietsch, 2008: 373, fig. (after Pietsch, 1994). Baker et al., 2009: 17, appendix 3 (South Australia). Arnold and Pietsch, 2012: 128 (molecular phylogeny, revised classification).

Rhycherus globeri: Lindberg et al., 1997: 213 (misspelling of specific name).

MATERIAL EXAMINED

Eleven specimens, 29 to 115 mm SL.

Holotype of *Rhycherus gloveri*: USNM 231754, 113 mm, Clifton Beach Rocks, Bunbuly, Western Australia, found dead in tidepool, Collette, 8 February 1970.

Paratypes of *Rhycherus gloveri*: NMV A.524, 1 (33 mm), off Port Lincoln, 22 October 1968. SAMA F.4378, 1 (100 mm), Grindle Island, near Thistle Island, South Austra-

Fig. 186. *Rhycherus filamentosus* (Castelnau). Rye Jetty, Mornington Peninsula, Victoria, Australia. Photo by Brian Mayes.

Fig. 187. *Rhycherus filamentosus* (Castelnau). Blairgowrie Marina, Port Phillip, Victoria, Australia. Photo by James Peake.

Fig. 188. *Rhycherus filamentosus* (Castelnau). (**A–E**) Rye Jetty, Mornington Peninsula, Victoria, Australia. (**F**) Blairgowrie Jetty, Mornington Peninsula, Victoria, Australia. Photos by Brian Mayes.

lia, Wilkshire, 10 August 1978; SAMA F.2685, 1 (86 mm), Whyalla, South Australia, Foster, 21 September 1950; SAMA F.614, 1 (29 mm), Wallaroo, November 1919; SAMA F.1465, 1 (89 mm), Port Broughton, South Australia, Keast, 24 November 1930. WAM P.3114, 1 (115 mm), North Beach, Western Australia, Bond, 14 July 1946; WAM P.2315, 1 (87 mm), Cottesloe Beach, Western Australia, Glauert, 18 August 1941; WAM P.24867, 1 (100 mm), Googee Beach near Perth, Bruce, 7 August 1974; WAM P.5814, 1 (103 mm), Warnbro Beach, Hunt, 23 August 1963; WAM P.12111, 1 (111 mm), Ledge Point, Western Australia, Harley, 22 July 1964.

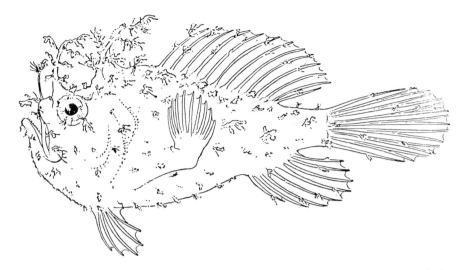

Fig. 189. *Rhycherus gloveri* Pietsch. SAMA F.614, 29 mm SL, juvenile, off Wallaroo, Spencer Gulf, South Australia. After McCulloch and Waite (1918, text-fig. 31).

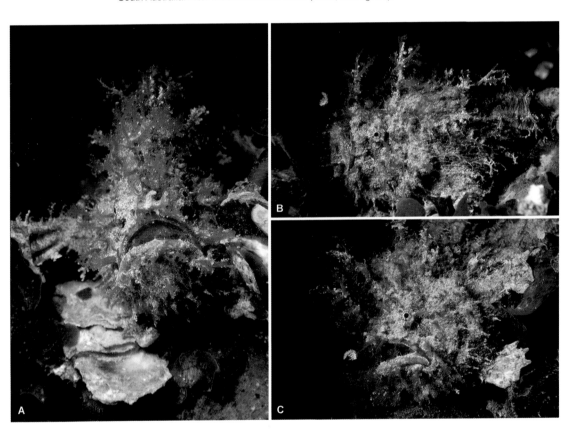

Fig. 190. *Rhycherus gloveri* Pietsch. (**A–C**) Port Hughes, South Australia. Photos by John Lewis.

DIAGNOSIS

A member of the genus *Rhycherus* unique in having a short illicium, 9.9 to 17.6% SL (Fig. 183) and an esca consisting of a single, tapering appendage (as long as 22% SL, WAM P.12111) bearing short filaments within a "V-shaped" depression along its inner margin, and a tight cluster of more elongate filaments arising from its base (Pietsch and Grobecker, 1987, fig. 102A) (Fig. 190).

DESCRIPTION

Pectoral-fin rays nine in South Australian population, 11 in Western Australian population (Table 16; see Geographic Variation, below).

Remaining description as given for the genus.

Largest specimen examined: 115 mm SL (WAM P.3114).

GEOGRAPHIC VARIATION

A distinct difference in the number of pectoral-fin rays within the known material of *R. gloveri* correlates with the curious disjunct distribution displayed by this species (see below). The six known specimens collected from Western Australia all have 11 pectoral-fin rays; the remaining five specimens, all from South Australia, all have nine (Table 16). Detailed examination of the known material, however, revealed no additional features to justify assigning separate species status to these two populations—molecular analysis may well prove otherwise.

DISTRIBUTION

Endemic to temperate Australia, *Rhycherus gloveri* is known from two general localities (Fig. 291, Table 22): of the 11 specimens examined by us, six were collected from the vicinity of Perth, Western Australia (captured between about 31° and 35° S latitude), and five specimens from the vicinity of Adelaide, South Australia (between 135° and 138° E longitude). The holotype is from Clifton Beach Rocks, Bunbuly, Western Australia. Depth of capture is recorded for only a single specimen, NMV A.524, taken off Port Lincoln, Spencer Gulf, in 20 m.

Genus *Porophryne* Arnold, Harcourt, and Pietsch

Porophryne Arnold, Harcourt, and Pietsch, 2014: 534 (type species *Porophryne erythrodactylus* Arnold, Harcourt, and Pietsch, 2014, by original designation and monotypy).

DIAGNOSIS

Porophryne is unique among antennariids in having a club-shaped second dorsal-fin spine (narrow at its proximal end, abruptly expanded and anteroposteriorly flattened distally; Fig. 191), and unusually large pectoral fins (some specimens with a conspicuous tuft of filaments, usually brightly colored, on the midline of the lip of the lower jaw that may serve as accessory lure, unknown in any other anglerfish). It differs further in having the following combination of character states: skin covered with close-set bifurcate dermal spinules, the length of the spines of each spinule not more than twice the distance between tips of spines; illicium sparsely covered with dermal denticles; esca distinct; pectoral-fin lobe broadly attached to side of body; caudal peduncle present, the membranous posteriormost margins of soft-dorsal and anal fins attached to body distinctly anterior to base of outermost rays of caudal fin; endopterygoid absent; pharyngobranchial

Fig. 191. *Porophryne erythrodactylus* Arnold, Harcourt, and Pietsch. Holotype, UW 118988, 54.5 mm SL, Kurnell, Botany Bay, New South Wales, Australia. Photo by David Harasti.

I present; epural absent; pseudobranch absent; swimbladder present; dorsal-fin rays 13; anal-fin rays seven; all caudal-fin rays bifurcate; pectoral-fin rays 10 or 11 (eight on one side of smaller paratype); vertebrae 20; some or all fins fringed with red pigmentation (Tables 14, 15).

In addition to the morphological evidence given above, the validity of *Porophryne* is further supported by molecular data (see Evolutionary Relationships, p. 378). An 8.4 to 9.5% divergence in the COI gene was found between *Phyllophryne* and *Echinophryne*, a 10.9 to 11.3% divergence was calculated between *Porophryne* and *Kuiterichthys*, and 100% identity was found between the three individuals of the type species *Porophryne erythrodactylus*.

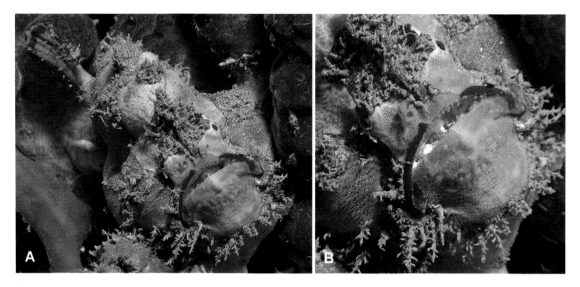

Fig. 192. *Porophryne erythrodactylus* Arnold, Harcourt, and Pietsch. (**A, B**) Bare Island, Botany Bay, New South Wales. Photos by Gary Dunnett.

DESCRIPTION

Esca red, gray, or brown, divided into two tufts of short, slender, filamentous appendages; illicium 6.3 to 7.7% SL, sparsely covered with dermal denticles, without groove alongside base of second dorsal-fin spine; second dorsal-fin spine 17.6 to 21.0% SL, without posterior membrane, sometimes with slender filamentous appendages distally, narrow for approximately one third of its length, abruptly expanded distally, forming an anteroposteriorly flattened, club-shaped structure; third dorsal-fin spine 27.4 to 35.3% SL, mobile, covered with thick skin; diameter of eyes 6.9 to 7.5% SL; dorsal-fin rays 13 (all bifurcate except anteriormost one or two); anal-fin rays seven (all bifurcate); pectoral fin exceptionally large, width across base of rays 18.3% SL, distance from distal end of third radial to tip of longest ray 38.5% SL; pectoral-fin rays 10 or 11 (eight on left side of smaller paratype; all simple); pelvic-fin rays five (posteriormost ray bifurcate); caudal-fin rays nine, all bifurcate. Body everywhere covered with dermal denticles, except within small black spots (see below); orange, yellow, pink, red, and white individuals usually with brown and red cutaneous appendages occurring in patches on body, especially around chin and opercle; occasionally a conspicuous cluster of fleshy filaments, often brightly pigmented, on midline of lip of lower jaw (possibly serving as an additional luring device); oral valve present, lining both upper and lower jaws; swimbladder present; ovaries paired, lobular; caudal peduncle present, posteriormost rays of dorsal and anal fins not connected by membrane to caudal peduncle and outermost rays of caudal fin.

Coloration in life: pale to dark gray, orange and red, orange-red and pink, white and pink, white and red, or yellow and red, with outer margin of some or all fins fringed with red; first and second dorsal-fin spines rarely red; upper jaw sometimes purple or bright red (Fig. 192); tiny light to dark gray spots on gray individuals, only visible on close inspection, giving the fish an overall light to dark gray color, with some individuals appearing nearly white; gray individuals usually with a variable number of naked, black, osculum-like spots scattered asymmetrically on head and body (Figs. 193, 194).

Coloration in preservation: dark to pale gray, fading ventrally, with larger diffuse gray spots (covered with dermal denticles) on abdomen; white where red pigment has leached

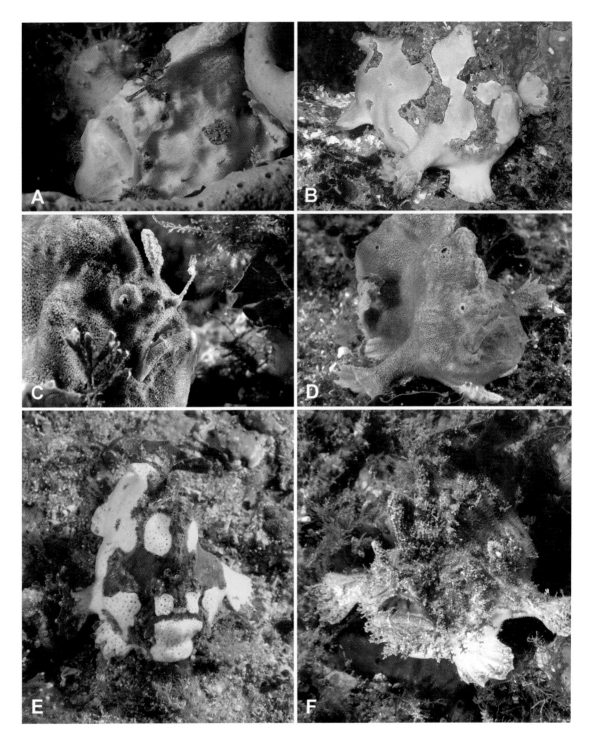

Fig. 193. *Porophryne erythrodactylus* Arnold, Harcourt, and Pietsch. (**A**) paratype, AMS I.44699, 74.5 mm SL, Bare Island, Botany Bay, New South Wales. Photo by Rob Harcourt. (**B**) Middle Ground, Jervis Bay, New South Wales. Photo by Kim Sebo. (**C–F**) Bass Point, Shellharbour, New South Wales. Photos by Michael McKnight.

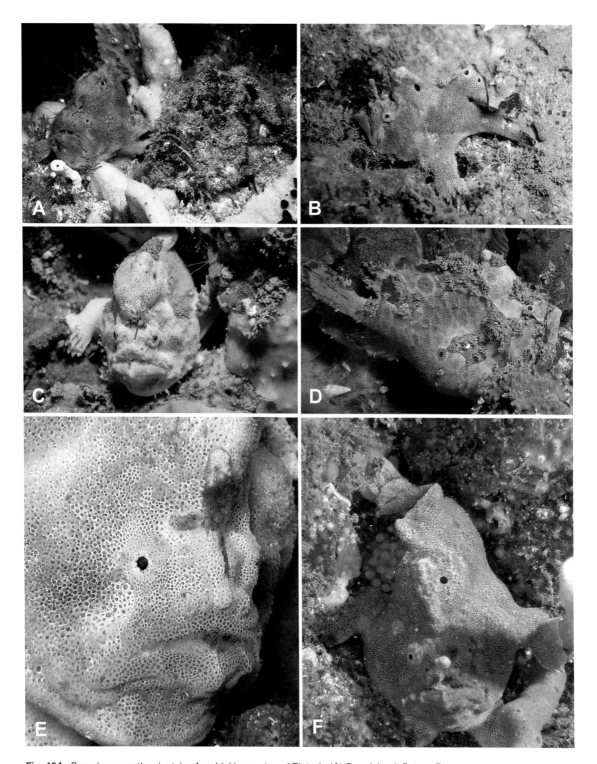

Fig. 194. *Porophryne erythrodactylus* Arnold, Harcourt, and Pietsch. (**A**) Bare Island, Botany Bay, New South Wales. Photo by David Harasti. (**B**) Kurnell, Botany Bay, New South Wales. Photo by Rachel J. Arnold. (**C**) Bare Island, Botany Bay, New South Wales. Photo by Nick Missenden. (**D**) Bare Island, Botany Bay, New South Wales. Photo by Gary Dunnett. (**E**) close-up showing tuft of filaments on symphysis of lower jaw that perhaps serves as an additional luring device, Bare Island, Botany Bay, New South Wales. Photo by Sheila Bowtle. (**F**) guarding a cluster of eggs, Bass Point, Shellharbour, New South Wales. Photo by Michael McKnight.

away; pigment of orange and red paratype (AMS I.44699) leached away (except for margins that were red in life, now light brown), patches of cutaneous appendages brown.

A single species.

Porophryne erythrodactylus Arnold, Harcourt, and Pietsch
Red-Fingered Frogfish

Figs. 191–194, 291; Tables 14, 15, 22

Antennariidae gen. et sp. nov.: Arnold and Pietsch, 2012: 117, 124, 125, 128, figs. 1G, 5, 7, table 1 (molecular phylogeny).

Porophryne erythrodactylus Arnold, Harcourt, and Pietsch, 2014: 534–539, figs. 1–3, table 1 (original description, three specimens, holotype UW 118988, Kurnell, Botany Bay, New South Wales, Australia). Schoer, 2015: 32 (after Arnold et al., 2014). Wheeler, 2015 ("new to nature"; after Arnold et al., 2014).

MATERIAL EXAMINED

Three specimens, 54.5 to 74.5 mm SL.

Holotype of *Porophryne erythrodactylus*: UW 118988, female, 54.5 mm, Kurnell, Botany Bay, New South Wales, Australia, scuba, 8 m, D. Harasti, 16 May 2009 (Fig. 191).

Paratypes of *Porophryne erythrodactylus*: AMS I.43749, female, 1 (68.5 mm), Bare Island, Botany Bay, New South Wales, Australia, approximately 33°59′ S, 151°13′ E, scuba, 13.7 m, R. Harcourt, 30 October 2005; AMS I.44699, female, 1 (74.5 mm), east side of Bare Island, 33°59′33″ S, 151°13′57″ E, 10.7 m, R. Harcourt, 18 January 2009 (Fig. 193A).

DIAGNOSIS

As given for the genus.

DESCRIPTION

Porophryne erythrodactylus appears to have two distinct color phases: gray individuals usually with naked black spots scattered asymmetrically on the head and body, and a near absence of cutaneous appendages on the body; and orange, yellow, pink, red, and white individuals, typically without naked black spots, but usually with many more cutaneous appendages on the body, the latter presumably to aid in camouflage or to enhance luring through mimicry of the substrate. Orange, yellow, pink, red, and white individuals are also usually larger than gray individuals: of the three type specimens, the orange and red paratype (AMS I.44699) is the largest at 74.5 mm (Fig. 193A), while *in situ* photographs of reddish individuals found near gray individuals are generally larger (Fig. 194A). It was initially thought that *P. erythrodactylus* may be sexually dichromic or that there were two undescribed species, but dissections of the three specimens in question confirmed that all are female, and divergence within the COI gene was 0% (see Evolutionary Relationships, p. 378).

Remaining description as given for the genus.

Largest known specimen: 74.5 mm SL (AMS I.44699).

DISTRIBUTION

The type material of *Porophryne erythrodactylus* was collected from nearshore waters of Botany Bay, New South Wales, at depths of 8 to 13.7 m, but observations by divers have been made in as little as 6 m and as deep as 24 m (Fig. 291, Table 22). A number of additional, uncollected individuals of *P. erythrodactylus* have been reported from the Sydney region, especially around Bare Island in Botany Bay, but confirmed sightings have also been made as far south as Jervis Bay, New South Wales.

HABITAT

Porophryne erythrodactylus is generally found in subtidal rocky-reef habitat dominated by small filamentous and foliose algae of the genera *Zonaria, Corallina, Amphiroa,* and *Laurencia* (often with some *Sargassum* spp.). A few scattered larger seaweeds, including *Ecklonia radiata* and *Sargassum* spp., with large patches of filamentous or foliose algae, may also be present (Creese et al., 2009). Vertical or sloping walls on the deep edge of nearby reefs are inhabited by ascidians, corals, and sponges, the latter including *Spongia* sp., *Tedania anhelans, Ephydatia fluviatilis, Darwinella australiensis, Chondrilla australiensis, Mycale australis,* and *Holopsamma laminaefavosa.* More specifically, this species is most often found closely associated with small sponges, blending in with the gray coloration of *Psammocinia* sp. or the red or yellow coloration of *Darwinella* sp., or nestled next to fan-shaped *Echinoclathria leporina.* In addition, the black "osculum-like" spots typically present on gray color morphs of the new species presumably aid in the mimicry of sponges (Arnold, 2010a: 68; Arnold et al., 2014: 535, 538).

COMMENTS

This frogfish was first collected by David B. Grobecker in Sydney Harbour on 15 October 1980 but the specimen (UW 21020, 60 mm) was subsequently lost and the species remained undescribed for the next quarter of a century. Three specimens were collected on 30 October 2005, 18 January 2009, and 16 May 2009 by scuba divers near Kurnell and Bare Island in Botany Bay, New South Wales, Australia, and subsequently described as a new genus and species by Arnold et al. (2014). Although frequently misidentified by local divers as a member of the genus *Antennarius, Porophryne* more closely resembles *Kuiterichthys*; in fact, the two were recovered as sister taxa in our molecular analysis (Arnold and Pietsch, 2012; see Evolutionary Relationships, p. 378). At the same time, however, the two are remarkably distinct, differing in a host of ways, the most striking of which is the morphology of the first and second dorsal-fin spines and the distinctive red pigmentation on the distal margin of the fins. While a 10.9 to 11.3% divergence in the COI gene was found between *Porophryne* and *Kuiterichthys*, 100% identity was found between the three individuals of *Porophryne erythrodactylus* examined (Arnold et al., 2014).

Genus *Kuiterichthys* Pietsch

Kuiterichthys Pietsch, 1984b: 32, 37 (type species *Chironectes furcipilis* Cuvier, 1817b, by original designation and monotypy).

DIAGNOSIS

Kuiterichthys is unique among antennariids in having the following combination of character states: skin covered with close-set bifurcate dermal spinules, the length of the spines of each spinule not more than twice the distance between tips of spines (Fig. 26E); illicium naked, without dermal spinules; esca distinct; pectoral-fin lobe broadly attached to side of body; caudal peduncle present, the membranous posteriormost margins of soft-dorsal and anal fins attached to body distinctly anterior to base of outermost rays of caudal fin; endopterygoid absent; pharyngobranchial I present; epural absent; pseudobranch absent; swimbladder present; dorsal-fin rays 10 to 14 (usually 11 to 13); anal-fin rays seven or eight; outermost rays of caudal fin simple, the seven innermost bifurcate; pectoral-fin rays seven to 11 (usually eight to 10); vertebrae 20 or 21 (Tables 14, 15).

DESCRIPTION

Esca well developed, complex; illicium naked and unprotected, no naked groove along-side second dorsal-fin spine, the shallow naked depression between second and third dorsal-fin spines absent; illicium length, 9.7 to 24.6% SL (Fig. 195); anterior end of pterygiophore of illicium usually terminating at symphysis of upper jaw, but occasionally extending slightly anteriorly, beyond symphysis; illicium and second dorsal-fin spine widely spaced on pterygiophore, the distance between bases of spines 6.2 to 9.3% SL; second dorsal-fin spine narrow (about twice width of illicium), slightly tapering toward tip, strongly curved posteriorly, not connected to head by membrane, covered with close-set dermal spinules, and usually terminating in tight cluster of spinules and cutaneous appendages; length of second dorsal-fin spine 17.0 to 33.2% SL; third dorsal-fin spine curved posteriorly, tapering slightly toward base (width at distal end 3.7 to 5.1% SL), the distal end protruding considerably beyond membranous connection to head; length of third dorsal-fin spine 22.6 to 30.4% SL; eye more or less surrounded by ring of separate, relatively widely spaced clusters of dermal spinules, diameter 5.0 to 12.1% SL; only distal tip (about 25% of length) of maxilla naked and tucked beneath folds of skin, the remaining proximal portion covered directly with spinulose skin; a few, small cutaneous appendages scattered over head, body, and fins, but presence not conspicuous; wartlike patches of clustered dermal spinules absent; epibranchial I with single tooth-bearing plate; ceratobranchial I toothless; dorsal-fin rays 11 to 13 (rarely 10 or 14), at most only four posteriormost bifurcate; anal-fin rays seven or eight, as many as six posteriormost bifurcate; pectoral-fin rays eight to 10 (rarely seven or 11), all simple; all rays of pelvic fin simple; vertebrae 20 or 21, caudal centra 15 or 16 (Tables 15, 17).

Coloration in life: reddish-brown interspersed with brown mottling, fading to almost white on belly, paired fins, and distal portions of second and third dorsal-fin

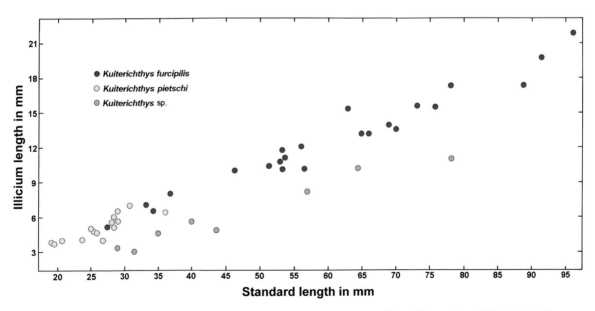

Fig. 195. Relationship between illicium length and standard length for species of *Kuiterichthys*. *K. furcipilis* (red), N = 23; *K. pietschi* (yellow), N = 15; *K.* sp. (blue), N = 8.

Table 17. Frequencies of Fin-Ray Counts of Species of the Genus *Kuiterichthys*

Species	Dorsal-fin rays					Anal-fin rays		Pectoral-fin rays*				
	10	11	12	13	14	7	8	7	8	9	10	11
K. furcipilis			3	31	2	4	32			12	58	2
K. pietschi	2	14				16		2	30			
K. sp.			5	2		5	2			4	10	
TOTALS	2	14	8	33	2	25	34	2	30	16	68	2

*Left and right sides combined.

spines; scattered dark brown to black spots on flank; conspicuous basidorsal spot usually present.

Coloration in preservation: beige, gray, pinkish-brown, yellow-brown, usually somewhat more lightly pigmented on lower portion of head and body, particularly on belly; usually a dark brown patch extending from nape of neck posteroventrally to behind and below pectoral-fin lobe and a similarly colored band extending across dorsalmost margin of soft-dorsal fin, continuing as broad, vertical bar across caudal peduncle; additional small, dark brown to black spots occasionally scattered over upper portion of head and body, but particularly on cheeks and second dorsal-fin spine; three specimens everywhere covered (except on belly) with a complex, almost white network of narrow, interconnecting lines, dividing darker ground coloration to form numerous, small, roughly circular (AMS IA.2957, NMV A.1583) to elongate and irregularly shaped (AMS IA.3614) patches; darkly pigmented basidorsal spot retained in some specimens; illicium without banding; no bars of pigment radiating from eye.

Two species (see Comments below), both restricted to the Indo-Australian Archipelago.

COMMENTS

Kuiterichthys was established by Pietsch (1984b: 37) to accommodate only *K. furcipilis*, a species originally placed by Cuvier (1817b) in the nominal genus *Chironectes* and later reallocated to *Antennarius* by Schinz (1822) and to *Trichophryne* (here considered a synonym of *Echinophryne*, following Pietsch and Kuiter, 1984) by Le Danois (1964). There are no known derived features uniquely shared by *Kuiterichthys* and *Antennarius*. Although *Kuiterichthys* shares four apomorphic features with *Echinophryne* (Table 14), none is unique to these two genera and three are loss character states that most likely do not reflect common ancestry. Instead, our molecular analysis supports a sister-group relationship between *Kuiterichthys* and *Porophryne* (Arnold and Pietsch, 2012; see Evolutionary Relationships, p. 378).

Key to the Known Species of the Genus *Kuiterichthys*

1. Esca with as many as nine (usually six to eight) appendages (short and thick to long and slender, often darkly pigmented) arising in a single plane from a common base; a variable number of short, slender escal filaments sometimes present arising from base (Fig. 198); dorsal-fin rays 13 (rarely 12 or 14); anal-fin rays eight (rarely seven); pectoral-fin rays nine or 10 (rarely 11) (Table 17) . . . .
. . . . *Kuiterichthys furcipilis* (**Cuvier, 1817b**), **p. 280**
Victoria, Tasmania, and New South Wales

2. Esca consisting of a single, compressed, tapering appendage bearing filaments along its lateral margins and within a shallow "V-shaped" depression along its inner margin, with a tight cluster of slightly longer filaments arising from its base; dorsal-fin rays 11 (rarely 10); anal-fin rays seven; pectoral-fin rays eight (rarely seven) (Table 17) . . . .
. . . . **Kuiterichthys pietschi Arnold, 2013, p. 283**
New South Wales

Kuiterichthys furcipilis (Cuvier)
Rough Frogfish

Figs. 15A, 26E, 195–199, 290; Tables 17, 22

Chironectes furcipilis Cuvier, 1817b: 429, pl. 17, fig. 1 (original description, single specimen, holotype MNHN A.4618, locality unknown; Fig. 196). Cuvier, 1829: 252 (after Cuvier, 1817b). Valenciennes, 1837: 423 (after Cuvier, 1817b). Valenciennes, 1842: 189 (after Cuvier, 1817b). Günther, 1861a: 184 (footnote reference to Cuvier, 1817b). Schultz, 1957: 73 [in synonymy of *Phrynelox scaber* (= *A. scaber*)]. Pietsch, 1985a: 85, fig. 24, table 1 (original manuscript sources for *Histoire Naturelle des Poissons* of Cuvier and Valenciennes). Pietsch et al., 1986: 139 (type catalog).

Antennarius furcipilis: Schinz, 1822: 500 (new combination; after Cuvier, 1817b).

Trichophryne furcipilis: Le Danois, 1964: 132, figs. 67, 68, 70a (new combination, description of holotype). Pietsch, 1981b: 419 (osteology). Last et al., 1983: 256, fig. 22.8 (description; common in D'Entrecasteaux Channel, Tasmania, on muddy bottoms between 15–50 m; in key).

Trichophryne mitchelli (not of Morton): Whitley, 1929a: 357 (AMS IA.3614, off Eden, New South Wales). Whitley, 1934, no. CLIV (reference to AMS IA.3614). Last et al., 1983: 256, fig. 22.8 (Tasmania; in key).

Kuiterichthys furcipilis: Pietsch, 1984b: 38 (new combination, genera of frogfishes). Pietsch, 1985a: 85, fig. 24, table 1 (original manuscript sources for *Histoire Naturelle des Poissons* of Cuvier and Valenciennes). Pietsch and Grobecker, 1987: 215, figs. 16E, 86–88, 129, pl. 41 (description, distribution, relationships). Paxton et al., 1989: 280 (Australia).

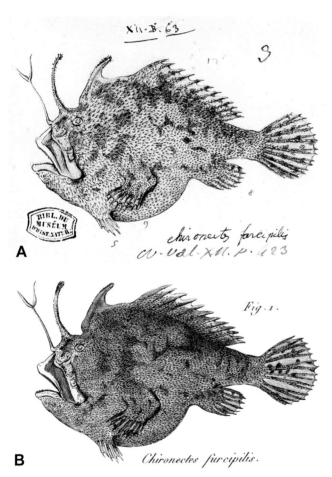

Fig. 196. *Kuiterichthys furcipilis* (Cuvier). Holotype, MNHN A.4618, 63 mm SL, locality unknown. (**A**) reproduced from the pencil original that served as the model for the engraving that accompanied the original description published by Cuvier. After Cuvier and Valenciennes, MS 504, XII.B, Bibliothèque Centrale, Muséum national d'Histoire naturelle, Paris; courtesy of Emmanuelle Choiseau, © Muséum national d'Histoire naturelle, Paris. (**B**) the published version of the above. After Cuvier (1817b, pl. 17, fig. 1).

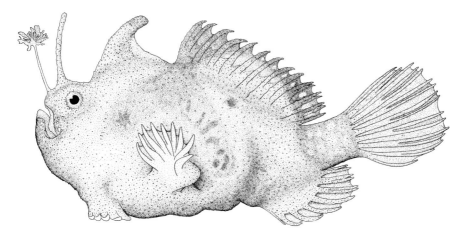

Fig. 197. *Kuiterichthys furcipilis* (Cuvier). NMV A.2848, 65 mm SL, Bass Strait, Victoria, Australia. Drawing by Cathy L. Short; after Pietsch and Grobecker (1987, fig. 86).

Fig. 198. Esca of *Kuiterichthys furcipilis* (Cuvier). NMV A.2848, 65 mm SL, Bass Strait, Victoria, Australia. Drawing by Cathy L. Short; after Pietsch and Grobecker (1987, fig. 87).

Kuiter, 1993: 50 (southeastern Australia). Pietsch, 1994: 292, fig. 260 (southern Australia). Kan et al., 1995: 23, table 2 (Yule Island, Gulf of Papua). Allen et al., 2006: 643 (southern Australia). Kuiter and Debelius, 2007: 124, color figs. (Tasmania). Pietsch, 2008: 370, fig. (after Pietsch, 1994). Arnold and Pietsch, 2012: 120, 125, 128, figs. 1H, 5, 7, table 1 (molecular phylogeny). Arnold, 2013: 496, fig. 2, table 1 (comparison with congener; color plate after Pietsch and Grobecker, 1987). Arnold et al., 2014: 534, table 1 (genetics, as an outgroup).

MATERIAL EXAMINED

Forty-four specimens, 27.5 to 96.5 mm SL.

Holotype of *Chironectes furcipilis*: MNHN A.4618, 63 mm, locality unknown (Figs. 15, 196).

Additional material: AMS, 19 (27.5–83 mm); CSIRO, 7 (45–89 mm); NMV, 14 (43–96.5 mm); UW, 3 (52.5–66 mm).

DIAGNOSIS

A member of the genus *Kuiterichthys* unique in having the following combination of character states: esca with as many as nine (usually six to eight) appendages (short and thick to long and slender, often darkly pigmented) arising in a single plane from a common base, with a variable number of short, slender filaments sometimes present arising from base (Figs. 197, 198); illicium length 18.8 to 24.6% SL (Fig. 195); second dorsal-fin spine length 17.8 to 29.8% SL (Figs. 196, 197); dorsal-fin rays 13 (rarely 12 or 14); anal-fin rays eight (rarely seven); pectoral-fin rays nine or 10 (rarely 11) (Table 17); vertebrae 21; adult body size relatively large, 27.5 to 96.5 mm SL (average of 44 specimens, 62 mm SL); basidorsal spot and complicated network of interconnecting white lines present or absent (Fig. 199).

DESCRIPTION

As given for the genus.

Largest known specimen: 96.5 mm SL (NMV A.1583).

DISTRIBUTION

Kuiterichthys furcipilis is restricted to coastal waters of Victoria, the east coast of Tasmania, and New South Wales (below 32° S) (Fig. 290, Table 22). The record of Kan et al. (1995: 23) from Yule Island, Gulf of Papua, is most probably erroneous. The type locality is unknown.

Depths of capture, known for 21 specimens, range from about 9 to 240 m; 96% of these specimens, however, were taken in less than 170 m, and 85% in less than 150 m. The average depth of capture for all known specimens is 97 m.

COMMENTS

Not knowing the locality of his only specimen, Cuvier (1817b: 430), and later Valenciennes (1837: 424), confused *Kuiterichthys furcipilis*, a subtropical Australian endemic, with a Western Atlantic frogfish. Both authors agreed that their *"furcipilis"* best resembled the figure of a Brazilian antennariid published by Johannes de Laet (1633: 574) under the name *Pira Vtoewah, forma monstrosa* (later appearing as *Guaperua Brasiliensibus* in Marcgrave, 1648: 150, and as *Guaperua Brasiliensibus Margravii,* "The American Toad-Fish," in Willughby, 1686; see title page illustration). It is difficult to understand how they came to this conclusion—the only real similarity between the two is perhaps the long illicium and long, free, second dorsal-fin spine—but, on the contrary, all evidence indicates that *Pira Vtoewah,* the earliest depiction of an antennariid, represents *Antennarius multiocellatus* (see Synonymy, p. 131).

Within the known material of *Kuiterichthys furcipilis* identified by Pietsch and Grobecker (1987: 217), there are seven specimens (labeled *Kuiterichthys* sp.) that do not conform to the diagnosis of either of the two species of the genus recognized here: AMS I.3945, 1(64.5 mm), off New South Wales; LACM 11537–1, 1(31 mm), SW of Kangaroo Island; NMV A.525, 1(57 mm), off Ninety Mile Beach, Victoria; UW 20986, 3(29–44 mm), Great Australian Bight; WAM P.28197–001, 1(35 mm), Rottnest Island, Western Australia. The length of the illicium of these individuals is significantly shorter: 9.7 to 15.5% SL compared to 18.8 to 24.6% for *K. furcipilis* and 17.8 to 22.6% for *K. pietschi* (Fig. 195). A thorough search among Australian collections resulted in one additional specimen (CSIRO H.2808, 78 mm SL). It is likely that these eight individuals represent a third, undescribed species of *Kuiterichthys*, but description is postponed until additional specimens and distinguishing characters are discovered. Collectors and divers are encouraged to send us specimens, tissues for DNA analysis, and any additional information pertaining to this form.

Kuiterichthys pietschi Arnold
Crowdy Head Frogfish

Figs. 195, 200, 290; Tables 17, 22

Kuiterichthys pietschi Arnold, 2013: 496, fig. 1, table 1 (original description, 20 specimens, holotype AMS I.33555-004, 26 mm, Crowdy Head, New South Wales, Australia).

MATERIAL EXAMINED

Twenty specimens, 19 to 37 mm SL.

Holotype of *Kuiterichthys pietschi*: AMS I.33555-004, 26 mm SL, female, Crowdy Head, New South Wales, 31.9° S, 152.9° E (Fig. 200).

Paratypes of *Kuiterichthys pietschi*: AMS I.32163-002, 2 (19–24 mm), off Crowdy Head, New South Wales, 31.9° S, 152.9° E, 89 m; AMS I.26229-001, 19.5 mm, southeast of Evans Head, off Iluka, New South Wales, 29.3° S, 153.6° E; AMS I.33548-002, 4 (25–32.5 mm), off Crowdy Head, New South Wales, 31.9° S, 152.9° E; AMS I.38410-004, 25.5 mm, Clarence River, New South Wales, 29.4° S, 153.6° E, 68 m; AMS I.38477-001, 26.5 mm, off Newcastle, New South Wales, 33.1° S, 151.8° E, 67 m; AMS I.43908-001, 21 mm, Broken Bay, New South Wales, 33.472° S, 151.542° E, 60 m; AMS I.33675-003, 30.5 mm, off Crowdy Head, New South Wales, 31.9° S, 152.8° E, 84 m; AMS I.26451-003, 27 mm, Broken Bay, New South Wales, 33.6° S, 151.7° E; AMS I.32166-002, 27.5 mm, off Crowdy Head, 32.0° S, 152.8° E, 86 m; AMS I.33551-006, 29 mm, off Yamba, New South

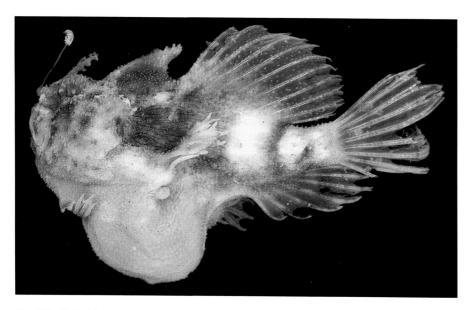

Fig. 200. *Kuiterichthys pietschi* Arnold. Holotype, AMS I.33555–004, 26 mm SL, Crowdy Head, New South Wales. Photo by Rachel J. Arnold; after Arnold (2013, fig. 1).

Wales, 29.4° S, 153.6° E; AMS I.33555-002, 2 (26.5–28 mm), off Crowdy Head, New South Wales, 31.9° S, 152.9° E; UW 151890, 23.5 mm, Clarence River, New South Wales, 29.4° S, 153.6° E, 68 m; UW 151891, 37 mm, Broken Bay, New South Wales, 33.6° S, 151.7° E; UW 151892, 25.5 mm, off Crowdy Head, New South Wales, 31.9° S, 152.8° E, 84 m.

DIAGNOSIS

A member of the genus *Kuiterichthys* unique in having the following combination of character states: esca consisting of a single, compressed, tapering appendage bearing filaments along its lateral margins and within a shallow "V-shaped" depression along its inner margin, with a tight cluster of slightly longer filaments arising from its base; illicium length 17.8 to 22.6% SL (Fig. 195); second dorsal-fin spine length 17.0 to 22.5% SL (Fig. 200); dorsal-fin rays 11 (rarely 10); anal-fin rays seven; pectoral-fin rays eight (rarely seven); vertebrae 20 (Table 17); adult body size relatively small, 19 to 37 mm SL (average of 20 specimens, 26 mm SL); basidorsal spot and complicated network of interconnecting white lines absent.

DESCRIPTION

Color in preservation: body light brown with darker brown marbling, lighter on ventral parts of head and body; a wide pale bar posterior to midpoint of body; similar pale bar across caudal peduncle; caudal fin with light brown spotting on rays, interradial membranes clear; paired fins light to dark brown proximally, lighter distally; dorsal, anal, and caudal fins of some specimens with pale bars (Fig. 200).

Remaining description as given for the genus.

Largest known specimen: 37 mm SL (UW 151891).

DISTRIBUTION

All known specimens of *Kuiterichthys pietschi* were collected off New South Wales, Australia, primarily from Crowdy Head, but ranging from southeast of Evans Head, off Iluka, to Broken Bay (Fig. 290, Table 22), at depths of 60 to 89 m (average 73 m).

Genus *Phyllophryne* Pietsch

Phyllophryne Pietsch, 1984b: 31, 39 (type species *Histiophryne scortea* McCulloch and Waite, 1918, by original designation).

DIAGNOSIS

Phyllophryne is unique among antennariids in having the anteriormost radial of the soft-dorsal fin supporting only a single fin ray (supporting two rays in all other antennariids examined). *Phyllophryne* is further distinguished from all other antennariids by having the following combination of character states: skin without dermal spinules (except for a single, tiny, crescent-shaped spinule associated with each pore of acoustico-lateralis system; Fig. 201); illicium naked, without dermal spinules; esca well developed (Fig. 202); pectoral-fin lobe broadly attached to side of body; caudal peduncle absent, the membranous posteriormost margin of soft-dorsal and anal fins attached to body at base of outermost rays of caudal fin; endopterygoid absent; pharyngobranchial I present; epural absent; pseudobranch absent; swimbladder present; dorsal-fin rays 14 to 16 (usually 15); anal-fin rays seven to nine (usually eight); outermost rays of caudal fin simple, the seven innermost bifurcate; pectoral-fin rays 10 to 12 (usually 11); vertebrae 20 to 22 (usually 21) (Tables 14, 15).

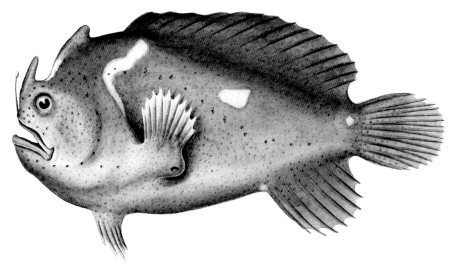

Fig. 201. *Phyllophryne scortea* (McCulloch and Waite). Holotype, SAMA F.618, 49.5 mm SL, Stansbury, Gulf St. Vincent, South Australia. Drawing by Phyllis Clarke; after McCulloch and Waite (1918, pl. 7, fig. 2); courtesy of Lea Gardam and the South Australian Museum, North Terrace, Adelaide.

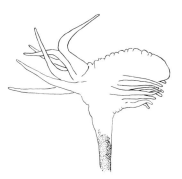

Fig. 202. Esca of *Phyllophryne scortea*. USNM 266613, 42 mm SL, Point Peron, south of Perth, thought to mimic a pontogeneiid amphipod of the genus *Gondogeneia* (see Comments, p. 289; and Pietsch and Grobecker, 1987: 244, pl. 46). Drawing by Cathy L. Short; after Pietsch and Grobecker (1987, fig. 100).

DESCRIPTION

Esca an oval-shaped, laterally directed appendage with usually six tapering filaments arising from posterior margin and a variable number of shorter, more slender filaments on anteroventral margin [strongly resembling a pontogeneiid amphipod of the genus *Gondogeneia*, perhaps *G. microdeuteropa* (Haswell); Fig. 202; see Comments, p. 289]; illicium, when laid back on head, fitting into narrow groove on either left or right side of second dorsal-fin spine, the tip of illicium (esca) coming to lie just posterior to base of second spine (esca perhaps capable of being concealed or covered by second dorsal-fin spine when spine is fully depressed), no obvious depression on head between second and third dorsal-fin spines; illicium about two-thirds length of second dorsal-fin spine, 7.1 to 12.9% SL; anterior end of pterygiophore of illicium terminating distinctly posterior to symphysis of upper jaw; illicium and second dorsal-fin spine relatively closely spaced on pterygiophore (illicium appearing to arise from base of second spine), the distance between bases of spines less than 5% SL; second dorsal-fin spine straight to slightly curved posteriorly, anteroposteriorly flattened, slightly concave on posterior surface, not connected to head by membrane, but enveloped by loose, puffy skin, length 12.0 to 21.7% SL; third dorsal-fin spine curved posteriorly, enveloped by loose, puffy skin, length 17.2

Fig. 203. *Phyllophryne scortea* (McCulloch and Waite). (A) Edithburgh Jetty, South Australia. Photo by Alexius Sutandio. (**B**) Edithburgh Jetty, South Australia. Photo by James Peake.

to 23.3% SL; eye diameter 3.8 to 6.8% SL; distal tip (about 20% of length) of maxilla tucked beneath folds of skin; head and anterior half of body with numerous, flattened cutaneous appendages (outer margin nearly always rounded, sometimes scalloped or fringed), the appendages particularly large and numerous on chin, corner of mouth, and cheek (Fig. 203); epibranchial I with a single tooth-bearing plate; ceratobranchial I toothless; vertebrae 21 or 22, caudal centra 16 or 17; dorsal-fin rays 15 (rarely 14 or 16), all simple; anal-fin rays seven to nine (usually eight), all simple; pectoral-fin rays 10 to 11 (rarely 12), all simple; all rays of pelvic fin simple (Tables 14, 15).

Coloration in life: overall cream, light yellow, yellow-green, yellow-brown, orange, red, dark-brown, to solid black, often covered with irregularly shaped, scabrous white patches especially well developed on face and anterior half of body; often two or three dark spots along base of dorsal fin and similar, but smaller and more diffuse, spots scattered over body and dorsal fin. Solid orange with irregularly shaped white blotches beneath eye and on head between second and third dorsal-fin spines; flattened cutaneous appendages of chin and corner of mouth white (Figs. 203–205). Esca sometimes blood-red (WAM P.26616-003). Ground coloration capable of sudden and dramatic change (Rudie H. Kuiter, personal communication, 15 March 1984).

Coloration in preservation: tan, yellow-brown, brown, to dark gray, often with cream-colored patches on either side of lower jaw, on head between second and third dorsal-fin spines and between third spine and soft-dorsal fin; cream-colored band extending from eye to corner of mouth sometimes present; cutaneous appendages often cream-colored, arising from center of similarly pigmented patch of skin; rarely covered everywhere with small dark, close-set spots (WAM P.7412); often two or three diffuse basidorsal spots; illicium sometimes darkly pigmented, but without banding; small bars of pigment radiating from eye sometimes present; darker-color phases with tips of all fin rays white.

A single species.

COMMENTS

Phyllophryne was established by Pietsch (1984b: 39) to accommodate a single species, *P. scortea* (McCulloch and Waite, 1918). Why this species was originally placed in the genus *Histiophryne* is difficult to imagine. Although four derived features are shared by

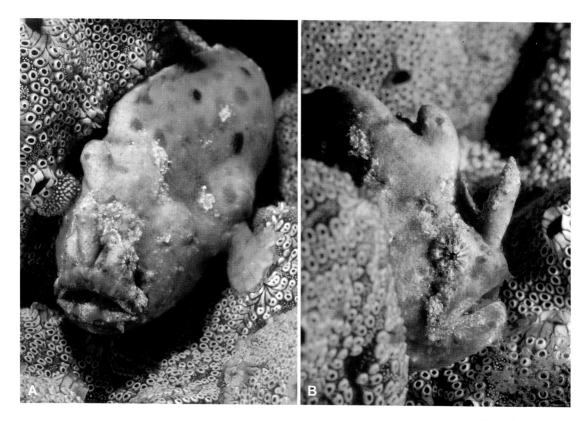

Fig. 204. *Phyllophryne scortea* (McCulloch and Waite). (**A**, **B**) Edithburgh, South Australia. Photos by David J. Hall.

Phyllophryne and *Histiophryne* (Table 14), none is by any means unique to these taxa, and three out of the four are loss character states that most likely do not reflect common ancestry. Our molecular analysis instead supports a hypothesis of sister-group relationship between *Phyllophryne* and *Echinophryne* (see Evolutionary Relationships, p. 378).

Phyllophryne scortea (McCulloch and Waite)
Smooth Frogfish

Figs. 201–205, 290, 344; Tables 14, 15, 22

Histiophryne scortea McCulloch and Waite, 1918: 74, pl. 7, fig. 2 (original description, three specimens, holotype SAMA F.618, Stansbury, Gulf St Vincent, South Australia; Fig. 201). Waite, 1921: 181, fig. 301 (after McCulloch and Waite, 1918; including *H. scortea*, var. *inconstans*). Waite, 1923: 209, fig. (after McCulloch and Waite, 1918, Waite, 1921). McCulloch, 1929: 409 (Gulf St Vincent). Schultz, 1957: 69, pl. 2, fig. D (synonymy, description after McCulloch and Waite, 1918). Mees, 1959: 10 (new to Western Australia). Scott et al., 1974: 297, fig. (description, Western and South Australia; in key). Glover, 1976: 170 (type catalog). Saunders, 2012: 308 (reference to original description).

Histiophryne scortea, **var.** *inconstans*: McCulloch and Waite, 1918: 75 (color variants; Gulf St Vincent, Kangaroo Island, South Australia). McCulloch, 1929: 409 (Gulf St Vincent, Kangaroo Island).

Phyllophryne scortea: Pietsch, 1984b: 40 (new combination, genera of frogfishes). Pietsch and Grobecker, 1987: 243, figs. 99, 100, 131, pls. 46, 47 (description, distribution, relationships). Paxton et al., 1989: 280 (southern Australia). Kuiter, 1993: 50 (southern Australia). Hutchins, 1994: 42, appendix (Western Australia). Pietsch, 1994: 293,

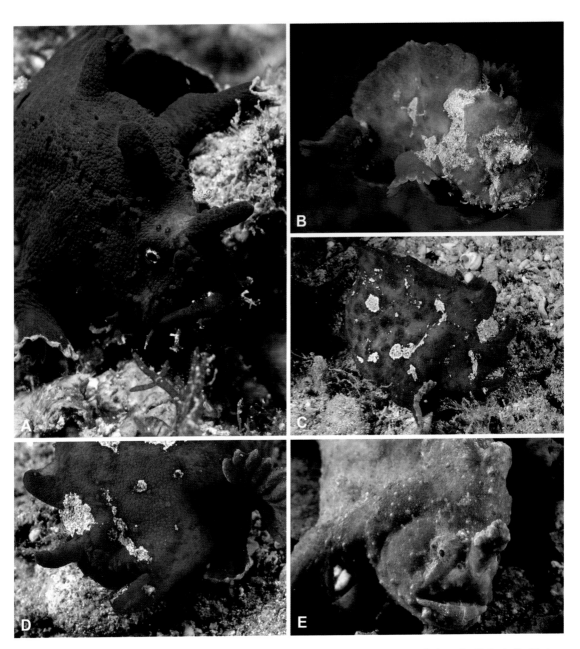

Fig. 205. *Phyllophryne scortea* (McCulloch and Waite). (**A**) Edithburgh Jetty, South Australia. Photo by James Peake. (**B–E**) Edithburgh Jetty, South Australia. Photos by John Lewis.

fig. 261 (southern Australia). Kuiter, 1997: 42 (southern Australia). Bavendam, 1998: 49, color fig. (natural history). Schleichert, 2000: 39, color fig. (natural history). Hutchins, 2001: 22 (Western Australia). Allen et al., 2006: 644 (southern Australia). Kuiter and Debelius, 2007: 125, color fig. (South Australia). Pietsch, 2008: 371, fig. (after Pietsch, 1994). Baker et al., 2009: 18, appendices 2, 3 (South Australia). Pietsch, 2009: 249 (aggressive mimicry). Currie and Sorokin, 2010: 203, 208 (Spencer Gulf, South Australia). Arnold and Pietsch, 2012: 120, 128, figs. 2A, 5, 7, table 1 (molecular phylogeny). Saunders, 2012: 308 (reference to original description). Arnold et al.,

2014: 534, table 1 (genetics, as an outgroup). Karplus, 2014: 36 (aggressive mimicry; after Pietsch and Grobecker, 1987).

MATERIAL EXAMINED

Ninety-two specimens, 16 to 78 mm SL.

Holotype of *Histiophryne scortea*: SAMA F.618, 49.5 mm, Stansbury, Gulf St Vincent, South Australia (Fig. 201).

Paratype of *Histiophryne scortea*: SAMA F.617, 48 mm, data as for holotype; SAMA F.619, 22 mm, data as for holotype.

Additional material: AMS, 8 (19–53 mm); BMNH, 1 (31 mm); NMV, 14 (16–54 mm); SAMA, 52 (17–78 mm); USNM, 1 (42 mm); UW, 1 (42 mm); WAM, 13 (18–55 mm).

DIAGNOSIS AND DESCRIPTION

As given for the genus.

Largest known specimen: 78 mm SL (SAMA F.3412).

DISTRIBUTION

Phyllophryne scortea is known only from South Australia, Victoria, and Tasmania, and the southwest corner of Western Australia. The type material was collected from Gulf St Vincent, South Australia (Fig. 290, Table 22). Depths of capture, recorded for only 12 specimens, range from 1 to 44 m, with an average depth of 13.7 m.

COMMENTS

The esca of *P. scortea* appears to mimic a pontogeneiid amphipod of the genus *Gondogeneia*, perhaps *G. microdeuteropa* (Haswell) (Pietsch and Grobecker, 1987: 244, pl. 46; Fig. 202). This discovery was made by Rudie H. Kuiter of Museums Victoria, Melbourne, based on aquarium observations of specimens collected at Edithburg Jetty, South Australia. According to Kuiter (personal communication, 26 August 1984) the esca is "a perfect copy of a swimming amphipod common along the south coast of Australia, but what makes this particularly interesting is that *P. scortea* feeds primarily on gobies of the genus *Nesogobius*, the several common South Australian species of which feed primarily on amphipods" (see Aggressive Mimicry, p. 411).

Genus *Echinophryne* McCulloch and Waite

Echinophryne McCulloch and Waite, 1918: 66 (type species *Echinophryne crassispina* McCulloch and Waite, 1918, by original designation).

Trichophryne McCulloch and Waite, 1918: 68 (type species *Antennarius mitchellii* Morton, 1897, by original designation).

DIAGNOSIS

Echinophryne is unique among antennariids in having the following combination of character states: skin covered with close-set, bifurcate dermal spinules, the length of spines of each spinule not more than twice the distance between tips of spines in *E. crassispina* and *E. reynoldsi* (Fig. 26F, H), about three to four times the distance between tips of spines in *E. mitchellii* (Fig. 26G); illicium thick, completely covered with close-set dermal spinules; distinct esca absent; pectoral-fin lobe broadly attached to side of body; caudal peduncle present, the membranous posteriormost margin of soft-dorsal and anal fins attached to body distinctly anterior to base of outermost rays of caudal fin; endopterygoid absent; pharyngobranchial I present; epural absent; pseudobranch absent; swimbladder present; dorsal-fin rays 13 to 16 (usually 14 or 15); anal-fin rays eight or nine (rarely 10); outermost rays of caudal fin simple, seven innermost bifurcate; pectoral-fin rays 10 or 11; vertebrae 21 to 23 (usually 22) (Tables 14, 15).

DESCRIPTION

Distinct esca absent, the spinule-covered illicium often terminating in a cluster of enlarged spinules and occasionally a small fleshy appendage (see species accounts below); illicium unprotected, no naked groove alongside second dorsal-fin spine, the shallow naked depression between second and third dorsal-fin spines absent; illicium about equal to or slightly longer than second dorsal-fin spine, 11.5 to 21.4% SL; anterior end of pterygiophore of illicium terminating at or slightly anterior to symphysis of upper jaw; second dorsal-fin spine not connected to head by membrane and covered with close-set dermal spinules, length 10.2 to 20.5% SL; third dorsal-fin spine curved posteriorly, length 12.5 to 28.8% SL; eye diameter 4.6 to 10.7% SL; only distal tip (about 20 to 25% of length) of maxilla naked and tucked beneath folds of skin, the remaining proximal portion covered directly with spinulose skin; wartlike patches of clustered dermal spinules absent; epibranchial I with a single tooth-bearing plate; ceratobranchial I with as many as four tooth-plates; dorsal-fin rays 14 or 15 (rarely 13 or 16), all simple (posteriormost occasionally bifurcate in E. mitchellii); anal-fin rays eight to 10 (usually nine), all simple; pectoral-fin rays 10 or 11 (usually 11), all simple; all rays of pelvic fin simple; vertebrae 21 to 23, caudal centra 16 to 18 (Tables 15, 18).

Three species, all restricted to temperate waters of Australia and Tasmania.

COMMENTS

Of the 13 antennariid genera recognized by Pietsch (1984b), two were characterized by having a spinulose illicium without a distinct esca: Echinophryne, containing at that time only E. crassispina McCulloch and Waite; and Trichophryne, containing only T. mitchellii (Morton). The two genera were distinguished from one another primarily by characters of the dermal spinules: relatively short, present in at least two size classes in Echinophryne; all more or less equal in size, but extremely long and narrow in Trichophryne. The argument against synonymizing the two was primarily the lack of uniqueness of the spinulose illicium, a condition also found in one of the two species of the genus Lophiocharon (Pietsch, 1984b: 39). However, the discovery of E. reynoldsi (Pietsch and Kuiter, 1984) created a dilemma. It too was characterized by having a spinulose illicium without an esca, but at the same time small dermal spinules, all more or less equal in size. On the evidence provided by the spinulose illicium alone, this new species appeared to

Table 18. Frequencies of Fin-Ray Counts of Species of the Genus Echinophryne

Species	Dorsal-fin rays				Anal-fin rays			Pectoral-fin rays*	
	13	14	15	16	8	9	10	10	11
E. crassispina			17	2	4	13	2	8	38
E. mitchelli	1	6			6	1			14
E. reynoldsi			8	1	6	3		17	
TOTALS	1	6	25	3	16	17	2	25	52

*Left and right sides combined.

be most closely related to *E. crassispina* or *T. mitchellii*. But in the absence of any known derived feature shared by only two of these taxa, a choice had to be made between erecting a new genus to contain the new species and synonymizing the genera to recognize only *Echinophryne* with three species. Rather than add still another monotypic genus to the family, Pietsch and Kuiter (1984) chose to synonymize *Trichophryne* McCulloch and Waite (1918: 68) with *Echinophryne* McCulloch and Waite (1918: 66). The rather narrow genetic divergence (6.49 to 6.65% within the COI gene) between *E. crassispina* and *E. mitchellii*, shown by the molecular analysis of Arnold and Pietsch (2012; see Phylogenetic Relationships, p. 380; Fig. 272), strongly supports this decision.

Key to the Known Species of the Genus *Echinophryne*

Like the other keys in this volume, the following works by progressively eliminating the most morphologically distinct taxon and for that reason should always be entered from the beginning. All features listed for each species must correspond to the specimen being keyed; if not, proceed to the next set of characters.

1. Skin covered with extremely long, narrow dermal spinules, length of spines of each spinule three to four times distance between tips of spines, giving skin a brush-like texture (Fig. 26G); tip of third dorsal-fin spine protruding beyond membranous connection to head; dorsal-fin rays 13 or 14; pectoral-fin rays 11 (Table 18) . . . . *Echinophryne mitchellii* (**Morton, 1897**), **p. 295**
    Victoria and Tasmania

2. Dermal spinules short, length of spines of each spinule not more than twice distance between tips of spines, giving skin a granular texture; spinules present in at least two distinct size classes (Fig. 26F); dorsal-fin rays 15 or 16; pectoral-fin rays 11 (rarely 10) (Table 18) . . . . *Echinophryne crassispina* **McCulloch and Waite, 1918, p. 291**
    South Australia, Victoria, Tasmania, and New South Wales

3. Dermal spinules short, length of spines of each spinule not more than twice distance between tips of spines (Fig. 26H); spinules all about equal in length (except for highly modified spinules associated with pores of acoustico-lateralis system); dorsal-fin rays 15 or 16; pectoral-fin rays 10 (Table 18) . . . .*Echinophryne reynoldsi* **Pietsch and Kuiter, 1984, p. 298**
    Western Australia, South Australia, and Victoria

Echinophryne crassispina McCulloch and Waite
Prickly Frogfish

Figs. 26F, 206–209, 289; Tables 18, 22

Echinophryne crassispina McCulloch and Waite, 1918: 67, pl. 6, fig. 2 (original description, three specimens, holotype SAMA F.609, Spencer Gulf, South Australia; paratypes apparently lost, Western Port, Victoria; Fig. 206). Waite, 1921: 182, fig. 302 (fig. after McCulloch and Waite, 1918). Waite, 1923: 208, fig. (after McCulloch and Waite, 1918). McCulloch, 1929: 408 (South Australia, Victoria). Schultz, 1957: 70, pl. 3, fig. A (after McCulloch and Waite, 1918). Scott et al., 1974: 298, fig. (South Australia, Victoria; in key). Glover, 1976: 170 (type catalog). Pietsch, 1981b: 419 (osteology). Last et al., 1983:

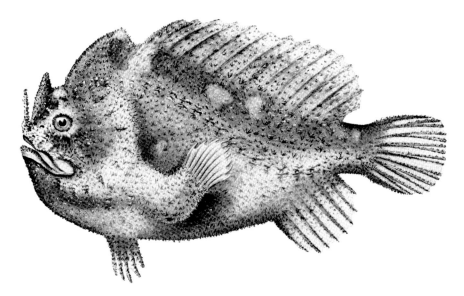

Fig. 206. *Echinophryne crassispina* McCulloch and Waite. Holotype, SAMA F.609, 36.5 mm SL, Spencer Gulf, South Australia. Drawing by Phyllis Clarke; after McCulloch and Waite (1918, pl. 6, fig. 2); courtesy of Lea Gardam and the South Australian Museum, North Terrace, Adelaide.

254, fig. 22.6 (Tasmania; in key). Pietsch, 1984b: 39, fig. 1C (genera of fishes). Pietsch and Grobecker, 1987: 236, figs. 16I, 97, 130, pl. 45 (description, distribution, relationships). Paxton et al., 1989: 279 (Australia). Kuiter, 1993: 51 (southeastern Australia). Hutchins, 1994: 42, appendix (southern Australia). Liem, 1994: 19, color photo (parental care). Pietsch, 1994: 287, fig. 255 (southern Australia). Kuiter, 1997: 42 (Australia). Liem, 1998: 19, color photo (parental care). Michael, 1998: 334 (identification, behavior, captive care). Watson et al., 2000: 120 (parental care). Hutchins, 2001: 22 (Western Australia). Kuiter and Debelius, 2007: 125, color fig. (South Australia, Victoria). Michael, 2007: 60 (dermal denticles). Pietsch, 2008: 366, fig. (after Pietsch, 1994). Baker et al., 2009: 18, fig. 10A, appendix 2 (South Australia). Arnold and Pietsch, 2012: 128, figs. 2B, 5, 7, table 1 (molecular phylogeny). Arnold et al., 2014: 534, table 1 (genetics, as an outgroup).

MATERIAL EXAMINED

Twenty-two specimens, 26 to 56 mm SL.

Holotype of *Echinophryne crassispina*: SAMA F.609, 36.5 mm, Spencer Gulf, South Australia (Fig. 206).

Additional material: AMS, 5 (24–54 mm); NMV, 8 (31–55.5 mm); SAMA, 6 (38.5–56 mm); UW, 2 (35.5–40 mm).

DIAGNOSIS

A member of the genus *Echinophryne* unique in having dermal spinules present in at least two distinct size classes (except for highly modified spinules associated with pores of acoustico-lateralis system, all are more or less equal in size in all other spinulose antennariids; compare dermal spinules shown in Fig. 26F–H).

Echinophryne crassispina differs further from its congeners in having the following combination of character states: dermal spinules relatively small, the length of spines of each spinule not more than twice distance between tips of spines (Fig. 26F); dorsal-fin rays 15 or 16; anal-fin rays eight to 10; pectoral-fin rays 10 or 11 (Table 18).

Spinule-covered illicium occasionally terminating in a small fleshy appendage (Fig. 206); illicium about equal to length of second dorsal-fin spine, 15.5 to 21.2% SL (distal tip probably missing in holotype); anterior end of pterygiophore of illicium terminating at symphysis of upper jaw; illicium and second dorsal-fin spine relatively closely spaced on pterygiophore, the distance between bases of spines less than 5% SL; second dorsal-fin spine conical, tapering to blunt point, length 14.1 to 18.2% SL; third dorsal-fin spine tapering toward base (width at distal end 6.8 to 8.5% SL), the full length connected to head by membrane; third dorsal-fin spine 22.2 to 28.8% SL; eye surrounded by ring of close-set dermal spinules, diameter 5.1 to 10.7% SL; dermal spinules of body posterior to pectoral-fin lobe distinctly larger than those of head, the spinules surrounding eye slightly larger than outlying spinules of face; tiny cutaneous appendage (usually forked at tip)

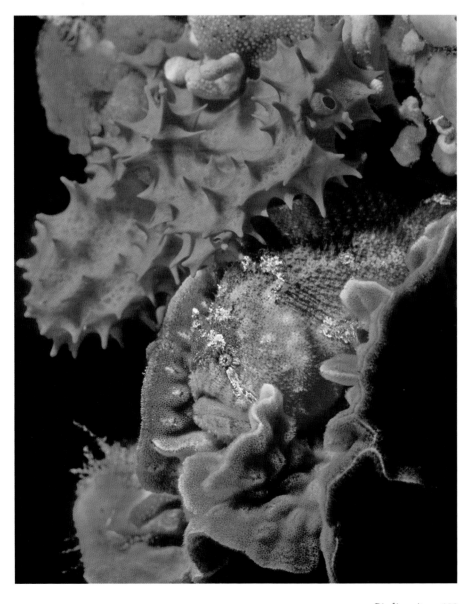

Fig. 207. *Echinophryne crassispina* McCulloch and Waite. Edithburgh, South Australia. Photo by David J. Hall.

Fig. 208. *Echinophryne crassispina* McCulloch and Waite. (**A–C**) Edithburgh, South Australia. Photos by Alexius Sutandio.

arising laterally from point of bifurcation of each of larger class of dermal spinules in largest known specimens (e.g., AMS I.19248-001); ceratobranchial I toothless; vertebrae 22 or 23, caudal centra 17 or 18; all fin rays simple (Table 18).

Coloration in life: red-orange with whitish patches on head between second and third dorsal-fin spines, and between third dorsal-fin spine and soft-dorsal fin; a similar patch on dorsal margin of caudal peduncle (Figs. 207–209).

Coloration in preservation: pinkish-yellow-brown to brown, the upper half of head and body somewhat darker than lower half, particularly on belly; dark brown pigment often present on head above eyes, nape of neck, anterior surface of dorsal-fin spines, space between third dorsal-fin spine and soft-dorsal fin, and on two or three anteriormost dorsal-fin rays; usually a dark saddle on caudal peduncle and occasionally a dark patch on cheek under eye; basidorsal spot absent; illicium without banding; small bars of dark pigment radiating from eye in some specimens.

Remaining description as given for the genus.

Largest known specimen: 56 mm SL (SAMA F.1726).

DISTRIBUTION

Echinophryne crassispina is restricted to the coastal waters of southern and eastern Australia, from the Recherche Archipelago, Western Australia, to Jervis Bay, New South Wales, including Tasmania (Fig. 289, Table 22). The type locality is Spencer Gulf, South Australia. Depths of capture, known for seven specimens, range from 5.0 to 18.3 m; the average depth was 12.7 m.

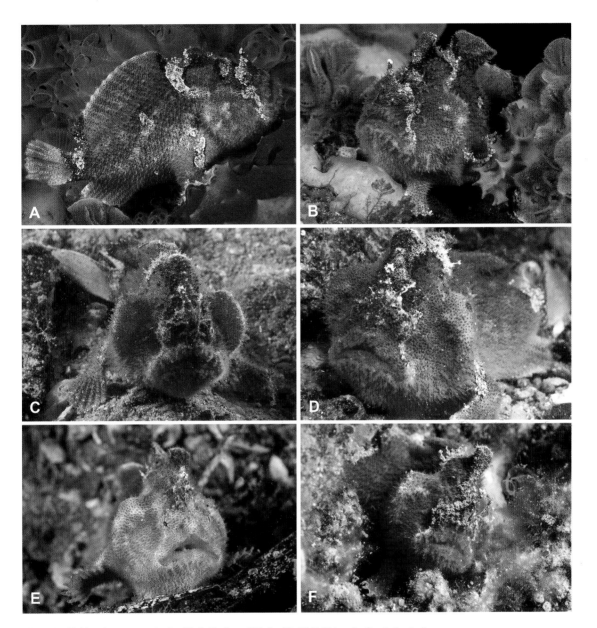

Fig. 209. *Echinophryne crassispina* McCulloch and Waite. (**A, B**) Edithburgh, South Australia. Photos by David J. Hall. (**C, D**) Rapid Bay, South Australia. Photos by Diana Fernie. (**E**) Edithburgh, South Australia. Photo by Diana Fernie. (**F**) Edithburgh, South Australia. Photo by Alexius Sutandio.

Echinophryne mitchellii (Morton)
Mitchell's Frogfish

Figs. 26G, 210, 211, 289; Tables 18, 22

Antennarius mitchellii Morton, 1897: xiv, 98 (original description, single specimen, holotype TMH D.136, east coast of Tasmania). Andrews, 1971: 3 (type catalog). Andrews, 1992: 110 (type catalog). Saunders, 2012: 186, 409 (after Morton, 1897).

Trichophryne mitchelli: McCulloch and Waite, 1918: 68, pl. 6, fig. 1 (new combination, description, additional material, two specimens, Brighton Beach, South Australia, and off Wilsons Promontory, Victoria; Fig. 210). Waite, 1921: 183, fig. 303 (after Mc-Culloch and Waite, 1918). Waite, 1923: 211, fig. (after Waite, 1921). McCulloch, 1929: 480 (Tasmania, South Australia, New South Wales). Schultz, 1957: 65, pl. 1, fig. C (synonymy, description after McCulloch and Waite, 1918). Last et al., 1983: 256, fig. 22.9 (Tasmania; in key). Pietsch, 1984b: 39, fig. 1D (genera of frogfishes). Pietsch, 2008: 374, fig. (southern Australia). Saunders, 2012: 186, 409 (as *T. mitchellii*; reference to original description).

Trichophryne mitchelli (not of Morton): Whitley, 1929a: 357 (AMS IA.3614 is *Kuiterichthys furcipilis*). Whitley, 1934, no. CLIV (after Whitley, 1929a).

Echinophryne mitchelli: Scott et al., 1974: 298, fig. (new combination; South Australia, Victoria, New South Wales, Tasmania; in key). Pietsch and Kuiter, 1984: 24 (*Trichophryne* synonymized; in key). Pietsch and Grobecker, 1987: 239, figs. 16J, 98, 131 (description, distribution, relationships). Paxton et al., 1989: 288 (southern Australia). Kuiter, 1993: 51 (southeastern Australia). Pietsch, 1994: 288, fig. 256 (southern Australia). Hutchins, 2001: 22 (Western Australia). Allen et al., 2006: 641 (Australia). Kuiter and Debelius, 2007: 124, color figs. (Tasmania). Arnold and Pietsch, 2012: 128, figs. 5, 7, table 1 (molecular phylogeny). Arnold et al., 2014: 534, table 1 (genetics, as an outgroup).

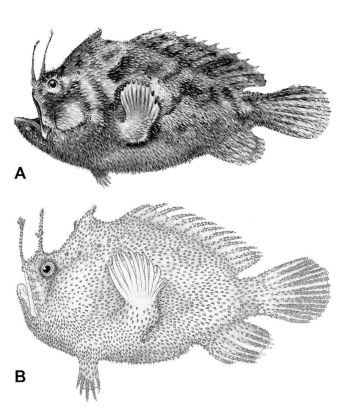

Fig. 210. *Echinophryne mitchellii* (Morton). (**A**) Brighton Beach, South Australia. Drawing by Phyllis Clarke; after McCulloch and Waite (1918, pl. 6, fig. 1); courtesy of Lea Gardam and the South Australian Museum, North Terrace, Adelaide. (**B**) AMS E.1261, 58.5 mm SL, Wilsons Promontory, Victoria. Drawing by Cathy L. Short; after Pietsch and Grobecker (1987, fig. 98).

MATERIAL EXAMINED

Fifteen specimens, 33 to 111 mm SL.

Holotype of *Antennarius mitchellii*: TMH D.136, 92 mm, Lisdillon, east coast of Tasmania, Mitchell, 1895.

Additional material: AMS, 1 (58.5 mm); NMV, 3 (80–111 mm); SAMA, 2 (89–94 mm); TMH, 8 (33–81 mm).

DIAGNOSIS

A member of the genus *Echinophryne* unique among antennariids in having skin covered with extremely long, narrow dermal spinules (some greater than 4% SL in large specimens), the length of spines of each spinule about three to four times distance between tips of spines (Fig. 26G). *Echinophryne mitchellii* differs further from its congeners in having the following combination of character states: dermal spinules all more or less equal in size (except for highly modified spinules associated with pores of acoustico-lateralis system); dorsal-fin rays 13 or 14; anal-fin rays eight or 9; pectoral-fin rays 11 (Table 18).

DESCRIPTION

Spinule-covered illicium terminating in a cluster of somewhat enlarged dermal spinules and occasionally a few, elongate cutaneous appendages (Figs. 210, 211); illicium about as long as second dorsal-fin spine, 17.0 to 21.4% SL; anterior end of pterygiophore of illicium terminating at or slightly anterior to symphysis of upper jaw; illicium and second dorsal-fin spine widely spaced on pterygiophore, the distance between bases of spines 5.9 to 11.1% SL; second dorsal-fin spine narrow (only slightly wider than illicium), covered with relatively widely spaced dermal spinules and usually terminating in a tight cluster of spinules and cutaneous appendages; length of second dorsal-fin spine 15.4 to 20.5% SL; third dorsal-fin spine equal in width throughout length, the distal end protruding beyond membranous connection to head; length of third dorsal-fin spine 12.5 to 21.4% SL; eye surrounded by ring of separate clumps of tightly clustered dermal spinules, diameter 4.6 to 6.7% SL; a tiny, cutaneous appendage (forked, or more complexly branched at tip) arising laterally from point of bifurcation of dermal spinules, especially spinules of chin and unpaired fins; ceratobranchial I with remnants of one or more tooth-plates; vertebrae 21 or 22, caudal centra 16 or 17; posteriormost ray of soft-dorsal fin occasionally bifurcate, all other rays simple (Table 18).

Coloration in preservation: beige, light pinkish-brown, yellow, to yellow-brown; upper part of head and body usually slightly darker than lower part, with irregular (but bilaterally symmetrical), dark brown to black mottling, especially on face and dorsal surface of head (face, dorsal surface of head, second and third dorsal-fin spines, and soft-dorsal fin gray in NMV A.1566); a conspicuous, dark basidorsal spot present in some specimens (well developed in NMV A.1566, and on left side of NMV A.540); illicium without banding; no pigment bars radiating from eye (Fig. 211).

Remaining description as given for the genus.

Largest known specimen: 111 mm SL (NMV A.1566).

DISTRIBUTION

All of the known material of *Echinophryne mitchellii* was collected in coastal waters of South Australia (off Newland Head), Victoria (Wilsons Promontory and Ninety Mile Beach), Flinders Island in the Bass Strait, and off the east coast of Tasmania as far south

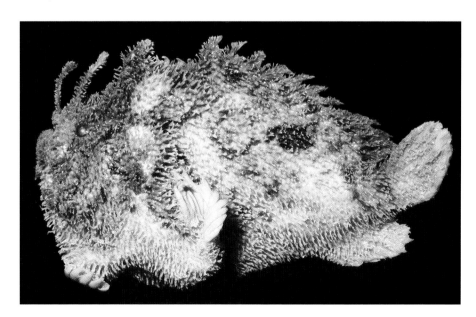

Fig. 211. *Echinophryne mitchellii* (Morton). NMV A.1566, 111 mm SL, near Flinders Island, Bass Strait, Tasmania. Photo by Rudie H. Kuiter.

as Bruny Island (Fig. 289, Table 22). The type locality is Lisdillon, Tasmania. The largest known specimen (NMV A.1566, 111 mm SL) was trawled from a depth of 69.5 m.

Echinophryne reynoldsi Pietsch and Kuiter
Sponge Frogfish

Figs. 26H, 212–214, 289; Tables 18, 22

Echinophryne reynoldsi Pietsch and Kuiter, 1984: 24, figs. 1–3 (original description, nine specimens, holotype NMV A.3212, Port Phillip Bay, Victoria; Figs. 212, 213). Pietsch and Grobecker, 1987: 234, figs. 16H, 96, 130, pls. 43, 44 (description, distribution, relationships). Paxton et al. 1989: 279 (southern Australia). Kuiter 1993: 51 (southeastern Australia). Pietsch, 1994: 289, fig. 257 (southern Australia). Hutchins 2001: 22 (Western Australia). Allen et al. 2006: 642 (southern Australia). Kuiter and Debelius, 2007: 124, color figs. (Victoria). Pietsch, 2008: 367, fig. (after Pietsch, 1994). Baker et al., 2009: 17, 18, fig. 10B, appendix 2 (South Australia). Arnold and Pietsch, 2012: 128 (molecular phylogeny, revised classification).

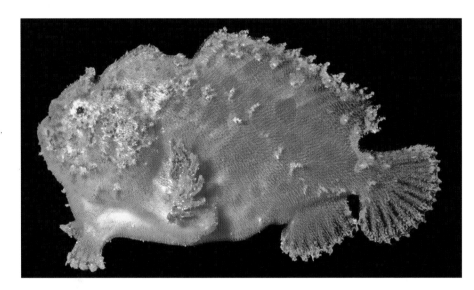

Fig. 212. *Echinophryne reynoldsi* Pietsch and Kuiter. Holotype, NMV A.3212, 52 mm SL, Portsea Hole, Port Phillip Bay, Victoria. Drawing by Cathy L. Short; after Pietsch and Grobecker (1987, fig. 96).

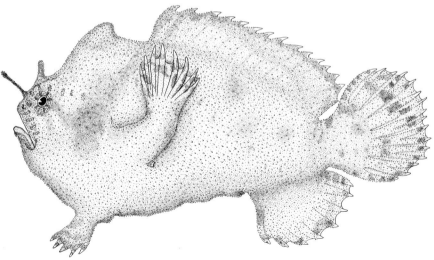

Fig. 213. *Echinophryne reynoldsi* Pietsch and Kuiter. Holotype, NMV A.3212, 52 mm SL, Portsea Hole, Port Phillip Bay, Victoria, found in a sponge at 20 m. Photo by Rudie H. Kuiter.

MATERIAL EXAMINED

Twelve specimens, 16.5 to 57.5 mm SL.

Holotype of *Echinophryne reynoldsi*: NMV A.3212, 52 mm, Portsea Hole, Port Phillip Bay, Victoria, in large orange sponge at 20 m, Reynolds, 4 August 1983 (Figs. 212, 213).

Paratypes of *Echinophryne reynoldsi*: NMV A.3064, 4 (16.5–31.5 mm, 28.5-mm and 31.5-mm specimens cleared and stained), Portsea, Victoria, quinaldine, rocky ledges from 3–15 m, Kuiter, January 1982; NMV A.2840, 1 (24.5 mm), Crawfish Rock, Victoria, 31 May 1970; NMV A.3209, 1 (37 mm), data as for holotype; NMV A.2843, 1 (57.5 mm), Westernport Bay, Victoria, Watson. SAMA F.3645, 1 (36.5 mm), Pondalowie Bay, Yorke Peninsula, South Australia, Walker, 19 February 1970.

Additional material: NMV, 1 (45 mm); SAMA, 1 (36 mm), WAM, 1 (29 mm).

DIAGNOSIS

A member of the genus *Echinophryne* unique in having the following combination of character states: dermal spinules small, the length of spines of each spinule not more than twice the distance between tips of spines (Fig. 26H); spinules all more or less equal in size (except for highly modified spinules associated with pores of acoustico-lateralis system); dorsal-fin rays 15 or 16; anal-fin rays eight or nine; pectoral-fin rays 10 (one side of one specimen, SAMA F.3645, with eight pectoral-fin rays, probably owing to injury) (Table 18).

DESCRIPTION

Spinule-covered illicium usually terminating in a cluster of slightly larger dermal spinules and a small, darkly pigmented cutaneous appendage; illicium slightly longer than second dorsal-fin spine, 11.5 to 17.1% SL; anterior end of pterygiophore of illicium terminating at symphysis of upper jaw; illicium and second dorsal-fin spine relatively closely spaced on pterygiophore, the distance between bases of spines less than 5% SL; second dorsal-fin spine short and stout, somewhat tapering, length 10.2 to 15.6% SL; third dorsal-fin spine tapering slightly toward base, its full length connected to head by membrane, length 17.4 to 22.0% SL; eye diameter 5.1 to 10.7% SL; tiny remnants of pharyngeal tooth-plates present (one on epibranchial I, two to four on ceratobranchial I, and one on ceratobranchials II and III); vertebrae 22, caudal centra 17 (Table 18).

Coloration in life: pale yellow to orange-brown over most of head, body, and fins, with ventral portion of head slightly lighter than rest of body; large, irregularly shaped, pale gray patches around eye, on cheek, and on margins of jaw; similar patches on body, one above base of pectoral fin, a second extending from behind head to base of anteriormost dorsal-fin rays, a third on dorsal margin of caudal peduncle; distal margin of dorsal, pectoral, and pelvic fins grayish-white, bordered in some specimens by a purplish-brown band on dorsal fin and a pale-brown band on paired fins; body and dorsal fin often with numerous pale to dark brown, eye-sized ocelli, each encircled with a white to yellow ring; occasionally a distinct white blotch just dorsal to base of pectoral fin (Fig. 214).

Coloration in preservation: cream to light gray, or pinkish-yellow-brown to brown, the upper half of head and body somewhat darker than lower; gray to black pigment usually present on margin of jaws, above eye, on cheek, and especially on anterior margin of second and third dorsal-fin spines; a band of dark gray pigment along distal margin of soft-dorsal fin, usually extending posteriorly as a dark saddle on caudal peduncle; flecks of dark gray to black on rays of pectoral and pelvic fins.

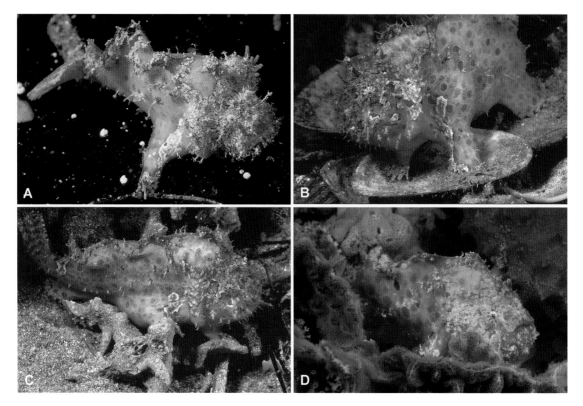

Fig. 214. *Echinophryne reynoldsi* Pietsch and Kuiter. (**A–C**) Edithburgh, South Australia. Photos by John Lewis. (**D**) Edithburgh, South Australia. Photo by David J. Hall.

Remaining description as given for the genus.
Largest known specimen: 57.5 mm SL (NMV A.2843).

DISTRIBUTION

Echinophryne reynoldsi was originally described from nine cataloged specimens and an unregistered tenth (now SAMA F.4781; Pietsch and Kuiter, 1984, fig. 3) collected at Yorke Peninsula, South Australia, and Port Phillip and Westernport bays, Victoria (Fig. 289, Table 22). At least two additional specimens have subsequently been identified, one from Edithburg Jetty, Yorke Peninsula, and a second that extends the known range of this species westward to Lucky Bay, Recherche Archipelago, Western Australia. The type locality is Port Phillip Bay. Depths of capture, recorded for six specimens, range from 5 and 20 m.

COMMENTS

Echinophryne reynoldsi is a reclusive, seldom observed species that appears to associate with yellow and orange encrusting sponges (Pietsch and Kuiter, 1984: 26). The holotype and one paratype (NMV A.3209) were collected from within a large orange sponge. Four additional paratypes (NMV A.3064) were collected when quinaldine was used within caves and under ledges where yellow sponges were abundant.

Genus *Tathicarpus* Ogilby

Tathicarpus Ogilby, 1907: 19 (type species *Tathicarpus butleri* Ogilby, 1907, by subsequent designation of McCulloch, 1929).

DIAGNOSIS

Tathicarpus is unique among antennariids in having a Y-shaped ectopterygoid (T-shaped in all other antennariids); epibranchial I with a row of six to 11 teeth borne directly on bone (not on separate tooth-plates as in other antennariids); proximal end of second pectoral radial reduced, not contributing to articulation of pectoral fin and girdle; all nine rays of caudal fin simple; and only six or seven pectoral-fin rays (eight or more in all other antennariids) (Tables 14, 15).

Tathicarpus is further distinguished from all other antennariids by having the following combination of character states: skin covered with close-set, bifurcate dermal spinules, the length of spines of each spinule not more than twice the distance between tips of spines (Fig. 26I); illicium naked, without dermal spines; esca distinct; pectoral-fin lobe detached from side of body for most of its length (Fig. 215); caudal peduncle present, the membranous posteriormost margin of soft-dorsal and anal fins attached to body distinctly anterior to base of outermost rays of caudal fin; endopterygoid absent; pharyngobranchial I present; epural absent; pseudobranch present; swimbladder absent; dorsal-fin rays 11 (rarely 10); anal-fin rays seven; all rays of caudal fin simple (extreme distal tip of dorsalmost ray bifurcate in some specimens); pectoral-fin rays seven (rarely six); vertebrae 18 (Tables 14, 15).

DESCRIPTION

Esca a thin, broad, membranous appendage bearing numerous fine, hairlike filaments along distal margin and one or two dark, eyelike pigment spots at base (Fig. 216); esca as long as 22.2% SL (KFRL FO-3887); illicium unprotected, no naked groove alongside second dorsal-fin spine, the shallow naked depression between second and third dorsal-fin spines absent; illicium length 24.0 to 46.9% SL, one and a half to nearly twice length of second dorsal-fin spine (Fig. 215); anterior end of pterygiophore of illicium extending beyond symphysis of upper jaw; illicium and second dorsal-fin spine relatively closely spaced on pterygiophore, the distance between bases of spines less than 5% SL; second dorsal-fin spine narrow, strongly curved posteriorly, the anterior and lateral surfaces covered with close-set dermal spinules, connected to head by thin, usually translucent membrane; length of second dorsal-fin spine 15.0 to 28.1% SL; third dorsal-fin spine curved posteriorly,

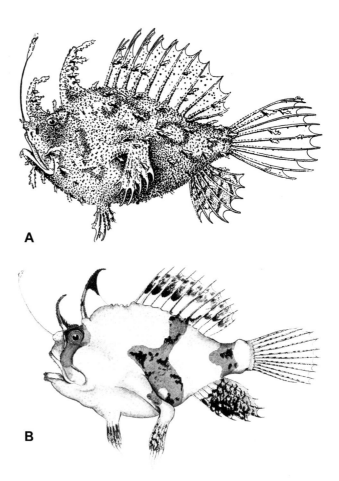

A

B

Fig. 215. *Tathicarpus butleri* Ogilby. (**A**) AMS I.13621, 56.5 mm SL, Shark Bay, Western Australia. Drawing by Allan R. McCulloch; after McCulloch (1915, pl. 37, fig. 4). (**B**) about 48 mm SL, Cairns, Queensland, Australia. Watercolor by George James Coates; courtesy of Rick Feeney and the Los Angeles County Museum of Natural History.

Fig. 216. Esca of *Tathicarpus butleri* Ogilby. WAM P.14663, 65 mm SL, Shark Bay, Western Australia. Drawing by Cathy L. Short; after Pietsch and Grobecker (1987, fig. 109).

equal in width throughout length and connected to head by thin, usually translucent membrane; length of third dorsal-fin spine 24.6 to 37.9% SL; eye with a few small clusters of dermal spinules on periphery, particularly along anterior margin, diameter 6.7 to 9.1% SL; only distal tip (about 20 to 25% of length) of maxilla naked and tucked beneath folds of skin, the remaining proximal portion covered directly with spinulose skin; opercular opening tubelike, extending considerably out from surface of body; large, fringed, cutaneous appendages scattered over surface of head, body, and fins; wartlike patches of clustered dermal spinules absent; distal one third to one half of pectoral-fin rays free, not interconnected by membrane, membrane interconnecting proximal portion of rays thin and translucent; all fin rays exceptionally long, the longest dorsal ray 30.8 to 42.1% SL, longest anal 30.6 to 39.6% SL, longest caudal 40.8 to 58.9% SL, longest pectoral 25.0 to 35.4% SL, longest pelvic 16.1 to 22.4% SL; epibranchial I with a row of eight to 11 teeth borne directly on bone (not on separate tooth-plates as in other antenariids); ceratobranchial I toothless; vertebrae 18, caudal centra 13; dorsal-fin rays 11 (rarely 10), as many as nine anteriormost rays bifurcate at tip (in all other antenariids examined, bifurcation of soft-dorsal and anal-fin rays proceeds from the posterior end rather than anterior); anal-fin rays seven, occasionally the anteriormost one or two rays bifurcate at tip; pectoral-fin rays seven (rarely six), all simple; all rays of pelvic fin simple (Tables 14, 15).

Coloration in life: almost white to light gray with large yellow-green bands through eye, on body extending from middle of soft-dorsal fin to belly, and across caudal peduncle; similar yellow-green coloration on second and third dorsal-fin spines, anal and paired fins, and pectoral-fin lobe; yellow-green markings of flank, caudal peduncle, and fins overlaid with irregularly shaped black reticulations; distal portion of third dorsal-

Fig. 217. *Tathicarpus butleri* Ogilby. Great Barrier Reef, Australia. Photo by Roger Steene.

fin spine black; black and brown markings on membranes between rays of soft-dorsal fin (Figs. 217, 218). Other variations from slate gray to brilliant orange (Ogilby 1907: 22).

Coloration in preservation: cream, beige, pinkish-brown, yellow-brown, to dark brown, everywhere densely speckled or mottled with darker brown to black; belly somewhat lighter-colored than head and body; dark mottling sometimes denser on fins than body; often a dark brown to black patch on anterior end of soft-dorsal and anal fins; a black, bilobed patch devoid of dermal spinules nearly always present just dorsal to base of pectoral-fin lobe; basidorsal spot absent; illicium usually banded; short bars of pigment occasionally present on periphery of eye.

A single species.

Tathicarpus butleri Ogilby
Butler's Frogfish

Figs. 26I, 215–218, 289; Tables 14, 15, 22

Tathicarpus butleri Ogilby, 1907: 20 (original description, single specimen, holotype QMB I.861, Port Curtis, Queensland). McCulloch, 1929: 405 (Queensland). Schultz, 1957: 64, pl. 1, fig. A (after

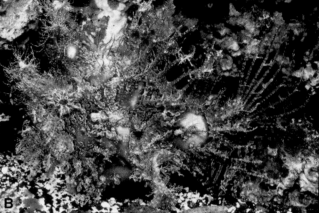

Fig. 218. *Tathicarpus butleri* Ogilby. (**A, B**) aquarium specimens, Brisbane, Australia. Photos by Scott W. Michael.

Ogilby, 1907). Whitley, 1964: 57, 60 [listed as synonym of *T. subrotundatus* (Castelnau)]. Kailola, 1975: 48 (Gulf of Papua). Kailola and Wilson, 1978: 26, 58 (after Kailola, 1975). Pietsch, 1981b: 419 (osteology). Pietsch, 1984b: 41 (genera of frogfishes). Pietsch and Grobecker, 1987: 259, figs. 108, 109, 132, pls. 54, 55 (description, distribution, relationships). Paxton et al., 1989 (Australia). Derbyshire and Dennis, 1990: 92 (prawn predation, Torres Strait). Blaber et al., 1994: 392 (Gulf of Carpentaria). Martin et al., 1995: 916 (Gulf of Carpentaria). Larson and Williams, 1997: 349 (Darwin Harbour, Northern Territory). Pietsch, 1999: 2015 (western central Pacific; in key). Hutchins, 2001: 22 (Western Australia). Allen and Adrim, 2003: 26 (Pulau Ujir, Moluccas). Allen et al., 2003: 367, color fig. (New Guinea, northern Australia). Shedlock et al., 2004: 134, 144, table 1 (molecular phylogeny). Allen and Erdmann, 2012: 155, color figs. (East Indies). Arnold, 2012: 63, fig. 4 (genetics, phylogeny). Arnold and Pietsch, 2012: 120, 128, figs. 2C, 5, 7, table 1 (molecular phylogeny). Larson et al., 2013: 54 (Northern Territory). Brandl and Bellwood, 2014: 32 (pair-formation). Moore et al., 2014: 177 (northwestern Australia).

Tathicarpus muscosus Ogilby, 1907: 22 [original description, single specimen, holotype lost (see Ogilby, 1912: 64), 98 mm, Port Curtis, Queensland]. Ogilby, 1912: 64 (description, additional material, QMB I.3448, Wide Bay, Queensland; holotype lost).

McCulloch, 1915: 277, pl. 37, fig. 4 (description, additional material, AMS I.13621, Shark Bay, Western Australia; Fig. 215A). McCulloch, 1929: 405 (Port Curtis, Queensland). Whitley, 1964: 60 [listed as synonym of *T. subrotundatus* (Castelnau)]. Marshall, 1964: 514, 516, pl. 64, fig. 497 (Queensland; Shark Bay, Western Australia; NW Australia).

Tathicarpus appeli Ogilby, 1922: 302, pl. 19, fig. 2 ("Scribbled Angler," original description, single specimen, holotype QMB I.3183, Wide Bay, SE Queensland). McCulloch, 1929: 406 (after Ogilby, 1922). Whitley, 1964: 60 [as synonym of *T. subrotundatus* (Castelnau)].

Tathicarpus mucosum: Schultz, 1957: 64, pl. 1, fig. A (in synonymy of *T. butleri*, misspelling of specific name).

MATERIAL EXAMINED

Ninety-three specimens, 26 to 102 mm SL.

Holotype of *Tathicarpus butleri*: QMB I.861, 63.5 mm, north end of Port Curtis Harbor, Queensland, Butler.

Holotype of *Tathicarpus appeli*: QMB I.3183, 58 mm, Wide Bay, Queensland, Appel.

Additional material: AMS, 17 (36–79 mm); KFRL, 3 (33–58.5); NMNZ, 8 (44–66 mm); QMB, 1 (44.5 mm); SAMA, 1 (64.5 mm); USNM, 1 (48 mm); UW, 2 (54–65.5 mm); WAM, 58 (26–102 mm).

DIAGNOSIS AND DESCRIPTION

As given for the genus.

Largest specimen examined: 102 mm SL (WAM P.5432).

DISTRIBUTION

As presently understood, *Tathicarpus butleri* ranges from the eastern Moluccan island of Ujir to Papua New Guinea and south to the coasts of Northern Territory (including the Gulf of Carpentaria and Torres Strait), Western Australia (as far south as about 33° S latitude), and Queensland (as far south as about 26° S latitude) (Fig. 289, Table 22). Depths of capture, known for 16 specimens, range from 7.3 to 146 m, with an average depth of 38 m; 14 of these records were made at depths of about 70 m or less, 11 in 45 m or less.

COMMENTS

Without explanation, Whitley (1964: 60) declared *T. butleri* to be a junior synonym of *T. subrotundatus* (Castelnau, 1875). How he came to this decision is a mystery, since Castelnau's description is so inadequate (glossing over the diagnostic features and providing no illustration) that it could not support such a hypothesis, and no type material seems to have survived (not found in any Australian institution or in the MNHN). A search through Whitley's literature card file (maintained in the Department of Ichthyology, Australian Museum, Sydney) shows that, at least up until 1957, when his reference to Schultz's (1957) revision of the Antennariidae was made, he recognized *T. butleri*, making no mention of *T. subrotundatus*. Having checked all possible sources, we have concluded that the basis for Whitley's (1964) statement cannot be found, and in the absence of any evidence to support his contention, Castelnau's (1875) name is listed as a *nomen dubium* (see p. 356). The name *T. butleri* (Ogilby, 1907) is therefore recognized as the oldest available name for this species, following the conclusion of Pietsch and Grobecker (1987).

Genus *Lophiocharon* Whitley

Lophiocharon Whitley, 1933: 104 [type species *Lophiocharon broomensis* Whitley, 1933 (= *Cheironectes trisignatus* Richardson, 1844), by original designation].

Plumantennarius Schultz, 1957: 89 [type species *Antennarius asper* Macleay, 1881 (= *Cheironectes trisignatus* Richardson, 1844), by original designation].

DIAGNOSIS

Lophiocharon is unique among antennariids in having the illicial bone and second dorsal-fin spine widely separated, the distance between bases of these elements about 50% of the length of pterygiophore of the illicium (less than about 35% of pterygiophore length in all other antennariids). This genus also differs from all other antennariid genera in having a row of two to four regularly spaced, more or less translucent (transparent in some specimens) ocelli on the membrane between each rays of the caudal fin (best observed by spreading the caudal rays), the illicium nearly always curved anteriorly and the anterior tip of the pterygiophore of the illicium distinctly upturned (Fig. 221).

Lophiocharon is further distinguished from all other antennariids in having the following combination of character states: skin covered with close-set, bifurcate dermal spinules, the length of spines of each spinule not more than twice the distance between tips of spines (Fig. 26J); illicium naked (*L. trisignatus*) or lightly covered throughout its length with dermal spinules (*L. hutchinsi* and *L. lithinostomus*); esca large and complex (*L. trisignatus*) or greatly reduced or absent (*L. hutchinsi* and *L. lithinostomus*); pectoral-fin lobe broadly attached to side of body; caudal peduncle absent, the membranous posteriormost margin of soft-dorsal and anal fins attached to body at base of outermost rays of caudal fin; endopterygoid absent; pharyngobranchial I absent; epural absent; swimbladder present; dorsal-fin rays 12 or 13 (usually 13); anal-fin rays six to eight (usually seven); all rays of caudal fin bifurcate; pectoral-fin rays eight or nine (usually nine); vertebrae 19 (Tables 14, 15).

DESCRIPTION

Esca a tuft of cylindrical to somewhat flattened, often multibranched appendages arising from common base (*L. trisignatus*) or greatly reduced (or absent) and scarcely differentiated from illicium (*L. hutchinsi* and *L. lithinostomus*); illicium unprotected, no naked groove alongside second dorsal-fin spine, the shallow naked depression between second and third dorsal-fin spines absent; illicium slightly greater than to more than twice length of second dorsal-fin spine, 9.6 to 36.4% SL (Fig. 219); upturned anterior tip of pterygiophore of illicium terminating slightly posterior to symphysis of upper jaw; illicium and second dorsal-fin spine widely spaced on pterygiophore, the distance between bases of spines 5.7 to 8.2% SL; second dorsal-fin spine slightly tapering toward base, strongly curved posteriorly, completely covered with close-set dermal spinules, connected to head and to proximal portion of third dorsal-fin spine by thick spinulose membrane; length of second dorsal-fin spine 6.8 to 17.2% SL; third dorsal-fin spine strongly curved posteriorly, slightly tapering toward base, fully attached to nape of neck and to anteriormost ray of soft-dorsal fin by thick spinulose membrane, length 19.8 to 24.7% SL; small, relatively widely spaced clusters of dermal spinules on periphery of eye; eye diameter 4.1 to 5.7% SL; only distal tip (about 25 to 30% of length) of maxilla naked and tucked beneath folds of skin, the remaining proximal portion covered directly with spinulose skin; numerous fringed cutaneous appendages scattered on head and body,

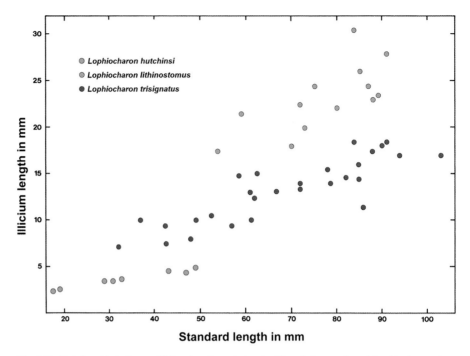

Fig. 219. Relationship between illicium length and standard length for species of *Lophiocharon*. *L. hutchinsi* (green), N = 8; *L. lithinostomus* (blue), N = 13; *L. trisignatus* (red), N = 28. Modified after Pietsch and Grobecker (1987, fig. 91).

Table 19. Frequencies of Fin-Ray Counts of Species of the Genus *Lophiocharon*

Species	Dorsal-fin rays		Anal-fin rays			Pectoral-fin rays*	
	12	13	6	7	8	8	9
L. hutchinsi		9		9			18
L. lithinostomus	1	12		13		2	24
L. trisignatus	3	53	2	52	2	1	111
TOTALS	4	74	2	74	2	3	153

*Left and right sides combined.

particularly conspicuous on chin and corner of mouth, some arising laterally from point of bifurcation of dermal spinules; wart-like patches of clustered dermal spinules absent; epibranchial I toothless; ceratobranchial I toothless; vertebrae 19, caudal centra 14; dorsal-fin rays 13 (rarely 12), all bifurcate; anal-fin rays seven (rarely six or eight), all bifurcate; pectoral-fin rays nine (rarely eight), all simple; all rays of pelvic fin simple (Tables 14, 15, 19).

Coloration in preservation: beige, yellow-orange, yellow-brown, brown, purplish, or reddish-brown to solid black. Light-color phases everywhere covered with dense mottling of darker brown; basidorsal spot and light-colored bar across base of caudal fin sometimes present; illicium with or without light banding. Dark-color phases usually with white to yellow-brown along outermost margin of soft-dorsal, anal, and caudal fins; illicium with or without dark banding; each ocellus of caudal fin usually encircled by a dark brown to black ring (Fig. 221); caudal ocelli faint or absent in solid-black individuals; dark pigment bars sometimes radiating from eye.

Three species, all restricted to the Indo-Australian Archipelago.

Key to the Known Species of the Genus *Lophiocharon*

This key, like the other keys in this volume, works by progressively eliminating the most morphologically distinct taxon and for that reason should always be entered from the beginning. All characters listed for each species must correspond to the specimen being keyed; if not, proceed to the next set of character states.

1.   Illicium naked, not covered with dermal spinules, length 13.4 to 24.8% SL (Figs. 219); esca well developed, morphologically complex (Fig. 226) . . . .
. . . .*Lophiocharon trisignatus* (**Richardson, 1844**), **p. 313**
Indo-Australian Archipelago

2.   Illicium long, length 21.6 to 36.4% SL, more or less covered from base to tip with dermal spinules (Figs. 219, 221); esca greatly reduced, scarcely if at all differentiated from illicium . . . .
. . . . *Lophiocharon lithinostomus* (**Jordan and Richardson, 1908**), **p. 309**
North Borneo and Sulu Archipelago to the Philippines

3.   Illicium short, length 9.6 to 12.9% SL, covered from base to tip with dermal spinules (Fig. 220); esca greatly reduced, scarcely if at all differentiated from illicium. . . . .
. . . .*Lophiocharon hutchinsi* **Pietsch, 2004a, p. 307**
Northern Australia and Aru Islands

Lophiocharon hutchinsi Pietsch
Hutchins's Frogfish

Figs. 219, 220, 286; Tables 19, 22

Antennarius caudimaculatus (not of Rüppell): Weber, 1913: 562 (misidentification, ZMA 116.513, Aru Islands).

Lophiocharon sp.: Pietsch and Grobecker, 1987: 224, 231, fig. 91 (five specimens, probably representing an undescribed species).

Lophiocharon hutchinsi Pietsch, 2004a: 160, figs. 1–2, pl. 1 (original description, nine specimens, holotype WAM P.27673-002, 43 mm, James Price Point, north of Broome, Western Australia; Fig. 220). Allen et al., 2006: 644 (Australia). Moore et al., 2009: 9 (type catalog). Arnold and Pietsch, 2012: 128 (molecular phylogeny, revised classification). Larson et al., 2013: 54 (Northern Territory). Moore et al., 2014: 177 (northwestern Australia).

MATERIAL EXAMINED

Nine specimens, 14 to 49 mm SL.

Holotype of *Lophiocharon hutchinsi*: WAM P.27673-002, 43 mm, James Price Point, 55 km north of Broome, Western Australia, 17°26′S 122°10′E, rotenone, 31 July 1982 (Fig. 220).

Paratypes of *Lophiocharon hutchinsi*: AMS I.15557-283, 2 (47–49 mm), Gulf of Carpentaria, Queensland, 17°00′S 140°16′E; WAM P.24486, 1 (33 mm), Exmouth Gulf, Western Australia, July 1973; AMS I.34780-001, 1 (14 mm), Lee Point, Darwin Country, Northern

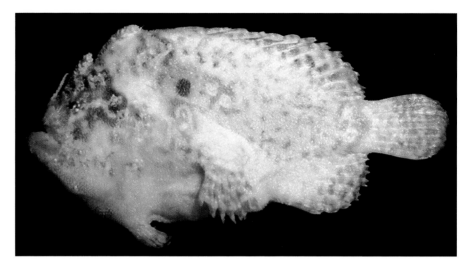

Fig. 220. *Lophiocharon hutchinsi* Pietsch. Holotype, WAM P.27673–002, 43 mm SL, James Price Point, 55 km north of Broome, Western Australia. Photo by T. W. Pietsch; after Pietsch (2004a, pl. 1).

Territory, 12°20′S 130°53′E, scuba, 3.0 m, 11 July 1993; WAM P.31884-001, 1 (31 mm), south of Bluff Point, Enderby Island, Dampier Archipelago, Western Australia, 20°40.9′S 116°33.2′E, box dredge on sandy mud, sponge, and seagrass, 9.0–9.2 m, 23 July 1999; WAM P.28416-015, 2 (19–29 mm), Gantheaume Point, Broome, Western Australia, 17°58′S 122°10′E, rotenone, 2–5 m, 13 September 1982; ZMA 116.513, 1 (18 mm), anchorage off Jedan Island, Aru Islands, Indonesia, Weber, 1899.

DIAGNOSIS

A member of the genus *Lophiocharon* unique in having the following combination of character states: illicium short, length 9.6 to 12.9% SL (Figs. 219, 220), covered from base to tip with dermal spinules; esca greatly reduced, scarcely if at all differentiated from illicium.

DESCRIPTION

Esca with a single tiny distal filament; length of second dorsal-fin spine 14.4 to 15.8% SL; length of third dorsal-fin spine 19.7 to 23.4% SL; distance between bases of illicium and second dorsal-fin spine 6.5 to 7.6% SL; diameter of eye 5.2 to 7.7% SL; dorsal-fin rays 13; anal-fin rays seven; pectoral-fin rays nine (Table 19).

Coloration in preservation: all known specimens in light-color phase: cream, beige, light yellow-brown to brown overall; dorsal and lateral surfaces, including fins, everywhere covered with speckles and mottling of darker brown, especially dense on face around eye; basidorsal spot and pale bar across base of caudal fin absent; illicium without banding; one or two dark circular spots on side of body above or slightly behind base of pectoral-fin lobe; dark streak sometimes radiating out from eye; caudal ocelli faint in preserved material, discernible only in largest known specimens (Fig. 220).

Additional description as given for the genus.

Largest known specimen: 49 mm SL (AMS I.15557-283).

DISTRIBUTION

Lophiocharon hutchinsi is restricted to northern Australian and southern New Guinean waters, with records extending from Western Australia, at Exmouth Gulf, to the Dampier Archipelago, and Broome to Northern Territory, at Lee Point, near Darwin; Queensland in the Gulf of Carpentaria; and the Aru Islands in the Arafura Sea (Fig. 286, Table 22). The type locality is James Price Point, north of Broome, Western Australia.

A relatively shallow-living species, all specimens for which depth data are known were collected in less than 10 m.

COMMENTS

Within the material of *Lophiocharon* examined by Pietsch and Grobecker (1987) there were five specimens that did not conform to the diagnosis of either of the two recognized species of the genus. Concluding that this material represented a third, undescribed species of *Lophiocharon*, but uncomfortable at the time about describing a new species based on only a few small individuals, the material was labeled *Lophiocharon* sp. and set aside pending the discovery of additional specimens (Pietsch and Grobecker, 1987: 231). The known material having nearly doubled in subsequent years, the species was described and named by Pietsch in 2004 in honor of Barry Hutchins, then curator of fishes at the Western Australian Museum, Perth.

Lophiocharon hutchinsi is clearly distinguished from *L. trisignatus* in lacking a distinct esca and in having a shorter, spiny illicium (Figs. 219, 220). On the other hand, the only feature that separates it from *L. lithinostomus* is its considerably shorter illicium (9.6 to 12.9% SL vs. 21.6 to 36.4% SL; Fig. 219). Because all nine known specimens of *L. hutchinsi* are small (less than 50 mm SL) and all known individuals of *L. lithinostomus* are relatively large (54 to 93 mm SL), it might be argued that *L. hutchinsi* simply represents juvenile specimens of the latter. However, if this were true it would necessitate an extremely rapid ontogenetic increase in illicial length, the evidence for which is lacking in all other lophiiform fishes for which adequate material has been studied. It should be pointed out also that *L. lithinostomus* and *L. hutchinsi*, as presently understood, are allopatric in distribution: the former ranges from North Borneo to the Sulu Archipelago and the Philippines, whereas the latter is known only from northern Australia and the Aru Islands (Fig. 286). The third known species of *Lophiocharon, L. trisignatus*, is sympatric with both its congeners, ranging from tropical Australia to the Philippines.

Lophiocharon lithinostomus (Jordan and Richardson)
Marble-Mouth Frogfish

Figs. 219, 221–224, 286, 338; Tables 19, 22

Antennarius lithinostomus Jordan and Richardson, 1908: 235, 286, fig. 12 (original description, single specimen, holotype SU 20204, Cuyo Island, Philippines; Fig. 221). Herre, 1934: 106 (Sitankai). Bôhlke, 1953: 148 (type at CAS). Herre, 1953: 854 (Sulu Province, Panay).

Antennarius chironectes (not of Latreille): Weber, 1913: 562 (in part, larger specimen, ZMA 101.875, Sulu Archipelago).

Lophiocharon (*Lophiocharon*) *caudimaculatus* (not of Rüppell): Schultz, 1957: 82 (in part, i.e., USNM 164243, 164360, North Borneo and Philippines). Schultz, 1964: 176, 178, table 1 (in part, i.e., SU 27872, North Borneo).

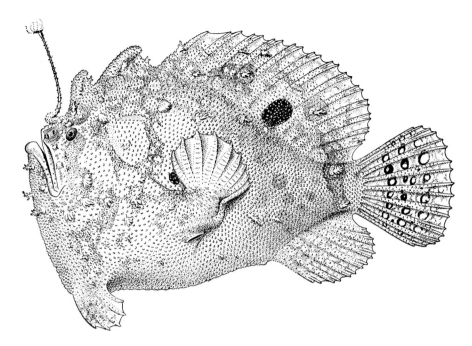

Fig. 221. *Lophiocharon lithinostomus* (Jordan and Richardson). Holotype, SU 20204, 85 mm SL, Cuyo Island, Philippines (note that a well-developed esca has been added by the artist; see Comments, p. 312). Drawing by William S. Atkinson; after Jordan and Richardson (1908, fig. 12).

Lophiocharon caudimaculatus (not of Rüppell): Le Danois, 1964: 111, figs. 54, 55g (in part, i.e., MNHN A.2286, Island Soulou).

Antennarius caudimaculatus (not of Rüppell): Pietsch and Grobecker, 1980: 551 (in part, i.e., USNM 164243, North Borneo; misidentification, parental care). Thresher, 1984: 42 (after Pietsch and Grobecker, 1980). Yoneda et al., 1998: 94 (misidentification, after Pietsch and Grobecker, 1980). Arnold et al., 2014: 534, table 1 (reference to Pietsch and Grobecker, 1980).

Lophiocharon lithinostomus: Pietsch, 1984b: 39 (new combination, genera of frogfishes). Pietsch and Grobecker, 1987: 229, figs. 91, 95, 130 (description, distribution, relationships). Michael, 1998: 334, 357, color fig. (identification, behavior, captive care). Pietsch, 1999: 2015 (western central Pacific; in key). Pietsch, 2004a: 159 (comparison with congeners; in key). Pietsch et al., 2009a: 37, 45, fig. 5 (genetics, as an outgroup). Arnold and Pietsch, 2011: 64, fig. 4 (genetics, phylogeny). Allen and Erdmann, 2012: 154, color figs. (East Indies). Arnold, 2012: 63, fig. 4 (genetics, phylogeny). Arnold and Pietsch, 2012: 120, 128, figs. 2D, 5, 7, table 1 (molecular phylogeny). Townsend, 2015: 4 (Sabah, Malaysia).

MATERIAL EXAMINED

Sixteen specimens, 54 to 93 mm SL.

Holotype of *Antennarius lithinostomus*: SU 20204, 85 mm, Cuyo Island, Philippines, McGregor (Fig. 221).

Additional material: BPBM, 1 (75 mm); FMNH, 1 (89 mm); MNHN, 1 (82 mm); SU, 7 (59–91 mm); UMMZ, 1 (81 mm); USNM, 2 (72–80 mm); UW, 1 (93 mm); ZMA, 1 (54 mm).

DIAGNOSIS

A member of the genus *Lophiocharon* unique in having the following combination of character states: illicium relatively long, length 21.6 to 36.4% SL (Figs. 219, 221), covered from base to tip with dermal spinules; esca greatly reduced, scarcely if at all differentiated from illicium (Fig. 221).

DESCRIPTION

Esca with or without one or more tiny distal filaments; length of second dorsal-fin spine 6.2 to 17.8% SL; length of third dorsal-fin spine 17.8 to 23.7% SL; distance between bases of illicium and second dorsal-fin spine 5.8 to 8.0% SL; diameter of eye 4.2 to 5.5% SL; dorsal-fin rays 13, rarely 12; anal-fin rays seven; pectoral-fin rays nine, rarely eight (Table 19).

Coloration in life: "Mottled and blotched with dark brown and gray; lips grass green; dorsal, caudal, and anal largely green; a pink spot above pectoral; the whole fish with the appearance of an alga-covered rock, even the interior of the mouth being mottled" (R. C. McGregor, in Jordan and Richardson, 1908: 287); mottled green with row of two to four large, white, dark-edged spots on membrane between each pair of caudal-fin rays; often resembling an algal-covered rock (Figs. 222–224).

Remaining description as given for the genus.

Largest known specimen: 93 mm SL (UW 115749).

DISTRIBUTION

The known material of *L. lithinostomus* was all collected from one general locality extending from Panay Island and the southern tip of the Sulu Archipelago (Philippines) to Brunei and Sabah (northern Borneo), and the Raja Ampat Islands (West Papua). The holotype is from Cuyo Island in the Philippines (Fig. 286, Table 22). Apparently a rather shallow-living frogfish, it is found nearly at the surface to a depth of about 10 m.

COMMENTS

Prior to the antennariid revision of Pietsch and Grobecker (1987), *Lophiocharon lithinostomus*, described by Jordan and Richardson (1908) from a single specimen, had not been recognized by subsequent authors, Herre (1934, 1953) being the only exception. Although clearly distinct from its congener, this species remained in synonymy for all this time owing largely to its rarity but also because its authors failed to mention, in their otherwise detailed description, the

Fig. 222. *Lophiocharon lithinostomus* (Jordan and Richardson). Aquarium specimen, Philippines. Photo by Scott W. Michael.

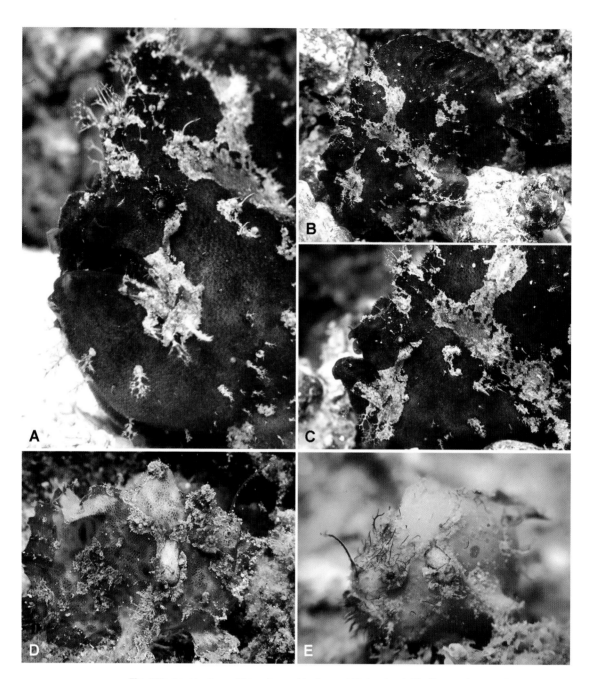

Fig. 223. *Lophiocharon lithinostomus* (Jordan and Richardson). (**A–C**) aquarium specimen, Philippines. Photos by Scott W. Michael. (**D, E**) Brunei. Photos by Cindy Tan.

diagnostic presence of dermal spinules along the length of the illicium. Moreover, Jordan and Richardson (1908, fig. 12) unfortunately (but understandably) failed to realize that their species differed from nearly all other antennariids in the severe reduction or absence of the esca. Assuming that the esca had been lost in their only specimen, they apparently instructed their artist to recreate this structure in their drawing of the holotype. Their figure was modified by Pietsch and Grobecker (1987: 231, fig. 95) to reflect this absence, but it is reproduced here in the original (Fig. 221).

Fig. 224. *Lophiocharon lithinostomus* (Jordan and Richardson). (**A**, **B**) Malapascua Island, Cebu, Philippines. Photos by Teresa Zuberbühler.

Lophiocharon trisignatus (Richardson)
Three-Spot Frogfish

Figs. 26J, 219, 225–230, 286, 339; Tables 19, 22

Cheironectes trisignatus Richardson, 1844: 15, pl. 9, fig. 1 (original description, single specimen, holotype, about 29 mm, lost, King George's Sound, Western Australia, after Macleay, 1881: 579; neotype designated by Pietsch and Grobecker, 1987: 229, WAM P.27274-013, 74 mm, Broome, Western Australia; Fig. 225A). Hutchins and Smith, 1991: 9 (type catalog). Moore et al., 2009: 9 (type catalog).

Chironectes caudimaculatus (not of Rüppell): Richardson, 1848: 125, pl. 60, figs. 8, 9 (description, additional material, Australia).

Antennarius urophthalmus Bleeker, 1851: 472, 488 (original description, single specimen, holotype lost, not in RMNH, 120 mm total length, Bintan, Riouw Archipelago; here synonymized on authority of Bleeker, 1865b: 15). Bleeker, 1859b: 130 (listed). Günther, 1861a: 192 (Singapore, Australia). Macleay, 1878: 356 (Port Darwin, Australia). Macleay, 1881: 578 (Port Darwin). McCulloch, 1929: 407 (Northern Territory, Australia).

Antennarius caudimaculatus (not of Rüppell): Bleeker, 1855a: 12 (new combination, eastern Australia). Bleeker, 1865b: 9, 15, pl. 197, fig. 6 (additional material; Banka, Biliton, Bintan, Singapore; Fig. 225B). Beaufort and Briggs, 1962: 207 [in part, synonymy includes references to Rüppell, "1835–1840" (= 1829, 1838), and Klunzinger (1871, 1884); Indonesia, Philippines, Australia; in key]. Pietsch and Grobecker, 1980: 551, figs. 1, 2 (in part, USNM 164243 is *L. lithinostomus*; parental care). Allen and Swainston, 1988: 38 (northwestern Australia).

Antennarius trisignatus: Bleeker, 1855a: 12 (new combination, eastern Australia). Macleay, 1881: 579 (after Richardson, 1844; type locality, King George's Sound, Western Australia). McCulloch, 1929: 407 (King George's Sound). Whitley, 1948: 31 (Western Australia).

Antennarius lindgreeni Bleeker, 1855b: 192 (original description, single specimen, holotype lost, not found at RMNH, no size given, Muntok, Banka Island, Sumatra; here synonymized on authority of Bleeker, 1865b: 15). Bleeker, 1855c: 425 (Banka). Günther, 1861a: 192 (after Bleeker, 1855b).

Chironectes trisignatus: Günther, 1861a: 184 (footnote reference to original description).

Antennarius candimaculatus (not of Rüppell): Günther, 1880: 295, figs. 109, 110 (figures after Richardson, 1848; misspelling of specific name).

Antennarius caudomaculatus (not of Rüppell): Günther, 1880: 474 (misspelling of specific name).

Antennarius asper Macleay, 1881: 580 (original description, single specimen, holotype AMS I.16427-001, Darnley Island, Torres Strait, Queensland). McCulloch, 1929: 407 (after Macleay, 1881). Whitley, 1941: 46, pl. 2, fig. 30 (additional specimen, AMS IA.3718, Murray Island, Queensland). Whitley, 1954: 28 (AMS IB.2419, Darwin, Northern Territory). Mees, 1960: 20 (new to Western Australia). Marshall, 1964: 514, 515 (Thursday Island, Queensland).

Lophiocharon broomensis Whitley, 1933: 104, pl. 25, fig. 1 (original description, single specimen, holotype AMS IA.5562, Broome, Western Australia). Whitley, 1948: 31 (Western Australia).

Lophiocharon (*Lophiocharon*) *caudimaculatus* (not of Rüppell): Schultz, 1957: 82, pl. 6, fig. A (new combination; in part, USNM 164243 and 164360 are *L. lithinostomus*; synonymy includes reference to Rüppell, 1835 (= 1829, 1838); North Borneo; in key). Schultz, 1964: 176, 178, table 1 (in part, SU 27872 is *L. lithinostomus*; Singapore, Philippines; *A. asper* Macleay a synonym, *Plumantennatus* a synonym of *Lophiocharon*).

Antennarius (*Plumantennatus*) *asper*: Schultz, 1957: 89, pl. 8, fig. C (synonymy, description; Singapore, west of Port Darwin; in key). Palmer, 1961: 550 (additional specimen, Monte Bello Islands, Western Australia).

A

B

Fig. 225. *Lophiocharon trisignatus* (Richardson). **(A)** holotype, specimen lost, about 29 mm SL, King George's Sound, Western Australia. After Richardson (1844, pl. 9, fig. 1). **(B)** Bleeker's *Antennarius caudimaculatus*, specimen apparently lost. Drawing by Ludwig Speigler under the direction of Pieter Bleeker; after Bleeker (1865b, pl. 197, fig. 6).

Lophiocharon caudimaculatus (not of Rüppell): Le Danois, 1964: 111, fig. 55g (in part, i.e., Red Sea locality of Rüppell, 1835 (= 1829, 1838), MNHN A.2286 is *L. lithinostomus*, MNHN 1954-25 is *A. coccineus*).

Uniantennatus caudimaculatus (not of Rüppell): Le Danois, 1970: 90, fig. 30D (material is *A. coccineus*).

Antennarius hispidus (not of Bloch and Schneider): Burgess and Axelrod, 1973a: 528, color pl. 485 (plate represents *L. trisignatus*). Wheeler, 1975, pl. 182 (plate represents *L. trisignatus*).

Lophiocharon trisignatus: Pietsch, 1984b: 39 (new combination, genera of frogfishes). Pietsch and Grobecker, 1987: 224, figs. 16G, 91–94, 130, pl. 42 (description, distribution, relationships). Paxton et al., 1989: 280 (Australia). Pietsch and Grobecker, 1990a: 99, fig. (aggressive mimicry). Pietsch and Grobecker, 1990b: 76, 80, fig. (after Pietsch and Grobecker, 1990a). Hutchins, 1994: 42, appendix (Western Australia). Allen, 1997: 62 (tropical Australia). Larson and Williams, 1997: 349 (Darwin Harbour, Northern Territory). Michael, 1998: 330, 333, 357, color fig. (identification, behavior, captive care). Pietsch, 1999: 2015 (western central Pacific; in key). Allen, 2000a: 62, pl. 12-14 (Indo-Australian Archipelago). Pietsch, 2000: 597 (South China Sea). Hutchins, 2001: 22 (Western Australia). Allen and Adrim, 2003: 25 (Kalimantan to Riau Islands, Indonesia). Allen et al., 2003: 368, color fig. (Asian Pacific). Pietsch, 2004a: 159 (comparison with congeners; in key). Ortiz-Ramirez et al., 2005: 23 (early life history). Kuiter and Debelius, 2007: 124, color figs. (Western Australia). Allen and Erdmann, 2009: 594 (West Papua). Pietsch et al., 2009a: 37, 45, fig. 5 (genetics, as an outgroup). Arnold and Pietsch, 2011: 64, fig. 4 (genetics, phylogeny). Allen and Erdmann, 2012: 143, 155, color figs. (East Indies). Arnold, 2012: 63, fig. 4 (genetics, phylogeny). Arnold and Pietsch, 2012: 128, figs. 5, 7, table 1 (molecular phylogeny). Kwik, 2012: 93 (Sentosa Island, Singapore). Larson et al., 2013: 54 (Northern Territory). Moore et al., 2014: 177 (northwestern Australia). Ng et al., 2015: 330 (eastern Johor Strait, Singapore). Toh et al., 2016: 86, 92, appendix 1 (Raffles Marina, Singapore).

MATERIAL EXAMINED

Ninety-nine specimens, 8.0 to 147 mm SL.

Neotype of *Cheironectes trisignatus*: WAM P.27274-013, 74 mm, Gantheaume Point, Broome, Western Australia, Sarti et al., 18 January 1981.

Holotype of *Antennarius asper*: AMS I.16427-001, 75 mm, *Chevert* Expedition, Darnley Island, Torres Strait, 1875.

Holotype of *Lophiocharon broomensis*: AMS IA.5562, 93 mm, Broome, Western Australia, Bourne, 1932.

Additional material: AMS, 7 (69.5–128 mm); ANSP, 2 (91–92 mm); BMNH, 10 (25.5–93 mm); CSIRO, 1 (64 mm); FMNH, 3 (62.5–90 mm); MCZ, 1 (56.5 mm); NMV, 1 (96 mm); NTM, 8 (8–121 mm); RMNH, 4 (78–114 mm); SAMA, 2 (82–123 mm); SU, 17 (38.5–116 mm); USNM, 1 (84 mm); UW, 5 (61–106 mm); WAM, 35 (32–147 mm); ZMA, 1 (88 mm); ZMUC, 2 (85–122 mm).

DIAGNOSIS

A member of the genus *Lophiocharon* unique in having the following combination of character states: illicium relatively short, length 13.4 to 24.8% SL (Fig. 219), naked, without dermal spinules; esca large, morphologically complex (Fig. 226).

Fig. 226. Esca of *Lophiocharon trisignatus* (Richardson). Neotype, WAM P.27274–013, 74 mm SL, Gantheaume Point, Broome, Western Australia. Drawing by Cathy L. Short; after Pietsch and Grobecker (1987, fig. 94).

DESCRIPTION

Esca consisting of a tuft of cylindrical to somewhat flattened, often multi-branched appendages arising from common base; length of second dorsal-fin spine 6.8 to 17.2% SL; length of third dorsal-fin spine 19.8 to 24.7% SL; distance between bases of illicium and second dorsal-fin spine 5.7 to 8.2% SL; diameter of eye 4.1 to 5.7% SL; dorsal-fin rays 13, rarely 12; anal-fin rays seven, rarely six or eight; pectoral-fin rays nine, rarely eight (Table 19).

Coloration in life: brown, greenish- or reddish-brown on upper half of head and body, gradually turning to yellow on mid-body, tan, cream to almost white on chin and belly; numerous dark brown to black reticulations scattered over entire head, body, and fins; each ocellus of caudal fin encircled with dark brown to black ring (Fig. 225); in some specimens (e.g. the neotype, WAM P.27274-013) three dark spots on side of body—the explanation for Richardson's (1844) name *trisignatus*—one spot basidorsal, a second just dorsal to base of pectoral-fin lobe, a third spot about halfway between base of pectoral-fin lobe and base of caudal fin (Fig. 226A). Dark reticulations of dorsal half of body mixed with orange-red blotches; dark brown patches on face, on flank immediately dorsal and ventral to pectoral-fin lobe, and on anterior portion of anal fin (Figs. 227–229).

Remaining description as given for the genus.

Largest known specimen: 147 mm SL (WAM P.16773).

Fig. 227. *Lophiocharon trisignatus* (Richardson). Off Exmouth, Western Australia. Photo by Warren Taylor.

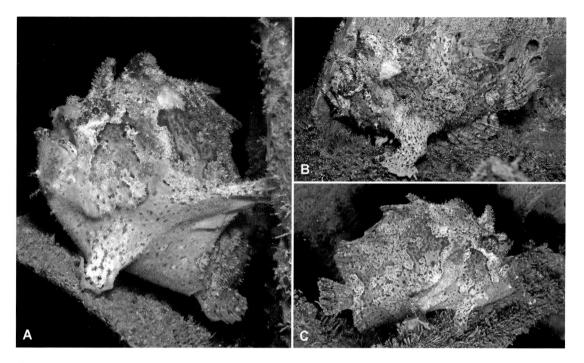

Fig. 228. *Lophiocharon trisignatus* (Richardson). (**A–C**) Raja Ampat, West Papua, Indonesia. Photos by Ned DeLoach.

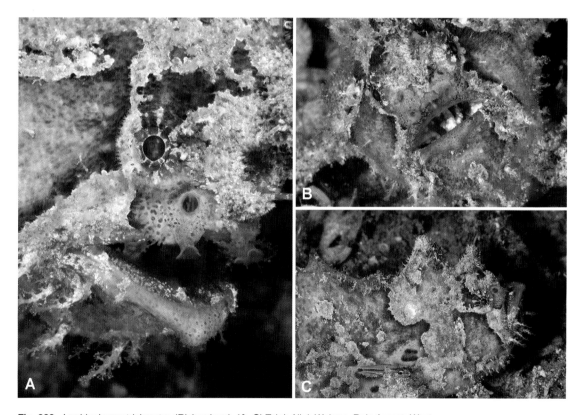

Fig. 229. *Lophiocharon trisignatus* (Richardson). (**A–C**) Teluk Aljui, Waigeo, Raja Ampat, West Papua, Indonesia. Note (**A**) the large elaborate nostrils, each with a double opening, that appear to mimic a tunicate (Robert Delfs, personal communication, 4 November 2017); notice also (**C**) the tiny solitary Red Spot cardinalfish (*Ostorhinchus parvulus*) swimming by its side. Photos by Robert Delfs.

DISTRIBUTION

Lophiocharon trisignatus is restricted to the Indo-Australian Archipelago, ranging from Singapore, Malaysia, the Philippines (Luzon and Bilatan Islands), and throughout Indonesia (including the Riau Islands and Kalimantan to West Papua) to Western Australia, Northern Territory, and Queensland). The original type locality (specimen now lost) is said to be King George's Sound, Western Australia (Macleay, 1881: 579), but we have not examined material collected along this coast south of Shark Bay (WAM P.25168-002). The neotype, designated by Pietsch and Grobecker (1987: 229), is from Broome, Western Australia (Fig. 286, Table 22). Depths range from about 2 to 20 m.

COMMENTS

Although remarkably accurate drawings of *L. trisignatus* are known from the turn of the 18th century (Fig. 230) and originally described as *Cheironectes trisignatus* from a single Western Australian specimen by Richardson (1844), this species was recognized for nearly 140 years under the specific name *caudimaculatus* (see Synonymy, p. 315), an epithet first attached to a frogfish from the Red Sea by Rüppell (1829; see also Rüppell, 1838). The cause of this confusion was probably Richardson (1848) himself, who, four years after his description of *C. trisignatus*, referred a second specimen of his species to *C. caudimaculatus* Rüppell. This decision no doubt prompted subsequent authors to assume that Richardson's (1844) *C. trisignatus* was a synonym of Rüppell's species described some nine years earlier. It turns out, however, that *C. caudimaculatus* Rüppell is a synonym of *Antennarius commerson* (Latreille, 1804; see Comments, p. 123, and Fig. 83); thus, the name *trisignatus*, resurrected by Pietsch and Grobecker (1987), is now accepted for this highly distinctive species. Because Richardson's (1844) holotype has not survived, a neotype was designated by Pietsch and Grobecker (1987: 229) to maintain nomenclatorial stability: WAM P.27274-013, 74 mm, from Broome, Western Australia.

Genus *Histiophryne* Gill

Histiophryne Gill, 1879b: 222 (type species *Chironectes bougainvilli* Valenciennes, 1837, by original designation).

Golem Whitley, 1957b: 70 (type species *Antennarius cryptacanthus* Weber, 1913, by original designation).

Xenophrynichthys Schultz, 1957: 81 (type species *Antennarius cryptacanthus* Weber, 1913, by original designation).

DIAGNOSIS

Histiophryne is unique among antennariids in having the second and third dorsal-fin spines bound down to the surface of the cranium by skin, emerging only as low protuberances on top of the head and nape of the neck; the posteriormost margin of the soft-dorsal and anal fins extending beyond the base of the caudal fin and connecting to the proximal portion of the outermost rays of the caudal fin (Tables 14, 15); and eggs carried in a "brood pouch" formed by the soft-dorsal, anal, and caudal fins.

Histiophryne is distinguished further from other antennariid genera by the following combination of character states: dermal spinules of skin usually absent (rough to the touch in about 45% of the material examined), but when present tiny and simple (Fig. 26K); illicium reduced (often discernable only with the aid of a microscope and sometimes completely hidden beneath the skin of the head), naked, without dermal spinules; esca present or absent, when present tiny and often difficult to distinguish from illicium, sometimes hidden beneath skin; pectoral-fin lobe broadly attached to side of body; caudal peduncle absent, the membranous posteriormost margin of soft-dorsal and anal fins broadly connected to proximal portion of outermost rays of caudal fin; endopterygoid absent; pharyngobranchial I absent; epural absent; pseudobranch absent; swimbladder present; dorsal-fin rays 13 to 16 (tropical endemics 13, rarely 14; Australian endemics 14 to 16); anal-fin rays seven to nine (tropical endemics seven, rarely eight; Australian endemics eight or nine, rarely seven); outermost rays of caudal fin simple, only seven innermost bifurcate; pectoral-fin rays eight or nine; vertebrae 19 to 23 (tropical endemics 20; Australian endemics 21 to 23) (Tables 14, 15).

DESCRIPTION

Illicium and esca, when present, laid back onto head, usually fitting into narrow cavity (sometimes hidden by folds of skin) on mid-dorsal line (cavity extending onto dorsal surface of second dorsal-fin spine in some specimens of *H. bougainvilli*), groove alongside second dorsal-fin spine absent, depression between second and third dorsal-fin spines absent; illicium short (length never more than 9% and usually less than 5% of SL; Fig. 231) to minute (evident only with aid of a microscope); esca, when present, a small oval or lanceolate appendage, the surface often covered with irregular, longitudinal folds of tissue; anterior end of pterygiophore of illicium terminating considerably posterior to sym-

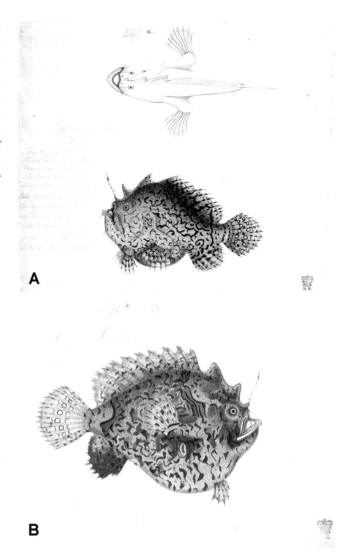

Fig. 230. The earliest known depictions of *Lophiocharon trisignatus* (Richardson). (**A**) BMNH 1982.3.8.32, 82 mm SL, dorsal and lateral views, "The Habits of this Fish in the Manner of taking its Food are very singular." (**B**) specimen unknown, about 57 mm SL. Drawings by unknown artists based on specimens observed at Singapore, taken from the voluminous collection of illustrations of Asiatic zoology amassed by Major-General Thomas Hardwicke (1756–1835) between approximately 1790 and 1820; bequeathed to the British Museum in 1835, these drawings are now part of the manuscript collections of the museum's Zoological Library. Courtesy of Stephan Atkinson and the Picture Library, Natural History Museum, London.

physis of upper jaw; eye diameter 3.2 to 7.8% SL; only distal tip (about 25% of length) of maxilla tucked beneath folds of skin; cutaneous appendages usually absent; thick, raised, wartlike patches of skin sometimes present on head and body (when present, remarkably symmetrical when sides of specimen are compared, particularly on head); rays of paired fins interconnected nearly to tip by membrane; epibranchial I toothless;

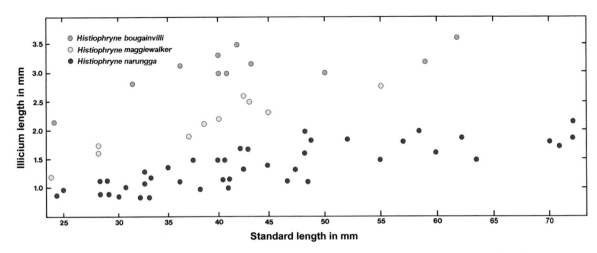

Fig. 231. Relationship between illicium length and standard length for three species of *Histiophryne*. *H. bougainvilli* (blue), N = 11; *H. maggiewalker* (yellow), N = 10; *H. narungga* (red), N = 43.

ceratobranchial I toothless; vertebrae 20 to 23, caudal centra 15 to 18; dorsal-fin rays 13 to 16, all simple; anal-fin rays seven to nine, all simple; pectoral-fin rays eight or nine, all simple; all rays of pelvic fin simple (Tables 14, 15).

Coloration: light beige, pale yellow, yellow-brown, pinkish-brown to dark-chocolate brown, the lower portion of head and body usually somewhat more lightly pigmented than upper; usually one or two small, light-colored patches on each side of body just dorsal to base of pectoral-fin lobe, and occasionally similar patches on dorsal margin of pectoral-fin lobe; basidorsal spot absent; illicium without banding; bars of pigment radiating from eye usually absent. Some specimens (about 38% of material examined) everywhere covered with small, somewhat darker-colored, close-set spots or ocelli; others with irregular, darkly pigmented scaby or wartlike patches on sides of head and body; still others everywhere covered with a dense "mosaic" pattern, or a swirling array of close-set, more-or-less concentric stripes (faded and inconspicuous in preserved material but still evident with the aid of a microscope).

Six species, all restricted to the Indo-Australian Archipelago.

COMMENTS

The genus *Histiophryne* differs most strikingly from other frogfishes in having greatly reduced dorsal-fin spines: the first (illicium) is unusually short to nearly absent, often completely hidden beneath the skin of the head; the second and third dorsal-fin spines are immobile, bound down to the surface of the cranium by skin, emerging only as low protuberances on top of the head. In addition, the skin is naked or only sparsely covered with tiny dermal spinules; and the posteriormost margins of the soft-dorsal and anal fins extend well beyond the base of the caudal fin and are connected broadly to the proximal portion of the outermost caudal-fin rays (Pietsch and Grobecker, 1987).

In addition to morphology, *Histiophryne* is unique in a host of other ways. Its mode of locomotion is notable, with its tail often curled to the left of its body, making it somewhat comical to watch as it crawls across the substrate or tumbles along the seafloor in the case of *H. psychedelica* (Pietsch et al., 2009). The peculiar way in which it holds its body is a reminder that it is the only genus within the family that carries its eggs within a "brood pouch" formed by the soft dorsal, anal, and caudal fins (Pietsch and Grobecker, 1987; see Parental Care, p. 481). The eggs of *Histiophryne*, which hatch as fully formed juveniles, lack a distinct pelagic larval stage (Arnold and Pietsch, 2012; Arnold, 2013; Arnold et al., 2014). Combined with this highly specialized mode of parental care and its limited swimming ability, it is hypothesized that *Histiophryne* and other genera within the subfamily Histiophryninae have greatly limited dispersal potential, resulting in relatively narrow geographic distributions (Arnold and Pietsch, 2012, 2018).

As determined by Arnold and Pietsch (2018), *Histiophryne* now contains six species, three of which are endemic to tropical waters of the western Pacific: *H. cryptacanthus* (Weber, 1913), represented in collections around the world by at least 37 individuals from localities ranging from Taiwan to southern Indonesia and Papua New Guinea; *H. pogonius* Arnold, 2012, six specimens ranging from the Philippines south to Lombok and Komodo Islands, Indonesia; and *H. psychedelica* Pietsch et al., 2009, known from three preserved specimens, all from Ambon, Indonesia, the only confirmed locality. The remaining three species are subtropical and temperate endemics: the type species, *H. bougainvilli* (Valenciennes, 1837), known from at least 15 individuals, all collected from off New South Wales; *H. maggiewalker* Arnold and Pietsch, 2011, 19 specimens, all from off Queensland; and *H. narungga* Arnold and Pietsch, 2018, 61 specimens, all from Western and South Australia. While five of these species appear to be rather well defined, the taxonomic limits of *H. cryptacanthus* are poorly understood, despite a recent reexamination of nearly all known material collected from throughout its currently recognized geographic range (Arnold and Pietsch, 2018). Striking differences in pigment pattern seem to justify the naming of one or more additional new species but such action is unsupported by preliminary molecular analysis (for further discussion, see Comments, p. 332).

Key to the Known Species of the Genus *Histiophryne*

This key should be considered tentative, perhaps lacking the best means to discriminate species and most likely underestimating the diversity within the genus—the reader should therefore proceed with caution (see Comments, above). Like the other keys in this volume, it works by progressively eliminating the most morphologically distinct taxon and for that reason should always be entered from the beginning. All characters listed for each species must correspond to the specimen being keyed; if not, proceed to the next set of character states. For an additional or alternative means of identification, all species of *Histiophryne* may be compared simultaneously by referring to Tables 20 and 21.

1. Dorsal-fin rays 14 or 15; anal-fin rays eight (rarely seven); pectoral-fin rays eight; vertebrae 21; illicium relatively long, length 5.4 to 8.9% SL (Fig. 231), lying within shallow groove on mid-dorsal line of snout, but not enveloped by folds of tissue and easily discernible without aid of a microscope; esca present, usually elongate, sometimes lanceolate, easily distinguished from illicium; skin naked, without dermal spinules (sometimes sparsely present on top of head, especially on bases of second and third dorsal-fin spines), cutaneous filaments and appendages absent; head, body, and fins nearly white (occasionally peppered with tiny dark spots) to beige, pale yellow, yellow-brown, pinkish brown or dark-chocolate brown . . . .
. . . . *Histiophryne bougainvilli* (**Valenciennes, 1837**), p. 323
New South Wales, Australia

Table 20. Comparison of Distinguishing Character States of Species of the Genus *Histiophryne* (measurements in percent SL)

Character	*H. bougainvilli*	*H. cryptacanthus*	*H. maggiewalker*	*H. narungga*	*H. pogonius*	*H. psychedelica*
Dermal spinules	weak/absent	absent	absent	weak/absent	absent	absent
Cutaneous appendages	absent	present/absent	absent	absent	present	absent
Illicium length	5.4–8.9	minute	2.5–4.7	tiny	minute	minute
Esca	present	minute	present	minute	absent	absent
Dorsal-fin rays	14–15	13 (14)	15 (14)	15–16	13 (14)	13
Anal-fin rays	8 (7)	7 (8)	8 (7)	8–9	7	7
Pectoral-fin rays	8	8 (9)	8	9 (8)	8	8
Vertebrae	21	20	21 (22–23)	22–23	20	20
Coloration	monochrome	monochrome/ geometric	monochrome/ spotted	monochrome/ spotted	spotted	striped
Distribution	New South Wales	Taiwan to Papua New Guinea	Queensland	Western and South Australia	Philippines to Papua New Guinea	Indonesia

Table 21. Frequencies of Fin-Ray and Vertebral Counts of Species of the Genus *Histiophryne*

Species	Dorsal-fin rays				Anal-fin rays			Pectoral-fin rays*		Vertebrae			
	13	14	15	16	7	8	9	8	9	20	21	22	23
Tropical endemics													
H. cryptacanthus	29	2			29	2		60	2	21			
H. pogonius	5	1			6			12		6			
H. psychedelica	3				3			6		3			
Australian endemics													
H. bougainvilli		8	7		1	14		30			12		
H. maggiewalker		3	14		2	15		32			11	1	2
H. narungga			37	15		35	19	12	84			28	19
TOTALS	37	14	58	15	41	66	19	152	86	30	23	29	21

*Left and right sides combined.

2.   Dorsal-fin rays 15 (rarely 14); anal-fin rays eight (rarely seven); pectoral-fin rays eight; vertebrae 21 (rarely 22 or 23); illicium relatively short, length 2.5 to 4.7% SL (Fig. 231), lying within shallow groove on mid-dorsal line of snout, sometimes at least partially enveloped by folds of tissue but easily discernible especially with aid of a microscope; esca simple, oval to elongate, expanded distally, sometimes difficult to distinguish from illicium; skin naked, lacking any trace of dermal spinules, cutaneous filaments and appendages absent; head, body, and fins nearly white with light brown spots to all dark-chocolate brown . . . . . . ***Histiophryne maggiewalker*** **Arnold and Pietsch, 2011, p. 332**

Queensland, Australia

3. Dorsal-fin rays 15 or 16; anal-fin rays eight or nine; pectoral-fin rays nine (rarely eight); vertebrae 22 or 23; illicium tiny, length not more than 4.2% SL, easily discernible with aid of a microscope; esca present, a small tuft of filaments, easily distinguished from illicium; skin covered with tiny dermal spinules, especially on head, cutaneous filaments and appendages absent; head, body, and fins typically white, often peppered with numerous small, close-set ocelli . . . .*Histiophryne narungga* **Arnold and Pietsch, 2018, p. 334**
Western and South Australia

4. Dorsal-fin rays 13; anal-fin rays seven; pectoral-fin rays eight; vertebrae 20; illicium minute, barely distinguishable even with aid of a microscope, usually lying completely hidden beneath skin; esca absent; skin naked, lacking any visible trace of dermal spinules, cutaneous filaments and appendages absent; head, body, and fins covered with a swirling pattern of more-or-less concentric white stripes on a yellow-brown or peach-colored background (faded and inconspicuous in preserved material but evident with aid of a microscope) . . . .*Histiophryne psychedelica* **Pietsch, Arnold, and Hall, 2009a, p. 340**
Ambon, Indonesia

5. Dorsal-fin rays 13 (rarely 14); anal-fin rays seven; pectoral-fin rays eight; vertebrae 20; Illicium minute, barely distinguishable even with aid of a microscope, usually lying completely beneath skin; esca absent; skin lacking any visible trace of dermal spinules, but chin and sides of head more or less covered with cutaneous appendages, presenting a bearded appearance; head, body, and fins creamy white to light pink, everywhere peppered with tiny dark close-set spots. . . . .*Histiophryne pogonius* **Arnold, 2012, p. 338**
Philippines to Eastern Indonesia

6. Dorsal-fin rays 13 (rarely 14); anal-fin rays seven (rarely eight); pectoral-fin rays eight (rarely nine); vertebrae 20; illicium minute, barely distinguishable even with aid of a microscope, sometimes lying completely hidden beneath skin; esca present but minute, often difficult to distinguish from illicium; skin lacking any visible trace of dermal spinules, cutaneous filaments and appendages absent; nearly white, often with dark coralline algae-like patches on chin and sides of head or everywhere covered with a dense mosaic pattern . . . .*Histiophryne cryptacanthus* **(Weber, 1913), p. 328**
Taiwan to Eastern Indonesia and Papua New Guinea

Histiophryne bougainvilli (Valenciennes)
Bougainville's Frogfish

Figs. 231–234, 288, 340; Tables 20–22

Chironectes bougainvilli Valenciennes, 1837: 431 (original description, several specimens, lectotype by subsequent designation of Le Danois, 1964, MNHN A.4546, whereabouts of paralectotypes unknown, *Mers des Indes*). Pietsch, 1985a: 94, table 1 (original manuscript sources for *Histoire Naturelle des Poissons* of Cuvier and Valenciennes). Pietsch et al., 1986: 139 (type catalog).

Antennarius bougainvillii: Günther, 1861a: 199 (new combination; description after Valenciennes, 1837).

Cheironectes bougainvillii: Gill, 1863: 90 (type species of *Histiophryne* Gill; emendation of generic name).

Histiophryne bougainvillii: Gill, 1879b: 222 (new combination).

Histiophryne bougainvilli: McCulloch and Waite, 1918: 72, pl. 7, fig. 1 (in part, illustrated specimen, AMS I.14241, New South Wales; see Comments below; Fig. 232). Waite, 1921: 181, fig. 300 (after McCulloch and Waite, 1918). Waite, 1923: 209, fig. (after McCulloch and Waite, 1918). Whitley, 1927b, no. CLIV (Shell Harbour, Port Stephens, New South Wales). McCulloch, 1929: 408 (New South Wales). Whitley, 1934, no. CLIV (after McCulloch, 1929). Schultz, 1957: 69, pl. 2, fig. C (in part, synonymy, description and figure after McCulloch and Waite, 1918). Le Danois, 1964: 106, fig. 52a (lectotype designated). Marshall, 1964: 513, pl. 64, fig. 494 (in part, New South Wales). Scott et al., 1974: 296, fig. (in part, New South Wales; figure after McCulloch and Waite, 1918; in key). Pietsch, 1984b: 40 (in part; genera of frogfishes). Pietsch, 1985a: 94, table 1 (original manuscript sources for Histoire Naturelle des Poissons of Cuvier and Valenciennes). Pietsch and Grobecker, 1987: 253, figs. 104, 133 (in part, confused with congeners; description, distribution, relationships). Paxton et al., 1989: 279 (in part, Australia). Pietsch, 1994: 290, fig. 258 (in part, New South Wales). Allen et al., 2006: 642 (in part, Australia). Kuiter and Debelius, 2007: 126, color fig. (eastern Australia). Pietsch et al., 2009a: 37, 38, 44, 45, fig. 5 (in part; comparison with congeners). Pietsch, 1999: 2015 (in part, western central Pacific; in key). Pietsch, 2008: 368, fig. (after Pietsch, 1994). Arnold, 2010a: 68 (locomotion). Arnold and Pietsch, 2011: 64, fig. 4 (in part, specimens from Nelson Bay, New South Wales; genetics, comparison with congeners). Arnold, 2012: 65, fig. 4 (in part; after Arnold and Pietsch, 2011).

Histiophryne bougainvilli (not of Valenciennes, 1837): Whitley, 1941: 46, fig. 31 (figure represents *H. maggiewalker*; Heron Island, Queensland). Mees, 1964: 52 (new to

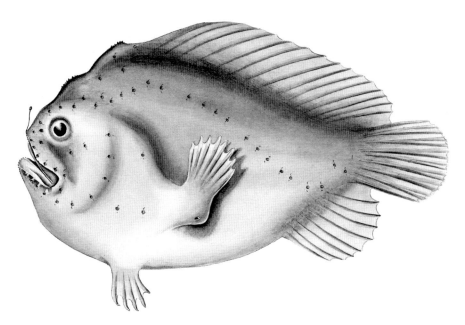

Fig. 232. *Histiophryne bougainvilli* (Valenciennes). AMS I.14241, 32 mm SL, locality unrecorded, but certainly New South Wales and probably Sydney Harbour, Australia. Drawing by Phyllis Clarke; after McCulloch and Waite (1918, pl. 7, fig. 1); courtesy of Lea Gardam and the South Australian Museum, North Terrace, Adelaide.

Western Australia). Pietsch, 1981b: 419 (osteology, NMV A.535 cleared and stained; Spencer Gulf, South Australia). Baker et al., 2009: 17 (South Australia).

MATERIAL EXAMINED

Fifteen specimens, 14.5 to 62.5 mm SL.

Lectotype of *Chironectes bougainvilli*: MNHN A.4546, 50 mm, *La Thétis, Mers des Indes* (i.e., New South Wales), Hyacinthe de Bougainville, 1825.

Additional material: AMS, 10 (14.5–62.5 mm); BMNH, 1 (54 mm); UW, 3 (24–40 mm).

DIAGNOSIS

A member of the genus *Histiophryne* unique in having the following combination of character states: illicium relatively long, length 5.4 to 8.9% SL (Fig. 231), lying within shallow groove on mid-dorsal line of snout, but usually not enveloped by folds of tissue and easily discernible without aid of a microscope; esca present, small, sometimes oval but usually elongate, sometimes lanceolate, easily distinguished from illicium (Fig. 232); skin appearing naked, without dermal spinules but sometimes sparsely present especially on head, cutaneous filaments and appendages absent; dorsal-fin rays 14 or 15, anal-fin rays eight (rarely seven), pectoral-fin rays eight, vertebrae 21 (Tables 20, 21); head, body, and fins nearly white to beige, pale yellow, yellow-brown, pinkish brown or dark-chocolate brown (Figs. 233, 234); genetic divergence from congeners in the nuclear recombination activation gene-2 (RAG2) and cytochrome oxidase-I (COI) genes of at least 6.4% (Arnold and Pietsch, 2011: 67, fig. 4).

Fig. 233. *Histiophryne bougainvilli* (Valenciennes). Nelson Bay, New South Wales, Australia.
Photo by David Harasti.

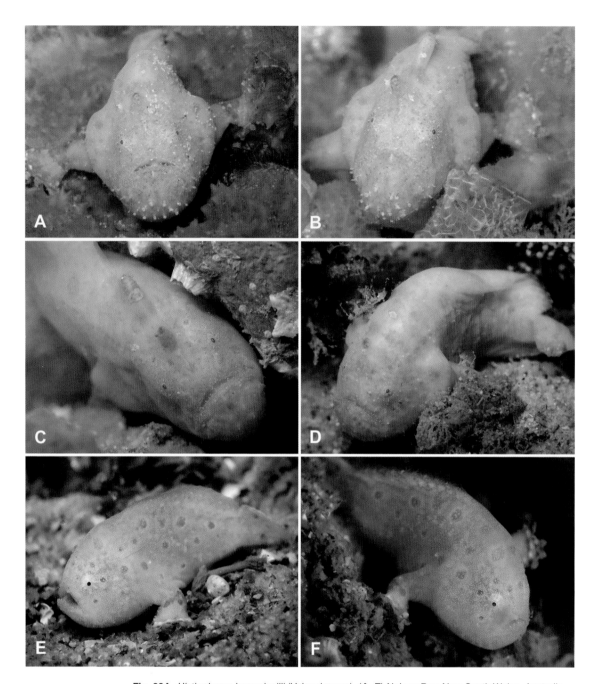

Fig. 234. *Histiophryne bougainvilli* (Valenciennes). (**A–F**) Nelson Bay, New South Wales, Australia. Photos by Brian Mayes.

DESCRIPTION

As given for the genus.

Largest known specimen: 62.5 mm SL (AMS IB.8375).

DISTRIBUTION

All known material of *Histiophryne bougainvilli* was collected in coastal waters of New South Wales, Australia, ranging from North Solitary Island south to Currarong in the

Shoalhaven District (Fig. 288, Table 22). The center of abundance appears to be Nelson Bay, where all recent observations and photographs have been made; in fact, according to David Harasti (personal communication, 22 January 2018) "Nelson Bay seems to be the only locality where this species now occurs in Australia."

The holotype was collected from some unrecorded locality in the *Mers des Indes* during the circumnavigation of Hyacinthe Yves de Bougainville (1781–1846), ship's captain in command of the frigate *La Thétis* and the corvette *L'Espérance* (1824–1826). However, judging from the original cruise track (Bougainville, 1837, pl. 56), it had to have been New South Wales where in 1825 the expedition made landfall in the region of present-day Sydney.

Specific depths of capture are unavailable but it appears that this species resides in relatively shallow waters, from just beneath the surface to about 6 m.

COMMENTS

Histiophryne bougainvilli as recognized by Pietsch and Grobecker (1987) and thought by them to range from Queensland to South Australia, has now been shown, largely through DNA analysis (Arnold and Pietsch, 2012, figs. 5, 7), to be a composite of three species: *H. maggiewalker* Arnold and Pietsch (2011), found only off Queensland and the Great Barrier Reef, and *H. narungga* Arnold and Pietsch (2018), from Western and South Australia, leaving *H. bougainvilli* sensu stricto confined to the coastal waters of New South Wales. The description and figure of *H. bougainvilli* published by McCulloch and Waite (1918: 72, pl. 7, fig. 1; later cited by Waite, 1921, fig. 300; 1923, unnumbered figure; and Schultz, 1957, pl. 2, fig. C)—in a paper entitled "Some New and Little-known Fishes from South Australia"—was partially to blame for this confusion. McCulloch and Waite (1918: 73) described their specimen as "well preserved," but from an "unknown locality." Their figure shows a well-developed first dorsal-fin spine (Fig. 232), which they described as "free, short and slender, with spinules at its base, and a fleshy knob at its tip," and "slightly longer than the eye." Judging from the figure, its length is about 6.8% SL, well within the range of that of *H. bougainvilli* (Fig. 231). But *H. bougainvilli*, as presently understood, is endemic to the coastal waters of New South Wales—so, if identified correctly, how is it that this specimen ended up in the South Australian Museum prior to 1918?

In their account of *H. bougainvilli*, McCulloch and Waite (1918: 73) mention two additional specimens of *Histiophryne*, both collected from Gulf St Vincent and still extant (*H. narungga*, SAMA F.615, F.616), which, although badly preserved, were described by them as "apparently identical with this species." It seems likely, therefore, that McCulloch and Waite used a specimen borrowed from the Australian Museum, Sydney, to better describe their South Australian specimens (Ralph Foster, personal communication, 23 May 2018). In fact, in their introduction, McCulloch and Waite (1918: 65) stated that their paper "is the result of an examination of some fishes preserved in the South Australian Museum, marked "Old Collection," but specimens in the Australian Museum have also been used for comparison." Initially assumed lost, the specimen has been found, surviving in relatively good condition (AMS I.14241, 32 mm SL; Fig. 232).

Pietsch and Grobecker (1987, figs. 105, 106A, B) caused additional confusion by publishing a drawing of *H. maggiewalker* (AMS I.17445-044), as well as figures of the luring apparatus of this species, misidentified as *H. bougainvilli*. Also, in this same publication (Pietsch and Grobecker, 1987, fig. 106C, D), figures of the luring apparatus labeled *H. cryptacanthus* depict that of *H. narungga*.

Histiophryne cryptacanthus (Weber)
Cryptic Frogfish

Figs. 235–238, 287; Tables 20–22

Antennarius cryptacanthus Weber, 1913: 564, pl. 3, fig. 2 (original description, two specimens, lectotype ZMA 101.874 by subsequent designation of Pietsch and Grobecker,
1987, Rotti Island off Timor, Indonesia; paralectotype ZMA 101.898, 52 mm, Karake-
land Island off Beo, Indonesia; Fig. 235A). Nijssen et al., 1982: 64 (type catalog).

Antennatus (Xenophrynichthys) cryptacanthus: Schultz, 1957: 81, pl. 6, fig. B (new combination; description after Weber 1913).

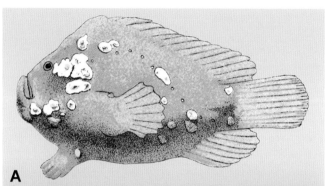

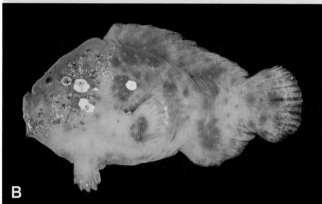

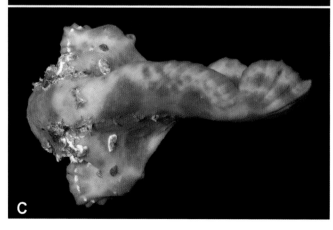

Histiophryne cryptacanthus: Beaufort and
Briggs, 1962: 220, fig. 49 (new combination; after Weber, 1913). Pietsch, 1984b:
40, fig. 1E (in part; genera of frogfishes).
Pietsch and Grobecker, 1987: 255, figs.
104, 107, 133, pls. 51 (in part, confused
with congeners; description, distribution, relationships). Michael, 1998: 334
(in part; identification, behavior, captive
care). Pietsch, 1999: 2015 (in part, western central Pacific; in key). Allen, 2000b:
96 (Calamianes Islands, Philippines). Pietsch, 2000: 597 (South China Sea). Allen
and Adrim, 2003: 25 (Timor and West
Papua, Indonesia). Allen et al., 2003: 367,
color fig. (Asian Pacific). Kwang-Tsao
et al., 2008: 243 (southern Taiwan). Allen
and Erdmann, 2009: 594 (West Papua).
Chen et al., 2009: 23 (Green Island, Taiwan). Pietsch et al., 2009a: 37, 38, 44, 45,
fig. 5 (in part; genetics, comparison with
congeners). Arnold and Pietsch, 2011: 64,
fig. 4 (in part; genetics, comparison with
congeners). Allen and Erdmann, 2012:
153, color figs. (East Indies). Arnold,
2012: 65, fig. 4 (in part; genetics, compar-

Fig. 235. *Histiophryne cryptacanthus* (Weber).
(**A**) lectotype, ZMA 101.874, 63.5 mm SL, Pelela
Bay, Roti Island, Indonesia. Drawing by Joan
François Obbes; after Weber (1913, pl. 3,
fig. 2). (**B**) BPBM 22731, 27 mm SL, Nixon
Rock, south end of Truan-Fan-Shih, Taiwan.
Photo by John E. Randall; courtesy of Lisa
Palmer and the Smithsonian Institution, NMNH,
Division of Fishes, Washington, DC. (**C**) Lombok,
Indonesia. Photo by Vincent Chalias.

ison with congeners). Arnold and Pietsch, 2012: 120, 128, figs. 5, 7, table 1 (in part; molecular phylogeny).

Golem cryptacanthus: Whitley, 1957b: 70 (new combination, description, Port Moresby, AMS IA.5720). Strasburg, 1966: 475, 477 (comparison with *Golem cooperae, nomen dubium*; see p. 354). Munro, 1967: 584, pl. 78, fig. 1089 (New Guinea, figure after Weber, 1913). Kailola, 1975: 47 (Kapa Kapa, central Papua).

Histiophryne sp.: Kailola, 1975: 48 (Gulf of Papua). Kailola and Wilson, 1978: 26, 58 (after Kailola, 1975).

Histiophryne cryptacanthus (not of Weber, 1913): Paxton et al., 1989: 279 (Australia). Hutchins, 1994: 42, appendix (Western Australia). Pietsch, 1994: 291, fig. 259 (photo represents *H. narungga*; southern Australia). Hutchins, 2001: 22 (Western Australia). Kuiter and Debelius, 2007: 126, color figs. (photos show *H. narungga*; Western and South Australia). Pietsch, 2008: 369, fig. (photo shows *H. narungga*; southern Australia). Currie and Sorokin, 2010: 203, 208 (Spencer Gulf, South Australia).

Histiophryne cryptacantha (not of Weber, 1913): Kuiter, 1993: 52 (southeastern Australia). Pietsch, 1994: 291, fig. 259 (southern Australia). Baker et al., 2009: 17, appendix 2 (South Australia).

MATERIAL EXAMINED

Thirty-seven specimens, 13.5 to 86 mm SL.

Lectotype of *Antennarius cryptacanthus*: ZMA 101.874, 63.5 mm, *Siboga* station 301, Pepela Bay, Roti Island, Indonesia, 27–45 m, Weber, 30 January–1 February 1900 (Fig. 235A).

Paralectotype of *Antennarius cryptacanthus*: ZMA 101.898, 62 mm, *Siboga* station 131, anchorage off Beo, Karakelang Island, Indonesia, Weber, 24–25 July 1899.

Additional material: AMS, 2 (37.5–71 mm); ASIZP, 4 (69–86 mm); BMNH, 1 (46 mm); BPBM, 1 (27 mm); FMNH, 2 (15.5–19 mm); KFRL, 4 (50–64.5 mm); RMNH, 1 (63 mm); USNM, 7 (13–45 mm); UW, 12 (13.5–67 mm, including four "mosaic" morphs); WAM, 1 (27 mm, "mosaic" morph).

DIAGNOSIS

A member of the genus *Histiophryne* unique in having the following combination of character states: illicium minute, barely distinguishable even with aid of a microscope, sometimes lying completely hidden beneath skin; esca present but minute, often difficult to distinguish from illicium; skin appearing naked, lacking any visible trace of dermal spinules, but chin and sides of head sometimes covered with cutaneous appendages, presenting a bearded appearance; dorsal-fin rays 13 (rarely 14), anal-fin rays seven (rarely eight), pectoral-fin rays eight (rarely nine), vertebrae 20 (Tables 20, 21); head, body, and fins nearly white, light beige, pale yellow, or yellow-brown, often with dark coralline algae-like patches on chin and sides of head, or rarely everywhere covered with a dense "mosaic" pattern (Figs. 235B, C, 236–238; genetic divergence from congeners in the nuclear recombination activation gene-2 (RAG2) and cytochrome oxidase-I (COI) genes of at least 5.5% (Arnold and Pietsch, 2011: 67, fig. 4).

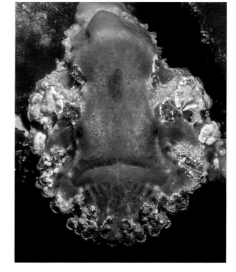

Fig. 236. *Histiophryne cryptacanthus* (Weber). Milne Bay, Papua New Guinea. Photo by Roger Steene.

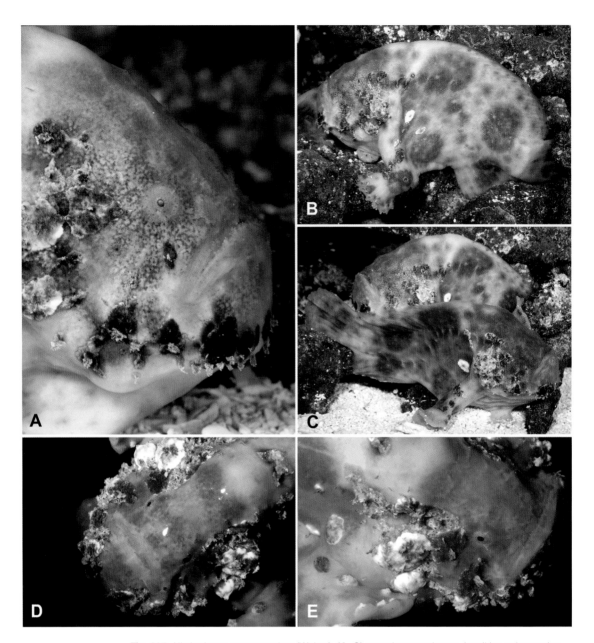

Fig. 237. *Histiophryne cryptacanthus* (Weber). (**A–C**) aquarium specimens, localities unknown but probably Indonesia. Photos by Scott W. Michael. (**D, E**) Lombok, Indonesia. Photos by Vincent Chalias.

DESCRIPTION

Description as given for the genus.

Largest known specimen: 86 mm SL (ASIZP 55187).

DISTRIBUTION

Heretofore thought to range south into Australian waters, *Histiophryne cryptacanthus* is now recognized as restricted to a much narrower region, extending from Taiwan south

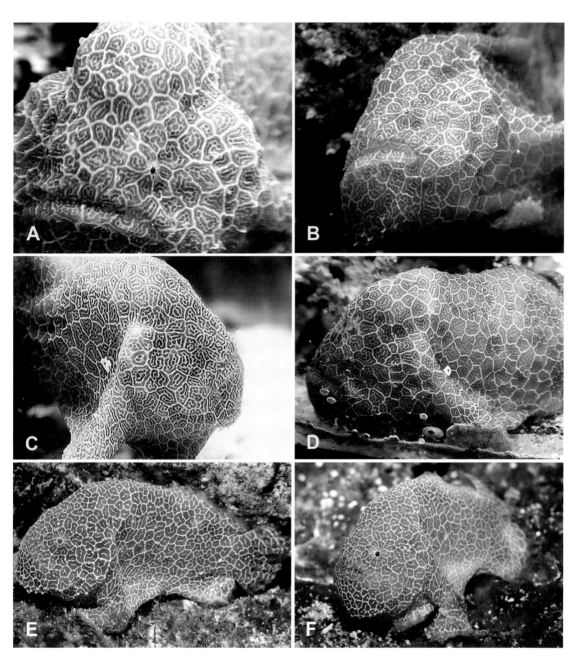

Fig. 238. *Histiophryne cryptacanthus* (Weber), "mosaic" morphs. (**A–D**) aquarium specimens, localities unknown but probably Indonesia. Photos by Bryan Gim. (**E**) Raja Ampat, Indonesia. Photo by Roger Steene. (**F**) West Papua, Indonesia. Photo by Gerry Allen.

to the Philippines, Indonesia (Roti and Timor to West Papua), and Papua New Guinea (Fig. 287, Table 22). The type locality is Roti Island, Indonesia.

Depths of capture, recorded for 10 specimens, range from just beneath the surface to 45 m, but most of these were captured in 20 m or less; the average depth for all known captures is 18 m.

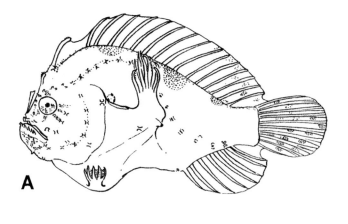

A

B

C

Fig. 239. *Histiophryne maggiewalker* Arnold and Pietsch. (**A**) AMS IB.771, 24 mm SL, Heron Island, Queensland. After Whitley (1941, fig. 31). (**B**) AMS I.17445–044, 45 mm SL, One Tree Island, Great Barrier Reef. Drawing by Cathy L. Short; after Pietsch and Grobecker (1987, fig. 105). (**C**) paratype, QMB I.37621, 60 mm SL, found washed up on the beach at Alexandria Bay, Noosa Heads, southeast Queensland, Australia. Photo by Rachel J. Arnold.

COMMENTS

This species was previously thought to extend south into Australian waters (Pietsch and Grobecker, 1987: 257, fig. 133), but now, thanks largely to DNA analysis (Arnold and Pietsch, 2011, 2012), its southern limit has been reestablished at approximately 11° S latitude along southernmost Indonesia and the coastal waters of Papua New Guinea. Australian members of the genus *Histiophryne* now include only three allopatric species—*H. maggiewalker* from Queensland, *H. bougainvilli* from New South Wales, and the recently described *Histiophryne narungga* from Western and South Australia (Arnold and Pietsch, 2018).

Although now much better understood, the possibility still remains that *H. cryptacanthus* represents a mosaic of two or more cryptic species that we have been unable to sort out despite a detailed comparative examination of all known preserved material collected from throughout its currently recognized geographic range. Of great interest are the so-called "mosaic" morphs of which there seems to be at least two (Fig. 238). Other than their unique pigment patterns, they bear no discernable distinguishing features from the much more common non-patterned forms, and no intermediates have been observed. Further puzzling, molecular analysis shows no divergence within the cytochrome oxidase I (COI) gene for "mosaic" morphs purportedly from Indonesia (Rachel J. Arnold, unpublished data).

Histiophryne maggiewalker Arnold and Pietsch
Maggie's Frogfish

Figs. 231, 239, 240, 287; Tables 20–22

Histiophryne bougainvilli (not of Valenciennes, 1837): Whitley, 1941: 46, fig. 31 (in part, figure represents *H. maggiewalker*; Heron Island, Queensland; Fig. 239A). Pietsch and Grobecker, 1987: 254 (in part, figs. 105, 106A, B, represent *H. maggiewalker*; Fig. 239B).
Histiophryne **sp. 3**: Arnold and Pietsch, 2012: 121, figs. 5, 7, table 1 (QMB I.38176, molecular phylogeny).

Histiophryne maggiewalker Arnold and Pietsch, 2011: 64, figs. 1–4 (original description, six specimens, holotype QMB I.38176, 55 mm, Alexandria Bay, Noosa Heads, southeast Queensland, Australia; Fig. 239C). Arnold, 2012: 65, fig. 4 (genetics, comparison with congeners). Arnold and Pietsch, 2012: 128, figs. (molecular phylogeny, revised classification).

MATERIAL EXAMINED

Nineteen specimens, 12.5 to 57 mm SL.

Holotype of *Histiophryne maggiewalker*: QMB I.38176, 52 mm, found washed up on beach at Alexandria Bay, Noosa Heads, southeast Queensland, Australia.

Paratypes of *Histiophryne maggiewalker*: QMB I.29079, 1 (16.5 mm), north side of Mujimba Island, Mujimba, Queensland, Australia; QMB I.37621, 1 male (51 mm), 1 female (57 mm), and two juveniles (12.5–14 mm), Shag Rock, off northeast end of North Stradbroke Island, Queensland, Australia (Fig. 239B).

Additional material: AMS, 9 (17–45 mm); CSIRO, 1 (55 mm); QMB, 3 (23–37 mm).

DIAGNOSIS

A member of the genus *Histiophryne* unique in having the following combination of character states: illicium relatively short, length 2.5 to 4.7% SL (Fig. 231), lying within shallow groove on mid-dorsal line of snout, sometimes at least partially enveloped by folds of tissue but easily discernible especially with aid of a microscope (Fig. 240; esca minute, simple, oval to elongate, usually slightly expanded distally, but sometimes difficult to distinguish from illicium; skin naked, lacking any visible trace of dermal spinules, cutaneous filaments and appendages absent; dorsal-fin rays 15 (rarely 14), anal-fin rays eight (rarely seven), pectoral-fin rays eight, vertebrae 21 (rarely 22 or 23) (Tables 20, 21); head, body, and fins creamy white, with small close-set, light-brown spots, to dark-chocolate brown, almost black; reticulate pattern sometimes evident with aid of a microscope; genetic divergence from congeners in the nuclear recombination activation gene-2 (RAG2) and cytochrome oxidase-I (COI) genes of at least 6.1% (see Arnold and Pietsch, 2011: 67, fig. 4).

DESCRIPTION

As given for the genus.

Largest specimen examined: 57 mm SL (QMB I.37621).

DISTRIBUTION

Histiophryne maggiewalker is known only from shallow inshore waters of Queensland, extending from South Molle Island to North Stradbroke Island, and from Heron Island, One Tree Island, and Swains Reefs on the Great Barrier Reef (Fig. 287, Table 22). The holotype was found washed up on the beach at Noosa Heads. All specimens for which depths are known were collected in three to 12 m of water.

Fig. 240. Illicia and escae of *Histiophryne maggiewalker* Arnold and Pietsch. (**A**) AMS I.20532, 40 mm SL, One Tree Island, Great Barrier Reef. (**B**) AMS I.17445–044, 43.5 mm SL, One Tree Island, Great Barrier Reef. Drawings by Cathy L. Short; after Pietsch and Grobecker (1987, fig. 106A, B).

A   B

Four specimens, a male, a non-gravid female, and two juveniles, were collected together at the same time at Shag Rock off the northeast end of North Stradbroke Island. While frogfishes are notorious for their insatiable appetite, including a tendency toward cannibalism, it is possible that these four were living together peacefully (see Comments below). We are tempted to speculate that the two juveniles were the offspring of one or both of the adults.

COMMENTS

Pietsch and Grobecker (1987: 254, figs. 105, 106A, B) failed to notice the uniqueness of *H. maggiewalker*, assigning most of the material known at the time to *H. bougainvilli* and publishing figures of misidentified specimens (see Comments, p. 327). It was discovered only much later that sequence data originating from tissues received from the Queensland Museum, Brisbane, did not match sequences of *H. bougainvilli* for which identification had been verified. Subsequent examination of the specimens revealed an esca that differed from anything described previously for *Histiophryne* (Arnold and Pietsch, 2011: 64, fig. 3; Fig. 240). Subsequent analysis of RAG2 and COI sequences, including 441 and 570 bps, respectively, recovered *H. maggiewalker* as sister of *H. bougainvilli*, with a posterior probability of 1.0. A clade containing *Histiophryne cryptacanthus* and *H. psychedelica*, two tropical species of this genus, was recovered as sister of *H. bougainvilli* and *H. maggiewalker*, with a posterior probability of 1.0. The intraspecific COI DNA sequence divergence between *H. cryptacanthus* and *H. psychedelica* was 6.1%, and between *H. bougainvilli* and *H. maggiewalker*, 7.1%.

Histiophryne narungga Arnold and Pietsch
Narungga Frogfish

Figs. 26K, 231, 241, 242, 287; Tables 20–22

Histiophryne bougainvilli (not of Valenciennes, 1837): McCulloch and Waite, 1918: 72 (in part, SAMA F.615, F616; Gulf St Vincent, South Australia). Waite, 1921: 181, fig. 300 (after McCulloch and Waite, 1918). Waite, 1923: 209, fig. (after McCulloch and Waite, 1918). Schultz, 1957: 69 (in part; description after McCulloch and Waite, 1918). Marshall, 1964: 513, pl. 64, fig. 494 (in part, South Australia). Mees, 1964: 52 (new to Western Australia). Scott et al., 1974: 296, fig. (in part, South Australia; in key). Pietsch, 1981b: 419 (osteology). Pietsch, 1984b: 40 (in part; genera of frogfishes). Paxton et al., 1989: 279 (in part, Australia). Pietsch, 1999: 2015 (in part, western central Pacific; in key). Allen et al., 2006: 642 (in part, Australia). Baker et al., 2009: 17 (South Australia). Pietsch et al., 2009a: 37, 38, 44, 45, fig. 5 (in part; genetics, comparison with congeners). Arnold, 2010a: 68 (locomotion). Arnold and Pietsch, 2011: 64, fig. 4 (in part; genetics, comparison with congeners). Arnold, 2012: 65, fig. 4 (in part; genetics, comparison with congeners).

Histiophryne cryptacanthus (not of Weber, 1913): Pietsch and Grobecker, 1987: 255, figs. 16K, 104, 106C, D, 133, pls. 52, 53 (in part, confused with congeners; description, distribution). Paxton et al., 1989: 279 (Australia). Hutchins, 1994: 42, appendix (Western Australia). Pietsch, 1994: 290, fig. 258 (southern Australia). Hutchins, 2001: 22 (Western Australia). Kuiter and Debelius, 2007: 126, color fig. (Western and South Australia). Pietsch, 2008: 369, fig. (southern Australia). Currie and Sorokin, 2010: 203, 208 (Spencer Gulf, South Australia).

Histiophryne cryptacantha (not of Weber, 1913): Kuiter, 1993: 52 (southeastern Australia). Pietsch, 1994: 291, fig. 259 (southern Australia). Baker et al., 2009: 17, appendix 2 (South Australia).

Histiophryne **sp. 2**: Arnold and Pietsch, 2012: 121, figs. 5, 7, table 1 (SAMA F.11719, molecular phylogeny).

Histiophryne narungga Arnold and Pietsch, 2018: 623, figs. 1–7, tables 1, 2 (original description, 60 specimens, holotype SAMA F.13936, 74 mm, Port Lincoln, Spencer Gulf).

MATERIAL EXAMINED

Sixty specimens, 13.5 to 75 mm SL.

Holotype of *Histiophryne narungga*: SAMA F.13936, 74 mm, Port Lincoln, Spencer Gulf, South Australia Research and Development Institute (SARDI), Spencer Gulf Prawn Bycatch Survey, Station BC31, 34° 33′ 39″ S, 136° 32′ 53″ E, 33 m, 16 February 2007.

Paratypes of *Histiophryne narungga* (34, 16–75 mm): AMS I.20222-011, 32.5 mm, Mondrain Island, Recherche Archipelago, 34° 12′ S, 122° 23′ E, Russell and Kuiter. NMV A.527, 30.5 mm, Cape Donington, Port Lincoln, Spencer Gulf, South Australia, 34° 44′ S, 136° 00′ E, Veitch, 8 April 1968; NMV A.535, 64 mm, Tiparra Reef, Spencer Gulf, South Australia; NMV A.2836, 3 (25–43 mm), Port Lincoln, Spencer Gulf, South Australia, 34° 44′ 25″ S, 135° 53′ 04″ E, 11–18 m, Veitch, 13 March 1969; NMV A.20798, 48 mm, Lucky Bay, Western Australia, 33° 59′ 36″ S, 122° 13′ 19″ E, Kuiter, 11–13 March 1986; NMV A.30651-002, 47.5 mm, Edithburgh Jetty, Gulf St Vincent, South Australia, 35° 05′ 06″ S, 137° 44′ 59″ E, Arnold, 15 June 2009; NMV A.3334, 74 mm, Edithburgh Jetty, Yorke Peninsula, 35° 05′ 05″ S, 137° 45′ 00″ E, 3–8 m, Kuiter, 11 March 1984. SAMA F.4568, 60 mm, off Cowell, Spencer Gulf, South Australia, 33° 44′ S, 136° 55′ E, Ebon, November 1980; SAMA F.5520, 63 mm, Lusby Island Rocks, Sir Joseph Banks Group, Spencer Gulf, South Australia, 34° 33′ S, 136° 16′ E, Church, 24 January 1986; SAMA F.5521, 53 mm, Nicholas Bay, Reevesby Island, Sir Joseph Banks Group, Spencer Gulf, South Australia, 34° 32′ S, 136° 17′ E, under rocks in shallow water, Phillips, 23 January 1986; SAMA F.7323, 40 mm, off reef north of Home Bay, Reevesby Island, Spencer Gulf, Sir Joseph Banks Group, South Australia, 34° 35′ S, 136° 13′ E, Holmes and Horsefall, 23 January 1968; SAMA F.11719, 55 mm, Smoky Bay, Zippel's oyster lease, South Australia, 32° 18′ S, 133° 50′ E, 30 June 2007 (DNA extracted; see Arnold and Pietsch, 2012, table 1); SAMA F.12085, 2 (64–70 mm), Shoalwater Point (between Cowell and Wallaroo), Spencer Gulf, South Australia, 38° 48′ 14″ S, 137° 27′ 43″ E, Kuiter, March 1982. UW 158282, 72 mm, 16.5 miles from Wallaroo outer stick, 6 miles from Yarraville Light, Spencer Gulf, South Australia, 33° 31′ 33″ S, 137° 34′ 10″ E, May 1982. WAM P.3808, 75 mm, Cottesloe, Western Australia, 32° 03′ S, 115° 44′ E, Wintman, 6 June 1955; WAM P.4352, 41 mm, north of Lancelin Island, Western Australia, 31° 00′ S, 115° 19′ E, 62 m, Weekes, 10 June 1958; WAM P.15872, 28.5 mm, north side of Cape Naturaliste, 33° 32′ S, 115° 01′ E, Wilson et al., 29 December 1967; WAM P.15873, 33 mm, north side of Cape Naturaliste, 33° 32′ S, 115° 01′ E, Wilson et al., 29 December 1967; WAM P.26004-010, 38 mm, island on east side of Lucky Bay, Recherche Archipelago, 34° 01′ S, 122° 15′ E, Hutchins et al., 17 March 1978; WAM P.26008-013, 42.5 mm, bay on west side of Mondrain Island, Recherche Archipelago, 13 m, Hutchins, 21 March 1978; WAM P.26009-010, 33.5 mm, Lucky Bay, Recherche Archipelago, 13 m, Hutchins and Kuiter, 22 March 1978; WAM P.26812-003, 30 mm, 3 km off Dunsborough, Geographe Bay, Western Australia, 16 m, Allen, 26 December 1978; WAM P.27902-001, 40 mm, found in craypot, Western Australia, Graham, 4 August 1972; WAM P.28276-001, 49 mm, northeast side of Cape Naturaliste, Western Australia, 33° 50′ S, 115° 00′ E, Wilson, 21 September 1969; WAM P.28293-008, 17 mm, unnamed island on east side of Lucky Bay, Western Australia, 5–7 m, 34° 00′ S, 122° 14′ E, Hutchins et al., 12 April 1984; WAM P.28296-011, 16 mm, headland on E side of Mondrain

Island, Recherche Archipelago, Western Australia, 34° 08′ S, 122° 15′ E, 5–6 m, Hutchins et al., 13 April 1984; WAM P.28509-001, 28 mm, Duke of Orleans Bay, Western Australia, 33° 56′ S, 122° 36′ E, 5 m, Longbottom, 1 March 1985; WAM P.28513-002, 36 mm, John Island, Duke of Orleans Bay, Western Australia, 33° 54′ S, 122° 37′ E, 8 m, Bryce and Haigh, 7 March 1985; WAM P.28514-001, 47 mm, Little Wharton Beach, Western Australia, 33° 58′ S, 122° 34′ E, 4–5 m, Longbottom, 8 March 1985; WAM P.28515-001, 41 mm, Lion Island, Recherche Archipelago, Western Australia, 33° 53′ S, 122° 02′ E, 4–6 m, Bryce and Haigh, 10 March 1985; WAM P.28517-001, 29 mm, 3 km north of Dunsborough, Western Australia, 33° 35′ S, 115° 06′ E, 15–17 m, Hutchins, 12 April 1985.

Additional material (25, 16–58 mm): AMS, 1 (45 mm); LACM, 1 (43 mm); NMV, 3 (21–41.5 mm); SAMA, 18 (23.5–58 mm); WAM, 2 (23–57 mm).

DIAGNOSIS

A member of the genus *Histiophryne* distinguished in having the skin more or less everywhere covered with tiny dermal spinules (scattered spinules sometimes present on top of head of *H. bougainvilli*). It is further unique in having the following combination of character states: illicium tiny (length not more than 4.2% SL) but easily discernible without aid of a microscope; esca present, a tiny tuft of filaments, easily distinguished from illicium; cutaneous filaments and appendages absent; dorsal-fin rays 15 or 16, pectoral-fin rays nine (rarely eight), vertebrae 22 or 23; head, body, and fins typically white to beige, light pink or gray, sometimes everywhere peppered with numerous small, close-set ocelli (dark spots each surrounded by a narrow white ring); genetic divergence from congeners in the nuclear recombination activation gene-2 (RAG2) and cytochrome oxidase-I (COI) genes of at least 8.9% (Arnold and Pietsch, 2018, fig. 6).

DESCRIPTION

Head, body, and fins nearly white to beige, light pink or light gray, sometimes peppered everywhere with numerous small, close-set ocelli (dark spots each surrounded by a narrow white ring); gray spots or patches occasionally present on sides of body and caudal

Fig. 241. *Histiophryne narungga* Arnold and Pietsch. Edithburgh Jetty, South Australia. Photo by Scott W. Michael.

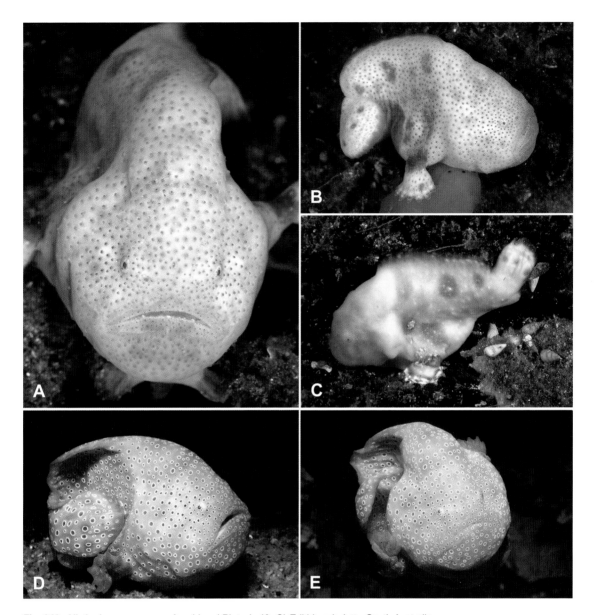

Fig. 242. *Histiophryne narungga* Arnold and Pietsch. (**A–C**) Edithburgh Jetty, South Australia. Photos by Scott W. Michael. (**D, E**) Edithburgh, South Australia. Photos by Rudie H. Kuiter.

fin; occasionally, dark, often greenish, coralline algae-like patches present on sides of body and pectoral-fin lobe (Figs. 241, 242).

The largest specimen examined: 75 mm SL (WAM P.3808), from Cottesloe, Western Australia.

DISTRIBUTION

Histiophryne narungga ranges from off Cape Cuvier, Western Australia, and south and east along the path of the Leeuwin Current, to Kangaroo Island and Aldinga Beach, South Australia, with the center of abundance in Spencer Gulf and Gulf St Vincent, along the shores of Yorke Peninsula, South Australia (Fig. 287, Table 22).

Depths of capture, recorded for 24 specimens, range from 3–130 m, but 83% of these individuals were collected in less than 20 m and 50% in less than 10 m; the average depth for all known captures is 13 m.

COMMENTS

This species has long been confused with *H. cryptacanthus* and made known only recently through DNA analysis (Arnold and Pietsch, 2011, 2012, 2018). *Histiophryne cryptacanthus* is now recognized as a tropical species ranging in distribution from Taiwan to southern Indonesia and Papua New Guinea, while *H. narungga* is a subtropical and temperate form, restricted to shallow coastal waters of Western and South Australia (Fig. 287).

Histiophryne pogonius Arnold
Bearded Frogfish

Figs. 243, 244, 288; Tables 20–22

Histiophryne **sp. 1**: Arnold and Pietsch, 2012: 121, figs. 5, 7, table 1 (UW 118820, molecular phylogeny).

Histiophryne **pogonius** Arnold, 2012: 63, figs. 1–4 (original description, four specimens, holotype UW 118820, 48 mm, procured through the live fish trade, reportedly from Cebu, Philippines).

MATERIAL EXAMINED

Six specimens, 37 to 54 mm SL.

Holotype of *Histiophryne pogonius*: UW 118820, 48 mm, procured through the live fish trade, reportedly from Cebu, Philippines.

Paratypes of *Histiophryne pogonius*: UW 119522, 1 (54 mm), female, Lombok, Indonesia; UW 119523, 1 (53.5 mm), female, Lombok, Indonesia; UW 119920, 1 (49.5 mm), female, Lombok, Indonesia (Fig. 243).

Additional material: AMS, 2 (37–49 mm).

DIAGNOSIS

A member of the genus *Histiophryne* unique in having the following combination of character states: illicium minute, barely distinguishable even with aid of a micro-

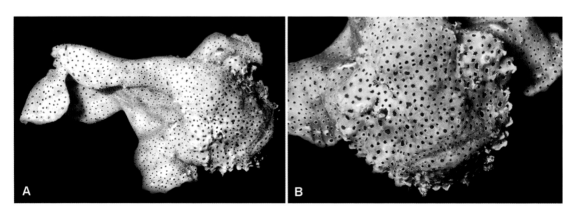

Fig. 243. *Histiophryne pogonius* Arnold. (**A**, **B**) paratype, UW 119920, 49.5 mm SL, Lombok, Lesser Sunda Islands, Indonesia. Photos by Vincent Chalias.

scope, usually lying completely beneath skin; esca absent; skin lacking any visible trace of dermal spinules, but chin and sides of head more or less covered with cutaneous appendages, presenting a bearded appearance; cheeks and chin with numerous small shallow depressions, giving a pitted appearance to head; dorsal-fin rays 13 (rarely 14), anal-fin rays seven, pectoral-fin rays eight, vertebrae 20 (Tables 20, 21); head, body, and fins creamy white to light pink, everywhere peppered with tiny dark close-set spots (Figs. 243, 244); genetic divergence from congeners in the cytochrome oxidase-I *c* subunit (COI) gene of at least 5.5% (see Arnold, 2012: 65, fig. 4; Arnold and Pietsch, 2012: 125, figs. 5, 7).

Fig. 244. *Histiophryne pogonius* Arnold. (**A–E**) near Bangka Island, North Sulawesi, Indonesia. Photos by Volker Mattke.

Color in preservation: small white patches, resembling coralline algae, always present on body behind pectoral fins, and sometimes present on cheeks. Pinkish coloration in life lost on preservation, but small dark spots that cover head and body, and white patches on cheeks and behind pectoral fins, still clearly visible.

Largest specimen examined: 54 mm SL (UW 119522).

DISTRIBUTION

Histiophryne pogonius is known only from shallow inshore waters of Lombok, Komodo, and Bangka (North Sulawesi) islands, Indonesia; off Port Moresby, Papua New Guinea; and perhaps nearshore waters surrounding Cebu, Philippines (Fig. 288, Table 22). The species appears to occur within reef habitat and may be more active at night.

COMMENTS

Despite numerous reports of aggression and cannibalism in aquarium-held frogfishes, especially *Histrio histrio* and a few members of the genus *Antennarius* (Mosher, 1954; Rasquin, 1958; Pietsch and Grobecker, 1987; see Voracity and Cannibalism, p. 451), two specimens of *H. pogonius* procured through the tropical fish trade lived peacefully together in an aquarium for many months (Arnold, 2012). Several specimens of the closely related *H. cryptacanthus* have also been reported to coexist peacefully in an aquarium (Bryan Gim, personal communication, 7 March 2007). Arnold and Pietsch (2011) speculated further that two adult and two juvenile *H. maggiewalker* collected together, might have been coexisting peacefully, suggesting that *Histiophryne* may not be cannibalistic or as aggressive as purported for frogfishes in general.

Histiophryne psychedelica Pietsch, Arnold, and Hall
Psychedelic Frogfish

Figs. 245–247, 288, 341–343; Tables 20–22

Histiophryne psychedelica Pietsch et al., 2009a: 38, figs. 1–3, 4A, 5, 6, 8 (original description, three specimens, holotype NCIP 6377, female, 87 mm, Ambon Bay, Ambon Island, Maluku Province, eastern Indonesia). Helfman et al., 2009: 326, 327, figs. (aggressive mimicry). Arnold and Pietsch, 2011: 67, fig. 4 (genetics, comparison with congeners). Arnold, 2012: 65, fig. 4 (genetics, comparison with congeners). Allen and Erdmann, 2012: 153, color figs. (Ambon). Arnold and Pietsch, 2012: 128, figs. 2E, 5, 7, table 1 (molecular phylogeny). Arnold et al., 2014: 538 (locomotion and defense). Ricciardi, 2018: 62 (vertical distribution).

MATERIAL EXAMINED

Three specimens, 65 to 87 mm SL.

Holotype of *Histiophryne psychedelica*: NCIP 6377, female, 87 mm, collected by scuba in 6.5 m of water, approximately 20 m offshore, at the Laha I dive site, near a commercial jetty in the harbor at Ambon City, Ambon Bay, Ambon Island, Maluku Province, eastern Indonesia, R. J. Arnold and party, 2 April 2008.

Paratypes of *Histiophryne psychedelica*: UW 22454, 1 male (65 mm), 1 female (76 mm), received by an importer in a shipment of marine fishes from Bali, Lesser Sunda Islands, Indonesia; died in the Dallas Aquarium, received at UW 23 June 1992.

DIAGNOSIS

A member of the genus *Histiophryne* unique in having the following combination of character states: skin naked, lacking any visible trace of dermal spinules or cutaneous filaments and appendages, everywhere thick and loose, forming conspicuous fleshy folds that envelope the unpaired fins; illicium minute, evident only with aid of a microscope; esca absent; dorsal-fin rays 13, anal-fin rays seven, pectoral-fin rays eight, vertebrae 20 (Tables 20, 21); head, body, and fins everywhere covered with a swirling pattern of more-or-less concentric white stripes on a yellow-brown or peach-colored background, radiating from eyes and continuing back to body and tip of caudal fin (faded and inconspicuous in preserved material but evident with the aid of a microscope); genetic divergence from congeners in the nuclear recombination activation gene-2 (RAG2), cytochrome oxidase-I (COI), and 16S rRNA genes (Pietsch et al., 2009a: 38, fig. 5; Arnold, 2012: 65, fig. 4; Arnold and Pietsch, 2012: 000, figs. 5, 7); and a set of unique behavioral traits (see below).

DESCRIPTION

Face broad, flat, surrounded by thick, fleshy, laterally expanded cheeks and chin, with eyes directed anteriorly; fleshy extensions of the cheeks and chin, appearing scalloped along the outer margins (Fig. 245); extremely gelatinous in life, quickly hardening and shrinking considerably during fixation and preservation (shortly after the holotype was placed in ethanol it measured 100 mm SL, but when measured again four months later it was only 87 mm SL), fleshy extensions of cheeks and chin becoming considerably less evident, eyes taking on a more lateral position, but everywhere thick fleshy skin still evident (see Pietsch et al., 2009a, fig. 4A).

Coloration: in life, head, body, and fins everywhere adorned with a swirling pattern of more-or-less concentric white stripes on a yellow-brown or peach-colored background, each individual stripe bisected by a somewhat darker, bluish-gray, central stripe (in sharp contrast to its rather bland congeners); stripes radiating out from eyes, but roughly parallel to each other on chin, forming small closed ovals or circles at various places on pectoral-fin lobes and posteroventral part of body; skin surrounding eyes, and along periphery of fins bright turquoise in life, enhanced remarkably when placed in for-

Fig. 245. *Histiophryne psychedelica* Pietsch, Arnold, and Hall. (**A**) adult specimen, about 110 mm SL, Ambon, Indonesia, shown "jet-propelling" in open water. (**B**) adult specimen, about 150 mm SL, propped up between bottom substrate and a coral head by its armlike pectoral fins. Photos by David J. Hall.

malin, but all pigment lost almost immediately in ethanol, leaving entire surface of head, body and fins creamy white (Pietsch et al., 2009a, fig. 4A); striped pattern, however, still readily discernible with aid of a dissecting microscope even after years in preservative (Figs. 245, 246).

Largest specimen examined: the holotype, 87 mm SL (NCIP 6377), but considerably larger, uncollected individuals have been observed, approaching 150 mm SL (Figs. 245, 246).

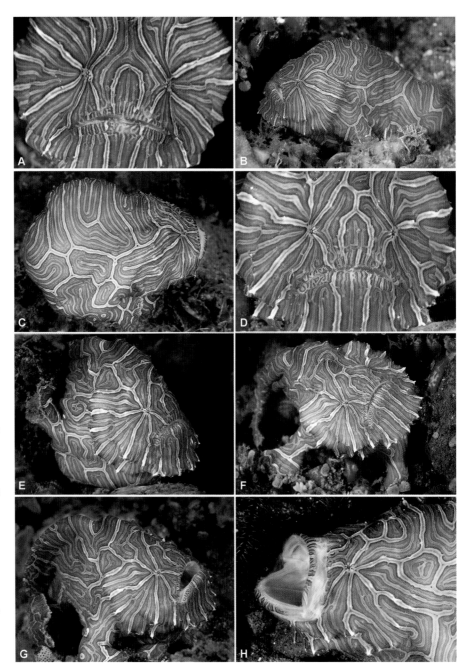

Fig. 246. *Histiophryne psychedelica* Pietsch, Arnold, and Hall. Ambon, Indonesia. (**A, B**) anterior and left lateral views, about 110 mm SL. (**C**) taking on spherical appearance while jet-propelling in open water, about 110 mm SL. (**D, E**) anterior and right lateral views, about 150 mm SL. (**F**) "walking," with armlike pectoral fins, about 150 mm SL. (**G**) right lateral view, about 150 mm SL. (**H**) about 150 mm SL specimen, "yawning," in right lateral view. Photos by David J. Hall.

DISTRIBUTION AND HABITAT

Histiophryne psychedelica is known only from shallow inshore waters of central and eastern Indonesia, the holotype collected at the Laha I dive site in Ambon Bay, Ambon Island, Maluku Province, by scuba in 6.5 m of water, approximately 20 m offshore (Fig. 288, Table 22). The two paratypes were received by the Dallas Aquarium in a shipment of marine fishes from Bali, Lesser Sunda Islands.

One or both of two adult individuals of *H. psychedelica* were observed and photographed on six occasions over a 10-day period from 6 to 15 March 2008. On each occasion, the fish were found at a depth of 5 to 7 m, within 100 m of the commercial jetty in the harbor at Ambon City (the "Laha I" dive site, previously well known to divers as the "Twilight Zone," because of the presence of many rare animals). Currents varied from virtually nonexistent to strong (at times almost too strong to swim against) and visibility varied between about 10 and 30 m. The location was immediately offshore from a small village, along a gently sloping beach. When both fish were seen on the same dive, they were always found within 10 to 15 m of one another. The sea floor in this location slopes gently near shore, but more steeply below 10 m, quickly reaching a depth of 100 m or more. Aside from coral rubble, the bottom was littered with a variety of human refuse. Both individuals were consistently found hidden among coarse coral rubble, usually encrusted with ascidians, sponges, and coralline algae.

COMMENTS

Histiophryne psychedelica was first recognized as "something different" in June 1992 by staff of the Dallas Aquarium, when two frogfish with an unusual pigment pattern were found in a shipment of live fishes from Bali; arriving "in very poor condition," the specimens soon died (David M. Schleser, personal communication, 23 June 1992). The specimens were then sent to TWP for identification; but, unfortunately, failing to make more than a superficial examination at that time, they were labeled *H. cryptacanthus* and placed on a shelf where they remained unnoticed for the next decade and a half.

The species was not seen again until 28 January 2008, when employees of a sport-dive operation located on Ambon Island, Indonesia, photographed a frogfish that had never been observed before in their 20 years of diving at that locality. Unable to identify the fish, after circulating photographs throughout the dive community, they sent pictures to TWP who recognized it as new: "I can say that in my 40 years studying frogfishes, and anglerfishes in general, I have never seen one quite like this. If I had to say what its closest relative might be, I'd suggest the genus *Histiophryne*, but this fish differs in a host of ways" (T. W. Pietsch to Andrew Shorten, personal communication, 21 February 2008). Excited that something important had been discovered, the University of Washington Office of News and Information issued a press release on 2 April 2008 ("New fish has a face even Dale Chihuly could love") that was picked up instantly and circulated for months by news agencies and various on-line periodicals around the world. Eventually a specimen was obtained that provided, in part, the basis for the original description.

Although the species is presently represented by only three preserved specimens, one from Ambon and two from an unrecorded locality, divers at the time located another dozen or so individuals at the same locality off Ambon, all within 40 m of each other and all remarkably similar in color and pigment pattern (Hall, 2008; Klein, 2008). Among these uncollected individuals were a spawned-out female, approximately 150 mm (the basis for comments on reproduction below); two small juveniles, each about 20 mm; and several tiny, newly hatched juveniles.

When observed in the field, the fish were often so well hidden that they could be located only by overturning rubble in a systematic search of an area approximately 100 m². Once a fish was uncovered, it was observed and photographed until it re-entered a crevice or hole, usually within 10 to 15 minutes. On several occasions, the fish were observed entering holes headfirst. The openings were often small enough that one or two minutes were required for the fish to work its way in (Fig. 247). Repeated observations of this behavior revealed rigorous twisting and turning of the body, together with active use of the pelvic fins, to facilitate entry. In one such example, the fish turned nearly upside down as it worked its way through the tight opening (Fig. 247D), appearing to rotate so that its pelvic fins would first make contact with sharp pointed coral that lined one section of the opening, perhaps in an attempt to protect the body from injury. Despite this vigorous activity against highly abrasive substrata, neither fish bore evidence of superficial scratches or wounds of any kind.

Locomotion: Once uncovered by divers, the fish did not remain in an exposed location for long. They traveled along the rubble-covered bottom primarily by using "jet propulsion" (forcefully expelling water through restricted, posteriorly directed opercular openings; see Pelagic Locomotion and Jet Propulsion, p. 459) as the dominant means of locomotion, eventually entering a hole or crevice, usually within 10 to 15 minutes following initial exposure. Jet propulsion is often used by other antennariids, but in a somewhat different manner. Members of the genus *Antennarius* typically move short distances by "walking," with the aid of pectoral and pelvic fins, or longer distances by swimming continuously from one location to the next, using a combination of jet propulsion and fin movements (typically with the fins erected and widely splayed out), without touching bottom until they reach their final destination. In *H. psychedelica*, however, jet propulsion in a sustained, long-distance swimming effort was never observed. On the contrary, both individuals moved consistently in a series of short "hops," the paired pelvic fins making frequent contact with, and appearing to push off, the bottom at each bounce. While moving in this way and when viewed from the side, the fish assumed a strikingly globose shape (Fig. 246C), with the dorsal fin bent to one side, the posterior part of the body and tail bent strongly forward to approach the head, the fleshy lateral extensions of the cheeks and chin pulled back, and the dorso-ventral diameter increased significantly. The overall visual impression, when viewed from the side, was reminiscent of an inflated rubber ball bouncing along the bottom (Hall, 2008).

Pigment pattern and camouflage: Most antennariids depend heavily on camouflage for both predation and defense. The circular shape assumed by *H. psychedelica* when moving about in open water may help to disguise its head and fins. The curvilinear, disruptive pigment pattern probably also provides some protection by allowing the animal to blend in with the complex coral-reef substrate. At the same time, the pigment pattern resembles the growth pattern and yellow-brown background coloration common to a number of widely distributed, Indo-West Pacific hermatypic corals (e.g., members of the families Agariciidae, Faviidae, Mussidae, and Pectiniidae; Fig. 248). It is therefore tempting to speculate that the fish is a mimic of hard corals.

A remarkably similar external pigment pattern has been observed in all known individuals of *H. psychedelica,* although subtle differences make it possible to distinguish and identify individuals in photographs (e.g., compare Figs. 245, 246). The individual pattern and coloration of each fish did not appear to change over the 10-day period of observa-

Fig. 247. *Histiophryne psychedelica* Pietsch, Arnold, and Hall. Adult specimen, about 150 mm SL, Ambon, Indonesia; a sequence of photographs taken over a two-minute period, showing rigorous twisting and turning of the body, together with active use of pectoral and pelvic fins, to facilitate entry into a small hole. (**A**) approaching a hole. (**B, C**) entering head first. (**D, E**) turning upside down and pushing off with pelvic fins. (**F, G, H**) additional squeezing and final disappearance. Photos by David J. Hall.

Fig. 248. Common Indo-West Pacific species of hermatypic coral, with curvilinear, concentric growth pattern and yellow-brown background coloration, reminiscent of the pigment pattern of *Histiophryne psychedelica*. (**A**) *Symphyllia sinuosa*, family Mussidae. (**B**) *Leptoseris explanata*, Agariciidae. (**C**) *Pachyseris rugosa*, Agariciidae. (**D**) *Platygyra ryukyuensis*, Faviidae. (**E**) *Pectinia lactuca*, Pectiniidae. (**F**) *Caulastrea furcata*, Faviidae. Photos by David J. Hall.

tion and was still present several months later as frequently reported by divers. Subsequent photographs of hatchlings and juveniles also displayed essentially the same pattern. Antennariids typically have a highly variable pigment pattern that changes over time to adapt to different surroundings (see Color and Color Pattern, p. 36). We are aware of no other antennariid with such a highly complex, fixed pigment pattern that is apparently common to all individuals regardless of habitat. Unfortunately, all pigmentation was lost almost immediately when the holotype from Ambon was placed in ethanol following formalin fixation, leaving the entire fish devoid of color (Pietsch et al., 2009a, fig. 4A). The distinctive striped pattern, however, is still readily discernible with aid of a dissecting microscope, even in the two paratypes from Bali that have been in ethanol since 1992.

Predation and defense: Antennariids are lie-in-wait predators that employ active mimicry to attract prey. An illicium equipped with a terminal esca is wriggled and waved about to simulate the movements of a small fish or invertebrate. Members of the family are typically found in exposed locations where the lure can be readily observed by potential prey. But, in sharp contrast, species of *Histiophryne* have a vestigial illicium with or without an esca, and while its congeners, especially *H. bougainvilli* and *H. cryptacanthus*, are commonly observed out in the open, our observations suggest that *H. psychedelica* spends most of its time hidden deep within holes and crevices. We suspect that prey may be trapped within these holes as *H. psychedelica* enters and blocks an escape route. This notion is consistent with our observations that *H. psychedelica* preferentially chooses holes with small openings that are effectively closed off as the fish passes through. Such a small opening likely also serves a defensive purpose by denying entry to larger potential predators. It is tempting to speculate further that the fleshy lateral extensions of the cheeks and chin, with highly flexible scalloped margins, are an adaptation to a life in the dark, functioning like cat's whiskers, enabling the frogfish to better detect objects and movement of predators and prey in the interstices of rock and coral.

Reproduction and parental care: Direct field observation by one of us (RJA) at Ambon, from 1 to 6 April 2008, shows that females of *Histiophryne psychedelica*, like those of its congeners, practice a form of egg-brooding in which a small cluster of relatively large eggs, attached to each other by acellular filaments, is completely hidden from view and thereby protected within a pocket formed by the body and pectoral, dorsal, and caudal fins (for more, see Parental Care, p. 482).

Relationships: Analyses of RAG2, COI, and 16S sequences, including 478, 701, and 603 bps, respectively, resulted in all three maximum-likelihood gene trees with the same topology and bootstrap support (Pietsch et al., 2009a: 38, fig. 5). A clade containing three individuals of *H. cryptacanthus* was resolved with bootstrap support of 100%. *Histiophryne psychedelica* formed a lineage distinct from *H. cryptacanthus* with bootstrap support of 100%. The intraspecific COI DNA sequence divergence for *H. cryptacanthus* was 0.0% and between *H. cryptacanthus* and *H. psychedelica*, 5.7%.

A psychedelic look-alike: Of considerable interest with regard to *H. psychedelica* is a putative undescribed member of the genus *Histiophryne*, dubbed the "Paisley Frogfish," so far known only from a few photographs taken by Barbara Auer at Halmahera, Maluku Islands, Indonesia, in April 2016 (personal communication, 14 April 2017; Fig. 249). Based on pigment pattern alone, it appears to be closely related to the Psychedelic Frogfish. Formal description awaits its rediscovery and the collection of specimens.

Fig. 249. *Histiophryne* sp., dubbed the "Paisley Frogfish." An uncollected specimen, about 80 mm SL, similar to *H. psychedelica* but clearly different in pigment pattern and most likely new to science. Observed and photographed by Barbara Auer at Halmahera, Maluku Islands, Indonesia, in April 2016; courtesy of Barbara Auer.

Genus *Allenichthys* Pietsch

Allenichthys Pietsch, 1984b: 32, 38 (type species *Echinophryne glauerti* Whitley, 1944a, by original designation and monotypy).

DIAGNOSIS

Allenichthys is unique among antennariids in having a reduced second dorsal-fin spine, its length less than 9% SL (comparable only to some specimens of the genus *Antennatus* (Fig. 250). *Allenichthys* is further distinguished by having the following combination of character states: skin covered with close-set, bifurcate dermal spinules, the length of the spines of each spinule not more than twice the distance between tips of spines (Fig. 26L); illicium naked, without dermal spinules; esca distinct; pectoral-fin lobe broadly attached to side of body; caudal peduncle absent, the membranous posteriormost margin of soft-dorsal and anal fins attached to body at base of outermost rays of caudal fin; endopterygoid absent; pharyngobranchial I present (with a single, tooth-bearing plate, but present only on right side of a single cleared-and-stained specimen examined); epural absent; pseudobranch absent; swimbladder present; dorsal-fin rays 15 or 16; anal-fin rays eight; all rays of caudal fin bifurcate; pectoral-fin rays nine to 11 (rarely nine); vertebrae 22 or 23 (usually 22) (Tables 14, 15).

DESCRIPTION

Esca a single, elongate appendage, everywhere covered with slender, sometimes fine, hairlike filaments (Fig. 251); illicium unprotected, no naked groove alongside second dorsal-fin spine, the shallow, naked depression between second and third dorsal-fin spines

Fig. 250. *Allenichthys glauerti* (Whitley). WAM P.19214, 90 mm SL, Green Head, Western Australia. Drawing by Cathy L. Short; after Pietsch and Grobecker (1987, fig. 89).

absent; illicium more than twice length of second dorsal-fin spine, 14.8 to 21.8% SL (Fig. 250); anterior end of pterygiophore of illicium terminating posterior to symphysis of upper jaw; illicium and second dorsal-fin spine widely spaced on pterygiophore, the distance between bases of spines 5.5 to 8.3% SL; second dorsal-fin spine unusually short (6.1 to 8.9% SL), constricted at tip, curved posteriorly, and covered with thick, spinulose skin, not connected to head by membrane; third dorsal-fin spine curved posteriorly, tapering toward base (width at distal end 3.9 to 6.3% SL), the full length connected to head by membrane, length 21.2 to 25.8% SL; eye not distinctly surrounded by separate clusters of dermal spinules, diameter 3.2 to 4.8% SL; only distal tip (about 20 to 25% of length) of maxilla naked and tucked beneath folds of skin, the remaining proximal portion covered directly with spinulose skin; a few, tiny, cutaneous appendages scattered over head and body, but not conspicuous; wartlike patches of clustered dermal spinules absent; epibranchial I with a single, elongate, tooth-bearing plate; ceratobranchial I with a single tooth-bearing plate (reduced on left side of single cleared-and-stained specimen examined); dorsal-fin rays 15 or 16, at most six posteriormost bifurcate; anal-fin rays eight, none to all eight bifurcate; pectoral-fin rays 10 or 11 (rarely nine), all simple; all rays of pelvic fin simple; vertebrae 22 or 23, caudal centra 17 or 18 (Tables 14, 15).

Coloration in preservation: light tan, light yellow, pinkish-brown, yellow-brown, to brown, usually somewhat more lightly pigmented on lower half of head and body, particularly on belly; usually dark brown pigment on top of head, from symphysis of maxilla (and sometimes extending down along surface of maxilla to corner of mouth) to space between second and third dorsal-fin spines; similar pigment between third dorsal-fin spine and soft-dorsal fin, and along outermost edge of all fins; dark patch at base of pectoral and pelvic lobe extending onto fin; broad dark bar across base of caudal fin; a few dark, roughly circular spots scattered over head and body; WAM P.8927 (and to a much lesser extent WAM P.19213) with numerous scattered, dark brown markings (about 14 on each side of body, one on each side of third dorsal-fin spine, three or four on outer surface of pectoral fin), each consisting of four (sometimes two or three) small, tightly spaced spots sometimes interconnected to form a ring; WAM P.26502 completely

Fig. 251. Esca of *Allenichthys glauerti* (Whitley), WAM P.5965, 103 mm SL, Beagle Islands, Western Australia. Drawing by Cathy L. Short; after Pietsch and Grobecker (1987, fig. 90).

covered with numerous tiny, tightly spaced, dark brown spots; basidorsal spot absent; illicium sometimes banded; no dark pigment bars radiating from eye (Figs. 252, 253).

A single species.

COMMENTS

Allenichthys was established by Pietsch (1984b: 38) to accommodate a single species, *A. glauerti*, originally placed by Whitley (1944a) in the genus *Echinophryne*. Although *Allenichthys* and *Echinophryne* share three derived features (Table 14), none is unique to these two genera and all are loss character states that most likely do not reflect common ancestry (see Evolutionary Relationships, p. 385).

Allenichthys glauerti (Whitley)
Glauert's Frogfish

Figs. 26L, 250–253, 288; Table 22

Echinophryne glauerti Whitley, 1944a: 272 (original description, two specimens, holotype WAM P.1459, paratype WAM P.1232, Cottesloe Beach, Western Australia). Whitley, 1944b: 28, fig. 5 (figured). Schultz, 1957: 70, pl. 3, fig. B (description after Whitley, 1944a, figure after Whitley, 1944b). Glover, 1968: 794, fig. 3 (additional specimen, SAMA F.3429, Pearson Island, southwest of Elliston, South Australia). Scott et al., 1974: 299, fig. (Western Australia, South Australia; in key). Hutchins and Smith, 1991: 9 (type catalog). Moore et al., 2009: 9 (type catalog). Saunders, 2012: 423 (reference to original description).

Antennarius glauerti (not of Whitley, 1944a, 1944b): Whitley, 1957a: 207, fig. (new combination; specimen described as new, AMS IB.2979, a typical specimen of *A. striatus*). Whitley, 1958: 50 (new to Queensland).

Fig. 252. *Allenichthys glauerti* (Whitley). Busselton Jetty, Busselton, Western Australia. Photo by Daniel Barker.

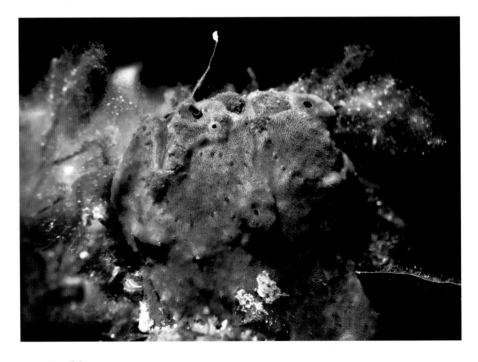

Fig. 253. *Allenichthys glauerti* (Whitley). (**A**, **B**) Busselton Jetty, Busselton, Western Australia. Photos by Ann Storrie. (**C–E**) Busselton Jetty, Busselton, Western Australia. Photos by Jacob Loyacano.

Allenichthys glauerti: Pietsch, 1984b: 38 (new combination, genera of frogfishes). Pietsch and Grobecker, 1987: 221, figs. 16F, 89, 90, 129 (description, distribution, relationships). Paxton et al., 1989: 277 (Western Australia). Pietsch, 1994: 286, fig. 254 (southern coast of Australia). Hutchins, 2001: 22 (Western Australia). Allen et al., 2006: 638 (Australia). Pietsch, 2008: 365, fig. (after Pietsch, 1994). Baker et al., 2009: 17 (South Australia). Arnold and Pietsch, 2012: 128, fig. 2F (molecular phylogeny). Saunders, 2012: 423 (reference to original description).

MATERIAL EXAMINED

Fourteen specimens, 65 to 142 mm SL.

Holotype of *Echinophryne glauerti*: WAM P.1459, 131 mm, Cottesloe Beach, Western Australia, Glauert, June 1932.

Paratype of *Echinophryne glauerti*: WAM P.1232, 61 mm, data as for holotype.

Additional material: SAMA, 1 (102 mm); WAM, 11 (65–142 mm).

DIAGNOSIS AND DESCRIPTION

As given for the genus.

Largest known specimen: 142 mm SL (WAM P.26835-001).

DISTRIBUTION

Thirteen of the 14 specimens of *Allenichthys glauerti* examined by us were collected along the coast of Western Australia, from Shark Bay to Busselton (approximately 25° to 33° S latitude; Fig. 288, Table 22). The single exception (SAMA F.3429), a 102-mm specimen captured in a craypot off Pearson Island, South Australia, on 9 January 1967, is likely the result of dispersal via the Leeuwin and Zeehan currents. The Leeuwin Current flows down and around the Western Australian coast where it meets the Zeehan Current, which in turn flows along the continental slope from the Great Australian Bight to Tasmania. The combination of the Leeuwin and Zeehan currents often bring tropical species well into southern coastal waters (Gomon et al., 2008).

Depth of capture is recorded for only three specimens: WAM P.19214 was trawled between 164 and 183 m; WAM P.19213, in 145 m; and SAMA F.3429, in 62 m; but divers have observed this species at depths as shallow as 7 or 8 m depending on the tide.

COMMENTS

Whitley's (1944a) original description of this species as *Echinophryne glauerti* should not be confused with his 1957 description of *Antennarius glauerti*, based on a typical specimen of *A. striatus* (see Synonymy, p. 350).

Species *Incertae Sedis*
Lophius bigibbus Latreille
nomen dubium

Antennarius (Bigibbus), nigro et griseo variegatus nobis: Commerson, MSS 889(1): 13, 891: 135 (observed by Commerson at Mauritius, about 1770; no extant material). Lacepède, 1798: 325 (footnote reference to Commerson, MSS).

La Lophie Double-Bosse: Lacepède, 1798: 302, 325 [invalid, not Latinized; description, and Latin polynomial in footnote, based on Commerson, MSS; no specimens; *voisine des côtes orientales de l'Afrique* (= Mauritius)]. Sonnini, 1803: 161, 192 (after Lacepède, 1798).

Lophius bigibbus Latreille, 1804: 73 (binomial for *La Lophie Double-Bosse* of Lacepède, 1798; constitutes original description, no known type material). Pietsch and Grobecker, 1987: 262 (*nomen dubium*).

Chironectes bigibbus: Cloquet, August 1817: 599 (new combination; binomial for *La Lophie Double-Bosse* of Lacepède, 1798).

COMMENTS

The specific name *bigibbus* originated in the manuscripts of Philibert Commerson [MSS 889(1): 13, 891: 135], based on specimens examined by him at Mauritius sometime between 1768 and 1770. However, the unpublished materials of Commerson that relate to this form contain only a single, brief descriptive polynomial: "*Antennarius (Bigibbus) Nigro et griseo variegatus nobis,*" a reference to body coloration that could easily apply to a number of different Indo-Pacific antennariids. Of the four drawings of antennariids left by Commerson (see Introduction, p. 9), none can be correlated with this description. Lacepède (1798: 326), in utilizing Commerson's unpublished materials, but without examining specimens, recognized Commerson's species under the vernacular name *La Lophie Double-Bosse*. Six years later, Latreille (1804), again without reference to specimens, fixed the name *bigibbus* in the scientific literature by being the first to provide a Latinized binomial (*Lophius bigibbus*) for Lacepède's *La Lophie Double-Bosse*. Because neither Lacepède nor Latreille provided any additional evidence that might help to identify Commerson's species, the name *Lophius bigibbus* must remain a *nomen dubium* (for further discussion, see Comments, p. 256).

Antennarius commersonii Var. β. *cantoris* Günther
nomen dubium

Antennarius commersoni (not of Latreille): Cantor, 1849: 1186 (description based on "a single individual . . . observed at Singapore in May 1840"). Cantor, 1850: 204 (after Cantor, 1849).

Antennarius commersonii (not of Latreille) **Var. β. *Cantoris*** Günther, 1861a: 193 (original description based on Cantor, 1849, 1850; no specimens). Pietsch and Grobecker, 1987: 263 (*nomen dubium*).

Antennarius cantori: Bleeker, 1865b: 10, 21 (after Cantor, 1849, 1850, and Günther, 1861a; no specimens).

COMMENTS

Antennarius commersonii Var. β. *cantoris* Günther was based solely on a description of a frogfish observed at Singapore and published by Danish naturalist Theodore Edward Cantor (1849: 204) under the name *A. commersoni*. Cantor's description is almost totally devoted to characters of color and color pattern that do not seem to fit any known antennariid; the combination of fin-ray counts and apparent lack of cutaneous appendages only add to the confusion. Günther (1861), having no specimen but realizing that the set of characters provided by Cantor could not be assigned to any known species, chose to name it as a variety of *A. commersonii*. Four years later, and again without reference to material, Bleeker (1865b) gave it full specific status. Because there is little possibility that the true identity of Cantor's (1848) observation will ever be discovered, the name *Antennarius cantoris* is set aside as a *nomen dubium*.

Golem cooperae Strasburg
nomen dubium

Golem cooperae Strasburg, 1966: 475, figs. 1, 2 (original description, single specimen, holotype USNM 199683, 139 mm, Lami, mangrove area 4.8 km NW of Suva, Viti Levu Island, Fiji, cement company wharf, stoned to death in 0.6 m, July 1964). Pietsch and Grobecker, 1987: 265 (*nomen dubium*).

COMMENTS

Golem cooperae Strasburg (1966) was based on a single aberrant specimen from Fiji in which the three cephalic dorsal-fin spines had been lost due to injury sustained prior to capture (despite this loss of the luring apparatus, the specimen has large, well-developed ovaries and otherwise appears to have been healthy). Healing of the damaged tissue of the dorsal surface of the head resulted in an antennariid that was mistaken by Strasburg (1966) as an undescribed species of the nominal genus *Golem* Whitley (a synonym of *Histiophryne* Gill, see p. 318), characterized by having all three dorsal-fin spines embedded beneath the skin of the head. There is no doubt that the holotype of *G. cooperae* represents either *Antennarius striatus* or *A. hispidus*, both known from Fiji; however, it is impossible at this time to determine with which of the two it should be synonymized.

Lophius marmoratus Shaw
nomen dubium

Fig. 254

Lophius marmoratus Shaw, 1794, pl. 176, lower fig. (*Lophius subcompressus lividus*, "The Marbled Lophius," original description, type material unknown, native of "the coast of New Holland and the Neighboring isles"; engraving by Frederick P. Nodder; Fig. 254). Shaw, 1804: 386, pl. 165, lower fig. ("subcompressed livid angler," after Shaw, 1794, no specimens, "observed about the coast of Otaheitee, etc."). Cuvier, 1816: 310 (after Shaw, 1794). Günther, 1861a: 184 (in footnote). Gill, 1879c: 226 ("incomprehensible, . . . must represent a factitious [sic] fish"). Swain, 1883: 308 (identified with *Antennarius*). Schultz, 1957: 103 (in synonymy of *Histrio histrio*). Pietsch and Grobecker, 1987: 261, pl. 56 (*nomen dubium*).

Lophius histrio (not of Linnaeus) **var. c. *marmoratus***: Bloch and Schneider, 1801: 141 (after Shaw, 1794). Schultz, 1957, 104 (in synonymy of *Histrio histrio*).

Chironectes marmoratus: Cuvier, 1829: 252 (new combination; in part, confused with *Histrio histrio*).

Antennarius marmoratus **var. ε. *marmorata***: Günther, 1861a: 187 (new combination; in part, confused with *Histrio histrio*).

Antennarius marmoratus: Borodin, 1930: 62 (after Bloch and Schneider, 1801).

COMMENTS

The earliest association of the name *marmoratus* with the family Antennariidae was made by Shaw (1794, pl. 176, lower figure), with the description of *Lophius marmoratus*, said to be based on material originating from "the coast of New Holland and the neighboring isles." The engraving accompanying his description, however, is clearly not recognizable as a frogfish or, for that matter, any known teleost (Fig. 254). Gill (1879c: 226) summarized Shaw's description as "incomprehensible, and, if the figure is at all correct, must

Fig. 254. George Shaw's *Lophius marmoratus*, *nomen dubium*, long confused with *Histrio histrio* (Linnaeus). Drawing and engraving by Frederick Polydore Nodder; after Shaw (1794, pl. 176, lower fig.); courtesy of the Natural History Museum, London.

represent a factitious [sic] fish." Rather than based on an actual specimen (even a damaged or deformed individual), his brief description ("subcompressed livid lophius, varied with whitish and ferruginous; with a single dorsal fin") seems likely to have been made directly from this strange drawing. In any case, if a specimen was available to Shaw, no trace of it has survived.

Soon after Shaw's description appeared, the name *marmoratus* became nearly inextricably associated with what we now know as *Histrio histrio*. This confusion was initiated by Bloch and Schneider (1801: 141) who recognized *Lophius marmoratus* (as described by Shaw) as a variety of *Lophius histrio* (i.e., *L. histrio* var. c. *marmoratus*). The problem was later compounded by Cuvier (1829: 252), who synonymized *L. histrio* var. c. *marmoratus* Bloch and Schneider and *L. raninus* Tilesius (both synonyms of *H. histrio*) under a new combination, *Chironectes marmoratus*. From that time on, a host of authors, from Lesson (1831) to Padmanabhan (1958), have routinely used *marmoratus*, in combination with either *Chironectes* or *Antennarius*, as the specific name for *Histrio histrio* (see synonymy of *H. histrio*, p. 182; on top of this confusion with *H. histrio*, Valenciennes (1837: 432) associated Shaw's *L. marmoratus* with *Histiophryne bougainvilli*; see Pietsch, 1985a: 68, fig. 6). Clearly, however, Shaw's (1794) name cannot reasonably by associated with *H. histrio* nor with any other known antennariid, and must therefore be set aside as a *nomen dubium*.

Antennarius commersoni Var. γ. *Musei britannici* Günther
nomen dubium

Antennarius commersoni (not of Latreille) **Var. γ. *Musei britannici*** Günther, 1861a: 193 (original description, single specimen, holotype lost, not found in BMNH, no locality). Pietsch and Grobecker, 1987: 264 (*nomen dubium*).

COMMENTS

This form, initially described by Günther (1861: 193) as a variety of *A. commersoni*, was based on a single "fine specimen" of unknown origin. The brief description, devoted almost solely to characters of color and color pattern, could be applied to almost any antennariid. Günther's specimen could not be found in the collections of the Natural History Museum (BMNH). Thus, without some further information this name cannot be identified and must be set aside as a *nomen dubium*.

Antennarius portoricensis Stahl
nomen nudum

Antennarius portoricensis: Stahl, 1882: 246 (authorship attributed to Felipe Poey; Puerto Rico). Evermann and Marsh, 1900: 335 [uncertainly placed in synonymy of *A. inops* (= *Histrio histrio*)]. Pietsch and Grobecker, 1987: 265 (*nomen nudum*).

COMMENTS

This name was listed by Puerto Rican physician and naturalist Agustín Stahl (1882) in a published catalog of his zoological collections, without application to a description or type. It must therefore be set aside as a *nomen nudum*.

Antennarius princeps Gill
nomen nudum

Antennarius princeps: Gill, 1863: 90 (designated type species of "Cheironectes Cuv. nec Illiger"; authorship attributed to Commerson). Pietsch and Grobecker, 1987: 264 (*nomen nudum*).

COMMENTS

This name was used by Gill (1863) in a synopsis of the major representatives of the families of the "Pediculati," without application to a description or type. It therefore constitutes a *nomen nudum*.

Chironectes subrotundatus Castelnau
nomen dubium

Chironectes subrotundatus Castelnau, 1875: 25 (original description, single specimen, existence of holotype unknown, not found in MNHN nor in any other institution; about 50 mm total length; Port Walcott, Western Australia). Whitley, 1933: 105 [description after Castelnau, 1875; comparison with *Lophiocharon broomensis* (= *L. trisignatus*)]. Schultz, 1957: 80 [in synonymy of *Antennatus bigibbus* (= *A. tuberosus*)]. Pietsch and Grobecker, 1987: 264 (*nomen dubium*).
Antennarius subrotundatus: Macleay, 1881: 580 (new combination; after Castelnau, 1875). McCulloch, 1929: 407 (after Castelnau, 1875).
Lophiocharon subrotundatus: Whitley, 1948: 31 (new combination, Western Australia).
Tathicarpus subrotundatus: Whitley, 1964: 57, 60 (new combination; erroneously listed as a senior synonym of *T. butleri, T. muscosus,* and *T. appeli*).

COMMENTS

Castelnau's (1875) *Chironectes subrotundatus* was based on a single specimen collected by a Mr. Bostock at Port Walcott, Western Australia. The description provided by Castelnau is inadequate and the type specimen has not survived (the latter concluded after a thorough search of the MNHN and all Australian institutions). Thus, this nominal form must be listed as a *nomen dubium*.

Antennarius tigrinus Thresher
nomen nudum

Antennarius tigrinus: Thresher, 1984: 41 (name given without authorship or locality).

COMMENTS

This name was mentioned by Thresher (1984), without application to a description or type, in reference to the production of egg rafts in antennariids. Mosher (1954) and Burgess (1976) are cited but neither of these authors mention the name. It is perhaps an error for *Antennarius tigris* (a junior synonym of *A. scaber*) but there is no way to know for sure. It must therefore be set aside as a *nomen nudum*.

The Fossil Record

Despite the discovery of many new fossils and reevaluation of poorly known extinct taxa, the evolutionary history of lophiiform fishes is still poorly understood. The extraordinary anatomical diversification of the various subunits of the order has made it difficult to decipher their origin (see Evolutionary Relationships, p. 364). Moreover, the age of divergence of the various lophiiform clades cannot be precisely established on the basis of paleontological data because representatives of these groups are still relatively rare in the fossil record (Carnevale and Pietsch, 2011, 2012). Although present in varying amounts and degrees of preservation, we do, however, have remains of all five lophiiform suborders. The Lophioidei, the basal clade of the order, is represented by disarticulated fragmentary bones (Lawley, 1876; Stefano, 1910; Leriche, 1910, 1926; Ray et al., 1968; Landini, 1977; Purdy et al., 2001; Schultz, 2006), otoliths (e.g., see Nolf, 1972, 1985), and teeth (Leriche, 1906, 1908; Hasegawa et al., 1988), but articulated skeletal remains are extremely rare (but see Pietsch and Carnevale, 2011; Figs. 255, 256; and Carnevale

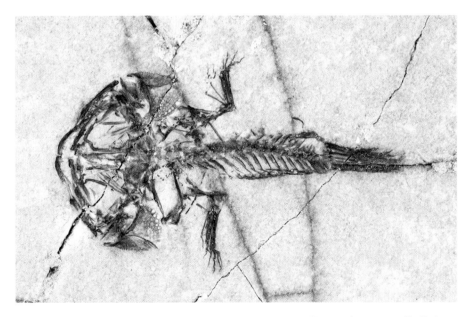

Fig. 255. †*Sharfia mirabilis* Pietsch and Carnevale. Holotype, MNHN Bol 38/39, 40 mm SL, Early Eocene, Late Ypresian, Monte Bolca, Pesciara cave site. Photo by Philippe Loubry; after Pietsch and Carnevale (2011, fig. 2).

Fig. 256. Hypothetical reconstruction of †*Sharfia mirabilis* Pietsch and Carnevale. Drawing by Joseph R. Tomelleri; after Pietsch and Carnevale (2011, fig. 3).

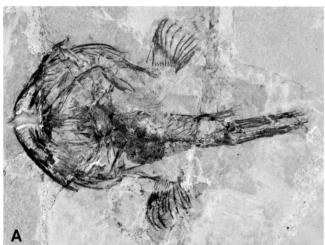

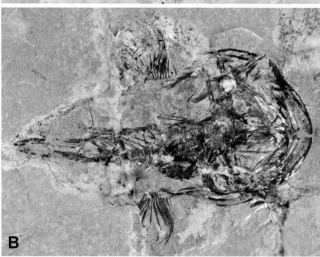

Fig. 257. †*Caruso brachysomus* (Agassiz). (**A**, **B**) lectotype, MNHN Bol 42/43, part and counterpart, 126 mm SL, Early Eocene, Late Ypresian, Monte Bolca, Pesciara cave site. Photos by Philippe Loubry; after Carnevale and Pietsch (2012, fig. 1).

and Pietsch, 2012; Fig. 257). Eocene material is restricted to a few whole specimens from the localities of Monte Bolca (e.g., Volta, 1796; Agassiz, 1833–1844; Zigno, 1874; Blot, 1980; for a review of the fossils from the famous collecting site of Monte Bolca, lying northeast of Verona, in the foothills of the Italian Alps, see Carnevale et al., 2014), and from Gornyi Luch, North Caucasus, Russia (Bannikov, 2004). Another, yet undescribed lophiid has been identified from the Oligocene clay of Grube Unterfeld (Frauenweiler), Upper Rhine Graben, Baden-Württemberg, Germany. While Neogene fossils of this family are also known, from Algeria (Arambourg, 1927), Azerbaijan (Sychevskaya and Prokofiev, 2010), and Italy (Sorbini, 1988), the Eocene diversity of these fishes suggests that the origin of the lophioid body plan, and therefore that of lophiiforms as a whole, is certainly much older.

The Eocene existence of ogcocephaloids is also well established, testified by the otolith-based taxon †*Halieutaea cirrhosa* from the Lutetian of southern England and Germany, and possibly also from the Bartonian of southern England (see Stinton, 1978; Schwarzhans, 2007). Ogcocephalid otoliths are known as well from the Oligocene and Miocene of Europe (e.g., Schwarzhans, 1994, 2010) and the Miocene of Australia (Schwarzhans,

Fig. 258. †*Tarkus squirei* Carnevale and Pietsch. (**A**, **B**) holotype, MCSNV T.158/159, part and counterpart, 121 mm SL, Early Eocene, Late Ypresian, Monte Bolca, Pesciara cave site. Photos courtesy of Roberto Zorzin, Anna Vaccari, and Museo Civico di Storia Naturale, Verona; after Carnevale and Pietsch (2011, fig. 1).

Fig. 259. †*Tarkus squirei* Carnevale and Pietsch. Paratypes, Early Eocene, Late Ypresian, Monte Bolca, Pesciara cave site. (**A**) MCSNV T.371, 138 mm SL. (**B**) MCSNV T.492, 162 mm SL. (**C**) MCSNV T.281, 126 mm SL. Courtesy of Roberto Zorzin, Anna Vaccari, and Museo Civico di Storia Naturale, Verona; after Carnevale and Pietsch (2011, fig. 2).

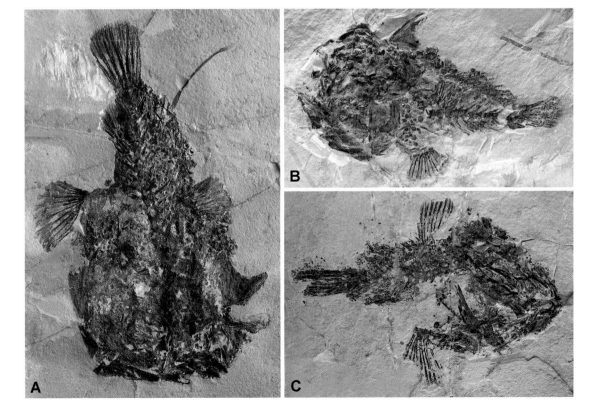

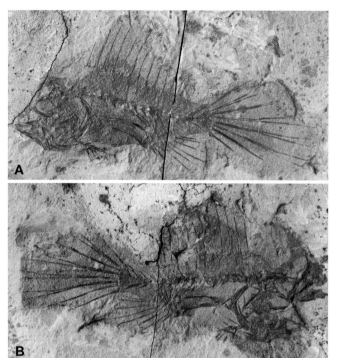

Fig. 260. †*Histionotophorus bassani* (De Zigno). (**A**, **B**) holotype MGPD 68487, part and counterpart, 55 mm SL, Early Eocene, Late Ypresian, Monte Bolca, Pesciara cave site. Photos by Stefano Castelli; after Carnevale and Pietsch (2010, fig. 1).

1985). But more significant is the recent description of †*Tarkus squirei*, based on complete articulated skeletons from Monte Bolca (Carnevale and Pietsch, 2011; Figs. 258, 259), providing the earliest evidence of whole fossil remains of the family Ogcocephalidae, thereby filling a large gap that existed in the skeletal record of the order, and suggesting also that the modern structural plan of batfishes originated before the early Eocene.

Turning now to antennarioid fossil remains, the earliest described taxon is †*Histionotophorus bassani* (Zigno, 1887) from the Eocene of Monte Bolca, traditionally placed in the Antennariidae (e.g. Eastman, 1904, 1905; Gill, 1904; Regan, 1912; Le Danois, 1964; Blot, 1980) but more recently reallocated to the antennarioid family Brachionichthyidae (Rosen and Patterson, 1969; Pietsch, 1981; Carnevale and Pietsch, 2010; Fig. 260). A second brachionichthyid genus and species, †*Orrichthys longimanus*, was described by Carnevale and Pietsch (2010) on the basis of two nearly complete skeletons, part and counterpart, from the Eocene Pesciara cave site of Monte Bolca (Fig. 261).

As for the family Antennariidae, the earliest described record is represented by the otolith-based †*Antennarius euglyphus* from the basal part of the London Clay Formation, southern England (Stinton, 1966). Additional fossil remains include a single otolith identified as that of a species of *Antennarius* from the Eocene of Sables de Lede, Belgium (Nolf, 1972), and the otolith-based taxa †*Antennarius excavatus*, from the Lutetian (Middle Eocene) Selsey Formation of Southern England (Stinton, 1978), and the † "genus *Antennariidarum*" *furcatus*, from the Bartonian (Middle Eocene) of southern England and the Priabonian (Upper Eocene) Marne di Possagno of northern Italy (see Girone and Nolf, 2008). Nearly complete articulated skeletons include †*Fowlerichthys monodi* (originally described as †*Antennarius monodi* by Carnevale and Pietsch, 2006; Fig. 262) from the Messinian (Upper Miocene) diatomites of the Chelif Basin, northeast Algeria; †*Eophryne barbutii* from the Early Eocene, late Ypresian, Pesciara locality of Monte Bolca (Carnevale and Pietsch, 2009b; Fig. 263); and another, as yet undescribed, antennariid, also from Monte Bolca.

The coeval existence of lophioids, antennarioids, and ogcocephaloids, as well as chaunacoids, which are represented by otoliths from the Eocene of New Zealand (see Schwarzhans, 1980), implies that the deep-sea ceratioids, represented by a dozen or so fossils from the Late Miocene Puente Formation of Southern California (see Pietsch and

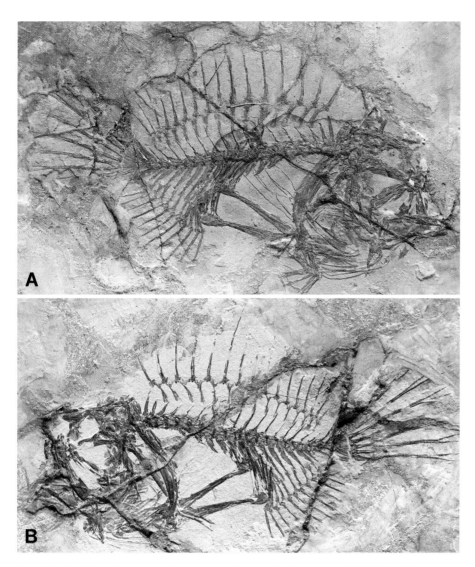

Fig. 261. †*Orrichthys longimanus* Carnevale and Pietsch. (**A**, **B**) holotype, MCSNV T.160/161, part and counterpart, 63 mm SL, Early Eocene, Late Ypresian, Monte Bolca, Pesciara cave site. Photos courtesy of Roberto Zorzin, Anna Vaccari, and Museo Civico di Storia Naturale, Verona; after Carnevale and Pietsch (2010, fig. 5).

Lavenberg, 1980; Pietsch and Shimazaki, 2005; Carnevale et al., 2008; Pietsch, 2009; Carnevale and Pietsch, 2009a; Fig. 264), were probably present at that time as well. Thus, Patterson and Rosen (1989), based on phylogenetic considerations and the puzzling lophiiform fossil record known at that time, postulated that all lophiiform lineages were already in existence and well diversified before the early Eocene.

That all major anglerfish clades were most likely present at the same time, makes it impossible, on the basis of the fossil record alone, to reconstruct the order of events in the phylogeny of lophiiform fishes and to estimate the minimum age of origin of the order as a whole (Carnevale and Pietsch, 2006, 2012). Recent molecular-clock analyses

Fig. 262. †*Fowlerichthys monodi* Carnevale and Pietsch. Holotype, MNHN ORA 388, 122 mm SL, Upper Miocene (Messinian) of Raz-el-Aïn, near Oran, northeastern Algeria (note a second fish evident within the stomach). Photo by Philippe Loubry; after Carnevale and Pietsch (2006, fig. 1).

Fig. 263. †*Eophryne barbutii* Carnevale and Pietsch. Holotype, MCSNV B.6513, 20.5 mm SL, Early Eocene, Late Ypresian, Monte Bolca, Pesciara cave site; the earliest evidence of articulated antennariid skeletal remains, demonstrating that the modern frogfish body plan was already in existence approximately 53 Mya. Courtesy of Roberto Zorzin, Anna Vaccari, and Museo Civico di Storia Naturale, Verona; after Carnevale and Pietsch (2009b, fig. 1).

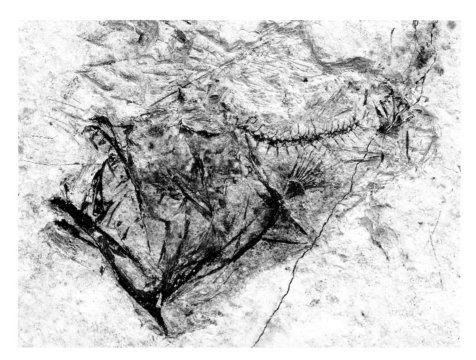

Fig. 264. †*Acentrophryne* sp., LACM-VP 117685, 28 mm SL, from the Late Miocene of Chalk Hill, western Puente Hills, Hacienda Heights (LACM locality 6908), Los Angeles County, California. Courtesy of Gary T. Takeuchi and the Department of Vertebrate Paleontology, Natural History Museum of Los Angeles County; after Pietsch and Lavenberg (1980, fig. 1) and Carnevale and Pietsch (2009a, fig. 1).

of divergence times have proposed two different hypotheses for the origin of these fishes. Alfaro et al. (2009) and Santini et al. (2009) suggested that the origin of lophiiforms is relatively consistent with the fossil record and must be searched for in the Paleocene. On the contrary, Miya et al. (2010) concluded that all the lophiiform clades originated in the Cretaceous during a relatively short time interval between 130 and 100 million years ago, thereby implying the existence of a minimum range of 50 million years.

5. Evolutionary Relationships

It is generally recognized that to move beyond guesswork and place questions of biological adaptation in a true scientific context, an understanding of the sequence of evolution of the major clades of a group of organisms is essential. A rigorously tested systematic hypothesis of relationships allows for a precise reconstruction of aspects of character evolution, which in turn casts light on broader evolutionary processes that have taken place within a taxon, in this case, the Antennariidae. The present chapter is devoted to that effort.

The morphological analysis of relationships undertaken here follows, in a general way, the phylogenetic approach first described in detail by Hennig (1950, 1965, 1966). Concisely stated, the Hennigian or cladistic methodology requires that taxa be monophyletic in the sense that they include all the descendants of a hypothetical ancestral species. Monophyletic units are defined on the basis of shared derived (synapomorphic) features. Shared primitive (symplesiomorphic) features, unshared derived (autapomorphic) features, degrees of difference, grades, and overall similarity are not utilized in forming hypotheses of common ancestry. Species or groups of species hypothesized to have had a common ancestor are called "sister-groups." An apomorphic (derived) feature used for the definition of a sister-group pair cannot serve for the definition of taxa contained within either sister-group, because it is plesiomorphic (primitive) at the level of the included subtaxa. The relative primitiveness of character states is identified by the procedure of outgroup comparison, as discussed by Eldredge and Cracraft (1980: 63). In the only significant departure from the Hennigian approach, we have not formally named all branching points in our phylogenetic diagrams (trees or cladograms), with the result that our classification is not strictly dichotomous (see Pietsch, 1981b, 1984d; Pietsch and Grobecker, 1987). The loss of convenience in discussing individual sister-groups by a single name is outweighed by the avoidance of adding a multiplicity of new taxonomic categories and names, as well as the necessity of altering names that are well-established in the scientific literature.

Ordinal Relationships

Anglerfishes have traditionally been allied with the toadfishes of the teleost order Batrachoidiformes (Fig. 265). This alignment dates back to Georges Cuvier's second edition of *Le Règne Animal* (1829: xi, 249), in which the then known lophioids, antennarioids, and ogcocephaloids were placed together with the batrachoidids in a group called the *Acanthoptérygiens à Pectorales Pédiculées*, a name proposed by Cuvier in reference to the footlike structure of the pectoral fins of these fishes. This phylogenetic arrangement, or a variation of it that placed anglerfishes and toadfishes side by side, was later followed by Valenciennes (1837: 335) and a host of others (e.g., Cope, 1872: 340; Jordan and Evermann, 1898: 2712; Jordan, 1902: 361; Boulenger, 1904: 717; Jordan, 1905: 542; Eaton et al., 1954: 216), including Albert Günther (1861a: 178) who was the first to use the name Pediculati, which he Latinized from the original French vernacular of Cuvier (1829). Gill (1863: 88), however, vehemently denied a close relationship between anglerfishes and toadfishes: "The genus *Batrachus*, referred to the Pediculati by Cuvier, has really little affinity to the true representatives of the group, and has been, by general consent, separated from them by all the more modern systematists." But just as sure of himself a few years later, Gill (1872: xli–xlii) changed his mind: "The natural character of the association of forms combined therein [the Pediculati] is obvious, and has never been questioned, and the comparatively slight affinity with them of the Batrachoids, which were formerly combined with them, is now universally conceded. . . . Their relations are most intimate with the Batrachoid and Blennioid forms, and doubtless they have descended from the same common progenitors."

Like Cuvier (1829) and his disciple Valenciennes (1837), Regan (1912: 278) initially believed anglerfishes and toadfishes to be so closely related that he included them as suborders of the Pediculati, a taxon recognized by him as an acanthopterygian order: "The Batrachoidea are here included

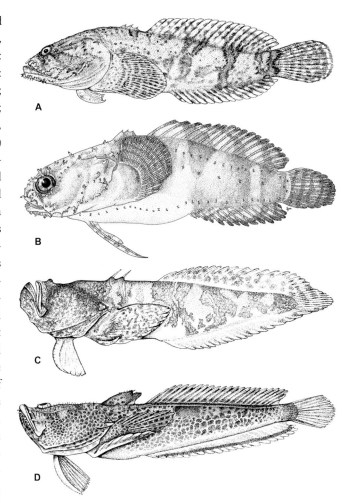

Fig. 265. Representative species of the teleost family Batrachoididae. Although contradicted by recent molecular evidence, this group has been recognized by nearly all ichthyologists as the most likely sister-taxon of lophiiform fishes. (**A**) *Potamobatrachus trispinosus*, holotype, 49 mm SL, MZUSP 4335. Drawing by Keiko H. Moore; after Collette, 1995. (**B**) *Halophryne hutchinsi*, 92 mm SL, USNM 150899. Drawing by Susan G. Monden; after Greenfield, 1998. (**C**) *Thalassophryne amazonica*, 68 mm SL, USNM 210867. Drawing by Mildred H. Carrington; after Collette, 1966. (**D**) *Daector schmitti*, paratype, 80 mm SL, CAS-SU 14949. Drawing by Mildred H. Carrington; after Collette, 1966, 1968). Courtesy of Karin Tucker and the University of California Press.

in the Pediculati rather than in the Percomorphi, for it can hardly be the case that the resemblance in osteological characters, especially in the structure of the pectoral arch, is not due to real affinity." Later, however, in his review of *The Pediculate Fishes of the Suborder Ceratioidea*, Regan (1926: 3) separated the Lophiiformes from the Batrachoidiformes (but kept them side-by-side, thereby implying a sister-group relationship), stating that "although the resemblances in the pectoral arch may be due to relationship, the differences in other characters are sufficient to keep them apart." Since that time, and up to only recently, Regan's (1926)

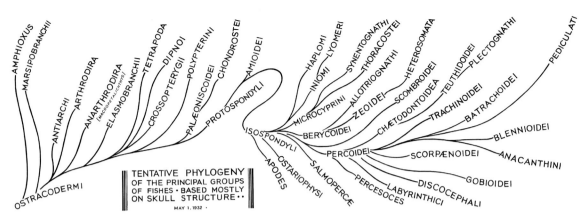

Fig. 266. William King Gregory's "Tentative Phylogeny of the Principal Groups of Fishes," one of many evolutionary hypotheses published throughout the twentieth century that supports a close relationship between the toadfishes (Batrachoidei) and the anglerfishes (Pediculati), indicated by a shared branching point at the far right of the diagram. After Gregory (1933, pl. 2); courtesy of Karin Tucker and the University of California Press.

revised opinion was almost universally accepted—the more significant studies reaching that conclusion were those of Regan and Trewavas (1932: 25), Gregory (1933: 386; Fig. 266), Gregory and Conrad (1936: 198, 200), Berg (1940: 498), Gregory (1951: 225), Eaton et al. (1954: 216), Monod (1960: 622), Greenwood et al. (1966: 388), Rosen and Patterson (1969: 438), and Lauder and Liem (1983: 152). For a return to Regan's 1912 proposal, see Gosline (1971: 173).

In a paper titled "The Paracanthopterygii Revisited: Order and Disorder," Patterson and Rosen (1989: 23) reaffirmed their earlier (see Rosen and Patterson, 1969: 438) contention that the Batrachoidiformes and Lophiiformes are sister-groups, citing new evidence derived from the dorsal gill-arch skeleton: in both these orders the first pharyngobranchial and the suspensory tip of the first epibranchial are reduced or lost, and, if present, both are withdrawn laterally away from the second and third pharyngobranchials. Patterson and Rosen (1989: 23) then listed two other "probable" synapomorphies: (1) the ventral gill arches converge on a very short copula, which is ossified very feebly or not at all; and (2) the prezygapophyses of the first vertebra insert into hollow exoccipital bony tubes that are secondarily elongated, extending to or beyond the basioccipital condyle (Rosen, 1985: 29).

More recent morphological evidence for a batrachoidiform-lophiiform relationship, however, has been rather limited. A notable exception is a detailed analysis of infrabranchial musculature (muscles that serve the ventral portion of the gill arches) by Datovo et al. (2014: 5–6, figs. 5B, 7) that identified two unique derived characters that clearly favor the hypothesis that the Batrachoidiformes and Lophiiformes are sister taxa: an extremely reduced *obliquus ventralis I* muscle and a complex insertion of the *rectus communis* on ceratobranchials 4 and 5 (Fig. 267). Datovo et al. (2014: 6) also found that most, but not all, batrachoidiforms and lophiiforms additionally share other specializations of the gill-arch system that are not found elsewhere among acanthopterygians, such as an extreme reduction of *obliquus ventralis II*, a "musculous" insertion of the *rectus communis* on ceratobranchial 5, and the anterior tip of ceratobranchial 1 positioned posterior to that of ceratobranchial 2 (see Pietsch and Orr, 2007: 15, fig. 12B). The authors (Datovo

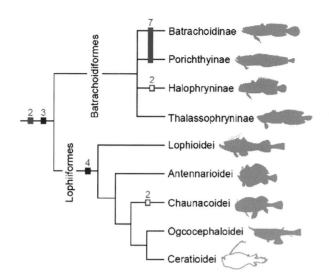

Fig. 267. Maximum parsimony optimization (ACCTRAN) of identified derived characters of the infrabranchial musculature, with characters superimposed on a cladogram of relationships among the Batrachoidiformes and Lophiiformes (for details see Datovo et al., 2014). After Datovo et al. (2014, fig. 5B); courtesy of Alessio Datovo.

et al., 2014: 6) warned, however, that "precise optimizations of such characters are problematic and demand a more encompassing taxonomic sampling of the Pediculati."

In summary, the Batrachoidiformes is the only group of fishes ever shown to bear convincing morphological evidence of a sister-group relationship with lophiiforms (Pietsch, 1981b: 413): "the latter are surely monophyletic, the former less surely, but if so, the two are sister groups" (Patterson and Rosen, 1989: 24). But more recent evidence of a different kind now indicates that perhaps anglerfishes have nothing at all to do with toadfishes, but are most closely related instead to triggerfishes, boxfishes, puffers, and their allies; that is, to all those fishes that make up the teleost order Tetraodontiformes. The evidence for this is molecular. In a comprehensive study of teleost relationships based on 100 complete mitochondrial DNA sequences, Miya et al. (2003: 131) showed that the commonly accepted position of the Lophiiformes among "relatively primitive groups within higher teleosts (Paracanthopterygii)" could not be supported. They found instead that lophiiforms were "confidently placed" within a crown group of teleosts that also included two tetraodontiform species (*Sufflamen fraenatus* and *Stephanolepis cirrhifer*), plus *Antigonia capros* (Zeiformes, Caproidae). Holcroft (2004: 8; see also Holcroft, 2005; and Yamanoue et al., 2007), in testing hypotheses of tetraodontiform relationships using the RAG1 gene, also suggested a more derived placement of lophiiforms, the one member of this order examined, *Lophius americanus*, clustering with *Siganus doliatus*; and these two taxa, along with *Antigonia capros*, forming the sister-group of an assemblage containing, among other groups, a monophyletic Tetraodontiformes.

Miya et al. (2005), in a follow-up study of the relationships of the Batrachoidiformes, this time inferring relationships from a partitioned Bayesian analysis of 102 whole mitochondrial genomic sequences, verified the earlier findings (Miya et al., 2003), showing once again that the Lophiiformes form a sister-group relationship with a clade comprising tetraodontiforms, plus *Antigonia*, supported by 99% to 100% posterior probabilities (Miya et al., 2005: 297). They concluded also that the least comprehensive monophyletic group that includes both Batrachoidiformes and Lophiiformes encompasses nearly the entire Percomorpha (Percomorpha minus Ophidiiformes), strongly suggesting that the two groups diverged relatively basally within the Percomorpha. Bayesian analysis of the three data sets found no tree topology that is congruent with the above two hypotheses, indicating that the probability of the Batrachoidiformes being both a member of the Paracanthopterygii and the sister-group of the Lophiiformes is less than 1/60,000 or 0.00002 in the Bayesian context (Miya et al., 2005: 297). While additional molecular support for a close relationship between lophiiforms and tetraodontiforms (along with various other taxa, including acanthuroids, caproids, siganids, and zeiforms) in a highly derived position within percomorph fishes has been reported by many others (e.g., Chen et al., 2003, 2007, 2014; Simmons and Miya, 2004; Dettaï and Lecointre, 2008; Holcroft and Wiley, 2008; Li et al., 2008; Santini et al., 2009, 2013; Miya et al., 2010; Near et al., 2012; Betancur-R et al., 2013, 2017; Arcila and Tyler, 2017; Nelson et al., 2017; Alfaro et al., 2018; for a good critique of some of these more recent classifications, see Britz, 2017), there is also some morphological evidence of a lophiiform-tetraodontiform relationship. In a remarkable study of the phylogenetic significance of color patterns in marine teleost larvae, Baldwin (2013: 523, 549, figs. 27, 51) pointed out the striking similarity of the larvae of some tetraodontiforms and lophiiforms in having the head and body enclosed in an inflated sac covered with one or more kinds of chromatophores. Chanet et al. (2013) provided evidence for a close phylogenetic relationship between tetraodontiforms and lophiiforms based on an analysis of soft anatomy: (1) a reduced gill opening; (2) rounded and anteriorly disposed kidneys; (3) a compact thyroid included in a blood sinus; (4) an abbreviated spinal cord; (5) an asymmetric liver; and (6) clusters of supramedullary neurons in the rostral part of the spinal cord. While these similarities may constitute synapomorphies, the authors acknowledged that their conclusions require confirmation through rigorous phylogenetic analysis of a much larger sample size of species of both orders.

In summary, considering the totality of evidence for a batrachoidiform versus a tetraodontiform relationship, no satisfactory conclusion can be reached. The Lophiiformes, like numerous other acanthomorph orders, is currently considered *incertae sedis* within percomorph fishes (Wiley and Johnson, 2010: 165). The contradictory morphological and molecular data remain for future phylogeneticists to resolve.

Subordinal and Familial Relationships

While agreeing with almost everything concerning the ordinal relationships of lophiiform fishes that had been proposed earlier by Regan (1912, 1926), Gregory (1933, 1951) and Gregory and Conrad (1936) provided an elaborate scenario of how members of this order might have evolved and diversified from shallow-water anglerfishes (antennarioids and lophioids). They illustrated their ideas with the first branching diagrams of anglerfishes ever published—"pictorial phylogenies of the pediculate fishes" that attempt to show the "principal lines of cleavage leading respectively to the true anglers (*Lophius*), the sea-bats (ogcocephalids), the sea-mice (antennariids) and the deep-sea anglers (ceratioids)" (Gregory and Conrad, 1936: 194, 198, figs. 3, 4); and the "divergent evolution" or "adaptive branching" of the deep-sea anglers (Gregory, 1933: 405, 1951: 312). In their scheme, an *Antennarius*-like ancestor is hypothesized as basal to everything else, with a divergent branch giving rise to lophiids and ogcocephaloids and another leading to all the deep-sea families (Fig. 268). Although based on little more than guesswork, and contrary to current thought, these ideas set the stage for the subsequent, more rigorous work of Rosen and Patterson (1969), Patterson and Rosen (1989), and others. Not surprisingly, however, up until this time, the order as a whole had not been adequately defined, and certainly not diagnosed in modern cladistic terms.

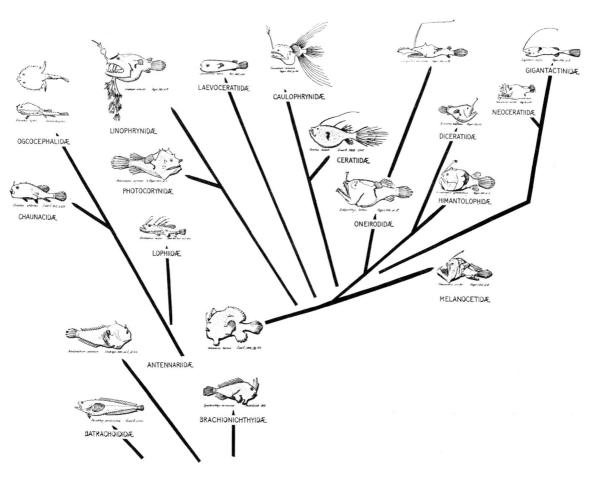

Fig. 268. "Pictorial phylogeny of the pediculate fishes": the earliest published branching diagram of relationships of anglerfish families, showing the Antennariidae giving rise on one hand to lophiids, chaunacids, and ogcocephalids and on the other to the families of the deep-sea Ceratioidei, based almost exclusively on the conclusions of Regan (1926) and Regan and Trewavas (1932). After Gregory and Conrad (1936, figs. 3, 4); courtesy of Judy Choi and the University of Chicago Press.

That the order Lophiiformes itself constitutes a natural assemblage seems certain, based on at least seven unique and morphologically complex, shared derived features (numbered 1–7 in Fig. 269; a modification of Pietsch, 1981b, 1984d, 2009; Pietsch and Grobecker, 1987: 268; see also Wiley and Johnson, 2010: 165):

1. Spinous dorsal fin primitively of six spines, the anteriormost three of which are cephalic in position, the first modified to serve as a luring apparatus, involving numerous associated specializations, including a medial depression of the anterior portion of the cranium, loss of the nasal bones (the nasal of Rosen and Patterson, 1969, is the lateral ethmoid) and supraoccipital lateral-line commissure, and modifications of associated musculature and innervation (Carnevale and Pietsch, 2012: 57, 68);

2. Epiotics separated from the parietals and meeting on the midline posterior to the supraoccipital (Pietsch, 1981b: 391, 397, figs. 3, 15–19; Carnevale and Pietsch, 2012: 54, 58, figs. 4, 9);

Order Lophiiformes

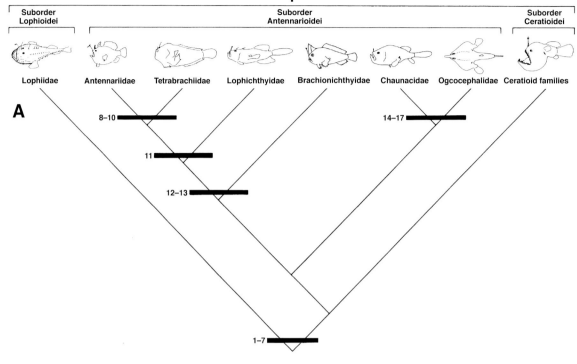

Order Lophiiformes

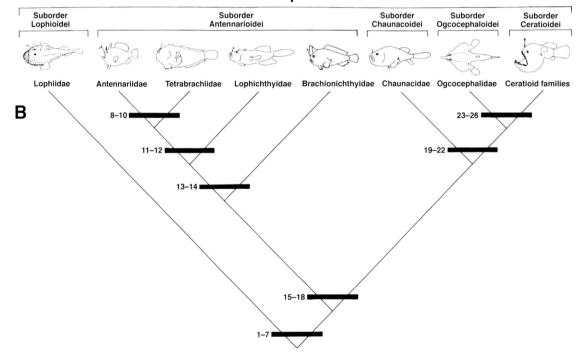

Fig. 269. Cladistic hypotheses of relationship of the families of lophiiform fishes. (**A**) three suborders recognized, with the Antennarioidei, containing six families, indicated as the sister-group of the Ceratioidei. (**B**) five suborders recognized, with the Antennarioidei, containing four families, indicated as the sister of a group containing the Chaunacoidei, Ogcocephaloidei, and Ceratioidei. Modified after Pietsch (2009, fig. 188); courtesy of Karin Tucker and the University of California Press.

3. Gill opening restricted to a small, elongate, tubelike opening situated immediately dorsal, posterior, or ventral to (rarely partly anterior to) pectoral-fin base (Carnevale and Pietsch, 2012: 68);

4. Urohyal absent, *rectus communis* muscle originating from the dorsal hypohyal (Datovo et al., 2014: 6, fig. 5B);

5. A single hypural plate (sometimes deeply notched posteriorly) emanating from a single, complex half-centrum (Rosen and Patterson, 1969: 441; Pietsch, 1981b: 396, 408, figs. 12, 33–35; Carnevale and Pietsch, 2012: 57, fig. 14);

6. Ventralmost pectoral radial considerably expanded distally (Pietsch, 1981b: 397, 411, figs. 14, 40; Carnevale and Pietsch, 2012: 57, 65, fig. 15);

7. Eggs spawned in an oval or double scroll-shaped mucous sheath (Rasquin, 1958: 340, figs. 1, 2, pl. 47; Ray, 1961: 230, fig. 1; Pietsch and Grobecker, 1987: 351, fig. 161, pl. 9; Arnold, 2010b: 17; Arnold and Pietsch, 2012: 128).

Since Regan (1912), three major lophiiform taxa of equal rank have been recognized by nearly all authors. These taxa, together with their currently recognized families (the 11 families of the bathypelagic Ceratioidei excluded; see Bertelsen, 1951: 29; and Pietsch, 1972: 18), are as follows:

Suborder Lophioidei
    Family Lophiidae
Suborder Antennarioidei
    Family Antennariidae
    Family Tetrabrachiidae
    Family Lophichthyidae
    Family Brachionichthyidae
    Family Chaunacidae
    Family Ogcocephalidae
Suborder Ceratioidei

Pietsch (1981b: 416) tested the validity of Regan's (1912) concept of three major lophiiform taxa using cladistic analysis (see Hennig, 1950, 1966). In that study, serious difficulty was encountered in efforts to establish monophyly for the six families of Regan's (1912) Antennarioidei. Although a number of synapomorphic features were found to support a sister-group relationship between the four families Antennariidae through Brachionichthyidae, and between the families Chaunacidae and Ogcocephalidae, no convincing synapomorphy was found to link these two larger subgroups (Fig. 269A).

In a later attempt, Pietsch (1984d: 323) proposed a new hypothesis, which, in resolving the former difficulties, differed significantly from that previously published (Pietsch, 1981b). In this revised cladogram (Fig. 269B), the suborder Antennarioidei is restricted to just four families: the Antennariidae, recognized as the sister-group of the Tetrabrachiidae, these two families together forming the sister-group of the Lophichthyidae, and this assemblage of three families forming the sister-group of the Brachionichthyidae. These relationships are supported by a total of seven synapomorphies, all previously described by Pietsch, 1981b (numbered 8 through 14 in Fig. 269B):

8. Posteromedial process of vomer emerging from ventral surface as a laterally compressed, keel-like structure, its ventral margin (as seen in lateral view) strongly convex (Pietsch, 1981b: 391, 397, figs. 4–6);

9. Postmaxillary process of premaxilla spatulate (Pietsch, 1981b: 393, 398, figs. 8, 20A);

10. Opercle similarly reduced in size (Pietsch, 1981b: 394, 401, figs. 9, 21);

11. Ectopterygoid triradiate, a dorsal process overlapping the medial surface of the metapterygoid (Pietsch, 1981b: 393, 400, figs. 9, 21, 22);

12. Proximal end of hypobranchials II and III deeply bifurcate (Pietsch, 1981b: 395, 407, figs. 11, 28, 29);

13. Interhyal with a medial, posterolaterally directed process that makes contact with the respective preopercle (Pietsch, 1981b: 393, 400, fig. 26);

14. Illicial pterygiophore and pterygiophore of the third dorsal spine with highly compressed, bladelike dorsal expansions (Pietsch, 1981b: 396, 410, figs. 13, 36).

The present interpretation of lophiiform relationships differs further from the previously proposed hypothesis of Pietsch (1981b) in considering the Antennarioidei (*sensu stricto*) to form the sister-group of a much larger group that includes the Chaunacoidei (Pietsch, 1984d: 323, fig. 166), Ogcocephaloidei (Pietsch, 1984d: 324, fig. 166), and Ceratioidei. The Ogcocephaloidei are in turn recognized as the sister-group of the Ceratioidei (Fig. 269B).

Monophyly for a group containing the suborders Antennarioidei, Chaunacoidei, Ogcocephaloidei, and Ceratioidei is supported by four synapomorphies, all previously identified by Pietsch, 1981b, 1984d (numbered as they appear in Fig. 269B):

15. Eggs and larvae small (at all stages the eggs are considerably less than 50% of the diameter of those of lophioids; the smallest larvae are certainly less than 50%, and probably less than 30%, of the size of those of lophioids; size at transformation to the prejuvenile stage is less than 60% that of lophioids; Pietsch, 1984d: 323, fig. 166);

16. Head of larvae proportionately large relative to body (always greater than 45% SL, compared to less than 30% in lophioids; Pietsch, 1984d: 324, fig. 166);

17. Pharyngobranchial IV absent (present and well-toothed in lophioids; Pietsch, 1981b: 395, 401, figs. 11, 28–32);

18. Number of dorsal-fin spines reduced from a primitive six in lophioids to three or less (Pietsch, 1981b: 409, figs. 36–38).

Monophyly for a group containing the suborders Chaunacoidei, Ogcocephaloidei, and Ceratioidei is supported by four synapomorphies (numbered as they appear in Fig. 269B):

19. Interhyal simple, cylindrical, without medial, posterolaterally directed process (present in lophioids and all antennarioids; Pietsch, 1981b: 400, fig. 26);

20. Gill filaments of gill arch I absent (but present on proximal end of ceratobranchial I of some ceratioids; Bradbury, 1967: 408; Pietsch, 1981b: 415);

21. Second dorsal-fin spine embedded beneath skin of head (Pietsch, 1981b: 410, fig. 38);

22. Third dorsal-fin spine embedded beneath skin of head (absent in ceratioids; Pietsch, 1981b: 410, fig. 38).

Monophyly for a group containing the Ogcocephaloidei and Ceratioidei is supported by four synapomorphies (numbered as they appear in Fig. 269B):

23. Posttemporal fused to cranium (attached to the cranium in batrachoidiforms and all other lophiiforms in such a way that considerable movement in an anterodorsal-posteroventral plane is possible; Pietsch, 1981b: 411, fig. 19);

24. Epibranchial I simple, without ligamentous connection to epibranchial II (in batrachoidiforms and all other lophiiforms, epibranchial I bears a medial process ligamentously attached to the proximal tip of epibranchial II; Pietsch, 1981b: 401, fig. 32);

25. Second dorsal-fin spine reduced to a tiny remnant (well-developed in the ceratioid family Diceratiidae, and in all other lophiiforms; Bertelsen, 1951: 17; Pietsch, 1981b: 410, fig. 38);

26. Third dorsal-fin spine and pterygiophore absent (present in all other lophiiforms; Bertelsen, 1951: 17; Bradbury, 1967: 401; Pietsch, 1981b: 410, fig. 38).

Of the possible cladograms that could be constructed from the morphological data provided by Pietsch (1981b, 1984d) and Pietsch and Grobecker (1987), the one shown in Fig. 269B is by far the most parsimonious (but see Pietsch, 1984d: 324, for a discussion of convergence or reversal of character states).

Key to the Major Subgroups of the Lophiiformes

Plesiomorphic and autapomorphic features of the major subgroups of the Lophiiformes are incorporated into the following analytical key:

1A. Postcephalic spinous dorsal-fin of one to three spines; pharyngobranchial IV present; cleithrum with prominent posterior spine; subopercle with large ascending process attached to anterior margin of ventral rami of opercle; pseudobranch well-developed; eggs and larvae large; head of larvae small relative to body . . . .**Suborder Lophioidei**

1B. Postcephalic spinous dorsal-fin absent; pharyngobranchial IV absent; cleithral spine absent; subopercle with ascending process absent or reduced to a small projection detached from opercle; pseudobranch greatly reduced or absent; eggs and larvae small; head of larvae large relative to body 2

2A. Spinous dorsal-fin of three spines emerging from dorsal surface of cranium; illicial pterygiophore and pterygiophore of third dorsal-fin spine with highly compressed, bladelike dorsal expansions; interhyal with a medial, posterolaterally directed process that comes into contact with the respective preopercle; interopercle flat and broad (**Suborder Antennarioidei**) 3

2B. Spinous dorsal-fin of two or three spines, but only anteriormost spine emerging from dorsal surface of cranium (spines II and III reduced and embedded beneath skin of head or lost); illicial pterygiophore and pterygiophore of third dorsal-fin spine without bladelike dorsal expansions; interhyal without a medial, posterolaterally directed process; interopercle elongate and narrow 6

3A. Parietals meeting on midline dorsal to supraoccipital; ectopterygoid roughly oval in shape or absent; ceratobranchials I through III with one or more tooth-plates; hypobranchial II simple, hypobranchial III absent; pectoral radials two; pelvic fin of one spine and four rays . . . . **Family Brachionichthyidae**

3B. Parietals well separated by supraoccipital; ectopterygoid triradiate, a dorsal process overlapping medial surface of metapterygoid; ceratobranchials I through IV toothless; hypobranchials II and III bifurcated proximally; pectoral radials three; pelvic fin of one spine and five rays 4

4A. Vomer wide, the width between lateral ethmoids nearly as great as that between lateral margins of sphenotics; vomer without posteromedial process; dorsal head of quadrate broad, the width equal to or greater than that of metapterygoid; postmaxillary process of premaxilla tapering to a point; opercle expanded posteriorly; pharyngobranchial and epibranchial of first arch toothed; bony connection between tips of haemal spines of 14th through 16th preural centra; pterygiophore of illicium elongate, greatly depressed and laterally expanded posteriorly . . . .**Family Lophichthyidae**

4B. Vomer narrow, the width between lateral ethmoids considerably less than that between lateral margins of sphenotics; posteromedial process of vomer emerging from ventral surface as a laterally compressed, keel-like structure, its ventral margin (as seen in lateral view) strongly convex; dorsal head of quadrate narrow, the width less than that of metapterygoid; postmaxillary process of premaxilla spatulate; opercle reduced in size; pharyngobranchial and epibranchial of first arch toothless; bony connection between tips of haemal spines absent; pterygiophore of illicium short, the posterior end cylindrical 5

5A. Body elongate, vertebral column more or less linear; eyes dorsal; dorsal-fin spines reduced; mouth small; fifth pectoral-fin ray membranously attached to side of body; posteriormost pelvic-fin ray membranously attached to pectoral-fin lobe . . . . **Family Tetrabrachiidae**

5B. Body short and deep, vertebral column sigmoid; eyes lateral; dorsal-fin spines well developed; mouth large; all pectoral-fin rays free, not membranously attached to side of body; posteriormost pelvic-fin ray free, not membranously attached to pectoral-fin lobe . . . .**Family Antennariidae**

6A. Second dorsal-fin spine elongate, embedded beneath skin of head; third dorsal-fin spine and pterygiophore present, embedded beneath skin of head; epibranchial I with a medial process ligamentously attached to proximal tip of epibranchial II . . . .**Suborder Chaunacoidei**

6B. Second dorsal-fin spine reduced to a tiny remnant embedded beneath skin of head and lying on, or fused to, dorsal surface of pterygiophore just behind base of illicial bone; third dorsal-fin spine and pterygiophore absent; epibranchial I simple, without ligamentous attachment to epibranchial II 7

7A. Palatine teeth usually present; ceratobranchial V toothed, expanded proximally; pelvic fins present; obvious sexual dimorphism absent, males not reduced . . . .**Suborder Ogcocephaloidei**

7B. Palatine teeth absent; ceratobranchial V toothless, reduced to a slender rod-shaped element; pelvic fins absent (except in larval caulophrynids); sexual dimorphism strongly developed, males reduced to a small fraction of size of females . . . . **Suborder Ceratioidei**

Interrelationships of Antennariid Genera and Species Groups
Morphological Evidence

In the earliest attempt to reconstruct the evolutionary history of frogfishes, Leonard P. Schultz (1957: 50) asked himself: "(1) How shall I evaluate the characters observed for the numerous species of Antennariidae, and (2) what level of generic interpretation should I give to each phyletic line?" He then added a surprisingly modern statement: "I believe that in evaluating each important character I should consider the evolutionary trend toward its more specialized condition." He went on to list what he interpreted as important characters of the Antennariidae, indicating in each case the character state he considered to be more specialized or derived:

1. The skin varies from highly dentigerous to almost naked, and in the most naked species some embedded prickles can be detected microscopically. Naked skin is assumed to be the most specialized condition.

2. The most specialized condition of the gill filaments on the first gill arch is that of greatest reduction, where only one-half the lower part of the first gill arch bears filaments.

3. The most primitive condition of the bait or lure at the tip of the first dorsal spine may be represented by a simple filament. The next step would be the development of a tuft of filaments, followed by specialization into bifid or trifid tentacles.

4. The more numerous and most complex development of dermal cirri all over the body would represent the more specialized condition. These cirri even replace the dermal denticles in one genus [*Rhycherus*].

5. Fin rays present somewhat of a problem but, in general, branched soft rays are assumed to be a more specialized condition than simple soft rays.

6. The more posterior position of the gill opening is considered to represent specialization.

7. The most movable condition of the dorsal spines should represent the more generalized condition, whereas embedded dorsal spines should represent the most specialized.

8. The distinct caudal peduncle is considered more primitive than when the median fins are membranously attached to the base of the caudal fin.

9. Adult antennariids in general have a sedentary habitat. *Histrio* has a sedentary habitat in seaweed but often the seaweed floats pelagically in the ocean. Thus, I consider *Histrio* to be more specialized in its "sedentary" pelagic habitat than the other antennariids.

Using the above characters, Schultz (1957: 53, fig. 1) constructed a branching diagram "suggesting the more important phyletic lines of evolution among the antennariids" (Fig. 270). But, unfortunately, Schultz's characters and his diagram are not very in-

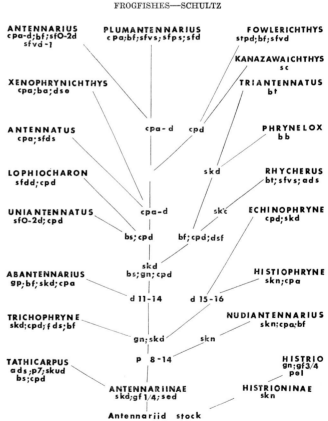

FROGFISHES—SCHULTZ

FIGURE 1.—Presumed phylogeny of the Antennariidae.

Fig. 270. Leonard P. Schultz's "Presumed phylogeny of the Antennariidae, suggesting the more important phyletic lines of evolution." For an explanation of abbreviations of anatomical characters, see Schultz (1957: 52, fig. 1). Courtesy of Lisa Palmer and the Smithsonian Institution, NMNH, Division of Fishes, Washington, DC.

formative. Polarization of many of these characters is questionable and except for character 6, the posterior displacement of the gill opening—clearly derived but found in separate states in each of only two members of the genus *Abantennarius*—and character 9, which is specific only to *Histrio histrio*, most are found in almost continuous variation throughout the family, while derived states of the others pop up multiple times within a number of disparate genera.

But our own attempts to provide morphological evidence of phylogenetic interrelationships of antennariids have not fared much better. Pietsch and Grobecker (1987) had only limited success. Monophyly for the genus *Antennarius*, then thought to contain 24 species, could not be established—despite considerable effort, they were unable to identify a single clearly derived feature shared by all included taxa. Thus, the genus was defined by a suite of what were believed to be primitive features. They hypothesized that *Antennarius* is the least derived member of the family—"all of the known characters of systematic importance found among the 12 genera [then recognized] (at least all those characters for which relative primitiveness of states could be reasonably hypothesized) appear to be present in *Antennarius* in the primitive state" (Pietsch and Grobecker, 1987: 275)—a conclusion contradicted in part by the molecular analysis of Arnold and Pietsch (2012: 126), which assigned this position to the genus *Fowlerichthys* (a clade formerly recognized as the *Antennarius ocellatus* group; see below). In any case, each of the remaining 11 genera were found by Pietsch and Grobecker (1987: 275) to possess at least two, and in some cases as many as nine, apomorphic character states (autapomorphic states included) that indicated its derived position relative to *Antennarius*.

Whereas a hypothesis of monophyly for each of the remaining 11 genera could be supported (six genera were monotypic at the time and none contained more than three species), their phylogenetic interrelationships as elucidated by cladistics could not be established. No group of two or more genera was found to possess any convincing synapomorphy that did not also occur within lophiiform taxa lying just outside the group in question, thus inviting the possibility of evolutionary convergence. For example, while still assuming that *Antennarius* is basal to all other antennariids and thus adequate to serve as an immediate outgroup to all the other genera, the following character states were considered apomorphic:

1. The skin lacks dermal spinules in *Rhycherus* and *Phyllophryne*; these are very much reduced in *Nudiantennarius* and tetrabrachiids, present or absent in *Histrio* and *Histiophryne*, and absent in all lophiids (numerous close-set dermal spinules cover the head and body of all other antennariid genera as well as lophichthyids and brachionichthyids);

2. The pectoral-fin lobe is detached from the side of the body for most of its length in *Histrio* and *Tathicarpus*, but this condition also occurs in lophiids, brachionichthyids, and ogcocephalids (attached to the side of the body for most of its length in *Antennarius* and in all other antennarioids);

3. The caudal peduncle is absent in four antennariid genera (Tables 1, 14); it is also absent in tetrabrachiids and some (but not all) members of *Antennatus* and the now defunct *Antennarius nummifer* group, the latter now recognized as *Abantennarius* (present in lophiids, *Antennarius*, and all other antennarioids);

4. The endopterygoid (mesopterygoid of Pietsch, 1981b; Pietsch and Grobecker, 1987) is absent in eight antennariid genera (nine genera when *Porophryne* is added; Tables 1, 14); it is also absent in tetrabrachiids, lophichthyids, and brachionichthyids

(Pietsch, 1981b: 400, table 2) [present in lophiids (but fused to the ectopterygoid; see Carnevale and Pietsch, 2012: 59, figs. 5, 11)], *Antennarius*, and all remaining antennariid genera);

5. Pharyngobranchial I is absent in *Lophiocharon* and *Histiophryne*; it is also absent in lophiids, brachionichthyids, and at least some ogcocephalids (e.g., *Dibranchus*; see Pietsch, 1981b: 401, table 2) (present in *Antennarius* and all remaining antennarioids);

6. An epural is absent in eight antennariid genera (nine genera when *Porophryne* is added; Tables 1, 14); it is also absent in tetrabrachiids and brachionichthyids, and greatly reduced or absent in lophichthyids (Pietsch, 1981b: 409, table 2) (present in lophiids, *Antennarius*, and all remaining antennariid genera);

7. A pseudobranch is absent in ten antennariid genera (Tables 1, 14); it is also absent in all other antennarioid families (Pietsch, 1981b: 405, table 2) (present in lophiids, *Antennarius*, and all remaining antennariid genera);

8. A swimbladder is absent in *Kuiterichthys* and *Tathicarpus*; it is also absent in lophiids and all other antennarioid families (Pietsch, 1981b: 405, table 2) (present in *Antennarius* and all remaining antennariid genera).

Two of these assumed derived features, the loss of the endopterygoid and epural, corroborate each other in defining the same group of eight genera (nine genera when *Porophryne* is added; Table 14). Nevertheless, Pietsch and Grobecker (1987: 276; see also Pietsch, 1984b: 42), although admitting the possibility, failed to formally recognize a monophyletic subgroup within the Antennariidae, which our molecular analysis has now confirmed (helping to define the subfamily Histiophryninae; Arnold and Pietsch, 2012). Setting aside the evidence, Pietsch and Grobecker (1987: 276) concluded: "the extent to which these and the remaining derived character states listed above have been independently acquired in the various taxa under consideration cannot now be determined. Consequently, it appears that new characters, most likely non-osteological ones, must be identified and analyzed before the interrelationships of antennariid genera can be resolved."

Molecular Evidence

With problems rampant in a morphological approach, we turn now to a molecular analysis, made possible almost solely through the efforts of the junior author (Arnold, 2010b; Arnold and Pietsch, 2012). In an effort to address the intergeneric relationships of the family and to test the validity of the species groups of *Antennarius* hypothesized by Pietsch (1984d) and Pietsch and Grobecker (1987), DNA sequences were assembled from 31 species, including two outgroup taxa and 47 individual antennariids from 10 of the 12 genera recognized at the time. *Porophryne erythrodactylus*, then an undescribed genus and species from New South Wales, Australia, was also included (identified as Antennariidae gen. et sp. nov.). *Allenichthys*, a monotypic genus endemic to Western Australia, was excluded for lack of available tissues. The monotypic genus *Nudiantennarius*, previously excluded from the analysis for lack of tissues has now been added. All polytypic genera, except *Antennarius*, were assumed to be monophyletic. Representatives of four of the six species groups of *Antennarius* recognized by Pietsch and Grobecker (1987), the *A. striatus*, *A. pictus*, *A. ocellatus*, and *A. nummifer* groups (the latter two now recognized as *Fowlerichthys* and *Abantennarius*, respectively), were included, but the *A. biocellatus* group, containing a single species, and the *A. pauciradiatus* group, with two species, were excluded, again for lack of available tissues (for a full list of tissue and voucher specimens, see Arnold and Pietsch, 2012: 121, table 1). A dataset of 1709 bp from the genes 16S, cytochrome oxidase c subunit 1 (COI), and recombination activation gene 2 (RAG2) were analyzed using Bayesian and maximum likelihood methods (for full details, see Arnold and Pietsch, 2012: 120–124).

In addition to a well-corroborated phylogeny of the family (described below), Arnold's work resulted in four major changes to the classification presented by Pietsch and Grobecker (1987: 41): (1) recognition that the family contains two monophyletic subunits, the Antennariinae and Histiophryninae (based as well on ovarian morphology and reproductive behavior, and nicely corroborated by geographic distribution); (2) resurrection of *Fowlerichthys* Barbour (1941) to contain the members of the former *Antennarius ocellatus* group; (3) expansion of the genus *Antennatus* Schultz (1957) to contain members of the former *Antennarius nummifer* group (but here resolved by resurrection of the genus *Abantennarius* Schultz, 1957); and (4) reallocation of *Antennarius indicus* to the *Antennarius striatus* group. Details of these modifications are given as follows:

Two major clades were recovered, the first composed of the subfamily Antennariinae (Fig. 271), containing *Fowlerichthys* (resurrected from the synonymy of *Antennarius*), *Antennarius, Nudiantennarius, Histrio*, and *Antennatus*; and the second, a sister clade, subfamily Histiophryninae (Fig. 272), containing the remaining genera *Rhycherus, Porophryne* (Antennariidae gen. et sp. nov. of Arnold and Pietsch, 2012), *Kuiterichthys, Phyllophryne, Echinophryne, Tathicarpus, Lophiocharon*, and *Histiophryne*. Within the Antennariinae (Fig. 271), *Antennarius* was rendered paraphyletic, requiring resurrection and recognition of *Fowlerichthys* Barbour, 1941, to contain former members of the *Antennarius ocellatus* group. At the same time, *Antennatus* clustered with the *Antennarius nummifer* group, thus requiring reallocation of all members of the latter (as recognized by Pietsch and Grobecker, 1987) to a much expanded *Antennatus*. To designate the distinction between these two subgroups, Arnold and Pietsch (2012) chose to recognize an *Antennatus tuberosus* group to encompass members of *Antennatus sensu stricto* and an *Antennatus nummifer* group to contain all the former members of the *Antennarius nummifer* group. As an alternative, we have chosen here to resurrect the genus *Abantennarius* Schultz, 1957, to accommodate the latter (see below, p. 383).

The *Antennarius pictus* group (as originally recognized by Pietsch, 1984d) was recovered as monophyletic, but the resurrected *Fowlerichthys* was found to be monophyletic only after exclusion of *A. indicus*, a species recovered as sister of the monophyletic *Antennarius striatus* group. The *A. pictus* group was found to be sister to the *A. striatus* group, these two taxa together forming the sister group of *Histrio*, plus *Nudiantennarius* in the somewhat expanded analysis presented here. The *A. pictus* and *A. striatus* groups, together with *Nudiantennarius* and *Histrio* comprise a clade sister to *Abantennarius* plus *Antennatus. Fowlerichthys* was recovered as sister to all the species-groups of *Antennarius*, plus *Nudiantennarius, Histrio, Abantennarius*, and *Antennatus*, and is thus recognized as the basal lineage of the subfamily Antennariinae.

The subfamily Histiophryninae, defined, in part, to include all members of the Antennariidae that have lost the endopterygoid and epural, was subdivided into two groups (Fig. 272), the first containing *Rhycherus, Porophryne, Kuiterichthys, Phyllophryne*, and *Echinophryne*; the second consisting of *Tathicarpus, Lophiocharon*, and *Histiophryne*. *Phyllophryne* and *Echinophryne* were recovered as sister taxa, as were *Porophryne* and *Kuiterichthys*; these two groups plus *Rhycherus* were recovered in a polytomy. Finally, *Tathicarpus* was recovered as sister of a clade containing *Lophiocharon* and *Histiophryne*.

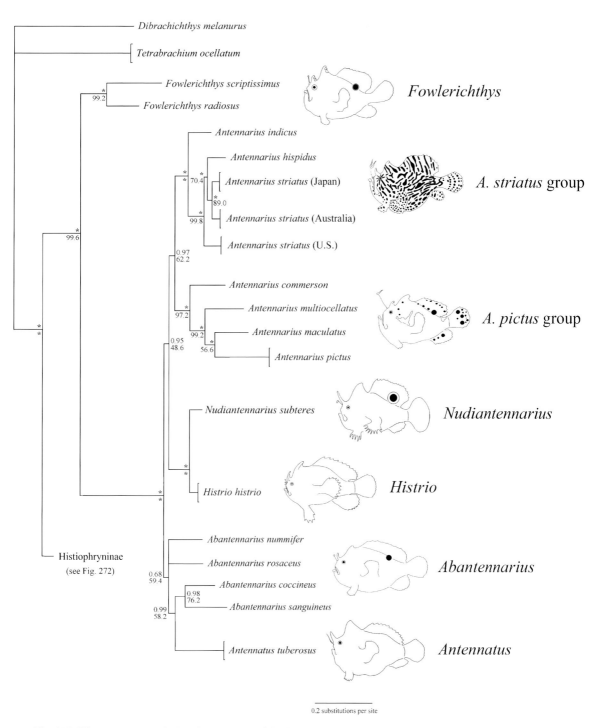

Fig. 271. Fifty percentage majority rule phylogeny of the Antennariinae, from trees sampled in the posterior, generated from Bayesian analyses without the saturated data (third codon position of COI). Branch lengths are measured in expected substitutions per site and are proportional to length. Numbers above nodes are posterior probabilities and numbers below nodes are bootstrap proportions from 500 pseudoreplicates used from maximum likelihood analysis; an asterisk indicates a posterior probability of 1.00 or a bootstrap percentage of 100. Modified after Arnold and Pietsch (2012, fig. 6).

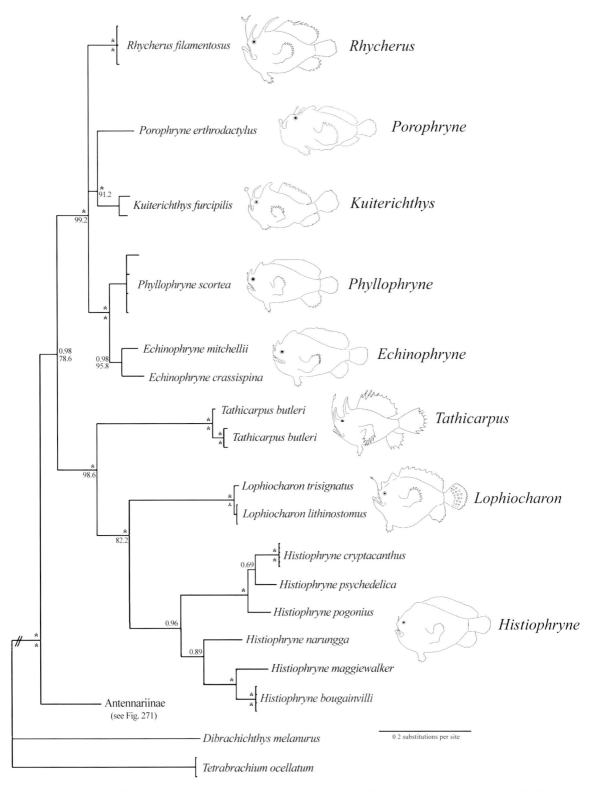

Fig. 272. Fifty percentage majority rule phylogeny of the Histiophryninae, from trees sampled in the posterior, generated from Bayesian analyses without the saturated data (third codon position of COI). Branch lengths are measured in expected substitutions per site and are proportional to length. Numbers above nodes are posterior probabilities and numbers below nodes are bootstrap proportions from 500 pseudoreplicates used from maximum likelihood analysis; an asterisk indicates a posterior probability of 1.00 or a bootstrap percentage of 100. Modified after Arnold and Pietsch (2012, fig. 7).

Reconciling the Evidence

While Pietsch (1984d) and Pietsch and Grobecker (1987) were unable to resolve relationships within the Antennariidae, a few morphological characters identified in their studies correspond nicely with clades recovered in our molecular analysis (Arnold and Pietsch, 2012). For instance, within the Antennariinae (Fig. 271), *Fowlerichthys* has a unique combination of 20 vertebrae and five bifurcate pelvic-fin rays, while its sister-group, all the remaining members of the subfamily Antennariinae (*Antennarius*, *Nudiantennarius*, *Histrio*, *Abantennarius*, and *Antennatus*), have 19 vertebrae and at most only one bifurcate pelvic-fin ray (Tables 1, 14). All members of the Histiophryninae (Fig. 272) have lost the endopterygoid and epural, a synapomorphy that led Pietsch (1984b: 41) and Pietsch and Grobecker (1987: 276) to hypothesize that this lineage likely forms a monophyletic group. *Lophiocharon* and *Histiophryne*, recovered as sister-groups in the molecular analysis, are the only antennariids that have lost pharyngobranchial I (this element is also absent in brachionichthyids and at least some ogcocephalids; see Pietsch, 1981b: 401, 1984b: 41; Pietsch and Grobecker, 1987: 275).

Phylogenetic Position of *Fowlerichthys*

Fowlerichthys (containing most of the previously recognized members of the *Antennarius ocellatus* group, *sensu* Pietsch, 1984d: 34; Pietsch and Grobecker, 1987: 111, 277) was recovered in the molecular analysis as the basal genus of the subfamily Antennariinae. The members of this assemblage, *Fowlerichthys avalonis*, *F. ocellatus*, *F. radiosus*, *F. scriptissimus*, and *F. senegalensis* (excluding *Antennarius indicus*), share a character unique among lophiiforms: all five soft rays of the pelvic fins are bifurcate, in contrast to all other antennariids, which have at most only a single bifurcate pelvic-fin ray. They also have 20 vertebrae while nearly all other antennariids examined within the Antennariinae have 19 (with only two exceptions, noted by Pietsch and Grobecker, 1987: 51, table 4): a single specimen each of *A. striatus* and *A. commerson* with 18 and 20 vertebrae, respectively.

Of interest also is an unusual early life-history stage called the "scutatus" prejuvenile form characterized by a pair of shield-like bony extensions of the cranium that reach beyond the level of the opercular bones, and an expansion of the anterior margin of the bones of the suspensorium (see Pietsch, 1984d: 322, fig. 165; Pietsch and Grobecker, 1987: 121, figs. 43, 44; Figs. 28–30). This early life-history stage is so unusual that it was originally described as a new genus and species, *Kanazawaichthys scutatus*, by Schultz (1957), and only later shown by Hubbs (1958) to be the postflexion larval stage of *Fowlerichthys radiosus* (Figs. 29, 30). Although a similar morphological adaptation in closely related species was expected, Pietsch (1984d: 322) and Pietsch and Grobecker (1987: 122) found nothing comparable on examination of the smallest individuals available at the time. While larval antennariids are not especially rare in collections, rapid metamorphosis from larva to adult—in *Histrio histrio*, Adams (1960) found that 10.0 mm SL marks the transition from postlarva to juvenile and by 15.5 mm SL most individuals have taken on full adult appearance—severely limits the number of prejuveniles available in collections. Nevertheless, tiny postflexion larval stages of *Fowlerichthys avalonis* (3.4 to 5.3 mm SL, significantly smaller than known "scutatus" stages of *F. radiosus*, which are typically between about 15 and 24 mm SL; see Hubbs, 1958: 285, table 2), with comparable morphological features, have since been described and figured by Watson (1998:

234; Fig. 30). The presence of this specialized "scutatus" larval stage in two members of the genus *Fowlerichthys*, while unknown everywhere else in the family, hints at the possibility that this feature might be unique to *Fowlerichthys* (see Watson et al., 2000). Only examination of additional collections will tell.

Phylogenetic Position of *Antennarius indicus*

In the molecular analysis, members of the *A. striatus* and *A. pictus* groups remained clustered as hypothesized by Pietsch (1984d; see also Pietsch and Grobecker, 1987: 276, fig. 112), with one exception: *Antennarius indicus*, previously recognized as the basal species of the *A. ocellatus* group (reallocated here to the genus *Fowlerichthys*), was recovered as sister to the *A. striatus* group (Arnold and Pietsch, 2012; Fig. 271). Morphologically, however, *A. indicus* is somewhat at odds with members of the latter group as previously diagnosed. For example, the membrane behind the second dorsal-fin spine of *Fowlerichthys* is usually divided into naked dorsal and ventral portions by a dense cluster of dermal denticles and at most it is only weakly divided in *A. indicus* (Pietsch and Grobecker, 1987: 117, fig. 38A; Figs. 27, 59), but this division is absent in the *A. striatus* group *sensu stricto*. In *Fowlerichthys*, the ventral portion of this membrane usually extends posteriorly, dividing the depressed area between the bases of the second and third dorsal-fin spines and nearly reaching the base of the third, but the membrane behind the second dorsal-fin spine of *A. indicus* terminates distinctly anterior to the third dorsal-fin spine, as it does in the other members of the *A. striatus* group.

The pigment pattern of *A. indicus* also lies somewhat between the two taxa (Arnold and Pietsch, 2012: 126). The species of *Fowlerichthys sensu stricto* nearly always have one to three darkly pigmented ocelli on each side of the body—*A. indicus* usually has two or three ocelli but members of the *A. striatus* group have none. Roughly parallel, darkly pigmented streaks and blotches on the head and body, with similar markings on the unpaired fins (Figs. 52, 61, 63, 69), are typical (but certainly not diagnostic) of the *A. striatus* group. While the head and body of *A. indicus* are generally devoid of such streaks, elongate dark blotches do often occur along the posterior margin of the dorsal and anal fins, occasionally with numerous darkly pigmented blotches on the pectoral and pelvic fins (Figs. 57, 60, 61). A similar pigment pattern is unknown in *Fowlerichthys* as here restricted.

On the other hand, there are some clear morphological similarities shared between *A. indicus* and the other three species now allocated to the *A. striatus* group, which were inexplicably and embarrassingly missed by Pietsch and Grobecker (1987). Most striking is the position of the pterygiophore of the illicium—this element in the newly formed *A. striatus* group nearly always extends anteriorly beyond the symphysis of the upper jaw in contrast to all other antennariids.

Meristic support for inclusion of *A. indicus* in the *A. striatus* group is present as well. Pietsch and Grobecker (1987: 277, fig. 113) recognized *A. indicus* as the "least specialized member" of the *A. ocellatus* group; the remaining five species were thought to form a monophyletic assemblage on the basis of having all five rays of the pelvic fin, and most or all of the rays of the dorsal fin, bifurcate, and in having relatively high vertebral and dorsal-fin ray counts. Indeed, most or all of the rays of the dorsal fin, and all five rays of the pelvic fin, are bifurcate in the newly reconstituted *Fowlerichthys* (the latter feature unique among the Lophiiformes), while at most, only four of the posteriormost dorsal rays are bifurcate in the *A. striatus* group, including *A. indicus*. All members of *Fowlerich-*

thys sensu stricto have 20 vertebrae, but *A. indicus* has only 19, as do all other members of the *A. striatus* group as well as those of *Antennarius* as a whole. Exceptions are extremely rare: among dozens of specimens radiographed, Pietsch and Grobecker (1987: 51, table 4) found only a single specimen of *A. striatus* with 18 vertebrae and a single specimen of *A. commerson* with 20 vertebrae.

Intrarelationships of the *Antennarius striatus* Group

In the molecular analysis, *Antennarius indicus* was found to be basal to the other members of the *A. striatus* group, while a single specimen of *Antennarius hispidus* fell between *A. striatus* and *A. scaber* (Arnold and Pietsch, 2012: 127; Fig. 271). Considering the extremely close morphological similarity between the latter two species (both characterized by having a wormlike esca), this was a surprising result, providing some support for the decision to resurrect *A. scaber*. Moreover, the rather large genetic distance between specimens of *A. striatus* from Australia and those from Japan (where *A. striatus* has been long recognized as *A. tridens*), adds further support for the possibility that the widely distributed *A. striatus* might constitute two or more species previously synonymized by Pietsch and Grobecker (1987; see Comments, p. 110). More evidence is needed, however, especially from faster evolving nuclear markers, before these populations can be definitively recognized as distinct (see Arnold and Pietsch, 2012: 127).

Phylogenetic Positions of the *Antennarius pauciradiatus* and *Antennarius biocellatus* Groups

The *Antennarius pauciradiatus* and *A. biocellatus* groups each have characters unique among antennariids. The monotypic *A. biocellatus* group is characterized by having the second dorsal spine free, not connected to the head by membrane, and nearly always straight and tapering distally to a point; conspicuous cutaneous appendages on the throat; an unusually prominent chin that protrudes far forward, well beyond the opening of the mouth, resulting in a unique bulldog appearance; and, most peculiar, the physiological ability (in contrast to all other antennariids) to occupy brackish and even freshwaters for long periods of time, if not permanently (first recorded by Bleeker, 1865b: 16; see also Pietsch and Grobecker, 1987: 177). Members of the *A. pauciradiatus* group are neotenic forms that reach a maximum known standard length of only 40 mm (more than 82% of the material examined was less than 20 mm; the average standard length was less than 16 mm); in addition, they have a broad membranous connection between the second and third dorsal spines and between the third dorsal spine and the soft dorsal fin (Pietsch and Grobecker, 1987: 177; Fig. 112). There are no known derived morphological characters that unite these two groups to each other or to any other antennariid. Vertebral counts (19) and ovarian morphology (see below), along with one bifurcate and four simple pelvic-fin rays suggest their placement somewhere within the lineage containing *Antennarius*, *Nudiantennarius*, *Histrio*, *Antennatus*, and *Abantennarius* (Fig. 271), but until adequate morphological and molecular data are analyzed, their placement within this lineage remains largely unknown.

Antennatus and the Resurrection of *Abantennarius*

In addition to a strikingly different physiognomy apparent at first glance, the morphological characters that differentiate members of the genus *Antennatus*, as first recognized by Schultz (1957), include: (1) skin thick and firm, everywhere covered with extremely close-set dermal spinules (in some cases, the spinules are tightly clustered to form slightly raised, wartlike patches); (2) absence of small naked areas between pores of the acoustico-lateralis system; (3) illicium more or less tapering to a fine point, an esca absent or only barely distinguishable; (4) third dorsal spine more or less immobile, bound down to surface of the cranium by skin of the head; and (5) the absence of a pseudobranch (Pietsch and Grobecker, 1987: 45, 186). This suite of characters is seemingly more than enough to distinguish it from *Antennarius* and for that matter all other antennariids. It was thus a real surprise in the molecular analysis to see *Antennatus* cluster with members of the *Antennarius nummifer* group (Arnold and Pietsch, 2012: 125; Fig. 271). Morphologically, *Antennatus* shares rather little with members of this group: besides a smaller maximum body size compared to most other species groups of *Antennarius* (the *A. pauciradiatus* group being a notable exception), the second dorsal-fin spine is free, not connected to the head by a membrane. Yet, in the molecular analysis, *Antennatus* and the *A. nummifer* group were recovered as monophyletic and sister to a clade consisting of *Nudiantennarius*, *Histrio*, and all the remaining species groups of *Antennarius*. This result prompted Arnold and Pietsch (2012) to recognize an expanded *Antennatus*, the members of which fell conveniently into two groups: an *Antennatus tuberosus* group, comprised of *A. flagellatus*, *A. linearis*, *A. strigatus*, and *A. tuberosus*, characterized by having relatively thick skin, short close-set dermal denticles, a reduced luring apparatus, the third dorsal-fin spine more or less immobile, bound down to surface of cranium by skin of head, and a shorter, smaller caudal fin; and an *Antennatus nummifer* group, containing *A. analis*, *A. bermudensis*, *A. coccineus*, *A. dorehensis*, *A. duescus*, *A. nummifer*, *A. rosaceus*, and *A. sanguineus* (as well as *A. drombus*, here resurrected from the synonymy of *A. coccineus*; see Comments, p. 218), with a darkly pigmented basidorsal spot nearly always present, the dorsal fin with as many as six (usually only three) posteriormost rays bifurcate, and the posterior surface of the second dorsal spine usually devoid of dermal denticles (Pietsch and Grobecker, 1987: 135). To reflect this dichotomy, we have here opted to do away with these two species groups and resurrect instead the genus *Abantennarius* Schultz, 1957, to accommodate all the former members of the *A. nummifer* group as first recognized by Pietsch (1984b: 36; see also Pietsch and Grobecker, 1987: 135, 278).

This seemingly nice solution to the problem, however, is diminished by the realization that *Abantennarius*, as recognized here, is probably paraphyletic. This paraphyly is not apparent morphologically, but our preliminary molecular analysis, based on only four of the nine contained species (*A. coccineus*, *A. nummifer*, *A. rosaceus*, and *A. sanguineus*), divides the genus into two clades and places the genus *Antennatus* in between (Fig. 271). *Antennatus*, however, is clearly distinct and well-defined—at least based on morphology—and to consider the contents of both genera as one, in an expanded *Antennatus*, seems unwarranted. On the other hand, to erect a new genus to accommodate a portion of the species of *Abantennarius* appears equally ill-advised considering how little we know about the interrelationships within the latter. At this point, it therefore seems best to maintain *Abantennarius* as recognized here until further molecular analysis becomes possible through additional collections of tissues and voucher specimens.

Phylogenetic Position of *Histrio*

In our earlier analysis, Arnold and Pietsch (2012: 127) retained the monotypic genus *Histrio* by abolishing the *Antennarius ocellatus* and *A. nummifer* groups (recognized now as

Fowlerichthys and *Abantennarius*, respectively). Several characters, including exceptionally long pelvic fins (length greater than 25% SL), two cutaneous cirri consistently present on the mid-dorsal line of the snout between the symphysis of the premaxilla and base of the illicium, and a pseudo-pelagic lifestyle in floating *Sargassum* (Pietsch and Grobecker, 1987: 197), require that *Histrio* be recognized as distinct from the *A. pictus*, *A. striatus*, *A. biocellatus*, and *A. pauciradiatus* groups.

Phylogenetic Position of *Nudiantennarius*

A combination of morphological and meristic characters, including the presence of an endopterygoid and epural, 19 vertebrae, and double scroll-shaped ovaries indicates that *Nudiantennarius* is a member of the subfamily Antennariinae. Morphological and molecular analyses further indicate a sister-group relationship with *Histrio* (see Pietsch and Arnold, 2017, fig. 5; Fig. 271). Shared morphological features include a short illicium, no more than half the length of the second dorsal-fin spine; the second dorsal-fin spine unusually long and narrow, without a posterior membrane; the illicium and second dorsal-fin spine closely spaced, the former appearing to emerge from the base of the latter; all rays of the pelvic fin simple; and greatly reduced dermal spinules. In addition, the pectoral-fin lobe of both species is detached from the side of the body, partially so in *Nudiantennarius* but free for most of its length in *Histrio* (Pietsch and Grobecker, 1987). Finally, an analysis of COI sequences recovered *N. subteres* as sister to *H. histrio*, with a posterior probability of 1.0 (Fig. 271). Genetic distances for *N. subteres* and *H. histrio* differed by 4.58–4.81%, and 1.1% between the two specimens of *N. subteres*.

Phylogenetic Position of *Allenichthys*

Although *Allenichthys* was not included in the molecular analyses, the lack of an endopterygoid and epural, simple ovarian morphology (see below), and endemism off Western and South Australia, below 20° S latitude, suggest its placement within the Histiophryninae (Fig. 272), in the clade containing *Tathicarpus*, *Lophiocharon*, and *Histiophryne* (see below). Beyond this, its position remains unresolved.

Phylogeny, Life History, and Geographic Distribution

The Antennariinae, containing *Fowlerichthys*, *Antennarius*, *Nudiantennarius*, *Histrio*, *Abantennarius*, and *Antennatus*, has a relatively wide geographic distribution, with all genera, except *Nudiantennarius*, found circumglobally throughout the tropics and subtropics. In sharp contrast, the Histiophryninae, consisting of *Rhycherus*, *Porophryne*, *Kuiterichthys*, *Phyllophryne*, *Echinophryne*, *Tathicarpus*, *Lophiocharon*, *Histiophryne*, and *Allenichthys* is restricted to the Indo-Australian Archipelago, a region that extends from Taiwan to Tasmania, including all of the inland seas and islands of the Philippines, Indonesia, New Guinea, the Solomons, and the continent of Australia. Within the Histiophryninae, a clade containing *Rhycherus*, *Porophryne*, *Kuiterichthys*, *Phyllophryne*, and *Echinophryne* is restricted to temperate waters of Australia and Tasmania, below approximately 30° S latitude. Within the remaining clade, *Tathicarpus* and *Lophiocharon* range from the Philippines to approximately 35° S latitude, while *Histiophryne* extends from Taiwan to the temperate waters of southern Australia.

Reproductive life history also differs considerably between the two subfamilies: the Antennariinae has double scroll-shaped ovaries described in detail by Rasquin (1958: 340, figs. 1, 2, pl. 47; see also Pietsch and Grobecker, 1987: 351, fig. 161), while the Histiophryninae has an entirely different ovarian morphology consisting of a pair of oval-shaped structures (Ray, 1961: 230, fig. 1). In addition, each ovarian type corresponds to a different life history: members of the Antennariinae are broadcast spawners and go through a distinct larval stage, while those of the Histiophryninae undergo direct development and display various degrees of parental care. For more on ovarian morphology, life history, and associated behavior, see Reproduction and Early Life History, p. 461.

6. Zoogeography

Frogfishes are widely distributed in tropical and subtropical waters of all four major marine faunal realms of the world (Fig. 273). By far the majority of the species, however, and all but five of the 15 recognized genera, are confined geographically to the Indo-Australian Archipelago. The family is well represented in the gulfs of California and Mexico, the Red Sea, and the Persian Gulf, but like all antennarioids it is unknown in the Mediterranean Sea. A summary of world distribution of all taxa is given in Table 22.

Fowlerichthys, Antennarius, and *Abantennarius*, with a combined total of 26 species and with the possible exception of *Histrio* (see p. 191), are the only circumglobal genera, each approximately overlapping the distribution of the other two. In the Western Atlantic they range from Long Island, New York, to the mouth of the Rio de la Plata, Uruguay, and in the Eastern Atlantic, from the Azores, Madeira, Canary Islands, Cape Verde, Ascension, and St. Helena to the coasts of Morocco to Namibia. In the Indo-Pacific they range from the tip of South Africa, the Red Sea, and Persian Gulf eastward to virtually all oceanic island groups of the Indo-Pacific Ocean (including the Hawaiian Islands as well as Pitcairn and Easter Islands), and from Japan to southern Australia (including Tasmania), and New Zealand. In the eastern Pacific Ocean, they occur off Southern California and throughout the Gulf of California to the Galápagos Islands and to Isla San Félix off the coast of central Chile (Fig. 273).

The monotypic genus *Histrio* is sympatric with the ubiquitous brown alga genus *Sargassum* (Phaeophyta) throughout the Atlantic and Indo-Pacific Oceans, with confirmed captures made as far east as the Hawaiian Islands, including Guam in the north and Tonga in the south (but for a discussion of possible Eastern Pacific records of *Histrio*, see p. 191). The genus *Antennatus*, embracing four species, occurs throughout the tropical Indo-Pacific and Eastern Pacific Oceans, from off South Africa to the Gulf of California and the coasts of Mexico and Central and South America as far south as Ecuador and the Galápagos Islands. *Nudiantennarius* is known from fewer than a dozen preserved specimens collected in the Philippines and in Indonesia from Bali to Ambon. *Tathicarpus* is restricted to the Moluccas, Papua New Guinea, and Australia about as far south as Perth in the west and Brisbane in the east. *Lophiocharon* ranges from the Malaysian coast, throughout the islands of the Philippines and the Moluccas to the northern coast of Western Australia, Northern Territory, the Gulf of Carpentaria, and the northeastern coast of Queensland. *Histiophryne* occurs from Taiwan, the Philippines, and the Moluccas to the southern coast of Australia. *Allenichthys* occurs only along the shores of Western Australia below 20° S latitude, with at least one record from off Port Lincoln, South Australia. All five remaining genera of the family are restricted to the subtropical and temperate coastal waters of Australia below about 30° S latitude.

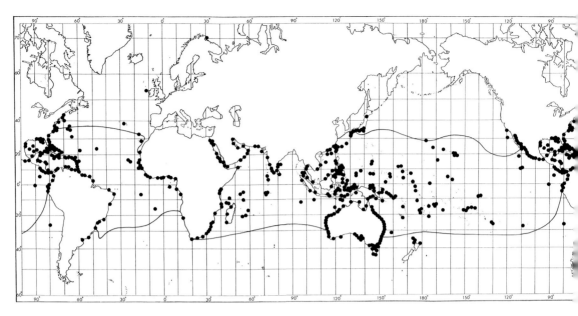

Fig. 273. World distribution of all 52 recognized species of the family Antennariidae, showing coincidence of northern and southern limits with the average annual surface isotherms for 20° C.

Horizontal Distribution

The geographic distribution of the taxonomic subunits of the Antennariidae is unequal around the world. As is characteristic of widely distributed tropical shorefish families (Goldman and Talbot, 1976: 129; Sale, 1980: 390; Springer, 1982: 117; Robertson et al., 2004), the number of antennariid species falls off sharply in either direction away from the Indo-West Pacific (Table 22). The 38 species that occur in the Indo-West Pacific are more than five times the number that are found in the Western Atlantic and nearly eight times the number in the Eastern Pacific and Eastern Atlantic (*Fowlerichthys radiosus* is most likely present in the Eastern Atlantic as a straggler; no breeding population is thought to exist; see Eastern Atlantic Endemics, p. 394).

Despite these numerical differences between major marine faunal realms of the world, many antennariid species have broad and relatively unrestricted geographic ranges (Table 22). One of the surprising outcomes of this investigation is that individual species of these sedentary fishes occur in three or more oceans of the world. It should not be forgotten, however, that pelagic eggs and larval stages (characteristic of the 32 species that make up the subfamily Antennariinae, nearly all of which have broad distributions), produced in large numbers by repeated spawnings during a lengthy breeding season, are subject to widespread dispersal. Five species are broadly distributed throughout the tropical waters of three or more oceans of the world. Of the 15 species that occur in the Western Indian Ocean (including the Red Sea; Table 22), all but two have distributions that continue eastward onto the Pacific Plate (see Springer, 1982); two of these (*Antennarius commerson* and *Abantennarius coccineus*) have distributions that extend beyond, into the Eastern Pacific.

On the other hand, not all antennariids display broad distributions. As might be expected, each of the four major marine-faunal regions of the world contains endemic species: six are restricted to the Western Atlantic, two to the Eastern Atlantic, 34 to the

Indo-Pacific, and three to the Eastern Pacific (Table 22). Of the 34 Indo-Pacific species, 22 are restricted to the Indo-Australian Archipelago, one (*Antennarius indicus*) is further confined to the Western Indian Ocean (and Red Sea), another (*Abantennarius drombus*) is found only at Johnston Atoll and the Hawaiian Islands, and yet another (*Antennatus flagellatus*), although currently known from only two specimens, may be endemic to Japan.

Although it is difficult, if not impossible, to explain why some antennariids have wide geographic ranges, whereas others are narrowly confined, the answer in some cases undoubtedly relates to differences in life history. The typical lophiiform mode of recruitment, employed by the 32 members of the subfamily Antennariinae, involves the production of large, egg-filled mucoid rafts or "veils" constructed to float freely at or near the surface and thus provide an excellent device for broadcasting a large number of small eggs over great geographical distances (see Ovarian Morphology and Egg-Raft Structure, p. 462). However, many antennariids employ an alternative reproductive strategy: the 20 members of the subfamily Histiophryninae retain and care for a small number of large eggs that hatch into relatively large, advanced young (as reflected in their ovarian morphology; see Parental Care, p. 479). Not surprisingly, these species have relatively restricted geographic ranges.

Wide-Ranging Species

Five species may be categorized as having wide geographic distributions; that is, they are found in three or more major oceans of the world (Table 22). Of these, *Histrio histrio* has the broadest longitudinal and latitudinal range. Its distribution is largely dependent on the dispersal capabilities of floating *Sargassum* (Dooley, 1972). In the Western Atlantic, it extends from the Gulf of Maine to the mouth of the Rio de la Plata, Uruguay, including the Gulf of Mexico and throughout the Caribbean (Fig. 274). On the eastern side of the Atlantic, it is apparently quite rare; we have seen specimens only from Madeira, the Azores, and off West Africa, and there is a recent record of five individuals observed at Cape Verde (Wirtz et al., 2017). The record from Vardø, northern Norway (Düben and Koren, 1846), is no doubt based on a straggler taken northward by the North Atlantic and Norwegian Currents.

In the Indian Ocean, *H. histrio* is known from the tip of South Africa eastward to India and Sri Lanka, with verified records from Oman, the Red Sea, Zanzibar, Madagascar, Réunion, and Mauritius. In the Western Pacific, it occurs from Hokkaido, Japan, to Australia (about as far south as Perth in the west and Sydney in the east), including Hong Kong, Taiwan, the Philippines and Moluccas, and Papua New Guinea, as well as rare but verified occurrences at Guam, the Marianas, Tonga, New Caledonia, the North Island of New Zealand, and the Hawaiian Islands (Fig. 274).

Although the *Sargassum* community has been described as a nearly worldwide, circumtropical phenomenon (but rare or perhaps absent in some parts of the central Pacific, especially in regions of low atolls; see Dooley, 1972; Tsuda, 1976; Phillips, 1995), with wind and current often carrying the weed and its associated fauna far into temperate waters (Adams, 1960), there are no verified records of *Histrio histrio* east of the Hawaiian Islands. Questionable occurrences of this species, however, have been reported from the Galápagos Islands (for further details of the geographic distribution of *H. histrio* in the Pacific, see Comments, p. 191).

As currently recognized, *Antennarius striatus* occurs in both the Atlantic and Indo-Pacific oceans. In the Eastern Atlantic (where it has been recognized as *A. occidentalis* by various authors) it is found off the West African coast from Senegal to Namibia, with

Table 22. World Distribution of Species of the Antennariidae

Species	Western Atlantic	Eastern Atlantic	Western Indian	Red Sea	Indo-West Pacific	Australia	Japan	Pacific Plate	Hawaiian Islands	Society Islands	Eastern Pacific	Text Figure
Wide-Ranging Species (5)												
Histrio histrio	X	X	X	X	X	X	X	X	X			274
Antennarius striatus		X	X	X	X	X	X	X	X	X		278
Abantennarius nummifer		X	X	X	X	X	X	X	X	X		277
Antennarius commerson			X	X	X	X	X	X	X	X	X	275
Abantennarius coccineus			X	X	X	X	X	X	X	X	X	276
Western Atlantic Endemics (6)												
Abantennarius bermudensis	X											276
Antennarius multiocellatus	X											281
Antennarius pauciradiatus	X											282
Antennarius scaber	X											278
Fowlerichthys ocellatus	X											280
Fowlerichthys radiosus	X											279
Eastern Atlantic Endemics (2)												
Antennarius pardalis		X										281
Fowlerichthys senegalensis		X										280
Western Indian Ocean Endemic (1)												
Antennarius indicus			X	X								278
Indo-West Pacific Endemic (1)												
Antennarius hispidus			X	X	X	X	X					280
Indo-Pacific Species (10)												
Fowlerichthys scriptissimus			X		X		X	X		X		279
Antennarius maculatus			X	X	X	X	X	X	X	X		275
Antennarius pictus			X	X	X	X	X	X	X	X		281
Abantennarius dorehensis			X		X	X	X	X		X		283
Antennatus linearis			X		X			X	X			285
Antennatus tuberosus			X	X	X		X	X	X	X		285
Abantennarius rosaceus				X	X	X	X	X				283
Antennarius randalli				X	X		X	X	X			282
Abantennarius analis					X			X	X	X		284
Abantennarius duescus					X			X	X			284

Species												Page
Indo-Australian Species (9)												
Antennarius biocellatus					X							282
Nudiantennarius subteres					X		X					274
Lophiocharon lithinostomus					X							286
Histiophryne cryptacanthus					X							287
Histiophryne pogonius					X							288
Histiophryne psychedelica					X							288
Lophiocharon hutchinsi					X	X						286
Lophiocharon trisignatus					X	X						286
Tathicarpus butleri					X	X						289
Southern Australian Endemics (13)												
Allenichthys glauerti						X						288
Echinophryne crassispina						X						289
Echinophryne mitchelli						X						289
Echinophryne reynoldsi						X						289
Histiophryne bougainvilli						X						288
Histiophryne maggiewalker						X						287
Histiophryne narungga						X						287
Kuiterichthys furcipilis						X						290
Kuiterichthys pietschi						X						290
Phyllophryne scortea						X						290
Porophryne erythrodactylus						X						291
Rhycherus filamentosus						X						291
Rhycherus gloveri						X						291
Japanese Endemic (1)												
Antennatus flagellatus							X					285
Hawaiian and Johnston Atoll Endemic (1)												
Abantennarius drombus								X	X			276
Eastern Pacific Endemics (3)												
Fowlerichthys avalonis											X	279
Abantennarius sanguineus											X	277
Antennatus strigatus											X	285
TOTAL NUMBER OF SPECIES	7	5	13	12	25	26	16	16	12	9	5	

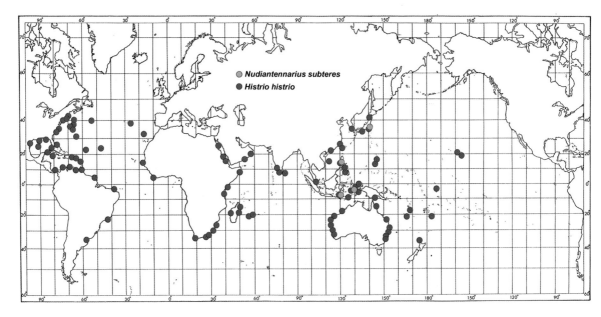

Fig. 274. Distribution of *Nudiantennarius subteres* (blue), Japan, Philippines, and Indonesia; and *Histrio histrio* (red), Atlantic and Indo-West Pacific.

records from St. Helena, Cape Verde, and the Canary Islands. In the Indo-Pacific it extends from the East African coast (including the Red Sea and the islands of Madagascar, Réunion, and Mauritius) to southern Japan, Taiwan, the Philippines, Sulawesi, the Moluccas, Australia, New Zealand, New Caledonia, and eastward to the Hawaiian, Society, and Marquesas Islands, with at least one record from Fiji.

Antennarius commerson and *Abantennarius coccineus* have similar distributions that range from South Africa, the Red Sea, and throughout the Indo-Pacific to the coasts of Central and South America (Figs. 275, 276). *Antennarius commerson*, rare in collections compared to *Abantennarius coccineus* (73 and 458 specimens examined, respectively), is known from off the east coast of South Africa, Mozambique, Oman, Réunion, the Seychelles, Ryukyu Islands, Taiwan, Malaysia, Bali to Papua New Guinea, Western Australia, the Andaman Sea, and New Caledonia. Farther east it has been reported from Fiji and the Mariana, Hawaiian, and Society Islands. Although its presence off Panama (including Cocos and the Revillagigedo Islands) and Colombia has now been verified, the single California record for this species (see Schultz, 1957: 92) remains doubtful (Fig. 275). Similarly, *Abantennarius coccineus* is well-known from off the east African coast, Mozambique, the Red Sea, Oman, and throughout the islands of the Western Indian Ocean and Pacific, including the Line and Marquesas Islands, Pitcairn and Easter Islands, as well as Clipperton, the Galápagos, Cocos, and Isla San Félix in the Eastern Pacific (Fig. 276). It is replaced in the Hawaiian Islands, however, by the endemic *A. drombus* (see Hawaiian and Johnston Atoll Endemic, p. 402).

Abantennarius nummifer has a typical Indo-Pacific distribution, ranging from the east coast of Africa to Réunion, the Red Sea, Oman, the Persian Gulf, and throughout the Indo-Australian Archipelago, including southern Japan, Taiwan, New Caledonia, New Zealand, and eastward to the Hawaiian and Society Islands (Fig. 277). It is also well represented at insular localities of the Eastern Atlantic, including the Azores, Madeira, Canaries, Cape Verde, and St. Helena.

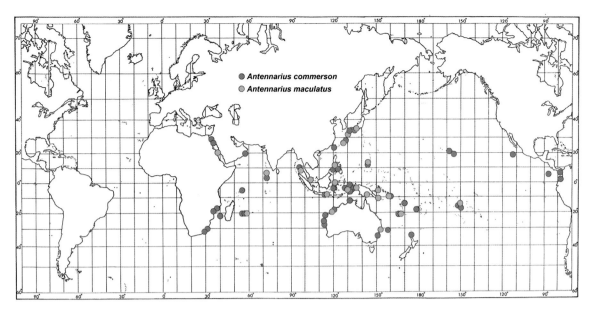

Fig. 275. Distribution of *Antennarius commerson* (red), Indo-Pacific; and *Antennarius maculatus* (blue), Indo-West Pacific.

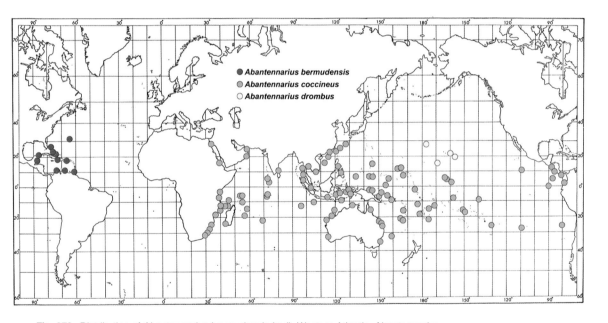

Fig. 276. Distribution of *Abantennarius bermudensis* (red), Western Atlantic; *Abantennarius coccineus* (blue), Indo-Pacific; and *Abantennarius drombus* (yellow), Hawaiian Islands.

Western Atlantic Endemics

Six antennariid species have distributions that are more or less restricted to the Western Atlantic Ocean (Table 22). *Antennarius scaber* ranges from Shinnecock Bay, Long Island, New York, to Florida, including Bermuda, the Bahamas, Cuba, the Gulf of Mexico, and throughout the island groups of the Caribbean to the southernmost coast of Brazil (Fig. 278). *Fowlerichthys radiosus* and *F. ocellatus* have similar distributions

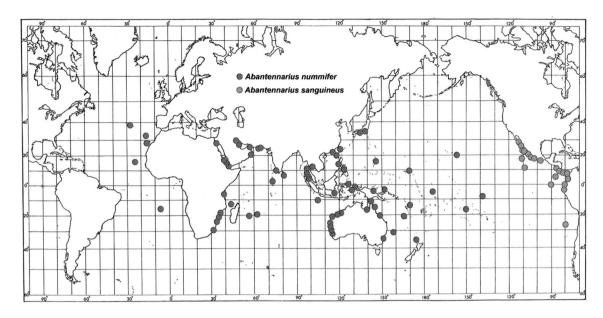

Fig. 277. Distribution of *Abantennarius nummifer* (red), Eastern Atlantic and Indo-West Pacific; and *Abantennarius sanguineus* (blue), Eastern Pacific.

(Figs. 279, 280). They occur mainly along continental margins, from off the mid-coastal states of the United States and throughout the Gulf of Mexico, and extending southward to Colombia and Venezuela; however, both are conspicuously absent from the islands of the Caribbean. In contrast, *Antennarius multiocellatus* (Fig. 281), *A. pauciradiatus* (Fig. 282), and *Abantennarius bermudensis* (Fig. 276) are insular species, all notably rare along continental margins and from the Gulf of Mexico, but present at Bermuda, the Bahamas, and most of the island groups of the Caribbean. Whereas, *Antennarius pauciradiatus* and *Abantennarius bermudensis* rarely occur farther south beyond the Caribbean, with fewer than a half dozen records from off Colombia and Venezuela, *Antennarius multiocellatus* is well-documented along the coasts of Central America, Colombia, Venezuela, and Brazil as far south as São Paulo (Fig. 281).

Two of these Western Atlantic species have distributions that very rarely extend into the eastern Atlantic: *Fowlerichthys radiosus* is known from a single prejuvenile specimen collected off Ireland (Wheeler, 1969) and two large adults from Madeira (Fig. 279); *Antennarius multiocellatus* occurs at Ascension Island in the mid-South Atlantic, at São Tomé in the Gulf of Guinea, and perhaps in the Azores (Fig. 281). It is assumed, however, that all these records represent expatriates and are not representative of breeding populations.

Eastern Atlantic Endemics

Two species are restricted in distribution to the Eastern Atlantic Ocean (Table 22). *Antennarius pardalis* ranges from off Senegal and the Cape Verde Islands southward and eastward to the Congo, with records from Príncipe and São Tomé, Gulf of Guinea (Fig. 281). In addition to several records from the Cape Verde Islands and at least one from the Azores, *Fowlerichthys senegalensis* has been collected from off Mauritania to Liberia (Fig. 280), but material has been reported from as far north as Morocco (Furnestin et al., 1958) and as far south as Cape Morro, Angola (Poll, 1959).

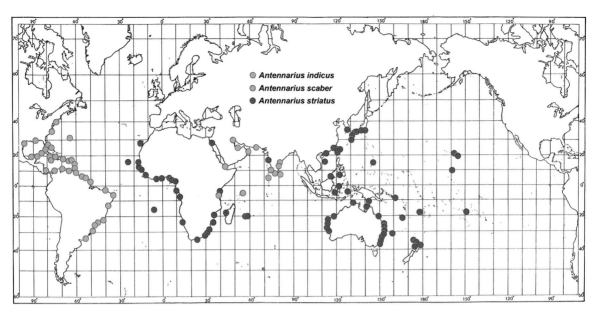

Fig. 278. Distribution of *Antennarius indicus* (blue), Western Indian Ocean; *Antennarius scaber* (green), Western Atlantic; and *Antennarius striatus* (red), Eastern Atlantic and Indo-West Pacific.

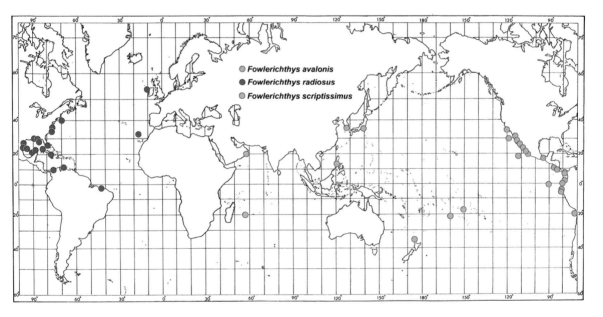

Fig. 279. Distribution of *Fowlerichthys avalonis* (green), Eastern Pacific; *Fowlerichthys radiosus* (red), Atlantic; and *Fowlerichthys scriptissimus* (blue), Indo-West Pacific. In this and subsequent charts a single symbol may indicate more than one capture.

Western Indian Ocean Endemic

Antennarius indicus is the only species of the family confined in distribution to the Western Indian Ocean (Fig. 278, Table 22). It extends from Zanzibar, Mozambique, and the Seychelles to the Gulf of Mannar, the Coromandel Coast of India, and Sri Lanka, including the Gulf of Aden, coast of Oman, Persian Gulf, Pakistani coast, and the Laccadive Archipelago (Jones and Kumaran, 1980).

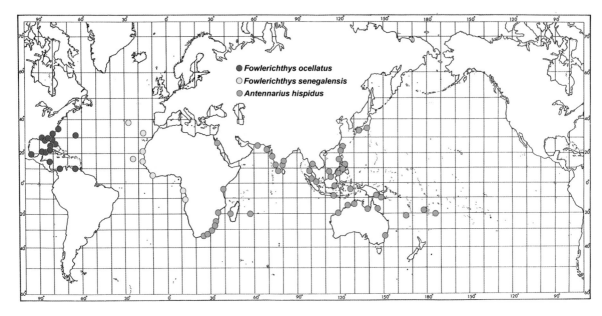

Fig. 280. Distribution of *Fowlerichthys ocellatus* (red), Western Atlantic; *Fowlerichthys senegalensis* (yellow), Eastern Atlantic; and *Antennarius hispidus* (blue), Indo-West Pacific.

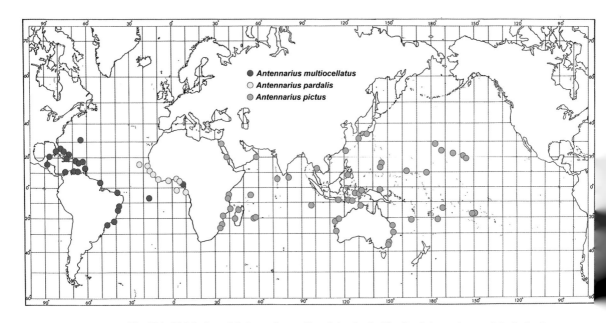

Fig. 281. Distribution of *Antennarius multiocellatus* (red), Atlantic; *Antennarius pardalis* (yellow), Eastern Atlantic; and *Antennarius pictus* (blue), Indo-West Pacific.

Indo-West Pacific Endemic

Only one species, *Antennarius hispidus*, is confined to the Indo-West Pacific Ocean (Table 22). It ranges from South Africa, Mozambique, the Red Sea, the Pakistani coast, India, and Malaysia to the Moluccas, and from southern Japan and Taiwan to northern Australia, with records from New Caledonia, Fiji, and Tonga (Fig. 280). Although rarely collected at oceanic islands of the Indian Ocean, Bleeker's (1859b: 128) records

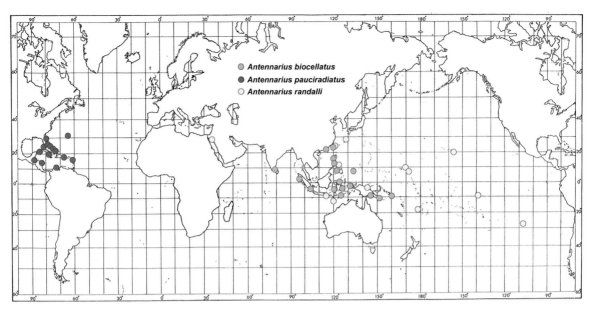

Fig. 282. Distribution of *Antennarius biocellatus* (blue), Indo-West Pacific; *Antennarius pauciradia-tus* (red), Western Atlantic; and *Antennarius randalli* (yellow), Western Pacific.

for Mauritius and Réunion, questioned by Pietsch and Grobecker (1987: 74), have been confirmed.

Indo-Pacific Species

Ten antennariids have distributions in the Indo-West Pacific that extend onto the Pacific Plate (Table 22). Most of the known material of *Abantennarius rosaceus* was collected in the Philippines and Indonesia (Java to West Papua, including Bali, Timor, and the Togian and Molucca Islands); additional localities include the Red Sea, Lord Howe Island, New Caledonia, Samoa, Bikini Atoll in the Marshalls, and Onotoa Atoll in the Gilberts (Fig. 283). *Antennarius randalli* has been documented at Taiwan, Cebu Island in the Philippines, Banda in the South Moluccas, Fiji, the Hawaiian Islands, Easter Island, and very recently off Nuweiba in the Gulf of Aqaba, Red Sea (Fig. 282). *Abantennarius analis* is known from numerous, widely scattered localities that range from Christmas Island and Rowley Shoals in the Eastern Indian Ocean to Okinawa, off northwestern Australia, and east to the Society and Hawaiian Islands, including the Marshalls and Marianas, and the islands of Palau, Yap, Enewetak, Guadalcanal, Tonga, Fiji, and Samoa (Fig. 284). *Abantennarius duescus*, previously recognized by Pietsch and Grobecker (1987: 295) as a possible Pacific Plate endemic (see Springer, 1982: 13)—then known only from the three types of specimens all collected in the Hawaiian Islands—has since been documented at Midway and Johnston Atolls; Flores in the Lesser Sunda Islands, Indonesia; off Madang, Papua New Guinea; and Ouvea Atoll in the Loyalty Islands, territory of New Caledonia (Fig. 284).

The six remaining species in this group of Indo-Pacific forms, *Fowlerichthys scriptissimus* (Fig. 279), *Antennarius maculatus* (Fig. 275), *Antennarius pictus* (Fig. 281), *Abantennarius dorehensis* (Fig. 283), *Antennatus linearis* (Fig. 285), and *Antennatus tuberosus* (Fig. 285), all have similar distributions that extend from East Africa and the Western Indian Ocean (with several records of *A. maculatus*, *A. pictus*, and *Antennatus tuberosus* in the Red Sea) to southern Japan, the Philippines, Indonesia, and onto the Pacific Plate. *Fowlerichthys*

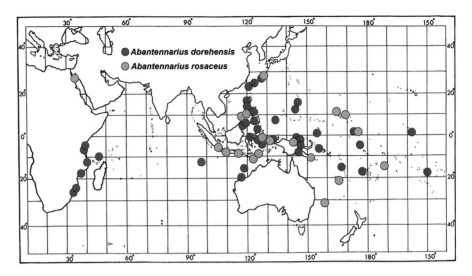

Fig. 283. Distribution of *Abantennarius dorehensis* (red), Indo-West Pacific; and *Abantennarius rosaceus* (blue), Indo-West Pacific.

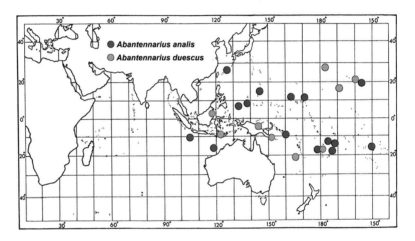

Fig. 284. Distribution of *Abantennarius analis* (red), Western Pacific; and *Abantennarius duescus* (green), Western Pacific.

scriptissimus, now known from about 30 specimens, has been collected from off Réunion, Oman, Korea, Japan, Manila Bay in the Philippines, the North Island of New Zealand, and the Cook and Society Islands (Fig. 279). All but *Antennarius randalli*, *Antennatus linearis*, *Abantennarius duescus*, and *Abantennarius rosaceus* are found in the Society Islands, and all but *Fowlerichthys scriptissimus*, *Abantennarius dorehensis*, and *Abantennarius rosaceus* occur in the Hawaiian Islands. *Antennarius pictus* extends as well to temperate Australia and *Antennatus tuberosus* is further represented at Pitcairn Island.

Indo-Australian Species

Five antennariid genera and nine species (excluding the southern and temperate Australian endemics described below) are confined to the Indo-Australian Archipelago, a region that extends from Taiwan to Tasmania, including all of the inland seas and islands of the Philippines, Indonesia, New Guinea, the Solomons, and the continent of Australia (Table 22).

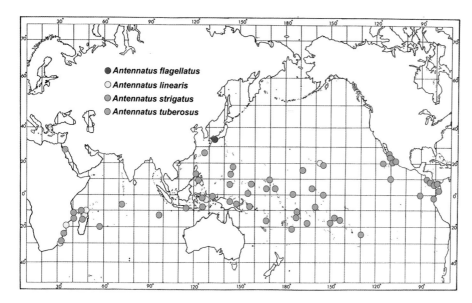

Fig. 285. Distribution of *Antennatus flagellatus* (red), Japan; *Antennatus linearis* (yellow), Indo-West Pacific; *Antennatus strigatus* (green), Eastern Pacific; and *Antennatus tuberosus* (blue), Indo-West Pacific.

Antennarius biocellatus is restricted to the shallow waters of Taiwan, the Philippines, Palau, Indonesia (Sumatra to West Papua), Papua New Guinea, and the Solomons, with at least one occurrence off Hong Kong and another from the Gulf of Mannar, India (Fig. 282, Table 22). *Nudiantennarius subteres* is known from fewer than a dozen specimens, at least two of which were collected off Luzon in the Philippines; the remaining material is from Flores in the Lesser Sunda Islands, and Ambon in the Moluccas. In addition, photographs have indicated its presence off North Sulawesi, Pantar and Alor Islands in the Alor Archipelago, and in Suruga Bay, Japan (Fig. 274). *Lophiocharon trisignatus* ranges from the Philippines to the western and northern coasts of Australia, with a single questionable record from some unknown locality on the Great Barrier Reef, Queensland. *Lophiocharon lithinostomus*, as presently known, is narrowly confined to an area extending from North Borneo and the southern tip of the Sulu Archipelago to Panay Island in the Philippines. The third member of this genus, *L. hutchinsi*, extends from southern Papua to the northern coast of Australia, including the Gulf of Carpentaria (Fig. 286, Table 22).

Histiophryne is found at various localities throughout this geographic area; three of the six recognized species are confined to tropical latitudes: *H. cryptacanthus* ranges from Taiwan and the Philippines to southern Indonesia and Papua New Guinea (Fig. 287); *H. pogonius* is known from the Philippines, eastern Indonesia, and Papua New Guinea (Fig. 288); and *H. psychedelica* is known only from Ambon, Maluku Islands, Indonesia (Fig. 288). *Tathicarpus butleri* ranges from Papua New Guinea (including the eastern Moluccan island of Ujir) and the coasts of Western Australia, Northern Territory (including the Gulf of Carpentaria and Torres Strait), and Queensland (Fig. 289).

Southern Australian Endemics

Seven antennariid genera and 13 species are endemic to subtropical and temperate waters of Australia, including Tasmania (Table 22). Within this restricted region, *Allenichthys*

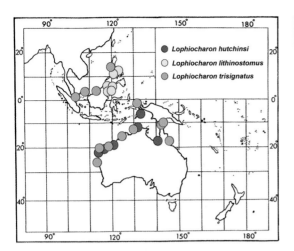

Fig. 286. Distribution of *Lophiocharon hutchinsi* (red), Indo-Australia; *Lophiocharon lithinostomus* (yellow), Philippines and Indonesia; and *Lophiocharon trisignatus* (blue), Indo-Australia.

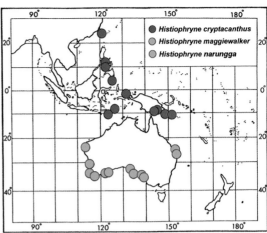

Fig. 287. Distribution of *Histiophryne cryptacanthus* (red), Indo-Australia; *Histiophryne maggiewalker* (blue), New South Wales; and *Histiophryne narungga* (green), Western and South Australia.

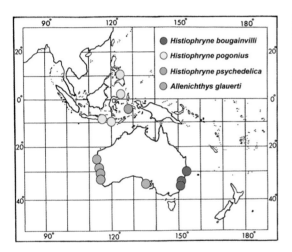

Fig. 288. Distribution of *Histiophryne bougainvilli* (red), eastern and southern Australia; *Histiophryne pogonius* (yellow), Philippines and Indonesia; *Histiophryne psyche-delica* (green), Ambon, Indonesia; and *Allenichthys glauerti* (blue), Western and South Australia.

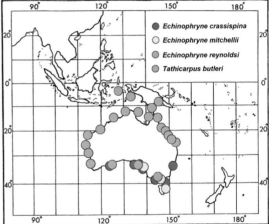

Fig. 289. Distribution of *Echinophryne crassispina* (red), southern Australia; *Echinophryne mitchellii* (yellow), southern Australia and Tasmania; *Echinophryne reynoldsi* (green), southern Australia; and *Tathicarpus butleri* (blue), Indo-Australia.

glauerti has been collected only along the coast of Western Australia (as far south as 35° S latitude, with a single record from Port Lincoln, South Australia (Fig. 288). *Echinophryne crassispina* ranges from Spencer Gulf, South Australia, to the coast of New South Wales (below about 32° S latitude), including Tasmania; *E. reynoldsi* extends from the Recherche Archipelago, Western Australia, to Victoria; and *E. mitchellii* is known from the coastal waters of Victoria, Flinders Island in the Bass Strait, and the east coast of Tasmania (Fig. 289).

The genus *Histiophryne* contains three allopatric subtropical and temperate species: the known material of *H. bougainvilli* (Fig. 288) and *H. maggiewalker* (Fig. 287) has been col-

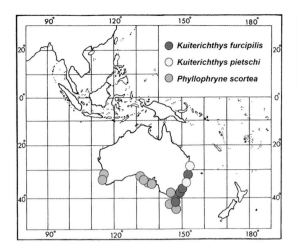

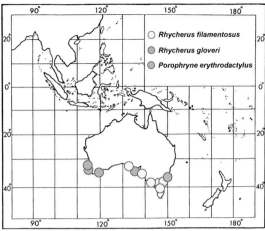

Fig. 290. Distribution of *Kuiterichthys furcipilis* (red), New South Wales, Victoria, and Tasmania; *Kuiterichthys pietschi* (yellow), New South Wales; and *Phyllophryne scortea* (blue), southern Australia and Tasmania.

Fig. 291. Distribution of *Rhycherus filamentosus* (yellow), southern Australia and Tasmania; *Rhycherus gloveri* (green), southern Australia; and *Porophryne erythrodactylus* (blue), New South Wales, Australia.

lected solely from off New South Wales and Queensland, respectively, while *H. narungga* extends from off Cape Cuvier, Western Australia, and south and east to Kangaroo Island and Aldinga Beach, South Australia (Fig. 287). *Kuiterichthys pietschi* and *K. furcipilis* are each found off New South Wales, while the latter extends farther south to the coasts of Victoria and Tasmania (Fig. 290). *Phyllophryne scortea* ranges from the southwest corner of Western Australia to South Australia, Victoria, and Tasmania (Fig. 290).

The two species of the genus *Rhycherus* are sympatric in the vicinity of Adelaide, South Australia; *R. gloveri*, however, is also well-documented off Western Australia (between 31° and 35° S latitude), whereas *R. filamentosus* extends farther east into the coastal waters of Victoria (Fig. 291). Finally, *Porophryne erythrodactylus* has only been collected from nearshore waters of Botany Bay, New South Wales, with confirmed sightings as far south as Jervis Bay, New South Wales (Fig. 291).

Japanese Endemic

Originally described from two specimens collected off Kashiwajima Island, Kochi Prefecture, southern Japan (Fig. 285), no additional material of *Antennatus flagellatus* has been collected since its initial discovery more than 25 years ago (Nobuhiro Ohnishi, personal communication, 1 September 2016). While it may eventually prove to be a Japanese endemic, the true boundaries of its distribution are presently unknown. Judging from the distributions of other antennariids, future collecting will probably show this species to have a considerably broader geographic range.

Williams (1989: 280) stated that "the Japan region harbors several distinct (endemic?) populations (species?) of frogfishes." But, aside from *Antennatus flagellatus*, which was not described until 1997, the only "population" that might qualify for this supposed endemism is the Japanese material of *Antennarius striatus* with a trifid esca, which several authors (e.g., Paulin, 1978: 490; Michael, 1998: 354, 2007: 58) have suggested should be recognized as *Antennarius tridens*, a name currently contained within the synonymy of *A. striatus*. But "*striatus*" specimens with a trifid esca are not at all limited to Japanese waters. On the contrary, 93% of the Indo-Pacific material of *A. striatus* examined (428 specimens from Japan to temperate Australia and from South Africa to the Hawaiian Islands) had a trifid esca (for more discussion, see Comments, p. 108).

Hawaiian and Johnston Atoll Endemic

Abantennarius drombus has one of the most restricted geographic distributions of any antennariid (Fig. 276). Now known from well over 50 specimens, all collected from Johnston Atoll and the Hawaiian Archipelago (including Midway Atoll), it is unlikely to be collected beyond these waters in the future.

Eastern Pacific Endemics

Three species are endemic to the Eastern Pacific Ocean (Table 22). *Fowlerichthys avalonis* is found throughout the Gulf of California and south along the coasts of Mexico and Central and South America to the port city of Iquique, northern Chile (at about 20° S latitude), with two vouchered records from off southern California, at least one specimen from Guadalupe Island, Mexico (Walther-Mendoza et al., 2013), and another from the Galápagos Islands (Grove and Lavenberg, 1997). The distributions of *Abantennarius sanguineus* and *Antennatus strigatus* are similar to that of *Fowlerichthys avalonis*, except that both are absent off southern California and in the northern half of the Gulf of California (Figs. 277, 285). In addition to occurring in the Galápagos Islands, both *Abantennarius sanguineus* and *Antennatus strigatus* are also found at Islas Tres Marías and Revillagigedo, Clipperton Island, Cocos and Caño Islands off Costa Rica, Gorgona Island, Colombia, and off coastal Ecuador. *Abantennarius sanguineus* has further been collected at Isla San Félix.

Vertical Distribution

With the single exception of *Histrio histrio*, which is pseudopelagic in *Sargassum* (and often found as well clinging to various inorganic floating debris such as discarded fishing nets)—and evidence that frogfish larvae are pelagic at considerable depths (scc Adams, 1960; Dooley, 1972; Boehlert and Mundy, 1996)—all antennariids are benthic, shallow to moderately deep-dwelling, marine fishes associated with flat, mud, or sand bottoms, as well as with eelgrass and rocky and coral reefs. *Antennarius biocellatus* is the only species commonly found in brackish or even fresh water (see Comments, p. 160).

Because depths of capture for most collections are often not recorded, vertical ranges for most species are difficult to define with precision. Generally speaking, however, frogfishes may be taken anywhere between the surface and about 350 m, with occurrences most common between about 20 and 100 m. Species of the genus *Fowlerichthys* are among the deepest-dwelling members of the family. All were taken at average depths greater than 50 m. Two species, *F. avalonis* and *F. radiosus*, averaged 88 and 90 m, respectively (*F. avalonis* has been trawled from depths in excess of 200 m); a 252-mm specimen of *F. scriptissimus* was taken by hook and line off Mahina, Tahiti, at a depth of 350 m. In contrast, however, off Destin, Florida, in the northern Gulf of Mexico, *Fowlerichthys ocellatus* is often found at depths as shallow as 15 m but it is more common at about 25 m and likely extends down to at least 50 m (Jim Garin, personal communication, 4 June 2017).

Within the genus *Antennarius*, depths of capture are usually similar for all species within a species-group. While two of the four species of the *A. striatus* group, *A. hispidus* and *A. scaber*, are often encountered within relatively shallow scuba depth, most preserved specimens in collections around the world were taken at considerably greater depths averaging 45 and 40 m, respectively. Depths of capture for the widely distributed *A. striatus*, recorded for 79 specimens, ranged from the surface to 219 m; about 86% of these, however, were taken in 30 m or more. But average depths for known captures varied widely depending on locality: Eastern Atlantic (30 records), 40 m; Western Indian Ocean (8 records), 43 m; Indo-Australian Archipelago (28 records), 73 m; and the Hawaiian Islands (13 records), 119 m. Although the average depth of known captures made in Australian waters exceeded 70 m, divers have routinely encountered this species in Sydney Harbour between about 10 and 25 m during the months of November to early December and March through April (Pietsch and Grobecker, 1987: 69). Similar occurrences are common at other popular dive sites throughout Indonesia and the Philippines. Depths of capture of *A. indicus*, the fourth member of this species group, are so poorly known that little can be concluded.

Members of the *A. pictus* group and the genus *Abantennarius* are relatively shallow-water species, nearly all having average depths of less than 25 m. An exception within the latter genus is the Eastern Atlantic population of *Abantennarius nummifer*, with six known depth records ranging between 44 and 293 m, and averaging 107 m. *Abantennarius dorehensis* is the shallowest-living antennariid, with all known captures (46 specimens) made in less than 2.5 m and 76% of the material taken in 1 m or less.

Known depths of capture for the only member of the *Antennarius biocellatus* group averaged 11 m, and those of the two species of the *Antennarius pauciradiatus* group, *A. pauciradiatus* and *A. randalli*, averaged 22 and 16 m, respectively. The monotypic genus *Histrio* is unique among lophiiform fishes in having a pseudopelagic association with *Sargassum*. Although Adams (1960) found that larval and postlarval individuals (up to 4 mm SL) occur at depths of 50 to 600 m, and Boehlert and Mundy (1996) provided additional evidence of the relatively deep occurrence of antennariid larvae, nearly all of the specimens examined in this study (for which depth data were available) were taken essentially at the surface.

Early collection data for the monotypic genus *Nudiantennarius* indicated a deep-water presence, at depths between 64 and 90 m (CAS 32765), 82 m (the holotype, USNM 70268), and 128 m (ZMUC P922045)—the reason why Pietsch and Grobecker (1987) called it the "Deepwater Frogfish." But numerous recent observations by scuba divers indicate a much shallower existence, from 3–30 m, although some hypothesize that most observations are based on juveniles, while adults occur at greater depths (Vincent Chalias, personal communication, 11 April 2017).

Allenichthys glauerti, represented by only two specimens for which depths of capture are known, appears to be rather deep-dwelling antennariid, trawled at 146 and 183 m. Members of the genus *Kuiterichthys* also appear to be deep-dwelling species: *K. furcipilis* ranged between 9 and 240 m, with an average capture depth of about 90 m; *K. pietschi* ranged from 60 to 89 m, with an average depth of 73 m. In contrast, *Echinophryne crassispina*, the monotypic *Phyllophryne*, and the four species of *Antennatus*, are all relatively shallow-water species, with average depths of capture between 10 and 14 m. Insufficient data are available to say much about depth preference for *Histiophryne bougainvilli*, but its congener *H. cryptacanthus* ranged from near the surface to 45 m, with an average depth of capture of 18 m. The remaining members of this genus were all taken at an average depth of 12 m. Finally, the monotypic genus *Tathicarpus*, taken between 7.3 and 146 m, had an average depth of 38 m. For all remaining antennariid taxa, vertical distributional data were either unavailable or insufficient to make any reasonable conclusions.

Ecological Zoogeography

A complete reconstruction of the ecological and historical factors responsible for the present-day distributional patterns shown by the Antennariidae is impossible: many species are so poorly represented in collections that their total geographic ranges are unknown, ecological data are unavailable for most species, and the fossil record for the family is minimal (see The Fossil Record, p. 357). But a more serious obstacle to zoogeographic explanation is the fact that the phylogenetic relationships among the genera and species are not fully understood (see Evolutionary Relationships, p. 374). For these reasons only a few broad generalizations can be made.

Temperature

The primary factor limiting the latitudinal range of the family is temperature. This can be illustrated by plotting the distributions of all antennariid species simultaneously on a chart constructed to show average annual surface isotherms (i.e., imaginary lines drawn through points of equal temperature). When this is done, it can be seen that the distribution of the family corresponds remarkably well to the broad tropical zone, a region limited by the average annual surface isotherm for 20° C (Fig. 273). One notable exception to this rule is the group of relatively cold-water species that are restricted to the southern coast of Australia (see Southern Australian Endemics, p. 399). Other, less-significant incursions into subtropical and temperate waters include the presence of members of the genus *Antennarius* and *Antennatus* in Japan and along the Atlantic coasts of the Americas. But rather than representing permanent features of total geographic ranges, these may well be the result of expatriated individuals taken north or south by currents during especially warm months of the year.

Continental versus Insular Species

In addition to temperature, there are obviously other, more complex, environmental parameters that help to shape the major patterns of distribution displayed by antennariids. The host of factors that contribute to the ecological differences between "continental" and "insular" habitats appear to be especially important. Continental species are those that are adapted to areas where broad muddy embayments are common and estuarine influence strong; these are typically forms that can withstand (or require) seasonal shifts in climates, changes in salinity and temperature due to runoff from large rivers, and turbidities caused by winds that stir rich bottom sediments (Robins, 1971: 251). By contrast, insular species are those that require (or can withstand) clear waters, buffered environmental conditions, and bottom sediments largely of calcium carbonate. Despite the fact that frogfishes are often included as part of coral-reef communities (which are typically well-developed in island localities), our investigation indicates that they are in no major way associated with coral (in agreement with Mead, 1970: 242). Although sufficient specific locality data are lacking to categorize all species, many appear to prefer a more continental habitat where reef-building corals are absent. For example, in the Western Atlantic, *Fowlerichthys radiosus* and *F. ocellatus* are distinctly continental in distribution, nearly all of the known material having been collected by commercial bottom trawling in the Gulf of Mexico; these species are notably absent from Bermuda, the Bahamas, and the island groups of the Caribbean (Figs. 279, 280). Similarly, the vast proportion of the known material of *Fowlerichthys senegalensis* and *Antennarius pardalis* in the Eastern Atlantic (Figs. 280, 281), *F. avalonis* in the Eastern Pacific (this species also occurring in the muddy-bottomed and turbid northernmost reaches of the Gulf of California; Fig. 279), and *Kuiterichthys furcipilis* of subtropical Australian waters (Fig. 290) was collected by trawls over flat, soft bottoms. The meager information available for *Antennarius indicus* of the Western Indian Ocean (Fig. 278), *Tathicarpus butleri* of New Guinea and Tropical Australia (Fig. 289), and *Allenichthys glauerti* of Western and South Australia (Fig. 288) also indicates a continental preference. *Antennarius hispidus*, restricted to the Indo-West Pacific, is conspicuously absent from most insular localities, except for a very few records from the high islands of New Caledonia, Fiji, and Tonga (Fig. 280). *Antennarius biocellatus* is strongly associated with estuarine and freshwater habitats (Fig. 282).

A preference for habitats more continental in nature probably also explains why a number of Indo-Pacific antennariids are widespread throughout the Indo-West Pacific and well established at the Hawaiian and Society Islands, but completely absent throughout the intervening low island groups of the Pacific Plate (e.g., *Antennarius striatus*, Fig. 278; *A. commerson*, Fig, 275; and *Abantennarius nummifer*, Fig. 277). The Hawaiian and Society Islands, in contrast to nearly all other island groups of the Pacific Plate (see Springer, 1982, fig. 2), are relatively high in elevation, and because of their consequent high rainfall, act to a large extent like continents, with permanent stream and river systems and well-developed muddy embayments. Why these same species are not also present at the nearby high islands of the Marquesas (or, for that matter, at other, more westerly high islands with similar habitats, such as Truk, Ponape, etc.) cannot be explained.

Not all antennariids, however, display continental distributional patterns. In the Western Atlantic, *Antennarius multiocellatus* (Fig. 281), *A. pauciradiatus* (Fig. 282), and *Abantennarius bermudensis* (Fig. 276) are distinctly insular, nearly all of the known material having been collected at Bermuda, throughout the Bahamas, and at island localities of the Caribbean. Similarly, the Indo-Pacific species *Antennarius maculatus*, *A. pictus*, *A. randalli*, *Abantennarius analis*, *A. coccineus*, *A. dorehensis*, *A. drombus*, *A. rosaceus*, and *Antennatus tuberosus* are all primarily associated with islands, including the low islands of the Pacific Plate (Figs. 275, 276, 281–285). *Abantennarius sanguineus* and *Antennatus strigatus* avoid the continental quality of the northern half of the Gulf of California (which is largely under the influence of the outflow of the Colorado River), but are well distributed in the lower, more insular portion of the Gulf (Figs. 277, 285); both species are present at Islas Tres Marías, Revillagigedos, Clipperton, Cocos, and the Galápagos Islands, but only *Abantennarius sanguineus* occurs also at Isla San Félix off Chile.

Three relatively well-known and widely distributed species, *Antennarius scaber*, *A. striatus*, and *Abantennarius nummifer*, cannot easily be categorized as being either continental or insular. All available ecological evidence indicates that *Antennarius scaber* and its sister-species *A. striatus* are strongly associated with eelgrasses and sponge beds on flat-sand, coral-rubble, or mud bottoms. In the Western Atlantic, *A. scaber* shows, for the most part, an insular distribution, being notably rare in the Gulf of Mexico, but present at Bermuda and throughout the Bahamas, the Florida Keys, and islands of the Caribbean (Fig. 278); at the same time, however, it occurs along the continental margin of South America, extending to southernmost Brazil. Except for a single known capture at the island of St. Helena, all known material of *A. striatus* from the Eastern Atlantic was taken by trawl on open soft bottoms. In the Indo-Pacific, *A. striatus* is conspicuously absent from all island groups of the Pacific Plate except for the high islands of the Hawaiian and Society groups.

In the Indo-Pacific, *Abantennarius nummifer* displays a distribution that is very similar to that of *Antennarius striatus* (compare Figs. 277 and 278). However, in the Eastern Atlantic, *Abantennarius nummifer* is found only at insular localities of the Azores, Madeira, the Canaries, and St. Helena, whereas *Antennarius striatus*, with the exception of a single record from St. Helena, is strongly associated with the continental margin of West Africa.

Historical Zoogeography

To provide some historical explanation for the present-day distributional patterns displayed by the family Antennariidae and its various subunits, the data were analyzed using the vicariant biogeographic methodology of Croizat (1958, 1964; Croizat et al., 1974; Rosen, 1975). Although this approach is now rather dated, especially when applied to marine organisms (see Evidence for Dispersal, p. 409), along with the concept of track analysis described below, we believe it is still instructive—we therefore present it here, with slight modifications following Pietsch and Grobecker (1987: 300).

The vicariant approach makes the assumption that on a global scale the general features of modern biotic distribution have been determined by a subdivision of ancestral biotas in response to changing geography; the events responsible for subdivision are termed "vicariant events." As stated by Rosen (1975: 432),

> the method . . . requires a comparison of the individual distributions (tracks) of species or monophyletic groups of species. A track is no more than a line on a map connecting the disjunct populations of a species or the disjunct species of a monophyletic group. Thus, a line may be drawn between the distribution of a monophyletic group of species and that of its sister group of one or more species. Plotting on a map the distributions of many different animal and plant assemblages from a certain region will demonstrate if commonality of distribution pattern occurs. If it does occur, the individual tracks will coincide to form a single pathway of massed tracks. This pathway, or generalized track, may be straight, curved, or irregular, but will be identifiable as a generalized track to the extent that all of its components share the same boundaries. Component tracks may occupy part or all of a generalized track. A generalized track which includes the individual tracks of many different monophyletic assemblages may connect adjoining or distant clusters of distributions. These clusters of distributions may be inferred to be fragments of the distribution of a parent biota. The reality of the parent biota may be tested repeatedly by the addition of other individual tracks to the system. The method of track construction is purely inductive, requiring only empirical knowledge of distribution of groups assumed to be monophyletic. An instance of lack of conformity of an individual track with a thoroughly documented generalized track may signify (1) that the individual track belongs to a different generalized track, (2) that the members of the track have broken away from the parent biota and have dispersed, or (3) that the track is based on a non-monophyletic group. The presence of distinct species clusters within a generalized track may then be theorized to have resulted at some time in the past from one or more vicariant events.

Track analysis of the distribution of the Antennariidae and its subunits revealed the presence of six distinct component tracks within the generalized track of the family (Fig. 292): an Atlantic-Indo-Pacific track, an Amphi-American track, a Trans-Atlantic track, an Indo-Pacific track, an Indo-Australian track, and a Southern-Australian track.

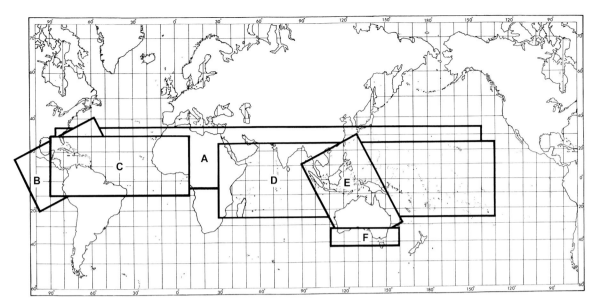

Fig. 292. Conceptual diagram of the six component tracks that characterize the present-day geographic distribution of the Antennariidae and its taxonomic subunits: (**A**) Atlantic-Indo-Pacific track. (**B**) Amphi-American track. (**C**) Trans-Atlantic track. (**D**) Indo-Pacific track. (**E**) Indo-Australian track. (**F**) Southern Australian track.

It should be emphasized that these component tracks of the Antennariidae are by no means unique to the family, but are in large part generalized biotic tracks for the entire marine shore fauna of the world. Similarly, most of the vicariant events responsible for the fragmentation of the ancestral antennariid biota are well-documented disruptive events that are responsible for the present-day distributional patterns displayed by wide-ranging tropical organisms of all kinds. The generalized track for the Antennariidae is thus only one component track of a generalized track of a worldwide tropical biota. For this reason, the conclusions reached here can be applied for the most part to a whole host of circumtropical marine organisms.

Generalized Track

The generalized track of the family largely coincides with the tropical zone of the world, as defined by the average annual surface isotherm for 20° C (see Temperature, p. 404; Fig. 273). Exceptions to this generalization include rare occurrences of stragglers in temperate waters and minor incursions into the subtropical waters of Japan and the Atlantic coast of the Americas; more significant, however, is the group of five genera and eight species endemic to southern Australia (Table 22). Members of two antennariid genera, *Fowlerichthys* and *Abantennarius*, have distributions that coincide with the generalized track of the family except that they fail to extend into southern Australian waters.

Atlantic-Indo-Pacific Track

This track is totally included within the generalized track of the family, but it does not extend eastward beyond the easternmost margin of the Pacific Plate (Fig. 292). It delimits the distributions of four antennariid taxa: the *Antennarius striatus* group, the *A. pictus* group (with the exception of *A. commerson*, represented by rare occurrences in the Eastern Pacific), the *A. pauciradiatus* group, and *Histrio histrio*. The strong biotic relationship between the Atlantic and Indo-Pacific, reflected here in the present-day distribution of the Antennariidae, is due to the fact that the faunas of these two regions were formerly parts of the same original fauna, that of the Tethys Sea (Ekman, 1953: 63). The existence of the Tethys Sea as a continuous seaway apparently ended at the time of the Oligocene-Miocene transition (about 25 MYBP), when as a result of plate tectonics, Africa-Arabia came into contact with Eurasia (for more discussion, see Middlemiss and Rawson, 1971; Pielou, 1979: 33, 151; Raven, 1979). The distribution of *Abantennarius nummifer* largely coincides with the Atlantic-Indo-Pacific track, except that it is absent in the Western Atlantic (Fig. 277). Although this distribution is unique among the Antennariidae, it is found in a number of other tropical shorefish taxa (see Rosenblatt, 1967: 582). As hypothesized by Rosen (1975: 443), the fact that Indo-Pacific elements in the Eastern Atlantic are not shared with the Western Atlantic may indicate that an Indo-Pacific-Eastern Atlantic track has been more recently superimposed on (and has perhaps partly replaced) the Eastern Atlantic components of an older, Trans-Atlantic generalized track. On the other hand, this track with respect to frogfishes is probably better explained by westward dispersal of organisms around the Cape of Good Hope by the Agulhas and Benguela currents (see Bowen et al., 2006; Floeter et al., 2008; Briggs and Bowen, 2013).

Amphi-American Track

This track interlocks with the Atlantic-Indo-Pacific track and includes the tropical Eastern Pacific as far west as the Galápagos Islands, the Gulf of California, and coastal regions of Central and South America as far south as Chile, and all of the tropical Western Atlantic (Fig. 292). As far as antennariids are concerned, it involves two sister-group relationships: the Eastern Pacific endemic, *Fowlerichthys avalonis*, is most probably the sister-group of a more derived group that includes the Atlantic sister-species pair, *F. ocellatus* and *F. senegalensis* (Fig. 280); the Western Atlantic endemic, *Abantennarius bermudensis*, as a member of a monophyletic group of three species (Fig. 276), is cladistically most closely related to the members of its genus that occur in the tropical Eastern Pacific (i.e., *A. coccineus* and *A. sanguineus*; see Evolutionary Relationships, p. 383). That the antennariid fauna of the tropical Eastern Pacific should show a strong affinity with that of the western Atlantic is no surprise. The strong biotic relationship between these two disjunct regions, a reflection of the connection that existed between these two oceans prior to the uplift of the Isthmus of Panama in the mid-Pliocene (some 6 MYBP; see Raven and Axelrod, 1974), has been extensively and precisely documented (see Ekman, 1953: 30, but especially Rosenblatt, 1967, and Rosen, 1975; and more recently Floeter et al., 2008, and Kulbicki et al., 2013).

Trans-Atlantic Track

This track interlocks with the Amphi-American track and is totally contained within the Atlantic-Indo-Pacific track (Fig. 292). It includes the tropical-coastal and island localities of the Western and Eastern Atlantic, and involves two sister-group relationships within the family: the Western Atlantic species, *Fowlerichthys ocellatus* and *Antennarius multiocellatus*, appear to be cladistically most closely related to the Eastern Atlantic endemics, *Fowlerichthys senegalensis* and *Antennarius pardalis*, respectively (Figs. 280, 281; see Evolutionary Relationships, p. 381). These examples among the Antennariidae represent only two of the numerous known components of a well-documented generalized track that describe a relationship between the two sides of the Atlantic (see Rosenblatt, 1967; Rosen, 1975; Floeter et al., 2008).

Indo-Pacific Track

This track is totally contained within the Atlantic-Indo-Pacific track and includes all of the tropical waters of the Indian and Pacific oceans as far east as the easternmost margin of the Pacific Plate (Fig. 292). The distributions of ten species in two genera (Table 22) are contained within the boundaries of this track, all of them overlapping and undoubtedly involving a complex history of vicariance and dispersal (see Evidence for Dispersal, below).

Indo-Australian Track

This track encompasses the tropical and subtropical waters of the entire Indo-Australian Archipelago, an area that extends from Taiwan to Australia and Tasmania, including all of the intervening island groups of the Western Pacific as far east as the westernmost margin of the Pacific Plate (Fig. 292). Five antennariid genera and 12 species have distributions that are confined to the boundaries of this track (Table 22). The high diversity among monophyletic groups of the Antennariidae in the Indo-Australian track is most likely due to the great number of vicariant events that have occurred in this area, especially in the region of the East Indies (Sale, 1980; Springer, 1982; Bellwood et al., 2005; Reaka et al., 2008; Gaither and Rocha, 2013; Sanciangco et al., 2013; Cowman et al., 2017). As in the Indo-Pacific track, most of these taxa have overlapping distributions.

Southern-Australian Track

This track is totally contained within the Indo-Australian track and includes the subtropical and temperate waters of Australia and Tasmania (Fig. 292). The five genera and eight species that are restricted to this area are conspicuously isolated from other members of the family (Figs. 287–291; Table 22). Their geographic ranges overlap only with those of two other antennariid taxa: the monotypic genus *Allenichthys*, found along the coast of Western and South Australia (Fig. 288), and *Histiophryne*, with a range that largely coincides with the Indo-Australian track described above (Figs. 287, 288). Ekman (1953: 197; see also Gomon et al., 2008; Eschmeyer et al., 2010) presented a host of data that confirm the distinctiveness of the marine coastal fauna of southern Australia, stating that the boundary between this and the fauna to the north was not only thermal but also hydrographic. Obviously, these subtropical endemics within the Antennariidae do not form a monophyletic group, but it is nevertheless apparent that the present-day distributions of frogfishes support the existence of a component track that unites the tropical Indo-Australian track with the subtropics of Australia and Tasmania.

Evidence for Dispersal

We recognize that nearly all more recent discussions of tropical inshore-fish zoogeography have focused on the role of dispersal rather than vicariance. There is now a massive amount of literature on the subject, based mostly on genetic research (e.g., Read et al., 2006; Rocha et al., 2007; Briggs and Bowen, 2013; Cowman and Bellwood, 2013; Bowen et al., 2016; Cowman et al., 2017). The missing component of generalized track analysis, as described above for antennariids, is time of vicariance. Generalized tracks can often be misleading because species or clades that have the same track may have arisen at different times. This can lead to misinterpretation of the origin of biogeographic patterns if the timing of vicariant and dispersal events is not included in the analysis. For example, a lack of congruence in the time of divergence may refute a common vicariant origin of species that is indicated by the generalized track alone, and instead supports dispersal as an important factor in the evolution of the species. Genetic methods of identifying times of divergence (i.e., cladogenesis) were not available when Croizat (1958, 1964), Croizat et al. (1974), Rosen (1975), and others developed and advocated the technique of generalized tracks. Although we have a good fossil-based starting date for the family as a whole—the earliest evidence of articulated antennariid skeletal remains described by Carnevale and Pietsch (2009b) shows that the modern frogfish body plan was already in existence in the Early Eocene, approximately 53 Mya (Fig. 263; see The Fossil Record, p. 357)—an analysis of frogfish divergence-time estimates is not yet available. Nevertheless, a complex history of dispersal among sister-group taxa of the Antennariidae is evidenced by the sympatry of members of the various monophyletic groups. This is especially apparent in the central Indo-West Pacific, where eight species of *Antennarius*, six species of *Abantennarius*, and nine genera of the family have broadly overlapping ranges (Table 22). Elsewhere, there are a number of instances where a member of a component track has apparently broken away from the parent biota and dispersed. For example, the genus *Antennatus* contains four species, two with broad Indo-Pacific distributions and one that is restricted to the tropical Eastern Pacific (the fourth member of this genus, *A. flagellatus*, is so far known only from Japan; Fig. 285). Rather than the result of bifurcation of a once-continuous Pacific fauna, this disjunct distribution is more likely the result of dispersal out of the Indo-Pacific and into the Eastern Pacific (by way of the North Equatorial Current system; see Hubbs and Rosenblatt, 1961; Robertson et al., 2004), followed by colonization and subsequent speciation. Two other fully documented examples of what appear to be relatively more recent dispersals of Indo-Pacific antennariids into Eastern Pacific waters are *Antennarius commerson*, with several confirmed Panamanian and Colombian records (Fig. 275), and *Abantennarius coccineus*, with more than 50 specimens collected from Clipperton, off Costa Rica, Cocos Island, Panama, Colombia, and at least one from Isla San Félix, Chile (Fig. 276).

7. Behavioral Ecology

The term "behavioral ecology" is here defined in a broad sense to include such diverse topics as feeding, defensive behavior, locomotion, and reproduction. The unique feeding behavior of frogfishes, a topic of great fascination for centuries, is complexly interrelated with a host of other behaviors and adaptations. Aggressive and defensive behavior among frogfishes, although poorly understood, appears to play an important role in avoiding competition and sharing trophic resources. Locomotion also plays an inseparable role in the acquisition of food, but the peculiar walking, stalking, and jet-propulsive behaviors of frogfishes, in many ways unique among fishes, are discussed separately. Finally, we summarize all available data regarding the unusual reproductive habits and early life history of frogfishes, including what little is known about several unique modes of parental care.

Feeding Dynamics

Here, we use the terms "aggressive mimicry" and "aggressive resemblance" to refer to those phenomena that are related directly to the luring of prey and establishing a foraging site, and "feeding behavior and biomechanics" primarily for activities that occur once the prey is positioned within the strike zone of the predator. While some argue that the use of a bodily part as a decoy to lure prey, as in the case of anglerfishes, should not be included under the term "aggressive mimicry"—the decoy being not a mimic in the strict sense but merely a structure that bears an alluring appearance, in contrast to a putative model that is "identifiable to a given species syntopic with the mimic" (Sazima, 2002: 38)—we define the term in the larger sense described below (following Wickler, 1968: 47; Ormond, 1980: 253; Randall and Kuiter, 1989: 51; Myrberg, 1991: 156; Randall, 2005b: 299; Jackson and Cross, 2013: 161).

Aggressive Mimicry

Luring as an aggressive mimetic device for acquiring energy is surprisingly widespread throughout both the plant and animal kingdoms. From the more than 600 known species of carnivorous plants—the flytraps, sun-dews, and pitchers, that attract flying insects either by displaying brightly colored shiny leaves or producing alluring smells or both—to members of numerous invertebrate phyla and all major groups of vertebrates, a vast array of relatively unrelated organisms conserve energy by using a feeding strategy of remaining motionless and offering enticement to would-be predators (e.g., Poulton, 1890; Cott, 1940; Wickler, 1968; Forbes, 2009; Shumaker et al., 2011; Alcock, 2013; Emlen, 2014; Stevens, 2016). From those organisms that lure with specialized structures that mimic food items or provide false sexual cues—whether visual, chemical, acoustic, or behavioral—to those that attract prey in more passive ways by offering what is mistaken by smaller organisms as suitable shelter or feeding substrate, the number of examples that could be cited is almost limitless. Among all of these, however, lophiiform fishes are surely the best-known "anglers" (Gill, 1909; Atz, 1950, 1951a; Shallenberger and Madden, 1973; Dawkins and Krebs, 1978; Pietsch and Grobecker, 1978, 1987; Gordoa and Macpherson, 1990; DeLoach and Humann, 1999; Randall, 2005b; Pietsch, 2009; Jackson and Cross, 2013: 161).

The unique feeding structures and remarkable behaviors of anglerfishes have attracted the attention of scientists, as well as the casual observer of nature, since classical times. Two species of monkfishes, *Lophius piscatorius* and *Lophius budegassa* (Fig. 293), both common in shallow waters throughout the Mediterranean and Black seas, were well-known to the ancients by such appropriate names as *Rana marina* and *Rana piscatrix*, the fishing frog (Waterman, 1939: 66). Aristotle's vivid and remarkably accurate account of how this sly fish uses its first dorsal-fin spine to attract prey, as told in his *Historia animalium* (c. 344 B.C.), is so precise and so original, it seems that he must have witnessed this behavior with his own eyes. Not only does he give the facts, but he attests to their credibility and provides further corroborating evidence (translation of Thompson, 1984, book 9 (37), lines 10–28):

> In marine creatures, also, one may observe many ingenious devices adapted to the circumstances of their lives. For the accounts commonly given of the so-called fishing-frog are quite true; as are also those given of the torpedo. The fishing-frog hunts little fish with a set of filaments that project in front of its eyes; they are long and thin like hairs, and are round at the tips; they lie on either side, and are used as baits. Accordingly, when the animal stirs up a place full of sand and mud and conceals itself therein, it raises the filaments, and, when the little fish strike against them, it draws them in underneath into its mouth. . . . That the creatures get their living by this means is obvious from the fact that, whereas they are peculiarly slow, they are often caught with mullets in their interior, the swiftest of fishes. . . . Furthermore, the fishing-frog is unusually thin when he is caught after losing the tips of his filaments.

But direct eyewitness accounts cannot be claimed for the writings of the several ancient Greek and Roman writers who came after Aristotle, and who made mention of the curious feeding habits of the anglerfish—none added anything new, and, in most cases, the accurate story first related by Aristotle becomes less exact and somewhat clouded. Of these accounts, taken roughly in chronological order (although, in most cases, the exact dates of when these authors prepared their manuscripts are unknown), the first is that of Cicero, who lived from 106 to 43 B.C. In his *De natura deorum* (On the Nature of the Gods), Cicero wrote that "Sea-frogs again are said to be in the habit of covering themselves with sand and creeping along at the water's edge, and then when fishes approach them thinking they are something to eat, these are killed and devoured by the frogs" (translation of Rackham, 1933: 243). Curiously, and although Cicero admits that most of his facts are taken from Aristotle, he provides none of the details of the act of luring—there is no mention of the "set of filaments that project in front of its eyes" and no indication that these filaments are baited at their tips and thus used to attract prey to the gaping mouth of the angler.

Pliny the Elder, said to be one of the most studious and most scholarly men of antiquity, born in A.D. 23 and famous, among other things, for having been killed during the great eruption of Vesuvius in A.D. 79, also glosses over the details in a somewhat corrupted abridgment of Aristotle's much more complete description. In Pliny's only extant

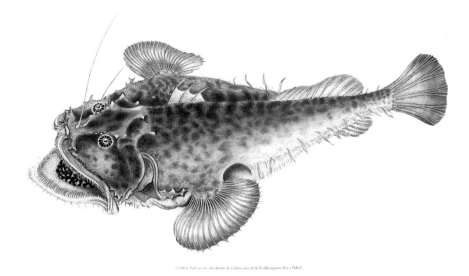

Fig. 293. *Lophius piscatorius* Linnaeus, "a fish of uncommon aspect and deformity, bearing a remote resemblance to the figure of the common frog in the tadpole state. Pliny and other Latin writers among the ancients call it *rana* and *rana marina*, in allusion to this similitude. The French, for the same reason, call it *Grenouille de mer*; the English fishermen the Fishing Frog, and Frog-fish; and the inhabitants of various other countries by names equally significant of its general resemblance to the tadpole of the common frog. The English name of Angler . . . is sufficiently expressive of the very singular arts employed by this curious fish in the capture of its prey." After Edward Donovan (1808, pl. 101).

work, the *Historia naturalis*, in thirty-seven books, first printed in Venice in 1469 (translation of Rackham, 1956: 261), he wrote: "The sea-frog called the angler-fish is equally cunning: it stirs up the mud and puts out the little horns that project under its eyes, drawing them back when little fishes frisk towards them till they come near enough for it to spring upon them . . . there are proofs of this cunning, because these fishes, though the slowest there are, are found with mullet, the swiftest of all fishes, in their belly."

Plutarch, whose dates are thought to be A.D. 46 to 120, likewise remarked, in *De sollertia animalium* (On the Intelligence of Animals), that "The so-called fishing-frog is known to many; it gets his name from his actions . . . for it throws out from its neck a filament that it extends to a distance in the manner of a line, letting it out and drawing it back at pleasure; this being done when it perceives any little fishes about it, it allows them to bite the extremity of this filament, being itself concealed beneath the sand or in the mud, when it gradually retracts the member until the fish is near enough to be swallowed by a quick motion" (translation of Helmbold, 1957: 435).

A much better account was left to us by Claudius Aelianus of Praeneste, born about A.D. 170, in his *De natura animalium* (translation of Scholfield, 1959: 245):

There is, it seems, a species of frog that bears the name of "angler," and is so called from what it does. It possesses baits above its eyes: one might describe them as elongated eyelashes, and at the end of each one is attached a small sphere. The fish is aware that nature has equipped it and even stimulated it to attract other fish by these means. Accordingly, it hides itself in spots where the mud is thicker and the slime deeper, and extends the afore-said hairs without moving. Now the tiniest fishes swim up to these eyelashes, imagining that the round, swinging objects at the end are edible;

meanwhile the angler lies in wait, never stirring, and when the little fishes are near to him, he withdraws the hairs towards himself (they are drawn in by some secret and invisible means), and the little fishes, whose gluttony has brought them close up, provide a meal for the aforesaid frog.

Oppian of Anazarbus, a Greek poet who flourished during the reign of Marcus Aurelius Antoninus (Emperor A.D. 161–180), provided another colorful example in his *Halieutica, of the Nature of Fishes and Fishing of the Ancients* (translation of Mair, 1928: 289):

The Fishing-frog again is likewise a sluggish and soft fish and most hideous to behold, with a mouth that opens exceeding wide. But for him also craft devises food for his belly. Wrapt himself in the slimy mud he lies motionless, while he extends aloft a little bit of flesh which grows from the bottom of his jaw below, fine and bright, and it has an evil breath. This he waves incessantly, a snare for lesser fishes which, seeing it, are fain to seize it. But the Fishing-frog quietly draws it again gently quivering within his mouth, and the fishes follow, not suspecting any hidden guile until, ere they know it, they are caught within the wide jaws of the Fishing-frog. As when a man, devising a snare for lightsome birds, sprinkles some grains of wheat before the gates of guile while others he puts inside, and props up the trap; the keen desire of food draws the eager birds and they pass within and no more is return or escape prepared for them, but they win an evil end to their banquet; even so the weak Fishing-frog deceives and attracts the fishes and they perceive not that they are hastening their own destruction.

This same passage was set to verse in the earliest English translation of Oppian's *Halieuticks* (Diaper and Jones, 1722: 65):

Hid in the Slime the Toad of form uncouth
(That fish is all one vast extended mouth)
Her tender Body wraps, on Prey intent,
And silent there concerts the great Event.
What softer Skin, and slower Pace deny,
With Foresight and successful Frauds supply
Within her jaws a fleshy Fibre lies,
Whose Whiteness, grateful Scent, and Worm-like Size
Attract the Shoals, and charm their longing Eyes.
She to allure oft shakes the tempting Bait;
They eager press, and hurry on their Fate.
But as they near approach, with subtle Art
The wily Toad contracts th' inviting Part;
Till giddy Numbers thus decoy'd she draws
Within the Circle of her widen'd Jaws.
    The Fowler thus the feather'd Race deceives,
And strows beneath his Snare the rifled Sheaves.
The busy Flocks peck up the scatter'd Seed,
Nor midst their Joy the fatal Engine heed;
Till with loud Clap the tilted Cover falls,
And the close Pit the flutt'ring Prey enthralls.

From the time of Oppian to the mid-sixteenth century, not a single mention of the anglerfish can be found, but the published works on fishes by the Renaissance ichthyologists and encyclopedists of natural history, including Pierre Belon (1551, 1553), Guillaume Rondelet (1554, 1555, 1558), Ippolito Salviani (1554–1558), and Conrad Gessner (1558), all speak of *Rana piscatrix*, and all contain passages describing similar luring behavior. Ulisse Aldrovandi, a professor of philosophy and medicine at Bologna, included an uncritical but encyclopedic synopsis of earlier accounts of "the fishing frog" in his *De Piscibus Libri V, et de Cetis Liber Unus* of 1613. In more recent times, however, these descriptions of the feeding habits of the anglerfish fell into disrepute as being no more than entertaining tales told by fishermen. In fact, many quoted Aristotle only to cast doubt on the accuracy of his observations. A quick inspection of the external anatomy of the angler will cause most anyone to admit that it must lure its prey by means of its baited first dorsal-fin spine, but ever since Aristotle, and not until very recently, had anyone been able to give a credible eyewitness account of this extraordinary behavior.

Writing in his *Natural History of Norway*, published in 1755, Erich Pontoppidan refused to believe that the first dorsal-fin spine is used for angling: some say that the "long and narrow bone that stands upon the snout of [*Rana piscatrix*], and hangs into the water, serves also as a bait to decoy the Fish: this may possibly be, tho' I should rather think that the creature used it to strike small Fish with." In the same paragraph, it is rather curious that Pontoppidan (1755: 152), while casting doubt on the angler's ability to lure with the dorsal spine, ascribes this very same function to the numerous highly branched, fleshy appendages that line the outer edge of the lower jaw of the angler:

> All round the under jaw-bone there hangs several slips, or false fins, of a gristly substance, about four inches long: these slips, before the fish is dried, look like so many worms. These the [*Rana piscatrix*] makes use of to decoy other fish with, when he wants to catch them. To this end he will get upon the edge of a rock, and open his jaws very wide: this vast mouth the other Fish, who are striving to get the supposed floating worms, take to be an opening or crack in the rock, so fall a prey to this Fish, and are devoured unawares.

With these statements, Pontoppidan was completely original. No one before had ever deduced that the hairy appendages surrounding the head and body of the angler might also serve as accessory attracting devices. However, British naturalist Jonathan Couch in his *History of the Fishes of the British Islands*, first published from 1862 to 1865, came up with the same idea: "These are not merely insensible doublings of the skin, but, although in a less degree, they perform the office commonly assigned to the fictitious bait suspended from the fishing-rod on the top of the head" (Couch, 1863: 206). But whether this notion came to Couch independent of Pontoppidan is unknown. In any case, the hypothesis that luring in lophiids is accomplished with the bodily tendrils, in addition to the function of the first dorsal-fin spine and bait, has not been addressed by others, yet many have described in some detail how the peripheral fringing aids in camouflaging the animal against the sea bottom.

Well over a century after Pontoppidan published his *Natural History*, William Saville-Kent (1874a, 1874b), who kept a four-foot angler alive for some days in the old Manchester Aquarium, in Manchester, England, also seriously doubted stories of luring activities. As part of a marvelous account of how the angler conceals itself among the animate and inanimate objects of the sea floor, Saville-Kent wrote (1874b: 511):

> As is universally known, this species derives its popular title of the "angler" or "fisher" from its supposed habit of deliberately angling or fishing for its prey, by aid of the single filamentous appendages that adorn its head. Pliny was one of the earliest writers to advocate this theory [but contradictory is the assertion by some] that the small-sized brain of the animal would scarcely seem to predicate its possession of so much sagacity.

Like, however, the long-cherished fiction of the sailing properties of the nautilus, the traditional fishing powers of the angler seem scarcely established on a basis sufficiently firm to withstand the rigid investigation of modern science.

Almost at the same time that Saville-Kent was watching his captive angler in a tank in Manchester, hoping to catch a glimpse of the fabled luring behavior, the Reverend Samuel James Whitmee, a missionary for the Anglican's London Missionary Society stationed on the island of Samoa in the South Pacific, was doing the same. But his was a very different kind of anglerfish—a member of the frogfish genus *Antennarius* that he kept alive for several days in a small aquarium in his study. Although it had been brought to him out of the water and "seemed somewhat exhausted," it soon recovered and Whitmee (1875: 543) was rewarded with some brief but spectacular behavior. While successful in observing angling behavior, apparently the first person to record such an event since Aristotle, he too was disappointed that no little fishes took the bait:

> It angled with the ciliated anterior dorsal for some of the small fish in the aquarium. I hoped to see it catch one; but they were too wary. There were seven fish, not too large for the *Antennarius*; but they had been some months in captivity, were quite at home in every nook and corner [of the aquarium], and knew too well the nature of the new inmate to allow themselves to be taken off their guard. I am accustomed to feed these with bread-crumbs, and I tried to entice them to the neighbourhood of the *Antennarius* by dropping some so as to fall immediately in front of it. But it was to no purpose; they kept at a safe distance. When one ventured to dash at a falling crumb rather nearer than usual, it immediately darted away again in evident fear.

Not long after these observations were published by Saville-Kent and the Reverend Whitmee, Austrian zoologist Richard Schmidtlein (1879: 14) reported similarly on anglers held in captivity in aquaria at the famous zoological station of Naples. Here is his account of yet another failure to witness successful angling behavior, later translated from the original German by American ichthyologist Theodore N. Gill (1905a: 506):

> *Lophius* embodies, so to speak, a living angling apparatus. Unfortunately, there is not much to record concerning its habits in captivity that might be considered as a contribution to the already known characters, for it is so peculiarly adapted for its dark mud-bottom, that it can never endure the confinement in our bright, well-lighted prisons with the clean sand for more than a few days. It lies for the most part on the bottom in perfect apathy without burying itself in the sand, and stares with its big dull, glazed eyes straight before it, while the jaws of the enormous mouth open a little and close at every breath, and the lobed barbels on the chin swing back and forth. At times it raises the "hooks" on the head and lets the terminal lappets play, or it yawns and changes the color of its dull mud dress into a lighter or darker shade. It never takes any food either voluntarily or by force. If it is made to feed it will spit out the morsel again.

George Brown Goode of the Smithsonian Institution, in a description of anglerfishes taken from deep water off the south coast of New England published in 1881, wrote disparagingly about the angler's ability to angle (p. 469): "It is the habit of many of the [anglerfish] family to lie hidden in the mud, with the long dorsal filament and its terminal soft expansion exposed. It has been imagined that the expansion is used as a bait to allure its prey, but it seems more likely that it is a sense organ intended to give notice of their approach." He repeats this idea a few years later in his *Food fishes of the United States* of 1884 (p. 174): "No one has ever seen the performance, and, although the theory is not altogether incredible, it seems more probable that the tops of these organs are intended by their sensitiveness to warn the fish of the approach of its prey than to act as allurements to attract other fishes." So sure of this alternative use of the lure, he repeats the claim all over again a decade later in *Oceanic Ichthyology* published in 1896 (p. 490) and coauthored with his colleague Tarleton H. Bean, then director of the New York Aquarium.

In a paper devoted to sensory structures of the angler published in 1891, Frédéric Guitel wrote that he kept a two-foot long *Lophius* for two months in an aquarium at the Laboratory of Arago. During all that time, however, he too never saw the lure in use—it was always folded down between the eyes on top of the head, out of the way. But looking back now on the kinds of marine research facilities that were available to Guitel and others prior to the turn of the nineteenth century, it is a wonder that they were able to keep anything alive at all, let alone observe and record animal behavior accurately, as it occurs in nature. Modern filtration systems, precise temperature and light controls, and full knowledge of water chemistry and fish health and nutrition, not to mention the dietary needs of individual species, were all unavailable. Saville-Kent and the Reverend Whitmee do not describe the conditions under which their animals were kept, but we can assume that the equipment was not very sophisticated. This was probably especially true for Whitmee, stationed on Samoa, where the hydraulics required for maintaining constantly running seawater and keeping it clean and well oxygenated were unavailable. It may thus seem surprising that Whitmee, unlike everyone else at the time, was at least able to observe luring behavior (but not prey engulfment), but we must remember that his observations were made on the much smaller, more easily transported and maintained, anglerfish genus *Antennarius*. A large, heavy, unwieldy specimen of *Lophius* presents a much more formidable set of problems: how to catch it and transport it to an aquarium facility without causing damage and minimizing trauma, and how to simulate a natural environment in a closed tank of finite size—both questions difficult to overcome.

The Laboratory of Arago was a state-of-the-art facility, almost brand new, when Frédéric Guitel made his observations on *Lophius* in the late 1880s. Founded by Henri de Lacaze-Duthiers in 1882, it is now the Oceanographic Observatory and Aquarium of Banyuls-sur-mer, located on the shores of the Mediterranean just south of Perpignan, southern France, close to the Spanish border. Certainly renovated to meet modern standards by now, the interior of the aquarium in 1890 would seem very primitive to us today. The aquaria themselves, arranged in a row amidst copies of Greek and Roman statuary, were too well lit and much too small to provide adequate space for even a small *Lophius*. Guitel admits as much when he says that fishes "never endure the confinement in our bright, well-lighted prisons with the clean sand for more than a few days." More than likely the specimen that Guitel kept was not at all happy—used to ample space and dark gloomy surroundings, with plenty of mud, weeds, and rocks to hide amongst, it is no surprise that it showed little interest in eating.

Facilities for maintaining marine animals in captivity were much improved by the 1920s, yet Ulric Dahlgren, then Professor of Biology at Princeton University, writing even as late as 1928, was still obliged to report that no one up to that time had ever actually seen an angler engulf a fish (but he was wrong by three years). He writes also that an angler has never been seen to eat when confined in an aquarium (p. 22): "He makes a poor aquarium subject, since he refuses all food in captivity, and hence does not live long." And further, "Since the fish seldom comes into water shallow enough for observation, but is usually caught at a depth of thirty or more feet, we are debarred from direct studies of its feeding habits." Although luring had been inferred—most notably by Parsons (1749), Pennant (1776), Tennent (1861), Day (1876), Agassiz (1882), Garman (1899), Gill (1905a, 1909)—no one had actually seen *Lophius*, or for that matter, any other kind of anglerfish, engulf prey that had been attracted by the luring apparatus. However, Dahlgren (1928) admits (p. 22):

> Probably it is true and we have one observation that seems to confirm it. One observer who had a fish in captivity touched the lure with a broom handle. The lure was moving back and forth and whenever the broom handle came in contact with the lure the fish gave a gulping snap with its huge mouth and accurately seized the stick at the region where it had been in contact with the lure. This would indicate a very active and delicate reflex.

A real experimentalist, Dahlgren (1928) demonstrated that the lure is attractive to small fishes. By severing the first dorsal-fin ray from an angler and dangling it in an aquarium containing small fishes, he showed rather conclusively how irresistible the device can be, stimulating numerous attacks from the would-be predators. He stated (p. 22) that the terminal lure itself is

> of a light yellow color, very different from the general color tone of all the rest of the body and would thus attract the eye. The "rod" or anterior dorsal fin ray on which the lure is placed is thin and stiff and at its base is provided with a series of muscles that move it back and forth from a posterior position in which it lies flat against the body, to one that is erect but leaning forward so that the lure hangs over the mouth. This would be its natural position in "fishing."

It was not until observations made during the first half of the twentieth century that the description provided by Aristotle some 2,300 years earlier was reconfirmed. Well-hidden to the average reader, in a large descriptive tome on the *Fishes of the Gulf of Maine* published in 1925, Henry B. Bigelow of Harvard University and his coauthor William W. Welsh casually mentioned (p. 528) the extraordinary observations of marine biologist William F. Clapp. It turns out that Clapp had many times watched the feeding habits of the anglerfish—or Goosefish (*Lophius americanus*) as he called it—at low tide in Duxbury Bay, Massachusetts, where this fish was often found in large numbers,

> lying perfectly motionless among the eelgrass, with the tag or "bait" on the tip of the first dorsal ray swaying to and fro over the mouth, either with the current or by some voluntary motion so slight as to be invisible. The only fish [Clapp] has seen them take are tomcod [*Microgadus tomcod*], and when one of these chances to approach, it usually swims close up to the "bait" but never actually touches it, for as soon as the victim is within a few inches the goosefish simply opens its vast mouth and closes it again, engulfing its victim instantaneously (Bigelow and Welsh, 1925: 528).

Bigelow and Welsh (1925: 528) went on to say that "These observations are the more welcome as no other recent student seems to have seen the feeding habits of this species in its natural surroundings, and they show that it depends mostly on such fish or Crustacea as chance to stray close enough to be snapped up from ambush or seized by a sudden rush." By using the word "recent" one really wonders if Bigelow and Welsh realized the true significance of these findings—that Clapp was the first irrefutable witness to the successful use of the anglerfish lure since Aristotle.

Clapp's observations were soon followed by others, notable among these being the brief but accurate account of luring in *Lophius piscatorius* published by Herbert Clifford Chadwick in 1929, based on observations made in aquaria at the Marine Biological Station at Port Erin on the Isle of Man in the Irish Sea. It seems that Chadwick had seen anglerfishes angling for their prey many times and only decided to write it down in published form on encouragement from colleagues (p. 337):

> It was our practice on many occasions to put into the tank containing the angler a few living young specimens of the coal-fish, *Gadus virens*. These would soon be noticed by the angler, which, while remaining stationary with closed mouth, raised the lure from its horizontal position along the back and jerked it to and fro. Suddenly, as the unsuspecting coal-fish hovered over the head of the angler and sampled the living and actively moving bait—I cannot say that I ever saw it touch the bait with its snout—the angler's mouth would open and as suddenly close upon its prey; the head of the coal-fish always disappearing first, while the tail projected from the tightly closed mouth. A few seconds later the tail would be drawn by a sort of suction into the still closed mouth and the angler would be ready for another meal.

Chadwick never saw the angler attempt to pursue its prey. The entire time, it stayed put, lying perfectly still on the bottom of the tank, the lure being actively jerked to and fro, but only when the smaller fishes were introduced. When no prey was in view, the lure was always laid back horizontally along the top of the angler's head.

In 1933, William Beebe, then director of the New York Zoological Society, wrote about successful luring in *Antennarius scaber*. While on a fish-collecting expedition in the West Indies, aboard the yacht *Antares*, a four-inch specimen of this species was captured and brought back alive to the New York Aquarium (p. 115):

> Its lure was, in size, shape, color and movement, a perfect imitation of a wriggling, grayish-white worm. While still on board the *Antares* it devoured three of its own cousins, and a beautiful blue-starred demoiselle. Twice I watched the process, and both times while the prospective prey was at least two inches away, the frogfish opened its mouth and with no apparent effort, created such a maelstrom, such an irresistible current, that the human eye could not see the fish disappear. It simply vanished from sight, the lure was tucked away and we imagined a gleam of satisfaction in the fishy eye.

Beebe went on to say (p. 115) that his *Antennarius* later shared an aquarium "with a herd of sea horses, but their armor proved no protection and three were forthwith devoured." Unfortunately, however, he failed to describe, in his usual colorful and meticulous way, the details of fishing, particularly the action of the lure. He later recalled, however, that the first dorsal spine was moved about actively, and the bait very much resembled a swimming or wriggling worm (see Gudger, 1945a: 111; 1945b: 544).

Additional observations, made by John Tee-Van and quoted by Eugene W. Gudger (1945a) in a paper entitled "The frogfish, *Antennarius scaber*, uses its lure in fishing" (p. 112), add some interesting details not provided by Beebe:

> Somewhere in the middle 1930s during the years of the Bermuda Oceanographic Expeditions, I recall being shown by Louis Mowbray, Sr., two specimens of a Frogfish that he had in one of the exhibition tanks of the Bermuda Aquarium. The two fish were resting among the interstices of the rocks at either side of the tank, probably five feet distant from each other. One of them was not more than 12 inches from the glass front of the tank, where it could be easily observed. This individual was headed toward the center of the tank and demonstrated a great deal of interest in some small specimens of so-called "Bermuda mangrove mullet" (*Fundulus bermudae*). Each time that one of these little fishes approached within a couple of feet of the Frogfish the latter alerted itself by pushing up the front of its body by tensing the pelvic fins, while at the same time it

moved its lure back and forth in line with the axis of its body. The so-called bait appeared to be considerably distended and somewhat larger than is normally seen in preserved specimens of *Antennarius*. The result was that as the antenna moved forward the lure trailed behind, and this bait reversed its movement when the tentacle moved back over the fishes' mouth. This trailing back and forth resembled the waving of a handkerchief held in the middle. Each time that one of the little killifishes came within a foot or so of the frogfish the action was repeated. Unfortunately, I was not able to witness the catching of the small fish, but judging by my observations of the feeding of another Antennariid, *Histrio histrio*, I would say that the Killies were in danger whenever they came any wise [sic] close to the frogfish.

All of these early accounts of luring behavior were summarized and greatly augmented by additional firsthand observations made by British zoologist Douglas P. Wilson and reported in full and precise detail in a 1937 paper entitled "The Habits of the Angler-fish, *Lophius piscatorius* L., in the Plymouth Aquarium." Wilson gave individual histories of three small specimens that lived in the tanks of the aquarium for months and were often seen to angle with the lure. Another individual was seen to hold the fishing rod over the closed mouth in readiness to fish but was not actually seen to catch anything. The entire article—which includes a wealth of information on general natural history of the anglerfish, from food and feeding to growth and respiration—is well worth reading in full, but here is what Wilson said specifically about angling (p. 487):

There is no doubt whatever that the first dorsal spine, or "rod," with its tag of skin, or "bait"—the whole being known as the "lure"—is used to attract fishes into reach of the mouth. This has not only been seen by Chadwick and others, but has been repeatedly confirmed by my own observations. An angler when hungry erects the lure immediately [when] any suitable fishes come anywhere near and endeavours to attract one of them close enough to be caught. The lure is quickly jerked to and fro and, as the rod is almost invisible, the bait (in my specimens always forked and "fly-like" not vermiform) simulates some tiny creature darting about. An attracted fish rushes up and endeavours to catch it; the bait is skillfully flicked out of its way just in time and, with a final cast, is dashed down in front of the mouth which may open very slightly. The intended victim, still following the bait, turns slightly head downward; it is now more or less directly head-on to the angler's mouth. The jaws snap faster than the eye can follow and the tail of the prey is next seen disappearing from sight through the firmly closed mouth. As far as I have been able to observe, the bait is not actually touched by the victim before it is caught, as has sometimes been supposed. Touching the bait with forceps does not cause a reflex snapping of the jaws.

The moving lure has a strong attraction for healthy hungry fishes of several kinds. Small pollack, whiting, pout, and bass were the species actually observed to be attracted to it, but probably most pelagic fish that capture moving prey by sight would at least swim up to investigate the darting object. Many fishes take a ready interest in moving things. A small angler crawling over the bottom of a tank into which it had just been placed, aroused, as a rule, the curiosity of nearby fishes, pollack and small dabs swimming towards it, keeping, however, at that time a safe distance away.

Besides the quick lashing motion of the lure just described, some anglers occasionally combine with it another movement. Every now and then the rod is depressed until the bait hangs just in front of and rather below the level of the lower jaw. It is then for a few minutes given a curious vibratory movement, after which the sharp flicking is again resumed, the bait being jerked over wide arcs in various directions.

Wilson explained further that the lure is not always used when the angler seeks its prey. If a fish happens to swim near the head of the angler, it may be engulfed. Anglers vary considerably in their use of the lure. Some individuals use it often, others only infrequently. When the lure is not in use, it is always laid back on top of the head between the eyes. Some individuals were observed to direct the lure forward horizontally across the mouth, keeping it quite still for hours at a time. Whether this behavior was the result of an especially hungry angler, with the fishing equipment ready for quick action if needed, or simply individual variation in behavior, is unknown.

In Wilson's description, as well as in those of others who have witnessed anglerfish feeding, the prey when struck is nearly always taken head-first just as it enters the space just above the level of the mouth, and the engulfment always occurs too rapidly for the human eye to follow. At one moment the prey is trying to seize the bait, in the next, the angler is sinking down into its sandy bed, with the tail of the prey projecting from its mouth, and in the final moment the tail is gone. In rare instances when the prey is caught broadside, it has to be worked around by the angler until the head goes in first.

Protective Resemblance

That mimicry in the form of protective resemblance—so closely allied with the concept of aggressive mimicry described above—plays an equally important role in the lives of anglerfishes was realized more than a century before the English naturalist Henry Walter Bates (1862) provided the first clear discussion of the phenomenon. In one of the earliest references to protective resemblance in the scientific literature, Pieter Osbeck, chaplain to a Swedish East Indiaman and a pupil of Linnaeus, described *Histrio histrio* as it clung to *Sargassum* in "The Grass-Sea" (taken from Forster's, 1771: 112, English translation of Osbeck, 1757): "Providence has clothed this fish with fulcra resembling leaves, that the fishes of prey might mistake it for sea-weed, and not entirely destroy the breed." Valenciennes (1837: 399) later wrote: "Osbeck thought the thin cutaneous appendages that cover the body of this [frog]fish might function to mimic the seaweed and thus make the fish less inclined to be attacked by other fishes or sea birds. We leave it up to the reader to judge the correctness of this hypothesis." Whitmee (1875: 544) further described his aquarium-held specimen of *Antennarius* as follows:

> It again fixed itself, in a vertical position with the head up, in an indentation in a coral block which pretty well matched its size. When attached it looked much like the block itself, the cutaneous tentacles and ocellated spots greatly resembling the fine seaweed and coloured nullipores [sic] with which the dead portions of corals and stones are more or less coated in these seas. As I watched it I could not help thinking that this fish presents us with what we now call (since Mr. Bates introduced the term) "mimicry."

To the present day, one cannot look at a frogfish without being struck with the notion that it must be a mimic of substrate and structure, encrusted with marine organisms (or a perfect resemblance to *Sargassum* weed, in the case of *Histrio histrio*; Fig. 128). Numerous authors have commented on the close morphological and chromatic similarity of benthic frogfishes to sponges (Fig. 294), tunicates, seagrasses, or coralline algae-encrusted rocks (Barbour, 1942: 28, 29; Breder, 1949: 95; Randall and Randall, 1960: 473; Böhlke and Chaplin, 1968: 719; Randall et al., 1990: 54; Michael, 1998: 330; Bavendam, 2000: 36; Schneidewind, 2003b: 86, 2005a: 23; Randall, 2005b: 300, 311; Helfman et al., 2009: 428; Arnold et al., 2014: 535; Stewart, 2015: 882), implying that this resemblance provides some protection against predation. Many more have written of the "perfect

Fig. 294. *Antennarius commerson* lurking behind a large fan sponge of the genus *Monanchora*, most likely either *M. ungiculata* or *M. viridis*: an incredible example of aggressive and protective resemblance. Photographed by Andrew Taylor at the Giants Castle dive site, Tofo, southern Mozambique, western Indian Ocean, at a depth of about 35 m; courtesy of Andrew Taylor.

camouflage" of *H. histrio* in *Sargassum* (Osbeck, 1757: 305; Forster, 1771: 112; Bouvier, 1888: 17; Ives, 1890; Gill, 1909: 603; Gregory, 1928: 408; Vignon, 1931: 131; Gordon, 1938: 14, 19; Atz, 1951b: 135; Gordon, 1955: 387; Adams, 1960: 79; Sisson, 1976: 194; Burnett-Herkes, 1986: 587; Randall et al., 1990: 56; Randall, 2005b: 300; Diamond and Bond, 2013: 19, 25). More recently, however, biologists have come to realize that mimicry in the form of protective resemblance is most likely secondary in importance to aggressive mimicry in the lifestyle of a "lie-in-wait" predator such as a frogfish (Wickler, 1968: 124; Drickamer and Vessey, 1982: 378, fig. 15-10; McFarland, 1982: 80; Sazima, 2002: 38; Randall, 2005b; for a thorough review and classification of mimetic systems, see Vane-Wright, 1976; Pasteur, 1982; see also Wiens, 1978; Brower, 1988; Ruxton et al., 2008).

Functional Morphology

In the 1930s, when Douglas Wilson was studying the habits of *Lophius piscatorius* at the Plymouth Aquarium, no one could have ever hoped to know just exactly what was taking place when anglerfishes gulp their food, nor could they imagine that the speed at which these movements take place might someday be measured. The technology required to do so had not yet been developed nor even imagined. We now have, however, a host of devices for studying what is called functional morphology, a relatively new branch of science that allows scientists to record and analyze animal movement in great detail and with high precision. High-speed video cameras, including x-ray video, and electromyography equipment, that detect the electrical potential generated by muscle cells and allow researchers to record muscle contractions while an animal is perform-

ing its normal movements, have provided a wealth of new information about how anatomical parts work together in living organisms (see Liem and Greenwood, 1981: 83).

While no one has yet fully applied these tools to investigate the functional morphology of feeding in anglerfishes, some related work on frogfishes was performed by Pietsch and Grobecker (1987). Their findings, described below, were the results of tests of
multiple hypotheses regarding the aggressive mimicry and feeding behavior of frogfishes
"as one of nature's most highly evolved examples of lie-in-wait predation" (Pietsch and
Grobecker, 1987: 309). In testing these ideas, they commented on the adaptive significance of luring as a feeding mode in high-diversity marine communities and made some
general predictions concerning the foraging strategy of these and other lie-in-wait predators. The data were gathered from preserved material, controlled laboratory experiments performed on captive individuals, and direct field observations of the organisms
in their natural habitat. They sought to maintain in captivity and observe in the field as
many species as could be made available, either as purchases from tropical wholesalers
and retailers or through their own collecting and diving efforts. In doing so, they were
able to gather unequal behavioral and ecological data for eight species of frogfishes: *Antennarius striatus*, *A. hispidus*, *A. maculatus*, *A. commerson*, *Abantennarius coccineus*, *A. sanguineus*, *Antennatus tuberosus*, and *Lophiocharon trisignatus*.

A number of related questions posed early on in their investigation involved the morphological specificity of the escae of frogfishes. Are the escae of frogfishes species-
specific, and do they mimic specific kinds of prey items? The answer to both of these
questions appears to be a qualified yes. With some exceptions (e.g., *Abantennarius coccineus*; see p. 208, Fig. 144), the morphological variation of escae within any given species is remarkably small, and again, with some exceptions (particularly within the more
species rich species-groups of *Antennarius*, and especially within the genus *Abantennarius*), species can be identified solely based on the structure of the esca. Although it is usually difficult, if not impossible, to correlate the structure of a specific esca with a particular organism that might serve as a model, a number of examples seem obvious. For
example, the wormlike baits of *Antennarius scaber*, *A. striatus*, and *Rhycherus filamentosus*
appears to mimic a polychaete (Figs. 64, 72, 187), that of *A. hispidus*, a feeding tube worm
(Fig. 53), that of *A. maculatus*, a small fish (Pietsch and Grobecker, 1978; Fig. 86), that of
A. commerson, a tiny shrimp (Fig. 78), and that of *Phyllophryne scortea*, an amphipod
(Pietsch and Grobecker, 1987, pl. 46; Fig. 202). Not only does the structure of escae appear to mimic smaller creatures, but the escae themselves are wriggled and otherwise
manipulated to simulate the natural movements of their models (Wickler, 1967: 547).

Assuming that escae mimic specific organisms, the question arises whether frogfishes
attract and consume a specific set of prey items (as once hypothesized for deep-sea ceratioid anglerfishes; see Pietsch, 1974: 98). For example, does a frogfish with a wormlike
esca have a diet that consists primarily of prey that normally feed on polychaetes or other
wormlike organisms? In an effort to test this idea, Pietsch and Grobecker (1987: 310) examined the diets of the most abundant prey species found in the stomachs of all available Hawaiian material of *Antennarius striatus* (17 specimens), *A. pictus* (52 specimens),
A. commerson (12 specimens), and *Antennatus tuberosus* (26 specimens). They then attempted to correlate this information with the specific escal morphology of the frogfish
from which each prey item was taken. The results failed to support the above hypothesis, but they revealed instead that the diets of frogfishes consist of a highly varied, overlapping assortment of prey types. (There does appear, however, to be a correlation between escal morphology and diet in *Phyllophryne scortea*; see the description of amphipod

mimicry provided by Rudie H. Kuiter, p. 289.) This wide diet must be due, at least in part, to the complexity of food acquisition in frogfishes. For example, not all prey are attracted by the esca. Many are simply stalked by the predator (see Feeding Behavior and Biomechanics, p. 437); others randomly enter the strike zone and are engulfed; still others are lured, not by the wriggling esca, but by the structure-like morphology of the animal itself, which appears to offer a suitable feeding substrate or shelter site (see Grobecker, 1983: 200; Aggressive Resemblance and Foraging-Site Selection, p. 431).

The esca may also attract prey items, particularly fishes, not because the esca is identified as a specific food item, but for competitive or defensive reasons. During active luring by *Antennarius maculatus* under laboratory conditions, damselfish (*Dascyllus aruanus*) were repeatedly observed to direct aggressive displays toward the fishlike esca of this species (Fig. 86A, B). On several occasions, in an overly aggressive attack, the damselfish entered the strike zone of the frogfish and was engulfed. In this situation the frogfish attracted prey that appeared to respond to the esca as a competing organism rather than as a food item. Finally, the esca may attract organisms for no other reason than that they are "curious" about any conspicuous, wriggling object (McFarland, 1982: 114).

LURING BEHAVIOR

The movement of the illicial apparatus during luring behavior can be divided into two components, primary and secondary, that are closely interrelated and apparently species-specific. The primary component of luring is defined as the pattern of movement of the illicium during a single luring sequence, while the secondary component is the motion of the esca itself. Both of these components of movement were examined by Pietsch and Grobecker (1987: 311) by video and from films made at 32 and 64 frames per second (see Approach and Procedures, p. 28).

Tracings made from films of the pathway within which the tip of the illicium was moved during a single luring sequence illustrate the specificity of this behavior in four species of *Antennarius* (Fig. 295): in *Antennarius striatus* the pattern of illicial movement is a rapid jerking motion that is confined within a relatively small area (Fig. 295A); *A. hispidus* moves the illicium in a roughly triangular-shaped pattern (Fig. 295B); in *A. maculatus* the movement is a rapid, circular, sweeping motion (Fig. 295C); and in *A. commerson*, the illicium describes a relatively simple pattern of strokes in the vertical plane (Fig. 295D).

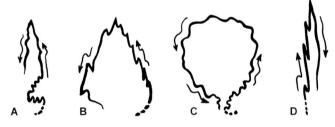

Fig. 295. Diagrams made from selected frames of 16-mm film showing primary component of luring behavior in four species of *Antennarius*. (**A**) *A. striatus*. (**B**) *A. hispidus*. (**C**) *A. maculatus*. (**D**) *A. commerson*. Arrows indicate direction of movement. After Pietsch and Grobecker (1987, fig. 135).

To quantify the primary component of luring behavior for each of the four species examined, the percent frequency of occurrence of the illicium was measured within six, 30-degree sectors radiating from the base of the illicium (see Tets, 1965: 27; Fig. 296). A comparison of percent differences further revealed specific illicial movements. From this analysis of primary luring behavior, it was possible to estimate the percent time that the tip of the illicium (esca) was positioned within the strike zone. These values were 95% for *Antennarius striatus* (Fig. 296A), 75% for *A. hispidus* (Fig. 296B), 65% for *A. maculatus* (Fig. 296C), and 80% for *A. commerson* (Fig. 296D).

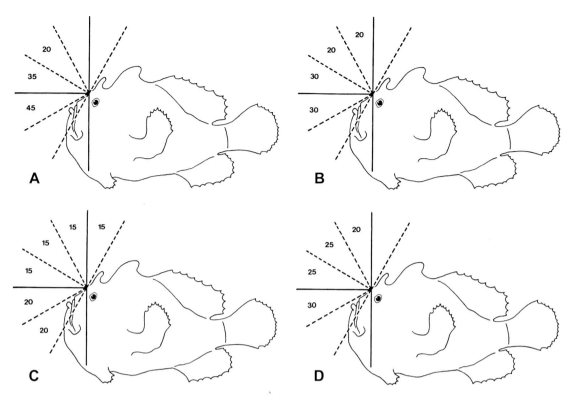

Fig. 296. Diagrammatic representation of percent frequency of occurrence of illicium within 30 sectors radiating from the base of the illicium in four species of *Antennarius*. (**A**) *A. striatus.* (**B**) *A. hispidus.* (**C**) *A. maculatus.* (**D**) *A. commerson.* After Pietsch and Grobecker (1987, fig. 136).

The secondary component of luring behavior, that is, the motion of the esca itself as it is moved about by the illicial apparatus, was analyzed in a similar way. Films made of luring behavior showed a specific motion of the esca for each of the four species examined: the wormlike esca of *A. striatus* wriggles in a sinusoidal pattern much like that of a swimming polychaete; the esca of *A. hispidus* makes a simple back-and-forth, swaying motion like that of a feeding tube worm; the fishlike esca of *A. maculatus* ripples as it is pulled through the water, appearing to mimic the lateral undulations of a swimming fish (Pietsch and Grobecker, 1978); and the esca of *A. commerson* is manipulated to simulate the quick, backward-darting motion of a swimming shrimp (Grobecker, 1981).

The two components of luring behavior are thus used simultaneously to attract potential prey to the vicinity of the predator and also to manipulate the prey into the strike zone. Whenever a prey item comes within close proximity to the angler (roughly one frogfish body length), but is not yet within the strike zone, the esca is positioned on the opposite side of the strike zone and is either held motionless or caused to vibrate slightly. This positioning of the esca creates a direct pathway from the prey item through the strike zone to the esca.

At the completion of luring activity, and just prior to a strike, the esca is contracted into a tight ball (accomplished through contraction of a complex network of smooth muscle fibers, which, in the absence of skeletal structures for muscle attachment, are "anchored" in an intricate pattern of collagen fibers; Fig. 297), and the illicium is laid back on top of the head alongside the base of the second dorsal-fin spine. In most cases, the esca in this non-luring or resting position is protected in some way from the potential nibblings and

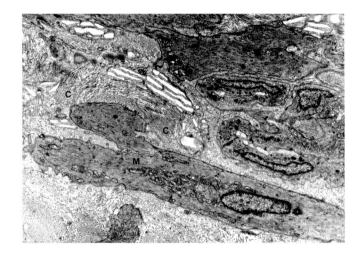

Fig. 297. Electron micrograph of cross-section through the esca of *Antennarius striatus*, showing "anchoring" of smooth muscle fibers (**M**) in a network of collagen fibers (**C**). After Pietsch and Grobecker (1987, fig. 137).

attacks of smaller fishes; in many species the contracted esca comes to lie within a shallow depression located between the second and third dorsal spines (Barbour, 1942: 25; Böhlke and Chaplin, 1968: 718; Pietsch and Grobecker, 1987: 29) and is covered either by the second spine or by membranous skin that connects the second spine to the dorsal surface of the head (see Illicial Apparatus and Associated Cephalic Spines, p. 30).

CHEMICAL ATTRACTION, BIOLUMINESCENCE, AND BIOFLUORESCENCE

Fine-structural, experimental, and behavioral observations indicate that *Antennarius striatus* incorporates an additional, chemical mechanism of attracting prey via secretory glands located in the esca (Pietsch and Grobecker, 1987: 313). A detailed fine-structural analysis revealed the presence of numerous secretory cells in the epidermal layer of the esca, each filled with numerous small granules (easily distinguished from larger melanophores; Fig. 298). As the granules mature, they approach a centrally located lumen into which the chemical attractant is apparently released (Fig. 299). Additional structures observed in these glands, which are usually associated with the secretory activities of cells (Downing and Novales, 1971), include Golgi bodies and rough-surfaced endoplasmic reticulum.

Controlled laboratory experiments were conducted by Pietsch and Grobecker (1987: 314) to test the hypothesis of chemical attraction. Water samples were siphoned from near the snout region of an actively luring individual of *Antennarius striatus* and presented to individual natural-prey items (*Dascyllus aruanus*) through a long, clear plastic tube. Attached to the distal end of the plastic tube was a fake esca of *A. striatus* constructed from a cream-colored, commercially produced rubber worm normally used for bait by sport fishermen. The controls were water samples taken from near the snout region of an individual of *A. striatus* from which the esca had been surgically removed. Recordings of prey approach to the rubber esca revealed a significant difference (p = .005) between the test and the control (for additional details of experimental procedure, see Approach and Procedures, p. 28).

It is to be expected that when a frogfish employs a chemical attractant, the majority of its prey would approach from down current. This is supported by observations made in Sydney Harbour, Australia (one of the few localities where *A. striatus* can be found at depths accessible to scuba), where seven of ten individuals observed were positioned in such a way that their strike zone was facing down current (Pietsch and Grobecker, 1987:

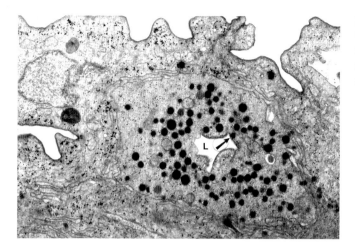

Fig. 298. Electron micrograph of cross-section through the esca of *Antennarius striatus*, showing lumen of granular cell (**L**). Arrow indicates granule interphasing with wall of lumen. After Pietsch and Grobecker (1987, fig. 138).

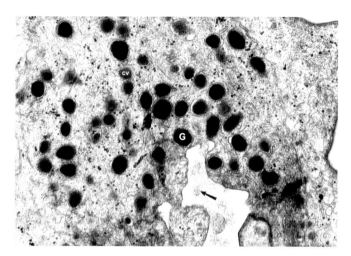

Fig. 299. Electron micrograph of longitudinal section through the esca of *Antennarius striatus*, showing portion of a granular cell. Note condensing vacuole (**CV**) and granule (**G**) located near wall of lumen. Arrow indicates discharged material. After Pietsch and Grobecker (1987, fig. 139).

315). It also seems likely that chemical attraction would be of great importance to a fish predator that spends a great deal of time foraging in waters of low-light intensity. Twenty-four-hour laboratory observations utilizing closed-circuit video revealed that *A. striatus* is particularly active nocturnally. Moreover, at least some populations of this species are comparatively deep-living; the mean depth of capture, for example, of all Hawaiian material of this species examined (n = 21) was 145 m. Pertinent here is Scott W. Michael's (1998: 330) report of aquarium-held individuals of *Abantennarius nummifer* actively luring in the dark. Using red fluorescent light to observe nocturnal behavior, he noticed also a distinct difference in luring behavior between night and day: a quick flicking movement of the illicium, along with rapid vibrations, was more often seen at night than during the day. He surmised that while visual cues may not be important in attracting prey at night, the rapid movement of the esca transmits a signal to the pressure-sensitive lateral-line system of prey. "Although a hungry cardinalfish or squirrelfish may not see the lure, it can feel it. So a dynamic luring pattern like the quick flick is more likely to be sensed by potential prey than one in which less movement of the bait is involved" (Michael, 1998: 330).

Of further interest with regard to luring in poorly lit environments is the discovery of bioluminescence in the esca of *Antennarius hispidus* by Ramaiah and Chandramohan (1992), an occurrence among anglerfishes heretofore known only in the deep-sea ceratioids (see Pietsch, 2009: 229). The escae of four living specimens collected at 50 m in coralline habitat off Bombay, India, were observed to glow when taken into a dark room. Later cultivation of extracts from the escae revealed the exclusive presence of the luminous bacterium *Photobacterium leiognathi*. Perhaps even more surprising is the recent paper by Brauwer and Hobbs (2016) describing biofluorescent baits in *A. striatus*. Using high-intensity blue LED torches and yellow filters while night diving in 5 m of water off Dauin, Philippines, the wormlike escae of three individuals were observed to display bright orange fluorescence. None of the three showed any fluorescence on the head and body, except for some limited red fluorescence caused by algal growth. Within 50 cm of the frogfishes, free-swimming unidentified worms of the same size as the escae exhibited the same orange fluorescence, prompting the authors to conclude that biofluorescence might well play a role in aggressive mimicry. Much more research is required to corroborate these findings and to assess the degree to which they are present among frogfishes in general, but it seems likely that luring in these fishes is mediated by a complex suite of cues that are not only visual, both day and night and beyond our visual spectrum, but chemical and tactile as well.

ESCAL ENLARGEMENT IN *ANTENNARIUS STRIATUS*

Antennarius striatus appears to further enhance its luring capabilities by enlarging the esca when actively luring (see Tee-Van, in Gudger, 1945a: 112). Indeed, the wormlike escal appendages of preserved specimens of this species (as well as those of its sister species *A. scaber*) show remarkable differences in size (without relationship to standard length), from short and stout (less than 4% SL) to extremely long and slender (27.8% SL). Films analyzed by Pietsch and Grobecker (1987: 315), as well as numerous underwater photos taken in the field by sport divers, revealed that both the length and the girth of the escal appendages can increase by some 35%. Cross-sections of escae reveal the presence of numerous desmosomes (gap junctions) in the epidermal layer (Fig. 300) that appear to facilitate escal enlargement by providing for structural maintenance, but despite careful examination by both light and electron microscopy, the exact mechanism responsible for this enlargement, a phenomenon not observed in any other antennariid examined, remains unknown.

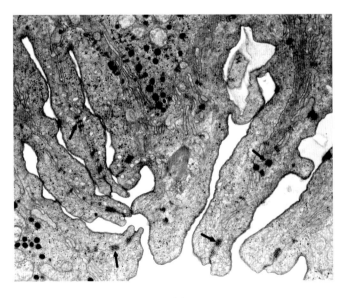

Fig. 300. Electron micrograph of cross-section through the esca of *Antennarius striatus*, showing the numerous desmosomes (indicated by arrows) that provide structural maintenance. After Pietsch and Grobecker (1987, fig. 140).

In anglerfishes, the illicium and esca act as a lure and are thus apparently highly suscep-tible to loss or damage by the attacks and nibblings of potential prey, as well as predators. Although we have examples indicating that the structure is not absolutely essential for sustained life, there seems little doubt that the luring apparatus is of prime importance in the life of the angler and that its loss in most instances would lead to serious conse-quences (for example, Aristotle remarked on the emaciated condition of individuals of *Lophius* that had lost the luring apparatus; see Thompson, 1984, book 9 (37), lines 27–28). The obvious question to be asked therefore is whether damaged or lost illicia and escae can be replaced. Assuming that the species-specificity of escal morphology has some important adaptive significance, one might further ask whether there is some regenera-tive mechanism that allows for the maintenance of this morphological specificity.

Although present throughout the entire animal kingdom, the ability to regenerate lost parts differs both in the extent to which, and the process by which, it occurs in vari-ous animal taxa. No other group of vertebrates exceeds the fishes in the variety of struc-tures capable of regeneration. Within the Teleostei the replacement of scales, fins, bar-bels, and various miscellaneous structures is well known (Goss, 1969; Kemp and Park, 1970; Saxena and Aggarwal, 1971; Bereiter-Hahn and Zylberberg, 1993; Akimenko et al., 2003; Nakatani et al., 2007; Zupanc and Sirbulescu, 2012). Of these structures, fins have the greatest potential for regeneration. Fins possess a generative zone along their outer margins that provides for their unlimited growth (Goss, 1969: 119; Akimenko et al., 2003; Nakatani et al., 2007; Singh et al., 2012). Thus, it is to be expected that, if lost or dam-aged, the escae of lophiiform fishes, lying at the terminal end of a modified dorsal-fin spine, will be replaced by tissue regeneration.

Günther (1861a: 184) seems to have been the first to suggest that the illicia and escae of antennariids are capable of regeneration: "the tentacle above the snout, which, being ten-der and delicate, is necessarily often injured, and probably reproduced." Somewhat later, but apparently based at least in part on Günther's earlier observations, Richard Owen (1866: 567) wrote "The modified dermoneurals [dorsal-fin spines] forming the cephalic tentacles of *Lophius* and *Antennarius* are as frequently reproduced as they are injured, to meet the particular use which these angling fishes make of them." Regan (1903: 278), in reference to lophiid anglerfishes, wrote that "the first ray [illicium] seems to become rela-tively longer during growth, but if, as frequently happens, it is broken off, a fresh flap de-velops at its end." Franz (1910: 111, pl. 10, fig. 12) described the internal structure of the esca of *Antennarius striatus* (as *A. tridens*) and suggested the possibility of regeneration. Wilson (1937: 494) provided further evidence that lost escae are regenerated in lophiids, and Pietsch (1974: 100) suggested that repair and replacement probably occur in deep-sea cera-tioid anglerfishes as well; but no one had rigorously tested a hypothesis of escal regenera-tion until the experimental work of Pietsch and Grobecker (1987: 319, figs. 141–143).

Although it is difficult to decide whether aberrant situations are genetic anomalies or the result of damage and subsequent regeneration, observations made on preserved an-tennariids provide some evidence that escae are replaced. Of 152 specimens of *Antennar-ius pictus* examined, one (BPBM 5805) appears to have lost the full length of the illicium, but a fully formed esca, one well within the morphological variation known for this spe-cies, has regenerated at the base (Fig. 301). Of 115 specimens of *A. hispidus* examined, a single individual (USNM 201569) appears to have sustained damage to the illicium at a point just below the esca, resulting in the development of a short piece of illicium that terminates in a second esca essentially identical to that diagnostic for the species (Fig. 302). Of 48 specimens of *Abantennarius drombus* examined, two individuals show similar escal anomalies: the illicium of BPBM 8471 is bifurcate at mid-length, each branch bearing a similar esca (Fig. 303A); that of BPBM 13332, a 52-mm specimen, is similar, but bifurcate from the base (Fig. 303B). Of 458 specimens of *A. coccineus* examined, the illicium of a specimen from the Red Sea (BPBM 19802, 55 mm) is also bifurcate from the base but the two branches of the illicium are connected by a thin membrane (Fig. 303C). Finally, it should be noted that at least one individual examined (the holotype of *Golem cooperae*; see Species *Incertae Sedis*, p. 354) appeared to have been surviving well (the body large and robust, and with well-developed ovaries) after having lost all three cephalic dorsal spines prior to capture; no evidence of any tissue regeneration was present.

Evidence for escal regeneration comes also from a series of laboratory experiments performed by Pietsch and Grobecker (1987: 319; see also Michael, 1998: 330) on antennariids in which the escae were surgically removed and allowed to regenerate. The es-

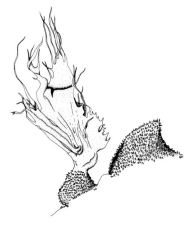

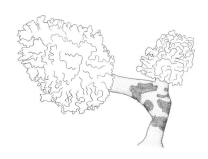

Fig. 302. Illicium and escae of *Antennarius hispidus*, USNM 201569, 88 mm SL, showing what appears to be damage at a point just below the esca, resulting in regeneration of a short extension of the illicium that terminates in a second esca (note that the second esca is smaller but essentially identical in morphology to the original terminally located esca). Drawing by T. W. Pietsch; after Pietsch and Grobecker (1987, fig. 142).

Fig. 301. Esca of *Antennarius pictus*, BPBM 5805, Kaawa, Oahu, Hawaiian Islands, 133 mm SL; the full length of the illicium has been lost and a new esca regenerated at its base. Drawing by Glen Higashi; after Pietsch and Grobecker (1987, fig. 141); courtesy of John E. Randall, Bernice P. Bishop Museum, Honolulu.

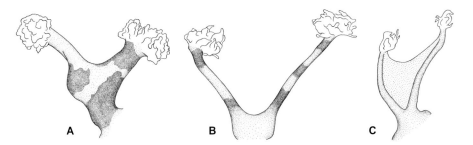

Fig. 303. Bifurcate illicia of three individuals of *Antennarius drombus*, anomalies apparently owing to damage and subsequent tissue regeneration. (**A**) BPBM 8471, 70 mm SL. (**B**) BPBM 13332, 52 mm SL. (**C**) BPBM 19802, 55 mm SL. Drawings by T. W. Pietsch; after Pietsch and Grobecker (1987, fig. 143).

cae of three individuals of *A. hispidus*, eight of *Antennarius striatus*, three of *Abantennarius coccineus*, one of *A. sanguineus*, and two of *Lophiocharon trisignatus* were amputated at various points below the base. The morphogenesis of regenerating escae was recorded from the time of initial surgery until growth and differentiation appeared complete. The gross morphology of the regenerated escae was then compared to the original amputated escae, as well as with the range of morphological variation found in the escae of preserved material of each species examined experimentally. In all cases, not only were the escae replaced, but the regenerated escae, with a single exception, were grossly identical in morphology to the original, non-regenerated escae. In one experiment, surgical removal of a single wormlike escal appendage of *A. striatus* was followed by regeneration of two similar filaments. The resulting regenerated esca, although different from the original non-regenerated esca, was still well within the expected morphological variation of the species (see Geographic Variation, p. 108).

Regeneration visible to the naked eye was apparent after only five days; in 42 days the escae were approximately one-third their original size; and in 123 to 240 days the escae attained from one-half to their full original size and structure.

The species-specificity of escal morphology is maintained in antennariids even in the event of loss or damage thus providing confirmation of the genetic basis of morphological species-specificity and strengthening confidence in its use as a character complex of considerable importance in anglerfish systematics. More significant from an ecological standpoint, however, is the reinforcement of the idea that there must be some strong, if as yet poorly understood, correlation between the often complex and specific embellishments of these structures and the efficiency of prey attraction. It could also be that escae, in mimicking different but specific kinds of smaller organisms (shrimps, copepods, amphipods, polychaetes, fishes, etc.), may attract a species-specific set of prey organisms, and thus avoid competition for food resources among closely related sympatric species (Pietsch, 1974: 98). However, an analysis of stomach contents revealed no significant differences between species in the kinds of prey taken; rather, it indicated that antennariids are generalists as far as diet is concerned. Nevertheless, the sample sizes of each species examined were small; it may be that a more detailed statistical analysis of stomach contents of larger samples would show differences in diet.

Aggressive Resemblance and Foraging-Site Selection

If an organism imitates a signal that is of interest to a signal-receiver, then this is a case of mimicry; but if generally uninteresting background or substrate is imitated, then protective resemblance or camouflage is involved (Poulton, 1890: 60; Cott, 1940: 115; Wickler, 1968: 238; Grobecker, 1983: 200; Randall, 2005b: 299). By imitating structure with the head, body, and fins, frogfishes signal uninteresting background to some signal-receivers, resulting in protective resemblance. Simultaneously, however, many lie-in-wait predators, including frogfishes, signal interesting background to other signal-receivers, displaying what is mistaken by potential prey as a possible shelter or foraging site. This is a functionally different kind of similarity between animal and object, called "aggressive resemblance," in which the concepts of protective resemblance (the imitation of uninteresting background) and mimicry (requiring a specific model) do not pertain (for a review and classification of mimetic systems, see Vane-Wright, 1976; Pasteur, 1982).

The signaling of interesting background by frogfishes, whether it be a sponge or a coralline algae-encrusted rock (i.e., attractive substrate), is enhanced by maintaining "open" or unoccupied space (other than by itself; see Grobecker, 1983: 200). This is accomplished simply by eating all fishes and invertebrates that enter that space. Since herbivorous fishes are eaten whenever the "open" space is invaded, algae (and other benthic organisms) are allowed to grow ungrazed, further enhancing the ambush site as a high-quality, "attractive" foraging site. It seems evident that frogfishes—as well as many other lie-in-wait predators, whether they be associated with high-diversity marine or terrestrial communities—gain a significant advantage by projecting a combination of signals that attract prey while, simultaneously, avoid being devoured.

Many, if not all, resident fishes on a coral reef have specific shelter sites (Smith and Tyler, 1972; Sale, 1991; Buchheim and Hixon, 1992; Almany, 2004; Komyakova et al., 2013). A shelter site provides refuge in which an organism may be relatively free from detection or attack by predators with little or no expenditure of energy. It has been suggested that space (shelter sites) is fully utilized on a reef; in fact, even the smallest patches of suitable substrate appear to be fully occupied (Smith and Tyler, 1972; Jones, 1991; Sale, 1991; Buchheim and Hixon, 1992; for an alternative view, see Doherty and Williams, 1988). Thus, there would appear to be a distinct advantage for a frogfish to prey on space-competing organisms (versus food-competing organisms), when space-competing prey would be in greater abundance over time (moving toward the carrying capacity) than would food-competing prey (MacArthur and Wilson, 1967: 149).

By selective or random removal of individual shelter-seeking prey organisms, frogfishes, in concert with other lie-in-wait predators in high-diversity marine communities, may allow for the coexistence of species that would be eliminated by competitive exclusion in the absence of such predators. Because the lie-in-wait feeding mode acts on the more abundant prey types, it may be frequency-dependent, thus promoting prey diversity (Pianka, 1978: 296).

Pietsch and Grobecker (1987: 320, pl. 5) found that *Antennarius striatus* in Sydney Harbour, Australia, maintains at least four distinct color phases, and thus appears to assume several forms of aggressive resemblance: a greenish yellow-brown color phase, with a dense growth of cutaneous filaments resembling algae-covered substrate; an orange phase closely resembling an orange sponge; a white phase resembling a white sponge; and a black color phase suggesting a black sponge (these sponges, at least the black and orange varieties, have been tentatively identified as *Spirastrella* spp.). Individuals of *A. striatus* observed in the field were each closely associated with the type of structure and coloration that their bodies resembled. Relevant here are the observations of Frank Schneidewind (2003b, 2005a) who described more recently a black phase of this species that greatly resembles a sea urchin, with which it associates in the Philippines (Fig. 304).

The presence of filamentous, algae-like cutaneous appendages characteristic of the greenish yellow-brown color phase of *A. striatus*, as well as the presence of real algae allowed to grow ungrazed within the ambush site, may be an important factor in luring herbivorous prey (see Grobecker, 1983: 200). The removal of herbivorous intruders through predation makes it possible for algae to grow unchecked and thus the attractiveness of the ambush site is enhanced.

Although algae in general are an important component of healthy reefs (e.g., see Vroom et al., 2006; Vroom and Braun, 2010; Vroom, 2011), larger filamentous and frondose species are conspicuously absent (Earle, 1972: 18; unhealthy reefs are another matter, see McCook et al., 2001; Bahr et al., 2015). Randall (1961, 1965, 1967) concluded that two factors may hold reef-dwelling macroalgae, as well as seagrasses, in check: strong competition for a foothold among the numerous sessile animals; but possibly more significant, heavy grazing by numerous algae-eating fishes. Of the 18 families of herbivorous

Fig. 304. (**A–E**) Black-phase individuals of *Antennarius striatus* and sea urchins, *Astropyga radiata*, off Sabang Beach, Mindoro Island, Philippines, about 50 m from shore at depths of 5 or 6 m, interpreted as "a form of aggressive and protective mimicry, as well as camouflage, previously unknown in frogfishes" (Schneidewind, 2005a: 23). Photos by Frank Schneidewind.

fishes listed by Randall (1967: 826), three, the Acanthuridae, Kyphosidae, and Scaridae, were almost entirely herbivorous (i.e., dependent on benthic algae as their primary food). For many other fish families, algae form an important part of the diet (i.e., among the 18 families identified by Randall, 59 species were found to be algae-eaters). Earle (1972: 33) found that the distribution of filamentous and frondose benthic algae in Lameshur Bay, St. John, Virgin Islands, was basically limited by light conditions, the availability of suitable substrate, and other physical factors in the environment. At the same time, however, she emphasized that the diversity and abundance of vegetation on and in the immediate vicinity of the reef is directly related to the occurrence of herbivorous fishes (see also Dawson et al., 1955: 14; Hiatt and Strasburg, 1960: 111; Stephenson and Searles, 1960: 265; Bakus, 1964: 6; Dart, 1972; Carpenter, 1986; Williams et al., 2001; Fong et al., 2016).

Further evidence that fishes strongly limit the abundance of macroalgae and seagrasses in high-diversity reef communities is provided by experiments designed to show the increase in algal growth within cages that exclude grazing fishes. In accord with the findings of Randall's (1961) now classic study, Earle (1972: 33) reported that the growth of algae within wire enclosures on a reef attained as much as 30 mm in height, whereas the algae outside averaged only 1 mm in height.

Other forms of aggressive resemblance are exemplified by the morphological and chromatic similarity of certain species of frogfishes and sessile invertebrates such as sponges and tunicates. In addition to observations of *Antennarius striatus* and sponges in Sydney Harbour (Pietsch and Grobecker, 1987: 320), Mowbray (in Barbour, 1942: 25) reported that the black color phase of *A. striatus* (identified as *A. nuttingi*) "usually lives cuddled up against or partly coiled about groups of black ascidians which abound about Bermuda." Describing *A. multiocellatus*, Barbour (1942: 28) wrote that dark individuals live among black tunicates. Breder (1949: 95) described an almost entirely black individual of this species resting on white sand that "showed a striking similarity to a very small logger-head sponge." Roessler (1977: 7) wrote that frogfishes "can sometimes be mistaken for nearby yellow, red, or brown sponges. I have seen black frogfish sit invisibly just inside the shadowed body cavity of a tube sponge."

Frogfishes that closely resemble sponges and tunicates (as well as perhaps other, as yet unidentified, marine organisms) probably attract and thus consume sponge- or tunicate-eating fishes and invertebrates (or sponge- and tunicate-dwelling organisms). (As an example of how extensive this behavior might be, Randall and Randall (1960: 473) listed juvenile *Antennarius pauciradiatus* as a mimic of dead *Udotea*, a benthic green alga.) That numerous fishes eat sponges has been shown by a number of workers. Hiatt and Strasburg (1960) listed numerous fishes from the Marshall Islands with sponge and tunicate fragments in their stomachs. Randall (1967: 828, 840) found that the stomachs of eight species of West Indian fishes contained more than 30% sponges by volume. Sixteen species fed at least in part on benthic tunicates; the stomachs of three of these, all species of trunkfishes (family Ostraciidae), contained 18% benthic tunicates by volume. Randall and Hartman's (1968: 225) stomach-content analysis of 212 species of West Indian reef and inshore fishes revealed sponge remains in 21 species. Potential prey may also be attracted to a sponge- or tunicate-like object when searching for a shelter site. Tyler and Böhlke (1972: 604) listed 39 species of primarily Caribbean fishes known to be either obligate or facultative sponge dwellers. They also listed a number of species whose occurrence in sponges is known to be fortuitous.

In examples like the observations of the four distinct color phases of *Antennarius striatus* in Sydney Harbour cited by Pietsch and Grobecker (1987: 320), it is intriguing to speculate that the proportions of the different color phases may reflect the relative efficiency of each form of aggressive resemblance in attracting prey, as well as the relative abundance of each type of structure. Of 90 specimens of *A. striatus* examined, 63% were of the greenish yellow-brown color phase; 12% the orange "sponge" phase; 2% the white "sponge" phase; and 23% the black "sponge" phase. These proportions perhaps indicate that, at least in Sydney Harbour, algae-covered structure is a more commonly occurring resource and one that is more eagerly sought by both food-seeking and shelter-seeking organisms.

Table 23. Relationship between Standard Length, Percent Vagility, and Percent Time Spent in Association with Structure in Two Specimens Each of Four Species of the Antennariidae

Species	Standard length (in mm)	Percent vagility	Percent time
Antennatus tuberosus*	15	12	96
Antennatus tuberosus*	23	7	89
Antennarius pictus	48	15	70
Antennarius commerson	66	16	52
Antennarius striatus	68	21	28
Antennarius striatus	73	32	24
Antennarius pictus	90	19	27
Antennarius commerson	140	15	16

*Values based on two 24-hour observation periods; all other values based on three 24-hour observation periods.

SIZE-DEPENDENT FORAGING

Foraging modes in frogfishes appear to depend primarily on body size but are also influenced by such variables as (1) amount of vagility, (2) degree of association with structure, and (3) depth preference. Following the experimental work of Pietsch and Grobecker (1987: 322), vagility is defined as the percent time the experimental animal was observed moving from one location to another, as opposed to moving about within a single location. Data on association with structure were recorded when an individual positioned its body within one (standard) body length of structure (such as coral or rock). The general trends, reflected in the data presented in Table 23, are presented in Figure 305.

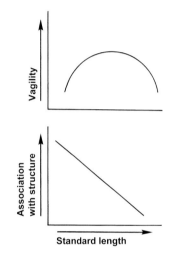

Fig. 305. Generalized relationship between size (standard length), amount of vagility, and relative time spent in association with structure determined by observations of aquarium-held individuals of four species of Antennariidae (Table 22). After Pietsch and Grobecker (1987, fig. 144).

Frogfishes of small size (less than 40 mm SL) usually have low vagility, and are most often found highly associated with structure in shallow water. Some of the advantages and consequences for these smaller individuals include (1) greater protection from predation, especially when positioned in the interstices of coral or rocks; (2) greater potential for attraction of prey, since structure in shallow water attracts small fishes and invertebrates that may serve as prey; and (3) reduced or obscured visibility of the esca to potential prey when associated with structure. For smaller frogfishes (with correspondingly smaller escae), luring may be of less value than for larger-sized frogfishes. For example, *Antennatus tuberosus*, a species that attains a maximum standard length of only about 70 mm and is strongly associated with coral rubble at depths that rarely exceed 20 m, lacks a distinct esca. Numerous individuals of this species have been maintained in aquaria for long periods of time, but luring behavior has never been observed.

As the body size of frogfishes increases, their association with structure decreases (Fig. 305, Table 23), a shift that may correlate with such factors as (1) lower predator risk, (2) increased consequences of aggressive resemblance, and (3) a less obstructed view of escae.

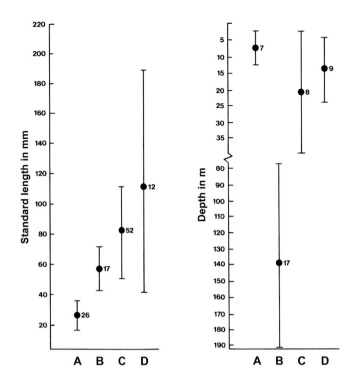

Vagility is highest for frogfishes of intermediate body size (roughly 70 to 80 mm SL). Frogfishes with a higher level of vagility may be able to take advantage of a wider range of foraging sites, making it possible to invade a wider range of depths, as exemplified by *Antennarius striatus* and *A. pictus* (Fig. 306). *Antennarius striatus* (and perhaps other frogfishes as well) incorporates an additional luring mechanism in the form of a chemical attractant (see Aggressive Mimicry, p. 411), which may make it possible for this species to occupy an even wider range of depths, especially greater depths where visibility is poor and visual signals inefficient. This species is common in surface waters, but has also been collected at depths that exceed 200 m (see Zoogeography, p. 402).

Fig. 306. Mean and standard deviation of variation in standard length (*left*) and depth of capture (*right*) in four species of the Antennariidae. (**A**) *Antennatus tuberosus*. (**B**) *Antennarius striatus*. (**C**) *Antennarius pictus*. (**D**) *Antennarius commerson*. After Pietsch and Grobecker (1987, fig. 145).

For intermediate to large-sized frogfishes (greater than 75 mm SL), vagility decreases (Fig. 305, Table 23), probably correlating with the higher cost of locomotion inherent in a larger animal, in this case a large animal with a globular, hydrodynamically poor body plan. The disadvantages of reduced vagility, however, are probably outweighed by the consequent advantages of having a larger, more conspicuous esca (aggressive mimicry) and a larger, structure-like body (aggressive resemblancc).

Being largely size-dependent, the foraging modes of frogfishes must shift throughout their life history as they grow larger. For individuals of species that attain a relatively small size (e.g., *Antennatus tuberosus*, less than 70 mm SL; Figs. 305, 306) there are few such modifications. In contrast, members of species that become relatively large (e.g., *Antennarius commerson*, in which the maximum known standard length is nearly 300 mm; Figs. 305, 306) probably undergo several modifications in foraging strategy during their lifetime. Each of the four species examined attains a different maximum standard length (Fig. 306), and each was found to pursue a different foraging mode. The differences may result in reduced interspecific competition for food, as well as for spatial resources.

THEORETICAL FORAGING-SITE SELECTION MECHANISM

A comparison of what has been learned about the foraging modes of frogfishes with previously published hypotheses (e.g., shell selection by the hermit crabs *Pagurus samuelis* and *Calcinus laevimanus* described by Reese, 1963; habitat selection by juvenile Convict Surgeonfish, *Acanthurus triostegus*, by Sale, 1969; and foraging behavior and habitat use in the Dash-and-dot Goatfish *Parupeneus barberinus* by Lukoschek and McCormick, 2001) have led to the following theoretical, foraging-site selection mechanism for frogfishes (Fig. 307).

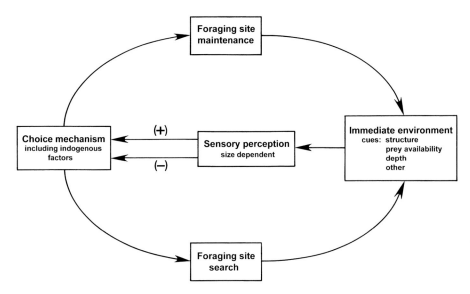

Fig. 307. Hypothesized foraging-site selection mechanism for a generalized antennariid (for a full explanation, see Foraging-Site Selection, p. 431). After Pietsch and Grobecker (1987, fig. 146).

In a frogfish, cues from the environment are constantly being monitored by sense organs that feed stimuli to a choice-making mechanism located in the central nervous system. These cues include the quality and availability of structure, availability of suitable prey organisms, water depth, and perhaps other, as yet unidentified, environmental signals. The magnitude and significance of the positive or negative stimuli perceived is dependent on the body size of the frogfish, small individuals requiring a high level of association with structure in shallow water, individuals of intermediate size displaying a reduced association with structure and occupying a wider range of water depths, and large individuals maintaining trophic efficiency with a reduced association with structure, within a narrower range of intermediate depths. The choice mechanism integrates perceived stimuli (including endogenous factors; see Lorenz, 1950: 221, 268), and in turn provides an output that initiates either a search for a new foraging site or the maintenance of the present foraging site. The foraging-site search response may be of two basic types, either random movement through the environment (Sale, 1969: 29) or a search for a site previously utilized (according to T. Nahacky, personal communication, 5 March 1982, individuals of *Antennarius pictus* periodically revisit specific foraging sites). The maintenance of a foraging site may include such activity as aggressive defensive displays (see Defensive Behavior, p. 451).

As in habitat selection by juvenile Convict Surgeonfish (*Acanthurus triostegus*; see Sale, 1969), foraging-site selection in antennariids can be considered a species-typical pattern of behavior. At the same time, however, the foraging-site selection mechanism described here is simplified to such a degree that it can be applied to a generalized antennariid, as well as to a host of other lie-in-wait predators.

Feeding Behavior and Biomechanics

Early comparative studies of the biomechanics of feeding in fishes were based solely on anatomical data (Gregory, 1933; Dobben, 1935; Tchernavin, 1953; Ballintijn and Hughes, 1965; and numerous others). More recently, however, anatomical data have been integrated with functional data obtained through the use of living material and cinematographic and electromyographic techniques (Osse, 1969; Liem, 1978, 1979, 1980; Wainwright and Lauder, 1986; Westneat, 1990; Ferry-Graham and Lauder, 2001; Konow and Bellwood, 2011; Westneat and Olsen, 2015, and numerous references therein). The addition of these relatively new techniques to anatomical analyses largely avoids the problems of extrapolating function from form and has provided a more accurate assessment of the role of individual bones, muscles, and ligamentous connections during feeding activity. In the past, most cinematographic analyses have been limited to film speeds of 13–250 frames per second (but see Nyberg, 1971; Grobecker and Pietsch, 1979; Grobecker, 1983; Pietsch and Grobecker, 1987; Van Wassenbergh et al., 2014, and numerous references therein). It has been suggested, however, that higher filming speeds might reveal that single feeding events in fishes occur at speeds greater than previously believed (Osse, 1969: 360).

In work described by Grobecker and Pietsch (1979: 1161) and Pietsch and Grobecker (1987: 326), prey capture was investigated through the use of high-speed cinematography at film speeds of 800 and 1,000 frames per second. Three antennariid species, *Antennarius hispidus*, *A. striatus*, and *A. maculatus*, were examined in detail, all providing the same general results. Anatomical analyses of the bones, muscles, and ligamentous connections of the feeding mechanism, integrated with functional data obtained from frame-by-frame analyses of films, revealed a complex process of prey capture that could be divided into three functionally distinct phases: (1) pre-strike behavior, consisting primarily of luring behavior, stalking behavior, and orientation to prey; (2) strike behavior, essentially the biomechanics of expansion and compression of the oral cavity; and (3) prey handling, or the manipulation and deglutition of prey.

PRE-STRIKE BEHAVIOR

When a prey item was introduced into the visual field of an individual of any of the three species of *Antennarius* examined, eye movement indicated that the prey was being followed visually. This behavior occurred 100% of the time. The next most commonly observed behavior was luring, occurring about 80% of the time (see Aggressive Mimicry, p. 411). A less common behavior, occurring about 15% of the time (and considerably more often in *A. striatus* than in any other species examined; see Vagility, p. 435), was a stalk toward the prey (a combination of jet propulsion and body and fin movement; see Locomotion, p. 455). Stalking behavior was initiated when a prey item came within a distance of about seven body lengths ("body length," when used to estimate distances between predator and prey, refers to frogfish standard length). As the frogfish began to move toward the prey, the body rose off the substratum and the dorsal fin became erect. The rate of approach was relatively rapid at first, but at a distance of less than three body lengths from the prey, the rate of locomotion decreased from about 0.30 to 0.10 standard lengths per second. During this slower phase of the stalk, the body of the frogfish became dorsoventrally depressed, taking on what may be described as a "crouching position"—at the same time the dorsal fin was laid back. This depression of the body apparently functions to reduce the size of the image of the oncoming predator, resulting in a less conspicuous approach to the prey. Individuals examined by us infrequently (less than 5% of the time) used the stalking and luring behaviors simultaneously.

Once the prey item was roughly one body length away, and just prior to prey seizure, orientation of the strike zone toward the prey took place. This strike-zone adjustment was accomplished in two ways: by a pitching-and-rolling movement of the body, produced by a lateral and antero-posterior movement of the pectoral and pelvic fins; and by a slight twisting movement of the body, causing the head of the frogfish to move to the side in a rolling-yawing motion (often functioning to adjust the strike zone in the transverse plane). Although each motion was capable of operating independently of the other, the two methods of orientation to prey were often used simultaneously.

STRIKE BEHAVIOR

Just prior (0.5–1.4 s) to oral expansion, two slight movements were observed: a ventral depression of the head and a slight opening of the mouth. These movements may be part of a way to build up potential energy in preparation for oral expansion. Although electromyographic data are unavailable for antennariids, the muscles most likely involved in these preparatory movements are the adductor mandibulae and geniohyoideus (Liem, 1978: 341).

In antennariids, the biomechanics involved in the expansion of the oral cavity and the protrusion of the upper jaw are similar to those employed by other teleosts (Ballintijn and Hughes, 1965; Alexander, 1967; Osse, 1969; Liem, 1970, 1978; Pietsch, 1978b; Motta, 1982; Grobecker, 1983; Wainwright and Lauder, 1986; Westneat, 1990; Ferry-Graham and Lauder, 2001; Wainwright et al., 2007; Konow and Bellwood, 2011). During prey seizure, contraction of the epaxial musculature lifts the head; at the same time, two biomechanical couplings are involved in the expansion of the oral cavity. The first is a direct coupling by way of the opercular apparatus (Figs. 308–310): contraction of the levator operculi muscle causes a lift of the opercle. This movement is translated to the lower jaw by way of the interopercle, causing the jaw to rotate around the articular-quadrate joint. In advanced teleosts, this is the primary mechanism for depressing the lower jaw (Lauder, 1979: 573).

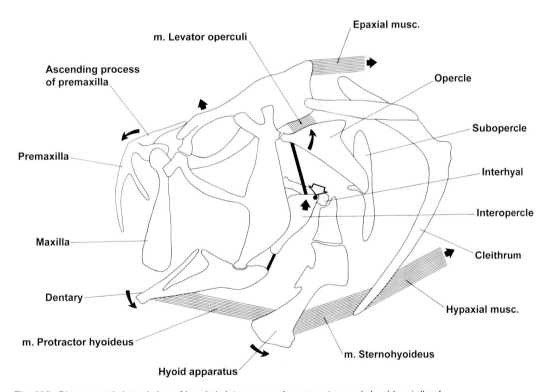

Fig. 308. Diagrammatic lateral view of head skeleton, opercular apparatus, and shoulder girdle of *Antennarius*, showing the major bony elements and muscles of the feeding mechanism. Closed arrows indicate direction of movement during the expansion phase of feeding; open arrow indicates ligamentous connection between interhyal and interopercle. After Pietsch and Grobecker (1987, fig. 147).

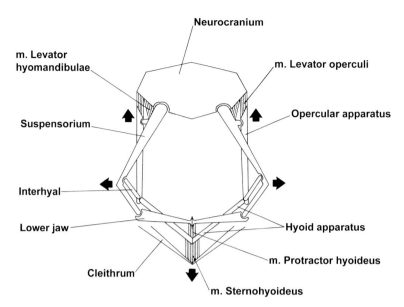

Fig. 309. Diagrammatic anterior view of head skeleton, opercular apparatus, and shoulder girdle of *Antennarius*, showing the major bony elements and muscles of the feeding mechanism. Arrows indicate direction of movement during the expansion phase of feeding. After Pietsch and Grobecker (1987, fig. 148).

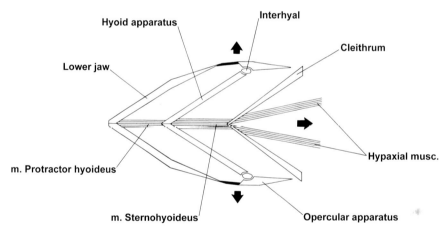

Fig. 310. Diagrammatic ventral view of portions of head skeleton, opercular apparatus, and shoulder girdle of *Antennarius*, showing the major bony elements and muscles of the feeding mechanism. Arrows indicate direction of movement during the expansion phase of feeding. After Pietsch and Grobecker (1987, fig. 149).

The second coupling is an indirect one, by way of the hyoid apparatus (Figs. 308–310). Contraction of the hypaxial musculature and the sternohyoideus pulls both the cleithrum and hyoid apparatus posterodorsally, a movement that is translated through the protractor hyoideus muscle to the lower jaw. The indirect coupling causes expansion of the oral, as well as the opercular, cavity in two ways: in the vertical plane, by lowering the floor of the oral cavity; and in the horizontal plane, by causing the sides of the head to expand laterally (see Ballintijn and Hughes, 1965: 355, fig. 4). In addition to the muscles already mentioned, the levator arcus palatini and the dilatator operculi are believed to be active during expansion of the oral cavity (Liem, 1978: 338).

These two couplings, the indirect hyoid coupling and the direct opercular coupling, are by no means functionally independent of each other. There is a strong, ligamentous connection between the interopercle and the interhyal (Fig. 308), and any posteriorly directed movement of the hyoid apparatus caused by contraction of the hypaxial musculature and the sternohyoideus is thus translated directly to the lower jaw.

The vast majority of teleosts, especially the spiny-rayed fishes (Acanthomorpha) and including nearly all lophiiform fishes (for exceptions, see Bertelsen and Struhsaker, 1977: 29; Bertelsen et al., 1981: 19; Pietsch, 2009: 270), are "gape-and-suck" feeders. They engulf prey by creating negative pressure (suction pressure) inside the mouth (Ballintijn and Hughes, 1965: 355; Liem, 1970: 145; Alexander 1970: 145; Grobecker and Pietsch, 1979). This negative pressure results from the large increase in volume produced by rapid expansion of the oral and opercular cavities. The amount of expansion (as well as the rate of expansion) of these cavities is crucial to the strike of a gape-and-suck feeder. In frogfishes, the amount of oral expansion is considerably greater than that of most, if not all, other teleosts examined (Whitmee, 1875: 544; Gudger, 1905: 842, 1945a: 111, 1945b: 544; Gill, 1909: 600; Beebe, 1933: 115; Barbour, 1942: 23, 25; Gordon, 1955: 389; Schultz, 1957: 52; Osse, 1969: 359; Elshoud-Oldenhave and Osse, 1976: 412; Kershaw, 1976: 231; Grobecker and Pietsch, 1979: 1162; Grobecker, 1983: 199; Pietsch and Grobecker, 1987: 330; Bergert and Wainwright, 1997: 568; Roos et al., 2009: 3490). To determine the magnitude of oral expansion, wax casts were made by injecting liquid paraffin into closed and fully expanded oral cavities. Comparison of these casts (Fig. 311) revealed that the three species of *Antennarius* examined are capable of creating oral volumes greater than 12 times that of the closed cavity (Table 24). In contrast, using a similar wax-cast technique, Osse (1969: 359) found that the European Perch, *Perca fluviatilis*, expands its mouth cavity about six times during a single feeding event.

During the expansion of the oral and opercular cavities, the premaxillae of the upper jaw of frogfishes undergo an extensive anteriorly directed protrusion. This movement, initiated just as the head begins to lift (Figs. 312A, 313Aa, 313Bb), is brought about

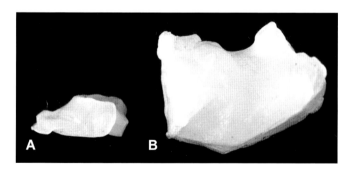

Fig. 311. Paraffin casts of closed (**A**) and fully expanded (**B**) oral cavity of *Antennarius hispidus*, UW 21015, 91 mm SL (Table 17). After Pietsch and Grobecker (1987, fig. 150).

Table 24. Weights of Paraffin Casts of Closed and Fully Expanded Oral Cavities, Volume Expansion of Oral Cavities, and Speed of Prey Engulfment for Three Species of *Antennarius*

| Species | Weight of cast (in g) | | | Speed of engulfment (in msec) | | |
	Closed	Fully expanded	Volume expansion	Trial 1	Trial 2	Trial 3
A. hispidus	1.2	16.4	13.7X	6.2	7.5	5.0
A. maculatus	2.1	25.4	12.1X	4.0*	10.0*	6.0*
A. striatus	2.3	29.9	12.8X	3.8	8.8	6.2

*Determined by high-speed cinematography at 1,000 fps; other values based on 800 fps.

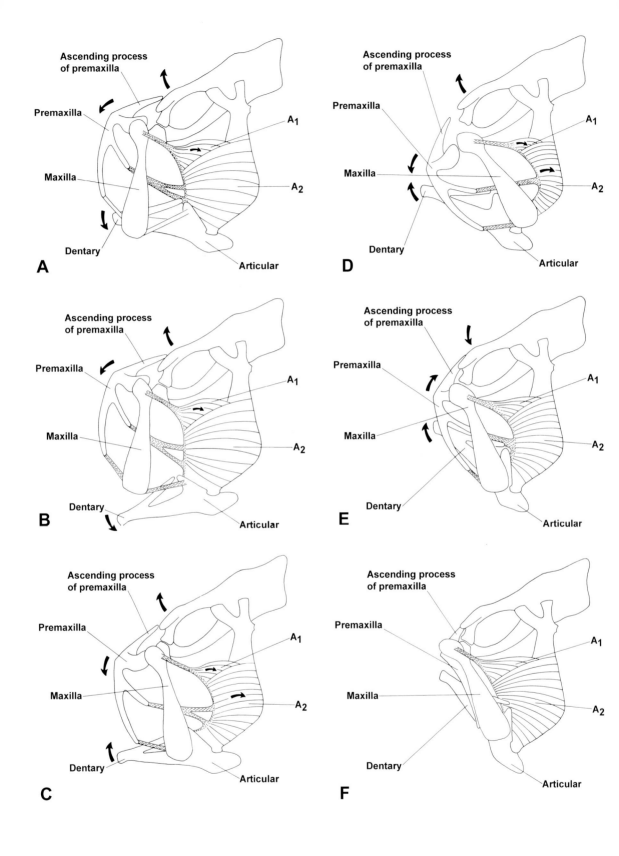

by a dual rotation of the maxillae: (1) an axial rotation and (2) an anteroventrally directed rotation of each maxilla about its articulation with the head of the palatine and ethmoid region of the cranium. These two kinds of maxillary movements are probably initiated simultaneously by a posteroventrally directed pull on the anterior maxillo-mandibular ligament caused by contraction of the A_1 portion of the adductor mandibulae (Figs. 312A, 312B, 313Bb, 313Cc; see Alexander, 1967: 47).

Protrusion of the upper jaw during a feeding event enhances the chances of prey capture in two ways: (1) it decreases the surface area of the mouth opening, thereby creating a greater pressure differential and increasing the velocity of water entering through the front of the mouth (following Poiseuille's law regarding the flow of a fluid through a tube; Osse, 1969: 359; Pietsch, 1978b: 259); and (2) it lengthens the oral cavity, thereby causing the suction force to be applied closer to the prey item (Osse 1969: 365).

Closure of the mouth and further protrusion of the premaxillae are presumably accomplished simultaneously by contraction of the entire adductor mandibulae (Figs. 312C, 313Dd). The force provided by this muscle complex rotates the shafts of the maxillae posterodorsally, thereby generating tension in a ligamentous sheath that forms the walls of the oral cavity and interconnects the distal portions of the maxillae and premaxillae. The resulting posteroventrally directed pull on the premaxillae causes a dramatic increase in the displacement of these bones, so that at maximum protrusion the premaxillae make contact with the nearly fully adducted lower jaw; at the same time, the proximal ends of the premaxillae become well separated from the proximal ends of the maxillae, and from the ethmoid region of the cranium (Figs. 312D, 313Ff).

This mechanism, allowing complete closure of the jaws while the oral and opercular cavities remain nearly fully expanded, functions to minimize water loss and the chances for prey escape through the front of the mouth. After the mouth closes, the oral and opercular cavities contract, forcing water out by way of the opercular openings (Figs. 312E, 313Gg). Simultaneously, the premaxillae are drawn back into the closed, nonfeeding position, primarily by the elasticity of the palatopalatine ligament (Motta, 1982: 308).

Perhaps more significant than the actual volume increase of the oral and opercular cavities during feeding is the speed at which this volume is increased during a single feeding event (e.g., see Gibb and Ferry-Graham, 2005). Analyses of high-speed films showing individual feeding sequences (Grobecker and Pietsch, 1979: 1162; Pietsch and Grobecker, 1987: 334) revealed that *Antennarius hispidus* (three individuals filmed at speeds of 800 frames per second) is capable of oral expansion and subsequent prey

---

Fig. 312. (opposite) Diagrammatic views of head of *Antennarius*, showing stages of oral expansion and contraction during a single feeding event, emphasizing the major bony elements, muscles, and ligaments involved in premaxillary protrusion. (**A**) cranium being lifted by contraction of epaxial musculature, premaxilla undergoing anteriorly directed protrusion, and maxilla undergoing dual rotation as a result of lower jaw depression and contraction of A1 portion of adductor mandibulae (corresponding to Fig. 313B). (**B**) more cranial elevation, maxillary rotation, and upper jaw protrusion; lower jaw nearly fully depressed (corresponding to Fig. 313C). (**C**) more cranial elevation and enhanced premaxillary protrusion as a result of contraction of entire adductor mandibulae; lower jaw undergoing adduction (corresponding to Fig. 313D). (**D**) premaxilla fully protruded, making contact with nearly fully adducted lower jaw (corresponding to Fig. 313E). (**E**) oral and opercular cavities contracting; premaxilla and cranium returning to nonfeeding position; lower jaw nearly fully adducted (corresponding to Fig. 313G). (**F**) elements of feeding mechanism in fully closed, nonfeeding position. After Pietsch and Grobecker (1987, fig. 151).

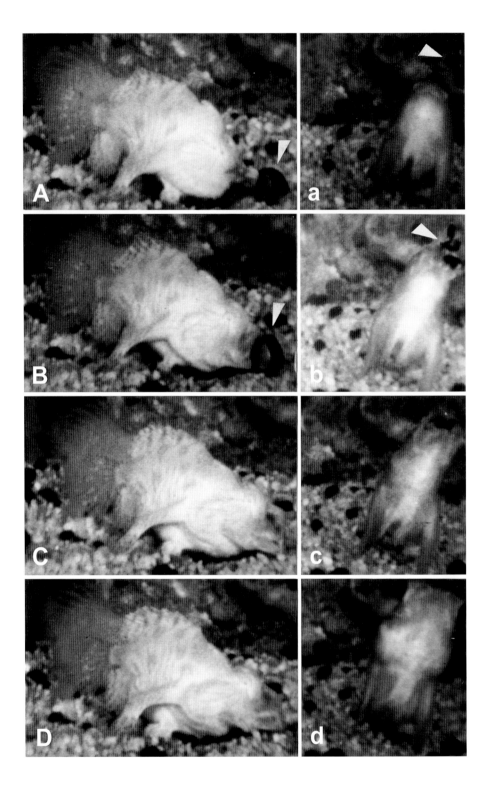

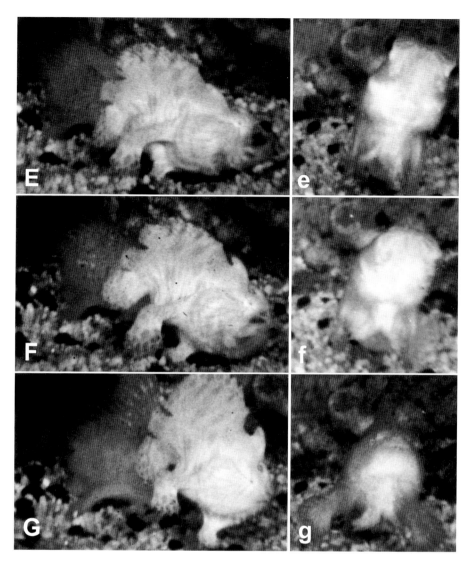

Fig. 313. (*above and opposite*) Selected 16-mm frame-to-print conversions of a single feeding event in *Antennarius hispidus*, filmed at 800 fps, right lateral views (**A–G**) and anteroventral views (**a–g**). (**Aa**) first stage of mouth opening. (**Bb**) prey being drawn toward expanding oral cavity (corresponding to Fig. 312A). (**Cc**) prey totally engulfed (corresponding to Fig. 312B). (**Dd**) slight adduction of lower jaw (corresponding to Fig. 312C). (**Ee**) lower jaw partially adducted and meeting protruded premaxillae (corresponding to Fig. 312D). (**Ff**) contraction of oral cavity, forcing engulfed water and prey into stomach (note expanded stomach in **d**). (**Gg**) release of engulfed water through opercular openings and mouth, causing a "post-feeding jump" (corresponding to Fig. 312E). Arrows indicate position of prey organism, the Whitetail Damselfish, *Dascyllus aruanus*. After Grobecker and Pietsch (1979, fig. 1); courtesy of *Science*, © 1979 American Association for the Advancement of Science.

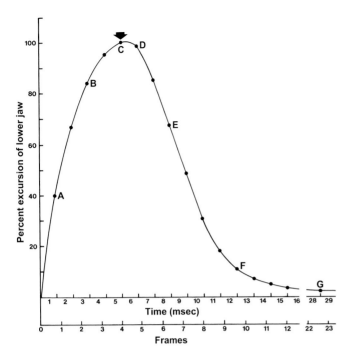

Fig. 314. Percent of excursion of lower jaw versus time (superimposed over cinematographic frames) of fastest single feeding event in *Antennarius hispidus* (Table 24). Letters indicate points on curve that correspond to individual cinematographic frame-to-print conversions shown in Fig. 303. Arrow indicates point at which prey organism is totally engulfed. After Grobecker and Pietsch (1979, fig. 2); courtesy of *Science*, © 1979 American Association for the Advancement of Science.

engulfment ranging from 5.0 to 7.5 msec (Fig. 314). Time sequences for total oral expansion in *A. striatus* (three individuals at 800 frames per second) and *A. maculatus* (three individuals at 1,000 frames per second) were similar, ranging from 3.8 to 10.0 msec, with an average speed for all three species combined of about 7 msec (Table 23). These speeds were unparalleled among the fish taxa that had been examined at the time Pietsch and Grobecker (1987: 334) summarized their findings: the Eurasian Ruffe, *Gymnocephalus cernua*, expands its oral cavity in about 250 msec (Elshoud-Oldenhave and Osse, 1976: 412); the European Perch, *Perca fluviatilis*, requires 40 msec (Osse, 1969: 359); the Freshwater Butterfly Fish, *Pantodon buchholzi*, 16 msec (Kershaw, 1976: 231); and the Stonefish, *Synanceia verrucosa*, 15 msec (Grobecker, 1983: 199). Since then, prey capture speeds comparable to those found in frogfishes have been recorded in only a few syngnathid fishes: an average of 5.8 msec in the Lined Seahorse, *Hippocampus erectus*; and 7.9 msec in the Dusky Pipefish, *Syngnathus floridae* (Bergert and Wainwright, 1997: 568; see also Roos et al., 2009).

In an attempt to understand the mechanism responsible for these extremely rapid feeding sequences in frogfishes, Pietsch and Grobecker (1987: 335) undertook a fine-structural examination of the muscles in *Antennarius striatus* presumed responsible for the major actions of the feeding mechanism, the A_1 and A_2 portions of the adductor mandibulae, the levator operculi, and the hypaxial musculature. The study revealed no significant structural differences in these muscles when compared to those of other vertebrates. An alternative explanation for such speed in feeding may be a sophisticated catapult mechanism in which muscles build up potential energy for a fast release (Alexander, 1982: 113; Roos et al., 2009; Van Wassenbergh et al., 2014). However, electromyographic data, coupled with high-speed cinematography, would be needed to test such a hypothesis.

Alexander (1967: 57) estimated that a teleost using a gape-and-suck feeding mode cannot engulf a prey item that is further from its mouth than about one-fourth of the length of the predator's head. However, Osse (1969: 360), using cinematography (at 32 frames per second) to record feeding in the European Perch (*Perca fluviatilis*), concluded that engulfment is possible when the prey is separated from the predator by a distance equal to one-half the head length. But high-speed cinematographic analyses of single feeding events in *Antennarius hispidus* (three individuals, 76–84 mm SL, at 800 frames per second) revealed that this species can remove prey from the surrounding water at a maximum distance slightly *greater* than the length of the predator's head.

Pietsch and Grobecker's (1987: 336) analyses of single feeding events filmed at high speed revealed that frogfishes utilize a functional repertoire of feeding modes that depends on prey location and/or size (Elshoud-Oldenhave and Osse, 1976: 420; Liem, 1978: 343, 1979: 115; Lauder and Liem, 1980: 387; Lauder, 1981: 158; Grobecker, 1983: 194). Two distinct feeding modes were distinguished: an exaggerated seizure and an abbreviated seizure. The exaggerated seizure was used when prey items were positioned in the outermost reaches of the strike zone (Fig. 315) or when the prey items were of a relatively large size (standard length of prey more than 50% that of the frogfish). This mode exhibited the most extreme movements of bony elements (i.e., cranial elevation, lower-jaw depression, maxillary swing, and premaxillary protrusion; Figs. 308, 312). During maximum depression of the lower jaw, the cranium was elevated about 30° relative to the body axis, and the maxillae were rotated through an arc of about 40° relative to the cranium.

The abbreviated seizure was used when smaller prey items (standard length of prey less than 50% that of the frogfish) were within closer proximity to the mouth (Fig. 316). The

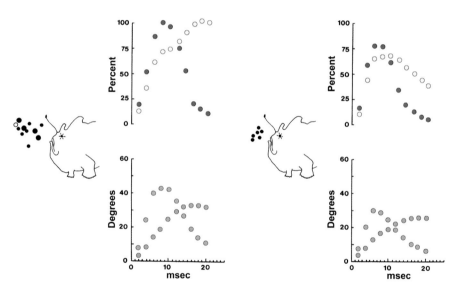

Fig. 315. Exaggerated feeding mode of *Antennarius hispidus*. In left lateral view of fish, large circles represent large prey items, small circles represent small prey items; closed circles represent position of prey at time of successful seizure, open circle represents unsuccessful attempt. Graphs show kinematic profiles of mouth gape (red), premaxillary protrusion (yellow), cranial elevation (blue), and anteroventral swing of maxillae (green). After Pietsch and Grobecker (1987, fig. 154).

Fig. 316. Abbreviated feeding mode of *Antennarius hispidus*. In left lateral view of fish, circles represent position of prey item at time of successful seizure. Graphs show kinematic profiles of mouth gape (red), premaxillary protrusion (yellow), cranial elevation (blue), and anteroventral swing of maxillae (green). After Pietsch and Grobecker (1987, fig. 155).

cranium was elevated only about 25° relative to the body axis, the maxillae were rotated only about 30° relative to the cranium, the mouth attained only about 75% of full gape, and the premaxillae attained only about 70% of their full protrusion. Thus, all four elemental movements were significantly less than in the exaggerated feeding mode (p = .05).

PREY HANDLING

Frogfishes have an unusual mechanism for passing a prey item into the stomach. As the oral and opercular cavities are contracted, a large quantity of the water surrounding the prey item is forced directly into the stomach, in an average of 16 msec (Pietsch and Grobecker, 1987: 336). This is accomplished by clamping the modified opercular openings shut with the pectoral fins (Fig. 313Aa–Ff). This closing of the opercular openings blocks escape routes for the engulfed water and sustains the positive pressure needed to force the volume of water back into the expandable, saclike stomach. Once the prey is in the stomach, the sphincter muscle of the esophagus presumably contracts to maintain the prey item inside the stomach, and the pectoral fins relax, allowing water to escape through the opercular openings and the front of the mouth (Fig. 313Gg).

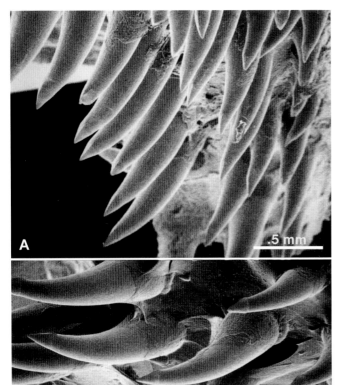

In the event that the prey was not fully engulfed, but held by the recurved, conical teeth of the dentaries, premaxillae, vomer, and palatines (Fig. 317A), the frogfish initiated a second feeding event, an abbreviated suction that resembles the movements observed in the abbreviated feeding mode described above. Once the prey is inside the mouth cavity, deglutition is achieved by a racking motion of the well-developed upper and lower pharyngeal teeth (Fig. 317B). Large prey items, sometimes larger than the frogfish itself (Fig. 318), are transported to the stomach in this more conventional way (see Grobecker, 1981).

After capture, rejection of prey occurred when the prey item was too large to be fully engulfed, or when it became lodged in the oral cavity. The rejection of prey was accomplished by a quick compression of opercular and oral cavities with the mouth agape.

The last behavior observed, often occurring just prior to a return to the nonfeeding, lie-in-wait position, was a series of cough-like movements, each resembling an abridged version of the movements used in the rejection of prey. This forcing of water out of the oral and opercular cavities functioned to expel bits of incidentally engulfed substrate, as well as loose scales from the

Fig. 317. Scanning electron micrographs of (**A**) premaxillary and (**B**) upper pharyngeal teeth of *Antennarius striatus*, UW 21016, 88 mm SL. After Pietsch and Grobecker (1987, fig. 156).

prey item. The movement of the cough may also aid in realigning the bony and soft-tissue elements employed during the strike phase.

All of the observed events associated with predation and prey handling in a generalized frogfish are summarized in Figure 319.

DISCUSSION

As in other prey-seizure mechanisms (e.g., those of the mantid, *Parastagmatoptera unipunctata*, Mittelstaedt, 1957; Common Cuttlefish, *Sepia officinalis*, Messenger, 1968; Largemouth Bass, *Micropterus salmoides*, Nyberg, 1971; and the Stonefish, *Synanceia verrucosa*, Grobecker, 1983), orientation to prey in antennariids is believed to be primarily a visually controlled, "closed-loop system" (e.g., see Ejaz et al., 2013). The movement of the prey is followed visually, the strike zone is sometimes adjusted, and in some cases the prey is manipulated toward the strike zone (see Luring Behavior, p. 424). But prey seizure in frogfishes is so rapid (mouth reaching full gape in as little as 4 msec, with no time for visual feedback of information regarding the position of the mouth parts) that it must be classed as an "open-loop system"; that is, a motor activity preprogrammed during orientation and controlled by an internal oscillator, as described in cichlids by Liem (1978: 356) and hypothesized for the Stonefish by Grobecker (1983: 199).

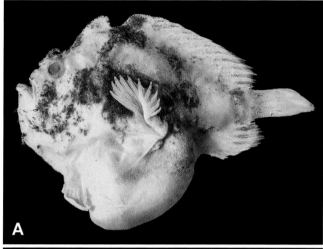

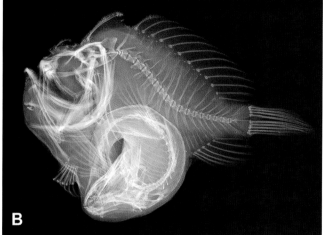

Fig. 318. *Abantennarius coccineus* (Lesson). (**A**, **B**) USNM 420800, 58 mm SL, Mururoa Atoll, French Polynesia, with swallowed prey, the body length of which is greater than that of the frogfish. Photo and radiograph by Sandra J. Raredon; courtesy of Lisa Palmer and the Smithsonian Institution, NMNH, Division of Fishes, Washington, DC.

In most predators, the overall rate of unsuccessful attacks has been estimated to be about 90% (Salt, 1967: 119). Yet the failure rates in the preprogrammed attacks of squid, mantids, Stonefish, and Largemouth Bass were extremely low, all around 10%—a value that compares well with that found in frogfishes (9%). Such a limitation, although quite small compared to that of most other predators, may be the result of the physiological limitations of exercising neurological control over relatively large bodily movements that occur within a duration of a few milliseconds (Nyberg, 1971: 140).

The exaggerated and abbreviated seizures are well-defined feeding modes that depend on prey location and/or size (see Strike Behavior, p. 439). The abbreviated seizure probably evolved as an energy-saving tactic that allows for capture of smaller prey items that are within close proximity; the more violent movements and greater energy consumption of the exaggerated seizure are apparently reserved for larger, more distant prey.

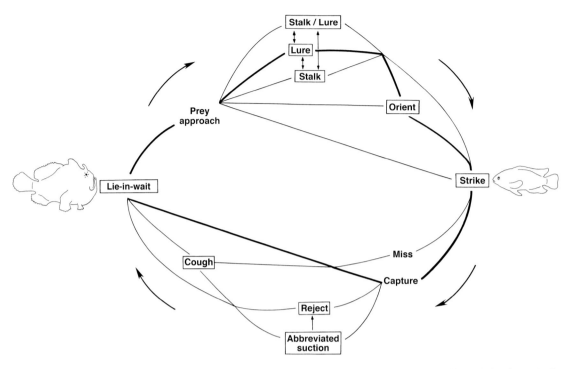

Fig. 319. Flow diagram showing sequence of possible events associated with predation (upper half of diagram) and prey handling (lower half of diagram) in a generalized antennariid. Thickened line represents the most often utilized series of behaviors and events. After Pietsch and Grobecker (1987, fig. 158).

Schoener (1971: 379) predicted that below a critical prey length, handling time should remain constant, and above the critical prey length, handling time should increase sharply. Analyses of prey length and the handling time of prey in frogfishes support Schoener's prediction. Small prey items were forced directly into the stomach in an average of 23 msec, whereas larger prey items were transported to the stomach by a more conventional mechanism (a racking motion of the tooth-bearing upper pharyngeal bones) in 3–40 sec, a 190- to 2,500-fold increase in handling time.

The three most important aspects of successful gape-and-suck prey capture—the amount of oral expansion, the rate of oral expansion, and the extent of protrusion of the upper jaw—are highly accentuated in lie-in-wait predators (Nyberg, 1971: 142; Liem, 1978: 341; Grobecker and Pietsch, 1979; Grobecker, 1983: 199). Unlike fast-swimming pursuit predators that incorporate body speed in engulfing prey, the lie-in-wait predator depends on ultrafast expansion of the mouth cavity and highly protrusible mouth parts in order to surprise and more successfully capture prey. Rapid prey capture provides the added advantage of visually obscuring or concealing the feeding activity of the predator to other potential prey; even fishes in close proximity to one another, for example those involved in schooling behavior, may overlook an attack on one of their members (Neill and Cullen, 1974: 551; Curio, 1976: 148).

The ability to rapidly expand and protrude the mouth parts in frogfishes is a major, required component of a highly successful feeding strategy based on predator immobility and aggressive mimicry (Pietsch and Grobecker, 1978; Grobecker and Pietsch, 1979; Grobecker, 1981). We now realize that this ability is widespread within a host of other, in many cases, relatively unrelated teleosts that use a similar mode of energy capture (e.g., Liem, 1970; Grobecker, 1983; Muller and Osse, 1984; Aerts et al., 1987; Aerts, 1989, 1990; Muller, 1989; Sazima, 2002; Sazima et al., 2005). It follows that an ultrafast feeding mechanism may in fact be a necessary prerequisite for the evolution of this kind of feeding. Since aggressive mimetic devices used for the purposes of capturing prey are believed to be widespread especially among acanthomorph teleosts (e.g., Liem, 1970; Ormond, 1980; Randall and Kuiter, 1989; Randall, 2005b; Cheney and Marshall, 2009; Catarino and Zuanon, 2010; Cheney, 2010), it seems evident that the evolution of ultrafast feeding mechanisms has been important in the proliferation of this largest and most morphologically diverse group of teleosts (Grobecker and Pietsch, 1979).

Voracity and Cannibalism

Frogfishes must be among the most voracious of fishes, and anyone who has maintained them in aquaria knows that they will eat nearly anything and everything that moves, from prey items much smaller than themselves to fishes considerably greater than their own standard length (Smith, 1898: 109; Gudger, 1905: 842; Gill, 1909: 606; Longley and Hildebrand, 1941: 303; Breder, 1949: 94; Straughan, 1954: 277; Schultz, 1957: 52; Randall, 1968: 291; Grobecker and Pietsch, 1979; Grobecker, 1981; Michael, 1998: 331, 336). Gill (1909: 606), in describing the feeding habits of *Histrio histrio*, wrote that if "careless or unlucky animals approach too near . . . the quiescent but hungry fish is stirred instantaneously into vigorous action. It leaps upon its prey as quickly as a tiger would upon its own." Straughan (1954: 277) observed individuals of this species "foolishly try to swallow a fish twice their size and then finally give up, spitting out the lifeless victim and swimming about the aquarium in a rage." Dooley (1972: 24) listed *H. histrio* as a major predator of the *Sargassum* complex, second only to jacks, filefishes, and triggerfishes. Our own experience with aquarium-held individuals of *Antennarius* has shown that once acclimated they will eat as much as, and whatever, is offered, bloating themselves to what appears to be dangerous levels (Straughan, 1954: 278). An individual of 100 mm SL can easily consume two or three goldfish (*Carassius auratus*) of 40 mm SL per day, for weeks on end with no ill effect.

Frogfishes are also such indiscriminate feeders that they will consume large numbers of their own kind. Reports of cannibalism in *Histrio histrio* are numerous (e.g., Smith, 1898; Gill, 1909: 607; Gordon, 1938: 20; Breder, 1949: 94; Mosher, 1954: 141; Straughan, 1954: 277; Gordon, 1955: 389; Böhlke and Chaplin, 1968: 717; Friese, 1973: 33). It is not at all unusual to find a dozen or more small individuals of *H. histrio* in the stomach of a larger one. If given half a chance, and especially if not kept well-fed, aquarium-held individuals of *Antennarius* will also eat each other; in one cannibalistic encounter, Pietsch and Grobecker (1987: 341) witnessed the demise of one of two nearly equal-sized individuals of *A. striatus* left unattended overnight in the same tank.

Defensive Behavior

To defend themselves, frogfishes turn usually to threats or aggression. Body inflation is an often-used response but retreat is apparently rare, and poisons and venoms are probably nonexistent.

Aggression

Under natural conditions, and except during periods of spawning, frogfishes are nongregarious, solitary predators. In the laboratory, individuals of many species, especially

Histrio histrio, will not tolerate the near approach of another. Thus, to avoid casualties and deaths through aggressive contests or loss through cannibalism (see Voracity and Cannibalism, above), either the frogfishes must be kept separately or their aquaria must be large and the animals kept well-fed.

A number of observers have commented on the aggressive behavior of frogfishes. Gudger (1905: 842) described daily combats between a pair of *Histrio histrio*, the smaller of the two suffering considerably, its filamentous appendages and the ends of its fins continually bitten off until the animal was finally killed by its tankmate. Gill (1909: 605) referred to *H. histrio* as, "in fact, a quarrelsome fish." Mosher (1954: 150) witnessed numerous ferocious attacks of males against females following courtship and spawning behavior, which often resulted in the death of the female. Finally, Gordon (1955: 389) wrote that captive specimens of *H. histrio* "bite and tear each other, their fleshy head and body ornaments are ripped to shreds. Their hand-like and foot-like fins become frayed and their delicate fin-rays protrude like broken bones."

Observations of defensive interaction, displayed by laboratory-maintained individuals of *Antennarius hispidus*, *A. striatus*, and *A. maculatus*, involved considerably less violence than that described for *Histrio histrio* (Pietsch and Grobecker, 1987: 342). In most instances, interspecific and intraspecific aggression seemed to be of equal intensity. On approach by another frogfish of similar size, the defender often made an attempt to position itself between its "foraging site" and the intruder. Whenever the intruder came within a distance roughly equal to its own body length, a conspicuous and complex lateral display was initiated by the defender. This display consisted of (1) elevation of the body by the defender pushing off the substrate with the pectoral fins; (2) extreme spreading of all median fins; (3) a blushing of coloration over the entire head, body, and fins; (4) the mouth set agape; and (5) a rapid quivering of the body while leaning in the direction of the intruder. In nearly all cases, the response from the intruder was rather passive, consisting of little more than slowly moving away from the defended site. However, in several instances involving intraspecific encounters, the intruder responded with a nearly identical display, resulting momentarily in a standoff until one or the other backed off and slowly moved away.

Body Inflation

Numerous authors have commented on the ability of frogfishes to expand their stomachs enormously by swallowing large quantities of air or water, an adaptation usually attributed only to the pufferfishes (families Diodontidae and Tetraodontidae; see Wainwright and Turingan, 1997: 507), the four species of the filefish genus *Brachaluteres* (Monacanthidae; Clark and Gohar, 1953: 46; Hutchins and Swainston, 1985: 57), a goby (*Sufflogobius bibarbatus*, Gobiidae; Smith, 1956: 714), and the swellshark (*Cephaloscyllium ventriosum*, family Scyliorhinidae; Compagno, 1984: 304). The increased size of the inflated body is believed to present a more formidable display to ward off potential predators, to make these slow-swimming fishes more difficult to swallow, or to defend a feeding or shelter site from intra- and interspecific competitors; or, as in the case of the swellshark, a defense mechanism by which it wedges itself into tight caves and crevices (Randall, 1967: 824; Keenleyside, 1979: 59; Myers, 1989: 266, 268; Wainwright et al., 1995: 614; Ferry-Graham, 1997: 1267; Wainwright and Turingan, 1997: 507).

The ability of frogfishes to inflate their bodies was first recognized by Marcgrave (1648: 150) who, in a description of *Guaperua* (i.e., *A. multiocellatus*), wrote: "while swimming it singularly spreads out its fins and inflates itself so that it appears round like a ball." This passage was later paraphrased by Linnaeus (1727–1730; Fig. 3A), and the observation was subsequently confirmed by Commerson (c. 1770; MSS 889, 891), Cuvier (1817b: 422), Valenciennes (1837: 389), Swainson (1838: 202), Günther (1861a: 184), Day (1876: 271), and a host of twentieth-century authors, including Jordan (1902: 367), Gregory (1933: 388), Gordon (1938: 20), Schmitt (in Longley and Hildebrand, 1941: 304, 305), Barbour (1942: 31), Schmitt and Longley (in Schultz, 1957: 52), Böhlke and Chaplin (1968: 714, 718), Randall (1968: 291), Halstead (1978: 354), and Michael (1998: 355). In the most significant of these many reports, Gordon (1938: 20) wrote that *Histrio histrio* sometimes "uses the quick gulping technique for self-defense. If it is attacked by a larger fish, [it] throws open its jaws, swallows water as it is on the point of being devoured, and instantly pumps itself up to an unexpected size. Thus, the swallower is forced to cough up the swallowee." In summary, a survey of the literature strongly suggests that body inflation is an often-used defensive response in frogfishes.

It is true that at least some species of frogfishes (e.g., *Fowlerichthys ocellatus, F. radiosus, Antennarius striatus, A. hispidus,* and *Histrio histrio*), either intentionally or accidentally, inflate themselves with air. Among a number of examples that could be cited, a convincing post on the Internet (Hendo, 2017) dated 5 January 2017, described recurring observations by Craig Hendo off Laurieton, New South Wales, Australia, of air inflation in *Antennarius striatus*: "The anglerfish was floating on the surface at night on a run out tide. When collected it floated in a bucket of water for a few minutes, then expelled air through its mouth and became negatively buoyant. I have observed this previously." Relevant here also are comments by Bruce A. Carlson (personal communication, 11 April 2018), former director of the Waikiki Aquarium and chief science officer of the Georgia Aquarium: "When collecting *A. commerson* in Hawaii for the Waikiki Aquarium, collectors were always warned not to take the animal out of the water or it would gulp air. We were never sure if this would harm the animal or not so we took precautions based on our past experiences."

Underwater photographer and author Scott W. Michael has witnessed body inflation in four species of frogfishes (*Antennarius hispidus, A. commerson, A. maculatus,* and *Abantennarius nummifer*), all while held in captivity (Scott W. Michael, personal communication, 14 May 2018):

> I kept a specimen of *A. hispidus* and a couple of *A. nummifer* for about 6 months while doing some research on luring behavior and ambush site selection. I observed *A. nummifer* inflate on three different occasions. This individual had been in the tank for several weeks and had acclimated well (i.e., it was feeding regularly and exhibiting a normal respiratory rate). I was recording data on 21 February 1985 when it inflated, just as it was beginning to stalk damselfish prey. In my notes, I wrote that it swallowed 10 times, inflating the stomach, and as it did so the damsels investigated the turgid frogfish. The whole bout, from the time of initial ingestion of water until the time when it was expelled from the stomach, lasted 32 seconds. After it had deflated, it immediately began to lure (the angling bout lasted 64 seconds). The second bout took place on 29 February 1985. This time, I simply reported that it sucked in water, did no luring while inflated, but underwent a color-change from an overall beige to a bright salmon (almost red) hue on top of the head while the body took on a darker tone. I observed the final bout on 2 April 1985. During this episode, the frogfish was in repose when it suddenly started to inflate. I wrote "inhales water into abdomen with successive gulps, interspersed with slight exhalations. The belly was about 1.75 inches in diameter and perfectly spherical." In my notes, I reported also that in all three cases,

there was no apparent harassment to illicit this behavior, neither from me or any tank-mate. In all cases, the fish showed no sign of stress.

Finally, the most interesting frogfish inflation episode I witnessed was when a captive *Antennarius pictus* attempted to ingest a smaller *A. maculatus*. It seized the head of the smaller frogfish, which immediately began to inflate with water. The pair ended up floating about the tank until the *A. pictus* spat out the inflated *A. maculatus*, which returned to the bottom of the tank and expelled the water from its stomach.

Surprisingly, however, with so much evidence to the contrary, we have never observed similar behavior. In our experience, initiating inflation when a frogfish is removed from the water nearly always takes a considerable amount of poking and manipulation on the part of the experimenter. Furthermore, in all our many years of maintaining living antennariids under laboratory conditions, often harassing the animals well beyond what might be expected under natural circumstances, we have not witnessed a single case of body inflation due to swallowing water. Nevertheless, the many witnesses of this phenomenon cannot be wrong—it does happen. One might conclude that this behavior occurs only under stress, brought on by captivity, but, in fact, irrefutable proof in the field has been provided by diver and photographer Stephane Bailliez. In a series of several dozen photographs taken on 23 April 2009 at Anilao, Batangas, Philippines, Bailliez recorded a small frogfish (*Nudiantennarius subteres*) desperately struggling (apparently successfully) to repel an attack from a hungry lizardfish (*Synodus jaculum*) by inflating its body well beyond the size of the predator's mouth. Further evidence comes from unpublished videos of *Antennarius maculatus* taken at the La Paz dive site, Southern Leyte, Philippines, in March 2015, and *A. striatus* made in Lembeh Strait, Sulawesi, in November 2016, by well-known underwater photographer Christa Holdt. In both cases, stress caused by the close proximity of the diver resulted in "hyperventilation" and "distinct swelling of the belly" (Christa Holdt, personal communication, 26 and 30 January 2017).

Günther (1861a: 184) went further than most by suggesting that body inflation provides a mechanism for dispersal: frogfishes are "enabled, by filling the spacious stomach with air, to sustain themselves on the surface of the water. They are therefore found in the open sea as well as near the coasts, and being bad swimmers, are driven with the currents into which they happen to fall. Thus it is a natural consequence that at least some of the species should have a very wide geographical range, not only over the Atlantic, but also over the Indian Ocean."

Jordan (1902: 367), in apparent reference to Günther (1861a), added that they are "therefore widely dispersed by the currents of the sea." Although some antennariids, particularly *Histrio histrio*, may drift on the surface in an inflated state for short distances, it seems very unlikely that geographic distributions have been altered substantially in this way.

Venom and Poison

There are scattered indications in the scientific literature that frogfishes might be venomous, or at least toxic. The Reverend Samuel James Whitmee (1875: 545), in a brief article on the habits of *Antennarius*, wrote that the natives of Samoa

frequently get "stung" by the third dorsal spine of this fish, when they happen to pick up a block to which it is attached, before they are aware of its presence. It causes very great agony, which usually lasts several hours, and sometimes two or three days.

Another fish, which I believe is also an *Antennarius*, but which I have not yet examined, produces effects much more alarming than this one. I have seen the hands and feet of natives swollen and greatly inflamed by a prick from the larger species, and have seen strong men weeping and groaning like children with the agony it caused. Sometimes the effect produced by a prick from this lasts for weeks.

As certain as Whitmee (1875) seemed to be in describing the harmful wounds inflicted by *Antennarius*, the dorsal spines of antennariids are not at all equipped to deliver such a "sting"; in all of our years of handling living frogfishes of many species we have never been injured by them in any way. As indicated by Gill (1909: 601), the guilty fish in Whitmee's (1875) description was probably not an anglerfish at all, but most likely a stonefish of the genus *Synanceia*.

As for toxicity, a number of workers have listed *Histrio histrio* as ciguatoxic (Phisalix, 1922: 582, 609; Maass-Berlin, 1937: 198; Bagnis et al., 1970: 88; Halstead, 1978: 348). These reports all appear to have originated from Phisalix (1922), who simply listed *H. histrio* with a number of other fish species suspected of causing ciguatera poisoning. However, if in fact antennariids are involved in ciguatoxicity, they could hardly have much impact on the health of human populations since they are rarely if ever consumed, even by local fishermen in remote areas. In addition to being small, nearly all of them rough and thick-skinned, and relatively rare in natural populations, antennariids do not form part of the diets of native peoples primarily because they are strongly associated with venomous scorpaeniform fishes. In fact, Samoan names for antennariids are generally the same as those given to scorpaenids; according to Wass (1984: 7), individuals less than 80 mm SL are called *la'otale*, larger individuals are called *nofu*.

Locomotion

Naturalists have long been intrigued by the unique tetrapod-like locomotion displayed by frogfishes. Myths concerning the amphibious nature of these animals date at least back to 1719 when Louis Renard, in his *Poissons, Ecrevisses et Crabes*, quoted the artist Samuel Fallours: "I kept it alive for three days in my house; it followed me everywhere with great familiarity, much like a little dog" (translated from the French; see Pietsch, 1984a: 64, 1995b, 1: 98; Fig. 3B). This story was later repeated by François Valentijn (1726: 370), and uncritically accepted nearly a century later by such eminent scientists as Philibert Commerson (MS 889, 891) (who was the first to recognize frogfishes as "espèces d'amphibies" in his manuscripts; see Historical Perspective, p. 9), Cuvier (1817b: 421), and Valenciennes (1837: 389). Valenciennes provided an appealing, but largely anecdotal, discussion of the locomotory behavior of antennariids:

The position of the paired fins gives them the appearance of having four feet, the pelvic fins in front, the pectoral fins behind. The tiny gill openings, represented only by a round aperture hidden in the axil of the pectoral-fin lobe, enable them to remain out of water for extended periods of time. This, in turn, enables them to crawl out over seaweed and mud, and thus they pursue their prey.

William Swainson (1838: 203), in his *Natural History and Classification of Fishes, Amphibians and Reptiles*, added that "their nature is so truly amphibious, that they can live out of water for two or three days; they are, in fact, so tenacious of life, that they have been transported alive from the Tropics to Holland, where they are sold as high as twelve ducats apiece."

Even as late as the second half of the nineteenth century, the alleged terrestrial habits of antennariids were retold. According to James Emerson Tennent, in his *Natural History of Ceylon* published in 1861 (p. 331), "the bones of the carpus form arms that support the pectoral fins, and enable these fishes to walk along the moist ground, almost like quadrupeds." Francis Day (1876: 271) later expanded on Tennent's remarks: "Their pediculated pectoral fins allow them to walk or hop over moist ground, or slimy rocks in quest of their prey, and even clasp pieces of wood or seaweeds, attached to which they often become carried away from the shore by currents, and are sometimes observed far out at sea."

Frogfishes do use their limb-like pectoral and pelvic fins to accomplish a kind of locomotion that resembles a tetrapod crawl, but their ability to remain out of water for any length of time, let alone move about without the buoyancy provided by their natural medium, is drastically limited. A living frogfish removed from an aquarium and placed on a flat surface cuts a rather poor figure, its body more or less immobile and spreading out pancake-like under its own weight.

Numerous authors have remarked on the resemblance of the paired fins of antennariids to tetrapod limbs (e.g., Swainson, 1838: 203; Whitmee, 1875: 545; Jones, 1879: 363; Gill, 1909: 600, 602, 606; Gregory, 1928: 409, 1933: 388; Gordon, 1938: 20; Barbour, 1942: 26; Whitley, 1949: 400; Gordon, 1955: 389; Friese, 1973: 30; Pietsch and Grobecker, 1985: 13; Edwards, 1989: 249, figs. 16–18; Schneidewind, 2005a: 23; Pietsch et al., 2009a: 41; Renous et al., 2011: 98; Arnold et al., 2014: 538). In one of the earliest of these descriptions, Swainson (1838: 203) wrote:

> The frogfishes have each of the pectorals supported by two [actually three] bones, analogous to the radius and ulna of the frogs, although, in reality, they belong to the carpus, and which, in this group, are longer than in any other. The ventrals [pelvic fins], again, are placed much before the pectorals, and stand, as it were, upon peduncles; they are thus enabled to perform the office of feet. The effect of this singular organization is, that these fishes can creep almost like small quadrupeds; the pectorals, from their position, performing the office of hind feet.

When at rest on open substrate, the proximal portion (the radials) of the pectoral fin (here referred to as the pectoral lobe) of a frogfish is held at a right angle to the body of the fish and parallel to the substrate, whereas the more distal, rayed portion of the fin is bent at a right angle to make contact with the substrate. The rays themselves bend at the point of contact with the substrate to form a division into "forearm" and "hand." The handlike distal portion of the fin is able to conform easily to the irregular surface of the substrate, and even has the ability to grasp solid objects (Gill, 1909: 606; Breder, 1949: 95; Le Danois, 1964: 81). Finally, to make the analogy complete, in many species (e.g., *Antennarius biocellatus*, *A. pauciradiatus*, *A. randalli*, *Abantennarius dorehensis*, *A. nummifer*, and *Nudiantennarius subteres*; Pietsch and Grobecker, 1987, fig. 77), the distal tips of the pectoral rays project beyond the skin-covered portion of the fin like short "fingers." Thus, these jointed, armlike and handlike pectoral fins are used in conjunction with the pelvic fins to "walk" in a typical quadruped-like fashion over flat bottoms, to clamber over and among rocks and coral (or, as in the case of *Histrio histrio*, among the fronds of *Sargassum*), and to brace or wedge the animals within holes and crevices provided by complex structure.

The following description and discussion of locomotion in frogfishes is the result of a thorough review of the literature on the subject, and Pietsch and Grobecker's (1987) osteological and myological studies and their behavioral observations of numerous aquarium-held individuals of eight species: *Antennarius hispidus*, *A. striatus*, *A. commerson*, *A. maculatus*, *Antennatus tuberosus*, *Abantennarius coccineus*, *A. sanguineus*, and *Lophiocharon trisignatus*. For a list of species examined osteologically, see Pietsch and Grobecker (1987: 22); myological data were obtained primarily from dissections of two species, *Antennarius hispidus* and *A. striatus*. The discussion has also benefited from personal communication with James L. Edwards, who generously shared results of his investigation of locomotor convergence of anglerfishes and tetrapods (see Edwards, 1977, 1989).

Functional Anatomy of the Pectoral Girdle and Fin

The osteology and myology of the lophiiform pectoral girdle and fin were described and figured in *Lophius americanus* by Eaton et al. (1954) and for *Fowlerichthys senegalensis* by Monod (1960: 648, figs. 24–31; see also Starks, 1930: 234, fig. 38, and Pietsch, 1981b: 396, 411, fig. 14; Edwards, 1989: 249, figs. 16–18). In antennariids the pectoral fin is supported by three pectoral radials. The two dorsalmost radials are similar in size and shape. The third or ventralmost radial is considerably larger; its expanded distal portion bears the articulating bases of the pectoral-fin rays, each ray associated with a small, cartilaginous distal radial. Proximally, the three radials converge at an unusually narrow point of articulation with an extremely small scapula and only slightly larger coracoid, the latter two elements united by cartilage (see Starks, 1930: 89; Pietsch, 1981b: 396, fig. 14). This narrow point of articulation provides for extreme rotational movement of the pectoral lobe relative to the pectoral girdle.

As noted by Winterbottom (1974: 271), the musculature of the pectoral fin of lophiiforms is considerably modified relative to the generalized state found among teleosts. Most neoteleosts possess only a single pair of muscles on either side of the pectoral girdle, two abductors and two adductors; many also have a coracoradialis, connecting the posterodorsal face of the coracoid to the ventralmost pectoral radial. In sharp contrast, antennariids have at least 14 well-differentiated pectoral muscles (Monod, 1960: 655, figs. 32–45): eight abductors (four superficial and four deep), five adductors (one superficial and four deep), and a coracoradialis.

When a frogfish is swimming rapidly in mid-water, its pectoral fins are held back against the body in the typical teleost position, but when it is associated with the bottom substrate, the side of the fin held against the body while swimming becomes the dorsal surface of the fin when "walking." Thus, during "walking," the muscles homologous to the adductors in other teleosts are located posterodorsally on the fin, the abductors anteroventrally. Although electromyological data are unavailable to verify proposed functions extrapolated purely from these anatomical observations, the abductors are at least in the correct position to protract the pectoral lobe, flex the proximal portion of the pectoral fin, and serve as adductors and opposers of the fin rays. The adductors are, on the other hand, in a position to function primarily as retractors of the pectoral lobe, extensors of the proximal portion of the pectoral fin, and abductors of the fin rays. The coracoradialis apparently functions to retract and rotate the pectoral lobe.

Benthic Locomotion and Tetrapod-Like Gaits

During frogfish locomotion across open substrate, the greater part of the propulsive force is created by cyclic movements of the pectoral fins, a distinct propulsive stroke (Fig. 320, frames 1, 19, and 49) followed by a recovery stroke (Fig. 320, frames 57, 69, and 94). The

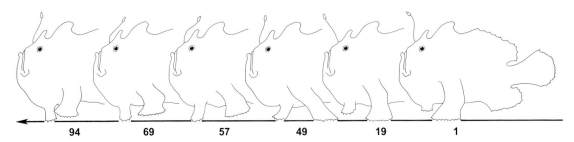

Fig. 320. Progressive changes in the position of pectoral and pelvic fins of *Antennarius striatus* moving forward by amphipedal progression, traced from selected frames of film exposed at 32 fps. Numbers correspond to individual frames of film, each frame representing a time lapse of 0.035 sec. After Pietsch and Grobecker (1987, fig. 159).

pelvic fins, positioned close together beneath the throat, create only a minor propulsive force, functioning primarily as a third point of support for the body.

Schematic fin-fall patterns made from analyses of films exposed at speeds of 32 frames per second illustrate two basic tetrapod-like gaits (Fig. 321A, B). The first, most commonly observed gait, may be categorized as amphipedal progression or "crutching," as described for *Periophthalmus koelreuteri* by Harris (1960: 123; Fig. 321A). Harris compared this kind of locomotion with that of a disabled person swinging along on a pair of crutches. The crutch-user swings his body forward on the crutches and, at the end of the stroke, transfers his weight to his weakened legs while the crutches are carried forward in synchrony to make ready for the next propulsion cycle. In a similar manner, the pectoral fins of frogfishes act as crutches to carry the body forward. At the end of the "crutching" stroke the weight of the body is transferred to the pelvic fins, which function in much the same way as the legs of the crutch-user. As measured in these analyses, the speed of forward progress using the amphipedal gait varies from less than 0.10 to 0.30 standard length per second.

The second tetrapod-like gait employed by benthic frogfishes may be called a "modified walk" (Fig. 321B). This method of locomotion approaches to some extent the characteristic ambulatory rhythm of tetrapods, in which the matching limbs are moved alternately in a definite time sequence (Gray, 1968: 262). But the role of the pelvic fins during the "walk," unlike that in amphipedal progression, is minimal; they contribute only a minor supporting function and no propulsive function (and thus are not represented in Fig. 321B). The "modified walk," one pectoral fin reaching forward, then the other, the pelvic fins contributing only to stability, is used primarily for clambering over rocks and coral, but is also often employed when stalking prey (see Feeding Behavior and Biomechanics, p. 437). Speeds reached during the "walk" vary from approximately 0.30 to 1.0 standard length per second.

It is important to point out that in these two approximations of tetrapod gaits, the sequence of fin movement and the fin-fall patterns are not as uniformly repeated as are the patterns of limb movement in terrestrial vertebrates (Hildebrand, 1974: 510). True tetrapods maintain precise, rhythmic movement in response to gravity. In an aquatic medium, frogfishes achieve near-neutral buoyancy, and thus are able to rely less on precise sequential locomotory patterns. Thus, one should not expect to see antennariids maintain either of these two locomotory modes for any great length of time; combinations, as well as modifications of the two basic gaits, are also often observed.

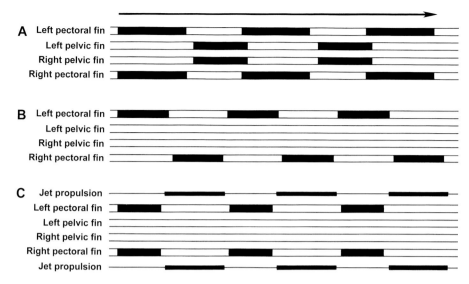

Fig. 321. Schematic fin-fall patterns of a generalized antennariid, showing three different modes of locomotion. (**A**) amphipedal progression or "crutching." (**B**) "modified walk" (the pelvic fins have only a minor supporting function, coming into contact with the substrate only randomly, and are thus not represented). (**C**) combination of amphipedal progression and jet propulsion (the pelvic fins have only a minor planing function and are therefore not represented). After Pietsch and Grobecker (1987, fig. 160).

Pelagic Locomotion and Jet Propulsion

Frogfishes display three distinct locomotory modes when moving through open water: (1) subcarangiform; (2) jet propulsion; and (3) a "kick-and-glide" (Videler, 1981: 10; Jayne and Lauder, 1996: 645; Flammang and Lauder, 2009: 279; Lauder, 2015: 522). Subcarangiform swimming, the mode most often observed in frogfishes, involves a simple series of rhythmic undulations of the body and caudal fin, with all other fins folded back flat against the body (Mowbray, in Barbour, 1942: 26). The movements are similar to those of a swimming perch or trout, the major difference being the low amplitude of the motion of the anterior portion of the body (for details and figures, see Webb, 1975: 40; Lauder, 2015: 522). The body of the frogfish is undulated between one-half and one full wavelength, propelling the animal forward at a speed of up to about one standard length per second.

Jet propulsion is a second form of swimming utilized by frogfishes. This mode of locomotion is surprisingly widespread among aquatic animals, and used for normal swimming as well as an escape response: for example dragon fly larvae of the genus *Aeshna* (Hughes and Mill, 1966; Mill and Pickard, 1975); the scallops *Pecten* and *Chlamys* (Moore and Trueman, 1971); the opisthobranch *Notarchus* (Martin, 1966); thaliacean tunicates *Pyrosoma* (Weihs, 1977) and *Doliolum* (Bone and Mackie, 1977); cephalopods *Sepia*, *Loligo*, and *Octopus* (Trueman and Packard, 1968; Ward and Wainwright, 1972; Trueman, 1975, 1980) and medusae and siphonophore bells, such as *Polyorchis*, *Cyanea*, and *Carybdea* (Gladfelter, 1972a, b; Trueman, 1980). In fishes, rapid contraction of the gill chambers is considered an important part of the thrust required for sudden takeoff from a resting position (as described for *Esox* by Lagler et al., 1977: 191; see also Brainerd et al., 1997: 1179).

A common principle applies to all animals using jet propulsion: the expulsion of water in one direction to propel the animal in the opposite direction. Accomplishing this requires a body form that allows a considerable quantity of water to be expelled during the power stroke, and that is then able to expand, inhaling water for the next power stroke. The effectiveness of a jet-propulsion mechanism depends largely on the velocity and mass of water ejected, which is in turn related to the strength of the musculature expelling the water, the cross-sectional area of the jet aperture, and the capacity of the chamber from which water is expelled (Trueman, 1980: 94; Brainerd et al., 1997: 1182). In frogfishes, the unique combination of enlarged, well-muscled oral and opercular cavities (see Feeding Behavior and Biomechanics, p. 437) and highly specialized, restricted opercular openings located behind the pectoral-fin bases admirably fulfill these requirements (Whitmee, 1875: 544; Gregory, 1928: 409, 1933: 388; Gordon, 1938: 20; Longley and Hildebrand, 1941: 303; Gordon, 1955: 389; Wickler, 1967: 546; Böhlke and Chaplin, 1968: 717).

The biomechanics of jet propulsion in frogfishes appear to be essentially the same as those involved in respiration, except that all of the movements are made with considerably more vigor. The propulsive force is generated by sucking a large quantity of water into the oral and opercular cavities and then forcing this water out through the tiny gill openings. A rapid, posteriorly directed discharge from these jetlike openings results in forward progress at speeds that vary between 0.20 and 1.25 standard lengths per second.

While the frogfish is jet-propelling in open water, its erected median and paired fins function as hydrofoils to control pitching, rolling, and yawing. The caudal and pectoral fins are also of considerable importance in maneuvering the body; the caudal functioning largely as a rudder by creating a yawing force on the body, the pectorals creating lift by acting as cambered plates operating at low Reynolds numbers (Webb, 1975: 26). During jet propulsion at low speeds (approximately 0.20 standard length per second), frogfishes are often observed to create additional propulsive forces by passing undulating waves in a posterior direction along the dorsal fin (a practice described in other fishes by Alexander, 1982: 27).

While antennariids typically move short distances by "walking," with the aid of pectoral and pelvic fins, or over longer distances by swimming continuously from one location to the next—using a combination of jet propulsion and fin movements, without touching bottom until they reach their final destination—the recently described Psychedelic Frogfish, *Histiophryne psychedelica*, employs a different approach. In this species, jet propulsion in a sustained, long-distance swimming effort has never been observed (Pietsch et al., 2009a: 41). On the contrary, individuals have been seen to move consistently in a series of short "hops," the paired pelvic fins making frequent contact with, and appearing to push off, the bottom at each bounce. While moving in this way and when viewed from the side, the fish assumed a strikingly globose shape (Fig. 247C), with the dorsal fin bent to one side, the posterior part of the body and tail bent strongly forward to approach the head, the fleshy lateral extensions of the cheeks and chin pulled back, and the dorsoventral diameter increased significantly. The overall visual impression, when viewed from the side, was reminiscent of an inflated rubber ball bouncing along the bottom (Hall, 2008: 16; Klein, 2008: 15; Pietsch et al., 2009a: 41).

The third swimming mode employed by frogfishes, regarded here as an escape response, is a kick-and-glide (Videler, 1981: 10; Jayne and Lauder, 1996: 645; Flammang and Lauder, 2009: 279). This movement is initiated by a "kick" composed of a combination of three to five rapid strokes of the caudal fin, a single rapid expulsion of water from the opercular openings (jet propulsion), and a rapid, posteriorly directed stroke of the pectoral fins. The initial result of this combination of propulsive forces is a rapid darting motion at speeds up to about five standard lengths per second. The glide segment is accomplished with the median and paired fins fully compressed and held tightly to the body to reduce drag, but at the same time the glide results in decreased stability and often erratic movement.

Finally, it should be noted that these benthic and pelagic locomotory modes are often used simultaneously. In a commonly observed form of locomotion on open bottoms, frogfishes combine jet-propulsion and amphipedal progression (Fig. 321C). The propulsive stroke provided by the pectoral fins moves the animal forward. During the recovery stroke that follows, a stream of water is forcibly ejected from the opercular openings. The alternation of pectoral movement and jet propulsion creates a smooth, continuous forward motion at speeds up to about 0.50 standard length per second. As in the "modified walk" (Fig. 321B), the pelvic fins play only a minor role in this form of locomotion, often acting as planing devices but providing no propulsive force (and are thus not represented in Fig. 321C).

Braking in frogfishes is accomplished by an extreme curvature of the caudal fin, which in this position functions as a "sea-anchor" (Breder, 1926: 208, fig. 58). Simultaneously, the pectoral fins are often widely spread and rotated into a vertical position to create drag.

Reproduction and Early Life History

Little is known about the reproduction and early life history of lophiiform fishes (Breder and Rosen, 1966: 598; Pietsch and Grobecker, 1980: 551, 1987: 350; Pietsch, 1984d: 320; Watson et al., 2000: 120; Arnold and Pietsch, 2012: 128; Pietsch et al., 2013: 663). Detailed information is available for only a few members of the Lophiidae and Antennariidae, and most ceratioid families (Pietsch, 2005, 2009). Scattered bits of published data are also available for the Tetrabrachiidae, Brachionichthyidae, Chaunacidae, and Ogcocephalidae, but nothing has been reported for the Lophichthyidae (Pietsch et al., 2009b: 491; Leis, 2015, table 3).

Unequal information on eggs and larvae is available for all four genera and at least 13 of the 27 recognized species of the Lophiidae (*Lophius americanus, L. budegassa, L. gastrophysus, L. litulon, L. piscatorius, L. vomerinus, Lophiodes reticulatus, L. beroe, L. caulinaris, L. monodi, L. spilurus, Lophiomus setigerus,* and *Sladenia shaefersi;* see Prince, 1891; Fulton, 1898; Bigelow and Welsh, 1925; Padoa, 1956; Martin and Drewry, 1978; Okiyama, 1988, 2014; Olivar and Fortuño, 1991; Watson, 1996a; Yoneda et al., 1998, 2001; Beltrán-León and Herrera, 2000; Duarte et al., 2001; Everly, 2002; Everly and Caruso, 2003, 2006; Maartens and Booth, 2005; Fariña et al., 2008; Ré and Meneses, 2008; Colmenero et al., 2017; and numerous additional references cited therein). Within the Antennariidae, information concerning early life-history stages is available for 10 of the 52 known species. These include rather complete descriptions of early egg and larval development of *Antennarius scaber* (Rasquin, 1958; Martin and Drewry, 1978: 380–384, figs. 200, 201) and *Histrio histrio* (Mosher, 1954; Rasquin, 1958; Martin and Drewry, 1978: 370–379, figs. 193–198; Okiyama, 1988: 345, figs.; Jackson, 2006a: 786–789, figs.); descriptions of the *"scutatus"* prejuvenile larval stages of *Fowlerichthys radiosus* (Schultz, 1957: 63; Hubbs, 1958; Pietsch, 1984d: 322, fig. 165; Jackson, 2006a: 786, fig. 1) and *F. avalonis* (Watson, 1996b: 559, fig. 1; 1998: 234, fig. 11; Beltrán-León and Herrera, 2000: 270, fig. 88; Figs. 28–30); detailed descriptions of early stages of *Abantennarius sanguineus* and *Antennatus strigatus* (Watson,

1998: 220, 231, figs. 1–10); and brief reports of parental care of eggs in *Lophiocharon trisignatus* (as *Antennarius caudimaculatus*; Pietsch and Grobecker, 1980) and four species of the genus *Histiophryne* (Pietsch and Grobecker, 1987: 361; Pietsch et al., 2009a; Arnold et al., 2014; Arnold and Pietsch, 2018: 627). Photos of larvae of three additional antennariid species, all questionably identified, were published by Maamaatuaiahutapu et al. (2006: 46; as *Antennarius commersonii*), Juncker (2007: 142, 144; as *A. coccineus* and *A. commerson*), and Waqalevu (2010: 18; as *A. commersonii*). Finally, Watson et al. (2000: 120–125, fig. 17) reviewed the reproduction and early development of antennariids and provided illustrations of five unidentified larvae.

For brachionichthyids, details of life history are unknown but some information is available for *Brachionichthys hirsutus* and *Brachiopsilus ziebelli*: females lay small clusters of large eggs (80 to 250 eggs per cluster, each egg measuring 1.8 to 2.0 mm in diameter) that are encapsulated and connected by filaments; the clusters are attached to the substrate by these filaments and there is evidence of parental care until they hatch; the young, resembling the adults in morphology, emerge from their egg cases and move by walking around on the substrate rather than swimming (Last et al., 1983: 250; Bruce et al., 1999; Last and Gledhill, 2009: 8, fig. 6A).

For chaunacids and ogcocephalids, aside from brief descriptions of ovarian structure (Rasquin, 1958: 343; Mead et al., 1964: 587), a limited (by available material) developmental series of three or four unidentified species of *Chaunax* have been described (Pietsch, 1984d: 322, fig. 164C, D; Okiyama, 1988: 345, fig.; Jackson, 2006b: 791–792, fig. 1), and somewhat more information on eggs and larvae are known for five species of ogcocephalids (*Zalieutes elater*, *Halieutichthys aculeatus*, *Ogcocephalus parvus*, and two unidentified species; Watson, 1996c: 563; Watson and Bradbury, 2000; Richards and Bradbury, 2005: 793–797, figs.; Okiyama, 2014). Finally, larvae, but not eggs, have been adequately described for most families of the Ceratioidei (see Bertelsen, 1951, 1984; Watson, 1996d; Pietsch, 2005, 2009).

Ovarian Morphology and Egg-Raft Structure

Probably the most striking characteristic of early ontogeny in lophiiform fishes is that eggs of most species are spawned encapsulated within a nonadhesive, balloon-shaped mucoid mass (Ray, 1961) or, more typically, a continuous, ribbonlike sheath of gelatinous mucous, often referred to as an "egg raft" or "purple veil" due to its purplish gray or brown coloration (Baird, 1871: 785; Collins, 1880: 8; Agassiz, 1882: 280; Fulton, 1898: 125; Gudger, 1937: 366; Pietsch and Grobecker, 1987: 351; for exceptions, see Parental Care, p. 479). Although gelatinous egg masses are also produced by some ophidiiforms (Sparta, 1929; Mito, 1962; Gordon et al., 1984; Fahay, 1992; Ambrose, 1996) and scorpaenoids (Barnhart, 1937; Orton, 1955; Phillips, 1957; Pearcy, 2011; Washington et al., 1984; Moser, 1996; Koya and Muñoz, 2007; Morris et al., 2011), the peculiar structure of lophiiform egg rafts differs considerably from that of any other ovarian product known in fishes (Rasquin, 1958; Breder and Rosen, 1966: 613). While serving, among other functions, as a device for broadcasting a large number of small eggs over great geographical distances, its positively buoyant quality provides for development in relatively productive surface waters (Gudger, 1905; Pietsch and Grobecker, 1980, 1987).

While there is some evidence that the unique manner of oviposition of anglerfishes was known to the ancient Greeks (see Gill, 1905a: 509)—in describing the habits of *Lophius*, Aristotle wrote that "the female lays her spawn all in a lump close in to shore" (translation of Thompson, 1984, book 6 (17), line 33)—the initial discovery of these epipelagic, egg-filled rafts has generally been attributed to Alexander Agassiz (1882: 280, pl. 16, fig. 2), apparently based on material received from Spencer Fullerton Baird in 1871 (see Gudger, 1937: 366). However, a report published by Joseph William Collins (1880: 8, 28) on the structure of the "egg raft" and its origin in the ovaries of *Lophius* preceded Agassiz by some two years:

> Some interesting experiments are being made in hatching the goose fish, monk fish or fishing frog (*Lophius piscatorius*), at the station of the United States Fish Commission . . . and much valuable information concerning the embryonic and early life of this natural angler has been obtained. One visiting the station can see the eggs in several stages of development, as well as the young fish that have been hatched. In the early stage the eggs are held together by a glutinous substance, which, floating in the water, looks like a thin sheet of jelly thickly dotted with small whitish beads. These sheets are from thirty to fifty feet long in their natural state, and float near the surface.

Although he cannot be given credit for its initial discovery, Alexander Agassiz (1882: 280, pl. 16, fig. 2) provided the first explicit description and illustration of this unique reproductive structure:

> The eggs of *Lophius* are laid embedded in an immense ribbon-shaped mucous band, from two to three feet broad and from twenty-five to thirty feet long. This gelatinous mass is often found floating on the surface of the sea during the last part of August. It looks at a short distance like an immense crape. The mucous mass is of a light violet gray color, and the black pigment spots of the young *Lophius*, still in the egg, give to the mass a somewhat blackish appearance. The eggs are laid in a single irregular layer through the mass, usually well separated by the mucus in which they float.

Following these initial descriptions, more detailed accounts of the egg rafts of *Lophius* were provided by Prince (1891), Fulton (1898), Bowman (1920), Procter et al. (1928), and others. According to Fulton (1898: 125), rafts of *Lophius* attain a length of nearly 11 m, weigh as much as 4.8 kg, and contain as many as 1,345,848 eggs. Connolly (1920: 10) reported the existence of even larger rafts with estimates of the number of eggs reaching 3,204,400.

That these rafts in *Lophius* are essentially identical to those of at least some antennariids (e.g., species of *Antennarius*; Fig. 322), and further that the rafts in both reflect the form and structure of the ovaries, was shown by Gill (1909: 570). In the most detailed account up until that time, Gill described the egg raft of *H. histrio* as "a soft jelly-like mass, quivering to the touch, but withal rather tenacious and [after full expansion in the water] 3 or 4 feet long by 2 to 4 inches or thereabouts in breadth. . . . The entire mass is thickly permeated with eggs, which appear to be in several irregular layers, or at least more than one" (Gill, 1909: 608).

Following Gill's (1909) thorough review, and with the exception of the detailed contributions of Mosher (1954) and Rasquin (1958), only scattered bits of additional information concerning egg-raft morphology became available. Noteworthy, however, are Hornell's (1921) description of a single raft of *Antennarius hispidus* (2.9 m long and 15.9 cm wide) and those of *A. scaber* described by Molter (1983) and *A. striatus* by Friese (as *Triantennatus zebrinus*, 1973, 1974). On the basis of some 50 spawnings that took place at the Taronga Zoo Aquarium during the summer of 1973, Friese (1973, 1974) reported that the width of the rafts varied little, ranging from about 9 to 11 cm. The length, however, varied enormously, from 40 to 90 cm, depending on the size and age of the female.

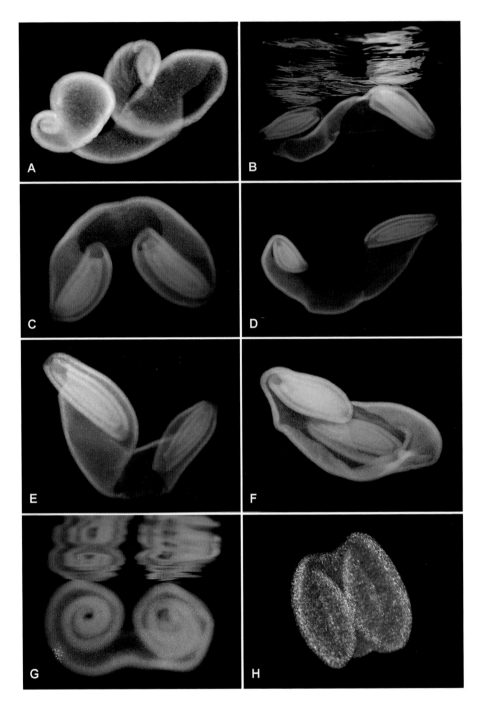

Fig. 322. Spawned egg rafts of species of *Antennarius* showing subtle interspecific differences in morphology. (**A**) *A. maculatus*, Anilao, Luzon, Philippines. Photo by Mike Bartick. (**B–F**) *A. multiocellatus*, Bonaire, Netherlands Antilles. Photos by Ellen Muller. (**G**) *A. pictus*, Dauin, Negros Island, Philippines. Photo by Daniel Geary. (**H**) *A. pauciradiatus*, Bonaire, Netherlands Antilles. Photo by Ellen Muller.

The only detailed study of the morphology of antennariid ovaries and egg rafts is that of Rasquin (1958). The following description is summarized from her paper:

The ovaries of *H. histrio* consist of a pair of glands fused on the midline over the region of the short oviduct. There is no septum dividing the two halves, the two thus forming a single organ. Each ovary forms a flattened, saclike structure, the long axis of which lies transversely in the abdominal cavity at nearly right angles to the anteroposterior axis of the body. Together the ovaries form a double scroll-shaped structure; the elongate, ribbonlike extension of each side is equally rolled inward from the distal tip toward the midline (Fig. 323; see also Rasquin, 1958: 336, pl. 47). Internally, the ovigerous tissue extends into the lumen of the ovary in the form of lamellae from only the inner wall of the scroll (Fig. 324). A considerable amount of smooth muscle is present in the ovarian walls (Rasquin, 1958, pls. 48–51).

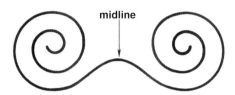

midline

Fig. 323. Diagrammatic representation of a cross-section through the double scroll-shaped ovaries of an antennariid; the elongate, ribbonlike extension of each side is rolled inward from its distal tip toward the midline. The structure of the mucoid egg raft is a replica of the internal surfaces of the ovaries; that is, it is formed by a mucoid impression of the epithelial surface of the lamellae of the ovarian walls (Figs. 325, 326). After Pietsch and Grobecker (1987, fig. 161).

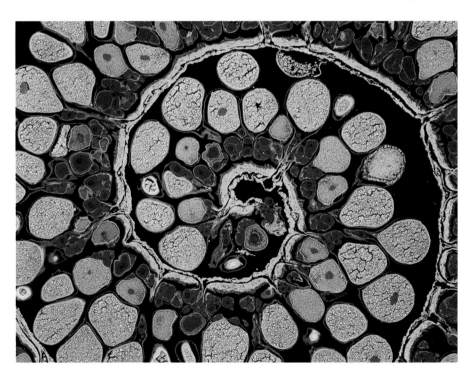

Fig. 324. A cross-section through the spiral-shaped ovary of an unidentified antennariid, with colors applied to differentiate the tissue types, using a combination of autofluorescence and inverse dark-field images. Generally, the yellow/orange is yolk, the green is connective tissue, and the red shows earlier-stage oocytes, all developing along the ovarian surface. Photomicrograph courtesy of James E. Hayden and The Wistar Institute, Philadelphia, Pennsylvania.

The gross morphology of the ovary of *A. scaber* is essentially the same as that of *H. histrio*, except that in *A. scaber* the straight sections on either side of the point of fusion of the two ovaries are longer, and there are more layers to the scrolled sections. The largest ovary of *H. histrio* examined by Rasquin (1958) had only three rows of ovarian tissue in the scroll, whereas those of *A. scaber* showed five rows. These differences are reflected in the expelled egg rafts of the two species: fresh rafts of *A. scaber* can be distinguished not only by their greater size, but also by the longer straight sections between the scrolled ends.

The structure of the egg raft is a replica of the internal surfaces of the ovary; that is, it is formed by a mucoid impression of the epithelial surface of the lamellae of the ovarian walls. After the oocytes are fully ovulated, the eggs and the mucoid raft are cast free into the ovarian lumen. This apparently does not occur until a short time before the raft is released from the body. The action of the smooth-muscle components of the ovarian wall and lamellae probably assists in releasing the mucoid material from its attachment to the lamellar epithelium. Late in the courtship period, strong contractions in the abdominal wall are a sign that the egg raft will be expelled from the body within a very few minutes. Pores in the raft are formed by spaces caused by the junction of the lamellae with the ovarian wall at a point where no mucous-producing epithelium is present (Rasquin, 1958, figs. 1, 2, pls. 51, 53; Figs. 325, 326). When the raft is expelled, sea-

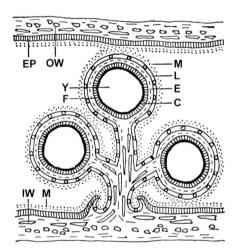

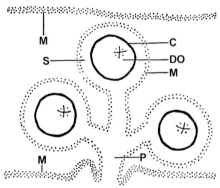

Fig. 325. Diagrammatic representation of a section through the ovary of *Histrio histrio*, including only one lamella, simplified for clarification. Both walls and lamellae contain collagenous and smooth muscle fibers. Abbreviations: **C**, chorionic membrane of the ripe ovum. **E**, endothelial layer. **EP**, epithelium lining the ovarian lumen. **F**, follicular layer. **IW**, inner wall of the ovary. **L**, lamellar epithelium. **M**, mucoid material produced by the lamellar epithelium and the epithelium lining the ovarian lumen. **OW**, outer wall of the ovary. **Y**, yolk of the ripe ovum. For a full explanation, see Rasquin (1958: 340–342, fig. 1). After Rasquin (1958, fig. 1); courtesy of Mai Reitmeyer and the American Museum of Natural History.

Fig. 326. Diagrammatic representation of the structure of the released egg raft of *Histrio histrio* for comparison with Figure 325. Abbreviations: **C**, chorionic membrane of the ovum. **DO**, developing ovum. **M**, mucoid material produced by the lamellar epithelium of the ovary and epithelium lining the ovarian lumen, which now forms the mucoid structure of the raft. **P**, pore open to the outside medium where seawater enters, probably carrying sperm. **S**, space between the ovulated ovum and mucoid wall filled with seawater in the released raft. For a full explanation, see Rasquin (1958: 340–342, fig. 2). After Rasquin (1958, fig. 2); courtesy of Mai Reitmeyer and the American Museum of Natural History.

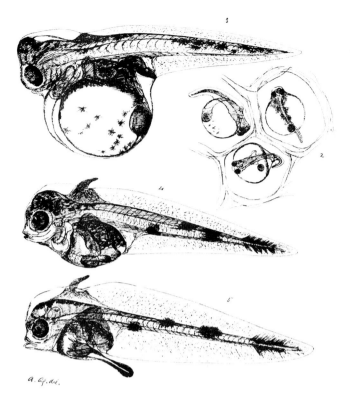

Fig. 327. Early development of *Lophius americanus*, showing similarity to that of *Histrio histrio* and other members of the antennariid subfamily Antennariinae. (**2**) three embryos, each lying within their own water-filled, gelatinous egg chamber. (**3**) larva removed from the egg just prior to hatching, showing large yolk sac. (**4**) larva just after hatching. (**5**) slightly older stage, showing depleted yolk sac. Drawings and description by Alexander Agassiz (1882, pl. 16).

water enters each pore, presumably carrying sperm with it. Each ovum is then confined within its own mucoid-walled compartment, free to float about in the seawater contained within the compartment (Figs. 326, 327). The raft remains afloat and intact for about three days prior to the time of hatching, when it begins to sink and disintegrate (Rasquin, 1958; see also Fujita and Uchida, 1959).

Ray (1961) described a peculiar variation of the typical lophiiform egg raft in a species said to be *Antennarius nummifer*. Over a period of some ten months, a single pair of individuals spawned 13 times, but instead of the elongate, scrolled rafts described by all previous workers, these took the form of a "perforated balloon" with a single entrance for water at its base. The mass was slightly positively buoyant and floated with the perforated base down. It was colorless, transparent, fairly resilient, and tough, and was composed of three layers of eggs, each egg in a separate mucoid chamber. A figure published by Ray (1961, fig. 1) shows a globular structure measuring roughly 9.7 cm in diameter and 11.8 cm in height (Fig. 328). Ray assumed that the ovary was globular as well, but our studies show that the ovaries of *A. nummifer* are indeed scrolled.

Fecundity

The surprising frequency with which females of *H. histrio* in captivity can produce egg rafts is well-documented. In nine spawnings observed by Breder (1949: 94; *H. histrio* identified as *H. gibba*), the intervals averaged 10.4 days (but ranged from 3 to 30 days). Mosher (1954: 143) recorded nine spawnings of two individuals, with intervals averaging 3.6 days (range 3 to 6 days). Rasquin (1958: 344, table 1) recorded 164 spawnings of 12 individuals, with intervals averaging 5.7 days (range 1 to 39 days). Walters (in Breder and Rosen, 1966: 600) observed 38 spawnings; on nine occasions two or more occurred on the same day, 22 of these multiple spawnings taking place within 1 to 111 minutes

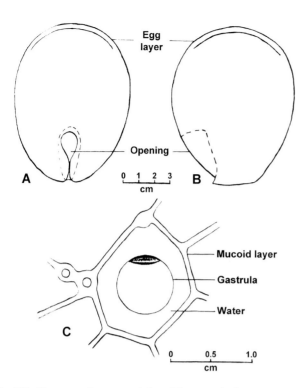

Fig. 328. Diagrammatic representation of the egg raft of an antennariid said to be *Antennarius nummifer.* (**A**) anterior view. (**B**) lateral view. (**C**) one of an estimated 48,800 water-filled, mucoid egg chambers containing a single developing egg. Dotted lines represent mucoid folds that are inverted inward, forming the borders of the opening at the base of the raft. After Ray (1961, fig. 1).

of one another, and in one instance five spawnings took place within 58 minutes. But despite many reports of prolific egg-raft production, very few estimates of the number of eggs per spawn have been made. Surprisingly, we have not been able to find any published fecundity estimates for *H. histrio.* Ray (1961) calculated that a single raft of his *Antennarius nummifer* (identification uncertain) contained about 48,800 eggs. For *A. striatus,* Friese (1973, 1974) gave the following approximations: a raft 44 cm long contained about 73,000 eggs, a 61-cm raft, about 180,000 eggs, and a 90-cm raft, about 288,000 eggs. Judging from these figures, mortality in nature is exceedingly high.

Sexual Size Differences

Lophiiform fishes are well known for size differences between conspecific males and females, the ultimate condition exemplified by the deep-sea Ceratioidei in which females may be more than 60 times the length and about a half-a-million times as heavy as the males (Pietsch, 2009: 3, figs. 3–7). Less extreme examples are found in lophiids and antennariids. In the former, size differences appear to be relatively small, but in all taxa available for examination (*Lophiodes* spp., *Sladenia shaefersi*), the largest individuals were female (see Pietsch et al., 2013, table 1).

Sexual size differences in antennariids are slightly more exaggerated than those observed in their lophiid relatives. Between June 2003 and May 2012, underwater photographer Ellen Muller of "Imagine Bonaire" documented 18 spawning events in several species of *Antennarius* in which the males in all cases were significantly smaller than the females (Pietsch et al., 2013: 663, fig. 4; Fig. 329). In many of these examples the male was observed resting on top of the female. Another example, recorded in August 2002 in Lembeh Strait, North Sulawesi, Indonesia, by Marna Zanoff and Chuck Boxwell of "Fish Tales Photography," shows an even greater size difference between male and female, in this case the male resting between her body and armlike pectoral fin (Fig. 329F). Many additional examples abound on the Internet.

Courtship and Spawning Behavior

Observations of courtship and spawning behavior have been reported for a few antennariids: *Antennarius scaber* (as *Phrynelox scaber* or *A. nuttingi*: Mowbray, in Barbour, 1942: 32; Krumholtz, in Martin and Drewry, 1978: 380; Molter, 1983: 66), *A. striatus* (as *Triantennatus zebrinus*: Friese, 1973, 1974), *A. hispidus* (Hornell, 1921), *A. multiocellatus* (Mosher, 1954: 142), *Histrio histrio* (Smith, 1898: 109; Gudger, 1905, 1937: 363; Breder, 1949: 94; Mosher, 1954: 142; Gordon, 1955; Rasquin, 1958: 344; Walters, in Breder and Rosen, 1966:

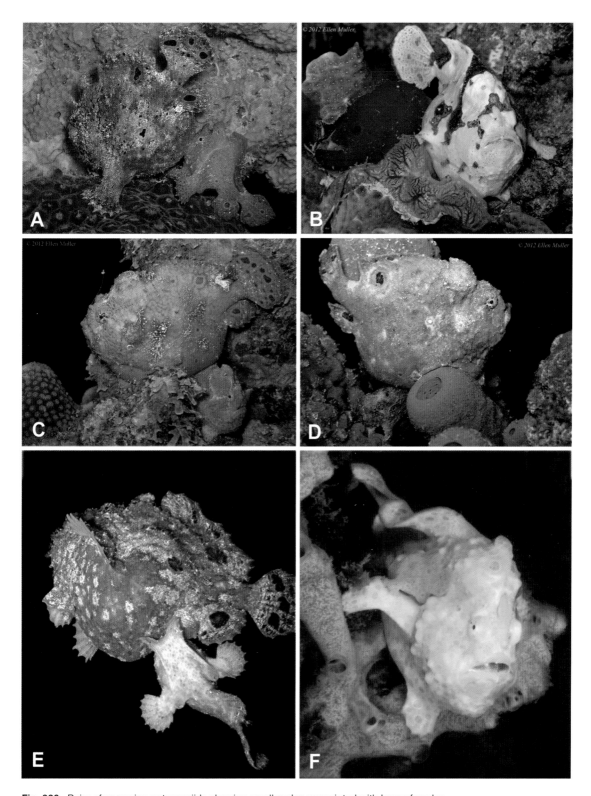

Fig. 329. Pairs of spawning antennariids showing small males associated with larger females. (**A**) male nudging female swollen with eggs. (**B**) about an hour before spawning. (**C**) moments before spawning. (**D**) tiny green male beneath gravid female, moments before spawning. (**E**) male prodding female, moments before spawning. (**F**) female with tiny male resting between her body and armlike pectoral fin. **A–E**: *Antennarius multiocellatus*, Bonaire, Netherlands Antilles. Photos by Ellen Muller. **F**: *Antennarius commerson*, Lembeh Strait, North Sulawesi, Indonesia. Photo by Chuck Boxwell.

600; Fujita and Uchida, 1959), and a species said to be *Antennarius nummifer* (Ray, 1961). But the full details of courtship, mating, and the production of a fertile egg raft were unavailable for any anglerfish prior to Carol Mosher's comprehensive study of *H. histrio* (1954). She provided not only detailed descriptions of courtship and spawning sequences, but also figures of egg rafts, typical courtship and spawning positions, and displays, some of which she took from selected motion-picture frames. Though the interested reader is encouraged to refer to Mosher's original work (as well as the subsequent study of Rasquin, 1958), the following is taken from her paper, as summarized by Gordon (1955):

On 3 January 1954 one of two individuals of *H. histrio* that had been maintained in solitary confinement for several months produced a single egg raft. When first spawned the raft was firm, about 89 mm long and tightly coiled at both ends; neither of the tightly rolled masses at the ends could be uncoiled. Later it softened and simultaneously swelled to about three times its original size. For some 48 hours it remained buoyant, but then began to sink to the bottom of the aquarium (had the fish spawned in its natural environment the egg raft would probably have become tangled in the complex branching of stems and bladders of *Sargassum*, which would have supported it and kept it afloat). Before spawning the female had swelled enormously, but after releasing a mass of eggs nearly equal to her own length, she returned to her normal shape.

In the meantime, the second of the two captive *H. histrio* had not spawned, nor had it undergone any change of shape. Acting on a hunch that the second individual might be a male, Mosher placed the two together in the same tank. From the beginning, neither showed any signs of the aggressive behavior so typical of *H. histrio* (see Defensive Behavior, p. 451). The supposed male constantly hovered close to the known female but made no attempt to attack. On 6 January the female retreated to a corner of the tank and remained there, clinging to *Sargassum* for about 8 hours, during which time her abdomen began to swell. As the swelling continued, her head came down and her posterior was uplifted so that she was more or less in a vertical position. Her coloration was a light tan, much like the *Sargassum* to which she attached herself. Despite her relative immobility she maintained her dorsal and caudal fins in a rigid manner. The male kept close by, circling and nudging her with his snout; his coloring became much darker, almost a chocolate brown, and his fins were withdrawn close to his body, but he would reach out and touch her repeatedly with his pectoral fin.

At about 1600 that same day, the female moved out from her corner into the open part of the tank. On the sandy bottom of the tank she began to "march" across the substrate, head down, "walking" by means of her pelvic fins. The male followed close behind, nudging her at the genital opening. The pair marched in closed formation back and forth across the length of the tank four times, accomplishing this in about 3 minutes. Then suddenly they turned upward and dashed to the surface, whereupon the female ejected an egg raft that seemed literally to burst out of her body fully formed. Simultaneously, the male shot by, presumably ejaculating spermatozoa. The spawning time could not have lasted more than a few seconds; so fast was the act of fertilization that the details could not be seen. After the spawning, the female appeared to be exhausted, swimming about erratically, but she soon recovered.

On 9, 12, 15, 18, and 22 January, the pair spawned again. Often between these sexual encounters the male became so aggressive, chasing the female about the tank, that they were placed in separate aquaria. Following the final spawning, the male grew increasingly and intolerably pugnacious, and before the female could be removed, he had attacked his mate so fiercely that she died of her injuries.

Not all males were as aggressive as this one in their courtship behavior. A second male, when presented with a female, immediately began to court. Its lower jaw trembled conspicuously whenever he approached her, and tremors passed through his entire body. While swimming back and forth with his fins outspread, he would occasionally brush against her, touching her with his pectoral fins. Several times he swam above her and settled back down on top of her, his pelvic fins straddling her back; he remained in this position for several hours. Their courtship continued throughout the day. On the next day the body of the female became much enlarged, and by evening she was almost spherical with ripened eggs and swollen mucous; in this condition she could hardly move about. By this time the male was pushing her about the tank, nudging her along with his snout. Occasionally she attempted to push him away with her pectoral fins. But his persistent pushing and biting continued, forcing her to stay off the bottom of the tank. An hour later they had spawned.

This same pair spawned again six days later, the female first leaving the *Sargassum* where she was resting. The male came up from below and shoved her. Then the two shot to the surface, the veil of eggs was explosively released, and they were fertilized. Immediately afterward both male and female quieted down, but several days later the male became so aggressive that he harassed and bit his mate to death, badly mutilating her body.

Friese (1973, 1974) described very similar courtship behavior in *Antennarius striatus* (as *Triantennatus zebrinus*), on the basis of repeated observations made at the Taronga Zoo Aquarium in Sydney, Australia:

> With the median fins firmly erect, the male will begin slowly but persistently to follow the female, "nudging" her gently with his head from below and slightly behind against the posterior area of the abdomen. Intermittently, there are brief periods of intensive trembling of the male, especially of the head region, which simultaneously assumes apparent rigidity. Similar trembling is displayed by the female, but not necessarily following that of the male in time, appearance, or intensity. These courting maneuvers may last for several hours and most of them take place in mid-water or just below the surface.

Although a significant amount of information is now available on courtship and spawning behavior in at least two antennariid species maintained in captivity, it is important to point out the difficulties in assuming that identical behavior occurs in the wild (Mosher, 1954: 151; Gordon, 1955: 393). Obviously the natural habitat of *H. histrio*, in the *Sargassum* "meadows" of the high seas, floating thousands of feet above the ocean bottom and continually buffeted by currents and agitated by winds, is very different from a laboratory aquarium. Mosher (1954: 151) suggested that in nature the male, once he has found a female on an adjacent clump of weed, follows her closely, never letting her get away; he does this by constantly touching, even grasping her with his mouth, or "riding" on her back for long periods of time. The well-defined prespawning march is no doubt considerably modified, and probably consists only of the male shifting in the rolling weed until he can position himself relative to the female for the final nuptial dash toward the surface.

Thanks to underwater photographers, courtship and spawning behavior of several species of *Antennarius* (e.g., *A. multiocellatus*, *A. scaber*, and *A. pauciradiatus*) has now been well-documented; numerous spectacular videos—among the very best are those of well-known diver and photographer Ellen Muller who has witnessed more than 65 spawnings at Bonaire—are readily available for viewing on the Internet. In these examples, the sequence of events, initiated nearly always just after sunset, is similar to that described for *H. histrio*, differing only in that it occurs near the ocean floor rather than near the surface in floating *Sargassum*: prior to spawning, the male and female "walk" along the bottom—or glide along just above the substrate using a combination of fin movements and jet-propulsion—with the female, bloated with eggs, in the lead, the smaller male following close behind, nudging her vent with his snout. Suddenly, without warning, the couple makes a dash toward the surface and the egg mass bursts forth, while milt is released by the male (Figs. 330, 331). Simultaneously and immediately following this rapid ascent, both parents (often more so by the male) undergo a violent spinning and somersaulting, functioning to aid in release of the eggs and to help uncurl and spread the mucous sheath, thereby facilitating the uptake of sperm-filled water and hydrating the eggs (Figs. 331–333). Once the dance begins, the entire process takes no more than a few seconds. Usually the act is performed one-on-one but occasionally two and sometimes even three or four males participate in fertilizing the eggs of a single female (Fig. 330B).

Descriptions of spawning and developing eggs in frogfish taxa other than *Histrio* and *Antennarius* are rare. A notable exception is Rudie H. Kuiter's (1993: 49) observations of *Rhycherus filamentosus*, which he described as "reasonably common" from Bass Strait to South Australia, on shallow weed-covered reefs to about 60 m, usually tucked into cracks or small hollows along ledges. In waters off Victoria, individuals congregate in October

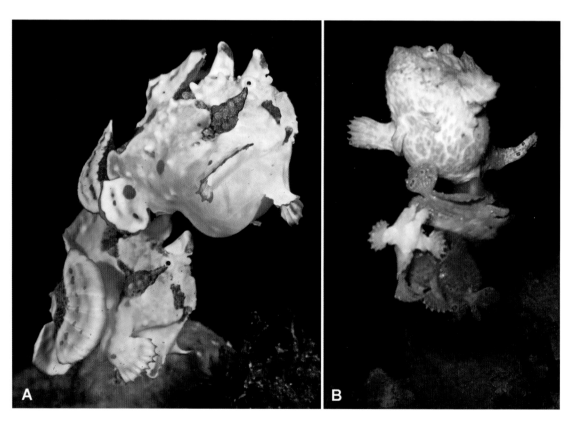

Fig. 330. Gravid females nudged off the bottom by courting males. (**A**) *Antennarius maculatus*, just prior to spawning, Anilao, Luzon, Philippines. Photo by Mike Bartick. (**B**) *Antennarius multiocellatus*, three males fertilizing the eggs of a single female (note the partially extruded egg raft), Bonaire, Netherlands Antilles. Photo by Ellen Muller.

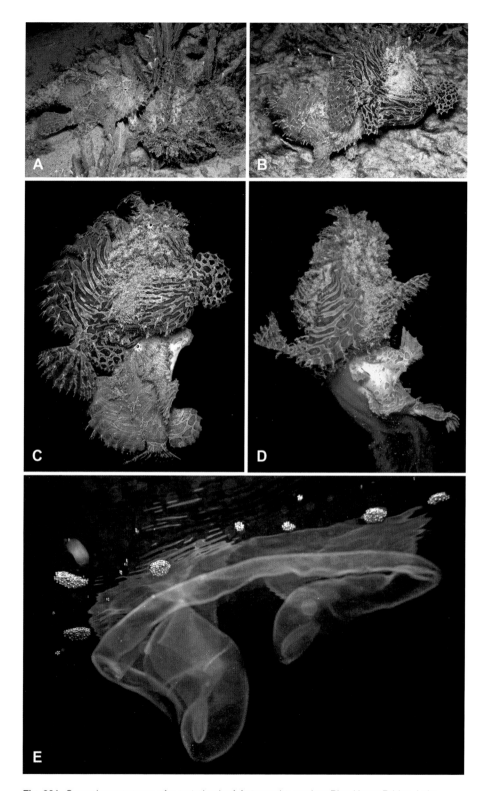

Fig. 331. Spawning sequence of a mated pair of *Antennarius scaber*; Blue Heron Bridge, Lake Worth Lagoon, Florida. (**A**) courting male on the left, gravid female on the right. (**B**) female nudged off the bottom by the male. (**C**) rising toward the surface, the male pushing from behind. (**D**) release of the egg raft. (**E**) spawned egg raft floating just beneath the surface. Photos by Ned DeLoach.

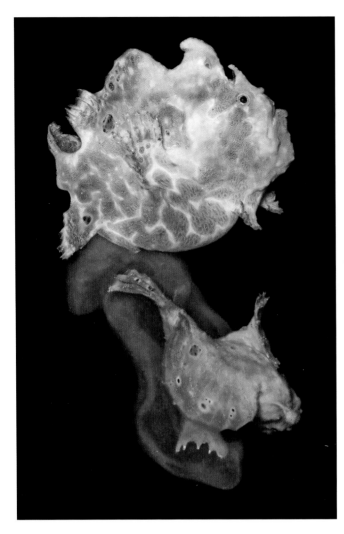

Fig. 332. *Antennarius multiocellatus*, a mated pair with egg raft, immediately after release of egg raft; Bonaire, Netherlands Antilles. Photo by Ellen Muller.

on shallow reefs to spawn, courting males outnumbering females three to one. Each female, greatly swollen with eggs, is surrounded by as many as four males, which are usually much smaller than the female and distinguished by the bright-blue breeding coloration of their lips (Fig. 334; see also Pietsch and Grobecker, 1987: 251, pl. 50). Stimulated by rapid vibration and lateral undulations of the males, the female eventually chooses a single mate and, pushing the other males aside, releases an egg mass that consists of numerous single-egg strands attached to a gelatinous disc-shaped structure measuring about 30 mm in diameter. The disc is laid first, followed rapidly by long strings of eggs, each attached in turn to a long sticky double filament (Fig. 335). As the male releases sperm, the female fans the eggs vigorously with her caudal fin and posterior part of her body, causing the egg mass to expand as it takes up sperm-filled water. The male is then chased away. The sticky threads become entangled, holding the egg mass together and often causing it to attach to surrounding algal growth on rocks. The female then huddles up against the mass guarding and protecting the eggs until hatching. The egg mass is surprisingly large, containing about 5,000 eggs, each about 5 mm in diameter. The eggs hatch after about 30 days and the young, still attached to a large bulbous yolk sac, but otherwise resembling miniature adults, sink rapidly to the bottom and quickly crawl into cracks and depressions in the surrounding substrate (Fig. 336).

Hybridization

The scientific literature contains a number of references to hybridization, either stated or implied, between what were thought to be distinct species of frogfishes (Krumholtz, in Schultz, 1957: 72; Ray, 1961; Friese, 1973, 1974; Michael, 1998: 334). Most of these reports may be explained by misidentifications. Louis Krumholtz, for example, observed spawning between what Schultz (1957: 72) identified as *Phrynelox scaber* and *P. nuttingi*, names that correspond to one and the same species; that is, the two extremes of coloration of *Antennarius scaber*. The "repeated hybridization" between what Friese (1973, 1974) called *Triantennatus zebrinus* (*A. striatus*) and *A. chironemus* (probably *A. pictus*) might also be attributed to intraspecific sexual encounters. Ray's (1961: 231) description of interaction between *H. histrio* and two unidentified individuals of *Antennarius* can most probably

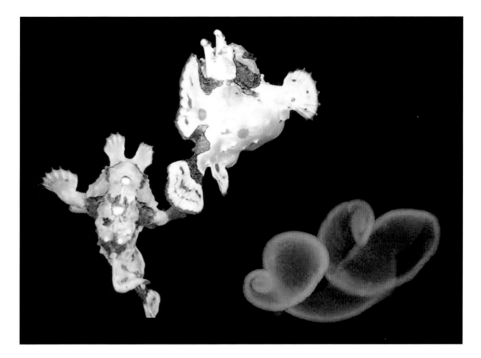

Fig. 333. A mated pair of *Antennarius maculatus* swimming away immediately after release of egg raft; Anilao, Luzon, Philippines. Photo by Mike Bartick.

Fig. 334. *Rhycherus filamentosus*, a gravid female surrounded by three well-camouflaged, courting males (note the bright blue breeding coloration of the lips of the males); Port Phillip Bay, Victoria, Australia. Photo by Rudie H. Kuiter.

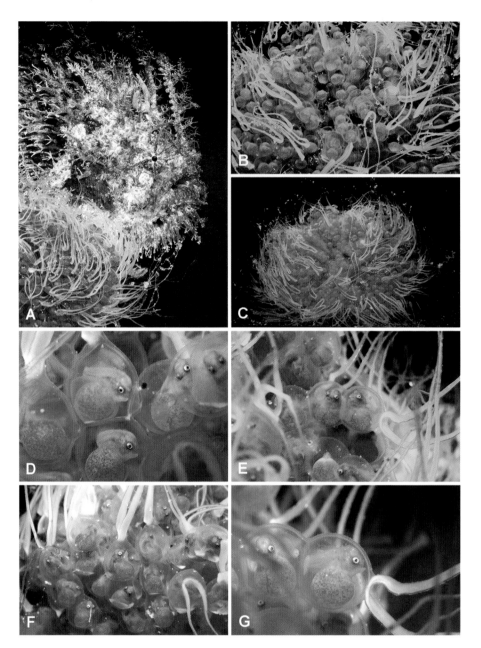

Fig. 335. *Rhycherus filamentosus*, showing the sequence of development of eggs spawned by specimens (shown in Figure 334) collected in 4 or 5 m, near Portsea Pier in Port Phillip Bay, Victoria, Australia, in October 1985 and held in an aquarium; spawning occurred on 23 October, hatching of eggs about 30 days later, and by 10 January 1986, when 30 to 35 mm long, the young were released in Port Phillip Bay (Rudie H. Kuiter, personal communication, 21 January 2018). (**A**) spent female with newly spawned egg mass, 23 October 1985. (**B**) egg mass on 23 October. (**C**) 26 October. (**D**) 8 November. (**E–G**) 11 November. Photos by Rudie H. Kuiter.

Fig. 336. *Rhycherus filamentosus*, juveniles resembling miniature adults, each 30 to 35 mm long, resulting from spawning of aquarium-held individuals and photographed just prior to their release in Port Phillip Bay (Figs. 334, 335). Photo by Rudie H. Kuiter.

be attributed to interspecific aggression (see Defensive Behavior, p. 451). But Scott W. Michael's (1998: 334) observations cannot be so easily set aside: "I kept a male Painted Frogfish [*A. pictus*] and a female Wartskin Frogfish [*A. maculatus*], which regularly crossbred both before and after I purchased them" (Fig. 337). He later added that egg rafts were produced on several occasions: "I saw it once and the person who gave me the pair saw it numerous times. I would suggest that when a conspecific mate is not available, as in the case with this aquarium-housed pair, they may mate with a close relative (of course, we see that in other fishes as well)" (Scott W. Michael, personal communication, 18 January 2018).

Egg and Larval Development

While general aspects of early development for the family were reviewed by Watson et al. (2000), detailed information on eggs and egg development, and the subsequent development of yolk-sac larvae through prejuvenile stages, is available only for *A. scaber* and *H. histrio*. The unfertilized eggs of *H. histrio* are colorless, almost transparent, and without oil droplets; the germinal disk lies partially sunk in a depression in the yolk, half its thickness being below the yolk surface, and the depression entirely filled with protoplasm; a single nucleus is visible (Martin and Drewry, 1978: 372). Fertilized eggs are extremely transparent and glassy, oval, or slightly elliptical, the major axis measuring 0.62 to 0.70 mm, the minor axis measuring 0.53 to 0.60 mm; they become spherical at the time of the second cleavage (Mosher, 1954: 148; Rasquin, 1958: 348; Fujita and Uchida, 1959: 280; Watson et al., 2000: 120).

Fig. 337. Aquarium-held specimens of *Antennarius pictus* (*above*) and *A. maculatus* (*below*), probably Indonesia, apparently in courtship. Photo by Scott W. Michael.

As development proceeds, the egg raft unrolls and expands to a length of 300 to 900 mm, a width of 51 to 76 mm, and a thickness of 8.2 to 16.4 mm (Smith, 1907: 400; Barbour, 1942: 22; Rasquin, 1958: 363). The membranes remain firm for an initial 48 hours or until about the sixth to eleventh myomere stages, then begin to deteriorate, the raft softening and expanding to about three times its original dimensions and finally beginning to sink (Mosher, 1954: 142; Fujita and Uchida, 1959: 280).

Mosher (1954: 148) published a detailed developmental schedule for eggs originating from Western Atlantic *H. histrio*. At water temperatures ranging from 21 to 23° C, hatching occurred in 108 hours. In a similar schedule of developmental steps based on *H. histrio* collected off Japan, Fujita and Uchida (1959: 280) found that eggs maintained at slightly higher temperatures (26.8 to 27.4° C) developed considerably faster, hatching in only 48 hours, 20 minutes. (For full descriptions and figures of egg development in *H. histrio*, see Mosher, 1954: 148; Fujita and Uchida, 1959: 280; Martin and Drewry, 1978: 372; and Watson et al., 2000: 120.)

Rasquin (1958: 348) compared the egg development of *H. histrio* to that of *A. scaber*, and to the developmental sequence described by Mosher (1954); only minor differences were found, and some additional data added (summarized by Martin and Drewry, 1978: 372).

Upon hatching, the yolk-sac larvae of *H. histrio* vary from 0.88 to 1.00 mm in total length; their length at the end of this stage, at least for laboratory-hatched larvae, is about 1.7 mm (Fujita and Uchida, 1959). The duration of the yolk-sac stage under laboratory conditions is about four days, regardless of whether the larvae are maintained at 26.8 to 27.4° C (Fujita and Uchida, 1959) or at about 30° C (Rasquin, 1958: 359). The developmental steps for the yolk-sac larvae of *H. histrio* and *A. scaber* are nearly identical, except for differences in pigmentation (Rasquin, 1958: 360; some additional minor differences were summarized by Martin and Drewry, 1978: 373–374).

Except for a few melanophores around the gut, laboratory hatchlings of *H. histrio* maintained at about 30° C had little pigment up to 95 hours after fertilization (23 hours after hatching; Rasquin, 1958: 359); yolk-sac larvae in the Japanese study, which hatched earlier (48.5 hours), had none (Fujita and Uchida, 1959: 281). At a lower temperature (21 to 23° C), "newly hatched" larvae (108 hours) were reported to have melanophores covering the head, and a solid band in the dorsal peritoneum across the top of the yolk sac (Mosher, 1954: 149). At 116 hours after fertilization, yolk-sac larvae maintained at about 30° C showed a considerable increase in pigmentation, especially around the gut and in the choroid region of the eye, but retinal pigment was still only weakly developed. On the head, where pigment spreads out behind the eyes, a Y-shaped pattern of melanophores appeared in *A. scaber* but not in *H. histrio*. Retinal pigmentation was completed at 168 hours, or about seven days after fertilization, three days after hatching (Rasquin, 1958: 360).

No exact information is available on the duration of the larval stage of *H. histrio* (complete when the minimum adult fin-ray complement is acquired), but a rough estimate based on length-frequency distribution of small samples from several years (Adams, 1960: 72; Dooley, 1972: 22; Fahay, 1975: 15) is one and a half to two months or more, depending on water temperature (Martin and Drewry, 1978: 375). The smallest larvae measure about 1.6 mm in total length; length at the completion of the larval stage is about 7.2 mm total length. The only estimates of pelagic larval duration in antennariids are 54 days for *Antennarius multiocellatus* and 47.5 days for *Antennatus tuberosus* (Luiz et al., 2013).

Postflexion larvae and small pelagic juveniles of *H. histrio* range from about 7.3 to 20.0 mm total length (5.0 to 15.2 SL) (Adams, 1960: 68); the duration of this stage, estimated from length-frequency data (Dooley, 1972: 22), appears to be about two months (Martin and Drewry, 1978: 376). Juvenile specimens range from about 15.5 to some unknown length beyond 30 mm SL. At 15.5 mm SL, the bifurcation of the posteriormost dorsal and all anal rays has appeared, the mottled markings of the typical adult color pattern have developed, and the midgut pigmentation characteristic of the earlier stages is barely discernible (Adams, 1960: 68, fig. 3). At about 28 mm SL, coloration is indistinguishable from that of the adult (Poll, 1959: 366).

Extremely little is known about growth beyond the juvenile stage, or age and size at maturity. Adams (1960: 69) showed that by 15 mm SL growth becomes relatively isometric. The length-frequency histogram for *H. histrio* collected under *Sargassum* in the Florida Current during 1966–1967 (Dooley, 1972: 22) suggests that growth from 5 to 45 mm SL took four to five months (Martin and Drewry, 1978: 379).

For considerably more detailed descriptions of early development in *H. histrio*, particularly the sequence of changes that occur in larvae, prejuveniles, and juveniles, you are referred to the publications of Mosher (1954), Rasquin (1958), Fujita and Uchida (1959), and Adams (1960), or to the concise summary provided by Martin and Drewry (1978; see also Watson et al., 2000). The little information that early life-history stages provide in formulating hypotheses of the relationship between the Antennariidae and the remaining families of the Lophiiformes has been summarized by Pietsch (1984d).

Parental Care

The typical lophiiform reproductive mode involves the production of a complex, pelagic egg raft that functions primarily to broadcast a large number of small eggs (see Egg and Larval Development, above) over great geographical distances, providing for development in relatively productive surface waters. A number of antennarioids, however, employ an alternative mode of recruitment that involves parental retention and care of a relatively small number of large eggs that hatch into relatively large, advanced young.

The earliest indication of an alternative mode of recruitment in anglerfishes was published by Pietsch and Grobecker (1980; see also Pietsch and Grobecker, 1987: 359, figs. 162, 163) who reported a single case of a male of *Lophiocharon trisignatus* (CAS 40369, 85 mm SL), with a cluster of eggs attached to the side of the body. The specimen was misidentified as *Antennarius caudimaculatus* and we strongly suspect that the sex of this individual was misidentified as well—the few subsequent examples of egg-carrying in this genus all appear to be female. In any case, the cluster contained about 650 spherical eggs, measuring 3.2 to 3.6 mm in diameter and securely attached to the left lateral surface just below the dorsal fin (Pietsch and Grobecker, 1987, fig. 162A). The eggs appeared to be in several layers, each layer being two or three eggs in thickness and separated from the adjacent layers by thin sheets of a transparent, acellular material. Within each layer the eggs were attached at two points (the radii from which consistently formed an

angle of about 140 degrees) to the surface of the transparent sheet and to each other by peculiar, double-stranded filaments. The eggs of the innermost layer were attached by these same double, threadlike structures to the spinulose epidermal surface of the fish (Pietsch and Grobecker, 1987, fig. 162B). The thin transparent sheet and the threadlike structures appear to be continuous with the outer surface of the egg, yet on the outermost eggs, and in areas where eggs had been torn away from each other, the threadlike extensions were left behind and formed closed loops (Pietsch and Grobecker, 1987, fig. 163). These loops apparently facilitate attachment to the surface of the fish.

Because of the strikingly different reproductive mode suggested by this single specimen of *Lophiocharon trisignatus*, the immediate reaction was that these attached eggs must be those of some other fish species deposited on the surface of the frogfish. Careful dissection of an individual egg, however, revealed a frogfish-like embryo having the characteristic crooked, "elbow-like" pectoral-fin lobe. Moreover, the low vertebral count (18 or 19 centra) compared well with that of lophiiforms while excluding most other teleost groups. Finally, eggs having the same double-stranded, threadlike structures on the outer surface were found inside the ovaries of a 72-mm SL female of *L. trisignatus*

Fig. 338. (A, B) *Lophiocharon lithinostomus*, with egg mass attached to its side, Raja Ampat, West Papua, Indonesia. Photo by Ned DeLoach.

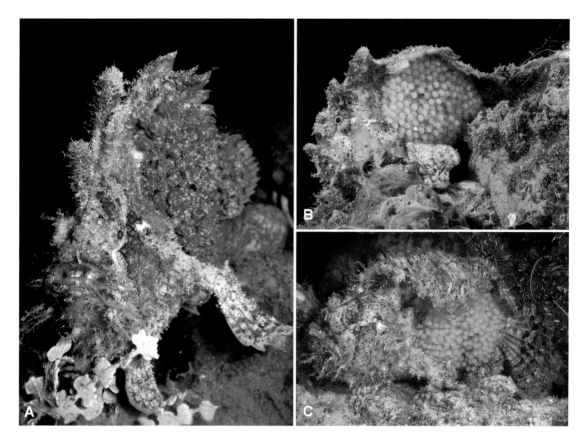

Fig. 339. *Lophiocharon trisignatus*, with egg mass attached to its side, Raja Ampat, West Papua, Indonesia. (**A**) Photo by Graham Abbott. (**B**) Photo by Frank Baensch. (**C**) Photo by Mary Jane Adams, courtesy of John and Suzanne Kelley.

(USNM 164243, about 320 eggs measuring 2.1 to 2.6 mm in diameter), which confirmed the conspecificity of eggs and parent. Further confirmation was later provided by Gerald R. Allen of the Western Australian Museum, Perth (personal communication, 9 February 1982), who brought our attention to a second example of egg-carrying in a specimen of *Lophiocharon trisignatus* obtained from a local aquarium shop. According to the owner of the shop, "a previous batch of eggs had successfully hatched in the aquarium and the newborn young were remarkably well developed, essentially miniature versions of the adults." Since that time a number of photographs of egg-carrying individuals of *Lophiocharon*, *L. lithinostomus* as well as *L. trisignatus*, have appeared on the Internet (some reproduced here; Figs. 338, 339), including a few videos that clearly show rapid rhythmic contractions of the muscles of the dorsal fin causing the egg mass to move up and down—a kind of pumping action that obviously serves to bring a continuous supply of well-oxygenated water to the developing embryos.

Additional examples of parental care of a relatively small number of large eggs are now known for other antennarioids. Neville Coleman, marine associate of the Australian Museum, Sydney (notes on file at the Australian Museum, dated 10 September 1973; Neville Coleman, personal communication, 7 December 1982) described a slightly different kind of egg-brooding in *Histiophryne cryptacanthus*, in which a cluster of eggs was completely hidden from view within a pocket formed by the pectoral fin, body, and caudal fin. A single specimen (AMS I.16170-001, 45 mm SL) was discovered while diving at

Fig. 340. (A–C) *Histiophryne bougainvilli*, mature females, each about 100 mm SL, with egg cluster protected within folds of body and fins; Nelson Bay, Port Stephens, New South Wales, Australia. Photos by David Harasti.

Cape Donington, South Australia, in December 1971:

It made short jerky movements that seemed to lack direction. Only the lower pectorals were used in a sort of grasping hand fashion. The tail was seen to be bent around to half the length of the fish and was not used at all. The fish was collected for identification, brought to the surface, and placed in a bucket of water for closer observation. In the bucket it made no attempt to move but rolled over onto the left side where the tail was curved. Closer examination revealed a hollow behind the left pectoral fin in which a number of eggs were being brooded, protected by the tail (Neville Coleman, personal communication, 7 December 1982).

In this case, the egg cluster, containing about 115 eggs measuring 3.6 to 4.2 mm in diameter, was not attached to the body of the fish. The eggs themselves were attached to each other by a single, flattened, acellular filament. This mode of parental care occurs also in *Histiophryne bougainvilli*; an egg cluster containing about 105 eggs, measuring 2.9 to 3.9 mm in diameter, was found associated with a 59-mm SL specimen (AMS I.17304-001) collected at Wollongong, New South Wales, in 1970 (Fig. 340).

Egg clusters are also well-documented in *Histiophryne psychedelica*. Direct field observations show that females of this species practice a similar, if not identical, form of egg-brooding in which a small cluster of relatively large eggs, attached to each other by acellular filaments, is completely hidden from view, and thereby protected, within a pocket formed by the body and pectoral, dorsal, and caudal fins (Pietsch et al., 2009a: 44; Figs. 341, 342). On 2 April 2008, a specimen of *H. psychedelica*, approximately 150 mm, was sighted in 6.4 m of water (Rachel J. Arnold, personal observation). The fish was apparently the same specimen that had been observed numerous times before in nearly the same spot, but it had now taken on a peculiar posture, with the body strongly curved to the left and the tail bent around in the same direction to almost meet the head (Fig. 341A, B). At the same time, the dorsal fin was folded to the left at a right angle relative to the body, the whole giving the fish a more or less circular appearance when viewed from the side. When disturbed by divers, an egg cluster containing about 220 spherical to slightly oval eggs, each approximately 3 to 4 mm in diameter, was discovered wrapped within the curve of the body and tail, behind the pectoral fin, and beneath the dorsal fin. Shortly thereafter, it became apparent that the eggs had been fertilized, and by 9 April 2008, the tiny developing embryos were clearly visible through the outer egg membranes. On the morning of 12 April 2008, the egg cluster was found abandoned (Fig. 341C), the female nowhere to be seen. About half the eggs by that time had hatched, with several newly hatched juveniles observed in the immediate area shortly thereafter (Fig. 343).

Additional examples of parental care in *Histiophryne psychedelica*, as well as similar modes of behavior in other members of the antennariid subfamily Histiophryninae, including *Porophryne erythrodactylus*, *Echinophryne crassispina*, and *Phyllophryne scortea*, have been recorded in recent years (e.g., see Liem, 1998; Figs. 194F, 344).

Tetrabrachium ocellatum Günther, one of two species that form the sister-family of the Antennariidae (see Pietsch, 1981b, and Evolutionary Relationships, p. 371), shows still another variation of egg-brooding. While curating the "old" collections, Helen K. Larson, of the Northern Territory Museum of Arts and Sciences, Darwin (Helen K. Larson, personal communication, 11 October and 9 November 1982), discovered a specimen of this species

Fig. 341. *Histiophryne psychedelica*, mature female, about 150 mm SL, Ambon, Indonesia. (**A**) with egg cluster protected within folds of the body and fins. (**B**) eggs emerging from protective folds. (**C**) abandoned egg cluster. Photos by Marcel Eckhardt.

with a tangled bunch of eggs hooked over [and wrapped around] two of its dorsal spines [the anteriormost soft-dorsal rays]. The eggs are on flattened flexible transparent filaments that are branched several times, with some branches bearing eggs and others

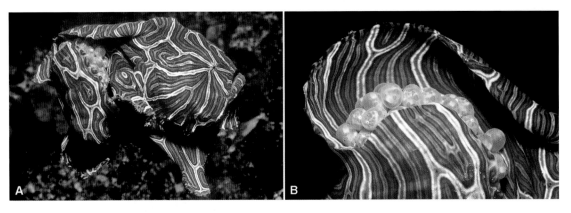

Fig. 342. (**A, B**) *Histiophryne psychedelica*, mature female, about 120 mm SL, Ambon, Indonesia, with egg cluster protected within folds of the body and fins. Photos by Roger Steene.

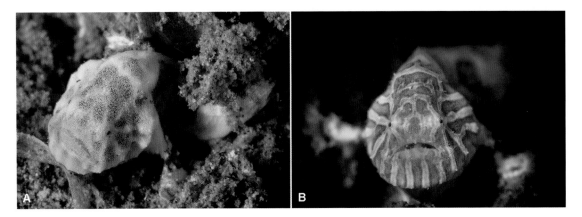

Fig. 343. *Histiophryne psychedelica*, juvenile specimens, Ambon Island, Indonesia. (**A**) Photo by Linda Ianniello. (**B**) Photo by Tony Cherbas.

ending in a pointed fork. Some of the egg cases are empty, and about a dozen have fallen off. The eggs are larger and less dense than those I've found in a few probably ripe females. . . . All of the ripe-looking *Tetrabrachium* I have looked at have been packed with dense eggs . . . [but] have no tendrils. *Tetrabrachium* have such odd hooked dorsal spines [i.e., rays; see Pietsch, 1981b: 390], it would be good to know that this was their purpose.

In all of these examples of parental care, the attached eggs are considerably larger and fewer in number than those spawned by related forms that employ the typical lophiiform reproductive mode. In *Lophiocharon trisignatus* the embryos are considerably larger, more highly developed, and supplied with a much greater amount of yolk than embryos of closely related forms (Pietsch and Grobecker, 1980: 552). For example, Rasquin (1958: 353, pls. 60, 69, fig. 3) described embryos of *A. scaber* at a stage just prior to hatching that measured less than one-third the length of those of *L. trisignatus*; the pectoral-fin bud consisted "of nothing more than a mass of undifferentiated, rapidly dividing cells" compared to the large pectoral-fin lobes of *L. trisignatus* (about 25% of total length); and the diameter of the yolk mass was considerably less than one-fifth that of *L. trisignatus*. In fact, the embryos of *L. trisignatus* appeared to be roughly compara-

Fig. 344. *Phyllophryne scortea* guarding a recently spawned egg mass; Edithburgh, South Australia. Photo by Fred Bavendam.

ble in development to the four-day-old, free-swimming larvae of *A. scaber*, yet even at this stage *A. scaber* larvae are still less than one-third the size of the embryos of *L. trisignatus* (Pietsch and Grobecker, 1987: 362). It seems apparent that these antennariid species are employing a strategy of recruitment that is directly opposed to that of most lophiiform fishes: retaining and caring for a small number of relatively large eggs that hatch into relatively large, advanced young—seeming to fall nicely within MacArthur and Wilson's (1967) and Pianka's (1970) theory of r versus K selection, with the two antennariid subfamilies on different sides of that gradient. Apparently as a result of this alternative reproduction mode, these three species have narrow geographic distributions compared to most other antennariids.

Parental care of developing eggs is well-documented in the species described above but by extrapolation we now know that this reproductive mode occurs in a host of additional, closely related species of the family. In the most significant contribution to our knowledge of antennariid biology to date, made possible through the efforts of the junior author (see Arnold, 2010b; Arnold and Pietsch, 2012; Arnold, 2013; Arnold et al., 2014), molecular analyses have provided overwhelming evidence that the family contains two monophyletic subunits, the Antennariinae and Histiophryninae, based on genetic divergence but also on ovarian morphology and reproductive behavior (corroborated as well by geographic distribution; see Zoogeography, p. 389). Reproductive life history differs considerably between the two subfamilies: the Antennariinae has double scroll-shaped ovaries, while the Histiophryninae has an entirely different ovarian morphology consisting of a pair of simple oval-shaped structures. In addition, each ovarian type corresponds to a different life history: members of the Antennariinae are broadcast spawners that go through a distinct larval stage, while those of the Histiophryninae undergo direct development and display various degrees of parental care. (For more, see Evolutionary Relationships, p. 385.)

It has been well-documented that frogfishes are efficient at luring prey by maintaining the immobile, inert appearance of substrate and structure while wriggling a highly conspicuous bait (see Aggressive Mimicry, p. 411). Since eggs provide excellent food for fishes, it seems apparent that having a visually conspicuous cluster of eggs attached to the substrate-like flank (in the case of *Lophiocharon trisignatus*), or dangling eggs at the tips of the dorsal-fin rays (as in *Tetrabrachium ocellatum*), would appreciably enhance the overall luring effect of the frogfish. If this is true, the unusual examples of parental care described here provide an unexpected trophic advantage. We speculate that this mode of egg carrying has evolved under a dual selective pressure to protect the offspring, on the one hand, and to enhance luring, on the other (Pietsch and Grobecker, 1980).

Alleged Nest Building in *Histrio histrio*

In 1872, Louis Agassiz—reporting on a discovery made at sea off St. Thomas, Virgin Islands, as a member of the 1871–1872 United States Coast Survey Deep-sea Dredging Expedition aboard the steamer *Hassler*—published a brief description of a globular egg-filled mass of floating *Sargassum*, which he identified as the "nest" of *Chironectes pictus* (i.e., *Histrio histrio*):

> The most interesting discovery of the voyage thus far is the finding of a nest built by a fish, floating on the broad ocean with its live freight. On the 13th of the month, Mr. Mansfield, one of the officers of the Hassler, brought me a ball of Gulf weed which he had just picked up, and which excited my curiosity to the utmost. It was a round mass of sargassum about the size of two fists, rolled up together. The whole consisted, to all appearance, of nothing but Gulf weed, the branches and leaves of which were, however, evidently knit together, and not merely balled into a roundish mass; for, though some of the leaves and branches hung loose from the rest, it became at once visible that the bulk of the ball was held together by threads trending in every direction, among the sea-weed, as if a couple of handfuls of branches of sargassum had been rolled up together with elastic threads trending in every direction. Put back into a large bowl of water, it became apparent that this mass of sea-weed was a nest, the central part of which was more closely bound up together in the form of a ball, with several loose branches extending in various directions, by which the whole was kept floating. . . . We had, no doubt, a nest before us, of the most curious kind: full of eggs too; the eggs scattered throughout the mass of the nest and not placed together in a cavity of the whole structure. (p. 154)

After describing the structure and its contents in greater detail and considering "What animal could have built this singular nest" (p. 154), Agassiz concluded: "It thus stands as a well authenticated fact that the common pelagic Chironectes of the Atlantic (named *Chironectes pictus* by Cuvier), builds a nest for its eggs in which the progeny is wrapped up with the materials of which the nest itself is composed; and as these materials are living Gulf weed, the fish-cradle, rocking upon the deep ocean, is carried along as an undying arbor, affording at the same time protection and afterwards food for its living freight" (p. 155).

Very soon after Agassiz's (1872) initial discovery, Jones (1872: 5), without speculating on its origin, described another floating "fish nest" of gulf weed "woven together by a maze of fine elastic threads, affording a raft from which depends the clustering mass of eggs, which I cannot illustrate better than by asking readers to imagine two or three pounds of No. 7 shot grouped together in bunches of several grains, and held in position by the elastic thread-work previously mentioned. . . . It is truly a wonderful speci-

Fig. 345. Alleged "nests" of *Histrio histrio* in *Sargassum*, now known to be formed by the sticky filamented eggs of flyingfishes. (**A**) Drawing by Sherman Foote Denton; after Agassiz (1888, fig. 210). (**B**) After Fowler (1928, fig. 82); courtesy of Tia Reber and the Bernice P. Bishop Museum, Honolulu, Hawaii. (**C**) After Breder (1938, fig. 48); courtesy of Rosemary Volpe and the Peabody Museum of Natural History, Yale University.

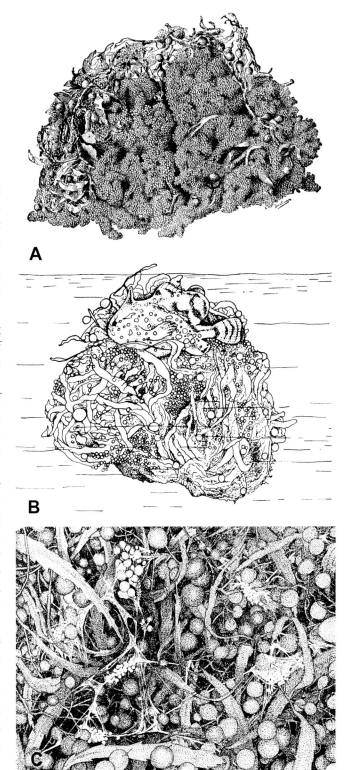

men of Nature's handiwork; a house built without hands, resting securely on the bosom of the rolling deep." Somewhat later, Jones (1879: 363) again described a nest, this time attributing it to the "Marbled Angler": "Here [in *Sargassum*] it makes its wonderful nest amidst the mass, suspended by means of those silk-like fibres, which prove amply strong enough to support the large bunches of eggs, which hang like grape clusters within their orbicular case."

In the years following, many more of these masses were discovered in *Sargassum*, and, accepting the identification of Agassiz, Vaillant (1887) explained in some detail how *Histrio histrio* (identified as *Antennarius marmoratus*) made the nest. Alexander Agassiz (1888: 31, fig. 210; Fig. 345A) published a figure of the nest, stating that "*Pterophryne* [i.e., *Histrio*], 'the marbled angler' of the Sargasso Sea, is especially adapted to live among the floating algae, to which it clings with its pediculated fins, and in which it intertwines its gelatinous clusters of eggs." Another figure of a nest showing *H. histrio* intertwined among the fronds of *Sargassum* was published by Fowler (1928: 476, fig. 82; Fig. 345B); Beebe (1932: 68–72, pls. 16, 17) provided two excellent photographs of a nest collected at Bermuda in 1931; and eggs clustered among *Sargassum* were depicted by Breder (1938: 110, fig. 48; Fig. 345C; see also Breder, 1934: 32).

Karl August Möbius (1894, figs. 1–3) cautiously accepted the idea that the eggs were probably those of *H. histrio*. He provided more details of the nest, and described and figured the eggs as having numerous bipolar filaments. After inspecting the ovarian eggs of *H. histrio* and finding no filaments (and also showing that the eggs were smaller than those found in the nest-like masses), Mö-bius concluded that the filaments must be acquired during the passage of the eggs through the oviduct. In 1905, Hugh Smith (in Gill, 1905b) noticed that the eggs expelled by individuals of *H. histrio* being maintained in an office aquarium had no filaments and were significantly smaller than those that had been described earlier by Möbius (1894). Almost simultaneously, Gudger (1905) reported on similar eggs that had been extruded from captive specimens of *H. histrio* in a long jellylike mass, much like those expelled by *Lophius*. It thus became obvious that the nest-like masses of *Sargassum* must be produced by a very different kind of fish. Gill (1905b) speculated, and later (1907, 1909) provided evidence, that these filamented eggs were those of a flyingfish (family Exocoetidae). He concluded, however, that the "masses of sargasso are not made by any fish at all, but by the eggs themselves. The eggs must be laid on the fronds of the weed and the long mo-tile tendril-like filaments clasp the finely cut branches of a frond till a globular mass is brought together" (Gill, 1907: 63). The question was settled once and for all by William Beebe (1932: 69, pl. 18), who, after collecting five egg-filled masses, managed to carry the ova through hatching and rearing to a stage of development that allowed him to identify them to the genus *Exonautes* (= *Hirundichthys*). Breder (1934: 37, figs.; see also Breder, 1938: 110, fig. 48) repeated Beebe's experiment, rearing the larvae into young flyingfish that he identified as *Exonautes rondeletii* (misspelled *rondoletti*), now known as *Hirundich-thys volador* (Jordan, 1884), the Atlantic Blackwing Flyingfish. (For a more complete his-torical account of the alleged nests of *H. histrio*, see Gudger, 1937.)

8. Tips for Aquarists and Divers and the Need for Conservation

With Scott W. Michael

We know that people have been keeping, or attempting to keep, frogfishes in captivity for centuries. Their unique froglike appearance, feeding behavior, and locomotor adaptations have made them highly sought after by naturalists, both amateur and professional, looking for something more behaviorally engaging to observe and study. To a great extent, they have become the favorite of a large, worldwide guild of marine aquarium hobbyists. Not only are frogfishes among the most interesting of aquatic animals, they are also ideally suited to the home aquarium. Their small size (most do not exceed 20 cm in total length), relative inactivity, and limited home range allows them to thrive in the small living space that an aquarium provides. That said, frogfishes do exhibit some special needs that must be met in order to ensure captive longevity.

The Frogfish Aquarium

Most frogfishes will do well in aquaria as small as 76 liters (20 gallons) and some of the dwarf species—like members of the *Antennarius pauciradiatus* group, which are unknown to exceed 40 mm SL—will thrive in smaller vessels. This is a positive feature because frogfishes pose a threat to, or may be threatened by, many potential tankmates (see below) and thus are best housed on their own. Smaller aquaria have inherent downsides—they are more prone to sudden changes in temperature and in water chemistry. As a result, more care must be taken when setting up a small tank, being attentive to where the tank is placed; for example, away from sources of heat or cold. The aquarist will also need to give the tank more regular attention, closely monitoring behavior and water parameters, and perform more frequent partial water changes. Fortunately, frogfishes do not require frequent feedings, and they produce easily detected "fecal packages" that can and should be removed with a dip net or siphon. This makes it easier to maintain optimal water quality. When setting up the aquarium, the aquarist should consider the rate of circulation and the direction of water movement. Frogfishes do not appreciate being buffeted by a strong current, so ensure that there are places in the tank sheltered from direct water flow. This is especially true if you have a small aquarium equipped with an external filter with rapid water return, or internal water pumps to increase water movement to benefit sessile invertebrates. Because they are not strong swimmers, install strainers or sponge filters attached to siphon tubes or power-head intakes. Many frogfishes have been mortally wounded when sucked into the intake of a submersible pump. While many, if not most, frogfishes can be housed in smaller aquaria, be aware that the adults of some of the larger species (e.g., *Fowlerichthys ocellatus* and *Antennarius commerson*) will require an aquarium of at least 285 liters (75 gallons). (This discussion and much of what follows that pertains to captive care and frogfishes in the field is taken almost verbatim from Scott W. Michael's landmark *magnum opus Reef Fishes* (1998: 329–339).

When it comes to aquarium aquascaping, most frogfishes are not particular as long as they have suitable perches on which to rest and crevices or caves in which to hide. Aquascaping material can include live rock, live or faux sponges and corals (avoid cnidarians that are strong-stingers), and PVC pipes or ceramic flower pots. If a frogfish is unable to find suitable refuge, it will likely become stressed, leading to possible fasting and health issues, such as a greater vulnerability to parasites. When dissatisfied with its surroundings, a frogfish will attempt to relocate. In the wild, this means moving to another section of the reef where shelter and water movement may be more to its liking or where prey items are more abundant. In an aquarium, a frogfish may not find the conditions it "prefers" and will end up moving about the tank incessantly. This is an obvious sign of stress and the aquarist must intervene. This can be done by providing more structure to the tank, including additional hiding places or more suitable perching sites, and checking to ensure that the frogfish is not being harassed by a tankmate—often a simple partial water change will eliminate the stress. If lack of structure is not the cause of hyperactivity, it may be the result of inadequate feeding—if not properly fed, a hungry frogfish will move about continuously in search of prey.

While many frogfishes are naturally found in coral or rocky-reef habitats, where caves and crevices provide shelter, there are some species (e.g., members of the *Antennarius striatus* group) that spend more time in less complex habitats, such as open sand or mud bottoms, with scattered sponges, algae, or large-polyped stony corals. These habitats can be replicated using light or black coral sand with scattered rubble piles, patches of macroalgae or filamentous microalgae, and large-polyped stony corals (of course, appropriate lighting and water parameters will be required to keep these organisms healthy). One antennariid with unique habitat preferences is the Sargassumfish, *Histrio histrio*, which is almost always found clinging to floating rafts of the brown macroalgae for which it is named. While it can be maintained without *Sargassum*, it does tend to retain its affinity for the upper layers of the water column, often perching on a taller piece of artificial coral or in the upper "canopy" of a grouping of artificial "plants." A Sargassumfish may even brace itself, using its armlike pectoral fins, in an upper corner of the aquarium. One way to better replicate the natural habitat of *H. histrio* is to encourage growth of macroalgae on the aquarium surface. Algae of the genera *Caulerpa*, *Chaetomorpha*, and *Gracilaria* will often thrive, if provided with enough illumination, on the water's surface and will serve as a more suitable habitat for a captive Sargassumfish.

Most frogfishes will do well in a "reef aquarium"; that is, an aquarium in which live corals and other sessile invertebrates are maintained. Because these aquarium environments tend to be replete with hiding places, it is more likely that the more secretive species may hide so much that they are rarely if ever observed—this is especially true if the aquarium is large. When adding a frogfish to a reef aquarium, one of the more conspicuous species should be selected—members of the *Antennarius striatus* or *A. pictus* groups are most highly recommended. These species tend to be more active and spend more time in the open. On the other hand, species that are best to avoid in tanks with complex topography (i.e., that contain lots of hiding places) include members of the *A. pauciradiatus* group and the genera *Abantennarius* and *Antennatus*. The latter genus contains especially cryptic species that do best in a smaller aquarium with limited rockwork to assure regular viewing. Members of the genus *Histiophryne* also tend to be highly secretive. In all cases, however, before introducing a frogfish, the aquarist should remember that any fish or invertebrate of a size that is possible to be swallowed is likely to get eaten. Frogfishes not only pose a serious threat to other fishes but to ornamental shrimps as well, including those species that are known cleaners (e.g., various species of the genus *Lysmata*), which are often "spared" by other predators.

Because frogfishes do not have to be fed frequently, they contribute relatively low levels of nitrogenous waste—a positive attribute in a closed system, making them a suitable addition to a small reef aquarium. They usually ingest all food items intended for them, and their feces are large and solid so they can be easily recognized and removed. The only drawback to keeping frogfishes with sessile invertebrates is that they sometimes perch on hard or soft corals, causing the polyps to stay retracted. This can be problematic if the frogfish remains on the same piece of coral for long periods of time. If this happens, carefully push the frogfish off its preferred spot with a piece of rigid airline tubing. If you do this every time you see it perching there, it will move to a new site. Feeding is usually not a problem in the reef aquarium—if you are feeding live food, the net method works particularly well (see below).

When adding a frogfish to a reef aquarium it is also important to realize that some invertebrates can pose a serious threat to the well-being of the antennariid. They should not be kept with invertebrates that have a potent sting, such as the Elegance Coral (*Catalaphyllia jardinei*), carpet anemones (*Stichodactyla* spp.), or fire anemones (*Actinodendron* spp.). These cnidarians can cause considerable damage to the skin of a frogfish when in contact with the tentacles; in the case of carpet anemones, a frogfish may even become ingested. Another potential frogfish predator is the Elephant Ear (*Amplexidiscus fenestrafer*)—some evidence indicates that this corallimorpharian cnidarian, closely related to stony or reef-building corals, may envelop and consume an unwary antennariid that settles on it. Other invertebrates that are a potential antennariid threat include hermit crabs (especially the larger, more omnivorous varieties), larger true crabs (Brachyura), spiny lobsters (family Palinuridae), and mantis shrimps (Stomatopoda). Some of the larger *Stenopus* shrimps (e.g., the Ghost Boxer Shrimp *Stenopus pyrsonotus*) have also been known to pester frogfishes. Finally, the piscivorous Green Brittle Star (*Ophiarachna incrassate*) has been known to capture more sedentary fishes and eat them, so they too are a threat to smaller antennariids.

Compatibility

The optimal captive environment for an antennariid is a "species aquarium"; that is, an aquarium devoted to a single species of fish. The primary reason for this is because frogfishes are so voracious (see Voracity and Cannibalism, p. 451). However, if the aquarist decides to keep a frogfish with other fishes, tankmate selection must be considered carefully. Because frogfishes can eat incredibly large prey items relative to their body size, any fish kept with them should be at least three times as long as the frogfish if they are of a slender build (e.g., wrasses), or slightly longer than the frogfish if they are deep-bodied forms (e.g., angelfishes, damselfishes). Even selecting tankmates that would appear to be too large for a frogfish to ingest may not stop a hungry antennariid from attempting to swallow it. There have been occasions when frogfishes have subdued (partially engulfed) a deep-bodied species (e.g., angelfish, butterflyfish), only to eventually spit out its now injured or even dead potential quarry. There are even occasions when the partial ingestion of a prey item has led to the demise of the frogfish because it could not release its quarry.

Conversely, a frogfish can be injured by its piscine tankmates. Fishes that browse on sessile invertebrates and algae may mistake a frogfish for a natural grazing surface. This can result in injury, infection, and even death. Although some representatives of these feeding guilds may not always bother their antennariid tankmates, the aquarist is taking a risk if these fishes are housed together, especially if the tank is small or is devoid of hiding places. Sessile-invertebrate feeding fishes that are likely to do damage to a frogfish include members of the following groups: spadefishes (family Ephippidae), angelfishes (Pomacanthidae), butterflyfishes (Chaetodontidae), triggerfishes (Balistidae), filefishes (Monacanthidae), trunkfishes (Ostraciidae), and pufferfishes (Tetraodontidae). While not a predator of sessile invertebrates, the porcupinefishes (Diodontidae) are also a threat to frogfishes. Herbivores like surgeonfishes (Acanthuridae), rabbitfishes (Siganidae), and blennies (Blenniidae) are potential frogfish nippers as well.

Fishes that make more suitable antennariid "neighbors" in captivity include soldierfishes and squirrelfishes (family Holocentridae), scorpionfishes (Scorpaenidae), groupers (Serranidae), larger hawkfishes (e.g., genus *Paracirrhites*, Cirrhitidae), snappers (Lutjanidae), grunts (Haemulidae), swallowtail angelfishes (genus *Genicanthus*, Pomacanthidae), pyramid butterflyfishes (genus *Hemitaurichthys*, Chaetodontidae), and monos (Monodactylidae). As mentioned above, it is important that these tankmates are large enough (usually at least the same length, in the case of deep-bodied species, to three times as long as the antennariid) to avoid predation by the frogfish. Also, be aware that smaller frogfishes are potential prey for certain piscivores like moray eels (Muraenidae), scorpionfishes (Scorpaenidae), groupers (Serranidae), and snappers (Lutjanidae).

Different species of frogfishes can be kept in the same tank, as long as the tank is large enough and the individual antennariids are similar in size. If there is sufficient size disparity between the individuals, the larger frogfish will not hesitate to eat its smaller relative (even conspecifics). One way to curb frogfish cannibalism is to keep them well-fed—they are less prone to eating one another if they are getting enough to eat.

Frogfishes may also behave aggressively toward each other, but they rarely inflict serious harm unless the aquarium is too small (see Defensive Behavior, p. 451). In most cases, agonistic interactions usually consist of lateral displays directed toward an intruder when it enters the ambush site of a resident individual: typically, the latter raises up on its pectoral fins while it spreads its dorsal, anal, and caudal fins; its color intensifies, and it undergoes a rapid shuddering of its entire body. In some cases, individuals will engage in mouth-gaping—the mouth is opened widely but not to the same degree as when they yawn. Usually the intruder will move off when confronted by these displays but, if the intruder is larger than the defender, it may begin to display as well and the level of aggression may escalate. The modified pectoral fins of antennariids come in handy during these melees—combatants will often put their "hands" on the head or body of their rival and attempt to push it away. If aggression occurs too often or escalates between captive frogfishes, they should be separated. The most aggressive member of the family is the Sargassumfish (*Histrio histrio*). During agonistic interactions, individuals of this species will bite each other, which often results in fins and skin flaps ripped to shreds (see Aggression, p. 451). These more serious injuries often lead to the death of the more subordinate individuals. The only time adult Sargassumfish will tolerate each other in the confines of an aquarium is during courtship and spawning. Even after spawning, females are best removed to avoid their being beaten to death by male suitors. Moreover, because of their belligerent nature, adult Sargassumfish cannot be trusted with other species of frogfishes.

Feeding

Frogfishes have a low metabolic rate and are extremely efficient at assimilating nutrients from their food. Therefore, it is only necessary to feed the adults twice a week, unless the aquarist is keeping more than one in the same aquarium. In the later situation, it is prudent for the aquarist to feed them three or four times a week in order to prevent a larger individual from eating a smaller relative. Also, juvenile frogfishes should be fed three or four times a week because of the greater metabolic demands associated with growth. It is not uncommon for adult frogfishes to consume two or three small feeder fish a day, but it is not necessary and may be harmful to feed them this much. Although it is entertaining to watch them eat live food, it is best to train a frogfish to consume a varied seafood diet to ensure its nutritional needs are met. The majority of individual antennariids can eventually be trained to take nonliving food items. However, if an individual refuses such offerings, it may be necessary to feed a captive frogfish live food. The best choice is live caridean shrimp (grass shrimp of the genus *Palaemonetes*). There is some risk of feeding a frogfish live feeder fish, such as guppies and mollies (Poeciliidae), damselfishes (Pomacentridae), and cardinalfishes (Apogonidae), because they can carry parasites into the display tank. Note that live freshwater feeder fish, in addition to being a possible parasite vector, are not a nutritious long-term alternative for frogfish fare. When feeding live food, it is important to "gut pack" these prey items before presenting them to the frogfish. This should be done by adding nutritious food to the holding tank where the prey species are kept, an hour or so before they are introduced to the display aquarium.

When feeding fresh or frozen foods, the best way to feed a frogfish is to impale the food item on the end of a rigid piece of clear airline tubing. To entice a frogfish to eat, the aquarist should mimic the movement of living prey by slowly moving it back and forth about one body length away from the frogfish. If the frogfish begins to lure, the food should be carefully moved into its strike zone. It is important to present food gradually and gently—abruptly sticking food in the face of the frogfish is more than likely to discourage feeding. If the frogfish is housed with other fishes, it may be necessary to fend off hungry tankmates to keep them from stealing food off the end of the feeding stick.

Frogfishes are quite methodical in their feeding behavior, often stalking prey for some time before ingesting it. If the aquarist is keeping an antennariid with aggressive feeders like groupers or snappers, it can be difficult to feed them, especially if the aquarist is simply introducing live food into the aquarium. When competitors are present, the best way to feed a frogfish live fare is to place the prey organism in a net and gradually move it toward the antennariid (keeping a grass shrimp or molly in a net is not difficult if you move the net slowly). The net should be positioned several body lengths from the frogfish and held still. In most cases, if patience is practiced, the frogfish will "walk" over to the net, crawl in, and capture the prey item. Soon the frogfish will learn to associate the net with food and will begin moving toward it as soon as it enters the tank. Frogfishes are often prompted to lure when a net is put up to the front of the aquarium—in one case, an individual was dubbed "Pavlov's frogfish" because it would lure frenetically anytime someone walked past the tank with a net (Michael, 1998: 338). The net method is also a great way to feed frogfishes in a reef aquarium, where live food is likely to quickly disappear into the reef structure.

When feeding an antennariid, it is extremely important not to present prey items greater than half of its total body length. It is better to feed it several smaller prey items than one large one. If a frogfish ingests a fish that is too large, the food may decay faster than it can be digested. The gas produced during decomposition often becomes trapped in the alimentary tract causing the overfed frogfish to float like a cork. Death typically occurs in two to three days after floating begins. When this happens, it is tempting to try to purge their stomachs with a syringe full of seawater or to pull the decaying prey item out of the digestive tract with tweezers, but these techniques are seldom successful. Even if the decaying food is removed, the event may still result in the death of the frogfish.

The fish species most frequently housed with, eaten by, and ending up killing a frogfish are anthias (Serranidae), wrasses (Labridae), and gobies (Gobiidae). This is because these fishes are elongate and flexible, making them easy to swallow and roll up within the stomach of a frogfish where they decay before they can be properly digested. Other fishes that have been known to cause the death of a "greedy" frogfish, even though they were considerably longer than the antennariid, include dottybacks (Pseudochromidae), tilefishes (Malacanthidae), certain blennies (Blenniidae), and sandperches (Pinguipedidae). Specific examples of death by "overeating" abound (Michael, 1998: 338): in one case, a frogfish swallowed a Comet (*Calloplesiops altivelis*, Plesiopidae) that was about 3 cm longer than itself; within 24 hours, the frogfish was floating around the aquarium and the next day it was dead. In another example, a frogfish attempted to eat a Whitespotted Bamboo Shark (*Chiloscyllium plagiosum*, Hemiscylliidae), at least twice as long as itself. When the antennariid died 24 hours later, the tail and posterior portion of the shark's body were still protruding from its mouth. Examples from nature exist as well: a Giant Frogfish (*Antennarius commerson*) was once discovered floating dead at the surface with an inflated porcupinefish (genus *Diodon*) in its stomach (Michael, 1998: 331).

Disorders and Diseases

Frogfishes have the potentially suicidal habit of ingesting air, especially during shipping or when they are lifted from the water. For this reason, it is extremely important to keep them submerged. When transferring an individual from one tank to another, it is imperative to use a container—not a net—to catch and move the frogfish. If they are being transported from the aquarium shop to a home tank, the water and frogfish should be carefully poured from the specimen container into a plastic bag. When arriving at home, the fish should be acclimated (slowly dripping water from the display tank into the bag is the preferred method), some of the water should be discarded from the bag, then the open bag should be submerged into the display tank and the frogfish should be released. Although it is usually not recommended to add water from a dealer's aquarium to the home display tank, in the case of a frogfish it is warranted. If a frogfish ingests air during the transfer, there is not much that can be done. It will often succeed in expelling the air on its own, but it is also not uncommon for an individual to float around until it dies.

Occasionally, a frogfish will damage the anterior edge of its lower jaw by rubbing up and down against the side of its plastic bag during shipping. This can abrade its skin causing redness and possible blistering. Although these wounds often heal without treatment, there is the risk of bacterial infection. Placing newspaper or black plastic around the outside of the bag may reduce stress and prevent the frogfish from injuring itself.

Another thing to consider when shipping frogfishes is that they should not be fed for at least a week before they are shipped. While closed up in a plastic bag, it is not unusual for a frogfish to regurgitate a meal up to five days after eating and the polluted water killing the fish while in transit. In most of these cases, the dealer was probably unaware that the antennariid had eaten a fish that had been placed unintentionally in the same holding tank.

Frogfishes are very sensitive to skin parasites so it is important to keep a close eye on them, especially newly acquired individuals. The latter, if intended to live with other fishes, should always be quarantined for a time (a quarantine minimum of three weeks) before being added to the display aquarium. It is often difficult to spot parasites on frogfishes because of their elaborate coloration, cutaneous appendages and filaments, and the scab-like ornamentation that helps them blend in with their surroundings. When they are ill, they look slightly swollen: their eyes appear sunken in, their skin appears smoother, and normal body contours disappear. The respiration of a sick frogfish is also very different from that of a healthy specimen. When a healthy frogfish respires, it is barely noticeable. Its mouth is left slightly open, water is pulled in and forced back past the gills where oxygen is extracted and carbon dioxide removed, and is eventually expelled through the tiny gill openings located at the base of the pectoral fins. When a frogfish is sick or greatly stressed, breathing is noticeably labored—rather than slowly pulling water into a barely open mouth, the frogfish gulps mouthfuls of water and the rate of intake and exhalation increases dramatically.

Treating a sick frogfish can be tricky. Some of the common anti-parasite medications may cause considerably more harm than good. Copper-based treatments are especially susceptible to misuse. The best way to deal with skin parasites is to prevent exposure in the first place. If the aquarist feeds a frogfish with live saltwater fishes, the latter need to be quarantined and prophylactically treated before introducing them into the display aquarium. If the frogfish is being kept in a community tank, all fishes introduced to the tank need to be quarantined. If a frogfish contracts an external stenohaline parasite (e.g., *Cryptocaryon irritans*, a species of ciliates that causes marine white spot disease or "marine ich") it can be treated by slightly reducing the relative salinity (amount of dissolved salts) by lowering the specific gravity in your aquarium from the recommended level of 1.023–1.025 to 1.010–1.011 for a couple of weeks. Beware that these lower specific gravity levels will kill most ornamental invertebrates. The specific gravity should be dropped over a period of several hours. When returned to a more "normal" range, this should be done over a longer time period (e.g., over two days). A UV sterilizer, when used correctly, can help prevent parasite problems in the first place. When it comes to their long-term health, the most important factor about frogfishes is that they are very susceptible to stress. While stress can cause problems for any marine fish, antennariids tend to have more difficulty bouncing back after stressful experiences.

Color Change

Frogfishes are able to change color in response to their surroundings. This can include radical chromatic shifts (e.g., from bright yellow to red or dark brown). Such color plasticity can cause much frustration and disappointment to the aquarist that has acquired (usually at a high cost) a brightly colored individual, only to have it transform to a dull muddy hue once established at home. One way to discourage chromatic change is to provide structure to the tank of the same color as the frogfish. For example, if you have a red Warty Frogfish (*Antennarius maculatus*) and want it to remain red, add some artificial or live red sponges for it to perch on. Do remember, however, that if the conditions around the feeding site are not conducive to frogfish perching (e.g., there is too much water flow), the frogfish may not use it. Unfortunately, this trick does not always work. You might provide a nice colorful background on which the frogfish rests, only to have it take on a less desirable hue over time. That said, occasionally a captive frogfish may exhibit a more desirable change of complexion. For example, I have seen a solid black frogfish turn orange—in this case it occurred in an aquarium without any colorful aquascaping.

Frogfishes in the Field

Because of their sedentary nature and tendency to tolerate the close presence of the careful diver, often not moving even when within inches, frogfishes are easy to observe and photograph in the field; that is, if you can find them. Being rather rare at most dive locations, they are usually solitary and situated far and few between—except, of course, when in reproductive mode. As masters of disguise, they are often overlooked by even the most observant and experienced diver. Moving slowly and exercising patience, however, they can be found at most popular dive sites within tropical latitudes of the world. One of the best places to look for frogfishes is among encrusting sponges on pier pilings. The island of Bonaire is one of the most popular snorkeling and scuba diving destinations in the Caribbean because of its multiple shore-diving sites and easy access to fringing reefs; the sponge and coral encrusted pillars of Salt Pier on the southern tip of the island, for example, are home to a large population of the Longlure Frogfish (*Antennarius multiocellatus*). Other Caribbean hot spots for frogfishes include St. Croix, Dominica, St. Lucia, and Barbados. The Lembeh Strait of North Sulawesi, Indonesia, is a prime site for observing and photographing frogfishes. Here you can commonly see as many as seven species on the coastal reefs and macroalgae-sponge beds on a dive trip lasting only several weeks. Other good sites to view these interesting creatures in the Western Pacific include the coastal reefs around Ambon and Bali (Indonesia), the reefs of Sipadan (Borneo), Batangas (Philippines), and the Izu Peninsula (Japan).

Purchasing underwater photography and video equipment can be daunting not to mention expensive—the options are almost unending, ranging from a simple point and shoot camera to an underwater housing for a compact camera, a single lens reflex (SLR) or video camera, a waterproof camera, underwater strobes, video lights, focus lights, and so on. While point and shoot cameras are fully capable of creating great images, serious amateurs and professionals shoot with digital single lens reflex (DSLR) cameras. DSLR cameras allow for use of a wide range of high-quality interchangeable lenses for specific shooting scenarios and subjects from super macro to extreme wide angle. Additionally, the viewfinder of a DSLR camera provides accurate composition through the actual lens. Different types of underwater photography, either macro or wide angle, require different lenses. Most divers interested in photographing frogfishes want to get up close, so a macro lens ranging from 60- to 105-mm is best. For more information on cameras, lenses, and accessories, several respectable "underwater photo guides" are available on the Internet.

It is important to practice good wildlife etiquette when photographing antennariids. You should never move a frogfish, even if it means getting a better photograph. The frogfish will perceive you as a potential predator not as an underwater photographer with good intentions. When you handle or prod a frogfish, its respiration rate will increase and it will burn valuable calories. These fishes have no sharp or poisonous spines, nor do they exude any distasteful toxins (see Venom and Poison, p. 454). They rely completely on camouflage to avoid becoming a meal; therefore, if they are cast up into the water column in an attempt to get a better shot or if they are displaced to a less suitable shelter site by a diver or photographer, they may end up as food for a hungry piscivore.

Equally important, care must always be taken to avoid damaging the habitat. There are many examples of patch reefs, once rich with marine life, including frogfishes, decimated by the regular visits of enthusiastic but thoughtless divers who clamber over the reef structure as they attempt to find a cryptic inhabitant or to get a better photograph.

Conservation

Now that the joys and relative ease of maintaining frogfishes in the home aquarium have been described with some enthusiasm, the contradictory downside must be addressed— the need to proceed with caution. Characterized by low population densities, restricted and patchy geographic distributions, small home ranges, and limited mobility, frogfishes, especially members of the subfamily Histiophryninae, are particularly vulnerable to population declines and risk of extinction due to competition from invasive species, over-collecting by aquarists and the tropical fish trade, climate change (global warming), and habitat degradation. Members of the subfamily Histiophryninae are especially susceptible because of their severely limited dispersal capabilities resulting from a reproductive strategy of producing relatively few large eggs that undergo direct development— hatching as juveniles that drop to the sea floor, thus bypassing a pelagic larval stage—mediated through elaborate and sustained parental care of eggs (see Reproduction, p. 461). Despite this vulnerability, there is a surprising lack of conservation concerns directed toward these fishes, a fact that is at odds with their economic and social importance (De Brauwer et al., 2018: 33; see below). Surveys of densities, population trends, and microhabitat preferences have apparently not been conducted for any antennariid, and few if any detailed assessments of threats to frogfishes in general have been made. Conservation publications that include antennariids invariably list them as unassessed, data deficient, or of least concern (De Brauwer et al., 2018: 33, 34, table 4). Of 14 species of antennariids listed by the International Union for Conservation of Nature (IUCN) Red List of Threatened Species (http://www.iucnredlist.org/search), all are indicated as "least concern"; population trends for 12 are indicated as "unknown" and two as "stable" (*Antennarius striatus* and *Histrio histrio*). All but three of the 14 (*Abantennarius analis*, *Abantennarius nummifer*, and *Histrio histrio*) are Atlantic or Eastern Pacific species; thus, the conservation status of the highly diverse Indo-Pacific frogfish fauna is essentially unknown.

In evaluating the future health of antennariids, it is important to consider the case of the Australian handfishes, members of the closely related anglerfish family Brachionichthyidae (see Evolutionary Relationships, p. 371; Fig. 269). Brachionichthyids, with five genera and 14 species, seven of which are found only in the Bass Strait and off Tasmania, are characterized by having modified pectoral fins that resemble human hands, much like those of frogfishes, but mounted on much longer "arms," providing enhanced flexibility—although unusually sedentary, "walking" is their primary mode of locomotion (Last and Gledhill, 2009). They are found on soft muddy and sandy bottoms where they form small localized populations. Spawning females attach gelatinous egg masses of up to 200 eggs onto stalked ascidians (i.e., sea squirts) and various marine vegetation and then guard the eggs for up to six weeks until they hatch (Barrett et al., 1996). Together, handfishes constitute the only lophiiform family with members listed as threatened and endangered (Bruce et al., 1998; Last and Gledhill, 2009). Three handfish species are now listed as endangered (Department of the Environment and Heritage, 2005; Commonwealth of Australia, 2016), but of greatest concern is the Spotted Handfish, *Brachionichthys hirsutus* (Barrett et al., 1996; Bruce, 1998; Bruce and Green, 1998; Bruce et al., 1997, 1998, 1999; Green and Bruce, 2002; Green et al., 2012; Lynch et al., 2015, 2016; Wong and Lynch, 2017; Fig. 1E). Once common throughout its range, in bays and estuaries of southeastern Tasmania at depths of 1 to 60 m, populations of the Spotted Handfish have suffered a significant decline in abundance, and individuals are now rarely encountered. The decline was first documented in the late 1980s when information was gathered on its basic biology and ecology (Barrett et al., 1996). Surveys conducted at the time failed to find any specimens in areas previously well-known for sightings, and, subsequently, only two specimens were reliably reported between 1990 and 1994 (Bruce and Green, 1998). More recent surveys of 50 sites covering its previously known range, located populations at only three localities, and isolated individuals at a further seven (Bruce and Green, 1998). In 1996, it became the first marine fish species to be listed as "critically endangered" in the IUCN Red List of Threatened Species. It is now also listed as "critically endangered" under the Commonwealth Environment Protection and Biodiversity Conservation Act of 1999, and "endangered" under Tasmania's Threatened Species Protection Act of 1995 (Department of the Environment and Heritage, 2005; Department of Primary Industries and Water, 2007). The factors that have made handfishes vulnerable to extinction are the very same that characterize members of the Histiophryninae: highly restricted and patchy distributions, low population densities, limited dispersal capabilities, and a reproductive strategy that involves parental care of a small number of large eggs that are highly susceptible to disturbance (Bruce and Green, 1998). Therefore, it is important that we heed the sad fate of handfishes—the same could happen to frogfish populations if not actively monitored and if remedial steps are not taken.

A seldom realized potential threat to frogfishes, as well as to those of many other marine organisms, is nature-based tourism, which constitutes an important source of income in both developing and developed countries (Balmford et al., 2009; De Brauwer and Burton, 2018). Global scuba diving, and reef-related tourism in general, is one of the most significant and fast-growing examples of nature-based tourism for a single ecosystem (Spalding et al., 2017). Coral reefs attract foreign and domestic visitors from around the world, and generate significant revenues in over 100 countries and territories (Spalding et al., 2017). Some 30% of the world's reefs are of value in the tourism sector, with a total estimated value of nearly US $36 billion, or over 9% of all coastal tourism value in the world's coral reef countries (Brander et al., 2007; Spalding et al., 2017). While reefs are of primary interest, a rapidly growing focus has been directed toward soft-sediment tropical habitats, where frogfishes in particular abound. Generally referred to as "muck diving" (Lew, 2013), this form of tourism, which focuses on finding and photographing cryptobenthic species, is currently thriving in suitable localities throughout the tropical Indo-Pacific and is estimated to be worth more than US $150 million annually

(De Brauwer and Burton, 2018). The habitats it depends on, however, do not benefit from any formal conservation activities and the taxa that primarily support tourism in soft-sediment habitats have been poorly studied (DeVantier and Turak, 2004; De Brauwer et al., 2017; De Brauwer and Burton, 2018).

In an effort to identify the species that are most important in attracting tourists, De Brauwer and Burton (2018) concluded that muck-diving tourism depends strongly on only a few taxa—perhaps not surprising, they found that the three (listed in order) most popular organisms sought by divers are "frogfishes, nudibranchs, and octopuses." (De Brauwer and Burton, 2018: 33). But the extent to which these and other popular organisms are impacted and what effects scuba divers have on the marine environment remains to be fully determined. To evaluate the environmental impacts, De Brauwer et al. (2018) observed 66 divers on coral reefs and soft-sediment habitats in Indonesia and the Philippines. They found that diver activity, specifically photography and associated interaction with animals, causes greater environmental disturbances than effects related to certification level, gender, dive experience, or age. Divers touched the substrate more often while diving on soft-sediment habitats than on coral reefs, and had a higher impact on the substrate and touched animals more frequently when observing or photographing the fauna. When using DSLR cameras (see Frogfishes in the Field, p. 496), divers spent up to five times longer interacting with organisms. Considering the unknown, long-term impacts on the cryptobenthic fauna and on soft-sediment habitats, the increasing popularity of underwater photography, and the growing economic value of muck-diving tourism, there is an urgent need for more research on the marine fauna of shallow soft sediment and other habitats important to tourism (De Brauwer and Burton, 2018). De Brauwer et al. (2018) concluded their study by arguing for the introduction of a muck-diving code of conduct.

Two additional, perhaps more obvious, potential threats to frogfishes are collecting by individual aquarists, and wholesale harvesting by the marine tropical fish trade. The global trade of aquatic organisms for home and public aquariums, along with associated equipment and accessories, has become a multibillion-dollar industry; although marine species make up less than 10% of the total volume of the ornamental trade, the percentage in terms of value is much higher and has increased significantly in recent years (Lin, 2017). Estimates of 18 to 30 million marine fishes, representing over 2,250 species, and hundreds of coral and other invertebrate species are traded globally every year (Wabnitz et al., 2003; Townsend, 2011; Rhyne et al., 2012; Baensch, 2014, 2018). The most recent global estimate indicates that more than 46 million individual marine organisms are collected and sold annually to approximately 2 million hobbyists worldwide, with a corresponding value exceeding US $300 million (Wabnitz et al., 2003; Palmtag, 2017). The Philippines and Indonesia, followed by the Solomon Islands, Sri Lanka, Australia, Fiji, the Maldives, and Palau, supply the overwhelming majority of livestock. Most of the specimens by far are imported by the United States (Fig. 346), with the remainder sent to Europe, Japan, and a handful of other countries (Bruckner, 2005; Palmtag, 2017; Rhyne et al., 2017b). Damselfishes, anemonefishes, and angelfishes constitute over 50% of the global volume; butterflyfishes, wrasses, blennies, gobies, triggerfishes, filefishes, hawkfishes, groupers, and basslets account for 31% of the trade, and the remaining 16% is represented by 33 additional families (Bruckner, 2005).

Scuba diving has opened up many habitats that until recently were inaccessible, providing numerous species heretofore unavailable to aquarists. Technological advances in underwater camera equipment have made it possible to more easily document the beauty

Fig. 346. Trade flow of marine aquarium fishes and invertebrates from source nations into the United States during 2008, 2009, and 2011. Numbers within circles denote percent contribution to total imports. Pie chart within the United States represents ports of entry (with the Midwest starting at 0 degrees, and clockwise, NE, SE, SW, and NW). Courtesy of Andrew Rhyne; after Rhyne et al. (2017a, fig. 5).

of marine organisms; in fact, photography, one of the most popular hobbies in the world, has greatly facilitated aquarium keeping, helping in turn to make it one of the most popular hobbies as well (Lin, 2017).

Unlike freshwater fishes of which more than 90 percent are produced through aquaculture, nearly all marine species are taken from the wild (Palmtag, 2017; Baensch, 2018). And it is not just the fishes—most of the live rock and invertebrates in the marine aquarium trade are also taken from the wild. Extensive harvesting and highly destructive collecting methods—such as the use of various toxic chemicals (including plant extracts, bleach, and most notably, cyanide; Rubec, 1986; Bruckner, 2001) and the common practice of breaking and demolishing coral (Pet-Soede and Erdmann, 1998; Bruckner, 2001; Wood, 2001), coupled with nonexistent or ineffective fisheries management—have resulted in serious depletion of target populations as well as irreparable damage to delicate coral reef ecosystems that are already under serious threats from global climate change, pollution, and other anthropogenic impacts (McManus et al., 1997; Palmtag, 2017).

While bad fishing practices and over-collecting of pelagic species—which constitute the largest proportion by far of the tropical marine fish fauna targeted by the aquarium trade (Rhyne et al., 2017b, table 3)—have a serious detrimental effect on their populations, the negative impact on benthic fishes, and especially cryptobenthic species such as antennariids, is even more severe. Like other lie-in-wait predators, frogfishes are not abundant. A local frogfish population could be decimated rather quickly if placed under heavy collection pressure. But antennariids are readily available in the aquarium trade—one to six individuals are generally available on a marine fish wholesaler price list every week. According to statistics provided by the Marine Aquarium Biodiversity and Trade Flow website (Aquarium Trade Data, 2018), a total of 19,225 antennariids were exported to various countries around the world during three years (2008, 2009, and 2011) of data collection. Of these 19,225, 11,098 came from the Philippines and 7,254 from Indonesia; the next highest number for any country was 344 specimens reported for Sri Lanka, followed by Kenya at 227, and a cluster of 13 other countries combined at 302. During these same years and based on the same analysis, a total of 20,909,681 ornamental marine fishes representing all families were exported to various countries around the world, of which 13,237,542 came from the Philippines and 6,630,442 from Indonesia; the next highest number for any country was 652,416 specimens reported for Sri Lanka, followed by Kenya at 389,281, and a cluster of 40 other countries combined at 2,617,739. Thus, the number of frogfishes collected and removed from their natural habitat is still quite small compared with exported fishes as a whole, amounting to only about 0.09%. Yet realizing the relative rarity of frogfishes in nature, even this seemingly limited amount of harvesting could have a serious detrimental effect over the long term.

The threat to the health of frogfishes, and to marine fish communities in general, can be mitigated by setting aside Marine Protected Areas (MPA) and Fish Replenishment Areas (FRA) in regions where fishes are most frequently collected for the aquarium trade (Rossiter and Levine, 2014; Bonaldo et al., 2017; Nagelkerken et al., 2017). An MPA is an area of coastline or reef where fishing and other activities that impact the habitat are managed, while an FRA is an area where no fishing is allowed at all. In these areas, resident adult fishes (brood stock) are left unmolested to supply tracts of safeguarded coastline with their progeny, theoretically ensuring the preservation of fish populations in and around the protected areas. A good example of the efficacy of FRAs in particular is the effort begun more than 20 years ago in the Hawaiian Islands (Williams et al., 2009; Rossiter and Levine, 2014). In 1998, in order to combat the degradation of Yellow Tang (*Zebrasoma flavescens*) populations on the west coast of Hawaii Island, FRAs were established prohibiting aquarium fishing along more than 30% of the coastline (Williams et al., 2009). Unlike other marine management approaches in Hawaii, which have largely been controversial, fraught with confusion over regulations, inadequately enforced, and lacking public support, these FRAs have been lauded as a marine conservation success, with wide-ranging support and evidence of rapid replenishment of the Yellow Tang population (Rossiter and Levine, 2014).

A potential alternative source of fishes for the marine aquarium trade is captive breeding programs (Calado, 2017a, 2017b; Calado et al., 2017; Fioravanti and Florio, 2017; Rhyne et al., 2017a). Culturing marine ornamentals is a lucrative and environmentally sound alternative to harvesting them from their reef habitat. The practice, however, still greatly lags behind the farming of freshwater tropicals because of the biological and economic constraints associated with culturing most species (Baensch, 2014, 2018). Many species have not been cultured because of their complex reproductive biology, others can only be raised a few individuals at a time, and for those that can be raised in large numbers, production is often not cost-effective compared to capturing their wild counterparts. Nevertheless, marine ornamental culture has made considerable progress in recent years. As of January 2018, 358 of the more than 2,250 traded marine aquarium fish species have been cultured, up from 170 species in 2000 (Sweet and Pedersen, 2018). Of these, about 65 species (4%) have been captive-bred commercially and about 30 of these are in production at any one time—mostly clownfishes, dottybacks, and sea horses, and a limited number of goby, blenny, cardinalfish, and angelfish species (Sweet and Pedersen, 2018). The remaining 260 marine aquarium species have been cultured for experimental purposes only (Baensch, 2018; Sweet and Pedersen, 2018). The list includes only a single frogfish species, *Rhycherus filamentosus*, based on work published more than 25 years ago by Rudie Kuiter (1993; see Courtship and Spawning Behavior, p. 468; Figs. 334, 335). The direct development of *Rhycherus filamentosus* and other members of the antennariid subfamily Histiophryninae certainly lend themselves nicely to captive breeding—species of *Lophiocharon* and *Histiophryne* especially seem like good candidates. The spawning of relatively few, larger eggs that hatch and produce relatively well-developed young are in marked contrast to the much more numerous, tiny eggs of the Antennariinae that undergo a prolonged larval stage. Aquaculture of members of the latter subfamily will prove to be much more difficult.

It should be noted that a recovery program for the Spotted Handfish was initiated in the mid-1990s, immediately after initial surveys indicated a drastic reduction in the size and geographic range of populations (Barrett et al., 1996; Bruce et al., 1997; Green et al., 2012). An effort to locate breeding colonies and to collect adults for the purpose of developing captive breeding protocols was made as early as 1997 (Bruce et al., 1997; Green et al., 2012). Publication of the first of several recovery plans quickly followed (Bruce and Green, 1998; Green and Bruce, 2002). Spotted Handfish juveniles were successfully reared in captivity and released in the late 1990s but problems attributed to technological constraints resulted in high mortalities (Bruce et al., 1997; Green and Bruce, 2002; Lynch et al., 2017; Wong and Lynch, 2017; Wong et al., 2018). By the end of 2017, however, better equipment and improved protocols produced a healthy captive population of 20 adults or subadults and 96 captive-bred juveniles, with zero mortality (Lynch et al., 2017; Wong et al., 2018). By all accounts, the handfish recovery program is working, and things look promising for the future.

Reallocation of Nominal Species of Frogfishes

Nominal Taxa

	Primary Senior Synonym
Antennarius albomaculatus Fowler, 1949	
[error for *albomarginatus*]	*Abantennarius dorehensis*
Antennarius albomarginatus Fowler, 1945	*Abantennarius dorehensis*
Antennarius altipinnis Smith & Radcliffe, in Radcliffe, 1912	*Abantennarius dorehensis*
Antennarius ampylacanthus Delais, 1951	
[error for *campylacanthus*]	*Antennarius pardalis*
Abantennarius analis Schultz, 1957	*Abantennarius analis*
Antennarius annulatus Gill, 1863	*Antennarius multiocellatus*
Tathicarpus appeli Ogilby, 1922	*Tathicarpus butleri*
Chironectes arcticus Düben & Koren, 1846	*Histrio histrio*
Antennarius argus Fowler, 1903	*Antennarius pictus*
Antennarius asper Macleay, 1881	*Lophiocharon trisignatus*
Antennarius astroscopus Nichols, 1912	*Antennarius multiocellatus*
Phrynelox atra Schultz, 1957	*Antennarius striatus*
Antennarius avalonis Jordan & Starks, 1907	*Fowlerichthys avalonis*
Chironectes barbatulus Eydoux & Souleyet, 1850	*Histrio histrio*
Antennarius bermudensis Schultz, 1957	*Abantennarius bermudensis*
Cheironectes bicornis Lowe, 1839	*Abantennarius nummifer*
Chironectes bifurcatus McCoy, 1886	*Rhycherus filamentosus*
Histiophryne bigibba Le Danois, 1964	*Antennatus tuberosus*
Lophius bigibbus Latreille, 1804	*nomen dubium*
Chironectes biocellatus Cuvier, 1817b	*Antennarius biocellatus*
Antennarius bivertex Schultz, 1957	*Antennarius commerson*
Antennarius pinniceps Var. β. *bleekeri* Günther, 1861a	*Antennarius striatus*
Chironectes bougainvilli Valenciennes, 1837	*Histiophryne bougainvilli*
Lophiocharon broomensis Whitley, 1933	*Lophiocharon trisignatus*
Tathicarpus butleri Ogilby, 1907	*Tathicarpus butleri*
Lophius calico Mitchill, 1818	*Histrio histrio*
Antennarius campylacanthus Bleeker, 1863	*Antennarius pardalis*
Antennarius candimaculatus Günther, 1880	
[error for *caudimaculatus*]	*Lophiocharon trisignatus*
Antennarius commersonii Var. β. *cantoris* Günther, 1861a	*nomen dubium*
Chironectes caudimaculatus Rüppell, 1829	*Antennarius commerson*
Antennarius caudomaculatus Günther, 1880	
[error for *caudimaculatus*]	*Lophiocharon trisignatus*
Lophius chironectes Latreille, 1804	*Antennarius pictus*
Antennarius chironemus Munro, 1967	*Antennarius pictus*
Chironectes chlorostigma Rüppell, 1838	
[emendation of *chlorostygma* Ehrenberg]	*Abantennarius nummifer*
Chironectes chlorostygma Valenciennes, 1837	*Abantennarius nummifer*

Nominal Taxa	Primary Senior Synonym
Chironectes coccineus Lesson, 1831	*Abantennarius coccineus*
Lophius cocinsinensis Shaw, 1811	*Histrio histrio*
Lophius commerson Latreille, March 1804	*Antennarius commerson*
Lophius commersonianus Lacepède, in Jordan, 1917b	*Antennarius striatus*
Lophius commersonii Shaw, June 1804	*Antennarius commerson*
Lophius concincinensis Valenciennes, 1837	
[error for *cocinsinensis*]	*Histrio histrio*
Golem cooperae Strasburg, 1966	*nomen dubium*
Antennarius corallinus Poey, 1865	*Antennarius multiocellatus*
Echinophryne crassispina McCulloch & Waite, 1918	*Echinophryne crassispina*
Antennarius cryptacanthus Weber, 1913	*Histiophryne cryptacanthus*
Antennarius cubensis Borodin, 1928	*Antennarius scaber*
Antennarius cunninghami Fowler, 1941	*Antennarius striatus*
Antennarius delaisi Cadenat, 1959	*Antennarius striatus*
Antennarius dorehensis Bleeker, 1859a	*Abantennarius dorehensis*
Antennarius drombus Jordan & Evermann, 1903	*Abantennarius drombus*
Antennarius duescus Snyder, 1904	*Abantennarius duescus*
Porophryne erythrodactylus Arnold, Harcourt, & Pietsch, 2014	*Porophryne erythrodactylus*
Antennarius pinniceps var. *fasciata* Steindachner, 1866	*Antennarius striatus*
Chironectes filamentosus Castelnau, 1872	*Rhycherus filamentosus*
Antennatus flagellatus Ohnishi, Iwata, & Hiramatsu, 1997	*Antennatus flagellatus*
Fowlerichthys floridanus Barbour, 1941	*Fowlerichthys radiosus*
Antennarius fuliginosus Smith, 1957	*Antennarius striatus*
Chironectes furcipilis Cuvier, 1817b	*Kuiterichthys furcipilis*
Lophius geographicus Quoy & Gaimard, 1825	*Histrio histrio*
Lophius gibba Cuvier, 1829	
[emendation of *gibbus*]	*Histrio histrio*
Peterophryne gibbo Connolly, 1920	
[emendation of *gibbus*]	*Histrio histrio*
Lophius gibbus Mitchill, 1815	*Histrio histrio*
Echinophryne glauerti Whitley, 1944a	*Allenichthys glauerti*
Antennarius glauerti Whitley, 1958	*Antennarius striatus*
Rhycherus gloveri Pietsch, 1984c	*Rhycherus gloveri*
Antennarius goramensis Bleeker, 1865a	*Antennarius commerson*
Antennarius guntheri Bleeker, 1865b	*Antennarius maculatus*
Lophius hispidus Bloch & Schneider, 1801	*Antennarius hispidus*
Lophius histrio Linnaeus, 1758	*Histrio histrio*
Antennarius horridus Bleeker, 1853a	*Antennarius pictus*
Lophiocharon hutchinsi Pietsch, 2004a	*Lophiocharon hutchinsi*
Antennarius immaculatus Le Danois, 1970	*Abantennarius coccineus*
Histiophryne scortea var. *inconstans* McCulloch & Waite, 1918	*Phyllophryne scortea*
Antennarius indicus Schultz, 1964	*Antennarius indicus*
Antennarius inops Poey, 1881	*Histrio histrio*
Batrachopus insidiator Whitley, 1934	*Antennarius striatus*
Histrio jagua Nichols, 1920	*Histrio histrio*
Antennarius japonicus Schultz, 1957	*Abantennarius nummifer*
Ostracion knorrii Walbaum, 1792	*Antennarius scaber*
Antennarius lacepedii Bleeker, 1856a	*Antennarius striatus*

Nominal Taxa	Primary Senior Synonym
Pterophryne laevigata Poey, 1881	
[emendation of *laevigatus*]	*Histrio histrio*
Lophius laevigatus Bosc, in Cuvier, 1816	*Histrio histrio*
Lophius laevis Latreille, 1804	
[senior primary homonym of *Lophius laevis*	
Lacepède, 1804, Brachionichthyidae]	*Histrio histrio*
Antennarius lateralis Tanaka, 1917	*Antennarius commerson*
Antennarius laysanius Jordan & Snyder, 1904	*Antennarius pictus*
Antennarius leopardinus Günther, 1864	*Abantennarius sanguineus*
Antennarius multiocellatus var. δ. *leprosa* Günther, 1861a	*Antennarius pictus*
Chironectes leprosus Eydoux & Souleyet, 1850	*Antennarius pictus*
Antennarius leucosoma Bleeker, 1854b	*Antennarius pictus*
Antennarius leucus Fowler, 1934b	*Abantennarius coccineus*
Antennarius lindgreeni Bleeker, 1855b	*Lophiocharon trisignatus*
Antennatus linearis Randall & Holcom, 2001	*Antennatus linearis*
Saccarius lineatus Günther, 1861a	*Antennarius striatus*
Antennarius lioderma Bleeker, 1865a	*Histrio histrio*
Antennarius lithinostomus Jordan & Richardson, 1908	*Lophiocharon lithinostomus*
Phrynelox lochites Schultz, 1964	*Antennarius striatus*
Chironectes lophotes Cuvier, 1817b	*Antennarius hispidus*
Antennarius lutescens Seale, 1906	*Antennarius commerson*
Phymatophryne maculata Le Danois, 1964	
[emendation of *maculatus*]	*Antennarius maculatus*
Chironectes maculatus Desjardins, 1840	*Antennarius maculatus*
Antennarius maculosus Myers & Shepard, 1980	
[error for *maculatus*]	*Antennarius maculatus*
Histiophryne maggiewalker Arnold & Pietsch, 2011	*Histiophryne maggiewalker*
Antennarius marmoratus Var. ε. *marmorata* Günther, 1861a	*Histrio histrio*
Lophius marmoratus Shaw, 1794	*nomen dubium*
Antennarius melas Bleeker, 1857	*Antennarius striatus*
Chironectes mentzelii Valenciennes, 1837	*Antennarius scaber*
Antennarius mentzelli Miranda-Ribeiro, 1915	
[error for *mentzelii*]	*Antennarius scaber*
Antennarius mitchellii Morton, 1897	*Echinophryne mitchellii*
Antennarius moai Allen, 1970	*Abantennarius coccineus*
Antennarius moluccensis Bleeker, 1855c	*Antennarius commerson*
Tathicarpus mucosum Schultz, 1957	
[error for *muscosus*]	*Tathicarpus butleri*
Antennarius multiocellatus Var. α. *multiocellata*	
Günther, 1861a	*Antennarius multiocellatus*
Chironectes multiocellatus Valenciennes, 1837	*Antennarius multiocellatus*
Antennarius mummifer Gregory, 1933	
[error for *nummifer*]	*Abantennarius nummifer*
Tathicarpus muscosus Ogilby, 1907	*Tathicarpus butleri*
Antennarius commersonii Var. γ. *Musei britannici*	
Günther, 1861a	*nomen dubium*
Histiophryne narungga Arnold and Pietsch, 2018	*Histiophryne narungga*
Abantennarius neocaledoniensis Le Danois, 1964	*Abantennarius coccineus*
Chironectes nesogallicus Valenciennes, 1837	*Histrio histrio*
Antennarius nexilis Snyder, 1904	*Abantennarius coccineus*
Chironectes niger Garrett, 1864	*Antennarius pictus*
Antennarius nigromaculatus Playfair, 1869	*Antennarius pictus*

Nominal Taxa	Primary Senior Synonym
Antennarius nitidus Bennett, 1827	*Histrio histrio*
Antennarius niveus Fowler, 1946	*Abantennarius dorehensis*
Antennarius notophthalmus Bleeker, 1853b	*Antennarius biocellatus*
Antennarius nox Jordan, 1902	*Antennarius striatus*
Chironectes nummifer Cuvier, 1817b	*Abantennarius nummifer*
Antennarius nuttingii Garman, 1896	*Antennarius scaber*
Antennarius occidentalis Cadenat, 1959	*Antennarius striatus*
Lophius histrio var. d. *ocellatus* Bloch & Schneider, 1801	*Fowlerichthys ocellatus*
Antennarius oligospilos Bleeker, 1857	*Antennarius maculatus*
Antennarius oligospilus Bleeker 1865b	
[emendation of *oligospilos*]	*Antennarius maculatus*
Antennarius pardal Fowler, 1936b	
[error for *pardalis*]	*Antennarius pardalis*
Chironectes pardalis Valenciennes, 1837	*Antennarius pardalis*
Antennarius pauciradiatus Schultz, 1957	*Antennarius pauciradiatus*
Chironectes pavoninus Valenciennes, 1837	*Antennarius multiocellatus*
Lophius pelagicus Banks, in Beaglehole, 1962	*Histrio histrio*
Chironectes peravok Montrouzier, 1856	*Antennarius pictus*
Antennarius phymatodes Bleeker, 1857	*Antennarius maculatus*
Antennarius marmoratus Var. α. *picta* Günther, 1861a	*Histrio histrio*
Lophius pictus Shaw, 1794	*Antennarius pictus*
Kuiterichthys pietschi Arnold, 2013	*Kuiterichthys pietschi*
Antennarius pinniceps Bleeker, 1856a	*Antennarius striatus*
Antennarius pleurophthalmus Gill, 1863	*Fowlerichthys ocellatus*
Histiophryne pogonius Arnold, 2012	*Histiophryne pogonius*
Antennarius polyophthalmus Bleeker, 1852b	*Antennarius pictus*
Antennarius portoricensis Stahl, 1882	*nomen nudum*
Antennarius princeps Gill, 1863	*nomen nudum*
Chironectes principis Valenciennes, 1837	*Antennarius multiocellatus*
Histiophryne psychedelica Pietsch, Arnold, & Hall, 2009a	*Histiophryne psychedelica*
Antennarius punctatissimus Fowler, 1946	*Abantennarius dorehensis*
Antennarius radiosus Garman, 1896	*Fowlerichthys radiosus*
Antennarius randalli Allen, 1970	*Antennarius randalli*
Antennarius marmoratus Var. γ. *ranina* Günther, 1861a	
[emendation of *raninus*]	*Histrio histrio*
Lophius raninus Tilesius, 1809	*Histrio histrio*
Antennarius reticularis Gilbert, 1892	*Antennatus strigatus*
Chironectes reticulatus Eydoux & Souleyet, 1850	*Antennatus tuberosus*
Echinophryne reynoldsi Pietsch & Kuiter, 1984	*Echinophryne reynoldsi*
Antennarius rosaceus Smith & Radcliffe, in Radcliffe, 1912	*Abantennarius rosaceus*
Chironectes rubrofuscus Garrett, 1863	*Antennarius commerson*
Lophius sandvicensis Bennett, 1840	*Antennarius pictus*
Antennarius sanguifluus Jordan, 1902	*Abantennarius nummifer*
Antennarius sanguineus Gill, 1863	*Abantennarius sanguineus*
Antennarius sarasa Tanaka, 1916	*Fowlerichthys scriptissimus*
Chironectes scaber Cuvier, 1817b	*Antennarius scaber*
Histiophryne scortea McCulloch & Waite, 1918	*Phyllophryne scortea*
Antennarius scriptissimus Jordan, 1902	*Fowlerichthys scriptissimus*
Kanazawaichthys scutatus Schultz, 1957	*Fowlerichthys radiosus*
Antennarius sellifer Borodin, 1928	
[error for *stellifer*]	*Antennarius multiocellatus*
Antennarius senegalensis Cadenat, 1959	*Fowlerichthys senegalensis*

Nominal Taxa	Primary Senior Synonym
Chironectes sonntagi Müller, 1864	*Histrio histrio*
Lophius spectrum Gray, 1854	*Antennarius scaber*
Antennarius spilopterus Friese, 1974	*Antennarius pictus*
Antennarius stellifer Barbour, 1905	*Antennarius multiocellatus*
Antennarius stigmaticus Ogilby, 1912	*Abantennarius coccineus*
Lophius striatus Shaw, 1794	*Antennarius striatus*
Antennarius strigatus Gill, 1863	*Antennatus strigatus*
Chironectes subrotundatus Castelnau, 1875	*nomen dubium*
Antennarius subteres Smith & Radcliffe, in Radcliffe, 1912	*Nudiantennarius subteres*
Antennarius tagus Heller & Snodgrass, 1903	*Abantennarius sanguineus*
Antennarius teleplanus Fowler, 1912	*Antennarius scaber*
Chironectes tenebrosus Poey, 1853	*Antennarius multiocellatus*
Antennarius tenerosus Borodin, 1928	
[error for *tenebrosus*]	*Antennarius multiocellatus*
Antennarius tenuifilis Günther, 1869b	*Antennatus strigatus*
Lophius thymelicus Meuschen, 1781	*Antennarius striatus*
Antennarius tigrinus Thresher, 1984	*nomen nudum*
Chironectes tigris Poey, 1853	*Antennarius scaber*
Chironectes tricornis Cloquet, 1817	*Antennarius striatus*
Chironectes tridens Temminck & Schlegel, 1845	*Antennarius striatus*
Cheironectes trisignatus Richardson, 1844	*Lophiocharon trisignatus*
Histiophryne tuberosa Le Danois, 1970	
[emendation of *tuberosus*]	*Antennatus tuberosus*
Chironectes tuberosus Cuvier, 1817b	*Antennatus tuberosus*
Pterophryne tumida Jordan, 1905	
[emendation of *tumidus*]	*Histrio histrio*
Lophius tumidus Osbeck, 1765	*Histrio histrio*
Antennarius unicornis Bennett, 1827	*Antennatus tuberosus*
Antennarius urophthalmus Bleeker, 1851	*Lophiocharon trisignatus*
Chironectes variegatus Rafinesque, 1814	*Histrio histrio*
Antennarius verrucosus Bean, 1906a	*Antennarius multiocellatus*
Chironectes verus Cloquet, 1817	*Antennarius pictus*
Cheironectes pictus var. *vittatus* Richardson, 1844	*Histrio histrio*
Antennarius vulgaris Osório, 1891	*Antennarius pardalis*
Rhycherus wildii Ogilby, 1907	*Rhycherus filamentosus*
Antennarius zebrinus Schultz, 1957	*Antennarius striatus*
Phrynelox zebrinus Schultz, 1957	*Antennarius striatus*
Antennarius ziesenhennei Myers & Wade, 1946	*Antennatus strigatus*

Glossary

Abdomen (adj. **abdominal**): referring to the belly.

Aberrant: unusual in form or behavior; abnormal.

Acanthopterygii: a superorder of derived actinopterygian fishes containing a huge assemblage of spiny-rayed teleosts, some 20 orders and 319 families, the interrelations of which are for the most part unknown.

Acoustico-lateralis system: or lateral-line system, a complicated network of sensory canals (containing mechanoreceptors, i.e., neuromasts) arranged on the head and body of fishes (and larval and permanently aquatic amphibians) that serves to transduce and relay water-borne vibrations to nerve cells, allowing for detection of disturbances in the surrounding environment.

Actinopterygii: the ray-finned fishes, a class or subclass of the bony fishes, the largest and most successful group of fishes, including more than half of all living vertebrates.

Additional material: specimens that are not part of the type series.

Aggressive mimicry: the imitation of a specific model in order to gain some advantage from the model; in the case of anglerfishes, to lure prey.

Aggressive resemblance: similarity of an animal to an object in its immediate surroundings (a plant, a sedentary animal such as a sponge or coral, or part of the substratum) that is of interest to potential prey (e.g., an animal can be mistaken for a possible shelter or foraging site).

Airbladder (**gasbladder, swimbladder**): a membranous, gas-filled organ located in the body cavity of bony fishes, lying between the vertebral column and the gut.

Alimentary canal: the passage through which food passes, where it is digested and absorbed and waste is excreted, including the esophagus, stomach, and intestine.

Allopatric: geographically separated, without overlap, usually referring to populations or species.

Anal: related to the anus or vent.

Anal fin: a rayed or partially spinous fin located between the anus and tail on the lower surface of the body.

Anal-fin origin: the anteriormost point of insertion of the anal fin.

Antennarioidei: a suborder of anglerfishes containing the frogfishes and handfishes, four families, 23 genera, and about 68 species of laterally compressed, shallow to moderately deep-water, benthic forms.

Anterior: directed toward or pertaining to the front or head end of an organism or structure.

Anus: the external, posterior opening of the digestive tract, synonymous with the vent.

Apomorphy: in cladistic analysis, a derived character state.

Autapomorphy: in cladistic analysis, a derived character state present in only a single taxon.

Axilla (pl. **axillae**): the area immediately behind a paired fin.

Bar: a contrasting vertical streak of pigment, often appearing in a parallel series.

Basibranchial teeth: the teeth on any (usually the most anterior) of the median bones located in the floor of the mouth.

Batrachoidiformes: one of some 46 orders of bony or ray-finned fishes (class Actinopterygii) that contains all the toadfishes of the world (a single family, 22 genera, and about 84 species), thought by some to be the sister-group of the Lophiiformes.

Benthic: phenomena and things that are strictly associated with the bottom, regardless of depth.

Benthopelagic: phenomena and things that are closely associated with the bottom, regardless of depth.

Bifurcate: forked, or divided, into two parts or branches.

Binomen: a combined generic and specific name to designate a species.

Binominal nomenclature: a combination of two names used to designate a species—a generic name plus a specific name.

Brackish: water with a salt concentration somewhere between that of freshwater and seawater.

Branchial: concerning the gills.

Branchiostegal ray: one of a series of elongate, flattened or cylindrical, riblike elements that support the membranous outer margin of the gill cover of teleost fishes; interconnected by muscles, the branchiostegals open and close relative to one another, thus increasing the efficiency of the pumping mechanism responsible for moving water across the gills; they are also partly responsible for feeding mechanisms that rely on the production of suction.

Buccal: pertaining to the mouth cavity.

Carnivore (adj. **carnivorous**): an animal that eats animals.

Carotenoids: fat-soluble yellow, orange, or red pigments.

Caudal: pertaining to the tail.

Caudal fin: the tail fin.

Caudal peduncle: the narrowest part of the body of a fish, nearly always located just anterior to the base of the caudal fin, responsible for transmitting the power of the lateral undulation of the locomotory trunk muscles to the tail.

Centrum (pl. **centra**): the cylindrical or spool-shaped part of a vertebra or unit of the backbone.

Cephalic: pertaining to the head.

Ceratioidei: a suborder of anglerfishes (order Lophiiformes) containing the seadevils, 11 families, 35 genera, and 168 currently recognized species of globose to elongate, mesopelagic, bathypelagic, and abyssal-benthic forms.

Chaunacoidei: a suborder of anglerfishes (order Lophiiformes) containing the gapers, coffinfishes, and sea toads, a single family, at least two genera, and perhaps as many as 29 species of globose, deep-water benthic forms.

Chromatophore: a modified cell of the skin that contains pigment and gives a fish its color.

Circumtropical: distributed throughout the tropics.

Cirrus (pl. **cirri**): a slender, flexible, cutaneous filament often found on various parts of the head and along the length of the lateral line.

Clade: a branch of a cladogram containing all the descendants of a common ancestor (*see* **Monophyletic**).

Cladistics: a method of phylogenetic analysis first described in detail by Willi Hennig (1950, 1966), in which monophyletic groups are identified by shared derived characters or synapomorphies.

Cladogenesis: the evolutionary splitting of a parent species to form two distinct species, forming a clade.

Cladogram: a branching diagram (tree) showing the sequence of evolutionary divergence of organisms through time, based on cladistic analysis; a product of systematic research.

Classification: the ordering of organisms into groups based on their similarity and relationship.

Classifications: concise lists of organisms, grouped or ranked according to the pattern of branching seen in a branching diagram (cladogram); a product of systematic research.

Cleithrum (pl. **cleithra**): a primary bone of the pectoral girdle that bears the pectoral fin and supports the posterior edge of the gill opening.

Compressed: flattened laterally, from side to side.

Congeneric: belonging to the same genus.

Conspecific: belonging to the same species.

Continental shelf: the region of the sea floor that extends from the continental coast down to a depth of about 200 m.

Continental slope: the region of the sea floor adjacent to the continental shelf that slopes down from a depth of about 200 m to about 2000 m.

Cosmopolitan: having a worldwide geographic distribution; describes an organism that is found everywhere.

Cranial: concerning the brain case or cranium.

Cranium: the skull.

Cryptic: pertaining to organisms that live among sheltering and concealing cover or that have protective morphology and coloration.

Cutaneous: pertaining to the skin.

Dentary: the primary bone of the lower jaw that usually bears teeth.

Dentition: the arrangement of teeth.

Depressed: flattened from above and below.

Depressible: not rigidly fixed, usually referring to spines or teeth.

Derived: modified relative to the primitive condition.

Dermal: pertaining to the skin.

Designation: the act of fixing (selecting or establishing) a type specimen; "original designation" refers to fixation of a type of a nominal taxon when the taxon was first established, whereas "subsequent designation" refers to fixation of a type subsequent to the date at which the taxon was established.

Dimorphism: of two shapes, often referring to differences in characters or in body proportions between the sexes, as in sexual dimorphism.

Distal: remote from the point of attachment.

DNA: deoxyribonucleic acid, the self-replicating material present in nearly all living organisms as the main constituent of chromosomes; the carrier of genetic information.

Dorsal: pertaining to the upper surface or back of an organism or structure.

Dorsal fin: a rayed or spinous fin located between the head and tail on the upper surface of the body.

Dorsal-fin origin: the anteriormost point of insertion of the dorsal fin.

Emarginate: an indented or notched distal margin, as a whole or between rays or spines.

Embedded: completely enveloped in skin with no free edge, usually referring to scales.

Emendation: an intentional change in the spelling of a name.

Endemic: in biogeography, a taxon found only in a particular geographic locality.

Epaxial: that portion of the body that develops dorsal to the axis of the body formed by the vertebral column.

Epipelagic: describes that part of the oceanic zone that extends from the surface to 200 m, a depth that corresponds on average to the margin of the continental slope, which in turn is approximately equivalent to the lower limit of photosynthesis, often called the Euphotic Zone.

Epural: a bony or cartilaginous element of the tail lying above the caudal vertebrae.

Erectile: capable of being raised or erected, usually referring to spines.

Esca (pl. **escae**): the Latin word for "bait," a fleshy structure borne on the distal tip of the illicium of anglerfishes; bioluminescent in females of nearly all deep-sea anglerfishes.

Estuarine: living primarily in estuaries.

Estuary: a partly enclosed coastal body of brackish water with one or more rivers or streams flowing into it and with a free connection to the open sea.

Etymology: the study of origins and derivations of words; from the Greek *etymos*, the true, original, literal root-meaning of a word often derived by analyzing its individual component parts.

Euryhaline: a wide physiological tolerance of salinity, usually referring to fishes that can move easily between saltwater and freshwater.

Expatriate: in the case of marine organisms, an individual of a species removed by chance dispersal from its established geographic range into a region unfavorable for reproduction.

Extinction: the loss of a species.

Extrinsic muscles: muscles that originate separately or away from, but insert on, the bony elements in question.

Family: a category of organismal classification containing one or more related genera.

Fauna: the community of animals in a specific geographic area.

Filiform: threadlike.

Fluvial: pertaining to rivers and organisms that live in rivers.

Frontal: a paired bone on top of the head lying between the orbits in front of the parietals.

Gape: the expanse of the open mouth.

Gasbladder (airbladder, swimbladder): a membranous, gas-filled organ located in the body cavity of bony fishes, lying between the vertebral column and the gut.

Genus (pl. **genera**): a category of organismal classification containing one or more related species.

Gill arches: the bony support of the gills.

Gill filaments: a series of cutaneous projections arising along the posterior margin of a gill arch and responsible for gas exchange.

Gill lamellae: *see* **Gill filaments**.

Gill membranes: thin membranous tissues, supported by branchiostegal rays, that cover the lower part of the gills.

Gill opening: the opening behind each opercle, leading to the gills.

Gill rakers: bony, toothlike structures located along the anterior margin of a gill arch that serve to strain or prepare food for swallowing.

Gonads: the sex organs that produce eggs and sperm.

Gravid: sexually ripe; having eggs or sperm ready for immediate spawning.

Hemibranch: a gill arch with only one row of gill lamellae.

Herbivore (adj. **herbivorous**): an animal that eats only plants.

Holobranch: a gill arch with two rows of gill lamellae, one row on each side.

Holotype: a single specimen designated as such by an author in an original description and set aside as the best example of the species in question at the time of description, thus serving as a reference standard for that taxon that cannot be arbitrarily changed.

Homology (adj. **homologous**): when characters in different taxa are structurally similar due to common evolutionary origin.

Homoplasy: non-homologous similarity due to convergent, parallel, or reversed evolution.

Hyoid apparatus: in teleost fishes, a series of bony elements (including the urohyal, dorsal and ventral hypohyals, epihyals, ceratohyals, and interhyals) that forms in part the floor of the mouth, supports the gills, and provides articulating surface for the branchiostegal rays.

Hypaxial: that portion of the body that develops ventral to the axis of the body formed by the vertebral column.

Hypural: one of a series of bony elements of the tail skeleton of a fish, which in anglerfishes are fused together to form a solid, roughly triangular bony plate (hypural plate) that supports the rays of the caudal fin.

Ichthyoplankton: fish eggs and larvae usually found in the epipelagic or photic zone of the water column, generally less than 200 m deep.

ICZN: International Commission on Zoological Nomenclature; the judicial body empowered to enforce and interpret the International Code of Zoological Nomenclature.

Illicium (pl. **illicia**): the luring apparatus of anglerfishes—the modified first dorsal-fin spine situated on the tip of the snout or on top of the head that bears the bait or esca.

Incertae sedis: of uncertain taxonomic placement.

Innervation: the distribution of nerves to an organ of the body.

Interspecific: between different species.

Intraspecific: within a single species.

Intrinsic muscles: muscles that both originate and insert on the bony elements in question.

Invalid: refers to a name or nomenclatural act that is not valid under the International Code of Zoological Nomenclature.

Isotherm: a line on a map or chart drawn to connect points of equal temperature at a given time or on average over a given period.

Jugular: pertaining to the throat.

Junior synonym: a scientific name that is superseded by a valid older name to designate the same species.

Juvenile: a young individual, usually similar in form to an adult but not yet sexually mature.

Lachrymal: the anteriormost bone of the infraorbital (circumorbital) series, located immediately in front of the orbit, often overlapping the maxilla when the mouth is closed.

Lamella (pl. **lamellae**): a flattened, leaflike sheet of tissue.

Lanceolate: spear-shaped, gradually tapering toward the extremity.

Larva (pl. **larvae**): a developmental stage between hatching (or birth) and attainment of full external meristic complements (fins and scales) and loss of temporary specializations for pelagic life.

Lateral: pertaining to the side.

Lateral-line system: or acoustico-lateralis system, a complicated network of sensory canals (containing mechanoreceptors, i.e., neuromasts) arranged on the head and body of fishes (and larval and permanently aquatic amphibians) that serves to transduce and relay waterborne vibrations to nerve cells, allowing for detection of disturbances in the surrounding environment.

Lectotype: a single specimen selected from a series of syntypes (or a specimen known to have been used by the original describer if no type was identified) by a later author to serve in place of a holotype.

Lingual: pertaining to the tongue.

Littoral zone: the near-shore area that extends from the high water mark to areas that are permanently submerged and where sunlight penetrates all the way to the bottom, allowing aquatic vegetation (macrophytes) to grow.

Lophiiformes: one of some 46 orders of bony or ray-finned fishes (class Actinopterygii) that contains all the anglerfishes of the world.

Lophioidei: a suborder of anglerfishes (order Lophiiformes) containing the goosefishes and monkfishes; a single family, four genera, and about 27 living species of shallow- to deepwater, dorso-ventrally flattened forms.

Mandible: the lower jaw.

Maxilla (pl. **maxillae**): the posteriormost of the two paired bones that form the upper jaw of teleost fishes.

Median (**medial**): pertaining to the middle.

Median fins: fins located on the midline, including the dorsal, anal, and caudal fins.

Melanin: a brown or black pigment.

Melanophore: a cell (chromatophore) containing brown or black pigment (melanin), usually capable of expansion and contraction that changes its size and shape.

Meristics: characters or features of fish anatomy, such as fin rays, branchiostegal rays, or vertebrae, that involve numbers or counts.

Metamorphosis: a stage in early development during which an animal goes through a relatively abrupt and radial change in structure and function.

Mimicry: visual, chemical, acoustic, or behavioral similarity of an organism (or part of an organism), the mimic, to another organism or inanimate object, the model, such that the mimic benefits in some way from the mistaken identity.

Monophyletic (n. **monophyly**): describes a natural group of organisms (i.e., a clade) that includes all the descendants of a common ancestor.

Monotypic (n. **monotypy**): describes a taxon above the species level that contains only a single valid species.

Morphology: the form and structure of an organism.

Myomere: a muscle segment of the body, separated from adjacent segments by connective tissue.

Naris (pl. **nares**): the nostril.

Nasal: pertaining to the nostrils.

Neoteny (adj. **neotenic, neotenous**): the retention of juvenile characters in adult life stages.

Neotype: a specimen designated as the name-bearing type of a nominal species or subspecies for which no holotype, lectotype, syntype, or prior neotype is believed to exist.

Neritic: describes the shallow pelagic zone over the continental shelf.

Neuromasts: mechanoreceptors of the acoustico-lateralis system.

Nomen dubium: a name representing a nominal species based on one or more unidentifiable specimens (e.g., damaged specimens that have lost all diagnostic characters), or a questionable taxon for which a type specimen or specimens have been lost or never retained; a name not certainly applicable to any known species.

Nomen nudum: a name appearing in the literature without application to a type, description, or illustration; a name without indication.

Nomen oblitum: a forgotten name.

Occipital: pertaining to the posterior part of the skull.

Ocellus (pl. **ocelli**): an eyelike pigment spot surrounded by a marginal ring.

Ogcocephaloidei: a suborder of anglerfishes (order Lophiiformes) containing the batfishes, a single family of 10 genera and about 77 species of dorso-ventrally flattened, deep-water benthic forms.

Omnivore (adj. **omnivorous**): an animal that eats both plant and animal material.

Ontogenetic: pertaining to developmental change through the life cycle of an organism.

Opercular apparatus: in teleost fishes, a series of bony elements (including the preopercle, opercle, subopercle, and interopercle) that forms a major part of the respiratory pump and suction feeding apparatus and that also serves to cover and protect the gills.

Oral: pertaining to the mouth.

Original description: the description of a nominal taxon when first established.

Osteology (adj. **osteological**): the study of bones.

Otolith (ear bone): a calcareous structure in the ear capsules of bony fishes.

Outgroup: a closely related taxon outside the study group, used in cladistic analysis to provide information about the direction of character state change.

Oviparity (adj. **oviparous**): expulsion of undeveloped eggs, rather than live young, sometimes fertilized before release but more often fertilized externally.

Paired fins: fins, such as the pectorals and pelvics, that do not insert on the midline.

Palate: the roof of the mouth.

Palatine: a paired, often tooth-bearing bone on the roof of the mouth, one on each side of the centrally located vomer.

Papilla (pl. **papillae**): a small fleshy projection.

Paracanthopterygii: a superorder of actinopterygian fishes containing the trout-perches, cods, cusk-eels, toadfishes, and anglerfishes; it is most likely an artificial, unnatural assemblage, but it is still recognized by some authorities.

Paralectotype: a syntype remaining after a lectotype has been selected from a type series by a subsequent describer.

Paratype: a specimen other than the specimen designated as the holotype used in the description of a new species.

Parsimony: in cladistic analysis, the principle that determines the choice of hypotheses requiring the fewest ad hoc assumptions about character convergence, parallelism, and reversal.

Pectoral fins: the anterior or uppermost of the paired fins, which correspond to the anterior limbs of tetrapods.

Pectoral girdle: in teleost fishes, bony elements (including the scapula, coracoid, cleithrum, supracleithrum, and one or more postcleithra) that support the pectoral fin.

Pectoral radials: in teleost fishes, those bony elements, consisting in anglerfishes of two to five separate ossifications, that form the pectoral-fin lobe and provide articular support for the rays of the pectoral fin.

Pediculati: now obsolete ordinal name first used by British ichthyologist Albert Günther (1861) to contain all the anglerfishes; a junior synonym of Lophiiformes.

Peduncle: usually a reference to the caudal peduncle, the narrow region of the body between the anal fin and the base of the caudal fin.

Pelagic: describing all those phenomena and things that lie within the water column, in contrast to the bottom.

Pelvic fins: the paired fins (sometimes called ventral fins) that correspond to the posterior limbs of tetrapods, which in antennariids are located far forward, beneath the throat.

Pelvic girdle: bony support for the pelvic fins.

Percomorpha: the most derived teleostean clade, containing all the perches and perchlike fishes, a huge assemblage of some 11 orders and 255 families, the interrelationships of which are mostly unknown.

Peritoneum: the smooth membrane that lines the gut cavity and is folded inward over the abdominal and pelvic viscera.

Pharyngeal: pertaining to the mouth, pharynx, or gills.

Pharyngeal teeth: in teleost fishes, teeth borne on upper and lower parts of the gill arches that permit the separation of grabbing (outer jaw) and chewing (pharyngeal jaws) functions of the mouth.

Pharynx: the throat or more generally the mouth.

Phylogeny (adj. **phylogenetic**): the evolutionary relationships among taxa, based on their descent from a common ancestor.

Piscivore (adj. **piscivorous**): a carnivorous animal that eats primarily fishes.

Plesiomorphy: in cladistic analysis, a primitive character state.

Polyphyletic (n. **polyphyly**): describes a group of organisms derived from more than one common evolutionary ancestor or ancestral group and therefore not suitable for placing in the same taxon.

Polytomy: in cladistic analysis, the branching of three or more clades from a single node; an unresolved branching point.

Posterior: directed toward or pertaining to the back or tail end of an organism or structure.

Premaxilla (pl. **premaxillae**): the anteriormost of the two paired bones that form the upper jaw of teleost fishes.

Preoccupied: at the generic level, describes a name predated by use of the same generic or subgeneric name for another taxon at an earlier date; at the species level, describes a binomen for another species or subspecies used for another taxon at an earlier date.

Primitive: unmodified relative to the derived condition.

Priority: seniority fixed by the date of publication; the earliest published has priority.

Protective resemblance: close similarity of an animal to an object in its immediate surroundings (a plant, a sedentary animal such as a sponge or coral, or part of the substratum) that is of no interest to a predator, thus providing concealment and therefore protection.

Protractile: capable of being drawn out or extended forward.

Protrusible: pertaining to a feeding mechanism in which the upper jaw projects forward when the mouth is opened.

Proximal: describes a region, margin, or point adjacent to the place of attachment of a projection or appendage.

Pseudobranch: a small structure situated on the inner wall of the opercular cavity of many fishes, composed of filaments like those of the gills, and apparently involved in providing oxygenated blood to the eye.

Pterygiophore: a bony element that supports a spine or ray of an unpaired fin.

Quadrate: a bone that forms part of the skeleton of the cheek and serves to support the lower jaw.

Ray: a paired, segmented, bony but usually flexible, fin support.

Recurved: curved backward, usually referring to teeth.

Riverine: pertaining to rivers.

Sargassum: a genus of brown seaweed (macroalgae of the order Fucales, class Phaeophyceae) containing numerous species, distributed throughout tropical and temperate oceans of the world, where it typically occurs in huge floating, island-like masses that provide food, refuge, and breeding grounds for an array of animals, including fishes, sea turtles, marine birds, crabs, and shrimp, many of which spend their entire lives in this habitat.

Senior synonym: the oldest available scientific name used to designate a species.

Sexual dichromatism: differences between the sexes in coloration.

Sexual dimorphism: the existence of structures that differ morphologically between the sexes of a species.

Sigmoid: the shape of a long drawn-out letter "S."

Sister-group: in cladistic analysis, a taxon most closely related to another by common ancestry.

Spatulate: broad, flat, and rounded in shape.

Speciation: the process by which species are formed.

Speciose: rich in the number of species.

Spine: a sharp, bony protuberance, usually on the head; or an unpaired, unsegmented, and unbranched, bony fin support, often stiff and sharp.

Spinule: a small spine or secondary spine.

Standard length: the length of a fish, excluding the rays of the tail or caudal fin, measured from the anteriormost tip of the upper jaw to the posteriormost margin of the bony structures that support the caudal-fin rays (except where noted, all fish lengths in this volume are standard length, abbreviated SL; notochord length, the length measured from the anteriormost tip of the upper jaw to the tip of the notochord before formation of the hypural plate, abbreviated NL, or less often total length, abbreviated TL, is occasionally used for small larvae that do not have a fully developed caudal skeleton).

Stenohaline: describes a narrow physiological tolerance of salinity, usually referring to fishes that are unable to move easily between salt and fresh waters.

Stripe: a contrasting, horizontal line or streak of pigment, often appearing in a parallel series.

Suspensorium: a functional unit of the teleost head skeleton (consisting of the hyomandibula, preopercle, symplectic, pterygoids, palatine, and quadrate) that functions primarily to support or "suspend" the upper and lower jaws.

Swimbladder (airbladder, gasbladder): a membranous, gas-filled organ located in the body cavity of bony fishes, lying between the vertebral column and the gut.

Sympatric: in biogeography, describes two or more taxa that occupy the same geographic locality.

Symplesiomorphy: in cladistic analysis, a shared primitive character.

Synapomorphy: in cladistic analysis, a shared derived character.

Synonym: one of two or more scientific names of the same rank that have been inadvertently assigned to the same taxon.

Synonymy: the relationship between synonymous names, or a list of synonymous names.

Syntype: one of a series of two or more specimens used in the description of a new species in which no single specimen is given holotype status.

Systematics: the study of biological diversity; or, more specifically, the ordering of the diversity of nature through construction of a classification that can serve as a general reference system.

Tautonymy: the use of the same word for the name of a genus-group taxon and for one of its included species or subspecies, such as *Histrio histrio*.

Taxon (pl. **taxa**): any of the formal categories used in classifying organisms.

Taxonomy: that branch of biological sciences that deals with the discovery, recognition, definition, and naming of groups of organisms.

Teleostei (teleost): a taxonomic unit containing most all of the bony or ray-finned fishes, well over 34,000 living species, more than all other vertebrate species combined.

Tetraodontiformes: one of some 46 orders of bony or ray-finned fishes (class Actinopterygii) that contains all the spikefishes, triggerfishes, boxfishes, puffers, and molas of the world (10 families, approximately 105 genera, and 436 species), thought by some to be closely related to the Lophiiformes.

Tetrapod: a four-legged vertebrate animal (i.e., an amphibian, reptile, bird, or mammal, including vertebrate taxa that have secondarily lost limbs).

Thoracic: pertaining to the chest.

Total length: the length of a fish, measured from the anteriormost tip of the upper jaw to the tip of the longest caudal-fin ray (except where noted, all fish lengths in this volume are standard length, abbreviated SL; total length, abbreviated TL, is occasionally used for small larvae that do not have a fully developed caudal skeleton).

Trophic: pertaining to nutrition or mode of feeding.

Type locality: the geographic location from which the type specimen of a species was collected.

Type species: the (nominal) species that is the name-bearing type of a genus or subgenus.

Type specimen: a specimen (or in some cases a group of specimens) of an organism to which the scientific name of that organism is formally attached; or, in other words, an example that serves to solidify the defining features of a species.

Unpaired fins: fins (the dorsal, anal, and caudal) that insert on the midline.

Vagility: the ability or degree to which an organism is able to move about freely or disperse within an environment.

Ventral: pertaining to the lower surface or belly of an organism or structure.

Ventral fins: a synonym of pelvic fins.

Vernacular name: a name proposed in a language used for general purposes, often referred to as a common name, as opposed to a name proposed only for zoological nomenclature.

Vertebra (pl. **vertebrae**): a cartilaginous or bony element of the spinal column.

Vertebrate: an animal with a backbone or spinal column.

Vestigial: reduced or very poorly developed.

Vicariance: the geographic splitting of a population by a physical or biotic barrier to gene flow or dispersal, resulting in a pair of closely related species.

Villiform: small, slender, and crowded together in patches or bands, usually pertaining to teeth.

Vomer: a median, unpaired, often tooth-bearing bony element of the cranium that forms the anteriormost part of the roof of the mouth.

Vomerine teeth: teeth borne on the vomer.

References

Acero, A., and J. Garzón. 1987. Peces arrecifales de la región de Santa Marta (Caribe Colombiano). Lista de especies y comentarios generales. *Acta Biologica Colombiana*, 1 (3): 83–105.

Adams, J. A. 1960. A contributionn to the biology and postlarval development of the sargassum fish, *Histrio histrio* (Linnaeus), with a discussion of the *Sargassum* complex. *Bulletin of Marine Science of the Gulf and Caribbean*, 10 (1): 55–82.

Adler, K. 1989. *Contributions to the History of Herpetology*. Contributions to herpetology, No. 5. Society for the Study of Amphibians and Reptiles, Oxford, Ohio, 202 pp.

Aerts, P. 1989. Mathematical biomechanics and the "what!" "how?" and "why?" in functional morphology. *Netherlands Journal of Zoology*, 40 (1): 153–172.

Aerts, P. 1990. Variability of the fast suction feeding process in *Astatotilapia elegans* (Teleostei: Cichlidae): A hypothesis on peripheral feedback control. *Journal of Zoology*, London, 220 (4): 653–678.

Aerts, P., J. W. M. Osse, and W. Verraes. 1987. Model of jaw depression during feeding in *Astatotilapia elegans* (Teleostei: Cichlidae): Mechanisms for energy storage and triggering. *Journal of Morphology*, 194 (1): 85–109.

Afonso, P., F. M. Porteiro, R. S. Santos, J. P. Barreiros, J. Worms, and P. Wirtz. 1999. Coastal Marine Fishes of São Tomé Island (Gulf of Guinea). *Arquipélago, Boletim da Universidade dos Açores*, 17A: 65–92.

Agassiz, A. 1882. On the young stages of some osseous fishes. Part III. *Proceedings of the American Academy of Arts and Sciences*, 9: 271–303.

Agassiz, A. 1888. *A Contribution to American Thalassography. Three Cruises of the United States Coast and Geodetic Survey Steamer "Blake" in the Gulf of Mexico, in the Caribbean Sea, and Along the Atlantic Coast of the United States, from 1877 to 1880*. Houghton, Mifflin, Boston & New York, Vol. 2, 220 pp.

Agassiz, L. 1833–1844. *Recherches sur les poissons fossiles*. Petitpierre, Neuchâtel, 1420 pp.

Agassiz, L. 1872. Fish-nest in the sea-weed of the Sargasso Sea. Extracts from a letter from Professor Agassiz to Prof. Peirce, Superintendent U.S. Coast Survey, dated Hassler Expedition, St. Thomas, December 15, 1871. *American Journal of Science and Arts*, Series 3, 3 (14): 154–156.

Aguilera, O. 1998. Los peces marinos del occidente de Venezuela. *Acta Biologica Venezuelica*, 18 (3): 43–57.

Akimenko, M.-A., M. Marí-Beffa, J. Becerra, and J. Géraudie. 2003. Old questions, new tools, and some answers to the mystery of fin regeneration. *Developmental Dynamics*, 226: 190–201.

Albuquerque, R. A. 1956. Peixes de Portugal e ilhas adjacentes. Chavas para a sua determinação. Part 2. *Portugaliae Acta Biologica*, Series B, 5: 561–1164.

Alcock, J. 2013. *Animal Behavior: An Evolutionary Approach*. Tenth edition. Sinauer Associates, Inc., Sunderland, Massachusetts, 522 pp.

Aldrovandi, U. 1613. *De piscibus libri V, et de cetis liber unus*. Bellagambam, Bononiae, ix + 732 + 26 pp.

Alexander, R. M. 1967. The functions and mechanisms of the protrusible upper jaws of some acanthopterygian fish. *Journal of Zoology*, London, 151: 43–64.

Alexander, R. M. 1970. Mechanics of the feeding action of various teleost fishes. *Journal of Zoology*, London, 162: 145–156.

Alexander, R. M. 1982. *Locomotion in Animals*. Blackie & Son, Glasgow & London, vii + 163 pp.

Alfaro, M. E., B. C. Faircloth, R. C. Harrington, L. Sorenson, M. Friedman, C. E. Thacker, C. H. Oliveros, D. Černý, and T. J. Near. 2018. Explosive diversification of marine fishes at the Cretaceous–Palaeogene boundary. *Nature Ecology and Evolution*, 98: 1–11.

Alfaro, M. E., F. Santini, C. Brock, H. Alamillo, A. Dornburg, D. L. Rabovsky, G. Carnevale, and L. J. Harmon. 2009. Nine exceptional radiations plus high turnover explain species diversity in jawed vertebrates. *Proceedings of the National Academy of Sciences*, 106: 13410–13414.

Al-Jufaili, S. M., G. Hermosa, S. S. Al-Shuaily, and A. Al Mujaini. 2010. Oman fish biodiversity. *Journal of King Abdulaziz University (JKAU), Marine Sciences*, 21 (1): 3–51.

Allen, G. R. 1970. Two new species of frogfishes (Antennariidae) from Easter Island. *Pacific Science*, 24: 517–522.

Allen, G. R. 1991. *Field Guide to the Freshwater Fishes of New Guinea*. Publication No. 9 of the Christensen Research Institute, Madang, Papua New Guinea, 268 pp.

Allen, G. R. 1997. *Marine Fishes of Tropical Australia and South-east Asia*. Western Australian Museum, Perth, 292 pp.

Allen, G. R. 2000a. *Marine Fishes of Tropical Australia and South-East Asia*. Fourth Edition. Western Australian Museum, Perth, 292 pp.

Allen, G. R. 2000b. Reef fishes recorded during the RAP survey of the Calamianes Islands. Rapid Assessment Program, *Bulletin of Biological Assessment*, 17: 95–125.

Allen, G. R., and M. Adrim. 2003. Coral Reef Fishes of Indonesia. *Zoological Studies*, 42 (1): 1–72.

Allen, G. R., N. J. Cross, D. J. Bray, and D. F. Hoese. 2006. Family Antennariidae. Pp. 637–646, In: *Zoological Catalogue of Australia*, Vol. 35, Fishes, pts. 1, 2.

Allen, G. R., and M. V. Erdmann. 2009. Reef fishes of the Bird's Head Peninsula, West Papua, Indonesia. *Check List*, 5 (3): 587–628.

Allen, G. R., and M. V. Erdmann. 2012. *Reef Fishes of the East Indies*. Tropical Reef Research, Perth, Australia, Vol. 1, x + pp. 1–424.

Allen, G. R., and M. V. Erdmann. 2013. *Reef Fishes of Bali, Indonesia*. Conservation International, RAP Bulletin of Biological Assessment, Bali Marine Rapid Assessment Program 2011, pp. 15–68.

Allen, G. R., D. F. Hoese, J. R. Paxton, J. E. Randall, B. C. Russell, W. A. Starck II, F. H. Talbot and G. P. Whitley. 1976. Annotated checklist of the fishes of Lord Howe Island. *Records of the Australian Museum*, 30 (15): 365–454.

Allen, G. R., and D. R. Robertson. 1994. *Fishes of the Tropical Eastern Pacific*. Crawford House Press, Bathurst, xx + 332 pp.

Allen, G. R., and D. R. Robertson. 1997. An annotated checklist of the fishes of Clipperton Atoll, tropical eastern Pacific. *Revista de Biologia Tropical*, 45 (2): 813–843.

Allen, G. R., R. C. Steene, P. Humann, and N. DeLoach. 2003. *Reef Fish Identification: Tropical Pacific*. New World Publications, Jacksonville, Florida; Odyssey Publishing, El Cajon, California, 457 pp. + index.

Allen, G. R., and R. Swainston. 1988. *The Marine Fishes of North-western Australia. A Field Guide for Anglers and Divers*. Western Australian Museum, Perth, vi + 201.

Almany, G. R. 2004. Does increased habitat complexity reduce predation and competition in coral reef fish assemblages? *Oikos*, 106 (2): 275–284.

Almeida, A. J., L. Amoedo, and L. Saldanha. 2001. Fish assemblages in the seagrass beds at Inhaca Island (Mozambique)—cold season. *Boletim do Museu Municipal do Funchal*, Supplement, 6: 111–125.

Alwany, M. A., M. H. Hanafy, M. M. Kotb, A. A.-F. A. Gab-Alla. 2007. Species diversity and habitat distribution of fishes in Sharm El-Maiya Bay, Sharm El-Sheikh, Red Sea. *Catrina*, 2 (1): 83–90.

Amaoka, K., H. Senou, and A. Ono. 1994. Record of the bothid flounder *Asterorhombus fijiensis* from the Western Pacific, with observations on the use of the first dorsal-fin ray as a lure. *Japanese Journal of Ichthyology*, 41 (1): 23–28.

Ambrose, D. A. 1996. Ophidiiformes. Pp. 512–545, In: H. Geoffrey Moser (editor), *Early Stages of Fishes in the California Current Region*, Atlas No. 33, Allen Press, Lawrence, Kansas, 1505 pp.

Anderson, C. 1926. Allan Riverstone McCulloch, 1885–1925. *Records of the Australian Museum*, 15 (2): 141–148.

Anderson, R. C., J. E. Randall, and R. H. Kuiter. 1998. Additions to the fish fauna of the Maldive Islands. Part 2: New records of fishes from the Maldive Islands, with notes on other species. *Ichthyological Bulletin of the J. L. B. Smith Institute of Ichthyology*, 67: 20–32.

Andrews, A. P. 1971. A catalogue of the type material (fishes) in the Tasmanian Museum. *Papers and Proceedings of the Royal Society of Tasmania*, 105: 1–3.

Andrews, A. P. 1992. A descriptive catalog of the type material (Chordates) in the Tasmanian Museum. *Papers and Proceedings of the Royal Society of Tasmania*, 126: 109–113.

Annandale, N., and J. T. Jenkins. 1910. Report on the fishes taken by the Bengal fisheries Steamer "Golden Crown." Part III, Plectognathi and Pediculati. *Memoires of the Indian Museum*, 3 (1): 7–21.

Anonymous. 1910. International Commission on Zoological Nomenclature. Opinion 24. *Antennarius* Commerson, 1798, and Cuvier, 1817, vs. *Histrio* Fischer, 1813. *Smithsonian Institution Publication*, 1938: 57–58.

Anonymous. 1925. International Commission on Zoological Nomenclature. Opinion 89. Suspension of the rules in the case of Gronow 1763, Commerson 1803, Gesellschaft Schauplatz 1775 to 1781, Catesby 1771, Browne 1789, Valmont de Bomare 1768 to 1775. *Smithsonian Miscellaneous Collections*, 73 (3): 27–33.

Anonymous. 1950. International Commission on Zoological Nomenclature. Meuschen's index to Gronovius, 1763–1781, "Zoophylacium Gronovianum." *Bulletin of Zoological Nomenclature*, 4: 502–504.

Anonymous. 1954. International Commission on Zoological Nomenclature. Opinion 261. Rejection for nomenclatorial purposes of the index to the "Zoophylacium Gronovianum" of Gronovius prepared by Meuschen (F. C.) and published in 1781. *International Commission on Zoological Nomenclature, Opinions and Declarations*, 5 (22): 281–296.

Anonymous. 1981. *Dictionary of Japanese Fish Names and Their Foreign Equivalents*. Ichthyological Society of Japan. Sanseido, Tokyo, vii + 834 pp.

Anonymous. 1982. *Antennarius hispidus*. *Revue Française d'Aquariologie et Herpetologie*, Nancy, 9 (1): front and back covers.

Araga, C. 1984. Family Antennariidae. Pp. 102–104, In: H. Masuda, K. Amaoka, C. Araga, T. Uyeno, and T. Yoshino (editors), *The Fishes of the Japanese Archipelago*, Tokai University Press, Tokyo, Japan.

Arai, T. 2015. Diversity and conservation of coral reef fishes in the Malaysian South China Sea. *Reviews in Fish Biology and Fisheries*, 25 (1): 85–101.

Arambourg, C. 1927. Les poissons fossiles d'Oran. *Matériaux pour la Carte Géologique Algérie, 1er Série Paléontologie*, 6: 1–218.

Arcila, D., and J. C. Tyler. 2017. Mass extinction in tetraodontiform fishes linked to the Palaeocene–Eocene thermal maximum. *Proceedings of the Royal Society B*, 284: 20171771, doi.org/10.1098/rspb.2017.1771

Aristotle. 1984. History of animals. P. 966, lines 10–19, In : J. Barnes, J. (editor), *The Complete Works of Aristotle, the Revised Oxford Translation*. Bollingen Series LXXI, Princeton University Press, Princeton, New Jersey, Vol. 2.

Arnold, R. J. 2010a. In with the new: a new anglerfish species is described in Botany Bay. *Australasia Scuba Diver*, 6 (5): 68–69.

Arnold, R. J. 2010b. Evolutionary history of frogfishes (Teleostei: Lophiiformes: Antennariidae): a molecular approach. Unpublished M.S. thesis, University of Washington, Seattle, 66 pp.

Arnold, R. J. 2012. A new species of frogfish of the genus *Histiophryne* (Teleostei: Lophiiformes: Antennariidae) from Lombok and Komodo, Indonesia. *Zootaxa*, 3252: 62–68.

Arnold, R. J. 2013. A new species of frogfish of the genus *Kuiterichthys* (Lophiiformes: Antennariidae: Histiophryninae) from New South Wales, Australia. *Zootaxa*, 3718 (5): 496–499.

Arnold, R. J., R. Harcourt, and T. W. Pietsch. 2014. A new genus and species of the frogfish family Antennariidae (Teleostei: Lophiiformes: Antennarioidei) from New South Wales, Australia, with a diagnosis and key to the genera of the Histiophryninae. *Copeia*, 2014 (3): 534–539.

Arnold, R. J., and T. W. Pietsch. 2011. A new species of frogfish of the genus *Histiophryne* (Teleostei: Lophiiformes: Antennariidae) from Queensland, Australia. *Zootaxa*, 2925: 63–68.

Arnold, R. J., and T. W. Pietsch. 2012. Evolutionary history of frogfishes (Teleostei: Lophiiformes: Antennariidae): a molecular approach. *Molecular Phylogenetics and Evolution*, 62 (2012): 117–129.

Arnold, R. J., and T. W. Pietsch. 2018. Fantastic beasts and where to find them: a new species of the frogfish genus *Histiophryne* Gill (Lophiiformes: Antennariidae: Histiophryninae) from Western and South Australia, with a revised key to conspecifics. *Copeia*, 2018 (4): 622–631.

Arruda, L. M. 1997. Checklist of the marine fishes of the Azores. *Arquivos do Museu Bocage* (new series), 3 (2): 13–162.

Atz, J. W. 1950. Strange animal lures. *Animal Kingdom, Bulletin of the New York Zoological Society*, 53 (4): 110–113.

Atz, J. W. 1951a. Strange fish lures. *Aquarium Journal*, 22 (4): 70–72.

Atz, J. W. 1951b. Fishes that look like plants. *Animal Kingdom, Bulletin of the New York Zoological Society*, 54 (5): 130–136.

Avendaño-Ibarra, R., G. Aceves-Medina, E. Godínez-Domínguez, R. De Silva-Dávila, S. P. A. Jiménez-Rosenberg, H. Urias-Leyva, and C. J. Robinson. 2014. Fish larvae from the Gulf of California to Colima, Mexico: an update. *Check List*, 10 (1): 106–121.

Ayala, D., L. Riemann, and P. Munk. 2016. Species composition and diversity of fish larvae in the Subtropical Convergence Zone of the Sargasso Sea from morphology and DNA barcoding. *Fisheries Oceanography*, 25 (1): 85–104.

Ayala-Bocos, A., M. Hoyos-Padilla, D. García-Benito, and V. Martínez-Castillo. 2015. New record of the frogfish *Fowlerichthys avalonis* (Actinopterygii, Antennariidae) at the oceanic Revillagigedo Archipelago, west Mexico. *Marine Biodiversity Records*, 8: 1–4.

Azevedo, C. J. C. de. 1971. Antenarídeos de Angola, contribuição para o seu estudo. *Ciências Biológicas da Faculdade de Ciências da Universidade de Luanda, Angola*, 1971: 89–92.

Azevedo, J. M. N., and P. C. Heemstra, 1995. New records of marine fishes from the Azores. Arquipélago, *Ciências Biológicas e Marinhas*, 13A: 1–10.

Baensch, F. 2014. The Hawaii larval fish project. *Coral, Reef and Marine Aquarium Magazine*, 11 (2): 64–77.

Baensch, F. 2018. *Fish Culture and Underwater Photography.* https://www.frankbaensch.com. Accessed September 2018.

Bagnis, R. A., F. Berglund, P. S. Elias, G. J. van Esch, B. W. Halstead, and K. Kojima. 1970. Problems of toxicants in marine food products. 1. Marine biotoxins. *Bulletin of the World Health Organization*, 42: 69–88.

Bahr, K. D., P. L. Jokiel, and R. J. Toonen. 2015. The unnatural history of Kāneʻohe Bay: coral reef resilience in the face of centuries of anthropogenic impacts. *PeerJ*, 3 (5), 3:e950; doi: 10.7717/peerj.950.

Baird, S. F. 1871. Spawning of the goose fish (*Lophius americanus*). *American Naturalist*, 5: 785–786.

Baker, J., S. Shepherd, H. Crawford, A. Brown, K. Smith, J. Lewis, and C. Hall. 2009. *Surveys of Uncommon, Rare and Cryptic Reef Fishes in South Australia.* Report to Commonwealth Department of the Environment, Water, Heritage and the Arts, Envirofund Project, Adelaide, South Australia, 37 pp.

Bakus, G. J. 1964. The effects of fish grazing on invertebrate evolution in shallow tropical waters. *Allan Hancock Foundation, Occasional Papers*, 27: 1–29.

Baldwin, C. C. 2013. The phylogenetic significance of colour patterns in marine teleost larvae. *Zoological Journal of the Linnean Society*, 168: 496–563.

Ballintijn, C. M., and G. M. Hughes. 1965. The muscular basis of the respiratory pumps in the trout. *Journal of Experimental Biology*, 43: 349–362.

Balmford, A., J. Beresford, J. Green, R. Naidoo, M. Walpole, and A. Manica. 2009. A global perspective on trends in nature-based tourism. *PLoS Biology*, 7 (6): e1000144, doi.org/10.1371/journal.pbio.1000144.

Bannikov, A. F. 2004. The first discovery of an anglerfish (Teleostei, Lophiidae) in the Eocene of the Northern Caucasus. *Paleontological Journal*, 38: 67–72.

Barbour, T. 1905. Notes on Bermudian fishes. *Bulletin of the Museum of Comparative Zoology*, 46 (7): 109–134.

Barbour, T. 1941. Notes on pediculate fishes. *Proceedings of the New England Zoological Club*, 19: 7–14.

Barbour, T. 1942. The northwestern Atlantic species of frog fishes. *Proceedings of the New England Zoological Club*, 19: 21–40.

Barnard, K. H. 1927. A monograph of the marine fishes of South Africa. Part II. *Annals of the South African Museum*, 21 (2): 419–1065.

Barnhart, P. S. 1937. Notes on the habits, eggs and young of some fishes of southern California. *Bulletin of the Scripps Institution of Oceanography, Technical Series*, 3: 87–99.

Barrett, N., B. D. Bruce, and P. R. Last. 1996. *Spotted Handfish Survey*. Report to the Australian Conservation Agency Endangered Species Program, Project No. 538, CSIRO, Hobart, Tasmania, 27 pp.

Bartels, C. E., K. S. Price, M. I. López, and W. A. Bussing. 1983. Occurrence, distribution, abundance and diversity of fishes in the Gulf of Nicoya, Costa Rica. *Revista de Biologia Tropical*, 31 (1): 75–101.

Bates, H. W. 1862. Contributions to an insect fauna of the Amazon Valley. Lepidoptera: Heliconidae. *Transactions of the Linnean Society*, London, 23: 495–566.

Bauchot, M. L. 1958. Sur *Antennarius pinniceps* Commerson (Téléostéen Lophiiforme) et sa signification taxonomique. *Bulletin du Muséum national d'Histoire naturelle*, Paris, Série 2, 30 (2): 139–143.

Bauchot, M. L., J. Daget, and R. Bauchot. 1997. Pp. 27–80, In: T. W. Pietsch and W. D. Anderson, Jr. (editors), *Collection Building in Ichthyology and Herpetology*. American Society of Ichthyologists and Herpetologists, Special Publication, 3.

Bauchot, M. L., P. J. P. Whitehead, and T. Monod. 1982. Date of publication of the fish names in Eydoux & Souleyet's zoology of *La Bonite*, 1841–1852. *Cybium*, 6 (3): 59–73.

Baughman, J. L. 1955. The oviparity of the whale shark, *Rhineodon typus*, with records of this and other fishes in Texas waters. *Copeia*, 1955 (1): 54–55.

Bavendam, F. 1998. Lure of the frogfish. *National Geographic*, 149 (1): 40–49.

Bavendam, F. 2000. Sneaky, freaky, frogfish. *Ranger Rick*, 34 (1): 34–39.

Beaglehole, J. C. (editor). 1962. *The Endeavour Journal of Joseph Banks, 1768–1771*. Trustees of the Public Library of New South Wales in association with Angus & Robertson, Sydney, Vol. 2, xvi + 406 pp.

Bean, T. H. 1897. Notes upon New York fishes received at the N.Y. Aquarium, 1895–1897. *Bulletin of the American Museum of Natural History*, 9 (24): 327–373.

Bean, T. H. 1903. Catalogue of the fishes of New York. *Bulletin of the New York State Museum*, Albany, 60, Zoology, 9, 784 pp.

Bean, T. H. 1906a. Descriptions of new Bermudian fishes. *Proceedings of the Biological Society of Washington*, 19: 29–34.

Bean, T. H. 1906b. A catalogue of the fishes of Bermuda, with notes on a collection made in 1905 for the Field Museum. *Field Columbian Museum, Zoological Series*, 7 (2): 21–89.

Bearez, P. 1996. Lista de los peces marinos del Ecuador continental. *Revista de Biologia Tropical*, 44 (2): 731–741.

Beaufort, L. F. de, and J. C. Briggs. 1962. *The Fishes of the Indo-Australian Archipelago*. Volume XI. Leiden: Brill, xi + 481 pp.

Beebe, W. 1932. *Nonsuch: Land of Water*. National Travel Club, New York, xv + 259 pp.

Beebe, W. 1933. On the *Antares* to the West Indies. Narrative of the fifteenth and seventeenth expeditions of the Department of Tropical Research. *Bulletin of the New York Zoological Society*, 36 (4): 97–115.

Beebe, W., and J. Tee-Van. 1928. The fishes of Port-au-Prince Bay, Haiti, with a summary of the known species of marine fish of the island of Haiti and Santo Domingo. *Zoologica*, New York, 10 (1): 1–279.

Beebe, W., J. Tee-Van. 1933. *Field Book of the Shore Fishes of Bermuda*. Putnam's Sons, New York and London, xiv + 337.

Bekkers, J. A. F. (editor). 1970. *Correspondence of John Morris with Johannes de Laet (1634–1649)*. Van Gorcum, Assen, xxvii + 256 pp.

Bellwood, D. R., T. P. Hughes, S. R. Connolly, and J. Tanner. 2005. Environmental and geometric constraints on Indo-Pacific coral reef biodiversity. *Ecology Letters*, 8: 643–651.

Belon, P. 1551. *L'histoire Naturelle des Estranges Poissons Marins, avec la Vraie Peincture & Description du Daulphin, & de plusieurs autres son espèce, observée par Pierre Belon du Mans*. Imprimerie de Regnaud Chaudiere, Paris.

Belon, P. 1553. *De aquatilibus, libri duo, cum conibus ad viuam ipsorum effigiem, quoad eius fieri potuit, expressis*. Carolum Stephanum, Paris.

Beltrán-León, B. S., and R. R. Herrera. 2000. *Estadios Tempranos de Peces del Pacifico Colombiano*. Republica de Colombia, Ministerio de Agricultura y Desarrollo Rural, Instituto Nacional de Pesca y Acuicultura (INPA), Buenaventura, Colombia, Vol. 1, pp. 1–359.

Bennett, E. T. 1827. Observations on the fishes contained in the collection of the Zoological Society. *Zoological Journal*, London, 3 (37): 371–378.

Bennett, F. D. 1840. *Narrative of a Whaling Voyage Around the Globe, from the Year 1833 to 1836*. Bentley, London, Vol. 2, 395 pp.

Bereiter-Hahn, J., and L. Zylberberg. 1993. Regeneration of teleost fish scales. *Comparative Biochemical and Physiology*, Part A, 105 (4): 625–641.

Berg, L. S. 1940. Classification of fishes, both recent and fossil. *Trudy Zoologitscheskogo Instituta, Akademiia Nauk SSSR*, 5: 87–517 [Reprint, J. W. Edwards, Ann Arbor, Michigan, 1947.]

Bergert, B. A., and P. C. Wainwright, 1997. Morphology and kinematics of prey capture in the syngnathid fishes *Hippocampus erectus* and *Syngnathus floridae*. *Marine Biology*, 127: 563–570.

Bertelsen, E. 1951. The ceratioid fishes. Ontogeny, taxonomy, distribution and biology. *Dana Report*, Carlsberg Foundation, 39, 276 pp.

Bertelsen, E. 1984. Ceratioidei: Development and relationships. Pp. 325–334, In: H. G. Moser, W. J. Richards, D. M. Cohen, M. P. Fahay, A. W. Kendall, Jr., and S. L. Richardson (editors), *Ontogeny and Systematics of Fishes*. American Society of Ichthyologists and Herpetologists, Special Publication No. 1, x + 760 pp.

Bertelsen, E. 1994. Anglerfishes. Pp. 137–141, In: J. R. Paxton and W. N. Eschmeyer (editors), *Encyclopedia of Fishes*, University of New South Wales Press, Sydney, Australia.

Bertelsen, E., and T. W. Pietsch. 1998. Anglerfishes. Pp. 137–141, In: J. R. Paxton and W. N. Eschmeyer (editors), *Encyclopedia of Fishes*, 2nd edition, Academic Press, San Diego, California.

Bertelsen, E., T. W. Pietsch, and R. J. Lavenberg. 1981. Ceratioid anglerfishes of the family Gigantactinidae: Morphology, systematics, and distribution. *Natural History Museum of Los Angeles County, Contributions in Science*, 332: 1–74.

Bertelsen, E., and P. J. Struhsaker. 1977. The ceratioid fishes of the genus *Thaumatichthys*: Osteology, relationships, distribution, and biology. *Galathea Report*, 14: 7–40.

Betancur-R, R., R. E. Broughton, E. O. Wiley, K. Carpenter, J. A. López, C. Li, N. I. Holcroft, D. Arcila, M. Sanciangco, J. C. Cureton, F. Zhang, T. Buser, M. A. Campbell, J. A. Ballesteros, A. Roa-Varon, S. Willis, W. C. Borden, T. Rowley, P. C. Reneau, D. J. Hough, G. Lu, T. Grande, G. Arratia, and G. Ortí. 2013. The tree of life and a new classification of bony fishes. *PLOS Currents Tree of Life*, Edition 1, doi: 10.1371/currents.tol.53ba26640df0ccaee75bb165c8c26288.

Betancur-R, R., E. O. Wiley, G. Arratia, A. Acero, N. Bailly, M. Miya, G. Lecointre, and G. Ortí. 2017. Phylogenetic classification of bony fishes. *BMC Evolutionary Biology*, 17, doi: 10.1186/s12862-017-0958-3.

Bianconi, G. G. 1855. *Specimina zoologica Mosambicana, quibus vel novae vel minus notae animalium species illustrantur*. Bononiae, Academiae Scientiarum, 10: 215–230.

Bibron, G. 1833. Batracoide. P. 399, In: F. E. Guérin-Méneville (editor), *Dictionnaire Pittoresque d'Histoire Naturelle*, Paris: Bureau de souscription [Impr. de Cossin], Vol. 1, 640 pp.

Bigelow, H. B., and W. C. Schroeder. 1936. Supplementary notes on fishes of the Gulf of Maine. *Bulletin of the United States Bureau of Fisheries*, 48 (20): 319–343.

Bigelow, H. B., and W. C. Schroeder. 1953. Fishes of the Gulf of Maine. *Fishery Bulletin, United States Fish and Wildlife Service*, 53 (74): 1–577.

Bigelow, H. B., and W. W. Welsh. 1925. Fishes of the Gulf of Maine. *Bulletin of the United States Bureau of Fisheries*, 40: 1–567.

Billberg, G. J. 1833. Om ichthyologien och beskrifning öfver några nya fiskarter af samkäks-slägtet *Syngnathus*. *Linnéska Samfundets Handlingar*, 1 (1832): 47–55.

Blaber, S. J. M., D. T. Brewer, and A. N. Harris. 1994. Distribution, biomass and community structure of demersal fishes of the Gulf of Carpentaria, Australia. *Australian Journal of Marine and Freshwater Research*, 45 (3): 375–396.

Blache, J. 1962. Liste des poissons signalés dans l'Atlantique tropico-oriental sud du Cap des Palmes à Mossamédès (Province Guinéo-Équatoriale). Office de la Recherche Scientifique et Technique Outre-mer (ORSTOM), Paris, Série Pointe-Noire, 2: 13–102.

Blache, J., J. Cadenat, and A. Stauch. 1970. Clés de détermination des poissons de mer signalés dans l'Atlantique Oriental (entre le 20e parallèle N. et le 15e parallèle S.). *Faune Tropicale*, 18, 479 pp.

Blanc, M. 1963. Travaux ichthyologiques et herpétologiques publiés par Achille Valenciennes. *Mémoires de l'Institut Français d'Afrique Noire*, 68: 71–75.

Bleeker, P. 1851. Bijdrage tot de kennis der ichthyologische fauna van Riouw. *Natuurkundig Tijdschrift voor Nederlandsch Indië*, 2: 469–497.

Bleeker, P. 1852a. Bijdrage tot de kennis der ichthyologische fauna van de Moluksche eilanden. Visschen van Amboina en Ceram. *Natuurkundig tijdschrift voor Nederlandsch Indië*, 3: 229–309.

Bleeker, P. 1852b. Nieuwe visschen van Banda-Neira. *Natuurkundig tijdschrift voor Nederlandsch Indië*, 3: 643–646.

Bleeker, P. 1853a. Bijdrage tot de kennis der ichthyologische fauna van Solor. *Natuurkundig tijdschrift voor Nederlandsch Indië*, 5: 67–96.

Bleeker, P. 1853b. *Antennarius notophthalmus*, eene nieuwe soort van de Meeuwenbaai. *Natuurkundig tijdschrift voor Nederlandsch Indië*, 5: 543–545.

Bleeker, P. 1853c. Nalezingen op de ichthyologie van Japan. *Verhandelingen van het Bataviaasch Genootschap van Kunsten en Wetenschapen*, 25: 1–56.

Bleeker, P. 1854a. Derde bijdrage tot de kennis der ichthyologische fauna van de Banda-eilanden. *Natuurkundig tijdschrift voor Nederlandsch Indië*, 6: 89–114.

Bleeker, P. 1854b. Bijdrage tot de kennis der ichthyologische fauna van het eiland Flores. *Natuurkundig tijdschrift voor Nederlandsch Indië*, 6: 311–338.

Bleeker, P. 1854c. Vijfde bijdrage tot de kennis der ichthyologische fauna van Amboina. *Natuurkundig tijdschrift voor Nederlandsch Indië*, 6: 455–508.

Bleeker, P. 1855a. Over eenige visschen van Diemensland. *Verhandelingen der Koninklijke Akademie van Wetenschappen te Amsterdam*, 2: 1–31.

Bleeker, P. 1855b. *Antennarius Lindgreeni*, eene nieuwe soort van Banka. *Natuurkundig tijdschrift voor Nederlandsch Indië*, 8: 192–193.

Bleeker, P. 1855c. Zesde bijdrage tot de kennis der ichthyologische fauna van Amboina. *Natuurkundig tijdschrift voor Nederlandsch Indië*, 8: 391–434.

Bleeker, P. 1856a. Beschrijvingen van nieuwe en weinig bekende vischsoorten van Amboina, verzameld op eene reis door den Moluksche Archipel, gedaan in het gevolg van den Gouverneur-Generaal Duymaer van Twist in September en October 1855. *Acta Societatis Regiae Scientiarum Indo-Neêrlandicae*, 1: 1–76.

Bleeker, P. 1856b. Nieuwe bijdrage tot de kennis der ichthyologische fauna van Bali. *Natuurkundig tijdschrift voor Nederlandsch Indië*, 12: 291–302.

Bleeker, P. 1857. Achtste bijdrage tot de kennis der vischfauna van Amboina. *Acta Societatis Regiae Scientiarum Indo-Neêrlandicae*, 2: 1–102.

Bleeker, P. 1858a. Vierde Bijdrage tot de kennis der ichthyologische fauna van Japan. *Acta Societatis Regiae Scientiarum Indo-Neêrlandicae*, 3: 1–46.

Bleeker, P. 1858b. Tiende bijdrage tot de kennis der vischfauna van Celebes. *Acta Societatis Regiae Scientiarum Indo-Neêrlandicae*, 3: 1–16.

Bleeker, P. 1858c. Vijfde bijdrage tot de kennis der ichthyologische fauna van de Kokos-eilanden. *Natuurkundig tijdschrift voor Nederlandsch Indië*, 15: 457–468.

Bleeker, P. 1859a. Bijdrage tot de kennis der vischfauna van Nieuw-Guinea. *Acta Societatis Regiae Scientiarum Indo-Neêrlandicae*, 6: 1–24.

Bleeker, P. 1859b. Enumeratio specierum piscium hucusque in Archipelago Indico observatarum, adjectis habitationibus citationibusque, ubi descriptiones earum recentiores reperiuntur, nec non speciebus Musei Bleekeriani Bengalensibus, Japonicis, Capensibus Tasmanicisque. *Acta Societatis Regiae Scientiarum Indo-Neêrlandicae*, 6: i–xxxvi + 1–276.

Bleeker, P. 1859c. Negende bijdrage tot de kennis der vischfauna van Banka. *Natuurkundig tijdschrift voor Nederlandsch Indië*, 18: 359–378.

Bleeker, P. 1860a. Over eenige vischsoorten van de Kaap de Goede Hoope. *Natuurkundig tijdschrift voor Nederlandsch Indië*, 21: 49–80.

Bleeker, P. 1860b. Elfde bijdrage tot de kennis der vischfauna van Amboina. *Acta Societatis Regiae Scientiarum Indo-Neêrlandicae*, 8: 1–14.

Bleeker, P. 1863. Mémoire sur les poissons de la côte de Guinée. *Natuurkundige verhandelingen van de Hollandsche Maatschappij der Wetenschappen*, 2, Verzameling Deel, 18: 1–136.

Bleeker, P. 1865a. Description de quelques espèces inédites de poissons de l'Archipel des Moluques. *Nederlandsch Tijdschrift voor de Dierkunde*, 2: 177–181.

Bleeker, P. 1865b. *Atlas ichthyologique des Indes Orientales Néêrlandaises, publié sous les auspices du Gouvernement Colónial Néêrlandais. V. Baudroies, Ostracions, Gymnodontes, Balistes.* Frédéric Muller, Amsterdam, 152 pp., pls. 194–231.

Bleeker, P. 1865c. Énumération des espèces de poissons actuellement connues de l'île d'Amboine. *Nederlandsch Tijdschrift voor de Dierkunde*, 2: 270–293.

Bleeker, P. 1873. Mémoire sur la faune ichthyologique de Chine. *Nederlandsch Tijdschrift voor de Dierkunde*, 4: 113–154.

Bloch, M. E. 1785. *Naturgeschichte der ausländischen Fische.* Berlin, Vol. 1, viii + 136 pp.

Bloch, M. E. 1787. *Ichthyologie, ou histoire naturelle générale et particulière des poissons. Avec des figures enluminées, dessinées d'après nature.* Berlin, Vol. 4, 136 pp.

Bloch, M. E., and J. G. Schneider. 1801. *M. E. Blochii systema ichthyologiae iconibus cx illustratum. Post obitum auctoris opus inchoatum absolvit, correxit, interpolavit J. G. Schneider, Saxo.* Sanderiano, Berlin, lx + 584 pp.

Blot, J. 1980. La faune ichthyologique des gisements du Monte Bolca (Province de Vérone, Italie). Catalogue systématique présentant l'état actuel des recherches concernant cette faune. *Bulletin du Muséum national d'Histoire naturelle, Paris*, C, 4: 339–396.

Blunt, W. 2001. *Linnaeus, the Compleat Naturalist.* With an introduction by William T. Stearns. Princeton University Press, Princeton and Oxford, 264 pp.

Boehlert, G. W., and B. C. Mundy. 1996. Ichthyological vertical distributions near Oahu, Hawaii, 1985–1986: data report. NOAA Technical Memorandum NMFS, NIOAA-TM-NMFS-SWFSC-235, 156 pp.

Boeseman, M. 1947. *Revision of the Fishes Collected by Burger and Von Siebold in Japan.* Brill, Leiden, viii + 242 pp.

Boeseman, M. 1983. Introduction. Pp. 1–22, In: P. Bleeker, *Atlas Ichthyologique des Indes Orientales Néêrlandaises.* Plates originally prepared for planned tomes XI–XIV. Smiths. Inst. Press, Washington, DC, 22 pp. + 152 pls.

Böhlke, E. B. 1984. *Catalog of Type Specimens in the Ichthyological Collection of the Academy of Natural Sciences of Philadelphia.* Special Publication, Academy of Natural Sciences of Philadelphia, 14, viii + 216 pp.

Böhlke, J. E. 1953. A catalogue of the type specimens of recent fishes in the Natural History Museum of Stanford University. *Stanford Ichthyological Bulletin*, 5: 1–170.

Böhlke, J. E., and C. C. G. Chaplin. 1968. *Fishes of the Bahamas and Adjacent Tropical Waters*. Livingston, Wynnewood, Pennsylvania, xxiii + 771 pp.

Bonaldo, R. M., M. M. Pires, P. R. Guimarães Jr., A. S. Hoey, and M. E. Hay. 2017. Small marine protected areas in Fiji provide refuge for reef fish assemblages, feeding groups, and corals. *PLoS ONE*, 12 (1): e0170638, doi.org/10.1371/journal.pone.0170638.

Bone, Q., and G. O. Mackie. 1977. Ciliary arrest potentials, locomotion and skin impulses in *Doliolum* (Tunicata: Thaliacea). *Rivista di Biologia Normale e Patologica*, 3: 181–191.

Bonnaterre, J. P. 1787. Poissons. In: *Encyclopédie Méthodique: Histoire naturelle des animaux*. Panckouche, Paris, Vol. 3, lx + 435 pp. [Anonymous; authorship on authority of W. Engelmann, 1846, *Bibliotheca historico-naturalis*, published by the author, Leipzig, viii + 786 pp.]

Bonnaterre, J. P. 1788. *Ichthyologie. Tableau encyclopédique et méthodique des trois règnes de la nature*. Panckoucke, Paris, lvi + 215 pp.

Bornbusch, A. H. 1989. Lacepède and Cuvier: A comparative case study of goals and methods in late eighteenth- and early nineteenth-century fish classification. *Journal of the History of Biology*, 22 (1): 141–161.

Borodin, N. A. 1928. Scientific results of the yacht "Ara" expedition during the years 1926 to 1928, while in command of William K. Vanderbilt: Fishes. *Bulletin of the Vanderbilt Oceanographic Museum*, 1 (1): 1–37.

Borodin, N. A. 1930. Scientific results of the yacht "Ara" expedition during the years 1926 to 1930, while in command of William K. Vanderbilt: Fishes (collected in 1929). *Bulletin of the Vanderbilt Oceanographic Museum*, 1 (2): 39–64.

Borodin, N. A. 1932. Scientific results of the yacht "Alva" world cruise, July 1931 to March 1932, in command of William K. Vanderbilt: Fishes. *Bulletin of the Vanderbilt Marine Museum*, 1 (3): 65–101.

Borodin, N. A. 1934. Scientific results of the yacht "Alva" Mediterranean cruise, 1933, in command of William K. Vanderbilt. *Bulletin of the Vanderbilt Marine Museum*, 1 (4): 103–123.

Bortone, S. A., P. A. Hastings, and S. B. Collard. 1977. The pelagic-*Sargassum* ichthyofauna of the eastern Gulf of Mexico. *Northeast Gulf of Mexico Science*, 1 (2): 60–67.

Bory de Saint-Vincent, G. J. B. M. 1822. Antennaria. *Dictionnaire Classique d'Histoire Naturelle*, 1: 411.

Bory de Saint-Vincent, G. J. B. M. 1826. Lophie. *Dictionnaire Classique d'Histoire Naturelle*, 9: 493–496.

Bosc, L. A. G. 1803. Lophie, *Lophius*, genre de poissons de la division des Branchiostèges. *Nouveau Dictionnaire d'Histoire Naturelle*, Ed. 1, 13: 310–14 (pl. E.30 in Vol. 12, opposite p. 409).

Bosc, L. A. G. 1817. Lophie, *Lophius*. *Nouveau Dictionnaire d'Histoire Naturelle*, Ed. 2, 18: 180–184 (pl. E.30 in Vol. 17).

Boschung, H. T. 1992 Catalogue of freshwater and marine fishes of Alabama. *Alabama Museum of Natural History Bulletin*, 14, xvi + 266 pp.

Bougainville, H. Y. P. P. 1837. *Journal de la navigation autour du globe, de la frégate La Thétis et de la corvette L'Espérance, pendant les années 1824, 1825 et 1826, publié par ordre du roi sous les auspices du Département de la marine par baron de Bougainville*. A. Bertrand, Paris, 56 pls.

Bougainville, L. A. de. 1772. *A Voyage Round the World. Performed by Order of His Most Christian Majesty, in the Years 1766, 1767, 1768, and 1769*. Translation by J. R. Forster. Nourse & Davies, London, xxviii + 476 pp.

Boulenger, G. A. 1887. An account of the fishes obtained by Surgeon-Major A. S. G. Jayakar at Muscat, east coast of Arabia. *Proceedings of the Zoological Society*, London, 43: 653–667.

Boulenger, G. A. 1897. A list of fishes obtained by Mr. J. Stanley Gardiner at Rotuma, South Pacific Ocean. *Annals and Magazine of Natural History*, 6 (20): 371–374.

Boulenger, G. A. 1904. Fishes (Systematic Account of Teleostei). Pp. 541–727, In: S. F. Harmer and A. E. Shipley (editors), *The Cambridge Natural History*. Vol. 7, Fishes, Ascidians, Etc. London: MacMillan, xvii + 760 pp.

Bouvier, E. L. 1888. Le mimetisme chez les poissons. *Naturaliste*, 10: 17–20.

Bowen, B. W., A. Muss, L. A. Rocha, and W. S. Grant. 2006. Shallow mtDNA coalescence in Atlantic pygmy angelfishes (genus *Centropyge*) indicates a recent invasion from the Indian ocean. *Journal of Heredity*, 97 (1): 1–12.

Bowen, B. W., M. R. Gaither, J. D. DiBattista, M. Iacchei, K. R. Andrews, W. S. Grant, R. J. Toonen, and J. C. Briggs. 2016. Comparative phylogeography of the ocean planet. *Proceedings of the National Academy of Sciences*, 113 (29): 7962–7969.

Bowman, A. 1920. The eggs and larvae of the angler (*Lophius piscatorius* L.) in Scottish waters. *Scientific Investigations of the Fishery Board for Scotland*, 2: 1–42.

Braam-Houckgeest, A. E. van. 1774. Bericht wegens der *Lophius histrio*. *Natuurkundige Verhandelingen van de Hollandsche Maatschappij der Wetenschappen te Haarlem*, 15: 20–26.

Bradbury, M. G. 1967. The genera of batfishes (family Ogcocephalidae). *Copeia*, 1967 (2): 399–422.

Bradbury, M. G. 1980. A revision of the fish genus *Ogcocephalus* with descriptions of new species from the Western Atlantic Ocean (Ogcocephalidae; Lophiiformes). *Proceedings of the California Academy of Sciences*, 42 (7): 229–285.

Bradbury, M. G. 1988. Rare fishes of the deep-sea genus *Halieutopsis*: A review with descriptions of four new species (Lophiiformes: Ogcocephalidae). *Fieldiana, Zoology*, 44: 1–22.

Bradbury, M. G. 1999. A review of the fish genus *Dibranchus*, with descriptions of new species and a new genus, *Solocisquama* (Lophiiformes: Ogcocephalidae). *Proceedings of the California Academy of Sciences*, 51 (5): 259–310.

Brainerd, E., B. Page, and F. Fish. 1997. Opercular jetting during fast-starts by flatfishes. *Journal of Experimental Biology*, 200 (8): 1179–1188.

Brander, L. M., P. Van Beukering, H. S. J. Cesar. 2007. The recreational value of coral reefs: a meta-analysis. *Ecological Economics*, 63 (1): 209–218.

Brandl, S. J., and D. R. Bellwood. 2014. Pair-formation in coral reef fishes: an ecological perspective. *Oceanography* and *Marine Biology Annual* Review, 52: 1–80.

Brandl, S. J., C. H. R. Goatley, D. R. Bellwood, and L. Tornabene. 2018. The hidden half: ecology and evolution of cryptobenthic fishes on coral reefs. *Biological Reviews*, 2018: 1–28, doi: 10.1111/brv.12423.

Brauwer, M. de, and J.-P. A. Hobbs. 2016. Stars and stripes: biofluorescent lures in the Striated Frogfish indicate role in aggressive mimicry. *Reef Sites*, doi: 10.1007/s00338-016-1493-1.

Breder, C. M. 1926. The locomotion of fishes. *Zoologica*, New York, 4 (5): 159–297.

Breder, C. M. 1934. The oceanographic vessel *Atlantis* in the West Indies. *Bulletin of the New York Zoological Society*, 37 (2): 31–39.

Breder, C. M. 1938. A contribution to the life histories of Atlantic flying fishes. *Bulletin of the Bingham Oceanographic Collection*, 6 (5): 1–126.

Breder, C. M. 1949. On the relationship of social behavior to pigmentation in tropical shore fishes. *Bulletin of the American Museum of Natural History*, 94 (2): 85–106.

Breder, C. M., and M. L. Campbell. 1958. The influence of environment on the pigmentation of *Histrio histrio* (Linnaeus). *Zoologica*, New York, 43: 135–144.

Breder, C. M., and D. E. Rosen. 1966. *Modes of Reproduction in Fishes*. Natural History Press, Garden City, New York, 941 pp.

Brienen, R. P. 2007. From Brazil to Europe: the zoological drawings of Albert Eckhout and Georg Marcgraf. Pp. 273–314, In: K. A. E. Enenkel and P. J. Smith (editors), *Early Modern Zoology: The Construction of Animals in Science, Literature and the Visual Arts*, Brill, Leiden and Boston.

Briggs, J. C. 1962. Restoration of the frogfish, *Antennarius reticularis* (Gilbert). *Copeia*, 1962 (2): 440.

Briggs, J. C., and B. W. Bowen. 2013. Marine shelf habitat: biogeography and evolution. *Journal of Biogeography*, 40: 1023–1035.

Briggs, P. T., and J. R. Waldman. 2002. Annotated list of fishes reported from the marine waters of New York. *Northeastern Naturalist*, 9 (1): 47–80.

Britz. R. 2017. Book review: *Fishes of the World*, fifth edition. *Journal of Fish Biology*, 90 (1): 451–459.

Brower, L. P. 1988. Avian predation on the monarch butterfly and its implications for mimicry theory. *American Naturalist*, 131: 54–56.

Browne, P. 1756. *The Civil and Natural History of Jamaica*. Osborne & Shipton, London, 503 pp.

Bruce, B. D. 1998. Progress on Spotted Handfish recovery, on the brink! *Threatened Species and Communities*, 11: 9.

Bruce, B. D., and M. A. Green. 1998. *The Spotted Handfish Recovery Plan 1999–2001*. Spotted Handfish Recovery Team, Report to Environment Australia, Project ESP 572, Commonwealth Scientific and Industrial Research Organization (CSIRO) Marine Research, Hobart, Tasmania, 52 pp.

Bruce, B. D., M. A. Green, and P. R. Last. 1997. *Developing Husbandry Techniques for Spotted Handfish,* Brachionichthys hirsutus, *and Monitoring the 1996 Spawning Season*. Final report to the Endangered Species Unit, Environment Australia, Project 538, CSIRO Division of Marine Research, Hobart, Tasmania, 29 pp.

Bruce, B. D., M. A. Green, and P. R. Last. 1998. Threatened fishes of the world: *Brachionichthys hirsutus* Lacépède, 1804 (Brachionichthyidae). *Environmental Biology of Fishes*, 52 (4): 418, doi .org/10.1023/A:1007415920088.

Bruce, B. D., M. A. Green, and P. R. Last. 1999. Aspects of the biology of the endangered Spotted Handfish, *Brachionichthys hirsutus* (Lophiiformes: Brachionichthyidae) off southern Australia. Pp. 369–380, In: B. Séret and J.-Y. Sire (editors). *Proceedings of the 5th Indo-Pacific Fish Conference*, Noumea, New Caledonia, 3–8 November 1997, Société Française d'Ichtyologie, Paris.

Bruckner, A. W. 2001. Tracking the trade in ornamental coral reef organisms: the importance of CITES and its limitations. *Aquarium Sciences and Conservation*, 3: 79–94.

Bruckner, A. W. 2005. The importance of the marine ornamental reef fish trade in the wider Caribbean. *Revista de Biologia Tropical*, 53 (Supplement 1): 127–138.

Buchheim, J. R., and M. A. Hixon. 1992. Competition for shelter holes in the coral-reef fish *Acanthemblemaria spinosa* Metzelaar. *Journal of Experimental Marine Biology and Ecology*, 164: 45–54.

Buettikofer, J. 1890. *Reisebilder aus Liberia. II. Die Bewohner Liberia's Thierwelt*. Brill, Leiden, 510 pp.

Burgess, G. H., S. H. Smith, and E. D. Lane. 1994. Fishes of the Cayman Islands. Pp. 199–228, In: M. A. Brunt and J. E. Davies (editors), *The Cayman Islands: Natural History and Biogeography*, Kluwer Academic, Dordrecht and Boston.

Burgess, W. E. 1976. Salts from the seven seas. *Tropical Fish Hobbyist*, 25 (3): 57–64.

Burgess, W. E., and H. R. Axelrod. 1972. *Pacific Marine Fishes, Book 1*. T.F.H. Publications, Hong Kong, 280 pp.

Burgess, W. E., and H. R. Axelrod. 1973a. *Pacific Marine Fishes, Book 2*. T.F.H. Publications, Hong Kong, pp. 281–560.

Burgess, W. E., and H. R. Axelrod. 1973b. *Pacific Marine Fishes, Book 3*. T.F.H. Publications, Hong Kong, pp. 561–839.

Burgess, W. E., and H. R. Axelrod. 1974. *Pacific Marine Fishes, Book 5*. T.F.H. Publications, Hong Kong, pp. 1111–1381.

Burgess, W. E., and H. R. Axelrod. 1976. *Fishes of the Great Barrier Reef. Pacific Marine Fishes, Book 7*. T.F.H. Publications, Hong Kong, pp. 1655–1925.

Burnett-Herkes, J. 1986. Class Osteichthyes (bony fishes). Pp. 571–650, In: W. Sterrer (editor), *Marine Fauna and Flora of Bermuda: A Systematic Guide to the Identification of Marine Organisms,* John Wiley & Sons, New York, xxx + 742 pp.

Burton, E. M. 1932. Some old and new records of the sargassum fish (*Antennarius ocellatus*). *Charleston Museum Quarterly*, 2 (1): 13.

Bussing, W. A., and M. I. López. 1994. Demersal and pelagic inshore fishes of the Pacific coast of lower central America. An illustrated guide. *Revista de Biologia Tropical*, Special Publication, 1: 1–164.

Bussing, W. A., and M. I. López. 2005. *Fishes of Cocos Island and Reef Fishes of the Pacific Coast of Lower Central America*. Editorial Universidad de Costa Rica, San José, Costa Rica, 192 pp.

Cadenat, J. 1937. Recherches systématiques sur les poissons littoraux de la côte occidentale d'Afrique récoltés par le navire Président Théodore Tissier, au cours de sa 5e croisière (1936). *Revue des Travaux de l'Office Scientifique et Technique des Peches Maritimes,* Paris, 10 (4): 423–562.

Cadenat, J. 1951. Poissons de mer du Sénégal: Initiations Africaines, III. *Bulletin* de l'*Institut* français d'*Afrique Noire,* Dakar, 1950 (1951), 345 pp.

Cadenat, J. 1959. Notes d'ichtyologie Ouest-Africaine. XIX. Les *Antennarius. Bulletin* de l'*Institut* français d'*Afrique noire,* Dakar, Série A, 21 (1): 361–394.

Cadenat, J. 1960. Notes d'ichtyologie Ouest-Africaine. XXX. Poissons de mer Ouest-Africains observés du Sénégal au Cameroun et plus spécialement au large des côtes de Sierra Leone et du Ghana. *Bulletin* de l'*Institut* français d'*Afrique noire,* Dakar, Série A, 22 (4): 1358–1420.

Calado, R. 2006. Marine ornamental species from European waters: a valuable overlooked resource or a future threat for the conservation of marine ecosystems? *Scientia Marina,* 70 (3): 389–398.

Calado, R. 2017a. The need for cultured specimens. Pp. 15–22, In: R. Calado, I. Olivotto, M. P. Oliver, and G. J. Holt (editors), *Marine Ornamental Species Aquaculture,* John Wiley & Sons Ltd, Chichester, West Sussex, United Kingdom.

Calado, R. 2017b. The role of public and private aquaria in the culture and conservation of marine ornamentals. Pp. 609–610, In: R. Calado, I. Olivotto, M. P. Oliver, and G. J. Holt (editors), *Marine Ornamental Species Aquaculture,* John Wiley & Sons Ltd, Chichester, West Sussex, United Kingdom.

Calado, R., I. Olivotto, M. P. Oliver, and G. J. Holt (editors). 2017. *Marine Ornamental Species Aquaculture.* John Wiley & Sons Ltd, Chichester, West Sussex, United Kingdom, xxxv + 677 pp.

Caldwell, D. K. 1957. Additional records of marine fishes from the vicinity of Cedar Key, Florida. *Quarterly Journal of the Florida Academy Science,* 20 (2): 126–128.

Caldwell, D. K. 1966. Marine and freshwater fishes of Jamaica. *Bulletin of the Institute of Jamaica,* Kingston, Science Series, 17: 1–120.

Camara, M. L., B. Mérigot, F. Leprieur, J. A. Tomasini, I. Diallo, M. Diallo, and D. Jouffre. 2016. Structure and dynamics of demersal fish assemblages over three decades (1985–2012) of increasing fishing pressure in Guinea. *African Journal of Marine Science,* 38 (2): 189–206.

Cantor, T. 1849. Catalogue of Malayan fishes. *Journal of the Asiatic Society,* Bengal, 18 (2): 983–1443.

Cantor, T. 1850. *Catalogue of Malayan Fishes.* Thomas, Baptist Mission Press, Calcutta, xii + 461 pp.

Capello, F. de Brito. 1871. Primeira lista dos peixes da Ilha da Madeira, Acores, e das possessões portuguezas d'Africa que existem no Museu de Lisboa. *Jornal de Sciências, Mathemáticas, Physicas e Naturaes,* Lisboa, 3 (11): 194–202.

Carnevale, G., A. F. Bannikov, G. Marramà, J. C. Tyler, and R. Zorzin. 2014. The Pesciara-Monte Postale Fossil-Lagerstätte: 2. Fishes and other vertebrates. Pp. 37–63, In: C. A. Papazzoni, L. Giusberti, G. Carnevale, G. Roghi, D. Bassi, and R. Zorzin (editors), *The Bolca Fossil-Lagerstätten: A window into the Eocene World,* Excursion guidebook CBEP 2014-EPPC 2014-EAVP 2014-Taphos 2014 Conferences, *Rendiconti della Società Paleontologica Italiana,* 4.

Carnevale, G., and T. W. Pietsch. 2006. Filling the gap: a fossil frogfish, genus *Antennarius* (Teleostei, Lophiiformes, Antennariidae), from the Miocene of Algeria. *Journal of Zoology,* 270: 448–457.

Carnevale, G., and T. W. Pietsch. 2009a. The deep-sea anglerfish genus *Acentrophryne* (Teleostei, Ceratioidei, Linophryne) in the Miocene of California. *Journal of Vertebrate Paleontology,* 29 (2): 372–378.

Carnevale, G., and T. W. Pietsch. 2009b. An Eocene frogfish (Lophiiformes: Antennariidae) from Monte Bolca, Italy: the earliest known skeletal record for the family. *Palaeontology,* 52 (4): 745–752.

Carnevale, G., and T. W. Pietsch. 2010. Eocene handfishes from Monte Bolca, with description of a new genus and species, and a phylogeny of the family Brachionichthyidae (Teleostei: Lophiiformes). *Zoological Journal of the Linnean Society,* 160 (4): 621–647.

Carnevale, G., and T. W. Pietsch. 2011. Batfishes from the Eocene of Monte Bolca. *Geological Magazine*, 147: 1–12.

Carnevale, G., and T. W. Pietsch. 2012. †*Caruso*, a new genus of anglerfishes from the Eocene of Monte Bolca, Italy, with a comparative osteology and phylogeny of the family Lophiidae (Teleostei: Lophiiformes). *Journal of Systematic Palaeontology*, 10 (1): 47–72.

Carnevale, G., T. W. Pietsch, G. T. Takeuchi, and R. W. Huddleston. 2008. Fossil ceratioid anglerfishes (Teleostei: Lophiiformes) from the Miocene of the Los Angeles Basin, California. *Journal of Paleontology*, 82 (5): 996–1008.

Carpenter, K. E., F. Krupp, D. A. Jones, and U. Zajonz. 1997. *FAO Species Identification Guide for Fishery Purposes. The Living Marine Resources of Kuwait, Eastern Saudi Arabia, Bahrain, Qatar, and the United Arab Emirates.* FAO Rome, 293 pp.

Carpenter, R. C. 1986. Partitioning herbivory and its effects on coral reef algal communities. *Ecological Monographs*, 56 (4): 345–364.

Caruso, J. H. 1976. A review of the lophiid angler fish genus *Sladenia*, with a description of a new species from the Caribbean Sea. *Bulletin of Marine Science*, 26 (1): 59–64.

Caruso, J. H. 1981. The systematics and distribution of the lophiid anglerfishes: I. A revision of the genus *Lophiodes*, with the description of two new species. *Copeia*, 1981 (3): 522–549.

Caruso, J. H. 1983. The systematics and distribution of the lophiid anglerfishes: II. Revisions of the genera *Lophiomus* and *Lophius*. *Copeia*, 1983 (1): 11–30.

Caruso, J. H. 1985. The systematics and distribution of the lophiid anglerfishes: III. Intergeneric relationships. *Copeia*, 1985 (4): 870–875.

Caruso, J. H. 1986. [Lophiidae]. Pp. 1362–1363, In: P. J. P. Whitehead, M.-L. Bauchot, J.-C. Hureau, J. Nielsen, and E. Tortonese (editors), *Fishes of the North-eastern Atlantic and Mediterranean.* Vol. 3, United Nations Educational Scientific and Cultural Organization, Paris.

Caruso, J. H. 1989a. Systematics and distribution of the Atlantic chaunacid anglerfishes (Pisces: Lophiiformes). *Copeia*, 1989 (1): 153–165.

Caruso, J. H. 1989b. A review of the Indo-Pacific members of the deep-water chaunacid anglerfish genus *Bathychaunax*, with the description of a new species from the Eastern Indian Ocean (Pisces: Lophiiformes). *Bulletin of Marine Science*, 45 (3): 574–579.

Caruso, J. H., and H. R. Bullis, Jr. 1976. A review of the lophiid angler fish genus *Sladenia*, with a description of a new species from the Caribbean Sea. *Bulletin of Marine Science*, 26 (1): 59–64.

Caruso, J. H., H.-C. Ho, and T. W. Pietsch. 2006. *Chaunacops* Garman, 1899, a senior objective synonym of *Bathychaunax* Caruso, 1989 (Lophiiformes: Chaunacoidei: Chaunacidae). *Copeia*, 2006 (1): 120–121.

Caruso, J. H., and R. D. Suttkus. 1979. A new species of lophiid angler fish from the Western North Atlantic. *Bulletin of Marine Science*, 29 (4): 491–496.

Castellanos-Galindo, G. A., E. A. Rubio Rincon, B. Beltrán-León, L. A. Zapata, and C. C. Baldwin. 2006. Check list of gadiform, ophidiiform and lophiiform fishes from Colombian waters of the tropical eastern Pacific. *Biota Colombiana*, 7 (2): 191–209.

Castelnau, F. 1872. Contribution to the ichthyology of Australia. II. Note on some South Australian fishes. *Proceedings of the Zoological and Acclimatisation Society of Victoria*, 1: 243–248.

Castelnau, F. 1873. Contribution to the ichthyology of Australia. No. IV. Fishes of South Australia. *Proceedings of the Zoological and Acclimatisation Society of Victoria*, 2: 59–82.

Castelnau, F. 1875. Researches on the fishes of Australia. *Official Record of the Victorian Intercolonial Exhibition, Intercolonial Exhibition Essays*, Melbourne, 2: 1–52.

Castro-Aguirre, J. L., H. Espinosa Pérez, and J. J. Schmitter-Soto. 1999. *Ictiofauna estuarino-Lagunar y vicaria de México.* Colección Textos Politécnicos, Serie Biotechnologías, 711 pp.

Catarino, M. F., and J. Zuanon. 2010. Feeding ecology of the leaf fish *Monocirrhus polyacanthus* (Perciformes: Polycentridae) in a terra firme stream in the Brazilian Amazon. *Neotropical Ichthyology*, 8 (1): 183–186.

Cervigón, F. 1966. Los peces marinos de Venezuela. *Estacion Investigaciones marinas de Margarita*, Caracas, 2 (12): 441–951.

Cervigón, F. 1991. Los peces marinos de Venezuela. *Fundación Científica Los Roques*, 1: 1–425.

Chadwick, H. C. 1929. Feeding habits of the angler-fish, *Lophius piscatorius. Nature*, 124: 337.

Chanet, B., C. Guintard, E. Betti, C. Gallut, A. Dettaï, and Guillaume Lecointre. 2013. Evidence for a close phylogenetic relationship between the teleost orders Tetraodontiformes and Lophiiformes based on an analysis of soft anatomy. *Cybium*, 37 (3): 179–198.

Chen, J. T. F., M. Liu, and S. Lee. 1967. A review of the pediculate fishes of Taiwan. *Biological Bulletin of the Tunghai University*, 33 (Ichthyological Series No. 7): 1–23.

Chen, J.-P., R.-Q. Jan, J.-W. Kuo, C.-H. Huang, and C.-Y. Chen. 2009. Fish fauna around Green Island. *Journal of National Park*, 19 (3): 23–45 [in Chinese].

Chen, Q.-C., Y.-Z. Cai, and X.-M. Ma (editors). 1997. Fishes from Nansha Islands to South China Coastal Waters. Science Press, Beijing, xx + 202.

Chen, W.-J., C. Bonillo, and G. Lecointre. 2003. Repeatability of clades as a criterion of reliability: a case study for molecular phylogeny of Acanthomorpha (Teleostei) with larger number of taxa. *Molecular Phylogenetics and Evolution*, 26: 262–288.

Chen, W.-J., R. Ruiz-Carus, and G. Ortí. 2007. Relationships among four genera of mojarras (Teleostei: Perciformes: Gerreidae) from the western Atlantic and their tentative placement among percomorph fishes. *Journal of Fish Biology*, 70: 202–218.

Chen, W.-J., F. Santini, G. Carnevale, J.-N. Chen, S.-H. Liu, S. Lavoué, and R. L. Mayden. 2014. New insights on early evolution of spiny-rayed fishes (Teleostei: Acanthomorpha). *Frontiers in Marine Science*, doi: org/10.3389/fmars.2014.00053.

Cheney, K. L. 2010. Multiple selective pressures apply to a coral reef fish mimic: a case of Batesian–aggressive mimicry. *Proceedings of the Royal Society* B, 277: 1849–1855.

Cheney, K. L., and N. J. Marshall. 2009. Mimicry in coral reef fish: how accurate is this deception in terms of color and luminance? *Behavioral Ecology*, 20: 459–468.

Chevey, P. 1932. Poissons des campagnes du "de Lanessan" (1925–1929). Ire Partie. *Travail de l'Institut Océanographique de Indochine*, Saigon, Mémoire 4, 155 pp.

Chi, E. Y., E. Ignácio, and D. Lagunoff. 1978. Mast cell granule formation in the beige mouse. *Journal of Histochemistry and Cytochemistry*, 26 (2): 131–137.

Chirichigno, N., and J. Vélez. 1998. *Clave para identificar los peces marinos del Peru*. Seguenda edición, revidada y actualizada. Instituto del Mar del Peru, Publicación especial, 496 pp.

Christie, B. L., P. Z. Montoya, L. A. Torres, and J. W. Foster. 2016. The natural history and husbandry of the walking batfishes (Lophiiformes: Ogcocephalidae). *Drum and Croaker*, 47: 7–40.

Clark, E., and H. A. F. Gohar. 1953. The fishes of the Red Sea: Order Plectognathi. Publications of the Marine Biology Station, Al Ghardaqa, University of Cairo, 8: 1–80.

Cloquet, H. 1817. Chironecte (Ichthyol.), *Chironectes. Dictionnaire des Sciences Naturelles*, Paris, 8: 597–599.

Coleman, W. 1964. *Georges Cuvier, Zoologist: A Study in the History of Evolution Theory*. Harvard University Press, Cambridge, Massachusetts, x + 212 pp.

Collett, R. 1875. Norges Fiske med Bemaerkninger om deres Udbredelse. *Forhandlinger i Videnskabs-selskabet i Kristiania*, 1874 (1875): 1–240.

Collett, R. 1896. Poissons provenant des campagnes du yacht l'Hirondelle (1885–1888). *Résultats des Campagnes Scientifiques* du *Prince Albert Ier*, 10, viii + 198 pp.

Collette, B. B. 1966. A review of the venomous toadfishes, subfamily Thalassophryninae. *Copeia*, 1966 (4): 846–864.

Collette, B. B. 1968. *Daector schmitti*, a new species of venomous toadfish from the Pacific coast of Central America. *Proceedings of the Biological Society of Washington*, 81: 155–160.

Collette, B. B. 1983. Mangrove fishes of New Guinea. Chap. 10, pp. 91–102, In: H. J. Teas (editor), *Tasks for Vegetation Science*. Junk, The Hague, Vol. 8.

Collette, B. B. 1995. *Potamobatrachus trispinosus*, a new freshwater toadfish (Batrachoididae) from Rio Tocantins, Brazil. *Ichthyological Exploration of Freshwaters*, 6 (4): 333–336.

Collette, B. B., and D. W. Greenfield. 2009. *Batrachus uranoscopus* Guichenot, 1866, supposedly from Madagascar, is not a threatened species of toadfish (Batrachoididae). *Cybium*, 33 (1): 79–80.

Collette, B. B., J. T. Williams, C. E. Thacker, and M. L. Smith. 2003. Shore fishes of Navassa Island, West Indies: a case study on the need for rotenone sampling in reef fish biodiversity studies. *Aqua, Journal of Ichthyology and Aquatic Biology*, 6 (3): 89–131.

Collins, J. W. 1880. Hatching the "angler" or "fishing frog." *Forest and Stream*, 15 (1): 8, 28.

Colmenero, A. I., V. M. Tuset, and P. Sánchez. 2017. Reproductive strategy of white anglerfish (*Lophius piscatorius*) in Mediterranean waters: implications for management. *United States Fishery Bulletin*, 115 (1): 60–73.

Commerson, P. MS 889. Manuscrits provenant de Philibert Commerson (1727–1773) et en partie de sa main. VI. Manuscrits sur l'histoire naturelle, principalement des poissons. Bibliothèque Centrale du Muséum national d'Histoire naturelle, Paris.

Commerson, P. MS 891. Manuscrits provenant de Philibert Commerson (1727–1773) et en partie de sa main. VIII, pp. 122–144, Acanthoptérygiens à pectorales pédiculées. Bibliothèque Centrale du Muséum national d'Histoire naturelle, Paris.

Commonwealth of Australia. 2016. *Recovery Plan for Three Handfish Species: Spotted Handfish* Brachionichthys hirsutus, *Red Handfish* Thymichthys politus, *Ziebell's Handfish* Brachiopsilus ziebelli. Department of the Environment and the Tasmanian Government, Hobart, Tasmania, 38 pp.

Compagno, L. J. V. 1984. *Sharks of the World: An Annotated and Illustrated Catalogue of Shark Species Known to Date*. FAO Species Catalogue 4, Food and Agriculture Organization of the United Nation, Rome, Part 2, Carcharhiniformes, x + pp. 251–655.

Connolly, C. J. 1920. Histories of new food fishes. III. The angler. *Bulletin of the Biological Board of Canada*, 3: 1–17.

Cope, E. D. 1871. Contribution to the ichthyology of the Lesser Antilles. *Transactions of the American Philosophical Society*, Philadelphia, 14 (3): 445–483.

Cope, E. D. 1872. Observations on the systematic relations of the fishes. *Proceedings of the American Association for the Advancment of Science*, August 1871, pp. 317–343.

Coston-Clements, L., L. R. Settle, D. E. Hoss, and F. A. Cross. 1991. *Utilization of the* Sargassum *habitat by marine invertebrates and vertebrates—a review*. NOAA Technical Memorandum, NMFS-SEFSC-296, 32 pp.

Cott, H. B. 1940. *Adaptive Coloration in Animals*. Methuen, London, xxxii + 508 pp.

Couch, J. 1863. *A History of the Fishes of the British Islands*. Groombridge and Sons, London, Vol. 2, iv + 265 pp.

Cowan, C. F. 1969. Cuvier's *Règne animal*, first edition. *Journal of the Society for the Bibliography of Natural History*, 5 (3): 219.

Cowman, P. F., and D. R. Bellwood. 2013. The historical biogeography of coral reef fishes: global patterns of origination and dispersal. *Journal of Biogeography*, 40: 209–224.

Cowman, P. F., V. Parravicini, M. Kulbicki, and S. R. Floeter. 2017. The biogeography of tropical reef fishes: endemism and provinciality through time. *Biological Reviews*, 92 (4): 2112–2130.

Creese, R. G., T. M. Glasby, G. West, and C. Gallen. 2009. *Mapping the habitats of NSW estuaries*. Industry and Investment NSW Fisheries Final Report Series 113. Nelson Bay, NSW, Australia, 1837–2112.

Croizat, L. C. M. 1958. *Panbiogeography*. Published by the author, Caracas, 1,018 pp.

Croizat, L. C. M. 1964. *Space, Time, Form: The Biological Synthesis*. Published by the author, Caracas, xix + 881 pp.

Croizat, L. C. M., G. Nelson, and D. E. Rosen. 1974. Centers of origin and related concepts. *Systematic Zoology*, 23 (2): 265–287.

Cruz-Agüero, J. de la, I. G. Lizárraga, G. Tiburcio-Pintos, C. Sánchez-García, and E. Acevedo-Ruiz. 2012. Mass stranding of fish in the Cape region of the Gulf of California. *Marine Biodiversity Records*, 5 (70): 1–4.

Cunxin, W. 1993. Ecological characteristics of the fish fauna of the South China Sea. Pp. 77–118, In: B. Morton (editor), *The Marine Biology of the South China Sea*, Proceedings of the First International Conference on the Marine Biology of Hong Kong and the South China Sea, 28 October–3 November 1990, Hong Kong University Press, Hong Kong, China.

Curio, E. 1976. *The Ethology of Predation*. Zoophys. Ecol., 7. Springer-Verlag, Berlin, x + 250 pp.

Currie, D. R., and S. J. Sorokin. 2010. The distribution and trophodynamics of demersal fish from Spencer Gulf. *Transactions of the Royal Society of South Australia*, 134 (2): 198–227.

Cuvier, G. 1816. *Le règne animal distribué d'après son organisation, pour servir de base à l'histoire naturelle des animaux et d'introduction à l'anatomie comparée*, 4 vols., Deterville, Poissons, Paris, 2b: 104–351.

Cuvier, G. 1817a. *Mémoires pour servir a l'histoire et à l'anatomie des mollusques*. VI. Mémoire sur la scyllée, l'eolide et le glaucus, avec des additions au Mémoire sur la tritonie. Deterville, Paris, 29 pp.

Cuvier, G. 1817b. Sur le genre *Chironectes* Cuv. (*Antennarius*. Commers.). *Mémoires du Muséum national d'Histoire naturelle,* Paris, 3: 418–435.

Cuvier, G. 1828. Tableau historique des progrès de l'Ichtyologie, depuis son origine jusqu' à nos jours. Pp. 1–270, In: G. Cuvier and A. Valenciennes, *Histoire naturelle des poissons*, Levrault, Paris & Strasbourg, Vol. 1.

Cuvier, G. 1829. *Le règne animal distribué d'après son organisation, pour servir de base à l'histoire naturelle des animaux et d'introduction à l'anatomie comparée*. Nouvelle édition, Deterville, Paris, Poissons, 2: 122–406.

Cuvier, G. 1845. *Histoire des sciences naturelles depuis leur origine jusqu'à nos jours, chez tous les peuples connus, commencée au collége de France par Georges Cuvier, ouvrage posthume, publié par M. Magdeleine de Saint-Agy*. Fortin, Masson, Paris, Vol. 5, 440 pp.

Cuvier, G., and A. Valenciennes. MS 504. Histoire Naturelle des Poissons, par Cuvier et Valenciennes. Materiaux de cet ouvrage publié de 1829 à 1849. Notes et nombreuses planches gravées ou originales en couleurs, etc. Les premiers volumes ont recu une numérotation. XII.B. Poissons à pectorales pédiculées. Bibliothèque Centrale du Muséum national d'Histoire naturelle, Paris.

Dahlgren, U. 1928. The habits and life history of *Lophius*, the angler fish. *Natural History*, 28 (1): 18–32.

Dall, W. H. 1916. Bibliographic memoir of Theodore Nicholas Gill, 1837–1914. *National Academy of Sciences, Biographical Memoirs*, 8: 313–343.

Dart, J. K. G. 1972. Echinoids, algal lawn and coral recolonisation. *Nature*, 239: 50–51.

Datovo A., M. C. C. de Pinna, and G. D. Johnson. 2014. The infrabranchial musculature and its bearing on the phylogeny of percomorph fishes (Osteichthyes: Teleostei). *PLoS ONE*, 9 (10): e110129. doi: 10.1371/journal.pone.0110129.

Daudin, F. M. 1816. Antennaire. *Dictionnaire des* Sciences Naturelles, 2nd edition, Paris, 2: 193.

Davies, M. R., and S. Piontek. 2016. The marine fishes of St. Eustatius. Pp. 73–82, In: B. W. Hoeksema (editor), *Marine Biodiversity of St. Eustatius, Dutch Caribbean. Preliminary Results of the Statia Marine Biodiversity Expedition, 2015*. Naturalis Biodiversity Center, Leiden, Netherlands.

Dawkins, R., and J. R. Krebs. 1978. Animal signals: information or manipulation? Pp. 282–309, In: J. R. Krebs and N. B. Davies (editors), *Behavioural Ecology: An Evolutionary Approach*. Blackwell Scientific Publications, Oxford.

Dawson, E. Y., A. A. Aleem, and B. W. Halstead. 1955. Marine algae from Palmyra Island with special reference to the feeding habits and toxicology of reef fishes. *Allan Hancock Foundation, Occasional Papers*, 17: 1–39.

Day, F. 1865. *The Fishes of Malabar*. Quaritch, London, 293 pp.

Day, F. 1876. *The Fishes of India, being a Natural History of the Fishes Known to Inhabit the Seas and Fresh Waters of India, Burma, and Ceylon*. Quaritch, London, Vol. 1, Pt. 2, pp. 169–368.

Day, F. 1889. *The Fauna of British India, including Ceylon and Burma*. Taylor & Francis, London, Vol. 2, Fishes, xiv + 509 pp.

Debelius, H. 1997. *Mediterranean and Atlantic Fish Guide*. IKAN-Unterwasserarchiv, Frankfurt, Germany, 305 pp.

De Brauwer, M., and M. Burton. 2018. Known unknowns: conservation and research priorities for soft sediment fauna that supports a valuable SCUBA diving industry. *Ocean and Coastal Management*, 160: 30–37.

De Brauwer, M., E. S. Harvey, J. L. McIlwain, J.-P. A. Hobbs, J. Jompa, and M. Burton. 2017. The economic contribution of the muck dive industry to tourism in southeast Asia. *Marine Policy*, 83: 92–99.

De Brauwer, M., B. J. Saunders, R. Ambo-Rappe, J. Jompa, E. S. Harvey, and J. L. McIlwain. 2018. Time to stop mucking around: impacts of underwater photography on cryptobenthic fauna found in soft sediment habitats. *Journal of Environmental Management*, 218: 14–22.

De La Cruz Agüero, J., M. A. Martínez, V. M. C. Gómez, and G. De La Cruz Agüero. 1997. *Catalogo de los peces marinos de Baja California Sur*. Instituto Politécnico Nacional, Centro Interdiciplinario de Ciencias Marinas, 346 pp.

DeKay, J. E. 1842. Zoology of New York, or the New York fauna. Comprising detailed descriptions of all the animals hitherto observed within the state borders. Class V. Fishes. *Natural History of New York Geological Survey*, Albany, 1 (3–4): 1–415.

De Laet, J. 1625. *Nieuvve wereldt, ofte, Beschrijvinghe van West-Indien, wt veelderhande schriften ende aen-teeckeninghen van verscheyden natien by een versamelt door Iean de Laet, ende met noodighe kaerten ende tafels voorsien*. In de druckerye van Isaack Elzeviet, Leiden, xii + 510 + [16] pages, [10] folded leaves of plates and maps.

De Laet, J. 1633. *Novus orbis, seu descriptionis Indiae Occidentalis, libri XVIII. Authore Joanne de Laet Antwerp. Novis tabulis geographicis et variis animantium, plantarum fructuumque iconibus illustrati*. Cum privilegio. Elsevier, Leiden, 690 pp.

De Laet, J. 1640. *L'histoire du nouveau monde ou description des Indes Occidentales, contenant dix-huict liures, par le Sieur Iean de Laet, d'Anuers; enrichi de nouvelles tables géographiques et figures des animaux, plantes et fruictes*. Elsevier, Leiden, 632 pp.

Delais, M. 1951. Notes d'ichthyologie Ouest-Africaine. I. Note sur les Antennariidés en collection au Laboratoire de Biologie Marine de l'I.F.A.N. à Gorée. *Bulletin de l'Institut Français d'Afrique Noire*, 13 (1): 145–150.

Del Moral-Flores, L. F., J. M. Gracian-Negrete, and A. F. Guzmán-Camacho. 2016. Peces del Archipélago de las islas Revillagigedo: una actualización sistemática biogeográfica. *Biología, Ciencia y Tecnología*, 9 (34): 596–619.

DeLoach, N., and P. Humann. 1999. *Reef Fish Behavior: Florida, Caribbean, Bahamas*. New World Publications, Inc., Jacksonville, Florida, 360 pp.

Delrieu-Trottin, E., J. T. Williams, P. Bacchet, M. Kulbicki, J. Mourier, R. Galzin, T. Lison de Loma, G. Mou-Tham, G. Siu, and S. Planes. 2015. Shore fishes of the Marquesas Islands, an updated checklist with new records and new percentage of endemic species. *Check List*, 11 (5): 1–13.

De Mello, J. A. G. 1967. Johannes de Laet e sua descrição do Novo Mundo. *Revista do Instituto Arqueológico Histórico e Geográfico Pernambuco*, 46: 135–161.

Dennis, G. D., and T. J. Bright. 1988. New records of fishes in the northwestern Gulf of Mexico, with notes on some rare species. *Northeast Gulf Science*, 10 (1): 1–18.

Dennis, G. D., W. F. Smith-Vaniz, P. L. Colin, D. A. Hensley, and M. A. Mcgehee. 2005. Shore fishes from islands of the Mona Passage, Greater Antilles, with comments on their zoogeography. *Caribbean Journal of Science*, 41 (4): 716–743.

Department of Primary Industries and Water. 2007. *Threatened Species List—Vertebrate Animals*. Department of Primary Industries and Water, State Government of Tasmania. http://www.dpiw.tas.gov.au/inter.nsf/WebPages/SJON-58K8WK?open. Accessed September 2018.

Department of the Environment and Heritage. 2005. *Recovery plan for the following species of handfish: Spotted Handfish* (Brachionichthys hirsutus)*, Red Handfish* (Brachionichthys politus)*, Ziebell's Handfish* (Sympterichthys sp. [CSIRO #T6.01])*, Waterfall Bay Handfish* (Sympterichthys sp. [CSIRO #T1996.01])*. Natural Heritage Trust, Australian Government, Department of the Environment and Heritage, Canberra. http://www.environment.gov.au/biodiversity/threatened/publications/recovery/4-handfish/pubs/4-handfish.pdf. Accessed September 2018.

Derbyshire, K. J., and D. M. Dennis. 1990. Seagrass fish fauna and predation on juvenile prawns. Pp. 83–97, In: J. E. Mellors (editor), *Torres Strait Prawn Project: A Review of Research 1986–88*. Queensland Department of Primary Industries, Brisbane, Information Series, Q190018.

Derouen, V. W. B. Ludt, H.-C. Ho, and P. Chakrabarty. 2015. Examining evolutionary relationships and shifts in depth preferences in batfishes (Lophiiformes: Ogcocephalidae). *Molecular Phylogenetics and Evolution*, 84: 27–33.

Desjardins, J. 1840. Description d'une nouvelle espèce de poisson de l'île Maurice, appartenant à la famille des pectorales pédiculées et au genre Chironecte. *Magasin de Zoology*, Poissons, 1840: 1–5.

Dettaï A., and G. Lecointre. 2008. New insights into the organization and evolution of vertebrate IRBP genes and utility of IRBP gene sequences for the phylogenetic study of the Acanthomorpha (Actinopterygii: Teleostei). *Molecular Phylogenetics and Evolution*, 48 (1): 258–269.

DeVantier, L., and E. Turak. 2004. *Managing Marine Tourism in Bunaken National Park and Adjacent Waters, North Sulawesi, Indonesia*. Project Report, Natural Resources Management Program, Jakarta, Indonesia, 113 pp.

Diamond, J., and A. B. Bond. 2013. *Concealing Coloration in Animals*. Harvard University Press, Cambridge, Massachusetts, x + 271 pp.

Diaper, W., and J. Jones. 1722. *Oppian's Halieuticks of the Nature of Fishes and Fishing of the Ancients in V. Books*. Translated from the Greek, with an account of Oppian's life and writings, and a catalogue of his fishes. Printed at the Theater, Oxford, Pt. 1, Of the nature of fishes, Bk. 2, pp. 59–107, lines 153ff.

Divita, R., M. Creel, and P. F. Sheridan. 1983. Foods of coastal fishes during brown shrimp, *Penaeus aztecus*, migration from Texas estuaries (June–July 1981). *United States Fishery Bulletin*, 81 (2): 396–404.

Dobben, W. H. van. 1935. Über den Kiefermechanismus der Knockenfische. *Archives Néerlandaises de Zoologie*, 2: 1–72.

Doherty, P. J., and D. M. Williams. 1988. The replenishment of coral reeffish populations. *Oceanography and Marine Biology Annual Review*, 26: 487–551.

Donaldson, T., and R. F. Myers. 2002. Insular freshwater fish faunas of Micronesia: patterns of species richness and similarity. *Environmental Biology of Fishes*, 65 (2): 139–149.

Donovan, E. 1808. *The Natural History of British Fishes, Including Scientific and General Descriptions of the Most Interesting Species, and an Extensive Selection of Accurately Finished Coloured Plates, Taken Entirely from Original Drawings, Purposely Made from the Specimens in a Recent State, and for the Most Part Whilst Living*. Printed for the Author and for F. C. and J. Rivington, London, Vol. 5, unpaged, 24 pls.

Dooley, J. K. 1972. Fishes associated with the pelagic sargassum complex, with a discussion of the sargassum community. *Contributions in Marine Science*, 16: 1–32.

Dooley, J. K., J. Van Tassell, and A. Brito. 1985. An annotated checklist of the shorefishes of the Canary Islands. *American Museum Novitates*, 2824, 49 pp.

Dor, M. 1984. CLOFRES: Checklist of the fishes of the Red Sea. *Israel Academy of Sciences and Humanities*, Jerusalem, xxii + 437 pp.

Downing, S. W., and R. R. Novales. 1971. The fine structure of lamprey epidermis. III. Granular cells. *Journal of Ultrastructure Research*, 35: 304–313.

Drickamer, L. C., and S. H. Vessey. 1982. *Animal Behavior. Concepts, Processes, and Methods*. Willard Grant Press, Boston, xvi + 510 pp.

Duarte, R., M. Azevedo, J. Landa, and P. Pereda. 2001. Reproduction of anglerfish (*Lophius budegassa* Spinola and *Lophius piscatorius* Linnaeus) from the Atlantic Iberian coast. *Fisheries Research*, 51: 349–361.

Duarte-Bello, P. P. 1959. Catalogo de peces Cubanos. *Universidad Católica de Santo Tomás de Villanueva, Laboratorio de Biologia Marina, Marianao, Monografia*, 6: 1–208.

Düben, M. W. von. 1845. Norriges Hafs-Fauna. *Kungliga Svenska Vetenskapsakademiens Handlingar*, Stockholm, 1844, 4 (5): 110–116.

Düben, M. W. von, and J. Koren. 1846. Ichtyologiska Bidrag. *Kungliga Svenska Vetenskapsakademiens Handlingar*, Stockholm, 1844 (1846): 27–120.

Dulvy, N. K., Y. Sadovy, and J. D. Reynolds. 2003. Extinction vulnerability in marine populations. *Fish and Fisheries*, 4 (1): 25–64.

Duméril, A. H. A. 1861. Reptiles et poissons de l'Afrique occidentale. Étude précédée de considérations générales sur leur distribution géographique. *Archives du Muséum d'Histoire Naturelle*, Paris, 10: 137–268.

Duncker, G., and E. Mohr. 1929. Die Fische der Südsee-Expedition der Hamburgischen Wissenschaftlichen Stiftung 1908–1909. Teil 3. *Mitteilungen aus dem Zoologischen Staatsinstitut und Zoologisches Museum,* Hamburg, 44: 57–84.

Dyer, B. S., and M. W. Westneat. 2010. Taxonomy and biogeography of the coastal fishes of Juan Fernández Archipelago and Desventuradas Islands, Chile. *Revista de Biología Marina y Oceanografía*, 45 (S1): 589–617.

Earle, S. A. 1972. The influence of herbivores on the marine plants of Great Lameshur Bay, with an annotated list of plants. Pp. 17–44, In: B. B. Collette and S. A. Earle (editors), Results of the Tektite Program: Ecology of Coral Reef Fishes. *Natural History Museum of Los Angeles County, Science Bulletin*, 14.

Eastman, C. R. 1904. Descriptions of Bolca fishes. *Bulletin of the Museum of Comparative Zoology,* 46: 1–36.

Eastman, C. R. 1905. Les types de poissons fossils de Monte Bolca au Muséum d'Histoire Naturelle de Paris. *Mémoires de la Société géologique de France, Paléontologie*, 13: 1–31.

Eaton, T. H., Jr., C. A. Edwards, M. A. McIntosh, and J. P. Rowland. 1954. Structure and relationships of the anglerfish, *Lophius americanus. Journal of the Elisha Mitchell Science Society*, 70 (2): 205–218.

Edwards, A. J. 1993. New records of fishes from the Bonaparte Seamount and Saint Helena Island, South Atlantic. *Journal of Natural History*, 27 (2): 493–503.

Edwards, A. J., and C. W. Glass. 1987. The shore fishes of St. Helena Island, South Atlantic Ocean. I. the shore fishes. *Journal of Natural History*, 21 (3): 617–686.

Edwards, J. L. 1977. The evolution of terrestrial locomotion. Pp. 553–577, In: M. K. Hecht, P. C. Goody, and B. M. Hecht (editors), *Major Patterns in Vertebrate Evolution*, Plenum Press, New York.

Edwards, J. L. 1989. Two perspectives on the evolution of the tetrapod limb. *American Zoologist*, 29: 235–254.

Ehrenbaum, E. 1901. Die Fische. *Fauna Arctica*, Jena, 2: 65–168.

Ehrenbaum, E. 1936. Naturgeschichte und wirtschaftliche Bedeutung der Seefische Nordeuropas. In: H. Lübbert and E. Ehrenbaum (editors), *Handbuch der Seefischerei Nordeuropas*, Schweizerbart'sche, Stuttgart, Vol. 2, x + 337 pp.

Ejaz, N., H. G. Krapp, and R. J. Tanaka, 2013. Closed-loop response properties of a visual interneuron involved in fly optomotor control. *Frontiers in Neural Circuits*, 7: doi: 10.3389/fncir.2013.00050.

Ekman, S. 1953. *Zoogeography of the Sea*. Sidgwick and Jackson, London, xiv + 417 pp.

Eldredge, N., and J. Cracraft. 1980. *Phylogenetic Patterns and the Evolutionary Process*. Columbia University Press, New York, viii + 349 pp.

Ellis, L. B., J. G. Landesman, S. L. Asato, and D. R. Zambrano. 1988. Second record of *Cubiceps paradoxus* and *Antennarius avalonis* from California. *California Fish and Game*, 74 (3): 174–176.

Elshoud-Oldenhave, M. J. W., and J. W. M. Osse. 1976. Functional morphology of the feeding system in the ruff, *Gymnocephalus cernua* (L. 1758) (Teleostei, Percidae). *Journal of Morphology*, 150 (2): 399–422.

Emlen, D. J. 2014. *Animal Weapons: The Evolution of Battle*. Henry Holt and Co., New York, xiii + 270 pp.

Endo, H., and G. Shinohara. 1999. A new batfish, *Coelophrys bradburyae* (Lophiiformes: Ogcocephalidae) from Japan, with comments on the evolutionary relationships of the genus. *Ichthyological Research*, 46 (4): 359–365.

Engel, H. 1937. The life of Albert Seba. *Svenska Linné-Sällskapets Årsskrift*, 20: 75–100.

Engel, H., K. H. Voous, and C. L. Hubbs. 1969. In memoriam: Professor Dr. L. F. de Beaufort. *Beaufortia*, 16 (221): 199–213.

Erisman, B. E., G. R. Galland, I. Mascareñas, J. Moxley, H. J. Walker, O. Aburto-Oropeza, P. A. Hastings, and E. Ezcurra. 2011. List of coastal fishes of Islas Marías Archipelago, Mexico, with comments on taxonomic composition, biogeography, and abundance. *Zootaxa*, 2985: 26–40.

Eschmeyer, W. N., R. Fricke, J. D. Fong, and D. A. Polack. 2010. Marine fish diversity: history of knowledge and discovery (Pisces). *Zootaxa*, 2525: 19–50.

Eschmeyer, W. N., and E. S. Herald. 1983. *A Field Guide to Pacific Coast Fishes of North America from the Gulf of Alaska to Baja California.* Peterson Field Guide Series, 28, Houghton-Mifflin Co., Boston, xii + 336 pp.

Espino, F., J. A. González, A. Boyra, C. Fernández, F. Tuya, and A. Brito. 2014. Diversity and biogeography of fishes in the Arinaga-Gando area, east coast of Gran Canaria (Canary Islands). *Revista de la Academia Canaria de Ciencias,* 26: 9–25.

Everly, A. W. 2002. Stages of development of the goosefish, *Lophius americanus,* and comments on the phylogenetic significance of the development of the luring apparatus in Lophiiformes. *Environmental Biology of Fishes,* 64: 393–417.

Everly, A. W., and J. H. Caruso. 2003. *Preliminary Guide to the Identification of the Early Life Stages of Lophiid Fishes of the Western Central North Atlantic.* NOAA Technical Memorandum NMFS-SEFSC-519, 14 pp.

Everly, A. W., and J. H. Caruso. 2006. Lophiidae: Goosefishes. Pp. 771–782, In: W. J. Richards (editor). *Early Stages of Atlantic fishes: An Identification Guide for the Western Central Atlantic,* CRC Press, Boca Raton, Florida, Vol. 1.

Evermann, B. W., and M. C. Marsh. 1900. The fishes of Porto Rico. *Bulletin of the United States Fish Commission,* 20 (1899): 49–350.

Eydoux, F., and L. Souleyet. 1850. *Voyage autour du Monde exécuté pendant les années 1836 et 1837 sur la Corvette La Bonite commandée par M. Vaillant.* Zoologie, Vol. 1, Pt. 2, Poissons, pp. 155–216 (Atlas plates 1–10).

Fahay, M. P. 1975. *An annotated list of larval and juvenile fishes captured with surface-towed meter net in the South Atlantic Bight during four RV Dolphin cruises between May 1967 and February 1968.* NOAA Technical Report NMFS, Special Scientific Report, Fisheries, 685, 39 pp.

Fahay, M. P. 1992. Development and distribution of cusk eel eggs and larvae in the Middle Atlantic Bight with a description of *Ophidion robinsi* n. sp. (Teleostei: Ophidiidae). *Copeia,* 1992 (3): 799–819.

Fariña, A. C., M. Azevedo, J. Landa, R. Duarte, P. Sampedro, G. Costas, M. A. Torres, and L. Cañás. 2008. *Lophius* in the world: a synthesis on the common features and life strategies. *ICES Journal of Marine Science,* 65: 1272–1280.

Farina, S. C., and W. E. Bemis. 2016. Functional morphology of gill ventilation of the goosefish, *Lophius americanus* (Lophiiformes: Lophiidae). *Zoology* (Jena), 119 (3): 207–215.

Fernholm, B., and A. Wheeler. 1983. Linnaean fish specimens in the Swedish Museum of Natural History, Stockholm. *Zoological Journal of the Linnean Society,* London, 78: 199–286.

Ferreira, C. E. L., J. E. A. Gonçalves, and R. Coutinho. 2001. Community structure of fishes and habitat complexity on a tropical rocky shore. *Environmental Biology of Fishes,* 61: 353–369.

Ferry-Graham, L. A. 1997. Feeding kinematics of juvenile swellsharks, *Cephaloscyllium ventriosum. Journal of Experimental Biology,* 200: 1255–1269.

Ferry-Graham, L. A., and G. V. Lauder. 2001. Aquatic prey capture in ray-finned fishes: A century of progress and new directions. *Journal of Morphology,* 248 (2): 99–119.

Fioravanti, M. L., and D. Florio. 2017. Early culture trials and an overview on U.S. marine ornamental species trade. Pp. 347–380, In: R. Calado, I. Olivotto, M. P. Oliver, and G. J. Holt (editors), *Marine Ornamental Species Aquaculture,* John Wiley & Sons Ltd, Chichester, West Sussex, United Kingdom.

Fischer von Waldheim, G. 1813. *Zoognosia tabulis synopticis illustrata in usum praelectionum Academiae Imperialis Medico-Chirurgicae Mosquensis,* 3rd ed. Vsevolosky, Moscow, Vol. 1, xii + 466 pp.

Fish, F. E. 1987. Kinematics and power output of jet propulsion by the frogfish genus *Antennarius* (Lophiiformes: Antennariidae). *Copeia,* 1987 (4): 1046–1048.

Fitch, J. E., and R. J. Lavenberg. 1975. *Tidepool and Nearshore Fishes of California.* University of California Press, Natural History Guide, Berkeley, 38, 156 pp.

Fitzinger, L. J. F. J. 1873. Versuch einer natürlichen Classification der Fische. *Sitzungsberichte der Kaiserlichen Akademie der Wissenschaften, Mathematisch-Naturwissenschaftliche Classe,* 67 (1): 5–58.

Flammang, B. E., and G. V. Lauder. 2009. Caudal fin shape modulation and control during acceleration, braking and backing maneuvers in bluegill sunfish, *Lepomis macrochirus*. *Journal of Experimental Biology*, 212: 277–286.

Floeter, S. R., L. A. Rocha, D. R. Robertson, J. C. Joyeux, W. F. Smith-Vaniz, P. Wirtz, A. J. Edwards, J. P. Barreiros, C. E. L. Ferreira, J. L. Gasparini, A. Brito, J. M. Falcón, B. W. Bowen, and G. Bernardi. 2008. Atlantic reef fish biogeography and evolution. *Journal of Biogeography*, 35: 22–47.

Fong, P., N. M. Frazier, C. Tompkins-Cook, R. Muthukrishnan, and C. R. Fong. 2016. Size matters: experimental partitioning of the strength of fish herbivory on a fringing coral reef in Moorea, French Polynesia. *Marine Ecology*, 37 (5): 933–942.

Forbes, P. 2009. *Dazzled and Deceived: Mimicry and Camouflage*. Yale University Press, New Haven, Connecticut, xv + 283 pp.

Forster, J. R. 1771. *A Voyage to China and the East Indies, by Peter Osbeck*. Translated from the German edition of 1765. White, London, Vol. 2, 367 pp.

Fourriére, M., H. Reyes-Bonilla, A. Ayala-Bocos, J. Ketchum, and J. C. Chávez-Comparan. 2016. Checklist and analysis of completeness of the reef fish fauna of the Revillagigedo Archipelago, Mexico. *Zootaxa*, 4150 (4): 436–466.

Fowler, H. W. 1903. Descriptions of several fishes from Zanzibar Island, two of which are new. *Proceedings of the Academy of Natural Sciences of Philadelphia*, 55: 161–176.

Fowler, H. W. 1912. Records of fishes for the middle Atlantic states and Virginia. *Proceedings of the Academy of Natural Sciences of Philadelphia*, 64: 34–40.

Fowler, H. W. 1927. Fishes of the tropical central Pacific. *Bulletin of the Bernice P. Bishop Museum*, 38: 1–32.

Fowler, H. W. 1928. The fishes of Oceania. *Memoirs of the Bernice P. Bishop Museum*, 10: 1–540.

Fowler, H. W. 1931. The fishes of Oceania. Supplement 1. *Memoirs of the Bernice P. Bishop Museum*, 11 (5): 313–381.

Fowler, H. W. 1934a. The fishes of Oceania. Supplement 2. *Memoirs of the Bernice P. Bishop Museum*, 11 (6): 385–466.

Fowler, H. W. 1934b. Fishes obtained by Mr. H. W. Bell-Marley chiefly in Natal and Zululand in 1929 to 1932. *Proceedings of the Academy of Natural Sciences of Philadelphia*, 86: 405–514.

Fowler, H. W. 1936a. South African fishes received from Mr. H. W. Bell-Marley in 1935. *Proceedings of the Academy of Natural Sciences of Philadelphia*, 87: 361–408.

Fowler, H. W. 1936b. The marine fishes of West Africa based on the collection of the American Museum Congo Expedition, 1909–1915. *Bulletin of the American Museum of Natural History*, 70 (2): 607–1493.

Fowler, H. W. 1938. A list of fishes known from Malaya. *Singapore Fisheries Bulletin*, 1: 1–268.

Fowler, H. W. 1940a. The fishes obtained by the Wilkes Expedition, 1838–1842. *Proceedings of the American Philosophical Society*, 82 (5): 733–800.

Fowler, H. W. 1940b. A collection of fishes obtained on the west coast of Florida by Mr. and Mrs. C. G. Chaplin. *Proceedings of the Academy of Natural Sciences of Philadelphia*, 92: 1–22.

Fowler, H. W. 1941. The George Vanderbilt Oahu survey: The fishes. *Proceedings of the Academy of Natural Sciences of Philadelphia*, 93: 247–279.

Fowler, H. W. 1944. The fishes. Results of the fifth George Vanderbilt Expedition, 1941. *Monographs of the Academy of Natural Sciences of Philadelphia*, 6: 57–529.

Fowler, H. W. 1945. Fishes from Saipan Island, Micronesia. *Proceedings of the Academy of Natural Sciences of Philadelphia*, 97: 59–74.

Fowler, H. W. 1946. A collection of fishes obtained in the Riu Kiu Islands by Captain Ernest R. Tinkham A.U.S. *Proceedings of the Academy of Natural Sciences of Philadelphia*, 98: 123–218.

Fowler, H. W. 1949. Fishes of Oceania. Supplement 3. *Memoirs of the Bernice P. Bishop Museum*, 12 (2): 37–186.

Fowler, H. W. 1959. *Fishes of Fiji*. Government of Fiji, Suva, v + 670 pp.

Francis, M. P. 1993. Checklist of the coastal fishes of Lord Howe, Norfolk, and Kermadec Island, southwest Pacific Ocean. *Pacific Science*, 47 (2): 136–170.

Franz, V. 1910. Die Japanischen Knochenfische der Sammlungen Haberer und Doflein. *Abhandlungen der Mathematisch-Physikalischen Klasse der Königlich Bayerischen Akademie der Wissenschaften*, 4 (Supplement 1): 1–135.

Franzén, O. 1977. Magnus Lagerström. Pp. 175, In: *Swedish Biographical Dictionary*, Swedish National Archives, Stockholm, 22.

Fricke, R. 1999. *Fishes of the Mascarene Islands (Réunion, Mauritius, Rodriguez): An Annotated Checklist, with Descriptions of New Species*. Koeltz Scientific Books, Oberreifenberg, Germany, *Theses Zoologicae*, 31, viii + 759 pp.

Fricke, R., G. R. Allen, S. Andréfouët, W.-J. Chen, M. A. Hamel, P. Laboute, R. Mana, H. H. Tan, and D. Uyeno. 2014. Checklist of the marine and estuarine fishes of Madang District, Papua New Guinea, western Pacific Ocean, with 820 new records. *Zootaxa*, 3832 (1): 1–247.

Fricke, R., P. Durville, G. Bernardi, P. Borsa, G. Mou-Tham, and P. Chabanet. 2013. Checklist of the shore fishes of Europa Island, Mozambique Channel, southwestern Indian Ocean, including 302 new records. *Stuttgarter Beiträge zur Naturkunde A (Biologie)*, Neue Serie, 6: 247–276.

Fricke, R., M. Kulbicki, and L. Wantiez. 2011. Checklist of the fishes of New Caledonia, and their distribution in the Southwest Pacific Ocean (Pisces). *Stuttgarter Beiträge zur Naturkunde A (Biologie)*, Neue Serie, 4: 341–463.

Fricke, R., T. Mulochau, P. Durville, P. Chabanet, E. Tessier, and Y. Letourneur. 2009. Annotated checklist of the fish species (Pisces) of La Réunion, including a Red List of threatened and declining species. *Stuttgarter Beiträge zur Naturkunde A (Biologie)*, Neue Serie, 2: 1–168;

Fries, B. F., C. U. Ekström, and C. J. Sundevall. 1893. *A History of Scandinavian fishes, revised and completed by F. A. Smitt*. Norstedt & Son, Stockholm & Paris, Vol. 1, 566 + viii pp.

Friese, U. E. 1973. Anglerfishes. *Marine Aquarist*, 4 (5): 29–36.

Friese, U. E. 1974. Anglerfishes. *Koolewong*, 3 (4): 7–11.

Fuentes, C. I., E. Acuña, and N. R. Hernández. 2015. Biogeography of continental shelf and upper slope fishes off El Salvador, Central America. *Journal of the Marine Biological Association of the United Kingdom*, 95 (3): 611–622.

Fujita, S., and K. Uchida. 1959. Spawning habits and early development of a sargassum fish, *Pterophryne histrio* (Linné). *Science Bulletin of the Faculty of Agriculture*, Kyushu University, 17 (3): 277–282.

Fulton, T. W. 1898. The ovaries and ovarian eggs of the angler or frog-fish (*Lophius piscatorius*) and of the John Dory (*Zeus faber*). *Annual Report of the Fisheries Board of Scotland*, 16 (3): 125–134.

Furnestin, J., J. Dardignac, C. Maurin, A. Coupé, and H. Boutière. 1958. Données nouvelles sur les poissons du Maroc atlantique. *Revue des Travaux de l'Institut des Pêches Maritimes*, Paris, 22 (4): 379–493.

Gaither, M. R., and L. A. Rocha. 2013. Origins of species richness in the Indo-Malay-Philippine biodiversity hotspot: evidence for the centre of overlap hypothesis. *Journal of Biogeography*, 40 (9): 1638–1648.

Galván-Magaña, F., L. A. Abitia-Cárdenas, J. Rodríguez-Romero, H. Perez-España, and H. Chávez-Ramos. 1996. Systematic list of the fishes from Cerralvo Island, Baja California Sur, Mexico. *Ciencias Marinas*, 22 (3): 295–311.

Galván-Villa, C. M., E. Ríos-Jara, D. Bastida-Izaguirre, P. A. Hastings, and E. F. Balart. 2016. Annotated checklist of marine fishes from the Sanctuary of Bahía Chamela, Mexico, with occurrence and biogeographic data. *ZooKeys*, 554: 139–157.

Garcia, J., M. F. Nóbrega, and J. E. L. Oliveira. 2015. Coastal fishes of Rio Grande do Norte, northeastern Brazil, with new records. *Check List*, 11 (3): 1–24.

Garman, S. 1896. Report on the fishes collected by the Bahama Expedition, of the State University of Iowa, under Professor C. C. Nutting in 1893. *Bulletin of the Laboratory of the Natural History*, State University of Iowa, 4 (1): 76–93.

Garman, S. 1899. Reports on an exploration off the west coasts of Mexico, Central and South America, and off the Galapagos Islands in charge of Alexander Agassiz, by the U.S. Fish Commission Steamer "Albatross" during 1891, Lieut. Commander Z. L. Tanner, U.S.A. commanding. XXVI. The fishes. *Memoires of the Museum of Comparative Zoology*, Harvard, 24: 1–431.

Garrett, A. 1863. Descriptions of new species of fishes. *Proceedings of the California Academy of Natural Sciences*, 3: 63–67.

Garrett, A. 1864. Descriptions of new species of fishes. II. *Proceedings of the California Academy of Natural Sciences*, 3: 103–107.

Garthe, C. 1837. *Zoologische Tabellen oder Systematische Uebersicht der Thierwelt nicht allein die Nominal Abtheilungen.* Pisces: Knorpelfische, Table 4. Renard & Dübnen, Köln & Berlin,

Gell, F. R., and M. W. Whittington. 2002. Diversity of fishes in seagrass beds in the Quirimba Archipelago, northern Mozambique. *Marine and Freshwater Research*, 53: 115–121.

Gerson, E. 2001. Longlure frogfish (*Antennarius mulliocellatus*) on yellow branching tube sponge (*Pseudoceratina crassa*). *American Journal of Roentgenology*, 176 (1): 240–240.

Gessner, C. 1558. De piscium et aquatilium animantium natura. Liber IV, In: *Historiae animalium*, Froschoverum, Zurich.

Gibb, A. C., and L. A. Ferry-Graham. 2005. Cranial movements during suction feeding in teleost fishes: Are they modified to enhance suction production? *Zoology*, 108 (2): 141–153.

Gilbert, C. H. 1892. Scientific results of explorations by the U.S. Fish Commission Steamer Albatross. No. 22. Descriptions of thirty-four new species of fishes collected in 1888 and 1889, principally among the Santa Barbara Islands and in the Gulf of California. *Proceedings of the United States National Museum*, 14: 539–566.

Gilbert, C. H., and E. C. Starks. 1904. The fishes of Panama Bay. *Memoirs of the California Academy of Sciences*, 4: 1–304.

Gilchrist, J. D. F. 1902. *Catalogue of fishes recorded from South Africa.* Cape Town, 179 pp.

Gilchrist, J. D. F., and W. W. Thompson. 1917. A catalogue of the sea fishes recorded from Natal. Part 2. *Annals of the Durban Museum*, 1 (4): 255–431.

Gill, T. N. 1861. Catalogue of the fishes of the eastern coast of North America, from Greenland to Georgia. *Proceedings of the Academy of Natural Sciences of Philadelphia*, 13: 1–63.

Gill, T. N. 1863. Descriptions of some new species of Pediculati, and on the classification of the group. *Proceedings of the Academy of Natural Sciences of Philadelphia*, 15: 88–92.

Gill, T. N. 1872. Arrangement of the families of fishes, or classes Pisces, Marsipobranchii, and Leptocardii. *Smithsonian Miscellaneous Collections*, 247, xlvi + 49 pp.

Gill, T. N. 1879a. Synopsis of the pediculate fishes of the eastern coast of extratropical North America. *Proceedings of the United States National Museum*, 1: 215–221.

Gill, T. N. 1879b. Note on the Antennariidae. *Proceedings of the United States National Museum*, 1: 221–222.

Gill, T. N. 1879c. On the proper specific name of the common pelagic antennariid *Pterophryne*. *Proceedings of the United States National Museum*, 1: 223–226.

Gill, T. N. 1883. Supplementary note on the Pediculati. *Proceedings of the United States National Museum*, 5: 551–556.

Gill, T. N. 1904. Extinct pediculate and other fishes. *Science*, 20: 845–846.

Gill, T. N. 1905a. The life history of the angler. *Smithsonian Miscellaneous Collections*, 47: 500–516.

Gill, T. N. 1905b. The sargasso fish not a nest builder. *Science*, 22: 841.

Gill, T. N. 1907. The work of *Pterophryne* and the flying-fishes. *Science*, 35: 63–64.

Gill, T. N. 1909. Angler fishes: their kinds and ways. *Smithsonian Institution, Annual Report*, 1908 (1909), pp. 565–615.

Girone, A., and D. Nolf. 2008. Fish otoliths from the Priabonian (Late Eocene) of north Italy and south-east France—their paleobiogeographical significance. *Revue de Micropaléontologie*, 52 (3): 195–218.

Gistel, J. 1848. *Naturgeschichte des Tierreichs für hühere Schulen.* Hoffman'sche, Stuttgart, xvi + 216 pp.

Gladfelter, W. B. 1972a. Structure and function of the locomotory system of the scyphomedusa *Cyanea capillata*. *Marine Biology*, 14: 150–160.

Gladfelter, W. B. 1972b. Structure and function of the locomotory system of *Polyorchis montereyensis* (Cnidaria, Hydrozoa). *Helgoländer Wissenschaftliche* Meeresuntersuchungen, 23: 38–79.

Glover, C. J. M. 1968. Further additions to the fish fauna of South Australia. *Records of the South Australia Museum*, Adelaide, 15 (4): 791–794.

Glover, C. J. M. 1976. Fishes. Pp. 169–175, In: Vertebrate type-specimens in the South Australian Museum. *Records of the South Australian Museum*, Adelaide, 17 (7–12): 169–219.

Gmelin, J. F. 1789. Pisces. Pp. 1126–1516, In: Caroli a Linné, *Systema Naturae per regna tria naturae, secundum classes, ordines, genera, species, cum characteribus, differentiis, synonymis, locis, ed. 13*. Beer, Leipzig, Vol. 1, Part 3.

Golani, D. 2006. An annotated list of types in the Hebrew University fish collection. *Haasiana: A Newsletter of the Biological Collections of the Hebrew University*, 2006 (3): 20–40.

Golani, D., and S. Bogorodsky. 2010. The fishes of the Red Sea—reappraisal and updated checklist. *Zootaxa*, 2463: 1–135.

Goldfuss, G. A. 1820. *Handbuch der Zoologie*. Schrag, Nürnberg, 2te Abt., xxiv + 510 pp.

Goldin, M. R. 2002. *Field Guide to the Sāmoan Archipelago: Fish, Wildlife, and Protected Areas*. Bess Press, Honolulu, Hawaii, xiv + 330 pp.

Goldman, B., and F. H. Talbot. 1976. Aspects of the Ecology of Coral Reef Fishes. Pp. 125–154, In: D. A. Jones and R. Endean (editors), *Biology and Geology of Coral Reefs*. Academic Press, New York, Vol. 3.

Gomon, M., D. Bray, and R. Kuiter. 2008. *Fishes of Australia's Southern Coast*. Reed, New Holland, and Chatswood, New South Wales, 928 pp.

González-Díaz, A. A., and M. Soria-Barreto. 2013. Preliminary systematic list of fishes the state of Nayarit, Mexico. *Revista Bio Ciencias*, 2 (3): 200–215.

Good, J. M., and O. Gregory. 1813. *Lophius*. In: *Pantologia, a New Cyclopaedia*. Kearsley, London, Vol. 7, unpaged.

Goode, G. B. 1876. Catalogue of the fishes of the Bermudas, based chiefly upon the collections of the United States National Museum. *Bulletin of the United States Natural Museum*, 1 (5): 1–82.

Goode, G. B. 1877. A preliminary catalogue of the reptiles, fishes, and leptocardians of the Bermudas, with descriptions of four species of fishes believed to be new. *American Journal of Science and Arts*, Series 3, 14 (82): 289–298.

Goode, G. B. 1881. Fishes from the deep water on the south coast of New England obtained by the United States Fish Commission in the summer of 1880. *Proceedings of the United States National Museum*, 3 (1880): 467–486.

Goode, G. B. 1884. The food fishes of the United States. pp. 163–682, In: *The Fisheries and Fishery Industries of the United States*, Section 1, Natural history of useful aquatic animals, with an atlas of two hundred and seventy-seven plates, Part 3, United States Commission of Fish and Fisheries, Government Printing Office, Washington, DC.

Goode, G. B., and T. H. Bean. 1882. A list of the species of fishes recorded as occurring in the Gulf of Mexico. *Proceedings of the United States National Museum*, 5: 234–240.

Goode, G. B., and T. H. Bean. 1896. Oceanic ichthyology, a treatise on the deep-sea and pelagic fishes of the world, based chiefly on the collections made by steamers Blake, Albatross and Fish Hawk in the northwestern Atlantic. *United States National Museum, Special Bulletin*, 2: 1–553.

Gordoa, A., and E. Macpherson. 1990. Food selection by a sit-and-wait predator, the monkfish, *Lophius upsicephalus*, off Namibia (South West Africa). *Environmental Biology of Fishes*, 27 (1): 71–76.

Gordon, D. J., D. F. Markle, and J. E. Olney. 1984. Ophidiiformes: development and relationships. Pp. 308–319, In: H. G. Moser, W. J. Richards, D. M. Cohen, M. P. Fahay, A. W. Kendall, Jr., and S. L. Richardson (editors), *Ontogeny and Systematics of Fishes*. American Society of Ichthyologists and Herpetologists, Special Publication No. 1, x + 760 pp.

Gordon, M. 1938. Animals of the sargasso merry-go-round. *Natural History*, New York, 42 (1): 12–20.

Gordon, M. 1955. *Histrio*: The fish on the Sargasso Sea merry-go-round. *Aquarium*, 24: 386–393.

Goren, M., and M. Dor. 1994. *An Updated Checklist of the Fishes of the Red Sea*. CLOFRES II. The Israel Academy of Sciences and Humanities, Jerusalem, xii + 120 pp.

Gosline, W. A. 1971. *Functional Morphology and Classification of Teleost Fishes*. University of Hawaii Press, Honolulu, Hawaii, ix + 208 pp.

Gosline, W. A., and V. E. Brock. 1960. *Handbook of Hawaiian Fishes*. University of Hawaii Press, Honolulu, ix + 372 pp.

Goss, R. J. 1969. *Principles of Regeneration*. Academic Press, New York, 287 pp.

Gotshall, D. W. 1998. *Sea of Cortez Marine Animals: A Guide to the Common Fishes and Invertebrates, Baja California to Panama*. Sea Challengers, Monterey, California, v + 110 pp.

Grace, M., M. Bahnick, and L. Jones. 2000. A preliminary study of the marine biota at Navassa Island, Caribbean Sea. *Marine Fisheries Review*, 62 (2): 43–48.

Grandcourt. E. 2012. Reef fish and fisheries in the gulf. Pp. 127–161, In: B. M. Riegl and S. J. Purkis, *Coral Reefs of the Gulf: Adaptation to Climatic Extremes. Coral Reefs of the World*, Vol. 3, Springer Science+Business Media B.V., Dordrecht, Netherlands.

Grant, E. M. 1965. *Guide to Fishes*. Department of Harbours and Marine Queensland, Brisbane, viii + 280 pp.

Grant, E. M. 1972. *Guide to Fishes*, 2nd ed. Department of Harbours and Marine, Brisbane, xxiv + 469 pp.

Grant, E. M. 1978. *Guide to Fishes*, 4th ed. Department of Harbours and Marine, Brisbane, 768 pp.

Gray, J. 1968. *Animal Locomotion*. Norton, World Nature Series, New York, xi + 479 pp.

Gray, J. E. 1854. *Catalogue of Fishes Collected and Described by Laurence Theodore Gronow, Now in the British Museum*. Trustees of the British Museum, London, vii + 196 pp.

Green M. A., and B. D. Bruce. 2002. *Spotted Handfish Recovery Plan 1999–2001: Year 3*. Final Report to Environment Australia, Endangered Species Program, CSIRO Hobart, Tasmania, 18 pp.

Green, M. A., R. D. Stuart-Smith, J. P. Valentine, L. D. Einoder, N. S. Barrett, A. T. Cooper, and M. D. Stalker. 2012. *Spotted Handfish Monitoring and Recovery Actions, 2011–2012*. Report for Derwent Estuary Program and Caring for Our Country. CSIRO Marine and Atmospheric Research and Institute of Marine and Antarctic Studies, 24 pp.

Greenfield, D. W. 1998. *Halophryne hutchinsi*: a new toadfish (Batrachoididae) from the Philippine Islands and Pulau Waigeo, Indonesia. *Copeia*, 1998 (3): 696–701.

Greenfield, D. W. 2003. A survey of the small reef fishes of Kaneohe Bay, Oahu, Hawaiian Islands. *Pacific Science*, 57 (1): 45–76.

Greenwood, P. H., D. E. Rosen, S. H. Weitzman, and G. S. Myers. 1966. Phyletic studies of teleostean fishes, with a provisional classification of living forms. *Bulletin of the American Museum of Natural History*, 131: 339–456.

Gregory, W. K. 1928. Studies on the body-form of fishes. *Zoologica*, New York, 8 (6): 325–421.

Gregory, W. K. 1933. Fish skulls: A study of the evolution of natural mechanisms. *Transactions of the American Philosophical Society*, 23 (2): 75–481.

Gregory, W. K. 1951. *Evolution Emerging: A Survey of Changing Patterns from Primeval Life to Man*. Macmillan, New York, Vol. l, 704 pp.

Gregory, W. K., and G. M. Conrad. 1936. The evolution of the pediculate fishes. *American Naturalist*, 70: 193–208.

Griffin, L. T. 1929. Studies in New Zealand fishes. *Transactions of the New Zealand Institute*, Wellington, 59: 374–388.

Grimsditch, G., A. Basheer, and D. E. P. Bryant. 2016. Extreme white colouration of frogfish *Antennarius maculatus* due to coral bleaching event. *Coral Reefs*, doi: 10.1007/s00338 -016-1500-6.

Grobecker, D. B. 1981. Steady as a rock, fast as lightning. *Natural History*, 90 (5): 50–53.

Grobecker, D. B. 1983. The "lie-in-wait" feeding mode of a cryptic teleost, *Synanceia verrucosa*. *Environmental Biology of Fishes*, 8 (3/4): 191–202.

Grobecker, D. B., and T. W. Pietsch. 1979. High-speed cinematographic evidence for ultrafast feeding in antennariid anglerfishes. *Science*, 205: 1161–1162.

Gronovius, L. T. 1754. *Museum Ichthyologicum, sistens piscium indigenorum et quorundam exoticorum, qui in Museo Laurenti Theodori Gronovii, J. U. D., adservantur, descriptiones ordine systematico, accedunt nonnullorum exoticorum piscium icones aeri incisae*. Haak, Leiden, 70 pp.

Gronovius, L. T. 1763. *Zoophylacii Gronoviani, fasciculus primus exhibens animalia quadrupeda, amphibia atque pisces, quae in museo suo adservat, rite examinavit, systematice disposuit, descripsis atque iconibus illustravit L. T. Gronovius, J. U. D.* Haak & Luchtmans, Leiden, 136 pp.

Gronovius, L. T. 1781. *Zoophylacii Gronoviani, fasciculus tertius exhibens animalia quadrupeda, amphibia atque pisces, quae in museo suo adservat, rite examinavit, systematice disposuit, descripsis atque iconibus illustravit L. T. Gronovius, J. U. D.* Haak & Luchtmans, Leiden, 139 pp.

Grove, J. S., and R. J. Lavenberg. 1997. *The Fishes of the Galápagos Islands.* Stanford University Press, Stanford, California, xliv + 863 pp.

Gruber, J. W. 1983. What is it? The echidna comes to England. *Archives of Natural History*, 11 (1): 1–15.

Gudger, E. W. 1905. A note on the eggs and egg laying of *Pterophryne histrio*, the gulfweed fish. *Science*, 22: 841–843.

Gudger, E. W. 1912. Natural history notes on some Beaufort, N.C., fishes, 1910–1911. *Proceedings of the Biological Society of Washington*, 25: 165–176.

Gudger, E. W. 1929. On the morphology, coloration and behavior of seventy teleostean fishes of Tortugas, Florida. *Papers from the Tortugas Laboratory of the Carnegie Institution of Washington*, 26: 149–204.

Gudger, E. W. 1937. Sargasso weed fish "nests" made by flying fishes not by sargasso fishes (antennariids): A historical survey. *American Naturalist*, 71: 363–381.

Gudger, E. W. 1945a. The frogfish, *Antennarius scaber*, uses its lure in fishing. *Copeia*, 1945 (2): 111–113.

Gudger, E. W. 1945b. The angler-fishes, *Lophius piscatorius* et *americanus*, use the lure in fishing. *American Naturalist*, 79 (785): 542–548.

Guérin-Méneville, F. E. 1844. *Iconographie du Règne Animal de G. Cuvier.* Baillière, Paris, Vol. 3, Poissons, 44 pp.

Guichenot, A. 1853. Peces. Pp. 145–255, In: R. de la Sagra (editor), *Historia física, política y natural de la Isla de Cuba.* Bertrand, Paris, Vol. 4.

Guichenot, A. 1863. Faune ichthyologique, Annexe C. Pp. C1–C32, In: L. Maillard, *Notes sur l'île de La Réunion (Bourbon)*, 2nd edition, Vol. 2, Dentu, Paris.

Guitart, D. J. 1978. Sinopsis de los peces marinos de Cuba. *Academia de Ciencias de Cuba, Instituto de Oceanología*, Habana, 4: 611–881.

Guitel, F. 1891. Recherches sur la ligne latérale de la baudroie (*Lophius piscatorius*). *Archives de Zoologie Expérimentale et Générale*, 2 (9): 125–184.

Günther, A. C. L. G. 1853. *Die Fische des Neckars, untersucht und beschrieben.* Verlag von Ebner & Seubert, Stuttgart, 136 pp.

Günther, A. C. L. G. 1861a. *Catalogue of the acanthopterygian fishes in the collection of the British Museum.* Trustees of the British Museum, London, Vol. 3, xxv + 586 pp.

Günther, A. C. L. G. 1861b. List of Ceylon fishes. Pp. 359–362, In: J. E. Tennent, *Sketches of the Natural History of Ceylon, with Narratives and Anecdotes Illustrative of the Habits and Instincts of the Mammalia, Birds, Reptiles, Fishes, Insects, &c., Including a Monograph of the Elephant and a Description of the Modes of Capturing and Training It, with Engravings from Original Drawings,* Longman, Green, Longman, and Roberts, London.

Günther, A. C. L. G. 1864. Report of a collection of fishes made by Messrs. Dow, Godman, and Salvin in Guatemala. *Proceedings of the Zoological Society of London*, 1864 (3): 144–151.

Günther, A. C. L. G. 1869a. Report of a second collection of fishes made at St. Helena by J. C. Melliss, Esq. *Proceedings of the Zoological Society of London*, 1869 (4): 238–239.

Günther, A. C. L. G. 1869b. An account of the fishes of the states of Central America, based on collections made by Capt. J. M. Dow, F. Godman, Esq., and O. Salvin, Esq. *Transactions of the Zoological Society of London*, 6 (14): 377–494.

Günther, A. C. L. G. 1876. Andrew Garrett's Fische der Südsee. *Journal des Museum Godeffroy*, 11 (5): 129–168.

Günther, A. C. L. G. 1880. *An Introduction to the Study of Fishes.* Black, Edinburgh, xvi + 720 pp.

Günther, A. E. 1980. *The Founders of Science at the British Museum, 1753–1900. A Contribution to the Centenary of the Opening of the British Museum (Natural History) on 18th April 1981.* Halesworth, Suffolk, England, ix + 219 pp.

Hall, D. J. 2008. Dive briefs: exposed, first reported frogfish sighting in Ambon. *Sportdiver.com*, 16: 16, July 2008.

Hallez, L. 1866. Muséum d'Histoire Naturelle. Cours de M. Lacaze-Duthiers. *Revue des Cours Scientifiques de la France et de l'Étranger*, 3 (23): 377–384.

Halstead, B. W. 1978. *Poisonous and Venomous Marine Animals of the World, revised edition*. Darwin Press, Princeton, New Jersey, xlvi + 1043 + 283 pp.

Han, S.-H., J. S. Kim, and C. B. Song. 2017. The first record of a frogfish, *Fowlerichthys scriptissimus* (Antennariidae, Lophiiformes), from Korea. *Fisheries and Aquatic Sciences*, 20 (2): 1–5.

Hanel, L., J. Plíštil, and J. Novák. 2009. Checklist of the fishes and fish-like vertebrates on the European continent and adjacent seas. *Bulletin Lampetra*, 6: 108–180.

Hanel, R., and H.-C. John. 2015. A revised checklist of Cape Verde Islands sea fishes. *Journal of Applied Ichthyology*, 31: 135–169.

Harasti, D., K. Martin-Smith, and W. Gladstone. 2014. Does a no-take marine protected area benefit seahorses? *PLoS ONE*, 9 (8): e105462. doi.org/10.1371/journal.pone.0105462.

Harris, V. A. 1960. On the locomotion of the mud-skipper *Periophthalmus koelreuteri* (Pallas): (Gobiidae). *Proceedings of the Zoological Society of London*, 134: 107–135.

Hasegawa, Y., Y. Tohida, N. Kohno, K. Ono, H. Nokariya, and T. Uyeno. 1988. Quaternary vertebrates from Shiriya Area, Shimokita Peninsula, Northeastern Japan. *Memoirs of the National Science Museum*, 21: 17–36 [in Japanese].

Hays, A. N. 1952. David Starr Jordan: A Bibliography of His Writings, 1871–1931. *Stanford University Publication, University Series, Library Studies*, Vol. 1, xiii + 195 pp.

Heemstra, E., P. C. Heemstra, M. J. Smale, T. Hooper, and D. Pelicier. 2004. Preliminary checklist of coastal fishes from the Mauritian island of Rodrigues. *Journal of Natural History*, 38: 3315–3344 pp.

Heemstra, P. C., and E. Heemstra. 2004. *Coastal Fishes of Southern Africa*. South African Institute for Aquatic Biodiversity, National Inquiry Services Centre (NISC), xxiv + 488 pp.

Helfman, G. S., B. B. Collette, and D. E. Facey. 1997. *The Diversity of Fishes*. Blackwell Science, Malden, Massachusetts, xii + 528 pp.

Helfman, G. S., B. B. Collette, D. E. Facey, and B. W. Bowen. 2009. *The Diversity of Fishes*. Wiley-Blackwell, Chichester, West Sussex, United Kingdom, xiv + 720 pp.

Heller, E., and R. E. Snodgrass. 1903. Papers from the Hopkins Stanford Galapagos Expedition, 1898–1899. XV. New fishes. *Proceedings of the Washington Academy of Sciences*, 5: 189–229.

Helmbold, W. 1957. *Plutarch, De sollertia animalium*, with an English translation by William Helmbold. Plutarch's *Moralia*, Loeb Classical Library, Harvard University Press, Cambridge, Massachusetts, Vol. 12, 12 + 590 pp.

Hemphill, A. H. 2005. Conservation on the high seas-drift algae habitat as an open ocean cornerstone. *Parks, High Seas Marine Protected Areas*, 15 (3): 48–56.

Hennig, W. 1950. *Grundzüge einer Theorie der Phylogenetischen Systematike*. Deutscher Zentralverlag, Berlin, 370 pp.

Hennig, W. 1965. Phylogenetic systematics. *Annual Review of Entomology*, 10: 97–116.

Hennig, W. 1966. *Phylogenetic Systematics*. University of Illinois Press, Urbana, 263 pp.

Herald, E. S. 1961. *Living Fishes of the World*. Hamilton, London, 303 pp.

Herre, A. W. C. T. 1934. *Notes on fishes in the Zoological Museum of Stanford University. 1. The fishes of the Herre 1931 Philippine Expedition with descriptions of 17 new species*. Newspaper Enterprise, Hong Kong, 106 pp.

Herre, A. W. C. T. 1945. Additions to the fish fauna of the Philippine Islands. *Copeia*, 1945 (3): 146–149.

Herre, A. W. C. T. 1953. *Check List of Philippine Fishes*. Fish and Wildlife Service, Research Report, 20, 977 pp.

Hiatt, R. W., and D. W. Strasburg. 1960. Ecological relationships of the fish fauna on coral reefs of the Marshall Islands. *Ecological Monographs*, 30 (1): 65–127.

Hildebrand, M. 1974. *Analysis of Vertebrate Structure*. Wiley, New York, London, Sydney & Toronto, xv + 510 pp.

Hildebrand, S. F. 1946. A descriptive catalog of the shore fishes of Peru. *Bulletin of the United States National Museum*, 189: 1–530.

Hindwood, K. A. 1970. The "Watling" drawings, with incidental notes on the "Lambert" and the "Latham" drawings. *Proceedings of the Royal Zoological Society of New South Wales*, 1968–1969 (1970): 16–32.

Hiyama, Y. 1937. *Marine Fishes of the Pacific Coast of Mexico*. T. Kumada (editor). Nisson Fisheries Institute, Odawara, Japan, 75 pp.

Hjortberg, G. F. 1768. Beschreibung einer Guaperva, die in den Seegewächse Sargazo gefangen worden. *Abhandlungen der Königlich Schwedischen Akademie der Wissenschaften*, 30: 353–355.

Ho, H.-C. 2013. Two new species of the batfish genus *Malthopsis* (Lophiiformes: Ogcocephalidae) from the Western Indian Ocean. *Zootaxa*, 3716 (2): 289–300.

Ho, H.-C., P. Chakrabarty, and J. S. Sparks. 2010. Review of the *Halieutichthys aculeatus* species complex (Lophiiformes: Ogcocephalidae), with descriptions of two new species. *Journal of Fish Biology*, 77: 841–869.

Ho, H.-C., T. Kawa, and F. Satria. 2015b. Species of the anglerfish genus *Chaunax* from Indonesia, with descriptions of two new species (Lophiiformes: Chaunacidae). *Raffles Bulletin of Zoology*, 63: 301–308.

Ho, H.-C., T. Kawai, and K. Amaoka. 2016. Records of deep-sea anglerfishes (Lophiiformes: Ceratioidei) from Indonesia, with descriptions of three new species. *Zootaxa*, 4121 (3): 267–294.

Ho, H.-C., and P. R. Last. 2013. Two new species of the coffinfish genus *Chaunax* (Lophiiformes: Chaunacidae) from the Indian Ocean. *Zootaxa*, 3710 (5): 436–448.

Ho, H.-C., and M. McGrouther. 2015. A new anglerfish from eastern Australia and New Caledonia (Lophiiformes: Chaunacidae: *Chaunacops*), with new data and submersible observation of *Chaunacops melanostomus*. *Journal of Fish Biology*, 86 (3): 940–951.

Ho, H.-C., R. K. Meleppura, and K. K. Bineesh. 2015a. *Chaunax multilepis* sp. nov., a new species of *Chaunax* (Lophiiformes: Chaunacidae) from the northern Indian Ocean. *Zootaxa*, 4103 (2): 130–136.

Ho, H.-C., C. D. Roberts, and K.-T. Shao. 2013b. Revision of batfishes (Lophiiformes: Ogcocephalidae) of New Zealand and adjacent waters, with description of two new species of the genus *Malthopsis*. *Zootaxa*, 3626 (1): 188–200.

Ho, H.-C., C. D. Roberts, and A. L. Stewart. 2013a. A review of the anglerfish genus *Chaunax* (Lophiiformes: Chaunacidae) from New Zealand and adjacent waters, with descriptions of four new species. *Zootaxa*, 3620 (1): 89–111.

Ho, H.-C., and K.-T. Shao. 2010. A new species of *Chaunax* (Lophiiformes: Chaunacidae) from the western South Pacific, with comments on *C. latipunctatus*. *Zootaxa*, 2445 (1): 53–61.

Hobbs, J.-P. A., S. J. Newman, G. E. A. Mitsopoulos, M. J. Travers, C. L. Skepper, J. J. Gilligan, G. R. Allen, H. J. Choat, and A. M. Ayling. 2014a. Checklist and new records of Christmas Island fishes: the influence of isolation, biogeography and habitat availability on species abundance and community composition. *Raffles Bulletin of Zoology*, 30: 184–202.

Hobbs, J.-P. A., S. J. Newman, G. E. A. Mitsopoulos, M. J. Travers, C. L. Skepper, J. J. Gilligan, G. R. Allen, H. J. Choat, and A. M. Ayling. 2014b. Fishes of the Cocos (Keeling) Islands: new records, community composition and biogeographic significance. *Raffles Bulletin of Zoology*, Supplement, 30: 203–219.

Hoese, H. D., and R. H. Moore. 1998. *Fishes of the Gulf of Mexico: Texas, Louisiana, and Adjacent Waters*. Second edition. Texas A & M University Press, College Station, Texas, xv + 422.

Holcroft, N. I. 2004. A molecular test of alternative hypotheses of tetraodontiform (Acanthomorpha: Tetraodontiformes) sister group relationships using data from the RAG1 gene. *Molecular Phylogenetics and Evolution*, 32: 749–760.

Holcroft, N. I. 2005. A molecular analysis of the interrelationships of tetraodontiform fishes (Acanthomorpha: Tetraodontiformes). *Molecular Phylogenetics and Evolution*, 34: 525–544.

Holcroft, N. I., and E. O. Wiley. 2008. Acanthuroid relationships revisited: a new nuclear gene-based analysis that incorporates tetraodontiform representatives. *Ichthyological Research*, 55 (3): 274–283.

Holm, A. 1957. Specimina Linnaeana. I. Uppsala berarade Zoologiska samlingar fran Linnés tid. *Uppsala Universitets Årsskrift*, 6: 1–68.

Holmes, A. 1978. *Holmes' Principles of Physical Geology*, 3rd edition. Wiley, New York, xvi + 730 pp.

Holthuis, L. B. 1959. Notes on pre-Linnean carcinology (including the study of Xiphosura) of the Malay Archipelago. Pp. 63–125, In: H. C. D. de Wit (editor), *Rumphius Memorial Volume*. Uitgeverij en Drukkerij Hollandia N.V., Baarn.

Holthuis, L. B. 1969. Albertus Seba's "Locupletissimi rerum Naturalium Thesauri . . ." (1734–1765) and the "Planches de Seba" (1827–1831). *Zoologische Mededelingen*, Leiden, 43 (19): 239–252.

Holthuis, L. B., and T. W. Pietsch. 2006. *Les Planches inédites de Poissons et autres Animaux marins de l'Indo-Ouest Pacifique d'Isaac Johannes Lamotius [Isaac Johannes Lamotius (1646–c. 1718) and His Paintings of Indo-Pacific Fishes and Other Marine Animals]*. Publications Scientifiques du Muséum, Muséum national d'Histoire naturelle, Paris, 292 pp., 93 color pls.

Hoover, J. P. 1993. *Hawaii's Fishes. A Guide for Snorkelers, Divers, and Aquarists*. Mutual Publishing, Honolulu, viii + 178 pp.

Hoover, J. P. 2003. *Hawaii's Fishes. A Guide for Snorkelers, Divers, and Aquarists*. Revised edition. Mutual Publishing, Honolulu, viii + 183 pp.

Hornell, J. 1921. The Madras Marine Aquarium. *Madras Fisheries Bulletin*, 14: 57–96.

Hornell, J. 1923. *Guide to the Madras Marine Aquarium*. Superintendent, Government Press, Madras, 45 pp.

Howell y Rivero, L. 1938. List of the fishes, types of Poey, in the Museum of Comparative Zoology. *Bulletin of the Museum of Comparative Zoology*, 82 (3): 169–227.

Hubbs, C. L. 1958. *Dikellorhynchus* and *Kanazawaichthys*: nominal fish genera interpreted as based on prejuveniles of *Malacanthus* and *Antennarius*, respectively. *Copeia*, 1958 (4): 282–285.

Hubbs, C. L. 1964a. History of ichthyology in the United States after 1850. *Copeia*, 1964 (1): 42–60.

Hubbs, C. L. 1964b. David Starr Jordan. *Systematic Zoology*, 13 (4): 195–200.

Hubbs, C. L., W. I. Follett, and L. J. Dempster. 1979. List of fishes of California. *Papers of the California Academy of Sciences*, 133: 1–51.

Hubbs, C. L., and R. H. Rosenblatt. 1961. *Effects of the equatorial currents of the Pacific on the distribution of fishes and other marine animals*. Tenth Pacific Science Congress, Abstract of Symp. Papers, Honolulu, pp. 340–341.

Hughes, G. M., and P. J. Mill. 1966. Patterns of ventilation in dragonfly larvae. *Journal of Experimental Biology*, 44: 317–333.

Hutchins, B. 1994. A survey of the nearshore reef fish fauna of Western Australia's west and south coasts—the Leeuwin Province. *Records of the Western Australian Museum*, Supplement, 46, 66 pp.

Hutchins, B. 2001. Checklist of the fishes of Western Australia. *Records of the Western Australian Museum*, Supplement, 63: 9–50.

Hutchins, J. B., and K. N. Smith. 1991. A catalogue of type specimens of fishes in the Western Australian Museum. *Records of the Western Australian Museum*, Supplement, 38, 1–56 pp.

Hutchins, J. B., and R. Swainston. 1985. Revision of the monacanthid fish genus *Brachaluteres*. *Records of the Western Australian Museum*, 12: 57–78.

Hutton, F. W. 1872. *Catalogue of New Zealand Fishes*. Hughes, Wellington, 133 pp.

Hutton, F. W. 1891. List of New Zealand fishes. *Transactions of the New Zealand Institute*, 1890 (1891), 22: 275–285.

Hylleberg, J., and C. Aungtonya. 2014. Biodiversity survey on Chalong Bay and surrounding areas, east coast of Phuket, Thailand. *Phuket Marine Biological Center*, Special Publication, 32: 81–112.

Ibarra, M., and D. J. Stewart. 1987. Catalogue of type specimens of Recent fishes in Field Museum of Natural History. *Fieldiana, Zoology*, 35: 1–112.

Illiger, J. K. W. 1811. *Prodromus systematis mammalium et avium; additis terminis zoographicis utriusque classis, eorumque versione germanica*. Salfeld, Berlin, xviii + 301 pp.

Iredale, T. 1958. History of New South Wales shells. Part III: The settlement years (cont.): Thomas Watling, artist. *Proceedings of the Royal Zoological Society of New South Wales*, 1956–1957: 162–169.

Ives, J. E. 1890. Mimicry of the environment in *Pterophryne histrio*. *Proceedings of the Academy of Natural Sciences of Philadelphia*, 1889: 344–345.

Jackson, R. R., and F. R. Cross. 2013. A cognitive perspective on aggressive mimicry. *Journal of Zoology*, London, 290 (3): 161–171.

Jackson, T. L. 2006a. Antennariidae. Pp. 783–789, In: W. J. Richards (editor), *Early Stages of Atlantic Fishes: An Identification Guide for the Western Central North Atlantic*, CRC Press, Boca Raton, Florida, Vol. 1.

Jackson, T. L. 2006b. Chaunacidae. Pp. 791–792, In: W. J. Richards (editor), *Early Stages of Atlantic Fishes: An Identification Guide for the Western Central North Atlantic*, CRC Press, Boca Raton, Florida, Vol. 1.

Jardine, W. 1846. Memoir of Cuvier. Pp. 17–58, In: *The Naturalist's Library*. Vol. XVI. Mammalia: Lions, Tigers, &c., &c. Lizars & Bohn, Edinburgh & London, 276 pp.

Jarocki, F. P. 1822. *Zoologiia czyli zwiérzetopismo ogólne podlug náynowszego systematu*, volume 4. Drusarni Laukiewicza, Warsaw, 464 + xxvii pp.

Jatzow, R., and H. Lens. 1898. Fische von Ost-Afrika, Madagaskar und Aldabra. *Abhandlungen der Senckenbergischen Naturforschenden Gesellschaft*, 21 (3): 497–531.

Jawad, L. A., and J. M. Al-Mamry. 2009. First record of *Antennarius coccineus* from the Gulf of Oman and second record of *Antennarius indicus* from the Arabian Sea coast of Oman. *Marine Biodiversity Records*, 2: 1–3.

Jawad, L. A., and S. Hussain. 2014. First record of *Antennarius indicus* (Pisces: Antennariidae), *Equulites elongatus* (Actinopterygiidae: Leiognathidae) and *Cheilinus lunulatus* (Actinopterygiidae: Labridae) from the marine waters of Iraq. *International Journal of Marine Science*, 4 (40): 1–5.

Jayne, B. C., and G. V. Lauder. 1996. New data on axial locomotion in fishes: how speed affects diversity of kinematics and motor patterns. *American Zoologist*, 36 (6): 642–655.

Jenkins, O. P. 1904. Report on collections of fishes made in the Hawaiian Islands, with descriptions of new species. *Bulletin of the United States Fisheries Commission*, 22: 417–511.

Jerdon, T. C. 1853. Ichthyological gleanings in Madras, communicated by Walter Elliot, Esq., Revenue Commissioner. *Madras Journal of Literature and Science*, 17 (6): 128–151.

Johnson, J. W. 1999. Annotated checklist of the fishes of Moreton Bay, Queensland, Australia. *Memoirs of the Queensland Museum*, 43 (2): 709–762.

Jones, G. P. 1991. Postrecruitment processes in the ecology of coral reef fish populations: a multifactorial perspective. Pp. 294–328, In: P. F. Sale (editor), *The Ecology of Fishes on Coral Reefs*, Academic Press, San Diego, California.

Jones, J. M. 1872. A pelagic floating fish nest. *Nature*, 5: 462.

Jones, J. M. 1879. The gulf-weed (*Sargassum bacciferum*) a means of migration for fishes and marine invertebrates. *Nature*, 19: 363.

Jones, S., and M. Kumaran. 1980. *Fishes of the Laccadive Archipelago*. Nature Conservation and Aquatic Sciences Services, Santinivas Kerala, India, xii + 760 pp.

Jonston, J. 1650. *Historiae naturalis*, Vol. 3, De piscibus et cetis libri V. Meriani, Frankfurt am Main, 228 pp.

Jordan, D. S. 1884. Notes on a collection of fishes from Pensacola, Florida, obtained by Silas Stearns, with descriptions of two new species (*Exocoetus volador* and *Gnathypops mystacinus*). *Proceedings of the United States National Museum*, 7: 33–40.

Jordan, D. S. 1885. List of fishes from Egmont Key, Florida, in the Museum of Yale College, with description of two new species. *Proceedings of the Academy of Natural Sciences of Philadelphia*, 1884: 42–46.

Jordan, D. S. 1890. Scientific results of explorations by the U.S. Fish Commission Steamer *Albatross*. No. IX. Catalogue of fishes collected at Port Castries, St. Lucia, by the Steamer *Albatross*, November 1888. *Proceedings of the United States National Museum*, 12: 645–652.

Jordan, D. S. 1902. A review of the pediculate fishes or anglers of Japan. *Proceedings of the United States National Museum*, 24: 361–381.

Jordan, D. S. 1905. *Guide to the Study of Fishes*. Holt, New York, 2 vols.: Vol. 1, xxvi + 624 pp.; Vol. 2, xxii + 559 pp.

Jordan, D. S. 1917a. Concerning Rafinesque's Précis des d'ecouvertes somiologiques, ou zoologiques et botaniques. *Proceedings of the Academy of Natural Sciences of Philadelphia*, 1917: 276–278.

Jordan, D. S. 1917b. The genera of fishes, from Linnaeus to 1920, with the accepted type of each. A contribution to the stability of scientific nomenclature. *Stanford University Publications, University Series, Biological Sciences*, 1: 1–161.

Jordan, D. S. 1922. *The Days of a Man, Being Memories of a Naturalist, Teacher and Minor Prophet of Democracy.* World Book Co., Yonkers, New York, Vol. 1, xxviii + 710 pp.; Vol. 2, xxviii + 906 pp.

Jordan, D. S., and C. H. Bollman. 1889. List of fishes collected at Green Turtle Cay, in the Bahamas, by Charles L. Edwards, with descriptions of three new species. *Proceedings of the United States National Museum*, 1888 (1889): 549–553.

Jordan, D. S., and B. W. Evermann. 1898. The fishes of North and Middle America, a descriptive catalogue of the species of fish-like vertebrates found in the waters of North America, north of the isthmus of Panama. *Bulletin of the United States National Museum*, 47 (2): xxx, 1241–2183; (3): xxiv, 2183–3136.

Jordan, D. S., and B. W. Evermann. 1903. Descriptions of new genera and species of fishes from the Hawaiian Islands. *Bulletin of the United States Fisheries Commission*, 22: 161–208.

Jordan, D. S., and B. W. Evermann. 1905. The aquatic resources of the Hawaiian Islands. Part 1. The shore fishes of the Hawaiian Islands, with a general account of the fish fauna. *Bulletin of the United States Fisheries Commission*, 1903 (1905), 23 (1): 1–574.

Jordan, D. S., and C. H. Gilbert. 1883. Synopsis of the fishes of North America. *Bulletin of the United States National Museum*, 16, lvi + 1018 pp.

Jordan, D. S., and C. L. Hubbs. 1925. Records of fishes obtained by David Starr Jordan in Japan, 1922. *Memoirs of the Carnegie Museum*, 10 (2): 93–346.

Jordan, D. S., and R. E. Richardson. 1908. Fishes from islands of the Philippine Archipelago. *Bulletin of the United States Bureau of Fisheries*, 27: 233–287.

Jordan, D. S., and A. Seale. 1906. The fishes of Samoa. Descriptions of the species found in the Archipelago, with a provisional check-list of the fishes of Oceania. *Bulletin of United States Bureau of Fisheries*, 1905 (1906), 25: 173–455.

Jordan, D. S., and A. Seale. 1907. Fishes of the islands of Luzon and Panay. *Bulletin of United States Bureau of Fisheries*, 1906 (1907), 26: 1–48.

Jordan, D. S., and J. O. Snyder. 1901. List of fishes collected in 1883 and 1885 by Pierre Louis Jouy and preserved in the United States National Museum, with descriptions of six new species. *Proceedings of the United States National Museum*, 23: 739–769.

Jordan, D. S., and J. O. Snyder. 1904. Notes on collections of fishes from Oahu Island and Laysan Island, Hawaii, with descriptions of four new species. *Proceedings of the United States National Museum*, 27: 939–948.

Jordan, D. S., and E. C. Starks. 1907. Notes on fishes from the island of Santa Catalina, southern California. *Proceedings of the United States National Museum*, 32: 67–77.

Jordan, D. S., S. Tanaka, and J. O. Snyder. 1913. A catalogue of the fishes of Japan. *Journal of the College of Science, Tokyo Imperial University*, 33 (1): 1–497.

Jordan, D. S., and W. F. Thompson. 1914. Record of fishes obtained in Japan, 1911. *Memoirs of the Carnegie Museum*, 6 (4): 205–313.

Joshi, K. K., M. P. Sreeram, P. U. Zacharia, E. M. Abdussamad, M. Varghese, M. Habeeb, K. Jayabalan, K. P. Kanthan, K. Kannan, K. M. Sreekumar, and G. George. 2016. Check list of fishes of the Gulf of Mannar ecosystem, Tamil Nadu, India. *Journal of the Marine Biological Association of India*, 58 (1): 34–54.

Juncker, M. 2007. *Jeunes Poissons Coralliens de Wallis et du Pacifique Central, Guide d'Identification.* Coral Reef Initiative for the South Pacific Programme (CRISP), Wallis & Futuna, Service de l'Environnement, Nouméa, Nouvelle Calédonie, 70 pp. [In French.]

Kailola, P. J. 1975. A catalogue of the fish reference collection at the Kanudi Fisheries Research Laboratory, Port Moresby. *Department of Agriculture, Stock and Fisheries, Research Bulletin*, Papua New Guinea, 16: 1–277.

Kailola, P. J., and M. A. Wilson. 1978. The trawl fishes of the Gulf of Papua. *Department of Primary Industry, Port Moresby, Research Bulletin*, 20: 1–85.

Kamohara, T. 1955. *Coloured Illustrations of the Fishes of Japan*. Hoikusha, Osaka, Japan, 135 pp.

Kan, T. T., J. B. Aitsi, J. E. Kasu, T. Matsuoka, and H. L. Nagaleta. 1995. Temporal changes in a tropical nekton assemblage and performance of a prawn selective gear. *Marine Fisheries Review*, 57 (3–4): 21–34.

Karplus, I. 2014. *Symbiosis in Fishes: The Biology of Interspecific Partnerships*. Wiley-Blackwell, Chichester, West Sussex, United Kingdom, x + 449 pp.

Keenleyside, M. H. A. 1979. *Diversity and Adaptation in Fish Behaviour*. Springer-Verlag, New York, xiii + 208 pp.

Kemp, N. E., and J. H. Park. 1970. Regeneration of lepidotrichia and actinotrichia in the tailfin of the teleost *Tilapia mossambica*. *Developmental Biology*, 22 (2): 321–342.

Kershaw, D. R. 1976. A structural and functional interpretation of the cranial anatomy in relation to the feeding of osteoglossoid fishes and a consideration of their phylogeny. *Transaction of the Zoological Society*, London, 33 (3): 173–252.

Klein, J. T. 1742. *Historiae piscium naturalis promovendae*. Part 3. De piscibus per branchias occultas spirantibus. Schreiber, Dantzig, 48 pp.

Klein, R. 2008. Dive log: Ambon Bay, Indonesia. *Scuba Diving*, 17: 15, September 2008.

Klunzinger, C. B. 1871. Synopsis der Fische des Rothen Meeres. II. *Verhandlungen der Zoologisch-Botanischen Gesellschaft in Wien*, 21: 441–668.

Klunzinger, C. B. 1884. *Die Fische des Rothen Meeres. Eine kritische Revision mit Bestimmungstabellen*. I. Theil. Acanthopteri veri. Stuttgart, 133 pp.

Kner, R. 1865. Fische. Pp. 1–272, In: *Reise der österreichischen Fregatte "Novara" um die Erde in den Jahren 1857–1859, unter den Befehlen des Commodore B. von Wullerstorf-Urbain*. Zoologischen Theil, Vol. 1.

Knorr, G. W. 1766–1767. *Deliciae naturae selectae, oder auserlesenes Naturalien-Cabinet welches aus den drey Reichen der Natur zeiget, was von curiösen Liebhabern aufbehalten und gesammlet zu werden verdienet, ehemahls herausgegeben von Georg Wolfgang Knorr berühmten Kupferstecher in Nürnberg, fortgesetzet von dessen Erben, beschrieben von Philipp Ludwig Statius Müller öffentlichen ordentlichen Lehrer der Weltweissheit auf der Friedrichs Universität zu Erlang, und in das Französische übersezet*. Matthaus Verdier de la Blaquiere, Nuremberg, 2 vols.

Koefoed, E. 1944. Pediculati from the "Michael Sars" North Atlantic Deep-Sea Expedition 1910. *Report of the Scientific Research of the "Michael Sars" Expedition*, 4, 2 (1): 1–18.

Koelreuter, J. T. 1766. Piscium rariorum e Museo Petropolitano exceptorum descriptiones continuatae. *Novi commentarii Academiae Scientiarum Imperialis Petropolitanae*, 10: 329–336.

Komyakova, V., P. L. Munday, and G. P. Jones. 2013. Relative importance of coral cover, habitat complexity and diversity in determining the structure of reef fish communities. *PLoS ONE*, 8 (12): e83178. doi: 10.1371/journal.pone.0083178.

Kon, T., and T. Yoshino. 1999. Record of the frogfish (Lophiiformes: Antennariidae), *Antennarius analis*, from Japan, with comments on its authorship. *Japanese Journal of Ichthyology*, 46 (2): 101–103.

Konow, N., and D. R. Bellwood. 2011. Evolution of high trophic diversity based on limited functional disparity in the feeding apparatus of marine angelfishes (f. Pomacanthidae). *PLoS ONE*, 6 (9): e24113. doi: 10.1371/journal.pone.0024113.

Kosaki, R. K., R. L. Pyle, J. E. Randall, and D. K. Irons. 1991. New records of fishes from Johnston Atoll, with notes on biogeography. *Pacific Science*, 45 (2): 186–203.

Kottelat, M. 2013. The fishes of the inland waters of southeast Asia: a catalogue and core bibiography of the fishes known to occur in freshwaters, mangroves and estuaries. *Raffles Bulletin of Zoology*, Supplement, 27: 1–663.

Kottelat, M., A. J. Whitten, S. N. Kartikasari, and S. Wirjoatmodjo. 1993. *Freshwater Fishes of Western Indonesia and Sulawesi*. Periplus Editions, Hong Kong, xxxviii + 259.

Kotthaus, A. 1979. Fische des Indischen Ozeans. Ergebnisse der ichthyologischen Untersuchungen während der Expedition des Forschungsschiffes "Meteor" in der Indischen Ozean,

Oktober 1964 bis Mai 1965. A. Systematischer Teil, 21, Diverse Ordnungen. *Meteor-Forschungsergebnisse*, Reihe D, 28: 6–54.

Koumans, F. P. 1953. Biological results of the Snellius Expedition. XVI. The pisces and leptocardii of the Snellius Expedition. *Temminckia*, 9: 177–275.

Koya, Y., and M. Muñoz. 2007. Comparative study on ovarian structures in scorpaenids: possible evolutionary process of reproductive mode. *Ichthyological Research*, 54: 221–230.

Krishnan, S., and S. S. Mishra. 1993. On a collection of fish from Kakinada-Gopalpur sector of the east coast of India. *Records of the Zoological Survey of India*, 93 (1–2): 201–240.

Kuiter, R. H. 1982. Fish in focus. *Scuba Diver*, October 1982: 34–39.

Kuiter, R. H. 1987. Diving under sargassum weed. *Sportdiving Magazine*, 3: 26–29.

Kuiter, R. H. 1993. *Coastal Fishes of South-Eastern Australia*. University of Hawaii Press, Honolulu, Hawaii, xxxi + 437 pp.

Kuiter, R. H. 1997. *Guide to Sea Fishes of Australia. A Comprehensive Reference for Divers and Fishermen*. New Holland Publishers, Frenchs Forest, New South Wales, Australia, xvii + 434.

Kuiter, R. H., and H. Debelius. 2007. *The World Atlas of Marine Fishes*. IKAN-Unterwasserarchiv, Frankfurt, Germany, vi + 720 pp.

Kulbicki, M., V. Parravicini, D. R. Bellwood, E. Arias-Gonzàlez, P. Chabanet, S. R. Floeter, A. Friedlander, J. McPherson, R. E. Myers, L. Vigliola, and D. Mouillot. 2013. Global biogeography of reef fishes: a hierarchical quantitative delineation of regions. *PLoS ONE*, 8 (12): e81847. doi:10.1371/journal.pone.0081847.

Kulbicki, M., and J. T. Williams. 1997. Checklist of the shorefishes of Ouvéa Atoll, New Caledonia. *Atoll Research Bulletin*, 444: 1–26.

Kwang-Tsao, S., H.-C. Ho, P.-L. Lin, P.-F. Lee, M-Y. Lee, C.-Y. Tsai, Y.-C. Liao, and Y.-C. Lin. 2008. A checklist of the fishes of southern Taiwan, northern South China Sea. *Raffles Bulletin of Zoology*, Supplement, 19: 233–271.

Kwik, J. T. B. 2012. Controlled culling of venomous marine fishes along Sentosa Island beaches: a case study of public safety management in the marine environment of Singapore. *Raffles Bulletin of Zoology*, Supplement, 25: 93–99.

Kyushin, K., K. Amaoka, K. Nakaya, H. Ida, Y. Tanino, and T. Senta (editors). 1982. *Fishes of the South China Sea*. Japan Marine Fishery Resource Research Center, 333 pp.

Laboute, P., and R. Grandperrin. 2000. *Poissons de Nouvelle-Calédonie*. Éditions Catherine Ledru, Nouméa, pp. 7–520.

Lacepède, B.-G.-E. de. 1788–1789. *Histoire des quadrupèdes, ovipares et des serpens*. Hôtel de Thou, Paris, Vol. 1, 18 + 651 pp.; Vol. 2, 20 + 527 pp.

Lacepède, B.-G.-E. de. 1798–1803. *Histoire naturelle des poissons*. Plassan, Paris, Vol. 1, 1798, ccxii + 532 pp.; Vol. 2, 1800, lxiv + 632 pp.; Vol. 3, 1802, 16 + lxvi + 558 pp.; Vol. 4, 1802, xliv + 728 pp.; Vol. 5, 1803, lxviii + 803 pp.

Lacepède, B.-G.-E. de. 1804. Memoire sur plusieurs animaux de la Nouvelle-Hollande dont la description n'a pas encore été publiée. *Annales du Muséum National d'Histoire naturelle*, Paris, 4: 184–211.

Lagler, K. F., J. E. Bardach, R. R. Miller, and D. R. M. Passino. 1977. *Ichthyology*, 2nd edition Wiley, New York, xv + 506 pp.

Lampe, M. 1914. Die Fische der Deutschen Südpolar-Expedition 1901–1903. III. Die Hochsee und Küstenfische. *Deutschen Südpolar-Expedition 1901–1903, Berlin*, 15 (2): 201–256.

Landini, W. 1977. Revisione degli "Ittiodontoliti pliocenici" della collezione Lawley. *Palaeontographia Italica*, 70: 92–134.

Larson, H. K., and R. S. Williams. 1997. Darwin Harbour fishes: a survey and annotated checklist. Pp. 339–380, In: J. R. Hanley, G. Caswell, D. Megirianand, and H. K. Larson (editors), *Proceedings of the Sixth International Marine Biological Workshop. The Marine Flora and Fauna of Darwin Harbour, Northern Territory*, Australia. Museums and Art Galleries, Northern Territory and Australian Scientific Association.

Larson, H. K., R. S. Williams, and M. P. Hammer. 2013. An annotated checklist of the fishes of the Northern Territory, Australia. *Zootaxa*, 3696 (1): 1–293.

Last, P. R., and D. C. Gledhill. 2009. A revision of the Australian handfishes (Lophiiformes, Brachionichthyidae), with descriptions of three new genera and nine new species. *Zootaxa*, 2252, 1–77.

Last, P. R., E. O. G. Scott, and F. H. Talbot. 1983. *Fishes of Tasmania*. Tasmanian Fisheries Development Authority, Hobart, viii + 563 pp.

Latreille, P. A. 1804. Tableaux Méthodiques des poissons. Pp. 71–105, In: *Nouveau Dictionnaire d'Histoire Naturelle*, 1st edition, Paris, Vol. 24.

Lauder, G. V. 1979. Feeding mechanisms in primitive teleosts and in the halecomorph fish *Amia calva*. *Journal of Zoology*, London, 187: 543–578.

Lauder, G. V. 1981. Intraspecific functional repertoires in the feeding mechanism of the characoid fishes *Lebiasina*, *Hoplias* and *Chalceus*. *Copeia*, 1981 (1): 154–168.

Lauder, G. V. 2015. Fish locomotion: recent advances and new directions. *Annual Review of Marine Science*, 7: 521–545.

Lauder, G. V., and K. F. Liem. 1980. The feeding mechanism andcephalic myology of *Salvelinus fontinalis*: form, function, and evolutionary significance. Pp. 365–390, In: E. K. Balon (editor), *Charrs: Salmonid Fishes of the Genus* Salvelinus. Junk, The Hague, ix + 919 pp.

Lauder, G. V., and K. F. Liem. 1983. The evolution and interrelationships of the actinopterygian fishes. *Bulletin of the Museum of Comparative Zoology*, 150 (3): 95–197.

Lawley, R. 1876. *Nuovi studi sopra ai pesci ed altri vertebrati fossili delle colline toscane*. Tipografia dell'Arte della Stampa, Firenze, 122 pp.

Leão de Moura, R., J. L. Gasparini, and I. Sazima, 1999. New records and range extensions of reef fishes in the Western South Atlantic, with comments on reef fish distribution along the Brazilian coast. *Revista Brasileira de Zoologia*, 16 (2): 513–530.

Le Danois, Y. 1964. Étude anatomique et systématique des Antennaires, de l'Ordre des Pédiculates. *Mémoires du Muséum national d'Histoire naturelle*, Paris, Série A, Zoologie, 31 (1): 1–162.

Le Danois, Y. 1970. Etude sur des poissons pediculates de la famille des Antennariidae recoltes dans la Mer Rouge et description d'une espece nouvelle. *Israel Journal of Zoology*, 19 (2): 83–94.

Le Danois, Y. 1974. Étude ostéo-myologique et revision systématique de la famille Lophiidae (Pédiculates Haploptérygiens). *Mémoires du Muséum national d'Histoire naturelle*, Paris, Série A, Zoologie, 91: 1–127.

Le Danois, Y. 1979. Révision systématique de la famille des Chaunacidae (Pisces Pediculati). *UO NO KAI* (Japanese Society of Ichthyologists), 30: 1–76.

Leis, J. M. 2015. Taxonomy and systematics of larval Indo-Pacific fishes: a review of progress since 1981. *Ichthyological Research*, 62: 9–28.

Lema, T. de. 1976. Ocorrência de várias espécies de peixes tropicais marinhos na costa do Estado de Santa Catarina, Brasil (Osteichthyes, Actinopterygii, Teleostei). *Iheringia, Série Zoologia*, Porto Alegre, 49: 39–65.

Leriche, M. 1906. Contribution à l'étude des poissons fossils du Nord de la France et des regions voisines. *Mémoires de la Société Géologique du Nord*, 5: 1–430.

Leriche, M. 1908. Note préliminaire due des poissons nouveaux de l'Oligocène Belge. *Bulletin de la Société Belge de Gèologie, Paléontologie, Hydrologie*, 21: 378–384.

Leriche, M. 1910. Les poissons oligocènes de la Belgique. *Mémoires du Musée Royal d'Histoire Naturelle de Belgique*, 5: 233–363.

Leriche, M. 1926. Les poissons Néogènes de la Belgique. *Mémoires du Musée Royal d'Histoire Naturelle de la Belgique*, 32: 365–472.

Lesson, R. P. 1831. Poissons. Pp. 66–238, In: L. I. Duperrey, *Voyage autour du monde sur la Corvette de la Magesté, La Coquille, pendant les années 1822, 1823, 1824 et 1825*. Zoologie, 2 (1).

Letourneur, Y., P. Chabanet, P. Durville, M. Taquet, E. Teissier, M. Parmentier, J.-C. Quéro, and K. Pothin. 2004. An updated checklist of the marine fish fauna of Reunion Island, South-Western Indian Ocean. *Cybium*, 28 (3): 199–216.

Lew, A. A. 2013. A World geography of recreational scuba diving. Pp. 29–51, In: G. Musa and K. Dimmock (editors), *Scuba Diving Tourism*, Routledge, Abingdon, England.

Li, C., G. Lu, and G. Orti. 2008. Optimal data partitioning and a test case for ray-finned fishes (Actinopterygii) based on ten nuclear loci. *Systematic Biology*, 57: 519–539.

Liem, K. F. 1970. Comparative functional anatomy of the Nandidae (Pisces: Teleostei). *Fieldiana, Zoology*, 56: 1–166.

Liem, K. F. 1978. Modulatory multiplicity in the functional repertoire of the feeding mechanism in cichlid fishes. I. Piscivores. *Journal of Morphology*, 158 (3): 323–360.

Liem, K. F. 1979. Modulatory multiplicity in the feeding mechanism in cichlid fishes, as exemplified by the invertebrate pickers of Lake Tanganyika. *Journal of Zoology*, London, 189: 93–125.

Liem, K. F. 1980. Adaptive Significance of intra- and interspecific differences in the feeding repertoires of cichlid fishes. *Integrative and Comparative Biology*, 20 (1): 295–314.

Liem, K. F. 1994. Introducing fishes. Pp. 14–19, In: J. R. Paxton and W. N. Eschmeyer (editors), *Encyclopedia of Fishes*, University of New South Wales Press, Sydney, Australia.

Liem, K. F. 1998. Introducing fishes. Pp. 14–19, In: J. R. Paxton and W. N. Eschmeyer (editors), *Encyclopedia of Fishes*, 2nd edition, Academic Press, San Diego, California.

Liem, K. F., and P. H. Greenwood. 1981. A functional approach to the phylogeny of the pharyngognath teleosts. *American Zoologist*, 21: 83–101.

Lieske, E., and R. F. Myers. 2004. *Coral Reef Guide: Red Sea to Gulf of Aden, South Oman*. Harper Collins, London, 384 pp.

Lilljeborg, W. 1884. *Sveriges och Norges Fiskar*. Schultz, Uppsala, Vol. 1, xxii + 782 pp.

Lin, J. 2017. Foreword. Pp. xxxi–xxxiv, In: R. Calado, I. Olivotto, M. P. Oliver, and G. J. Holt (editors), *Marine Ornamental Species Aquaculture*, John Wiley & Sons Ltd, Chichester, West Sussex, United Kingdom.

Lindberg, G. U., V. V. Fedorov, and Z. V. Krasyukova. 1997. *Fishes of the Sea of Japan and the adjacent parts of the Sea of Okhotsk and Yellow Sea*. Part 7. Handbook on the Identification of Animals, Zoological Institute of the Russian Academy of Sciences, 168, 350 pp.

Lindberg, G. U., A. S. Heard, and T. R. Rass. 1980. *Multilingual Dictionary of Names of Marine Food-Fishes of World Fauna*. Academy of Sciences of the U.S.S.R., 564 pp.

Linnaeus, C. 1727–1730. *Caroli Linnaei Manuscripta Medica Tom. 1, in quo continetur Historie naturalis fragmenta, ab anno 1727 Mens: Aug: ad 1730 initium* [from August 1727 to the beginning of 1730]. Linnean Collections, Linnean Society of London, Unit ID: GB-110/LM/LP/BIO/3/2, 538 folios, http://linnean-online.org/61326. Accessed January 2018.

Linnaeus, C. 1735. *Systema naturae, sive Regna tria naturae, systematice proposita per classes, ordines, genera, et species*. Theodor Haak, Leiden, 12 pp.

Linnaeus, C. 1747. *Västgöta-resa, på riksens högloflige ständers befallning förrättad år 1746, med anmärkningar uti oeconomien, naturkunnogheten, antiquiteter, inwånarnes seder och lefnads-sätt, med tilhörige figure*. Laurentii Salvii, Stockholm, 284 pp.

Linnaeus, C. 1754a. *Specimen academicum, sistens Chinensia Lagerströmiana, quod, annvente nob: a facultate medica, in R. Academia Upsal. Praeside viro nobilissimo et experientissimo D:n D. Carolo Linnaeo . . . publicae disquisitioni submittit Johannes Laurentius Odhelius*. Merckell, Stockholm, 36 pp.

Linnaeus, C. 1754b. *Museum S:ae R:ae M:tis Adolphi Friderici Regis Svecorum . . . in quo animalia rariora, imprimis, et exotica: quadrupedia, aves, amphibia, pisces, insecta, vermes describuntur et determinantur, Latine et Svetice, cum iconibus, jussu Sac. Reg. Maj:tis a Car. Linnaeo, Equ*. Momma, Stockholm, 96 pp.

Linnaeus, C. 1758. *Systema naturae per regna tria naturae, secundum classes, ordines, genera, species, cum characteribus, differentiis, synonymis, locis*. Editio decima, reformata. Laurentius Salvius, Stockholm, Vol. 1, [4] + 824 pp.

Linnaeus, C. 1759. Chinensia Lagerströmiana, praeside D. D. Car. Linnaeo, proposita a Johann Laur. Odhelio, W. Gotho. Upsaliae 1754, Decembr. 23, In: *Amoenitates Academicae*, 4 (61): 230–260.

Linnaeus, C. 1766. *Systema naturae per regna tria naturae, secundum classes, ordines, genera, species, cum characteribus, differentiis, synonymis, locis*. Editio duodecima, reformata. Laurentii Salvii, Stockholm, Vol. 1, Regnum animale, 532 pp.

Lloris, D., J. Rucabado, and H. Figueroa. 1991. Biogeography of the Macaronesian Ichthyofauna (the Azores, Madeira, the Canary Islands, Cape Verde and the African enclave). *Boletim do Museu Municipal do Funchal*, 43 (234): 191–241.

Longley, W. H., and S. F. Hildebrand. 1941. Systematic catalogue of the fishes of Tortugas, Florida, with observations on color, habits, and local distribution. *Papers from the Tortugas Laboratory of the Carnegie Institution of Washington*, 34: 1–331.

Lönnberg, E. 1896. Linnean type-specimens of birds, reptiles, batrachians and fishes in the Zoological Museum of the R. University in Upsala. *Bihang till Kongliga Svenska Vetenskaps-Akademiens Handlingar*, 22, 4 (1): 1–45.

Lorenz, K. Z. 1950. The comparative method in studying innate behavior patterns. *Symposia of the Society for Experimental Biology*, 4: 221–268.

Lowe, R. T. 1839. A supplement to a synopsis of the fishes of Madeira. *Proceedings of the Zoological Society of London*, 7: 76–92 [also published in *Transactions of the Zoological Society of London*, 1842, 3 (1): 1–20].

Lozano, S., and F. A. Zapata, 2003. Short-term temporal patterns of early recruitment of coral reef fishes in the tropical eastern Pacific. *Marine Biology*, 142 (2): 399–409.

Lucas, A. H. S. 1890. A systematic census of indigenous fish, hitherto recorded from Victorian waters. *Proceedings of the Royal Society of Victoria*, New Series, 2 (2): 15–47.

Lugendo, B. R., I. Nagelkerken, N. Jiddawi, Y. D. Mgaya, and G. Van Der Velde. 2007. Fish community composition of a tropical nonestuarine embayment in Zanzibar, Tanzania. *Fisheries Science*, 73 (6): 1213–1223.

Luiz, O. J., A. Carvalho-Filho, C. E. L. Ferreira, S. R. Floeter, J. L. Gasparini, and I Sazima. 2008. The reef fish assemblage of the Laje de Santos Marine State Park, Southwestern Atlantic: annotated checklist with comments on abundance, distribution, trophic structure, symbiotic associations, and conservation. *Zootaxa*, 1807: 1–25.

Luiz, O. J., A. P. Allen, D. R. Robertson, S. R. Floeter, M. Kulbicki, L. Vigliola, R. Becheler, and J. S. Madin. 2013. Adult and larval traits as determinants of geographic range size among tropical reef fishes. *Proceedings of the National Academy of Sciences*, 110 (41): 16498–16502.

Lukoschek, V., and M. I. McCormick. 2001. Ontogeny of diet changes in a tropical benthic carnivorous fish, *Parupeneus barberinus* (Mullidae): relationship between foraging behaviour, habitat use, jaw size, and prey selection. *Marine Biology*, 138 (6): 1099–1113.

Lütken, C. F. 1863. Nogle nye Krybdyr og Padder. *Videnskabelige Meddelelser Naturhistorisk Forening i København*, Series V, 2 (4): 292–311.

Lütken, C. F. 1878. Til Kundskab om to arktiske slaegter af Dybhavs-Tudesfiske: *Himantolophus* og *Ceratias*. *Danske Videnskabernes Selskab Skrifter*, 5. Raekke, *Naturvidenskabelig og Mathematisk Afdeling*, 11 (5): 309–388.

Lynch, T. P., M. A. Green, and C. Davies. 2015. Diver towed GPS increases both statistical power to estimate densities and functionality to observe behaviours of sparsely distributed Spotted Handfish (*Brachionichthys hirsutus*, Lacépède 1804). *Biological Conservation*, 191 (1): 700–706.

Lynch, T. P., L. Wong, T. Fountain, and C. Devine. 2017. *Establishment of Captive Breeding Populations of Spotted Handfish*. Report to the National Environmental Science Programme, Marine Biodiversity Hub, CSIRO, 64 pp.

Lynch, T. P., L. Wong, and M. A. Green. 2016. *Direct Conservation Actions for Critical Endangered Spotted Handfish*. Final report to the Threatened Species Commissioner. CSIRO Oceans and Atmosphere, Hobart, Tasmania, 22 pp.

Maamaatuaiahutapu, M., G. Remoissenet, and R. Galzin. 2006. *Guide d'Identification des Larves de Poissons Récifaux de Polynésie Française*. Coral Reef Initiative for the South Pacific Programme (CRISP), Éditions Téthys, Nouméa, Nouvelle Calédonie, 104 pp. [In French.]

Maartens, L., and A. J. and Booth. 2005. Aspects of the reproductive biology of monkfish *Lophius vomerinus* off Namibia. *African Journal of Marine Science*, 27: 325–329.

Maass-Berlin, T. A. 1937. Gift-tiere. In: W. Junk (editor), *Tabulae Biologicae*. Van de Garde, Drukkerij, Zaltbommel, Holland, Vol. 13, 272 pp.

MacArthur, R. H., and E. O. Wilson. 1967. *The Theory of Island Biogeography*. Princeton University Press, Princeton, New Jersey, 203 pp.

Macleay, W. 1878. The fishes of Port Darwin. *Proceedings of the Linnean Society of New South Wales*, 2: 344–367.

Macleay, W. 1881. Descriptive catalogue of the fishes of Australia. Part II. *Proceedings of the Linnean Society of New South Wales*, 5: 510–629.

Mair, A. W. 1928. *Oppian, Colluthus, Tryphiodorus, with an English translation by A. W. Mair*. Loeb Classical Library, Harvard University Press, Cambridge, Massachusetts, lxxx + 515 pp.

Major, R. H. 1859. *Early Voyages to Terra Australis, Now Called Australia: A Collection of Documents, and Extracts from Early Manuscript Maps, Illustrative of the History of Discovery on the Coasts of That Vast Island, from the Beginning of the Sixteenth Century to the Time of Captain Cook*. Edited with an introduction by R. H. Major. Printed for the Hakluyt Society, London, 25, cxix + 214 pp.

Manilo, L. G., and S. V. Bogorodsky. 2003. Taxonomic composition, diversity and distribution of coastal fishes of the Arabian Sea. *Journal of Ichthyology*, 43: S75–S149.

Marcgrave, G. 1648. Historiae rerum naturalium Brasiliae, libro octo: quorum, Tres priores agunt de plantis. Quartus de piscibus. Quintus de avibus. Sextus de quadrupedibus & serpentibus. Septimus de insectis. Octavus de ipsa regione, & illius incolis . . . Ioannes de Laet, Antwerpianus. In: G. Piso and G. Margrave, *Historia naturalis Brasiliae*, Part 2, Haack & Elsevier, Leiden, 293 pp. [Reprint 1942, São Paulo, in Portuguese.]

Marshall, T. C. 1964. *Fishes of the Great Barrier Reef and Coastal Waters of Queensland*. Angus & Robertson, Sydney, xvi + 566 pp.

Martin, F. D., and G. E. Drewry. 1978. *Development of Fishes of the Mid-Atlantic Bight: An Atlas of Eggs, Larvae and Juvenile Stages*. Vol. 6. Stromateidae through Ogcocephalidae. Fish and Wildlife Service, United States Department of the Interior, 416 pp.

Martin, R. 1966. On the swimming behavior and biology of *Notarchus punctatus* Phillipi (Gastropoda, Opisthobranchia). *Pubblicazioni della Stazione Zoologica di Napoli*, 35: 61–75.

Martin, T. J., D. Brewer, and S. J. M. Blaber. 1995. Factors affecting distribution and abundance of small demersal fishes in the Gulf of Carpentaria, Australia. *Australian Journal of Marine and Freshwater Research*, 46 (6): 909–920.

Martínez-Muñoz, M. A., D. Lloris, A. Gracia, R. Ramírez-Murillo, S. Sarmiento-Nafáte, S. Ramos-Cruz, and F. Fernández. 2016. Biogeographical affinities of fish associated to the shrimp trawl fishery in the Gulf of Tehuantepec, Mexico. *Revista de Biologia Tropical*, 64 (2): 683–700.

Mathews, G. M., and T. Iredale. 1915. On the ornithology of the Dictionnaire des sciences naturelles (Levrault). *Australian Avian Records*, 3 (1): 5–20.

Maul, G. E. 1959. *Aulostomus*, a recent spontaneous settler in Madeiran waters. *Bocagiana, Museu Municipal do Funchal*, 1: 1–18.

McAllister, D. E., 1990. A list of the fishes of Canada. *Syllogeus*, 64: 1–310.

McCook, L. J., J. Jompa, and G. Diaz-Pulido. 2001. Competition between corals and algae on coral reefs: a review of evidence and mechanisms. *Coral Reefs*, 19: 400–417.

McCosker, J. E., and R. H. Rosenblatt. 1975. Fishes collected at Malpelo Island. *Smithsonian Contributions to Zoology*, 176: 91–93.

McCosker, J. E., and R. H. Rosenblatt. 2010. The fishes of the Galápagos Archipelago: an update. *Proceedings of the California Academy of Sciences*, Series 4, 61 (11): 167–195.

McCoy, F. 1886. *Chironectes bifurcatus* (McCoy). The two-pronged toad-fish. *Prodromus of the Zoology of Victoria*, 2 (13): 87–89.

McCulloch, A. R. 1915. Notes on, and descriptions of, Australian fishes. *Proceedings of the Linnean Society of New South Wales*, 40: 259–277.

McCulloch, A. R. 1916. Ichthyological items. *Memoirs of the Queensland Museum*, Brisbane, 5: 58–69.

McCulloch, A. R. 1922. Check-list of the fish and fish-like animals of New South Wales. Part III. *Australian Zoologist*, 2 (3): 86–130.

McCulloch, A. R. 1929. A check-list of the fishes recorded from Australia. *Memoirs of the Australian Museum*, Sydney, 5 (3): 329–436.

McCulloch, A. R., and E. R. Waite. 1918. Some new and little-known fishes from South Australia. *Records of the South Australian Museum*, Adelaide, 1 (1): 39–78.

McEachran, J. D., and J. D. Fechhelm. 1998. *Fishes of the Gulf of Mexico*. Volume 1: Myxiniformes to Gasterosteiformes, University of Texas Press, Austin, 1112 pp.

McFarland, D. (editor). 1982. *The Oxford Companion to Animal Behavior*. Oxford University Press, Oxford & New York, xii + 657 pp.

McManus, J. W., R. B. Reyes Jr, and C. L. Nanola Jr. 1997. Effects of some destructive fishing methods on coral cover and potential rates of recovery. *Environmental Management*, 21: 69–78.

Mead, G. W. 1970. A history of South Pacific fishes. Pp. 236–251, In: W. S. Wooster (editor), *Proceedings of a Symposium held during the Ninth General Meeting of the Scientific Committee on Oceanic Research, June 18–20, 1968, at Scripps Institution of Oceanography, La Jolla, California*, National Academy of Sciences, Washington, DC.

Mead, G. W., E. Bertelsen, and D. M. Cohen. 1964. Reproduction among deep sea fishes. *Deep Sea Research*, 11: 569–596.

Meek, S. E., and S. F. Hildebrand. 1928. The Marine fishes of Panama. Part III. *Field Museum of Natural History, Zoological Series*, 15 (3): 709–1045.

Mees, G. F. 1959. Additions to the fish fauna of Western Australia. 1. *Fisheries Bulletin of Western Australia*, 9 (1): 5–11.

Mees, G. F. 1960. Additions to the fish fauna of Western Australia. 2. *Fisheries Bulletin of Western Australia*, 9 (2): 13–21.

Mees, G. F. 1964. Additions to the fish fauna of Western Australia. 4. *Fisheries Bulletin of Western Australia*, 9 (4): 31–55.

Mejía-Ladino L. M., A. Acero, L. S. Mejía, and A. Polanco. 2007. Taxonomic revision of the family Antennariidae including a new record of *Antennarius* for the Colombian Caribbean (Pisces: Lophiiformes). *Boletín de Investigaciones Marinas y Costeras*, 36 (1): 269–305.

Melliss, J. C. 1875. *St. Helena: A Physical, Historical, and Topographical Description of the Island, including its Geology, Fauna, Flora, and Meteorology*. Reeves, London, 14 + 426 pp.

Menezes, N. A. 2003. Family Antennariidae. P. 64, In: N. A. Menezes, P. A. Buckup, J. L. de Figueiredo, and R. L. de Moura (editors), *Catálogo das Espécies de Peixes Marinhos do Brasil*, Museu de Zoologia da Universidade de São Paulo.

Messenger, J. B. 1968. The visual attack of the cuttlefish, *Sepia officinalis*. *Animal Behavior*, 16: 342–357.

Metzelaar, J. 1919. Report on the fishes collected by Dr. J. Boeke in the Dutch West Indies, 1904–1905, with comparative notes on marine fishes of tropical West Africa. In: J. Boeke, *Rapport Betreffende een Voorloopig Onderzoek naar den Toestand van de Visserij en de Industrie van Zeeproducten in de Kolonie Curacao*. Schinkel, The Hague, 315 pp.

Meuschen, F. C. 1781. Index, continens nomina generica specierum propria, trivialia ut et synonyma. Part 3, Pisces. Following p. 380, 4 pp., unpaged, In: L. T. Gronovius, *Zoophylacii Gronoviani fasciculus tertius*. Haak & Luchtmans, Leiden.

Meyer, A. B. 1885. Catálogo de los peces recolectados en el archipiélago de las Indias orientales durante los años 1870 á 1873. *Anales de la Sociedad Española de Historia Natural*, 14: 5–49.

Michael, S. W. 1998. *Reef fishes, Volume 1: A Guide to Their Identification, Behavior, and Captive Care*. Microcosm Limited, Shelburne, Vermont, 624 pp.

Michael, S. W. 2006. Frogfish behavior: notes from the field. *Coral, Reef and Marine Aquarium Magazine*, 3 (5): 34–43.

Michael, S. W. 2007. New frogfish? Two unusual species that you may encounter at your local fish store. *Freshwater and Marine Aquarium*, July 2007, pp. 56–64.

Middlemiss, F. A., and P. F. Rawson. 1971. Faunal provinces in space and time: Some general considerations. Pp. 199–210, In: F. A. Middlemiss and P. F. Rawson (editors), *Faunal Provinces in Space and Time*. Seel House Press, Liverpool.

Mill, P. J., and R. S. Pickard. 1975. Jet propulsion in anisopteran dragonfly larvae. *Journal of Comparative Physiology, A*, 97: 329–338.

Milne-Edwards, A. 1867. Éloge de M. Valenciennes. *Journal de Pharmacie et de Chimie*, Paris, Série 4, 5: 5–17.

Miranda-Ribeiro, A. de. 1915. Fauna Brasiliense-peixes. *Arquivos do Museu* Nacional *do* Rio de Janeiro, 17, Lophiidae, 8 pp.

Mishra, S., R. Gouda, L. Nayak, and R. C. Panigrahy. 1999. A check list of the marine and estuarine fishes of south Orissa, east coast of India. *Records of the Zoological Survey of India*, 93 (3): 81–90.

Mishra, S. S., and S. Krishnan. 2003. Marine fishes of Pondichery and Karaikal. *Records of the Zoological Survey of India, Occasional Papers*, 216: 1–53.

Mitchill, S. L. 1815. The fishes of NewYork, described and arranged. *Transactions of the Literary and Philosophical Society of New York*, 1 (5): 355–492.

Mitchill, S. L. 1818. Dr. Mitchill's memoir on the fishes of New York. *American Monthly Magazine and Critical Review*, 2 (5): 321–328.

Mito, S. 1962. Pelagic fish eggs from Japanese waters—V. *Callionymina* and *Ophidiina*. *Science Bulletin of the Faculty of Agriculture, Kyushu University*, 19: 377–380.

Mittelstaedt, H. 1957. Prey capture in mantids. Pp. 51–71, In: B. T. Scheer (editor), *Recent Advances in Invertebrate Physiology*, University of Oregon Publication, Eugene.

Miya, M., T. W. Pietsch, J. W. Orr, R. J. Arnold, T. P. Satoh, A. M. Shedlock, H.-C. Ho, M. Shimazaki, M. Yabe, and M. Nishida. 2010. Evolutionary history of anglerfishes (Teleostei: Lophiiformes): a mitogenomic perspective. *BMC Evolutionary Biology*, 10 (58): 1–27.

Miya, M., T. P. Satoh, and M. Nishida. 2005. The phylogenetic position of toadfishes (order Batrachoidiformes) in the higher ray-finned fish as inferred from partitioned Bayesian analysis of 102 whole mitochondrial genome sequences. *Biological Journal of the Linnean Society*, 85: 289–306.

Miya, M., H. Takeshima, H. Endo, N. B. Ishiguro, J. G. Inoue, T. Mukai, T. P. Satoh, M. Yamaguchi, A. Kawaguchi, K. Mabuchi, S. M. Shirai, and M. Nishida. 2003. Major patterns of higher teleostean phylogenies: a new perspective based on 100 complete mitochondrial DNA sequences. *Molecular Phylogenetics and Evolution*, 26: 121–138.

Möbius, K. 1894. Über Eiernester pelagischer Fische aus dem Mittelatlantischen Ozean. *Sitzungsberichte der Königlich-Preussische Akademie der Wissenschaften*, Berlin, 1: 1203–1210.

Mohsin, A. K. M., and M. A. Ambak. 1996. *Marine Fishes and Fisheries of Malaysia and Neighbouring Countries*. Universiti Pertanian Malaysia Press, Serdang, xxxvi + 744 pp.

Molter, T., Jr. 1983. A spawning of the Atlantic anglerfish *Antennarius scaber* Cuvier. *Freshwater and Marine Aquarium*, 6 (1): 34–35, 66, 69.

Monod, T. 1927. Contribution à la faune du Cameroun. Pisces. *Faune des Colonies Françaises*, 1: 643–742.

Monod, T. 1960. A propos du pseudobrachium des *Antennarius* (Pisces, Lophiiformes). *Bulletin de l'Institut Français d'Afrique Noire*, 22, Série A, 2: 620–698.

Monod, T. 1963. Achille Valenciennes et l'Histoire Naturelle des Poissons. *Mémoires de l'Institut Français d'Afrique Noire*, 68: 9–45.

Monod, T., and Y. Le Danois. 1973. Antennariidae. Pp. 661–664, In: J. C. Hureau and T. Monod (editors), *Check-list of the fishes of the north-eastern Atlantic and of the Mediterranean*. UNESCO, Paris, Vol. 1.

Monteiro-Neto, C., Á. A. Bertoncini, L. de C. T. Chaves, and R. Noguchi. 2013. Checklist of marine fish from coastal islands of Rio de Janeiro, with remarks on marine conservation. *Marine Biodiversity Records*, 6: 1–13.

Montrouzier, P. X. 1856. Essai sur la faune de l'île de Woodlark ou Moiou. *Annales de la Société Impériale d'Agriculture, d'Histoire Naturalle et des Arts Utiles de Lyon*, Série 2, 8: 417–504.

Moore, G. I., J. B. Hutchins, K. N. Smith, and S. M. Morrison. 2009. Catalogue of type specimens of fishes in the Western Australian Museum (Second Edition). *Records of the Western Australian Museum*, Supplement, 74, vii + 69 pp.

Moore, G. I., S. M. Morrison, J. B. Hutchins, G. R. Allen, and A. Sampey. 2014. Kimberley marine biota. Historical data: fishes. *Records of the Western Australian Museum*, Supplement, 84: 161–206.

Moore, J. D., and E. R. Trueman. 1971. Swimming of the scallop *Chlamys opercularis* (L.). *Journal of Experimental Marine Biology and Ecology*, 6: 179–185.

Mora, C., and F. A. Zapata, 2000. Effects of a predatory site-attached fish on abundance and body size of early post-settled reef fishes from Gorgona Island, Colombia. *Proceedings of the 9th International Coral Reef Symposium*, Bali, Indonesia, 1: 23–27.

Mori, A., and S. Yamato. 1993. *Caprella simia* Mayer, 1903 (Crustacea: Amphipoda), collected from the body surface of a frogfish *Antennarius striatus* (Shaw & Nodder, 1794). *Nanki-Seibutu*, 35: 41–46.

Morris, J. A., Jr., C. V. Sullivan, and J. J. Govoni. 2011. Oogenesis and spawn formation in the invasive lionfish, *Pterois miles* and *Pterois volitans*. *Scientia Marina*, 75 (1): 147–154.

Morton, A. 1897. *Antennarius mitchellii* sp. nov. *Papers and Proceedings of the Royal Society of Tasmania*, 1896 (1897): 98.

Moser, H. G. 1996. Scorpaeniformes. Pp. 732–871, In: H. Geoffrey Moser (editor), *Early Stages of Fishes in the California Current Region*, Atlas No. 33, Allen Press, 1505 pp.

Mosher, C. 1954. Observations on the behavior and the early larval development of the Sargassum fish *Histrio histrio* (Linnaeus). *Zoologica*, New York, 39: 141–152.

Motomura, H., and M. Aizawa. 2011. Illustrated list of additions to the ichthyofauna of Yaku-shima Island, Kagoshima Prefecture, southern Japan: 50 new records from the island. *Check List*, 7 (4): 448–628.

Motomura, H., K. Kuriiwa, E. Katayama, H. Senou, G. Ogihara, M. Meguro, M. Matsunuma, Y. Takata, T. Yoshida, M. Yamashita, S. Kimura, H. Endo, A. Murase, Y. Iwatsuki, Y. Sakurai, S. Harazaki, K. Hidaka, H. Izumi, and K. Matsuura. 2010. Annotated checklist of marine and estuarine fishes of Yaku-shima Island, Kagoshima, southern Japan. Pp. 65–247, In: H. Motomura and K. Matsuura (editors), *Fishes of Yaku-shima Island, A World Heritage Island in the Osumi Group, Kagoshima Prefecture, Southern Japan*. National Museum of Nature and Science, Tokyo.

Motta, P. J. 1982. Functional morphology of the head of the inertial suction feeding butterflyfish, *Chaetodon miliaris* (Perciformes, Chaetodontidae). *Journal of Morphology*, 174: 283–312.

Müller, J. W. 1864. *Reisen in der Vereinigten Staaten, Canada und Mexico*. Brockhaus, Leipzig, Vol. 1, xiv + 394 pp.

Muller, M. 1989. A quantitative theory of expected volume changes of the mouth during feeding in teleost fishes. *Journal of Zoology*, London, 217 (4): 639–661.

Muller, M., and J. W. M. Osse. 1984. Hydrodynamics of suction feeding in fish. *Transactions of the Zoological Society of London*, 37: 51–135.

Mundy, B. C. 2005. Checklist of the fishes of the Hawaiian Archipelago. *Bishop Museum Bulletin in Zoology*, 6: 1–703.

Mundy, B. C., R. Wass, E. Demartini, B. Greene, B. Zgliczynski, R. E. Schroeder, and C. Musberger. 2010. Inshore Fishes of Howland Island, Baker Island, Jarvis Island, Palmyra Atoll, and Kingman Reef. *Atoll Research Bulletin*, 585: 1–133.

Munro, I. S. R. 1955. *The Marine and Fresh-water Fishes of Ceylon*. Department of External Affairs, Canberra, xvi + 351 pp.

Munro, I. S. R. 1958. The fishes of the New Guinea region. *Papua New Guinea Agricultural Journal*, 10 (4): 97–369.

Munro, I. S. R. 1964. Additions to the fish fauna of New Guinea. *Papua New Guinea Agricultural Journal*, 16 (4): 141–186.

Munro, I. S. R. 1967. *The Fishes of New Guinea*. Department of Agriculture, Stock and Fisheries, Port Moresby, New Guinea, xxxvii + 650 pp.

Murase, A., A. Angulo, Y. Miyazaki, W. Bussing, and M. López. 2014. Marine and estuarine fish diversity in the inner Gulf of Nicoya, Pacific coast of Costa Rica, Central America. *Check List*, 10 (6): 1401–1413.

Murray, J., and J. Hjort. 1912. *The Depths of the Ocean*. MacMillan, London, xx + 821 pp.

Myers, G. S. 1951. David Starr Jordan, ichthyologist, 1851–1931. *Stanford Ichthyological Bulletin*, 4 (1): 2–6.

Myers, G. S., and C. B. Wade. 1946. New fishes of the families Dactyloscopidae, Microdesmidae, and Antennariidae from the West Coast of Mexico and the Galapagos Islands, with a brief account of the use of rotenone fish poisons in ichthyological collecting. *Allan Hancock Pacific Expedition 1932–1940*, Los Angeles, 96: 151–178.

Myers, R. F. 1989. *Micronesian Reef Fishes, a Practical Guide to the Coral Reef Fishes of Tropical Central and Western Pacific*. Coral Graphics, Barrigada, Territory of Guam, vi + 298 pp.

Myers, R. F. 1999. *Micronesian reef fishes. A Comprehensive Guide to the Coral Reef Fishes of Micronesia*. 3rd revised edition. Coral Graphics, Barrigada, Territory of Guam, vi + 330 pp.

Myers, R. F., and T. J. Donaldson. 2003. The fishes of the Mariana Islands. *Micronesica*, 35–36: 594–648.

Myers, R. F., and J. W. Shepard. 1980. New records of fishes from Guam, with notes on the ichthyofauna of the Southern Marianas. *Micronesica*, 16 (2): 305–347.

Myrberg, A. A., Jr. 1991. Distinctive markings of sharks: ethological considerations of visual function. *Journal of Experimental Zoology*, 256 (supplement 5): 156–166.

Myrberg, A. A., Jr. 1997. Underwater sound: its relevance to behavioral functions among fishes and marine mammals. *Marine and Freshwater Behaviour and Physiology*, 29: 3–21.

Nagelkerken, I., K. B. Huebert, J. E. Serafy, M. G. G. Grol, M. Dorenbosch, and C. J. A. Bradshaw. 2017. Highly localized replenishment of coral reef fish populations near nursery habitats. *Marine Ecology Progress Series*, 568: 137–150.

Nakabo, T. (editor). 2000. *Fishes of Japan with Pictorial Keys to the Species*. Second edition. Tokai University Press, Vol. 1, lvi + 866 pp.

Nakabo, T. (editor). 2002. *Fishes of Japan with Pictorial Keys to the Species*, English edition. Tokai University Press Tokyo, Vol. 1, lxi + 866 pp.

Nakabo, T. 2013. *Fishes of Japan, With Pictorial Keys to the Species*. Third edition. Tokai University Press, Kanagawa, Japan, Vol. 1, xlix + pp. 1–864; Vol. 2, xxxii + pp. 865–1748; Vol. 3, xvi + pp. 1749–2428.

Nakatani, Y., A. Kawakami, and A. Kudo. 2007. Cellular and molecular processes of regeneration, with special emphasis on fish fins. *Development Growth and Differentiation*, 49 (2): 145–154.

Navarro, F. de P. 1943. La pesca de arrastre en los fondos del Cabo Blanco y de Banco Arguin (Africa Sahariana). *Trabaja Instituto Español de Oceanografía*, Madrid, 18: 1–225.

Near, T. J., R. I. Eytan, A. Dornburg, K. L. Kuhn, J. A. Moore, M. P. Davis, P. C. Wainwright, M. Friedman, and W. L. Smith. 2012. Resolution of ray-finned fish phylogeny and timing of diversification. *Proceedings of the National Academy of Sciences*, 109 (34): 13698–13703.

Neill, S. R., and J. M. Cullen. 1974. Experiments on whether schooling by their prey affects the hunting behaviour of cephalopods and fish predators. *Journal of Zoology*, London, 172: 549–569.

Nelson, J. S., T. C. Grande, and M. V. H. Wilson. 2017. *Fishes of the World*, fifth edition. John Wiley & Sons, Inc., Hoboken, New Jersey, 752 pp.

Ng, H. H., H. H. Tan, K. K. P. Lim, W. B. Ludt, and P. Chakrabarty. 2015. Fishes of the eastern Johor Strait. 2015. *Raffles Bulletin of Zoology*, Supplement, 31: 303–337.

Ni, I.-H., and K-Y. Kwok. 1999. Marine Fish Fauna in Hong Kong Waters. *Zoological Studies*, 38 (2): 130–152.

Nichols, J. T. 1912. Notes on West Indian fishes. I. *Antennarius astroscopus*, a new frog-fish from Barbadoes. *Bulletin of the American Museum of Natural History*, 31 (11): 109–110.

Nichols, J. T. 1920. A contribution to the ichthyology of Bermuda. *Proceedings of the Biological Society of Washington*, 33: 59–64.

Nichols, J. T. 1939. A frogfish, *Antennarius tagus* Heller and Snodgrass, from Ecuador. *Copeia*, 1939 (1): 48.

Nichols, J. T., and R. C. Murphy. 1944. A collection of fishes from Panama Bight, Pacific Ocean. *Bulletin of the American Museum of Natural History*, 83 (4): 217–260.

Nijssen, H., L. van Tuijl, and I. J. H. Isbrücker. 1982. A catalogue of the type-specimens of Recent fishes in the Institute of Taxonomic Zoology (Zoölogisch Museum), University of Amsterdam, The Netherlands. *Verslagen en Technische Gegevens, Instituut voor Taxonomische Zoölogie, Universiteit van Amsterdam*, 33: 1–173.

Nilsson, S. 1855. *Skandinavisk Fauna*. Fjerde Delen: Fiskarna. Lund: Gleerup, xxxiv + 768 pp.

Nolf, D. 1972. Deuxième note sur les téléostéens des Sables de Lede (Éocène Belge). *Bulletin de la Société Belge de Géologie, de Paléontologie et d'Hydrologie*, 81 (1–2): 95–109.

Nolf, D. 1985. Otolithi Piscium. *Handbook of Paleoichthyology*, 10: 1–146.

Norman, J. R. 1935. Coast fishes. Part I, The South Atlantic. *Discovery Report*, 12: 1–58.

Norman, J. R. 1939. Fishes. *Scientific Reports of the John Murray Expedition*, London, 7 (1): 1–116.

Norris, J. E., and J. D. Parrish. 1988. Predator-prey relationships among fishes in pristine coral reef communities. *Proceedings of the 6th International Coral Reef Symposium*, Townsville, Australia, 2: 107–113.

Nutting, C. C. 1895. Narrative and preliminary report of the Bahama Expedition. *Bulletin of the Laboratory of Natural History, State University of Iowa*, 3 (1–2): 1–252.

Nybelin, O. 1963. Zur morphologie und terminologie des Schwanzskelettes der Actinopterygier. *Arkiv för Zoologi*, 15 (35): 485–516.

Nyberg, D. W. 1971. Prey capture in the largemouth bass. *American Midland Naturalist*, 86 (1): 128–144.

O'Connell, M. T., A. M. U. O'Connell, and R. W. Hastings. 2009. A meta-analytical comparison of fish assemblages from multiple estuarine regions of southeastern Louisiana using a taxonomic-based method. *Journal of Coastal Research, Special Issue*, 54: 101–112.

Obura, D., G. Stone, S. Mangubhai, S. Bailey, A. Yoshinaga, C. Holloway, and R. Barrel. 2011. Baseline marine biological surveys of the Phoenix Islands, July 2000. *Atoll Research Bulletin*, 589: 1–61.

Ochiai, A. 1964. An aberrant frogfish from Japan. *Bulletin of the Misaki Marine Biology Institute*, 5: 39–41.

Ochiai, A., and F. Mitani. 1956. A revision of the pediculate fishes of the genus *Malthopsis* found in the waters of Japan (family Ogcocephalidae). *Pacific Science*, 10: 271–285.

Ogilby, J. D. 1907. Some new pediculate fishes. *Proceedings of the Royal Society of Queensland*, 20: 17–25.

Ogilby, J. D. 1912. On some Queensland fishes. *Memoirs of the Queensland Museum*, Brisbane, 1: 26–65.

Ogilby, J. D. 1922. Three new Queensland fishes. *Memoirs of the Queensland Museum*, Brisbane, 7: 301–304.

Ohnishi, N., A. Iwata, and W. Hiramatsu. 1997. *Antennatus linearis* (Teleostei: Antennariidae), a new species of frogfish from southern Japan. *Ichthyological Research*, 44 (2): 213–217.

Okada, Y. 1938. *A Catalogue of Vertebrates of Japan*. Maruzen, Tokyo, iv + 412 pp.

Okada, Y. 1955. *Fishes of Japan: Illustrations and Descriptions of Fishes of Japan*. Maruzen, Tokyo, 434 + 28 pp.

Okada, Y., and K. Matsubara. 1938. *Keys to the Fishes and Fish-like Animals of Japan*. Tokyo and Osaka, 584 pp.

Okada, Y., K. Uchida, and K. Matsubara. 1935. Color Atlas of Fishes of Japan. Sanseido Co. Ltd., Tokyo, 5 + 425 + 46 pp., 166 pls.

Okamura, O., and K. Amaoka. 1997. *Sea Fishes of Japan*. Yama-Kei Publishing Company, Tokyo, Japan, 784 pp. [in Japanese].

Okiyama, M. (editor). 1988. *An Atlas of Early Stage Fishes in Japan*. Tokai University Press, Tokyo, Japan, xii + 1154 pp. [In Japanese; English translation of selected pages, 1989, https://escholarship.org/uc/item/9284j82s.]

Okiyama, M. (editor). 2014. *An Atlas of Early Stage Fishes in Japan*. Second edition, Tokai University Press, Kanagawa, Japan, Vol. 2, pp. i–xiii + 977–1639.

Olivar, M.-P., and J.-M. Fortuño. 1991. Guide to the ichthyoplankton of the southeast Atlantic (Benguela Current Region). *Scientia Marina*, 55 (1): 1–383.

Oliver, S. P. 1909. *The Life of Philibert Commerson D.M., Naturaliste du Roi: An Old-world Story of French Travel and Science in the Days of Linnaeus*. G. F. S. Elliot (editor). Murray, London, xvii + 242 pp.

Ormond, R. F. G. 1980. Aggressive mimicry and other interspecific feeding associations among Red Sea predators. *Journal of Zoology*, London, 191 (2): 247–262.

Ortiz, M., and R. Lalana. 2005. *Marine Biodiversity of the Cuban Archipelago: An Overview*. Center for Marine Research, University of Havana, Cuba, 20 pp.

Ortiz-Ramirez, F. A, L. M. Mejia-Ladino, and A. P. Arturo. 2005. Description of eggs and early larval stages of the frogfish *Antennarius striatus* (Shaw, 1794) in captivity, with notes on its mechanism of reproduction. *Revista de Biología Marina y Oceanografía*, 40 (1): 23–31.

Orton, G. L. 1955. Early developmental stages of the California scorpionfish, *Scorpaena guttata*. *Copeia*, 1955 (3): 210–214.

Osbeck, P. 1757. *Dagbok öfwer en Ostindisk resa åren 1750, 1751, 1752. Med anmårkningar uti naturkunnigheten främmande folkslags språk, seder, hushållning, m. m. På fleras åstundan utgifwen af Pehr Osbeck . . . Jåmte 12 tabeller och afledne skepps-predikanten Toréns bref*. Grefing, Stockholm, 376 pp.

Osbeck, P. 1765. *Reise nach Ostindien und China, nebst O. Toreens Reise nach Suratte und C. G. Ekebergs Nachricht von der Landwirtschaft der Chineser*. Koppe, Rostock, xxiv + 552 pp.

Osório, B. 1891. Estudos ichthyologicos ácerca da fauna dos domínios portuguezes na Africa. 3ª Nota.—Peixes maritimos das ilhas de S. Thomé, do Princípe e ilho das Rolas. *Jornal do Sciências Mathemáticas, Physicas e Naturaes*, Série 2, 2 (6): 97–139.

Osório, B. 1894. Estudos ichthyologicos acerca da fauna dos dominios portuguezes da Africa. 3a nota: Peixes maritimos da Ilhas de S. Thomé, do Principe e Ilheo das Rolas. *Jornal do Sciências Mathemáticas, Physicas e Naturaes*, Série 2, 3: 136–140, 173–183.

Osório, B. 1898. Da distribuicão geographica dos peixes e crustaceos colhidos nas possessões portuguezas d'Africa occidental e existentes no Museu Nacional de Lisboa. *Jornal do Sciências Mathemáticas, Physicas e Naturaes*, Série 2, 5 (19): 185–202.

Osório, B. 1909. Peixes colhidos nas vishinhancas do Archipelago de Cabo Verde. *Memorias do Museu Bocage*, Lisboa, 1 (2): 51–77.

Osse, J. W. M. 1969. Functional morphology of the head of the perch (*Perca fluviatilis* L.): An electromyographic study. *Netherlands Journal of Zoology*, 19 (3): 289–392.

Owen, R. 1866. *On the Anatomy of Vertebrates, Volume I: Fishes and Reptiles*. Longmans, Green, and Co., London, xxxvi + 650 pp.

Padmanabhan, K. G. 1958. Early stages in the development of the toad fish, *Antennarius marmoratus* Bleeker. *Bulletin of the Central Research Institute, University of Travancore, Natural Sciences*, 5C(1): 85–92.

Padoa, E. 1956. Triglidae, Peristediidae, Dactylopteridae, Gobiidae, Echneidae, Jugulares, Gobiesocidae, Heterosomata, Pediculati. Pp. 627–888, In: *Uova, larve, e studi giovanili di Teleostei*. Fauna Flora Golfo Napoli, 38 [in Italian].

Page, L. M., H. Espinosa-Pérez, L. T. Findley, C. R. Gilbert, R. N. Lea, N. E. Mandrak, R. L. Mayden, and J. S. Nelson. 2013. *Common and Scientific Names of Fishes from the United States, Canada, and Mexico, 7th Edition*. American Fisheries Society, Special Publication, 34, 243 pp.

Palacios-Salgado, D. S., A. Ramírez-Valdez, A. A. Rojas-Herrera, J. G. Amores, and M. A. Melo-García. 2014. Marine fishes of Acapulco, Mexico (Eastern Pacific Ocean). *Marine Biodiversity*, 44 (4): 471–490.

Palmer, G. 1960. The first record of a frogfish (*Antennarius*) from Irish waters. *Annals and Magazine of Natural History*, Series 13, 3: 149–151.

Palmer, G. 1961. New records of fishes from the Monte Bello Islands, Western Australia. *Annals and Magazine of Natural History*, Series 13, 4: 545–551.

Palmer, G. 1970. New records, and one new species, of teleost fishes from the Gilbert Islands. *Bulletin of the British Museum of Natural History, Zoology*, 19: 231–234.

Palmtag, M. R. 2017. The marine ornamental species trade. Pp. 3–14, In: R. Calado, I. Olivotto, M. P. Oliver, and G. J. Holt (editors), *Marine Ornamental Species Aquaculture,* John Wiley & Sons Ltd, Chichester, West Sussex, United Kingdom.

Parenti, P., and T. W. Pietsch. 2003. *Ostracion knorrii* Walbaum, 1792, a senior synonym of the striated frogfish *Antennarius striatus* (Shaw and Nodder, 1794) invalidated by "reversal of precedence." *Copeia*, 2003 (1): 187–189.

Parin, N. V. 1968. *Ikhtiofauna Okeanskoi Epipelagiali*. Trudy Institute of Okeanology [English translation, I.P.S.T., Jerusalem, Ichthyofauna of the epipelagic zone. Institute of Oceanology, Academy of Sciences of the U.S.S.R., 206 pp.]

Parra, A. 1787. *Descripcion de diferentes piezas de historia natural, las mas del ramo maritimo, representadas en setenta y cinco laminas.* Published by the author, Havana, 195 pp.

Parrish, J. D., J. E. Norris, M. W. Callahan, J. K. Callahan, E. J. Magarifuji, and R. E. Schroeder. 1986. Piscivory in a coral reef fish community. Pp. 285–298, In: C. A. Simenstad and G. M. Cailliet (editors), *Developments in Environmental Biology of Fishes,* Vol. 7, Contemporary studies on fish feeding: the proceedings of GUTSHOP '84, Papers from the fourth workshop on fish food habits held at the Asilomar Conference Center, Pacific Grove, California, U.S.A., December 2–6, 1984.

Parsons, J. 1749. Some account of the *Rana Piscatrix. Philosophical Transactions of the Royal Society,* London, 46: 126–131.

Pasteur, G. 1982. A classificatory review of mimicry systems. *Annual Reviews of Ecology and Systematics,* 13: 169–199.

Patterson, C. and Rosen, D. E. 1989. The Paracanthopterygii revisited: order and disorder. Pp. 5–36, In: D. M. Cohen (editor), *Papers on the Systematics of Gadiform Fishes,* Natural History Museum of the Los Angeles County, 32.

Paugy, D. 1992. Antennariidae. Pp. 569–574, In: C. Lévêque, D. Paugy, and G. G. Teugels (editors), *Faune des poissons d'eaux douces et saumâtres de l'Afrique de l'Ouest.* Musée Royal de l'Afrique Centrale Tervuren, Belgique, Vol. 2.

Paulin, C. D. 1978. New records of anglerfishes (Antennariidae) from New Zealand. *New Zealand Journal of Zoology,* 5: 485–491.

Paulin, C. D., and C. D. Roberts. 1992. *The Rockpool Fishes of New Zealand. Te ika aaria o Aotearoa.* Museum of New Zealand *Te Papa Tongarewa,* Wellington, xii + 177 pp.

Paulin, C. D., A. Stewart, C. D. Roberts, and P. J. McMillan, 1989. New Zealand fish, a complete guide. *National Museum of New Zealand, Miscellaneous Series,* 19, xiv + 279 pp.

Paxton, J. R., D. F. Hoese, G. R. Allen, and J. E. Hanley. 1989. *Zoological Catalogue of Australia. Pisces, Petromyzontidae to Carangidae.* Australian Government Publishing Service, Canberra, Vol. 7, xii + 665 pp.

Pearcy, W. G. 1962. Egg masses and early developmental stages of he scorpaenid fish, *Sebastolobus. Journal of the Fisheries Research Board of Canada,* 19 (6): 1169–1173.

Pellegrin, J. 1914. Mission Gruvel sur la côte occidentale d'Afrique (1905–1912). Poissons. *Annales de l'Institut océanographique,* Monaco, 6: 1–99.

Pennant, T. 1776. *British Zoology,* 4th edition. White, London, Vol. 3, 425 pp.

Pequeño, G. 1989. Peces de Chile. Lista sistemática revisada y comentada. *Revista de Biologia Marina,* Valparaiso, 24 (2): 1–132.

Pequeño, G., and J. Lamilla. 2000. The littoral fish assemblage of the Desventuradas Islands (Chile) has zoogeographical affinities with the western Pacific. *Global Ecology and Biogeography Letters,* 9: 431–437.

Pereira, M. A. M. 2000. *Preliminary Checklist of Reef-associated Fishes of Mozambique.* Ministry for the Coordination of Environmental Affairs, Mozambique Coral Reef Management Programme, Maputo, Mozambique, iv + 21 pp.

Pereira, M. A. M., E. J. S. Videira, and K. G. S. Abrantes. 2004. Peixes associados a recifes e zonas litorais do sul de Moçambique. *Jornal de Investigação e Advocacia Ambiental,* 1 (1): 1–7.

Pereira-Guimarães, A. R. 1884. Lista dos peixes da Ilha da Madeira, Acores e das possessões d'Africa, que existun no Museu de Lisboa. *Jornal de Sciencias,* Lisboa, 37: 11–28.

Peters, W. 1877. Übersicht der während der von 1874 bis 1876 unter dem Commando des Hrn. Kapitän 2. S. Freiherrn von Schleinitz ausgeführten Reise S.M.S. "Gazelle" gesammelten und von der Kaiserlichen Admiralität der Königlichen Akademie der Wissenschaften übersandten Fische. *Monatsberichte der Akademic der Wissenschaft zu Berlin,* 1876 (1877): 831–854.

Petiver, J. 1702. *Gazophylacii Nature and Arts.* Decas secunda. London, pls. 11–20.

Pet-Soede, L., and M. Erdmann. 1998. An overview and comparison of destructive fishing practices in Indonesia. *SPC Live Reef Fish Information Bulletin,* 4: 28–36.

Phillipps, W. J. 1927. A check list of the fishes of New Zealand. *Journal of the Pan-Pacific Research Institute,* Honolulu, 2 (1): 9–16.

Phillips, J. B. 1957. A review of the rockfishes of California (family Scorpaenidae). *California Department of Fish and Game, Fish Bulletin*, 104, 158 pp.

Phillips, N. 1995. Biogeography of *Sargassum* (Phaeophyta) in the Pacific basin. Pp. 107–144, In: I. A. Abbott (editor), *Taxonomy of Economic Seaweeds with Reference to Some Pacific Species Vol. V.* California Sea Grant College, La Jolla, California.

Phisalix, M. 1922. *Animaux Venimeux et Venins.* Masson, Paris, Vol. 1, xxv + 656 pp.

Pianka, E. R. 1970. "On r and K selection." *American Naturalist*, 104 (940): 592–597.

Pianka, E. R. 1978. *Evolutionary Ecology*, 2nd edition. Harper & Row, New York, xii + 397 pp.

Pielou, E. C. 1979. *Biogeography.* Wiley, New York, ix + 351 pp.

Pietsch, T. W. 1972. A review of the monotypic deep-sea anglerfish family Centrophrynidae: Taxonomy, distribution and osteology. *Copeia*, 1972 (1): 17–47.

Pietsch, T. W. 1974. Osteology and relationships of ceratioid anglerfishes of the family Oneirodidae, with a review of the genus *Oneirodes* Lütken. *Natural History Museum of Los Angeles County, Science Bulletin*, 18: 1–113.

Pietsch, T. W. 1978a. Antennariidae. In: W. Fischer (editor), *FAO Species Identification Sheets for Fishery Purposes. Western Central Atlantic (fishing area 31)*, Vol. 1, unpaged.

Pietsch, T. W. 1978b. The feeding mechanism of *Stylephorus chordatus* (Teleostei: Lampridiformes): functional and ecological implications. *Copeia*, 1978 (2): 255–262.

Pietsch, T. W. 1979. Systematics and distribution of ceratioid anglerfishes of the family Caulophrynidae with the description of a new genus and species from the Banda Sea. *Natural History Museum of Los Angeles County, Contrutions in Science*, 310: 1–25.

Pietsch, T. W. 1981a. Antennariidae. In: W. Fischer, G. Bianchi, and W. B. Scott (editors), *FAO Species Identification Sheets for Fishery Purposes. Eastern Central Atlantic (fishing areas 34, 47 in part)*. Ottawa: Department of Fisheries and Oceans, Vol. 1, unpaged.

Pietsch, T. W. 1981b. The osteology and relationships of the anglerfish genus *Tetrabrachium*, with comments on lophiiform classification. *United States Fishery Bulletin*, 79 (3): 387–419.

Pietsch, T. W. 1984a. Louis Renard's fanciful fishes. A collection of grossly inaccurate scientific illustrations has some merit after all. *Natural History*, 93 (1): 58–67.

Pietsch, T. W. 1984b. The genera of frogfishes (Family Antennariidae). *Copeia*, 1984 (1): 27–44.

Pietsch, T. W. 1984c. A review of the frogfish genus *Rhycherus* with the description of a new species from Western and South Australia. *Copeia*, 1984 (1): 68–72.

Pietsch, T. W. 1984d. Lophiiformes: Development and relationships. Pp. 320–325, In: H. G. Moser, W. J. Richards, D. M. Cohen, M. P. Fahay, A. W. Kendall, Jr., and S. L. Richardson (editors), *Ontogeny and Systematics of Fishes.* American Society of Ichthyologists and Herpetologists, Special Publication No. l, x + 760 pp.

Pietsch, T. W. 1984e. Antennariidae. 2 pp., unpaged, In: *FAO Species Identification Sheets for Fishery Purposes. Western Indian Ocean (Fishing Area 51).* Food and Agriculture Organization of the United Nations, Rome, Vol. I.

Pietsch, T. W. 1985a. The manuscript materials for the *Histoire Naturelle des Poissons*, 1828–1849: Sources for understanding the fishes described by Cuvier and Valenciennes. *Archives of Natural History*, 12 (1): 59–106.

Pietsch, T. W. 1985b. The functional morphology of the feeding mechanism of shallow water anglerfishes of the family Antennariidae. *National Geographic Society, Research Reports*, 18: 593–600.

Pietsch, T. W. 1986a. Family no. 102: Antennariidae. Pp. 366–369, In: M. M. Smith and P. C. Heemstra (editors), *Smiths' Sea Fishes*, Macmillan South Africa Ltd., Johannesburg, South Africa.

Pietsch, T. W. 1986b. Family Antennariidae (including Pterophrynidae). Pp. 1364–1368, In: P. J. P. Whitehead, M.-L. Bauchot, J.-C. Hureau, J. Nielsen, and E. Tortonese, *Fishes of the Northeastern Atlantic and Mediterranean* (FNAM), Vol. 3, UNESCO.

Pietsch, T. W. 1986c. The original manuscript sources for the *Histoire Naturelle des Poissons*, 1828–1849: Keys to understanding the fishes described by Cuvier and Valenciennes. *Copeia*, 1986 (1): 216–219.

Pietsch, T. W. 1990. Antennariidae. Pp. 481–488, In: *Check-list of the Fishes of the Eastern Tropical Atlantic* (CLOFETA), UNESCO, Vol 1.

Pietsch, T. W. 1994. Family Antennariidae. Pp. 285–297, In: M. F. Gomon, D. J. M. Glover, and R. H. Kuiter (editors), *The Fishes of Australia's South Coast*, State Printer, Adelaide, Australia.

Pietsch, T. W. (editor). 1995a. *Historical Portrait of the Progress of Ichthyology, from Its Origins to Our Own Time*. Edited and annotated by T. W. Pietsch, translated from the French by A. J. Simpson. Johns Hopkins University Press, Baltimore, xxiv + 366 pp., 67 figures.

Pietsch, T. W. 1995b. *Fishes, Crayfishes, and Crabs: Louis Renard and His Natural History of the Rarest Curiosities of the Seas of the Indies*. Johns Hopkins University Press, Baltimore, Vol. 1, Commentary, xxii + 214 pp.; Vol. 2, Facsimile, 224 pp., 100 color pls.

Pietsch, T. W. 1999. Antennariidae: frogfishes (also sea mice, anglerfishes). Pp. 2013–2015, In: *FAO Species Identification Sheets for Fishery Purposes. Western Central Pacific (Fishing Area 71 and the southwestern part of Area 77)*. Food and Agriculture Organization of the United Nations, Rome. Vol. 3, Batoid fishes, Chimaeras, and Bony Fishes Part 1 (Elopidae to Linophrynidae).

Pietsch, T. W. 2000. Family Antennariidae (frogfishes). P. 597, In: J. E. Randall and K. K. P. Lim (editors), A checklist of the fishes of the South China Sea, *Raffles Bulletin of Zoology*, Supplement, 8: 569–667.

Pietsch, T. W. 2001. Charles Plumier (1646–1704) and his drawings of French and American fishes. *Archives of Natural History*, 28 (1): 1–57.

Pietsch, T. W. 2002a. Antennariidae, Frogfishes (sea mice, anglerfishes. Pp. 1050–1051, In: *FAO Species Identification Guide for Fishery Purposes. The Living Marine Resources of the Western Central Atlantic (Fishing Area 31)*. Food and Agriculture Organization of the United Nations, Rome. Vol. 2, Bony fishes, pt. 1 (Acipenseridae to Grammatidae).

Pietsch, T. W. 2002b. Frogfishes, family Antennariidae. Pp. 270–272, In: B. B. Collette and G. Klein-MacPhee (editors), *Bigelow and Schroeder's Fishes of the Gulf of Maine*, 3rd edition, Smithsonian Institution Press, Washington and London.

Pietsch, T. W. 2004a. A new species of the anglerfish genus *Lophiocharon* Whitley (Lophiiformes: Antennariidae) from Australian waters. *Records of the Australian Museum*, 56 (2): 159–162.

Pietsch, T. W. 2004b. Codfishes, anglerfishes, and allies, pp. 228–237, In: *The New Encyclopedia of Aquatic Animals*, Vol. 2, Fishes and Aquatic Mammals, A. Campbell and J. Dawes (editors), Facts On File, Inc., New York.

Pietsch, T. W. 2005. Dimorphism, parasitism, and sex revisited: Modes of reproduction among deep-sea ceratioid anglerfishes (Teleostei: Lophiiformes). *Ichthyological Research*, 52 (3): 207–236.

Pietsch, T. W. 2008. Family Antennariidae. Pp. 364–374, In: M. F. Gomon, D. Bray, and R. H. Kuiter (editors), *Fishes of Australia's Southern Coast*, Revised Edition, Reed New Holland, an imprint of New Holland Publishers, Sydney, Australia.

Pietsch, T. W. 2009. *Oceanic Anglerfishes: Extraordinary Diversity in the Deep-sea*. University of California Press, Berkeley and Los Angeles, xii + 557 pp., 310 figs.

Pietsch, T. W. 2011. Plumier's passion: foregoing the smallest of worldly pleasures, a seventeenth-century French monk compiled a meticulous record of plants and animals. His illustrations of fishes are just one of his neglected legacies. *Natural History*, 119 (7): 30–36.

Pietsch, T. W. 2016. Antennariidae, Frogfishes. Pp. 2051–2053, In: *FAO Species Identification Guide for Fishery Purposes. The Living Marine Resources of the Eastern Central Atlantic (Fishing Area 31)*. Food and Agriculture Organization of the United Nations, Rome. Vol. 3, Bony fishes, pt. 1 (Elopiformes to Scorpaeniformes).

Pietsch, T. W. 2017. *Charles Plumier (1646–1704) and His Drawings of French and Caribbean Fishes [Charles Plumier (1646–1704) et Ses Dessins de Poissons de France et des Antilles]*. Publications Scientifiques du Muséum, Muséum national d'Histoire naturelle, Paris, 408 pp., 46 figs., 121 pls.

Pietsch, T. W., and R. J. Arnold. 2017. The "Lembeh Frogfish" identified: redescription of *Nudiantennarius subteres* (Smith and Radcliffe, in Radcliffe, 1912) (Teleostei: Lophiiformes: Antennariidae). *Copeia*, 2017 (4): 659–665, cover photo.

Pietsch, T. W., R. J. Arnold, and D. J. Hall. 2009a. A bizarre new species of frogfish of the genus *Histiophryne* (Lophiiformes: Antennariidae) from Ambon and Bali, Indonesia. *Copeia*, 2009 (1): 37–45.

Pietsch, T. W., M. L. Bauchot, and M. Desoutter. 1986. Catalogue critique des types de Poissons de Muséum national d'Histoire naturelle. (Suite) Ordre des Lophiiformes. *Bulletin du Muséum national d'Histoire naturelle*, Paris, Série 4, Sec. A, 8: 131–156.

Pietsch, T. W., and G. Carnevale. 2011. †*Sharfia mirabilis,* A New Genus and Species of Anglerfish (Teleostei: Lophiiformes: Lophiidae) from the Eocene of Monte Bolca, Italy. *Copeia*, 2011 (1): 64–71.

Pietsch, T. W., J. H. Caruso, C. R. Fisher, S. W. Ross, and M. G. Saunders. 2013. *In-situ* observations of the deep-sea goosefish *Sladenia shaefersi* Caruso and Bullis (Lophiiformes: Lophiidae), with evidence of extreme sexual dimorphism. *Copeia*, 2013 (4): 660–665.

Pietsch, T. W., and D. B. Grobecker. 1978. The compleat angler: Aggressive mimicry in an antennariid anglerfish. *Science*, 201: 369–370.

Pietsch, T. W., and D. B. Grobecker. 1980. Parental care as an alternative reproductive mode in an antennariid anglerfish. *Copeia*, 1980 (3): 551–553.

Pietsch, T. W., and D. B. Grobecker. 1981. [Antennariid anglerfishes: Aggressive mimics of the reef]. *Anima*, 105 (12): 11–15 (in Japanese).

Pietsch, T. W., and D. B. Grobecker. 1985. Frogfishes: Aggressive mimics of the reef. *Freshwater and Marine Aquarium*, 8 (4): 10–16.

Pietsch, T. W., and D. B. Grobecker. 1987. *Frogfishes of the World: Systematics, Zoogeography, and Behavioral Ecology*. Stanford University Press, Stanford, California, xxii + 420 pp.

Pietsch, T. W., and D. B. Grobecker. 1990a. Frogfishes: masters of aggressive mimicry, these voracious carnivores can gulp prey faster than any other vertebrate predator. *Scientific American*, 262 (6): 96–103, June 1990.

Pietsch, T. W., and D. B. Grobecker. 1990b. Fühlerfische—getarnte Angler: Diese Meister der Angriffsmimikry saugen ihr ahnungsloses Opfer schneller ins Maul, als irgendein anderes räuberisches Wirbeltier zuzuschnappen vermag. Schwimmen können sie mit Düsenantrieb, aber mit den Flossen auch klettern, schreiten und galoppieren. *Spektrum der Wissenschaft*, pp. 74–82, August 1990.

Pietsch, T. W., D. B. Grobecker, and B. Stockley. 1992. The sargassum frogfish, *Histrio histrio* (Linnaeus) (Lophiiformes: Antennariidae), on the Pacific Plate. *Copeia*, 1992 (1): 247–248.

Pietsch, T. W., J. Johnson, and R. J. Arnold. 2009b. A new genus and species of the shallow-water anglerfish family Tetrabrachiidae (Teleostei: Lophiiformes: Antennarioidei) from Australia and Indonesia. *Copeia*, 2009 (3): 483–493.

Pietsch, T. W., and R. H. Kuiter. 1984. A new species of frogfish of the genus *Echinophryne* (Family Antennariidae) from southern Australia. *Revue Française d'Aquariologie et Herpetologie*, Nancy, 11 (1): 23–26.

Pietsch, T. W., and Lavenberg R. J. 1980. A fossil ceratioid anglerfish from the Late Miocene of California. *Copeia*, 1980 (4): 906–908.

Pietsch, T. W., and J. W. Orr. 2007. Phylogenetic relationships of deep-sea anglerfishes of the suborder Ceratioidei (Teleostei: Lophiiformes) based on morphology. *Copeia*, 2007 (1): 1–34.

Pietsch, T. W., and Shimazaki, M. 2005. Revision of the deep-sea anglerfish genus *Acentrophryne* Regan (Lophiiformes: Ceratioidei: Linophrynidae), with the description of a new species from off Peru. *Copeia*, 2005: 246–251.

Pietschmann, V. 1909. Über zwei stark variante Exemplare von *Antennarius tridens* (Schlegel) mit Bemerkungen über die Variabilität von *Antennarius*. *Annalen des Naturhistorischen Hofmuseums*, 23 (1): 1–5.

Pietschmann, V. 1930. Remarks on Pacific fishes. *Bulletin of the Bernice P. Bishop Museum*, 73: 3–24.

Playfair, R. L. 1866. *The Fishes of Zanzibar. Acanthopterygii*. Van Voorst, London, xiv + 153 pp.

Playfair, R. L. 1867. The fishes of Seychelles. *Proceedings of the Zoological Society of London*, 1867 (6): 846–872.

Playfair, R. L. 1869. Further contributions to the ichthyology of Zanzibar. *Proceedings of the Zoological Society of London*, 1869 (5): 239–241.

Plumier, C. MS 25. *Poissons d'Amérique, dessinés par le Père Plumier*. –90 feuillets, presque tous avec des dessins, en partie coloriés. Bibliothèque Centrale du Muséum national d'Histoire naturelle, Paris.

Poey y Aloy, F. 1853. XVI. Quironectos cubanos. Género de peces llamados vulgarmente pescadores. *Memorias sobre la historia natural de la Isla de Cuba*, Habana, 1 (16): 214–221.

Poey y Aloy, F. 1861. L. Conspectus Piscium Cubensium. *Memorias sobre la historia natural de la Isla de Cuba*, Habana, 2 (3): 357–404.

Poey y Aloy, F. 1865. Synopsis Piscium Cubensium. Pp. 1–412, In: *Repertorio físico-natural de la Isla de Cuba*, Habana, Vol. 1.

Poey y Aloy, F. 1868. Synopsis Piscium Cubensium. Pp. 279–484, In: *Repertorio físico-natural de la Isla de Cuba*, Habana, Vol. 2.

Poey y Aloy, F. 1876. Enumeratio Piscium Cubensium. Part 2. *Anales de la Sociedad Española de Historia Natural*, 4: 89–224.

Poey y Aloy, F. 1881. Peces. Pp. 317–350, In: D. J. Gundlach, Apuntes para la fauna Puerto-Riqueña. Part 3. *Anales de la Sociedad Española de Historia Natural*, 10 (2).

Poll, M. 1949. Résultats scientifiques des croisières du Navire-École Belge "Mercator." Vol. 4, Poissons des XIe, XIVe et XVIIe croisières. *Mémoires de l'Institut Royal des Sciences Naturelles de Belgique*, Série 2, 33: 173–269.

Poll, M. 1959. Résultats Scientifiques Expédition Oceanographique Belge dans les eaux côtières Africaines de l'Atlantique Sud, 1948–1949. Poissons IV. Téléostéens Acanthoptérygiens (Part 2). *Mémoires de l'Institut Royal des Sciences Naturelles de Belgique*, 4 (3B): 1–417.

Pontoppidan, E. 1755. *The Natural History of Norway, containing a particular and accurate account of the temperature of the air, the different soils, waters, vegetables, metals, minerals, stones, beasts, birds, and fishes; together with the dispositions, customs, and manner of living of the inhabitants; interspersed with physiological notes from eminent writers, and transactions of academics, in two parts*. A. Linde, London, pt. 1, xxiv + 206 pp.; pt. 2, viii + 303 pp.

Porteiro, F. M., and P. Afonso. 2007. The singlespot frogfish *Antennarius radiosus* (Lophiiformes, Antennariidae), a valid member of the ichthyofauna of the Azores. *Life and Marine Sciences*, 24A: 57–60.

Potter, B. 1906. *The Tale of Mr. Jeremy Fisher*. Frederick Warne & Co, London and New York, 85 pp.

Poulton, E. B. 1890. *The Colours of Animals, Their Meaning and Use, Especially Considered in the Case of Insects*. Kegan Paul, Trench, Trübner, & Co. Ltd., London, xvi + 360 pp.

Prakash, S., J. Balamurugan, T. T. A. Kumar, and T. Balasubramanian. 2012. Invasion and abundance of reef inhabited fishes in the Vellar estuary, southeast coast of India, especially the lionfish *Pterois volitans* Linnaeus. *Current Science*, 103 (8): 941–944.

Prihadi, D. J. 2015. Keberadaan ikan kodok (*Antennarius maculates*, Desjardins 1840) di Pulau Nusa Penida Provinsi Bali [The existence of frogfish (*Antennarius maculates*, Desjardins 1840) on the island of Nusa Penida, Bali Province]. *Jurnal Akuatika*, 6 (2): 187–197.

Prince, E. E. 1891. Notes on the development of the anglerfish *Lophius piscatorius*. *Annual Report of the Fishery Board for Scotland*, 9: 343–348.

Procter, W., H. C. Tracy, E. Helwig, C. H. Blake, J. E. Morrison, and S. Cohen. 1928. Fishes: A contribution to the life history of the angler (*Lophius piscatorius*). Pp. 1–29, In: *Biological Survey of the Mount Desert Region, Part 2*. Wistar Institute of Anatomy and Biology, Philadelphia.

Prokofiev, A. M. 2014. New species and new records of deepsea anglerfish of the family Oneirodidae. *Voprosy Ikhtiologii*, 54 (5): 611–616 [In Russian, English translation in *Journal of Ichthyology*, 54 (8): 602–607].

Psomadakis, P. N., H. B. Osmany, and M. Moazzam. 2015. *Field Identification Guide to the Living Marine Resources of Pakistan*. FAO Species Identification Guide for Fishery Purposes. Food and Agriculture Organization of the United Nations, and Marine Fisheries Department, Ministry of Ports and Shipping, Government of Pakistan, Rome, x + 386 pp.

Puentes, V., N. Madrid, and L. A. Zapata. 2007. Catch composition of the deep sea shrimp fishery (*Solenocera agassizi* Faxon, 1893; *Farfantepenaeus californiensis* Holmes, 1900, and *Farfantepenaeus brevirostris* Kingsley, 1878) in the Colombian Pacific Ocean. *Gayana*, 71 (1): 84–95.

Purdy, R. W., V. P. Schneider, S. P. Applegate, J. H. McLellan, R. L. Meyer, and B. H. Slaughter. 2001. The Neogene shark, ray, and bony fishes from Lee Creek Mine, Aurora, North Carolina. *Smithsonian Contributions to Paleobiology*, 90: 71–202.

Quoy, J. R. C., and P. Gaimard. 1825. Zoologie, Poissons. Pp. 183–401, In: L. de Freycinet, *Voyage autour du monde, entrepris par ordre du roi, sous le ministère et conformément aux instructions de S. Exc. M. le Vicomte du Bouchage, secrétaire d'état au Département de la Marine, exécuté sur les corvettes de S. M. l'Uranie et la Physicienne, pendant les années 1817, 1818, 1819 et 1820.* Imprimerie Royale, Paris, 712 pp.

Rackham, H. 1933. *Cicero, De Natura Deorum Academica*, with an English translation by H. Rackham. Harvard University Press, Cambridge, Massachusetts, xxiv + 664 pp.

Rackham, H. 1956. *Pliny, Natural History*, with an English translation by H. Rackham. Harvard University Press, Cambridge, Massachusetts, Vol. 3, 624 pp.

Radcliffe, L. 1912. Scientific results of the Philippine Cruise of the Fisheries Steamer "Albatross," 1907–1910. No. 16. New pediculate fishes from the Philippine Islands and contiguous water. *Proceedings of the United States National Museum*, 42: 199–214.

Rafinesque, C. S. 1814. *Précis des d'ecouvertes et travaux somiologiques de Mr. C. S. Rafinesque-Schmaltz entre 1800 et 1814; ou choix raisonné de ses principales d'ecouvertes en zoologie et en botanique pour servir d'introduction à ses ouvrages futurs.* Published by the author, Palermo, Italy, 55 pp.

Rafinesque, C. S. 1815. *Analyse de la nature, ou tableau de l'univers et des corps organisés.* Published by the author, Palermo, Italy, 224 pp.

Ramaiah, N., and D. Chandramohan. 1992. Occurrence of *Photobacterium leiognathi*, as the bait organ symbiont in frogfish, *Antennarius hispidus*. *Indian Journal of Marine Science*, 21: 210–211.

Randall, J. E. 1961. Overgrazing of algae by herbivorous marine fishes. *Ecology*, 42 (4): 812.

Randall, J. E. 1965. Grazing effects on sea grasses by herbivorous reef fishes in the West Indies. *Ecology*, 46 (3): 255–260.

Randall, J. E. 1967. Food habits of reef fishes of the West Indies. *Studies in Tropical Oceanography*, Miami, 5: 665–847.

Randall, J. E. 1968. *Caribbean Reef Fishes.* T. F. H. Publications, Hong Kong, 318 pp.

Randall, J. E. 1986. 106 new records of fishes from the Marshall Islands. *Bulletin of Marine Science*, 38 (1): 170–252.

Randall, J. E. 1995. *Coastal Fishes of Oman.* University of Hawaii Press, Honolulu, i–xvi + 1–439.

Randall, J. E. 1996. *Shore Fishes of Hawaii.* Natural World Press, Vida, Oregon, 216 pp.

Randall, J. E. 1998. Zoogeography of shore fishes of the Indo-Pacific region. *Zoological Studies*, 37 (4): 227–268.

Randall, J. E. 1999. Report on fish collections from the Pitcairn Islands. *Atoll Research Bulletin*, 461: 1–36 + 14 unnumbered pp.

Randall, J. E. 2005a. *Reef and Shore Fishes of the South Pacific. New Caledonia to Tahiti and the Pitcairn Islands.* University of Hawaii Press, Honolulu, xii + 707 pp.

Randall, J. E. 2005b. A review of mimicry in marine fishes. *Zoological Studies*, 44 (3): 299–328.

Randall, J. E. 2007. *Reef and Shore Fishes of the Hawaiian Islands.* Sea Grant College Program, University of Hawaii, Honolulu, xiv + 546 pp.

Randall, J. E. 2010. *Shorefishes of Hawaii*, revised edition. University of Hawaii Press, Honolulu, vi + 234 pp.

Randall, J. E., G. R. Allen, and R. C. Steene. 1990. *Fishes of the Great Barrier Reef and Coral Sea.* University of Hawaii Press, Honolulu, Hawaii, xx + 507 pp.

Randall, J. E., G. R. Allen, and R. C. Steene. 1997a. *Fishes of the Great Barrier Reef and Coral Sea.* Second edition, revised and expanded. University of Hawaii Press, Honolulu, and Crawford House Press, Bathurst, New South Wales, xx + 557 pp.

Randall, J. E., P. Bacchet, R. Winterbottom, and L. Wrobel. 2002. Fifty new records of shore fishes from the Society Islands and Tuamotu Archipelago. *Aqua, Journal of Ichthyology and Aquatic Biology*, 5 (4): 153–166.

Randall, J. E., and J. L. Earle. 2000. Annotated checklist of the shore fishes of the Marquesas Islands. *Occasional Papers of the Bernice P. Bishop Museum of Polynesian Ethnology and Natural History*, 66: 1–39.

Randall, J. E., J. L. Earle, T. Hayes, C. Pittman, M. Severns, and R. J. F. Smith. 1993a. Eleven new records and validations of shore fishes from the Hawaiian Islands. *Pacific Science*, 47 (3): 222–239.

Randall, J. E., J. L Earle, R. L. Pyle, J. D. Parrish, and T. Hayes. 1993b. Annotated checklist of the fishes of Midway Atoll, Northwestern Hawaiian Islands. *Pacific Science,* 47 (4): 356–400.

Randall, J. E., and W. D. Hartman. 1968. Sponge-feeding fishes of the West Indies. *Marine Biology*, 1 (3): 216–225.

Randall, J. E., and R. R. Holcom. 2001. *Antennatus linearis*, a new Indo-Pacific species of frogfish (Lophiiformes: Antennariidae). *Pacific Science*, 55 (2): 137–144.

Randall, J. E., H. Ida, K. Kato, R. L. Pyle, and J. L. Earle. 1997b. Annotated checklist of the inshore fishes of the Ogasawara Islands. *National Science Museum Tokyo, National Science Museum Monographs*, 11: 1–74.

Randall, J. E., and R. H. Kuiter. 1989. The juvenile Indo-Pacific grouper *Anyperodon leucogrammicus*, a mimic of the wrasse *Halichoeres purpurescens* and allied species, with a review of the recent literature on mimicry in fishes. *Revue Française d'Aquariologie et Herpetologie*, Nancy, 16: 51–56.

Randall, J. E., and H. A. Randall. 1960. Examples of mimicry and protective resemblance in tropical marine fishes. *Bulletin of Marine Science of the Gulf and Caribbean*, 10 (4): 444–480.

Randall, J. E., J. T. Williams, D. G. Smith, M. Kulbicki, G. Mou Tham, P. Labrosse, M. Kronen, E. Clua, and B. S. Mann. 2004. Checklist of the shore and epipelagic fishes of Tonga. *Atoll Research Bulletin*, 502: 1–35.

Rasquin, P. 1958. Ovarian morphology and early embryology of the pediculate fishes *Antennarius* and *Histrio*. *Bulletin of the American Museum of Natural History*, 114 (4): 331–371.

Raven, P. H. 1979. Plate tectonics and southern hemisphere biogeography. Pp. 3–24, In: K. Larsen and L. B. Holm-Nielsen (editors), *Tropical Botany*. Academic Press, London & New York, viii + 453 pp.

Raven, P. H., and D. I. Axelrod. 1974. Angiosperm biogeography and past continental movements. *Annals of the Missouri Botanical Garden*, 61: 539–673.

Ray, C. 1961. Spawning behavior and egg raft morphology of the ocellated fringed frogfish, *Antennarius nummifer* (Cuvier). *Copeia*, 1961 (2): 230–231.

Ray, C. E., A. Wetmore, D. Dunkle, and P. Drez. 1968. Fossil vertebrates from the marine Pleistocene of Southern Virginia. *Smithsonian Miscellaneous Collection*, 153: 1–25.

Ray, J. 1713. *Synopsis Methodica Piscium*. London: Innys, 166 pp. + index.

Ré, P., and I. Meneses. 2008. *Early Stages of Marine Fishes Occurring in the Iberian Peninsula*. Instituto Português do Mar e da Atmosfera (IPIMAR/IMAR), Lisbon, Portugal, 282 pp.

Rea, P. M. 1909. Notes from the museum. *Bulletin of the Charleston Museum*, 5 (6): 53–55.

Read, C. I., D. R. Bellwood, and L. van Herwerden. 2006. Ancient origins of Indo-Pacific coral reef fish biodiversity: a case study of the leopard wrasses (Labridae: *Macropharyngodon*). *Molecular Phylogenetics and Evolution*, 38: 808–819.

Reaka, M. L., P. J. Rodgers, and A. U. Kudla. 2008. Patterns of biodiversity and endemism on Indo-West Pacific coral reefs. *Proceedings of the National Academy of Sciences*, 105 (supplement 1): 11474–11481.

Reese, E. S. 1963. The behavioral mechanisms underlying shell selection by hermit crabs. *Behaviour*, 21: 78–126.

Regan, C. T. 1903. A revision of the fishes of the family Lophiidae. *Annals and Magazine of Natural History*, Series 7, 11 (34): 277–285.

Regan, C. T. 1912. The classification of the teleostean fishes of the order Pediculati. *Annals and Magazine of Natural History*, Ser. 8, 9 (28): 277–289.

Regan, C. T. 1918. Further additions to the fish fauna of Natal. *Annals of the Durban Museum*, 2 (2): 76–77.

Regan, C. T. 1926. The pediculate fishes of the suborder Ceratioidea. *Dana Oceanographic Report*, 2, 45 pp.

Regan, C. T., and E. Trewavas. 1932. Deep-sea anglerfish (Ceratioidea). *Dana Report*, 2, 113 pp.

Reichenbach, A. B. 1840. *Die Naturgeschichte in getreuen Abbildungen und mit ausführlicher Beschreibung derselben Fische.* Eisenach, Leipzig, 160 pp.

Renard, L. [1719]. *Poissons, ecrevisses et crabes de diverses couleurs et figures extraordinaire, que l'on trouve autour des Isles Moluques, et sur les côtes des Terres Australes: peints d'après nature durant la regence de Messieurs Van Oudshoorn, Van Hoorn, Van Ribeek & Van Zwoll, successivement gouverneurs-généraux des Indes Orientales pour la Compagnie de Hollande. Ouvrage, auquel on a employé pres de trente ans, & qui contient un très-grand nombre de poissons les plus beaux & les plus rares de la Mer des Indes: Divisé en deux tomes, dont le premier a été copié sur les originaux de Monsr. Baltazar Coyett, ancien gouverneur des isles de la Province d'Amboine, présentement directeur des dites isles & president des Commissaires à Batavia; le second tome a été formé sur les recueils de Monsr. Adrien van der Stell, gouverneur regent de la dite Province d'Amboine, avec une courte description de chaque poisson. Le tout muni de certificats & attestations authentiques. Histoire des plus rares curiositez de la Mer des Indes.* Louis Renard, Amsterdam, 2 vols. in 1: Vol. 1, 4 pp. + 43 pls.; Vol. 2, 2 pp. + 57 pls.; index.

Renard, L. 1754. *Poissons ecrevisses et crabes, de diverses couleurs et figures extraordinaire, que l'on trouve autour des Isles Moluques, et sur les côtes des Terres Australes: peints d'après nature durant la régence de Messieurs Van Oudshoorn, Van Hoorn, Van Ribeek & Van Zwoll, successivement gouverneurs-généraux des Indes Orientales pour la Compagnie de Hollande. Ouvrage, auquel on a employé près de trente ans, & qui contient un très-grand nombre de poissons les plus beaux & les plus rares de la Mer des Indes: Divisé en deux tomes, dont le premier a été copié sur les originaux de Monsr. Baltazar Coyett, ancien gouverneur & directeur des isles de la Province d'Amboine, & president des Commissaires à Batavia. Le second tome a été formé sur les recueils de Monsr. Adrien van der Stell, gouverneur regent de la dite Province d'Amboine, avec une courte description de chaque poisson. Le tout muni de certificats & attestations authentiques. Donné au public par Mr. Louis Renard, agent de S. M. Brit. à Amsterdam, & augmenté d'une préface par Mr. Arnout Vosmaer.* Reinier and Josué Ottens, Amsterdam, 2 vols. in 1: Vol. 1, 12 pp. + 43 pls.; Vol. 2, 2 pp. + 57 pls.; index.

Renard, L. 1782. *Natuurlyke historie Indische Zeeën; behelsende de visschen, kreeften en krabben van verschillende kleuren en buitengewoone gedaanten, van de Moluksche Eilanden en op de kusten der Zuidlyke Landen: door wylen den Wel-Edelen Heere L. Renard, agent van zyne Groot-Britannische Majesteit, te Amsteram, met eene voorreden van den Wel-Edelen Heere A. Vosmaer, opzichter der kabinetten van natuurlyke historie van zyne doorl. hoogheid den Heere Prinse van Oranje, enz. enz. enz., en beschrijvingen door den Wel-Edelen Heere P. Boddaert, M.D. oud-raad der Stad Vlissingen, lid van de Keizerlyke Academie der Natuur-Onderzoekers, van de Maatschappyen der Wetenschappen te Haarlem en Vlissingen, van het Provinciaal Genootschap te Utrecht, der Natuurbeschouwers te Berlin en Halle, en van de Maatschappy van Ober-Lauznits.* Abraham van Paddenburg & Willem Holtrop, Utrecht & Amsterdam, 2 vols. in 1: Vol. 1, 56 pp. + 43 pls.; Vol. 2, 57 pls.

Renous, S., J. Davenport, and V. Bels. 2011. To move on immersed and emersed substrata: adaptive solutions of extant "fishes." pp. 91–128, In: V. L. Bels (editor), *How Vertebrates Moved onto Land*, Publications Scientifiques du Muséum, Muséum national d'Histoire naturelle, Paris.

Rhyne, A. L., M. F. Tlusty, P. J. Schofield, L. Kaufman, J. A. Morris, and A. W. Bruckner. 2012. Revealing the appetite of the marine aquarium fish trade: the volume and biodiversity of fish imported into the United States. *PLoS ONE*, 7 (5): e35808, doi.org/10.1371/journal.pone.0035808

Rhyne, A. L., M. F. Tlusty, and J. T. Szczebak. 2017a. Early culture trials and an overview on U.S. marine ornamental species trade. Pp. 51–70, In: R. Calado, I. Olivotto, M. P. Oliver, and G. J. Holt (editors), *Marine Ornamental Species Aquaculture,* John Wiley & Sons Ltd, Chichester, West Sussex, United Kingdom.

Rhyne, A. L., M. F. Tlusty, J. T. Szczebak, and R. J. Holmberg. 2017b. Expanding our understanding of the trade in marine aquarium animals. *PeerJ*, Aquatic Biology Section, 5: e2949, doi.org/10.7717/peerj.2949.

Ricciardi, F. 2018. Ambon: the hidden biodiversity of a spice island. *Tropical Fish Hobbyist Magazine*, 67 (1): 60–65.

Richards, W. J., and M. G. Bradbury. 2005. Ogcocephalidae. Pp. 793–797, In: W. J. Richards (editor), *Early Stages of Atlantic Fishes: An Identification Guide for the Western Central North Atlantic*, CRC Press, Boca Raton, Florida, Vol. 1.

Richardson, J. 1844. Ichthyology. Pp. 1–16, In: J. Richardson and J. E. Gray (editors), *The Zoology of the Voyage of H.M.S. Erebus & Terror, under the Command of Captain Sir James Clark Ross, R.N., F.R.S., during the Years 1839–1843*. Jansen, London, Vol. 2.

Richardson, J. 1846. Report on the ichthyology of the seas of China and Japan. *Report of the British Association for the Advancement of Science, 15th Meeting*, 1845: 187–320.

Richardson, J. 1848. Ichthyology. Pp. 75–139, In: J. Richardson and J. E. Gray (editors), *The Zoology of the Voyage of H.M.S. Erebus & Terror, under the Command of Captain Sir James Clark Ross, R.N., F.R.S., during the Years 1839–1843*. Jansen, London, Vol. 2.

Ricker, K. E. 1959. Fishes collected from the Revillagigedo Islands during the 1954–1958 cruises of the "Marijean." *Institute of Fisheries, University of British Columbia, Museum Contributions*, 4: 1–10.

Robertson, D. R., and G. R. Allen. 1996. Zoogeography of the shorefish fauna of Clipperton Atoll. *Coral Reefs*, 15: 121–131.

Robertson, D. R., J. S. Grove, and J. E. McCosker. 2004. Tropical transpacific shore fishes. *Pacific Science*, 58 (4): 507–565.

Robins, C. R. 1971. *Distributional Patterns of Fishes from Coastal and Shelf Waters of the Tropical Western Atlantic*. Symposium on Progress in Marine Research in the Caribbean and Adjacent Regions, Papers on Fisheries Research, FAO, Rome, pp. 249–255.

Robins, C. R., and G. C. Ray. 1986. *A Field Guide to Atlantic Coast Fishes of North America*. Houghton Mifflin Co., New York, 354 pp.

Rocha, L. A., M. T. Craig, and B. W. Bowen. 2007. Phylogeography and the conservation of coral reef fishes. *Coral Reefs*, 26 (3): 501–512.

Roessler, C. 1977. Color control: Multihued fishes. *Oceans*, 10 (5): 6–13.

Rogers, C. S., T. W. Pietsch, J. E. Randall, and R. J. Arnold. 2010. The Sargassum Frog-fish (*Histrio histrio* Linnaeus) observed in mangroves in St. John, U.S. Virgin Islands. *Coral Reefs*, 29 (3): 577.

Rohde, F. C., R. G. Arndt, J. W. Foltz, and J. M. Quattro. 2009. *Freshwater Fishes of South Carolina*. University of South Carolina Press, Columbia, South Carolina, xxv + 430 pp.

Román, B. 1979. Peces marinos de Venezuela, claves dicotómicas de los géneros y las especies. *Ciencias Naturales La Salle*, 39 (111/112): 1–408.

Rondelet, G. 1554. *Libri de piscibus marinis, in quibus verae piscium effigies expressae sunt*. Matthiam Bonhomme, Lyon.

Rondelet, G. 1555. *Universae aquatilium historiae pars altera, cum veris ipsorum imaginibus*. Matthiam Bonhomme, Lyon.

Rondelet, G. 1558. *L'histoire entiere des poissons, composée premierement en Latin par Maistre Guilaume Rondelet, avec leurs pourtraits au naif*. Matthiam Bonhomme, Lyon.

Roos, G., S. Van Wassenbergh, A. Herrel, and P. Aerts. 2009. Kinematics of suction feeding in the seahorse *Hippocampus reidi*. *Journal of Experimental Biology*, 212: 3490–3498.

Rosen, D. E. 1975. A vicariance model of Caribbean biogeography. *Systematic Zoology*, 24 (4): 431–464.

Rosen, D. E. 1985. An essay on euteleostean classification. *American Museum Novitates*, 2827, 57 pp.

Rosen, D. E., and C. Patterson. 1969. The structure and relationships of the Paracanthopterygian fishes. *Bulletin of the American Museum of Natural History*, 141: 357–474.

Rosenblatt, R. H. 1963. Differential growth of the illicium and second dorsal spine of *Antennatus strigatus* (Gill) and its bearing on the validity of *A. reticularis* (Gilbert). *Copeia*, 1963 (2): 462–464.

Rosenblatt, R. H. 1967. The zoogeographic relationships of the marine shore fishes of tropical America. *Studies in Tropical Oceanography*, Miami, 5: 579–592.

Rosenblatt, R. H., J. E. McCosker, and I. Rubinoff. 1972. Indo-west Pacific fishes from the Gulf of Chiriqui, Panama. *Natural History Museum of Los Angeles County, Contributions in Science*, 234: 1–18.

Rossiter, J. S., and A. Levine. 2014. What makes a "successful" marine protected area? The unique context of Hawaii's fish replenishment areas. *Marine Policy*, 44: 196–203.

Roule, L., and F. Angel. 1933. Poissons provenant des campagnes du Prince Albert Ier de Monaco. *Résultats des Campagnes Scientifiques* du *Prince Albert Ier*, 86: 1–115.

Roux, C. 1976. On the dating of the first edition of Cuvier's Règne Animal. *Journal of the Society for the Bibliography of Natural History*, 8 (1): 31.

Roux, C., and J. Collignon. 1957. Poissons marins, 2e partie, Clef pour la d'etermination des principaux poissons marins fréquentant les côtes de l'A.E.F. Pp. 255–369, In: J. Collignon, M. Rossignol, and C. Roux (editors), *Mollusques, Crustacés, Poissons marins des côtes d'A.E.F. en collection au Centre d'Océanographie de l'Institut d'Études Centrafricaines de Pointe-Noire*, Office de la Recherche Scientifique et Technique Outre-mer (ORSTOM), Paris.

Rubec, P. J. 1986. The effects of sodium cyanide on coral reefs and marine fish in the Philippines. Pp. 297–302, In: J. L. Maclean, L. B. Dizon, and L. V. Hosillos (editors), *Proceedings of the First Asian Fisheries Forum, Manila, Philippines, 26–31 May 1986*, Asian Fisheries Society, Manila, Philippines.

Rüppell, E. 1829. Fische des Rothen Meeres. Pp. 27–94, In: *Neue Wirbelthiere zu der Fauna von Abysinien gehörig*. Siegmund Schmerber, Frankfurt am Main, Atlas, pls. 7–24.

Rüppell, E. 1838. Fische des Rothen Meeres. Pp. 81–148, In: *Neue Wirbelthiere zu der Fauna von Abysinien gehörig*. Siegmund Schmerber, Frankfurt am Main, Vol. 4.

Russell, P. 1803. *Descriptions and Figures of Two Hundred Fishes; Collected at Vizagapatam on the Coast of Coromandel*. Bulmer, London, Vol. 1, 83 pp.

Ruxton, G. D., D. W. Franks, A. C. V. Balogh, and O. Leimar. 2008. Evolutionary implications of the form of predator generalization for aposematic signals and mimicry in prey. *Evolution*, 62 (11): 2913–2921.

Ruysch, H. 1718. Collectio nova piscium amboinensium partim ibi ad vivum delineatorum partim Ex Museo Henrici Ruysch M.D. XX tabulis comprehensa. Pp. 1–40, In: *Theatrum universale omnium animalium: piscium, avium, quadrupedum, exanguium, aquaticorum, insectorum, and angium, CCLX. tabulis ornatum ex scriptoribus tam antiqua quam recentioribus . . . & aliis maxima cura to J. Jonstonio collectum, ac plus quam trecentis piscibus nuperrime ex Indiis orientalibus allatis, ac nunquam antea his terris visis, locupletatum cum enumeratione morborum, quibus medicamina ex his animalibus petuntur, ac notitiâ animalium, ex quibus vicissim remedia præstantissima possunt capi cura Henrici Ruysch . . . VI. partibus, tomis bus, comprehensum*. R. & G. Wetstenios, Amsterdam, Vol. 1.

Sadovy, Y., and A. S. Cornish. 2000. *Reef Fishes of Hong Kong*. Hong Kong University Press, Aberdeen, Hong Kong, xi + 321 pp.

Salas, E., C. Sánchez-Godínez, and A. Montero-Cordero. 2014. Marine fishes of Caño Island Biological Reserve: reef fish community structure and updated list for the coastal fish. *Revista de Biologia Tropical*, 63 (1): 97–116.

Sale, P. F. 1969. A suggested mechanism for habitat selection by the juvenile manini *Acanthurus triostegus sandvicensis* Streets. *Behaviour*, 35: 27–44.

Sale, P. F. 1980. The ecology of fishes on coral reefs. *Oceanography and Marine Biology, Annual Review*, 18: 367–421.

Sale, P. F. 1991. Reeffish communities: open nonequilibrial systems. Pp. 564–598, In: P. F. Sale (editor), *The Ecology of Fishes on Coral Reefs*, Academic Press, San Diego, California.

Salt, G. W. 1967. Predation in an experimental protozoan population (*Woodruffia-Paramecium*). *Ecological Monographs*, 37: 113–144.

Salviani, I. 1554–1558. *Aquatilium animalium historiae, liber primus, cum eorumdem formis, aere excusis*. Hippolyto Salviano typhernate, Romae medicinam profitente auctore, Rome.

Sampaio, F. D. F., and A. Ostrensky. 2013. Brazilian environmental legislation as tool to conserve marine ornamental fish. *Marine Policy*, 42: 280–285.

Sanciangco, J. C., K. E. Carpenter, P. J. Etnoyer, and F. Moretzsohn. 2013. Habitat availability and heterogeneity and the Indo-Pacific warm pool as predictors of marine species richness in the tropical Indo-Pacific. *PLoS ONE*, 8 (2): e56245. doi:10.1371/journal.pone.0056245.

Sanders, A. E., and W. D. Anderson, Jr. 1999. *Natural History Investigations in South Carolina from Colonial Times to the Present*. University of North Carolina Press, Columbia, North Carolina, xxxix + 333 pp.

Santini, F., L. J. Harmon, G. Carnevale, and M. E. Alfaro. 2009. Did genome duplication drive the origin of teleosts? A comparative study of diversification in ray-finned fishes. *BMC Evolutionary Biology*, 9: 164–178.

Santini, F., L. Sorenson, and M. E. Alfaro. 2013. A new phylogeny of tetraodontiform fishes (Tetraodontiformes, Acanthomorpha) based on 22 loci. *Molecular Phylogenetics and Evolution*, 69: 177–187.

Santos, R. S., F. M. Porteiro, and J. P. Barreiros. 1997. Marine Fishes of the Azores: Annotated Checklist and Bibliography. A Catalogue of the Azorean Marine Ichthyodiversity. *Bulletin of the University of the Azores*, Supplement 1, xxiii + 242 pp.

Satapoomin, U. 2011. The fishes of southwestern Thailand, the Andaman Sea—a review of research and a provisional checklist of species. *Phuket Marine Biological Centre Research Bulletin*, 70: 29–77.

Saunders, B. 2012. *Discovery of Australia's Fishes: A History of Australian Ichthyology to 1930.* CSIRO Publishing, Collingwood, Victoria, Australia, xii + 491 pp.

Saville-Kent, W. 1874a. An angler or fishing frog (*Lophius piscatorius*) at the Manchester aquarium. *Zoologist*, 2 (9): 4264–4266.

Saville-Kent, W. 1874b. Angler-fish at the Manchester Aquarium. *The Field, The Country Gentleman's Newspaper*, 14 November 1874, pp. 511–512 [excerpts published in *The Zoologist*, 2 (9): 4264–4266, 1874].

Saxena, P. K., and S. Aggarwal. 1971. Structure and regeneration of barbels in *Heteropneustes fossilis* (Bloch). *Anatomischer Anzeiger*, 128: 354–364.

Sazima, I. 2002. Juvenile snooks (Centropomidae) as mimics of mojarras (Gerreidae), with a review of aggressive mimicry in fishes. *Environmental Biology of Fishes*, 65 (1): 37–45.

Sazima, I., J. P. Krajewski, R M. Bonaldo, and C. Sazima. 2005. Wolf in a sheep's clothes: juvenile coney (*Cephalopholis fulva*) as an aggressive mimic of the brown chromis (*Chromis multilineata*). *Neotropical Ichthyology*, 3 (2): 315–318.

Schinz, H. R. 1822. *Das Thierreich eingetheilt nach dem Bau der Thiere als Grundlage ihrer Naturgeschichte und der vergleichender Anatomie, von dem Herrn Ritter von Cuvier, aus dem Französischen frei übersetzt und mit vielen Zusätzen versehen.* Cotta, Stuttgart und Tübingen, Fische, 2: 189–553.

Schinz, H. R. 1836. *Naturgeschichte und Abbildungen der Fische.* Brodtmann, Leipzig, 312 + viii pp.

Schleichert, E. 2000. Sneaky, freaky frogfish. What? These weird-looking blobs with legs are FISH? Yup—and awesome ones at that! *Ranger Rick, National Wildlife Federation*, 34 (1): 34–39.

Schmeltz, J. D. E. 1869. *Museum Godeffroy Catalog*, Hamburg, 4, xxxix + 139 pp.

Schmeltz, J. D. E. 1877. *Museum Godeffroy Catalog*, Hamburg, 6, vi + 108 pp.

Schmeltz, J. D. E. 1879. *Museum Godeffroy Catalog*, Hamburg, 7, viii + 99 pp.

Schmidtlein, R. 1879. Beobachtungen über die Lebensweise einiger Seethiere innerhalb der Aquarien der Zoologischen Station Mittbeilungen. *Mitteilungen aus der Zoologischen Station zu Neapel*, 1: 1–27.

Schmitter-Soto, J. J., L. Vásquez-Yeomans, A. Aguilar-Perera, C. Curiel-Mondragón, and J. A. Caballero-Vázquez. 2000. Lista de peces marinos del Caribe mexicano. *Anales del Instituto de Biología, Universidad Nacional Autónoma de México*, Serie Zoologia, 71 (2): 143–177.

Schneider, W., and R. J. Lavenberg. 1995. Antennariidae. Pp. 854–857, In: W. Fischer, F. Krupp, W. Schneider, C. Sommer, K. E. Carpenter, and V. H. Niem (editors), *Guía FAO para la identificación para los fines de la pesca. Pacifico centro-oriental.* Organización de Las Naciones Unidas Para la Agricultura y la Alimentación, Rome, Vol. 2, Vertebrados, pt. 1.

Schneidewind, F. 2003a. Anglerfische. *Aquaristik Fachmagazin und Aquarium Heute*, 35 (5): 82–88.

Schneidewind, F. 2003b. Neue Anglerfisch-Mimikry. *Aquaristik Fachmagazin und Aquarium Heute*, 35 (6): 86–88.

Schneidewind, F. 2005a. A frogfish (*Antennarius* sp.) as a mimic of sea urchins: a new form of mimicry in the family Antennariidae. *Aqua, Journal of Ichthyology and Aquatic Biology*, 10 (1): 23–28.

Schneidewind, F. 2005b. Ein Anglerfisch als Seeigel-imitator. *Datz, Die Aquarien- und Terrarienzeitschrift*, 58: 14–17.

Schneidewind, F. 2005c. Die Wiege der Anglerfische. *Aquaristik Fachmagazin und Aquarium Heute,* 37 (5): 84–88.

Schneidewind, F. 2006a. Frogfishes. *Coral, Reef and Marine Aquarium Magazine,* 3 (5): 18–25.

Schneidewind, F. 2006b. Frogfishes—the family Antennariidae. *Coral, Reef and Marine Aquarium Magazine,* 3 (5): 26–33.

Schneidewind, F. 2006c. Overview of frogfishes in the family Antennariidae. *Coral, Reef and Marine Aquarium Magazine,* 3 (5): 46–47.

Schneidewind, F. 2006d. Hairy discovery: the striated frogfish (*Antennarius striatus*). *Coral, Reef and Marine Aquarium Magazine,* 3 (5): 61.

Schoener, T. W. 1971. Theory of feeding strategies. *Annual Review of Ecology and Systemtics,* 2: 369–404.

Schoer, G. 2015. Celebrating Sydney's marine environment: from research to action. *Nature New South Wales,* 59 (1): 32–33.

Scholfield, A. F. 1959. *Aelian on the Charcteristics of Animals,* with an English translation by A. F. Scholfield. Loeb Classical Library, Harvard University Press, Cambridge, Massachusetts, Vol. 3, vii + 445 pp.

Schultz, L. P. 1957. The frogfishes of the family Antennariidae. *Proceedings of the United States National Museum,* 107: 47–105.

Schultz, L. P. 1958. Correction for "The Frogfishes of the family Antennariidae" by Leonard P. Schultz, No. 3383, Proc. U.S. National Museum, 1957. *Copeia,* 1958 (2): 147.

Schultz, L. P. 1964. Three new species of frogfishes from the Indian and Pacific Oceans, with notes on other species (Family Antennariidae). *Proceedings of the United States National Museum,* 116: 171–182.

Schultz, L. P., L. P. Woods, and E. A. Lachner. 1966. Fishes of the Marshall and Marianas Islands. Vol. 3. Families Kraemeriidae through Antennariidae. *Bulletin of the United States National Museum,* 202: 1–176.

Schultz, O. 2006. An anglerfish, *Lophius* (Osteichthyes, Euteleostei, Lophiidae), from the Leitha Limestone (Badenian, Middle Miocene) of the Vienna Basin, Austria (Central Paratethys). *Beitrage zür Päläontologie,* 30: 427–435.

Schwartz, F. J. 2005. Frogfishes of North Carolina. *Journal of the North Carolina Academy of Sciences,* 121 (3): 145–148.

Schwarzhans, W. 1980. Die Tertiäre Teleosteer-Fauna Neuseelands, rekonstruiert anhand von Otolithen. *Berliner Geowissenschaftliche Abhandlungen,* A, 26: 1–211.

Schwarzhans, W. 1985. Tertiäre Otolithen aus South Australia und Victoria (Australien). *Paleo Ichthyologica,* 3: 1–60.

Schwarzhans, W. 1994. Die Fisch-Otolithen aus dem Oberoligozän der Niederrheinischen Bucht. Systematik, Palökologie, Paläobiogeographie, Biostratigraphie und Otolithen-Zonierung. *Geologisches Jahrbuch, Reihe* A, 140: 1–248.

Schwarzhans, W. 2007. The otoliths from the middle Eocene of Osteroden near Bramsche, north-western Germany. *Neues Jahrbuch für Geologie und Paläontologie, Abhandlungen,* 244 (3): 299–369.

Schwarzhans, W. 2010. *The Otoliths from the Miocene of the North Sea Basin.* Backhuys Publishers, Leiden, 352 pp.

Scott, T. D., C. J. M. Glover, and R. V. Southcott. 1974. *The Marine and Freshwater Fishes of South Australia,* 2nd edition. Government Printing Department, Netley, South Australia, 392 pp.

Scott, W. B., and M. G. Scott. 1988. *Atlantic Fishes of Canada.* Canadian Bulletin of Fisheries and Aquatic Sciences, University of Toronto Press, 219, xxx + 730 pp.

Seale, A. 1906. Fishes of the South Pacific. *Occasional Papers of the Bernice P. Bishop Museum,* 4 (1): 1–89.

Seale, A. 1935. The Templeton Crocker Expedition to Western Polynesian and Melanesian Islands, 1933. Fishes. *Proceedings of the California Academy of Sciences,* 21 (27): 337–378.

Seale, A. 1940. Report on fishes from Allan Hancock Expeditions in the California Academy of Sciences. *Allan Hancock Pacific Expedition,* 9 (1): 1–46.

Seba, A. 1734. *Locupletissimi rerum naturalium thesauri accurata descriptio, et Iconibus artificiosissimis expressio, per universam physices historiam, etc.* Janssonius van Waesberge, Wettstein & Smith, Amsterdam, Vol. 1, 178 pp.

Senou, H., M. Hayashi, and S. Yokoyama. 1994. First record of a frogfish *Antennatus tuberosus* (Pisces: Antennariidae) from Japan. *I. O. P. Diving News*, 5 (12): 2–3. [In Japanese, English abstract.]

Senou, H., and T. Kawamoto. 2002. First record of a frogfish, *Antennarius randalli*, from Japan. *Izu Oceanic Park Diving News*, 13 (4): 2–6.

Senou, H., Y. Kobayashi, and N. Kobayashi. 2007. Coastal fishes of the Miyako Group, the Ryukyu Islands, Japan. *Bulletin of the Kanagawa Prefectural Museum, Natural Sciences*, 36: 47–74.

Senou, H., K. Matsuura, and G. Shinohara. 2006. Checklist of fishes in the Sagami Sea with zoogeographical comments on shallow water fishes occurring along the coastlines under the influence of the Kuroshio Current. *Memoires of the National Science Museum*, Tokyo, 41: 389–542.

Shallenberger, R. J., and W. D. Madden. 1973. Luring behavior in the scorpionfish, Iracundus signifer. *Behaviour*, 47: 33–47.

Shaw, G. 1794. *The Naturalist's Miscellany, or Coloured Figures of Natural Objects, Drawn and Described from Nature.* Nodder, London, 5, pls. 147–182.

Shaw, G. 1804. *General Zoology, or Systematic Natural History by George Shaw, M.D., F.R.S., etc., with Plates from the First Authorities and Most Select Specimens Engraved Principally by Mr. Hedath.* Kearsley, London, 5 (2): 251–463.

Shaw, G. 1811. *The Naturalist's Miscellany, or Coloured Figures of Natural Objects, Drawn and Described from Nature.* Nodder, London, 23, pls. 973–1020.

Shedlock, A. M., T. W. Pietsch, M. G. Haygood, P. Bentzen, and M. Hasegawa. 2004. Molecular systematics and life history evolution of anglerfishes (Teleostei: Lophiiformes): Evidence from mitochondrial DNA. *Steenstrupia*, Copenhagen, 28 (2): 129–144.

Sherborn, C. D. 1922. *Index animalium sive index nominum quae ab A. D. MDCCLVIII generibus et speciebus animalium imposita sunt, sectio secunda, a kalendis ianuariis MDCCCI usque ad finem Decembris MDCCCL.* Part I. Introduction, Bibliography, and Index A-Aff. Typographio Academico, Cambridge, England, cxxxi + 128 pp.

Sherborn, C. D., and B. B. Woodward. 1901. Notes on the dates of publication of the natural history portions of some French voyages. Part I. *Annals and Magazine of Natural History*, London, Series 7, 7: 388–392.

Sherborn, C. D., and B. B. Woodward. 1906. Notes on the dates of publication of the natural history portions of some French voyages.—"Voyage autour du Monde . . . sur . . . la *Coquille* pendant . . . 1822–1825. . . . Par L. J. Duperry," &c.—A correction. *Annals and Magazine of Natural History*, London, Series 7, 17: 335–336.

Shimazaki, M., and K. Nakaya. 2004. Functional anatomy of the luring apparatus of the deep-sea ceratioid anglerfish *Cryptopsaras couesii* (Lophiiformes: Ceratiidae). *Ichthyological Research*, 51 (1): 33–37.

Shimizu, T. 2001. An annotated list of the coastal fishes from Iyo City, Ehime Prefecture, Japan. *Bulletin of the Tokushima Prefectural Museum*, 11: 17–99.

Shinohara, G., H. Endo, K. Matsuura, Y. Machida, and H. Honda. 2001. Annotated Checklist of the Deepwater Fishes from Fishes from Tosa Bay, Japan. *Monographs of the National Science Museum*, Tokyo, 20: 283–343.

Shumaker, R. W., K. R. Walkup, and B. B. Beck. 2011. *Animal Tool Behavior: The Use and Manufacture of Tools by Animals*, Revised and Updated Edition. Johns Hopkins University Press, Baltimore, xvi + 282 pp.

Siegel, S. 1956. *Nonparametric Statistics for the Behavioral Sciences.* New York, Toronto, McGraw-Hill, London, xvii + 312 pp.

Sielfeld, W. 2010. *Antennarius avalonis* (Antennariidae, Lophiiformes) in the southeast Pacific. *Revista de Biología Marina y Oceanografía*, 45 (1): 757–760.

Simmons, M. P., and M. Miya. 2004. Efficiently resolving the basal clades of a phylogenetic tree using Bayesian and parsimony approaches: a case study using mitogenomic data from 100 higher teleost fishes. *Molecular Phylogenetics and Evolution*, 31: 351–362.

Simonini, G. 2018. Daniel Weiman and Libri Picturati A 16–31. *Archives of Natural History*, 45 (1): 40–53.

Singh, S. P., J. E. Holdway, and K. D. Poss. 2012. Regeneration of amputated zebrafish fin rays from de novo osteoblasts. *Developmental Cell*, 22 (4): 879–886.

Sisson, R. F. 1976. Adrift on a raft of sargassum. *Natural Geographic*, 149 (2): 188–199.

Smith, C. L., and J. C. Tyler. 1972. Space resource sharing in a coral reef fish community. Pp. 125–170, In: B. B. Collette and S. A. Earle (editors), Results of the Tektite Program: Ecology of Coral Reef Fishes. *Natural History Museum of Los Angeles County, Science Bulletin*, 14.

Smith, C. L., J. C. Tyler, W. P. Davis, R. S. Jones, D. G. Smith, and C. C. Baldwin. 2003. Fishes of the Pelican Cays, Belize. *Atoll Research Bulletin*, 497: 1–88.

Smith, G. B., H. M. Austin, S. A. Bortone, R. W. Hastings, and L. H. Ogren. 1975. Fishes of the Florida Middle Ground, with comments on ecology and zoogeography. *Florida Marine Research Publication*, 9: 1–14.

Smith, H. M. 1898. The fishes found in the vicinity of Woods Hole. *Bulletin of the United States Fish. Commission*, 1897 (1898), 17 (3): 85–111.

Smith, H. M. 1907. The fishes of North Carolina. *North Carolina Geological and Economic Survey*, 2: 1–453.

Smith, H. M., and T. E. B. Pope. 1906. List of fishes collected in Japan in 1903, with descriptions of new genera and species. *Proceedings of the United States National Museum*, 31: 459–499.

Smith, J. L. B. 1949. *The Sea Fishes of Southern Africa*. Central News Agency, Capetown, 550 pp.

Smith, J. L. B. 1956. Self-inflation in a gobioid fish. *Nature*, 177: 714.

Smith, J. L. B. 1957. Four interesting new fishes from South Africa. *South African Journal of Science*, 53 (8): 219–222.

Smith, J. L. B. 1958. The fishes of Aldabra. Part X. *Annals and Magazine of Natural History*, London, Ser. 13, 1 (1): 57–63.

Smith, J. L. B., and M. M. Smith. 1963. *The Fishes of Seychelles*. Department of Ichthyology, Rhodes University, Grahamstown, South Africa, 215 pp.

Smith, J. L. B., and M. M. Smith. 1966. *Fishes of the Tsitsikama Coastal National Park*. National Parks Board of Trustees, Republic of South Africa. Swan Press, Johannesburg, 161 pp.

Smith, K. L. 1973. Energy transformations by the sargassum fish, *Histrio histrio* (L.). *Journal of Experimental Marine Biology and Ecology*, 12: 219–227.

Smith M. M., and P. C. Heemstra (editors). 1986. *Smiths' Sea Fishes*. Macmillan South Africa Ltd., Johannesburg, South Africa, xx + 1047 pp.

Smith-Vaniz, W. F., and B. B. Collette. 2013. Fishes of Bermuda. *Aqua, Journal of ichthyology and Aquatic Biology*, 19 (4): 4–25.

Smith-Vaniz, W. F., B. B. Collette, and B. E. Luckhurst. 1999. *Fishes of Bermuda: History, Zoogeography, Annotated Checklist, and Identification Keys*. American Society of Ichthyologists and Herpetologists, Special publication 4, x + 424 pp.

Smith-Vaniz, W., and H. L. Jelks. 2014. Marine and inland fishes of St. Croix, U. S. Virgin Islands: an annotated checklist. *Zootaxa*, 3803 (1): 1–120.

Snyder, J. O. 1904. A catalogue of the shore fishes collected by the Steamer Albatross about the Hawaiian Islands in 1902. *Bulletin of the United States Fish Commission*, 1902 (1904), 22: 513–538.

Snyder, J. O. 1912. Japanese shore fishes collected by the United States Bureau of Fisheries Steamer "Albatross" Expedition of 1906. *Proceedings of the United States National Museum*, 1909 (1912), 42: 399–450.

Solander, D. MS Z1. Pisces Australiae (i.e., New Zealand); Pisces & c. Novae Hollandiae; Pisces & Anim. caetera, Oceani Pacifici; Animalia Javanensia & Capensia; Pisces Islandici. Zoological Library, British Museum of Natural History, London.

Solander, D. MS Z8. Manuscript descriptions of animals written on slips and systematically arranged in accordance with Linné's Systema Naturae. . . . Editio duodecima reformata. Amphibia, Vol. 1, Zoological Library, British Museum of Natural History, London.

Sonnini, C. S. 1803. *Histoire naturelle, générale et particulière, des poissons; ouvrage faisant suite à l'histoire naturelle, générale et particulière, composée par Leclerc de Buffon, et mise dans un nouvel ordre par C. S. Sonnini, avec des notes et des additions*. Dufart, Paris, Vol. 4, 427 pp.

Sorbini, L. 1988. Biogeography and climatology of Pliocene and Messinian fossil fish of Eastern-central Italy. *Bollettino del Museo Civico di Storia Naturale di Verona*, 14: 1–85.

Spalding, M., L. Burke, S. A. Wood, J. Ashpole, J. Hutchison, and P. zu Ermgassen. 2017. Mapping the global value and distribution of coral reef tourism. *Marine Policy*, 82: 104–113.

Sparta, A. 1929. Contributo alla conoscenza di uova e larve negli Ofididi *Ophidium vassali* Risso ed *O. barbatum* L. *Memorie del Regio Comitato Talassografico Italiano*, 149: 1–11.

Springer, V. G. 1982. Pacific Plate biogeography, with special reference to shore fishes. *Smithsonian Contributions to Zoology*, 367, 182 pp.

Springer, V. G. 1987. Leonard Peter Schultz, 1901–1986. *Copeia*, 1987 (1): 271–272.

Stahl, A. 1882. *Catalógo del gabinete zoológico del Dr. A. Stahl en Bayamón (Pto.-Rico): precedido de una clasificación sistemática de los animales que corresponden á esta fauna*. Imprenta del Boletin Mercantil, San Juan, Puerto Rico, 248 pp.

Starks, E. C. 1930. The primary shoulder girdle of the bony fishes. *Stanford University Publications, University Series, Biological Sciences*, 6 (2): 149–239.

Steene, R. C. 2005. *Oceanic Wilderness: Mysteries of the Deep*. Firefly Books Ltd., Richmond Hill, Ontario, Canada, 339 pp.

Stefano, G. de. 1910. Osservazioni sulla ittiofauna pliocenica di Orciano e San Quirico in Toscana. *Bollettino della Società Geologica Italiana*, 28: 539–648.

Steindachner, F. 1866. Zur Fischfauna von Port Jackson in Australien. *Sitzungsberichte der Akademie der Wissenschaften,* Wien, 53: 424–480.

Steindachner, F. 1895. Note I. Die Fische Liberia's. *Notes from the Leyden Museum*, 16: 1–96.

Steindachner, F. 1903. Fische. Pp. 409–464, In: W. Kükenthal (editor), *Ergebnisse Zoologischen Forschungsreise*, 2, Teil, Bd. 3. Abhandlungen der Senckenbergischen Naturforschenden Gesellschaft, 25.

Steindachner, F. 1906. Zur Fischfauna der Samoa-Inseln. *Sitzungsberichte der Akademie der Wissenschaften, Wien*, 115 (1): 1369–1425.

Steindachner, F., and L. Döderlein. 1885. Beiträge zur kenntniss der Fische Japan's (III). *Denkschriften der Akademie der Wissenschaften*, Wien, 49: 171–212.

Stephens, L. D. 2000. *Science, Race, and Religion in the American South: John Bachman and the Charleston Circle of Naturalists, 1815–1895*. University of North Carolina Press, Chapel Hill, North Carolina, xx + 338 pp.

Stephenson, W., and R. B. Searles. 1960. Experimental studies of the ecology of intertidal environments at Heron Island. I. Exclusion of fish from beach rocks. *Australian Journal of Marine and Freshwater Research*, 11 (2): 241–267.

Stevens, M. 2016. *Cheats and Deceits: How Animals and Plants Exploit and Mislead*. Oxford University Press, Oxford, xvi + 300 pp.

Stewart, A. L. 2015. Family Antennariidae. Pp. 882–887, In: C. D. Roberts, A. L. Stewart, and C. D. Struthers (editors), *The Fishes of New Zealand*, 3: 577–1152.

Stinton, F. C. 1966. Fish otoliths from the London Clay. Pp. 404–464, In: E. Casier (editor), *Faune ichthyologique du London Clay*, Trustees of the British Museum of Natural History, London.

Stinton, F. C. 1978. Fish otoliths from the English Eocene. *Palaeontographical Society Monographs*, 3: 127–189.

Storer, D. H. 1839a. A report on the fishes of Massachusetts. *Proceedings of the Boston Society of Natural History*, 2 (3–4), 12: 289–558.

Storer, D. H. 1839b. Fishes of Massachusetts. Pp. 1–202, In: *Reports on the Fishes, Reptiles and Birds of Massachusetts*. Dutton & Wentworth, Boston, 426 pp.

Storer, D. H. 1846. A synopsis of the fishes of North America. *Memoirs of the American Academy of Arts and Science*, 2: 253–550.

Storer, D. H. 1855. A history of the fishes of Massachusetts. *Memoirs of the American Academy of Arts and Science*, 5 (2): 257–296.

Strack, H. L. 1993. Results of the Rumphius Biohistorical Expedition to Ambon (1990). Part 1. General account and list of stations. *Zoologische verhandelingen*, Leiden, 289: 1–72.

Strasburg, D. W. 1966. *Golem cooperae*, a new antennariid fish from Fiji. *Copeia*, 1966 (3): 475–477.

Straughan, R. P. L. 1954. The Sargassum fish, *Histrio pictus*. *Aquarium*, 23: 277–279.

Suzuki, K. 1964. Results of Amami-Expedition 2. Fishes. *Report of the Faculty of Fisheries, Prefectural University of Mie*, 5 (1): 153–188.

Swain, J. 1883. An identification of the species of fishes described in Shaw's General Zoology. *Proceedings of the Academy of Natural Sciences of Philadelphia*, 1882 (1883): 303–309.

Swainson, W. 1838. On the Natural History and Classification of Fishes, Amphibians and Reptiles or Monocardian Animals. Longman, Orme, Brown, Green & Longmans, London, Vol. 1, 368 pp.

Swainson, W. 1839. On the Natural History and Classification of Fishes, Amphibians and Reptiles or Monocardian Animals. Longman, Orme, Brown, Green & Longmans, London, Vol. 2, 448 pp.

Sweet, T., and M. Pedersen. 2018. A *Coral* special report: the state of the marine breeder's art, 2018. *Coral, Reef and Marine Aquarium Magazine*, 15 (2): 38–47.

Sychevskaya, E. K., and Prokofiev, A. M. 2010. On the occurrence of anglerfishes (Lophiidae) in the Lower Miocene of Transcaucasia. *Journal of Ichthyology*, 50: 205–210.

Tanaka, S. 1916. Three new species of Japanese fishes. *Dobutsugaku Zasshi (Zoological Magazine, Tokyo)*, 28 (330): 141–144 (in Japanese).

Tanaka, S. 1917. *Antennarius lateralis* n. sp. *Dobutsugaku Zasshi (Zoological Magazine, Tokyo)*, 29 (345): 200 (in Japanese).

Tanaka, S. 1918. *Figures and Descriptions of the Fishes of Japan, Including Riukiu Islands, Bonin Islands, Formosa, Kurile Islands, Korea and Southern Sakhalin*. Tokyo Printing Co., Tokyo, 21–30: 371–557.

Tanaka, S. 1930. *Figures and Descriptions of the Fishes of Japan, Including Riukiu Islands, Bonin Islands, Formosa, Kurile Islands, Korea and Southern Sakhalin*. Tokyo Printing Co., Tokyo, 47: 925–944.

Taylor, W. R. 1967. An enzyme method of clearing and staining small vertebrates. *Proceedings of the United States National Museum*, 122: 1–17.

Tchernavin, V. V. 1953. *The Feeding Mechanisms of a Deep Sea Fish* Chauliodus sloani *Schneider*. Trustees of the British Museum, London, 101 pp.

Temminck, C. J., and H. Schlegel. 1845. Pisces. Pp. 113–172, In: P. F. von Siebold, *Fauna Japonica*. Müller, Amsterdam, Parts 7–9.

Tennent, J. E. 1861. *Sketches of the Natural History of Ceylon, with Narratives and Anecdotes Illustrative of the Habits and Instincts of the Mammalia, Birds, Reptiles, Fishes, Insects, &c., Including a Monograph of the Elephant and a Description of the Modes of Capturing and Training It, with Engravings from Original Drawings*. Longman, Green, Longman, and Roberts, London, xxiii + pp.

Tets, G. G. van. 1965. A comparative study of some social communication patterns in the Pelecaniformes. *Ornithological Monographs*, 2: 1–88.

Thaler, E. 2006. Notable experiences with frogfishes and thoughts on their care in the aquarium. *Coral, Reef and Marine Aquarium Magazine*, 3 (5): 44–45.

Thompson, d'A. W. 1984. History of animals. Pp. 774–993, In: J. Barnes (editor), *The Complete Works of Aristotle*, The Revised Oxford Translation, Bollingen Series 71, part 2, Princeton University Press, Princeton, New Jersey, Vol. 1.

Thompson, W. W. 1918. Catalogue of fishes of the Cape Province. *Marine Biological Report*, South Africa, 4: 75–177.

Thomson, C. W. 1878. *The Voyage of the "Challenger." The Atlantic: A Preliminary Account of the General Results of the Exploring Voyage of H.M.S. "Challenger" During the Year 1873 and the Early Part of the Year 1876, by Sir C. Wyville Thomson*. Harper, New York, Vol. 1, 29 + 424 pp., Vol. 2, 14 + 396 pp.

Thomson, D. A., L. T. Findley, and A. N. Kerstitch. 1979. *Reef Fishes of the Sea of Cortez: The Rocky-Shore Fishes of the Gulf of California*. Wiley-Interscience, New York, Chichester, Brisbane & Toronto, xvii + 302 pp.

Thomson, D. A., L. T. Findley, and A. N. Kerstitch. 2000. *Reef Fishes of the Sea of Cortez: The Rocky-shore Fishes of the Gulf of California*. Revised edition. University of Texas Press, Austin, Texas, xx + 353 pp.

Thomson, D. A., and C. E. Lehner. 1976. Resilience of a rocky intertidal fish community in a physically unstable environment. *Journal of Experimental Marine Biology and Ecology*, 22: 1–29.

Thresher, R. 1984. *Reproduction in Reef Fishes*. T.F.H. Publications, Inc. Ltd., Neptune City, New Jersey, 399 pp.

Tilesius, W. G. 1809. Description de quelques poissons observés pendant son voyage autour du monde. *Mémoires de la Société des naturalistes de Moscou*, 2: 212–249.

Tilesius, W. G. 1812. Opisanie. *Recherches Théorique Académie Impériale des Sciences de Saint Pétersbourg*, 3: 278–303 (in Russian and Latin).

Tinker, S. W. 1978. *Fishes of Hawaii: A Handbook of the Marine Fishes of Hawaii and the Central Pacific Ocean*. Hawaiian Service, Honolulu, xxxx + 532 + xxxvi pp.

Tobón-López, A., E. A. Rubio, and A. Giraldo. 2008. Composition and taxonomic analysis of the fish fauna in the Gulf of Tribugá, northern Colombian Pacific. *Latin American Journal of Aquatic Research*, 36 (1): 93–104.

Toh, K. B., C. S. L. Ng, W.-K. G. Leong, Z. Jaafar, and L. M. Chou. 2016. Assemblages and diversity of fishes in Singapore's marinas. *Raffles Bulletin of Zoology*, Supplement, 32: 85–94.

Tokioka, T. 1961. Record of an unusual fish stranding in winter, with the list of stranded fishes identified by Prof. K. Matsubara. *Publications of the Seto Marine Biological Laboratory*, 9 (2): 447–450.

Torruco, D., E. A. Chávez, and A. González. 2007. Spatio-temporal variation of the structural organization of demersal communities in the Southwestern Gulf of Mexico. *Revista de Biologia Tropical*, 55 (2): 509–536.

Townsend, D. 2011. Sustainability, equity and welfare: a review of the tropical marine ornamental fish trade. *SPC Live Reef Fish Information Bulletin*, 20: 2–12.

Townsend, D. 2015. First comprehensive list of the coral reef fishes of Tunku Abdul Rahman Park, Sabah, Malaysia (Borneo). *Check List*, 11 (7): 1–12, doi: dx.doi.org/10.15560/11.5.1762.

Tremeau de Rochebrune, A. 1883. *Fauna de la Sénégambie. Poissons*. Douin, Paris & Bordeaux, 166 pp.

Triay-Portella, R., J. G. Pajuelo, P. Manent, F. Espino, R. Ruiz-Díaz, J. M. Lorenzo, and J. A. González. 2015. New records of non-indigenous fishes (Perciformes and Tetraodontiformes) from the Canary Islands (north-eastern Atlantic). *Cybium*, 39 (3): 163–174.

Trueman, E. R. 1975. *The Locomotion of Soft-bodied Animals*. Edward Arnold, London, viii + 200 pp.

Trueman, E. R. 1980. Swimming by jet propulsion. Pp. 93–105, In: H. Y. Elder and E. R. Trueman (editors), *Aspects of Animal Movement*. Society for Experimental Biology Seminar Series 5, Cambridge University Press.

Trueman, E. R., and A. Packard. 1968. Motor performance of some cephalopods. *Journal of Experimental Biology*, 49: 495–507.

Tsuda, R. T. 1976. Occurrence of the genus *Sargassum* (Phaeophyta) on two Pacific atolls. *Micronesica*, 12: 279–282.

Tweddle, D., and M. E. Anderson. 2007. A collection of marine fishes from Angola, with notes on new distribution records. *Smithiana Bulletin*, 8: 3–24.

Tyler, J. C. 1963. A critique of Y. Le Danois' work on the classification of the fishes of the order Plectognathi. *Copeia*, 1963 (1): 203–206.

Tyler, J. C., and J. E. Böhlke. 1972. Records of sponge-dwelling fishes, primarily of the Caribbean. *Bulletin of Marine Science*, 22 (3): 601–642.

Ueno, T. 1966. Fishes of Hokkaido. 22. Anglerfishes, frogfishes, and deepsea anglerfishes. *Hokusuishi Geppo*, 23 (11): 532–541.

Ueno, T. 1971. List of the marine fishes from the waters of Hokkaido and its adjacent regions. *Scientific Report of the Hokkaido Fisheries Experimental Station*, 13, 46 (3): 61–102.

Ueno, T., and K. Abe. 1966. On rare or newly found fishes from the waters of Hokkaido (II). *Japanese Journal of Ichthyology*, 13 (4/6): 229–236.

Uyarra, M. C., and I. M. Côté. 2007. The quest for cryptic creatures: Impacts of species-focused recreational diving on corals. *Biological Conservation*, 136 (1): 77–84.

Uyeno, T., and M. Aizawa. 1983. Family Antennariidae. Pp. 246–248, In: T. Uyeno, K. Matsuura, and E. Fujii (editors), *Fishes Trawled off Suriname and French Guiana*. Japan Marine Fishery Resource Research Center, 519 pp.

Vaillant, L. 1887. Remarques sur la construction du nid l'*Antennarius marmoratus* Less. et G., dans la Mer des Sargasses. *Comptes Rendus des Séances de la Société de Biologie*, Paris, 4: 732–733.

Valenciennes, A. 1837. Des Chironectes (*Chironectes*, Cuv., *Antennarius*, Comm.). Pp. 389–437, In: G. Cuvier and A. Valenciennes, *Histoire Naturelle des Poissons,* Levrault, Paris & Strasbourg, Vol. 12.

Valenciennes, A. 1842. Poissons. In: G. Cuvier, *Le Règne Animal distribué d'après son organisation, pour servir de base à l'histoire naturelles des animaux et d'introduction à l'anatomie comparée*, ed. 3 (Éd. des Disciples). Fortin, Masson, Paris, Vol. 4, 392 pp.

Valentijn, F. 1726. Omstandig verhaal van de geschiedenissen en zaaken het kerkelyke ofte den godsdienst betreffende, zoo in Amboina, als in alle de eylanden, daar onder behoorende, van de oudste tyden af tot nu toe, benevens een fraaye verhandeling der boomen, planten, heesters, enz. Als ook der land-dieren, vogelen, visschen, horenkens, en zeegewasschen, in en by dezelve eylanded vallende; mitsgaders een naaukeurige beschryving van Banda, en de eylanden, onder die landvoogdy begrepen, als ook der eylanden Timor, en Solor, Celebes, ofte Macassar, Borneo, en Bali, mitsgaders van de koningryken Tonkin, Cambodia, en Siam, benevens een verhaal der zaaken, in de voornoemde eylanden, en Koningryken, tot nu toe voorgevallen; met zeer nette prentverbeeldingen, en landkaarten verrykt door François Valentyn, onlangs bedienaar des Goddelyken woords in Amboina, Banda, enz. Pp. 336–515, In: *Oud en nieuw Oost-Indiën, vervattende een naaukeurige en uitvoerige verhandelinge van Nederlands Mogentheyd in die Gewesten*, Van Braam & Gerard onder de Linden, Dordrecht & Amsterdam, Vol. 3.

Van der Hoeven, J. 1855. *Handboek der Dierkunde. Tweede, verbeterde en vermeerderde uitgave,* 2nd edition. J. C. A. Sulpke, Amsterdam, Vol. 2, xxviii+1068.

Van Dolah, R. F., P. P. Maier, G. R. Sedberry, C. A. Barans, F. M. Idris, and V. J. Henry. 1994. *Distribution of bottom habitats on the continental shelf off South Carolina and Georgia.* South Carolina Department of Natural Resources, Southeast Area Monitoring and Assessment Program, South Atlantic Committee, Final Report, viii+45 pp.

Van Wassenbergh, S., B. Dries, and A. Herrel. 2014. New insights into muscle function during pivot feeding in seahorses. *PLoS ONE*, 9 (10): e109068; doi.org/10.1371/journal.pone.0109068.

Vane-Wright, R. I. 1976. A unified classification of mimetic resemblances. *Biological Journal of the Linnean Society*, 8: 25–56.

Videler, J. J. 1981. Swimming movements, body structure andpropulsion in cod *Gadus morhua.* Pp. 1–27, In: M. H. Day (editor), *Vertebrate Locomotion*, Academic Press, London & New York.

Vignon, M. P. 1931. Le mimetisme chez les animaux marins. *Terre et la Vie*, 1: 131–150.

Villarreal-Cavazos, A., H. Reyes-Bonilla, B. Bermúdez-Almada, and O. Arizpe-Covarrubias. 2000. Los peces del arrecife de Cabo Pulmo, Golfo de California, México: Lista sistemática y aspectos de abundancia y biogeografía. *Revista de Biologia Tropical*, 48 (2/3): 413–424.

Volta, G. S. 1796. *Ittiolitologia Veronese del Museo Bozziano ora annesso a quello del Conte Giovambattista Gazola e di altri Gabinetti Fossili Veronesi.* Stamperia Giuliani, Verona, 323 pp.

Vroom, P. S. 2011. Coral dominance: a dangerous ecosystem misnomer? *Journal of Marine Biology*, Article ID 164137, 8 pp., doi: 10.1155/2011/164127.

Vroom, P. S., and C. L. Braun. 2010. Benthic composition of a healthy subtropical reef: baseline species-level cover, with an emphasis on algae, in the Northwestern Hawaiian Islands. *PLoS One*, 5 (3), e9733; doi:10.1371/journal.pone.0009733.

Vroom, P. S., K. N. Page, J. C. Kenyon, and R. E. Brainard. 2006. Algae-dominated reefs. *American Scientist*, 94: 429–437, doi: 10.1511/2006.61.430.

Wabnitz, C., M. Taylor, E. Green, and T. Razak. 2003. *From Ocean to Aquarium: The Global Trade in Marine Ornamental Species.* United Nations Environment Programme World Conservation Monitoring Centre (UNEP-WCMC), Cambridge, United Kingdom.

Wagner, H. 1939. Biologische Beobachtungen an *Antennarius marmoratus* Gthr. *Zoologischer Anzeiger*, 126: 285–297.

Wainwright, P. C., A. M. Carroll, D. C. Collar, S. W. Day, T. E. Higham, and R. A. Holzman. 2007. Suction feeding mechanics, performance, and diversity in fishes. *Integrative and Comparative Biology*, 47 (1): 96–106.

Wainwright, P. C., and G. V. Lauder. 1986. Feeding biology of sunfishes: patterns of variation in the feeding mechanism. *Zoological Journal of the Linnean Society*, 88 (3): 217–228.

Wainwright, P. C., and R. G. Turingan. 1997. Evolution of pufferfish inflation behavior. *Evolution*, 51 (2): 506–518.

Wainwright, P. C., R. G. Turingan, and E. L. Brainerd. 1995. Functional morphology of pufferfish inflation: mechanism of the buccal pump. *Copeia*, 1995 (3): 614–625.

Waite, E. R. 1901. Additions to the fish fauna of Lord Howe Island, No. 2. *Records of the Australian Museum*, 4: 36–47.

Waite, E. R. 1912. Notes on New Zealand fishes: No. 2. *Transactions and Proceedings of the New Zealand Institute*, Wellington, 44 (10): 194–202.

Waite, E. R. 1921. Illustrated catalogue of the fishes of South Australia. *Records of the South Australian Museum*, 2 (1): 1–208.

Waite, E. R. 1923. *The Fishes of South Australia*. British Scientific Guild, Adelaide, 243 pp.

Walbaum, J. J. 1792. *Petri Artedi Sueci Genera Piscium, in quibus systema totum ichthyologiae proponitur cum classibus, ordinibus, generum characteribus, specierum differentiis, observationibus plurimis, redactis speciebus 242 ad genera 52*. Ichthyologiae Pars III. Anton Ferdinand Röse, Grypeswaldiae, 723 pp.

Walsh, W. J. 1985. Reef fish community dynamics on small artificial reefs: the influence of isolation, habitat structure, and biogeography. *Bulletin of Marine Science*, 36 (2): 357–376.

Walsh, H. J., K. E. Marancik, and J. A. Hare. 2006. Juvenile fish assemblages collected on unconsolidated sediments of the southeast United States continental shelf. *United States Fishery Bulletin*, 104 (2): 256–277.

Walther-Mendoza, M., A. Ayala-Bocos, M. Hoyos-Padilla, and H. Reyes-Bonilla. 2013. New records of fishes from Guadalupe Island, northwest Mexico. *Hidrobiológica*, 23 (3): 410–414.

Waqalevu, V. P. 2010. *Capture, Identification and Culture Techniques of Coral Reef Fish Larvae (French Polynesia)*. Training course report, Component 2A—Project 2A1 PCC Development, University of the South Pacific, Laucala, Fiji, and Coral Reef Initiative for the South Pacific Programme (CRISP), Nouméa, Nouvelle Calédonie, [58 pp].

Ward, D. F., and S. A. Wainwright. 1972. Locomotory aspects of squid mantle structure. *Journal of Zoology*, London, 167: 437–449.

Washington, B. B., H. G. Moser, W. A. Laroche, and W. J. Richards. 1984. Scorpaeniformes: development. Pp. 405–428, In: H. G. Moser, W. J. Richards, D. M. Cohen, M. P. Fahay, A. W. Kendall, Jr., and S. L. Richardson (editors), *Ontogeny and Systematics of Fishes*. American Society of Ichthyologists and Herpetologists, Special Publication No. 1, x + 760 pp.

Wass, R. C. 1984. *An Annotated Checklist of the Fishes of Samoa*. United States Department of Commerce, NOAA Technical Report, NMFS SSRF-781, v + 43 pp.

Waterman, T. H. 1939. Studies of deep-sea angler-fishes (Ceratioidea). I. An historical survey of our present state of knowledge. *Bulletin of the Museum of Comparative Zoology*, 85 (3): 65–81.

Watson, W. 1996a. Lophiidae: Goosefishes. Pp. 553–557, In: H. G. Moser (editor), *The Early Stages of Fishes in the California Current Region*, California Cooperative Oceanic Fisheries Investigations, La Jolla, California, Atlas 33.

Watson, W. 1996b. Antennariidae: Frogfishes. Pp. 559–561, In: H. G. Moser (editor), *The Early Stages of Fishes in the California Current Region*, California Cooperative Oceanic Fisheries Investigations, La Jolla, California, Atlas 33.

Watson, W. 1996c. Ogcocephalidae: Batfishes. Pp. 563–565, In: H. G. Moser (editor), *The Early Stages of Fishes in the California Current Region*, California Cooperative Oceanic Fisheries Investigations, La Jolla, California, Atlas 33.

Watson, W. 1996d. [Ceratioid families]. Pp. 568–595, In: H. G. Moser (editor), *The Early Stages of Fishes in the California Current Region*, California Cooperative Oceanic Fisheries Investigations, La Jolla, California, Atlas 33.

Watson, W. 1998. Early stages of the Bloody Frogfish, *Antennarius sanguineus* Gill 1863, and the Bandtail Frogfish, *Antennatus strigatus* (Gill 1863) (Pisces: Antennariidae). *California Cooperative Oceanic Fisheries Investigations (CalCOFI) Report*, 39: 219–235.

Watson, W., and M. G. Bradbury. 2000. Ogcocephalidae (Batfishes). Pp. 126–130, In: J. M. Leis and B. M. Carson-Ewart (editors), *The Larvae of Indo-Pacific Coastal Fishes: An Identification Guide to Marine Fish Larvae*, Fauna Malesiana Handbook, Vol. 2, Brill, Leiden.

Watson, W., J. M. Leis, and T. Trnski. 2000. Antennariidae (anglerfishes, frogfishes). Pp. 120–125, In: J. M. Leis and B. M. Carson-Ewart (editors), *The Larvae of Indo-Pacific Coastal Fishes: An Identification Guide to Marine Fish Larvae*, Fauna Malesiana Handbook, Vol. 2, Brill, Leiden.

Webb, P. W. 1975. Hydrodynamics and energetics of fish propulsion. *Bulletin of the Fisheries Research Board of Canada*, 190: 1–158.

Weber, M. 1913. Die Fische der Siboga-Expedition. *Siboga-Expedition Monographs*, 57, xii + 710 pp.

Weber, M., and L. F. de Beaufort. 1911. *The Fishes of the Indo-Australian Archipelago. I. Index of the Ichthyological Papers of P. Bleeker.* Brill, Leiden, xi + 410 pp.

Weihs, D. 1977. Periodic jet propulsion of aquatic creatures. *Fortschritte der Zoologie*, 24: 171–175.

Weis, J. 1968. Fauna associated with pelagic sargassum in the Gulf Stream. *American Midland Naturalist*, 80 (2): 554–558.

Westneat, M. W. 1990. Feeding mechanics of teleost fishes (Labridae; Perciformes): A test of four-bar linkage models. *Journal of Morphology*, 205 (3): 269–295.

Westneat, M. W., and A. M. Olsen. 2015. How fish power suction feeding. *Proceedings of the National Academy of Sciences*, 112 (28): 8525–8526.

Wheeler, A. C. 1958. The Gronovius fish collection: A catalogue and historical account. *Bulletin of the Britrish Museum Natural History, Historical Series*, 1 (5): 185–249.

Wheeler, A. C. 1969. *The Fishes of the British Isles and North-West Europe.* Macmillan, London, Melbourne & Toronto, xvii + 613 pp.

Wheeler, A. C. 1975. *Fishes of the World, an Illustrated Dictionary.* Macmillan, New York, xiv + 366 pp.

Wheeler, A. C. 1979. The Sources of Linnaeus's Knowledge of Fishes. Svenska Linnésällskapets Årsskrift, Uppsala, pp. 156–211.

Wheeler, A. C. 1991. The Linnaean fish collection in the Zoological Museum of the University of Uppsala. *Zoological Journal of the Linnean Society*, 103 (2): 145–195.

Wheeler, A. C., and M. J. P. Van Oijen. 1985. The occurrence of *Sphoeroides pachygaster* (Osteichthyes-Tetraodontiformes) off north-west Ireland. *Zoologische Mededelingen,* Leiden, 59 (11): 101–107.

Wheeler, Q. 2015. New to nature No. 141: *Porophryne erythrodactylus*: this newly classified frogfish, which inhabits the subtidal waters of New South Wales, has two quite distinct colour phases and an atypical defense strategy. *The Guardian, Zoology: New to Nature*, 17 May 2015: theguardian.com/science/2015/may/17/new-to-nature-no-141-porophryne-erythrodactylus-frogfish.

Whitehead, P. J. P. 1967. The dating of the 1st edition of Cuvier's *Le Règne animal distribué d'après son organisation. Journal of the Society for the Bibliography of Natural History*, 4 (6): 300–301.

Whitehead, P. J. P. 1976. The original drawings for the *Historia naturalis Brasiliae* of Piso and Marcgrave (1648). *Journal of the Society for the Bibliography of Natural History*, 7 (4): 409–422.

Whitehead, P. J. P. 1979. Georg Marcgraf and Brazilian zoology. Pp. 424–471, In: E. van den Boogaart (editor), *Johan Maurits of Nassau-Siegen: A Humanist Prince in Europe and Brazil.* Johan Maurits van Nassau Stichting, The Hague.

Whitehead, P. J. P. 1982. The treasures at Grüssau. *New Scientist*, 22 April 1982, pp. 226–231.

Whitehead, P. J. P., and P. K. Talwar. 1976. Francis Day (1829–1889) and His Collections of Indian Fishes. *Bulletin of the British Museum (Natural History), Historical Series*, 5 (1): 1–189.

Whitley, G. P. 1927a. A check-list of fishes recorded from Fijian waters. *Journal of the Pan-Pacific Research Institute*, Honolulu, 2 (1): 3–8.

Whitley, G. P. 1927b. Additions to the check-list of the fishes of New South Wales, unpaged. In: A. R. McCulloch, *The Fishes and Fish-like Animals of New South Wales*, 2nd edition. Royal Zoological Society of New South Wales, Sydney, xxvi + 104 pp.

Whitley, G. P. 1929a. Additions to the check-list of the fishes of New South Wales, No. 2. *Australian Zoologist*, 5 (4): 353–357.

Whitley, G. P. 1929b. Studies in ichthyology. No. 3. *Records of the Australian Museum*, 17 (3): 101–143.

Whitley, G. P. 1929c. Names of fishes in Meuschen's index to the "Zoophylacium Gronovianum." *Records of the Australian Museum*, 17: 297–307.

Whitley, G. P. 1931. New names for Australian fishes. *Australian Zoologist*, 6 (4): 310–334.

Whitley, G. P. 1933. Studies in ichthyology. No. 7. *Records of the Australian Museum*, 19 (1): 60–112.

Whitley, G. P. 1934. Supplement to the check-list of the fishes of New South Wales. 12 pp., In: A. R. McCulloch, *The Fishes and Fish-like Animals of New South Wales*, 3rd edition. Royal Zoological Society of New South Wales, Sydney, xxvi + 104 pp.

Whitley, G. P. 1935. Some fishes of the Sydney District. *Australian Museum Magazine*, 5 (9): 291–304.

Whitley, G. P. 1936. Ichthyological genotypes: Some supplementary remarks. *Australian Zoologist*, 8 (3): 189–192.

Whitley, G. P. 1941. Ichthyological notes and illustrations. *Australian Zoologist*, 10 (1): 1–50.

Whitley, G. P. 1944a. New sharks and fishes from Western Australia. Part 1. *Australian Zoologist*, 10 (3): 252–273.

Whitley, G. P. 1944b. Illustrations of some Western Australian fishes. *Proceedings of the Royal Zoological Society of New South Wales*, 1943–1944: 25–29.

Whitley, G. P. 1945. New sharks and fishes from Western Australia. Part 2. *Australian Zoologist*, 11 (1): 1–42.

Whitley, G. P. 1948. A list of the fishes of Western Australia. *Fishery Bulletin*, Western Australian Fisheries Department, Perth, 2, 35 pp.

Whitley, G. P. 1949. The handfish. *Australian Museum Magazine*, 9 (12): 398–403.

Whitley, G. P. 1954. New locality records for some Australian fishes. *Proceedings of the Royal Zoological Society of New South Wales*, 1952–1953: 23–30.

Whitley, G. P. 1957a. A new angler fish. *Western Australian Naturalist*, 5 (7): 207–209.

Whitley, G. P. 1957b. Ichthyological illustrations. *Proceedings of the Royal Zoological Society of New South Wales*, 1955–1956: 56–71.

Whitley, G. P. 1958. Descriptions and records of fishes. *Proceedings of the Royal Zoological Society of New South Wales*, 1956–1957: 28–51.

Whitley, G. P. 1959. Ichthyological snippets. *Australian Zoologist*, 12 (4): 310–323.

Whitley, G. P. 1964. Presidential address. A survey of Australian ichthyology. *Proceedings of the Linnean Society of New South Wales*, 89 (1): 11–127.

Whitley, G. P. 1968a. A check-list of the fishes recorded from the New Zealand Region. *Australian Zoologist*, 15 (1): 1–100.

Whitley, G. P. 1968b. Some fishes from New South Wales. *Proceedings of the Royal Zoological Society of New South Wales*, 1966–1967: 32–40.

Whitley, G. P. 1970. Early history of Australian zoology. *Proceedings of the Royal Zoological Society of New South Wales*, 1970: 1–75.

Whitmee, S. J. 1875. On the habits of the fishes of the genus *Antennarius*. *Proceedings of the Zoological Society*, London, 35 (7): 543–546.

Wickler, W. 1967. Specialization of organs having a signal function in some marine fish. *Studies in Tropical Oceanography*, Miami, 5: 539–548.

Wickler, W. 1968. *Mimicry in Plants and Animals*. McGraw-Hill, World University Library, New York, 255 pp.

Wiens, D. 1978. Mimicry in plants. Pp. 365–403, In: M. K. Hecht, W. C. Steere, and B. Wallace (editors), *Evolutionary Biology*, Vol. 11, Springer, Boston, Massachusetts.

Wiley, E. O., and G. D. Johnson. 2010. A teleost classification based on monophyletic groups. Pp. 123–182, In: J. S. Nelson, H.-P. Schultze, and M. V. H. Wilson (editors), *Origin and Phylogenetic Interrelationships of Teleosts*, Verlag Friedrich Pfeil, München, Germany.

Williams, I. D., N. V. C. Polunin, and V. J. Hendrick. 2001. Limits to grazing by herbivorous fishes and the impact of low coral cover on macroalgal abundance on a coral reef in Belize. *Marine Ecology Progress Series*, 222: 187–196.

Williams, I. D., W. J. Walsh, J. T. Claisse, B. N. Tissot, and K. A. Stamoulis. 2009. Impacts of a Hawaiian marine protected area network on the abundance and fishery sustainability of the Yellow Tang, *Zebrasoma flavescens*. *Biological Conservation*, 142: 1066–1073.

Williams, J. T. 1989. [Review of] *Frogfishes of the World: Systematics, Zoogeography, and Behavioral Ecology* by Theodore W. Pietsch and David B. Grobecker. *National Geographic Research*, 5 (3): 277–280.

Williams, J. T., K. E. Carpenter, J. L. Van Tassell, P. Hoetjes, W. Toller, P. Etnoyer, and M. Smith. 2010. Biodiversity assessment of the fishes of Saba Bank Atoll, Netherlands Antilles. *PLoS ONE*, 5 (5): e10676. doi.org/10.1371/journal.pone.0010676.

Willughby, F. 1686. *De historia piscium libri quatuor, Jussu & Sumptibus Sociatatis Regiae Londinensis editi. Totum opus recognovit, coaptavit, supplevit, librum etiam primum et secundum integros adjecit Johannes Raius.* Theatro Sheldoniano, Oxonii, 343 pp.

Wilson, D. P. 1937. The habits of the angler-fish, *Lophius piscatorius* L., in the Plymouth Aquarium. *Journal of the Marine Biological Association of the United Kingdom*, 21 (2): 477–496.

Winterbottom, R. 1974. A descriptive synonymy of the striated muscles of the Teleostei. *Proceedings of the Academy of Natural Sciences of Philadelphia*, 125 (12): 225–317.

Winterbottom, R., A. R. Emery, and E. Holm. 1989. *An Annotated Checklist of the Fishes of the Chagos Archipelago, Central Indian Ocean.* Royal Ontario Museum, Life Science Contributions, 145, 226 pp.

Wirtz, P., J. Bingeman, J. Bingeman, R. Fricke, T. J. Hook, and J. Young. 2014. The fishes of Ascension Island, central Atlantic Ocean—new records and an annotated checklist. *Journal of the Marine Biological Association of the United Kingdom*, 2014 (6): 1–16.

Wirtz, P., A. Brito, J. M. Falcón, R. Freitas, R. Fricke, V. Monteiro, F. Reiner, and O. Tariche. 2013. The coastal fishes of the Cape Verde Islands—new records and an annotated checklist. *Spixiana*, 36 (1): 113–142.

Wirtz, P., E. d'Oliveira, and G. Bachschmid. 2017. One fish and seven invertebrate species new for the marine fauna of the Cape Verde Islands. *Arquipelago, Life and Marine Sciences*, 34: 51–54.

Wirtz, P., C. E. L. Ferreira, S. R. Floeter, R. Fricke, J. L. Gasparini, T. Iwamoto, L. Rocha, C. L. S. Sampaio, and U. K. Schliewen. 2007. Coastal fishes of São Tomé and Príncipe islands, Gulf of Guinea (Eastern Atlantic Ocean)—an update. *Zootaxa*, 1523: 1–48.

Wirtz, P., R. Fricke, and M. J. Biscoito. 2008. The coastal fishes of Madeira Island—new records and an annotated check-list. *Zootaxa*, 1715: 1–26.

Wong, L., T. Bessell, and T. Lynch. 2018. *Spotted Handfish Ambassador Fish Program: Captive Fish Studbook.* Report to the National Environmental Science Programme, Marine Biodiversity Hub, CSIRO, 24 pp.

Wong, L., and T. Lynch. 2017. *Monitoring of Spotted Handfish* (Brachionichthys hirsutus) *Populations and On Ground Conservation Actions.* Report to the National Environmental Science Programme, Marine Biodiversity Hub, CSIRO, 19 pp.

Wood, E. M. 2001. *Collection of Coral Reef Fish for Aquaria: Global Trade, Conservation Issues and Management Strategies.* Marine Conservation Society, Ross-On-Wye, United Kingdom, 56 pp.

Wright, J. J., R. E. Schmidt, and B. R. Weatherwax. 2016. New and previously overlooked records of several fish species from the marine waters of New York. *Northeastern Naturalist*, 23 (1): 118–133.

Wu, H. W. 1931. Notes on the fishes from the coast of Foochow region and Ming River. *Contributions from the Biological Laboratory of the Science Society of China, Zoological Series*, 7 (1): 1–64.

Yamakawa, T. 1979. *Studies on the Fish Fauna Around the Nansei Islands, Japan. 1. Check list of fishes collected by Toshiji Kamohara and T. Yamakawa from 1954 to 1971.* Reports of the Usa Marine Biology Institute, Kochi University, Supplement to No. 1, 47 pp.

Yamanoue, Y., M. Miya, K. Matsuura, N. Yagishita, K. Mabuchi, H. Sakai, M. Katoh, and M. Nishida. 2007. Phylogenetic position of tetraodontiform fishes within the higher teleosts: Bayesian inference based on 44 whole mitochondrial genome sequences. *Molecular Phylogenetics and Evolution*, 45: 89–101.

Yanong, R. P., E. W. Curtis, S. P. Terrell, and G. Case. 2003. Atypical presentation of mycobacteriosis in a collection of frogfish (*Antennarius striatus*). *Journal of Zoo and Wildlife Medicine*, 34 (4): 400–407.

Yokota, M., M. Hayashi, and Y. Shimamura. 1992. First record of a frogfish *Antennarius rosaceus* (Pisces: Antennariidae) from Japan. *Izu Oceanic Park Diving News*, 3 (9): 4–5.

Yokota, M., and H. Senou. 1991. A review of the frogfishes of Japan. *Izu Oceanic Park Diving News*, 2 (6): 2–5 (in Japanese).

Yoneda, M., H. Miura, M. Mitsuhashi, M. Matsuyama, and S. Matsuura. 2000. Sexual Maturation, Annual Reproductive Cycle, and Spawning Periodicity of the Shore Scorpionfish, *Scorpaenodes littoralis*. *Environmental Biology of Fishes*, 58 (3): 307–319.

Yoneda, M., M. Tokimura, H. Fujita, N. Takeshita, K. Takeshita, M. Matsuyama, and S. Matsuura. 1998. Ovarian structure and batch fecundity in *Lophiomus setigerus*. *Journal of Fish Biology*, 52 (1): 94–106.

Yoneda, M., M. Tokimura, H. Fujita, N. Takeshita, K. Takeshita, M. Matsuyama, and S. Matsuura. 2001. Reproductive cycle, fecundity, and seasonal distribution of the anglerfish *Lophius litulon* in the East China and Yellow Seas. *United States Fishery Bulletin*, 99 (1): 356–370.

Yoshida, T., H. Motomura, P. Musikasinthorn, and K. Matsuura (editors). 2013. *Fishes of Northern Gulf of Thailand*. National Museum of Nature and Science, Tsukuba, Research Institute for Humanity and Nature, Kyoto, and Kagoshima University Museum, Kagoshima, viii + 239 pp.

Youn, C.-H. 2002. *Fishes of Korea, With Pictorial Key and Systematic List*. Academy Publications, Seoul, Korea, 747 pp.

Zigno, A. de. 1874. Catalogo ragionato dei pesci fossili del calcare eoceno di Monte Bolca. *Atti del Reale Istituto Veneto di Scienze, Lettere ed Arti*, 4: 1–215.

Zigno, A. de. 1887. Nuove aggiunte alla ittiofauna dell'epoca Eocena. *Memorie del Reale Istituto Veneto di Scienze, Lettere ed Arti*, 23: 9–33.

Zug, G. R., V. G. Springer, J. T. Williams, and G. D. Johnson. 1988. The vertebrates of Rotuma and surrounding waters. *Atoll Research Bulletin*, 316: 1–25.

Zugmayar, E. 1911. Poissons provenant des campagnes du Yacht "Princesse Alice." *Résultats des Campagnes Scientifiques du Prince Albert Ier*, 35, 174 pp.

Zupanc, G. K. H., and R. F. Sirbulescu. 2012. Teleost fish as a model system to study successful regeneration of the central nervous system. Pp. 193–233, In: E. Heber-Katz and D. L. Stocum (editors), *New Perspectives in Regeneration, Current Topics in Microbiology and Immunology*, Springer-Verlag, Berlin, Vol. 367.

Illustration Credits

Sources for the illustrations used in this volume are usually given in the legends for the figures themselves; where more detailed credits are required, they are given below, cited by figure number. Illustrations for which no attribution is cited, either here or in the legends, originate with the authors or reside in the public domain.

Figures 1A, 1F: images courtesy of the NOAA *Okeanos Explorer* Program, NOAA Office of Ocean Exploration and Research, National Oceanic and Atmospheric Administration, Department of Commerce, http://www.photolib.noaa.gov.

Figures 1B, 1H, 4, 13, 14, 17, 21, 28, 122–127, 191, 193, 194, 199, 231, 241, 242, 245–248, 253B, 255, 256, 264, 269, 328, 329, 341: images that appear in publications of the American Society of Ichthyologists and Herpetologists; courtesy of Prosanta Chakrabarty and the American Society of Ichthyologists and Herpetologists. Used with permission.

Figure 3A: folio 166, In: *Caroli Linnaei Manuscripta Medica Tom. 1, in quo continetur Historie naturalis fragmenta*, a bound volume of notes and various writings on mineralogy, botany, zoology, and miscellanea, dating from August 1727 to the beginning of 1730; courtesy of Isabelle Charmantier, Andrea Deneau, and the Linnean Society of London. Used with permission.

Figures 7, 8, 10, 14, 16, 17, 19, 20, 104E: images in the public domain, courtesy of Wikimedia Commons.

Figure 9: images from the *Collection des Vélins du Muséum national d'Histoire naturelle*, Paris, 91, figs. 62–64; courtesy of Joyce Faust, © RMN–Grand Palais / Art Resource, Inc., New York. Used with permission.

Figures 12A, 12B, 129, 230A, 230B: images courtesy of Stephan Atkinson and the Picture Library, Natural History Museum, London. Used with permission.

Figures 17, 25: photographs courtesy of Godard Tweehuysen and the Naturalis Biodiversity Center, University of Amsterdam, Netherlands. Used with permission.

Figure 21: photograph courtesy of Mark McGrouther and Claire Vince, Australian Museum, Sydney. Used with permission.

Figure 22: photograph courtesy of Penny Haworth, © NRF-SAIAB, National Research Foundation, South African Institute for Aquatic Biodiversity, Grahamstown. Used with permission.

Figure 23: after Smith, 1949, *The Sea Fishes of Southern Africa*, pl. 98; courtesy of Penny Haworth, © NRF-SAIAB, South African Institute for Aquatic Biodiversity. Used with permission.

Figures 24, 31, 57, 64, 113A, 135A–B, 137A, 145A–D, 147B, 150A–B, 152A–D, 155, 159A–F, 161, 163, 172, 181A–C, 235B, 270, 318A–B: images courtesy of Lisa Palmer, Division of Fishes, National Museum of Natural History, Smithsonian Institution, Washington, DC. Used with permission.

Figures 26–29, 33, 37, 38, 41, 42, 46, 50, 58, 59, 70, 75, 78, 82, 83, 91, 92, 98C, 100, 103A–B, 107, 112A–B, 113B, 114, 118, 119, 123, 130, 137B, 139, 140, 141, 144A–I, 151A–B, 158A–C, 162, 169, 170A–B, 176, 177A–C, 183, 197, 198, 202, 210B, 212, 216, 219, 226, 239B, 240A–B, 250, 251, 295–302, 303A–C, 305–312, 315–317, 319–321, 323: photos and drawings originating from Pietsch and Grobecker, 1987; all rights reverted to the senior author.

Figures 30A–E: after Watson, 1996b, fig. 1, and 1998, fig. 11; publications of the *California Cooperative Oceanic Fisheries Investigations*; drawings courtesy of Barbara Sumida McCall and William Watson, NOAA Fisheries, Southwest Fisheries Science Center, La Jolla, California. Used with permission.

Figure 45C: after Okada et al., 1935, pl. 158; courtesy of Hiroki Tabo and Sanseido Co. Ltd., Tokyo, Japan. Used with permission.

Figures 48A–C: photographs courtesy of Hiroshi Senou and the Image Database of Fishes (Fish-Pix), Kanagawa Prefectural Museum of Natural History, Odawara, Kanagawa, Japan. Used with permission.

Figure 49: after Poll, 1959, *Mémoires de l'Institut Royal des Sciences Naturelles de Belgique*, 4 (3B), fig. 124; courtesy of Françoise Antonutti and Olivier Pauwels, © Royal Belgian Institute of Natural Sciences, Brussels. Used with permission.

Figures 52, 67A, 89, 96, 97, 156, 196A: images from Cuvier and Valenciennes, MS 504, XII.B.34, 43, 50, 54, 56, 60, 61, 63, 64, Bibliothèque Centrale, Muséum national d'Histoire naturelle, Paris; courtesy of Emmanuelle Choiseau, © Bibliothèque Centrale, Muséum national d'Histoire naturelle, Paris. Used with permission.

Figures 56, 62: original drawings for Francis Day's *Fishes of India*; courtesy of Sarah Broadhurst, Ann Sylph, and the Zoological Society of London. Used with permission.

Figure 67B: taken from Plumier MS 25, 90B, Bibliothèque Centrale, Muséum national d'Histoire naturelle, Paris; courtesy of Emmanuelle Choiseau, © Bibliothèque Centrale, Muséum national d'Histoire naturelle, Paris. Used with permission.

Figure 86: after Pietsch and Grobecker, 1978, *Science*, 201, fig. 1; courtesy of *Science*, © 1978, American Association for the Advancement of Science. Used with permission.

Figures 157B, 215B: images from the George James Coates Collection; courtesy of Rick Feeney and the Department of Ichthyology, Natural History Museum of Los Angeles County, Los Angeles, California. Used with permission.

Figures 185, 189, 201, 206, 210A, 232: after McCulloch and Waite, 1918, *Records of the South Australian Museum*, 1 (1), pls. 6, 7; courtesy of Lea Gardam and the South Australian Museum, Adelaide. Used with permission.

Figure 239A: after Whitley, 1941, *Australian Zoologist*, 10 (1), fig. 31; courtesy of Mark McGrouther, Vanessa Finney, and Patricia Egan, Australian Museum, Sydney. Used with permission.

Figures 265A–D, 266, 269: after Pietsch, 2009, *Oceanic Anglerfishes: Extraordinary Diversity in the Deep-sea*, figs. 186, 187, 188; courtesy of Karin Tucker, University of California Press, Berkeley and Los Angeles. Used with permission.

Figure 267: after Datovo et al., 2014, *PLoS ONE*, 9 (10), fig. 5B; courtesy of Alessio Datovo, University of São Paulo, Brazil. Used with permission.

Figure 268: after Gregory and Conrad, 1936, *American Naturalist*, 70, figs. 3, 4; courtesy of Judy Choi, University of Chicago Press. Used with permission.

Figures 313, 314: after Grobecker and Pietsch, 1979, *Science*, 205, figs. 1, 2; courtesy of *Science*, © 1979 American Association for the Advancement of Science. Used with permission.

Figures 325, 326: after Rasquin, 1958, *Bulletin of the American Museum of Natural History*, 114 (4), figs. 1, 2; courtesy of Mai Reitmeyer, American Museum of Natural History, New York. Used with permission.

Figure 345B: after Fowler, 1928, *Memoirs of the Bernice P. Bishop Museum*, 10, fig. 82; courtesy of Tia Reber, Bernice Pauahi Bishop Museum, Honolulu, Hawaii. Used with permission.

Figure 345C: after Breder, 1938, *Bulletin of the Bingham Oceanographic Collection*, 6 (5), fig. 48; courtesy of Rosemary Volpe and the Peabody Museum of Natural History, Yale University, New Haven, Connecticut. Used with permission.

Figure 346: after Rhyne et al., 2017a, *PeerJ*, 5:e2949, https://doi.org/10.7717/peerj.2949; courtesy of Andrew L. Rhyne. Used with permission.

Index

A page number in boldface type indicates the first page of a major description or critical discussion. An "f" following a page number denotes a figure; a "k" denotes a key to the identification of a taxon; an "m" indicates a map of the geographic distribution of a species; and a "t" denotes a table. Antennariid genera, species, and common names that are considered valid taxa are indicated in boldface to distinguish them from taxa that are regarded as junior synonyms. Personal names are indexed as they appear in the text, but names of those that appear only as authors of taxa, collectors of specimens, or authors in the References are not indexed.